D0820572

Mechanics of Wave Forces on Offshore Structures

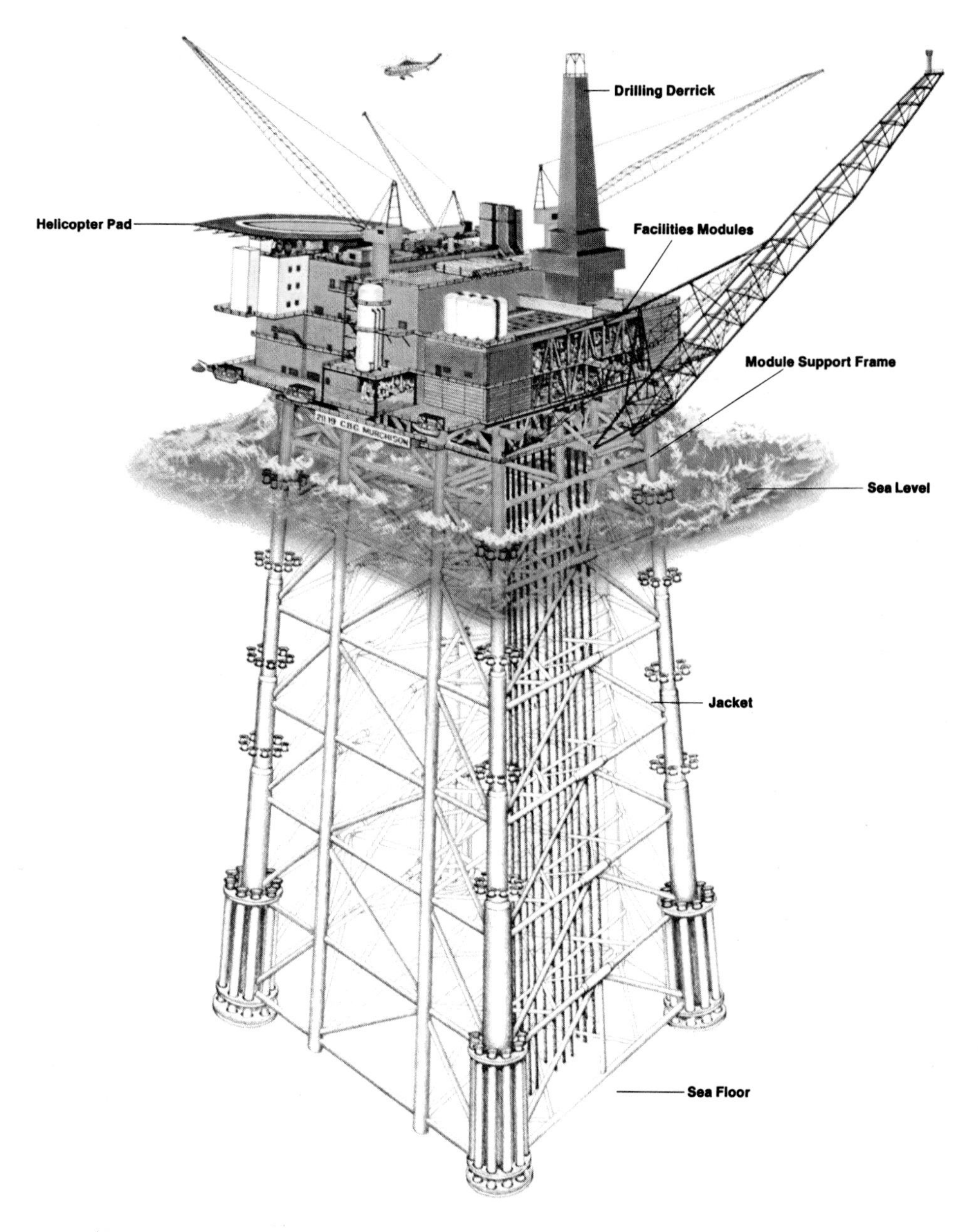

(From an original by D. Muller, and reproduced with the permission of Conoco Inc.)

Mechanics of Wave Forces on Offshore Structures

Turgut Sarpkaya

Michael Isaacson

VNR VAN NOSTRAND REINHOLD COMPANY
NEW YORK CINCINNATI ATLANTA DALLAS SAN FRANCISCO
LONDON TORONTO MELBOURNE

Van Nostrand Reinhold Company Regional Offices:
New York Cincinnati Atlanta Dallas San Francisco

Van Nostrand Reinhold Company International Offices:
London Toronto Melbourne

Library of Congress Catalog Card Number: 80-20237
ISBN: 0-442-25402-4

Manufactured in the United States of America

Published by Van Nostrand Reinhold Company
135 West 50th Street, New York, N.Y. 10020

Published simultaneously in Canada by Van Nostrand Reinhold Ltd.

15 14 13 12 11 10 9 8 7 6 5 4 3 2 1

Library of Congress Cataloging in Publication Data

Sarpkaya, Turgut, 1928–
Mechanics of wave forces on offshore structures.

Includes bibliographical references and indexes.
1. Offshore structures–Hydrodynamics. 2. Ocean waves. I. Isaacson, Michael, 1949– joint author. II. Title.
TC1650.S26 627′.98 80-20237
ISBN 0-442-25402-4

Preface

Offshore technology has experienced extremely rapid development since the 1940 s, and a thorough understanding of the interaction of waves with offshore structures has now become a vital factor in the safe and economical design of such structures. There has been a corresponding increase in research efforts to meet this need, but results are widely scattered throughout literature.

The present text is a modest effort in response to the clear need to assemble and organize the wide ranging research efforts pertinent to the central topic of wave forces on offshore structures. However, the intention is specifically not to present a compendium of experimental data and theoretical results. Rather, emphasis is placed on describing the vitally important physical concepts and underlying principles. Observations, laboratory and field experiments and theory have been kept continually in mind in the selection of topics and in their exposition. This is essential if the reader is to deal with a novel problem which might not entirely overlap with presently available results. In fact, in many instances the understanding of the limitations of the theoretical and experimental results and a sound judgment are the designer's most important recipes.

The text is intended to be both of fundamental interest to researchers, scientists and graduate students, as well as of immediate practical value to engineers involved in the design and construction of offshore structures. It may serve as a convenient text for graduate courses relating to wave forces, as well as for self-study by engineers interested in problems of wave forces on offshore structures.

A good background in mathematics and fluid dynamics is assumed. Even so, for sake of completeness, the fundamental concepts and governing equations of fluid motion are reviewed in Chapter 2. Subsequent chapters deal in turn with flow separation and time-dependent flows, wave theories, wave forces on small bodies, wave forces on large bodies, spectral methods, dynamic response and hydroelastic oscillations, and modelling of offshore structures. Each chapter is reasonably self-contained and the reader should find no serious difficulty in approaching any one chapter independently of the others. A comprehensive list of references is provided at the end of each chapter. As such, the book can serve also as a reference tool and perhaps as a point of departure for research.

Units of quantities are referred to fairly infrequently, but whenever this is the case the British System has been adopted in view of its widespread use in the United States.

Because of the wide range of topics covered, it has not been possible to maintain a consistent and distinct set of notation throughout the text without some overlap of symbols. However, the notation within any single chapter should be reasonably consistent and is defined wherever first encountered.

A text of this sort, which attempts to help bridge the gap between theoreticians and practicing engineers, cannot hope to fulfill the needs or expectations of all those within this wide spectrum. Indeed, not all topics related to ocean wave interaction with structures could be or are treated. More specifically, the text does not deal with such topics as wave interaction with breakwaters, seawalls and other coastal structures, coastal processes and scour, harbor design, and wave interaction with ships and ship-like vessels. Furthermore, highly specialized topics such as the discrete vortex model are not described in detail.

The contents of the 9 chapters reflect the collective experience in teaching, research and consulting of the authors and their assessment of the relevance of the material treated. Although most of the text describes material that is available in the technical literature, a number of original results and interpretations have been included.

There are many people to thank for aiding us in this effort. Professors Charles Dalton, of the University of Houston; John H. Nath, of Oregon State University; and Dr. Wayne W. Jamieson, of the National Research Council of Canada, each read the manuscript and gave helpful criticisms. Many graduate students have worked with us on this subject, notably Neil J. Collins, Neil MacKenzie, Farhad Rajabi, and Ray L. Shoaff. Many others devoted countless hours in carrying out experiments and evaluating data. Mr. Jack McKay's ingenuity permeated the design and construction of many research equipment which was invaluable in obtaining some of the results presented here. Sincere appreciation is also expressed here for the extensive research support and willing cooperation extended through the years by the representatives of the federal agencies, particularly the Office of Naval Research, the National Science Foundation and the Civil Engineering Laboratory of the Naval Construction Battalion Center (Port Hueneme, California) and the National Science and Engineering Council of Canada. Special appreciation is also extended to our families for their support.

This book is dedicated to the designers and builders of offshore structures and to researchers in this field. Their concern for the advancement of the state of the art motivated our work. We sincerely hope that our efforts, modest relative to their monumental achievements, will meet with their approval.

T. S. Monterey, California
M. I. Vancouver, Canada

Contents

Mechanics of Wave Forces on Offshore Structures

1
Introduction

Offshore technology has experienced a remarkable growth since the late 1940's, when offshore drilling platforms were first used in the Gulf of Mexico. At the present time a wide variety of offshore structures are being used, even under severe environmental conditions. These are predominantly related to oil and gas recovery, but they are also used in other applications such as in harbor engineering, in ocean energy extraction, and so on. Difficulties in design and construction are considerable, particularly as structures are being located in ever increasing depths and are subjected to extremely hostile environmental conditions. The discovery of major oil reserves in the North Sea has accelerated such advances, with fixed platforms in the North Sea now being located in water depths up to about 600 feet and designed to withstand waves of heights up to 100 feet.

The potential of major catastrophic failures, both in terms of human safety as well as economy loss, underlines the critical importance of efficient and reliable design. Two tragic failures are cases in point: In January 1961, the collapse of Texas Tower No. 4 off the New Jersey coast involved the loss of 28 lives. And more recently in March 1980, the structural failure and capsizing of the mobile rig Alexander Kielland in the Ekofisk field in the North Sea involved the loss of over 100 lives.

1.1 CLASSES OF OFFSHORE STRUCTURES

Before proceeding to describe the content of this text and its relation to the safe and economic design of offshore structures, it is appropriate at the outset

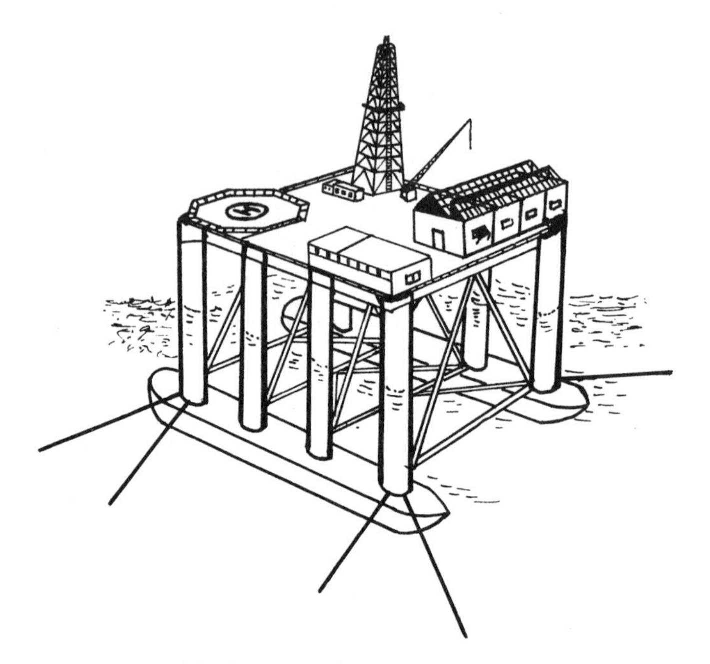

Fig. 1.1a. A semisubmersible platform.

to provide some perspective to what follows, by classifying briefly the wide variety of offshore structures that are in current use or that have been seriously proposed. The major offshore structures used in the various stages of oil recovery include both mobile and fixed drilling platforms, as well as a variety of supply, work and support vessels.

The various offshore structures presently in use have been described in detail in the trade and technical literature. Mention is made of Bruun (1976) who summarizes the recent offshore rigs used in the North Sea, and Watt (1978) who reviews the design and analysis requirements of fixed offshore structures used in the oil industry. Ships and shiplike marine vessels are also used extensively, but they are treated within the field of naval architecture and are not of primary consideration in this text.

Exploratory drilling is usually carried out with mobile drilling rigs. These include submersible platforms, which are limited to relatively shallow water, jackup platforms, drill ships or drill barges, and semisubmersible drilling platforms. A sketch of a semisubmersible platform is shown in Fig. 1.1a. Such platforms are capable of operating at depths up to about 1000 feet and are able to withstand severe weather conditions relatively well.

Development and production activities at an offshore site are primarily carried out with fixed platforms. The jacket or template structures, and ex-

Fig. 1.1b. A jacket-type platform.

tensions to them, are the most common platforms in use. A jacket platform comprises of a space frame structure, with piles driven through its legs. Extensions to this concept include the more recent space frame structures which employ skirt piles or pile clusters. Some platforms may contain enlarged legs to provide for self-buoyancy during installation. Jacket platforms are located throughout the world, including the North Sea where they may be exposed to waves with heights approaching 100 feet. Figure 1.1b shows a sketch of such a platform. The largest platform to date (1980) is one installed in the Cognac field off the Louisiana coast in a water depth of just over 1000 feet. This plat-

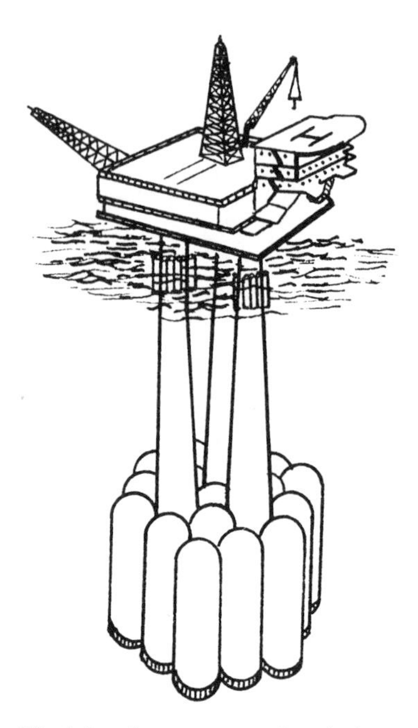

Fig. 1.1c. A concrete gravity platform.

form contains 59,000 tons of steel and was fabricated in three sections which were joined under water on site.

Gravity platforms are another class of fixed structures which are in use, and which depend on excessive weight, rather than on piles, for their stability. They are thus suited to sites with over-consolidated soils, and have been used primarily in the North Sea. Gravity platforms are usually constructed from concrete, although steel or concrete and steel platforms are also in use or have been proposed. The most usual gravity platforms comprise of a large base which has the capacity for significant oil storage and which supports a few columns as sketched in Fig. 1.1c. Examples include the Condeep, Sea Tank and Andoc platforms in use in the North Sea. The Ekofisk platform differs from the others by incorporating a perforated outer wall which surrounds a central column extending up to the water surface. In addition to their being located in depths of several hundred feet, gravity platforms are characterized by large horizontal dimensions. For example, a typical platform may be 600 feet high, its base may have a diameter of 300 to 400 feet, and have the capacity to store 1 million barrels of oil.

Compliant platforms are designed to withstand environmental loads by their ability to deflect from an equilibrium position when subjected to such loads. Examples of such platforms include the guyed tower, which is pivoted at the seabed and moored, and the tension leg platform.

Apart from drilling platforms and marine vessels, other offshore structures include storage tanks, submarine pipelines, risers extending from a surface vessel or platform deck down to the seabed, piles used in warfs and other coastal structures, caissons for ship berthing, articulated moorings for tankers, and so on. One particular example of some importance, but which is not directly related to the oil industry, pertains to Ocean Thermal Energy Conversion (OTEC) systems. The major structural component of an OTEC structure is a very long cold water pipe, typically about 30 feet in diameter, which extends from the water surface or near it to a depth of over 3,000 feet.

1.2 THE ROLE OF OFFSHORE ENGINEERING RESEARCH

We now turn to the role of research relating to offshore structures and the part this plays in the design process. The ultimate objective of such research is to develop methods of design and construction which will help to produce structures which are safe, functional, economical and able to resist the forces induced by man and environment over a required period of time. In order to achieve this goal, it is generally necessary to conduct research both in the laboratory and in the field, and to integrate fully these two complementary methods of investigation. The intended end result would then include the development of mathematical models, design rules and common-sense recommendations to the development of platform design and construction.

The environmental loads to be accounted for include those due to wind, waves, currents, ice, earthquakes and soil movement. An ideal experiment would be to subject an actual structure to such effects, to measure the structural loads and response, and to identify relationships between measured forces and response, and environmental conditions giving rise to those forces. Indeed, it is ideally desirable to provide selected prototype structures with permanently installed instrumentation systems, which would not only facilitate specific tests of such structures during and after construction, but also remain as a permanent monitoring system for the various environmental loads. This will continuously help to refine the calibration of platform loading models. However, this approach is not generally feasible because of the high cost of providing adequate instrumentation, and the rather low probability that one would be able to correlate the data obtained to achieve a reasonable degree of power of prediction.

In order to advance the study of loading and response mechanisms, it is necessary, then, to resort also to laboratory experiments with idealized conditions. Once a reliable model of the loading and response is established, one is then in a position to extend the model towards the prototype situation by considering the effects of additional parameters on the idealization and empiricism of the adopted model.

An intermediate and extremely valuable source of information is the erection and detailed instrumentation of model offshore platforms in major or judiciously chosen sites. The Ocean Test Structure (Haring and Spencer 1979) is one of such rigs, highly instrumented to measure wind and wave forces. It is a 20 × 40 × 120 feet platform installed in 66 feet of water in the Gulf of Mexico at south Timbalier Block 67. The results of such large scale investigations, taken together with those of controlled small-scale laboratory experiments, emphasize the importance of effective communication between researchers and design engineers in the attempt to improve overall design procedures.

The present text is concerned more specifically with wave-induced loads and has been prepared to provide a state-of-the-art account of wave forces on offshore structures. The sequence of calculation procedures needed to establish the structural loading generally involves some or all of the following steps:

(a) establishing the wave climate in the vicinity of a structure, either on the basis of recorded wave data, or by hindcasting from available meteorological data;
(b) estimating design wave conditions for the structure;
(c) selecting and applying a wave theory to determine the corresponding fluid particle kinematics;
(d) using a wave force formulation to determine the hydrodynamic forces on the structure;
(e) calculating the structural response; and
(f) calculating the structural loading, which includes base shear and moment, stresses and bending moments.

These steps are summarized in Fig. 1.2, but it is emphasized that they serve only as a rough indicator. For example, wave runup is an important consideration in many situations but is omitted in the above. In some cases, step (c) above may be omitted, such as in direct physical modelling, or is straightforward when linear wave theory is applied, such as when random loading is considered.

There are two distinct approaches whereby one can follow the above procedure. One is a deterministic approach, which itself may be either pseudo-static or time-dependent, and the second is a stochastic approach. In the deterministic pseudo-static method, which is the simplest to use, maximum loads are calculated and applied to a static analysis of the structure, whereas in the deterministic time-dependent method, a time record of the free surface elevation is used to calculate time-dependent loads, taking into account the dynamic response of the structure. In the stochastic approach, which complements the deterministic approach, the wave motion is treated as a random process and is described by its spectral density. A corresponding description of the loads on, and response of, the structure is obtained. Thus, in the stochastic approach all calculations are carried out in the frequency domain, whereas in the deter-

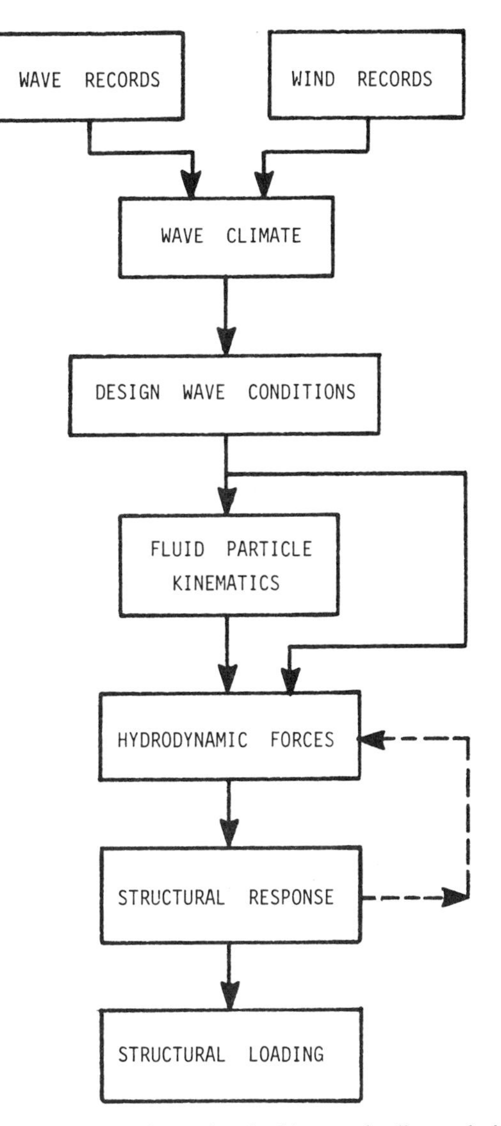

Fig. 1.2. Fundamental steps involved in wave loading analysis.

ministic approach they are carried out in the time domain. The different approaches outlined above have been described by Bea and Lai (1978) and summarized by them as indicated in Fig. 1.3. Finally, the above analyses may be carried out for extreme conditions, in order to calculate immediate structural failure, or for nominal conditions in order to calculate fatigue failure or structural response under operational conditions.

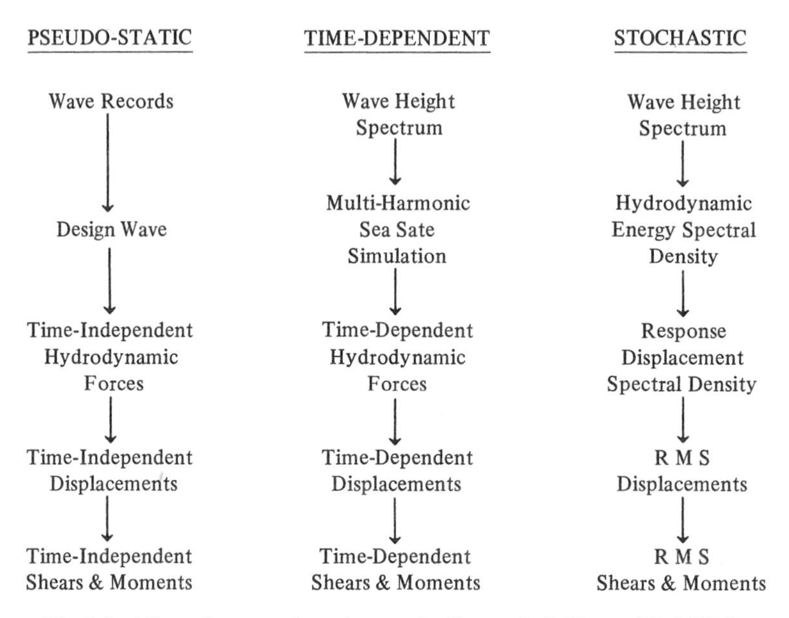

Fig. 1.3. Alternative procedures to wave loading analysis (Bea and Lai 1978).

1.3 HISTORICAL DEVELOPMENT

Having considered the role of wave force research in the overall design process, it may be appropriate at this point to pass on to a brief outline of the historical development of the subject of wave loading. There are two areas of background necessary for a thorough understanding of wave loading on structures. The first concerns the fundamental fluid mechanics of steady and unsteady flows past bodies. The second concerns the various wave theories which might be employed to provide the details of the fluid motion due to the waves. These two background areas were for the most part investigated quite independently since the last century, and it is only since about 1950 that the central topic of wave loading itself has been developed.

The theory of ideal fluid motion was formulated quite comprehensively in the last century and the earlier part of the present century. Thompson and Tait (1879) presented a general theory of the motions of solids in ideal fluids. The text by Lamb (1945), first published in 1879, is an invaluable reference relating to this kind of work. The fundamental result of interest concerns the hydrodynamic force, termed an inertia force, which is exerted on a body held in a uniformly accelerating ideal fluid. This is proportional to the acceleration du/dt and is expressed as

$$F_i = C_m \rho \forall \, du/dt \tag{1.1}$$

where ρ is the density of fluid, $\forall$ is the body volume and C_m is termed an inertia coefficient. This inertia force is directly related to the force on the body in the complementary flow corresponding to that body accelerating in a fluid otherwise at rest. In this case, the force is often expressed in terms of the added mass of the body, which adds on to its own mass when expressing the overall force needed to accelerate the body.

The uniform, steady, two-dimensional flow of a real (viscous) fluid pass a circular cylinder began to be understood only in the present century, following Prandtl's discovery of the boundary layer in 1904. The wake structure and associated vortex shedding, and drag and lift forces have now been described quite comprehensively, and since the 1960's have been simulated to varying degrees by numerical computation.

We now turn to the development of theoretical descriptions of surface wave behavior. The linear wave theory, also called Airy theory, small amplitude wave theory or sinusoidal wave theory, was first introduced by Airy (1845) and was treated in some detail by Stokes (1847, 1880). Stokes also introduced a perturbation procedure for obtaining higher order theories as successive approximations. These are the Stokes finite amplitude wave theories, and are valid in deep water or in intermediate depths.

On the other hand, the extreme shallow water case of a solitary wave was first described by Russell (1844) and first treated theoretically by Boussinesq (1871). Cnoidal wave theory, which describes periodic waves in the shallow water range, was introduced by Korteweg and de Vries (1895), but was not developed in more detail until much later (Keller 1948, Laitone 1961 and after).

Within the last few years, significant advances have been made, on the basis of numerical work requiring a computer, in accurately predicting the behavior of even very steep waves. At the present time, the properties of a regular wave train of any wave height and wave length can be described with considerable accuracy.

A method of calculating wave forces on piles was first proposed by Morison, Johnson, O'Brien and Schaaf (1950) and this provided a landmark in wave forces prediction methods. The formulation they proposed, the so-called Morison equation, forms the basis of most wave force calculations up to the present time. Morison and his co-workers suggested that the force acting on a section of a pile due to wave motion is made up of two components: a drag force, analogous to the drag on a body subjected to a steady flow of a real fluid associated with wake formation behind the body; and an inertia force, analogous to that on a body subjected to a uniformly accelerated flow of an ideal fluid. For the usual case of a circular section of diameter D, the Morison equation is thus expressed as

$$F' = 0.5\rho D C_d U|U| + 0.25\rho\pi D^2 C_m \, du/dt \tag{1.2}$$

where F' is the force per unit length; U, the incident (undisturbed) flow velocity taken at the section center; C_d, the drag coefficient; and C_m, the inertia coefficient.

The superposition of the drag and inertia components for the general case of wave motion–which constitutes an unsteady, nonuniform flow of a viscous fluid–may only be adopted if it is recognized that the corresponding drag and inertia coefficients no longer retain their values deriving from the two reference flows, but are now treated as empirical and taken to depend on the various parameters characterizing the flow.

There are strong similarities between the wave flow past a section of a pile and a two-dimensional sinusoidal flow past a circular section. The latter reference flow is conveniently obtained under laboratory conditions and may be used to generate a variety of pertinent experimental results. Keulegan and Carpenter (1958) investigated this flow and the characteristics of the drag and inertia coefficients corresponding to it. One of the important parameters of the flow is the amplitude of fluid motion relative to cylinder size, which is usually expressed as the Keulegan-Carpenter number $K = U_m T/D$, where U_m and T are respectively the velocity amplitude and period of the flow. The magnitude of K indicates the relative importance of drag and inertia forces. However, it is only relatively recently (Sarpkaya 1976 and later) that comprehensive, reliable data of the force coefficients have been obtained for the oscillatory flow, especially at relatively high Reynolds numbers, and the dependence of C_d and C_m on both the Keulegan-Carpenter number and the Reynolds number has been demonstrated.

A variety of additional effects have since been investigated and/or are subjects of continuing research. These include the proximity effects, wave slamming, cylinder roughness, structural flexibility (dynamic response), yaw to the incident flow, effects of combined steady and oscillatory flows (simulating the combined wave-current field), and so on.

The application of the Morison equation carries the implicit assumption that the body size is small relative to the wave length so that the incident flow is virtually uniform in the vicinity of the body. When the body is relatively large this is no longer true, the incident wave train itself being altered (diffracted) by the body, and an alternative formulation is required. However, since flow separation is now usually unimportant, a theoretical approach may be used in which the governing equations of the flow are solved to a first approximation, bearing in mind the additional boundary conditions introduced by the presence of the body. The wave forces predicted by this diffraction problem were first obtained for the special case of a large vertical circular cylinder by MacCamy and Fuchs (1954). Within the last decade bodies of more general shape have been treated by numerical schemes using a computer, and this approach forms the basis of present prediction methods for large offshore structures.

In describing ocean waves and wave forces as realistically as possible, one is faced with attempting to deal with the random nature of the waves. The statistical theory of random waves and wave forces paralleled the other developments since the 1950's so far described. The earliest results were obtained for waves possessing a narrow-band spectrum, after Longuet-Higgins (1952) first applied the statistical theory of noise in electrical circuits (Rice 1944–1945) to the case of ocean waves. Descriptions of random waves have since been extended to waves with spectra of arbitrary shape and to descriptions of the directional spreading of random waves.

The consequential random loads on structures may be calculated by application of the Morison equation, and have been considered by Borgman in a series of papers (see Borgman 1972). The most direct approach is to linearize the equation such that force components at any one frequency arise only from wave components at that same frequency. Extensions to deal with loads predicted from the complete (nonlinear) Morison equation, and with loads derived from directional wave spectra are relatively recent.

Finally, it is important to consider the response of a structure and its members to wave loading, particularly since the problem is generally coupled such that the hydrodynamic forces themselves depend on the motion of a structure. The dynamic analysis of offshore structures is generally carried out as part of the design process, although the wave loading under conditions of dynamic response has still not been adequately investigated.

1.4 OUTLINE OF THE TEXT

The content of this text largely follows the sequence of the outline just given. More specifically, Chapter 2 describes the fundamental concepts and governing equations of motion. Because of its central importance in fluid mechanics and relevance to the present text, the flow about a cylinder and the resulting forces are given special consideration.

Chapter 3 presents a detailed account of both steady and unsteady separated flows about cylinders, with particular reference to various two-dimensional, uniform flows past a circular cylinder. These include a steady flow, an impulsively-started flow, and a flow with constant acceleration. These lead to the important case of a harmonically oscillating flow past a circular section, which is closely related to the wave motion past a section of a cylindrical structural element.

Chapter 4 outlines the development of available wave theories, which express the details of the flow in terms of the overall characteristics of a wave train. The theories described include linear wave theory, Stokes finite amplitude wave theories, nonlinear shallow wave theories, and the more recent theories based on numerical computations. The relationship between the wave motion past a body and the two-dimensional sinusoidal flow described in Chapter 3 forms a vital step in most wave force calculations and is given special consideration.

The remainder of Chapter 4 summarizes briefly other aspects of wave behavior, including the mass-transport velocity, wave refraction, diffraction, attenuation, breaking and so on.

Chapter 5 concerns the calculation of wave loads on slender bodies, with particular reference to circular cylindrical structural members. It describes the principal factors of analysis and design and provides a detailed description of the selection of the appropriate force-transfer coefficients. Particular topics that are treated include the effects of cylinder roughness, cylinder orientation, combined wave-current flows and wall proximity. The chapter concludes with a detailed discussion of wave slamming and the analysis of marine risers.

Chapter 6 deals with very large offshore structures which are now an established aspect of offshore technology. It describes the available theoretical methods, summarizes various results, and presents some available comparisons with experimental measurements. Extensions to deal with the cases of floating bodies, steep (nonlinear) waves and currents superposed with the wave motion are summarized.

Chapter 7 treats the case of random waves and random forces. These topics are contained in a separate chapter since the techniques employed are fairly distinct from the remaining material covered in the text. Chapter 7 begins with a brief summary of necessary statistical concepts, and then describes various theoretical results concerning the probabilistic and spectral properties of waves and wave forces.

Chapter 8 describes the methods of analysis of the dynamic response of offshore structures and emphasizes the need for specifying hydrodynamic forces adequately as forcing functions in the equations of structural motion that are employed. The chapter continues with a detailed discussion of the hydroelastic oscillations of flexible cylindrical members, including both in-line and transverse oscillations occurring in both steady and oscillatory incident flows.

The final chapter, Chapter 9, describes the underlying principles of model laws, modelling of offshore structures, and instrumentation techniques used in wave and wave-force measurements, both in the laboratory and in the field. The chapter concludes with a brief discussion of the experience with actual structures.

1.5 REFERENCES

Airy, G. B. 1845. Tides and Waves. *Encyc. Metrop.*, Art. 192, pp. 241–396.

Bea, R. G. and Lai, N. W. 1978. Hydrodynamic Loadings on Offshore Structures. *Offshore Tech. Conf.*, Houston, Paper No. OTC 3064, pp. 155–168.

Borgman, L. E. 1972. Statistical Models for Ocean Waves and Wave Forces. *Advances in Hydroscience*, Vol. 8, pp. 139–181.

Boussinesq, J. 1871. Théorie de l'Intumescence Liquide, Appelée Onde Solitaire ou de

Translation se Propageant Dans un Canal Rectangulaire. *Comptes Rendus Acad. Sci.*, Paris, Vol. 72, pp. 755–759.

Bruun, P. 1976. North Sea Offshore Structures. *Ocean Eng.*, Vol. 3, No. 5, pp. 361–373.

Haring R. E. and Spencer, L. P. 1979. The Ocean Test Structure Data Base. *Civil Engineering in the Oceans IV, ASCE*, Vol. II, pp. 669–683.

Keller, J. B. 1948. The Solitary Wave and Periodic Waves in Shallow Water. *Commun. Appl. Math.*, Vol. 1, pp. 323–339.

Keulegan, G. H. and Carpenter, L. H. 1958. Forces on Cylinders and Plates in an Oscillating Fluid. *J. Res. Nat. Bureau of Standards*, Vol. 60, No. 5, pp. 423–440.

Korteweg, D. J. and De Vries, G. 1895. On the Change of Form of Long Waves Advancing in a Rectangular Canal, And on a New Type of Long Stationary Waves. *Phil. Mag.*, 5th Series, Vol. 39, pp. 422–443.

Laitone, E. V. 1961. The Second Approximation to Cnoidal and Solitary Waves. *JFM*, Vol. 9, pp. 430–444.

Lamb, H. 1945. *Hydrodynamics.* 6th ed., Dover, New York; also Camb. Univ. Press, 1932.

Longuet-Higgins, M. S. 1952. On the Statistical Distribution of the Heights of Sea Waves. *J. Mar. Res.*, Vol. 11, pp. 245–266.

MacCamy, R. C. and Fuchs, R. A. 1954. Wave Forces on Piles: A Diffraction Theory. U.S. Army Corps of Engineers, *Beach Erosion Board*, Tech. Memo. No. 69.

Morison, J. R., O'Brien, M. P., Johnson, J. W. and Schaaf, S. A. 1950. The Force Exerted by Surface Waves on Piles. *Petrol. Trans., AIME,* Vol. 189, pp. 149–154.

Rice, S. O. 1944–1945. Mathematical Analysis of Random Noise. *Bell System Tech. J.*, Vol. 23, pp. 282–332, Vol. 24, pp. 46–156; also in *Noise and Stochastic Processes*, ed. N. Wax, Dover, New York, 1954, pp. 133–294.

Russell, J. S. 1844. Report on Waves. *14th Meeting Brit. Assoc. Adv. Sci.*, pp. 311–390.

Sarpkaya, T. 1976. Vortex Shedding and Resistance in Harmonic Flow About Smooth and Rough Cylinders at High Reynolds Numbers. *Naval Postgraduate Sch.*, Monterey, Rept. No. NPS-59SL76021.

Stokes, G. G. 1847. On the Theory of Oscillatory Waves. *Trans. Camb. Phil. Soc.*, Vol. 8, pp. 441–455. Also *Math. Phys. Papers*, Vol. 1, Camb. Univ. Press, 1880.

Thompson, W. (Lord Kelvin) and Tait, P. G. 1879. Treatise on Natural Philosophy, Vol. 1, 1879, Vol. 2, 1883. *Camb. Univ. Press.*

Watt, B. J. 1978. Basic Structural Systems—A Review of Their Design and Analysis Requirements. In *Numerical Methods in Offshore Engineering*, eds. O. C. Zienkiewicz, R. W. Lewis and K. G. Stagg, Wiley, Chichester, England, pp. 1–42.

2
Review of the Fundamental Equations and Concepts

2.1 EQUATIONS OF MOTION

The equations of motion for an incompressible Newtonian fluid may be written as

$$\frac{Du}{Dt} = \frac{\partial u}{\partial t} + u\frac{\partial u}{\partial x} + v\frac{\partial u}{\partial y} + w\frac{\partial u}{\partial z} = X - \frac{1}{\rho}\frac{\partial p}{\partial x} + \nu\left(\frac{\partial^2 u}{\partial x^2} + \frac{\partial^2 u}{\partial y^2} + \frac{\partial^2 u}{\partial z^2}\right) \tag{2.1a}$$

$$\frac{Dv}{Dt} = \frac{\partial v}{\partial t} + u\frac{\partial v}{\partial x} + v\frac{\partial v}{\partial y} + w\frac{\partial v}{\partial z} = Y - \frac{1}{\rho}\frac{\partial p}{\partial y} + \nu\left(\frac{\partial^2 v}{\partial x^2} + \frac{\partial^2 v}{\partial y^2} + \frac{\partial^2 v}{\partial z^2}\right) \tag{2.1b}$$

$$\frac{Dw}{Dt} = \frac{\partial w}{\partial t} + u\frac{\partial w}{\partial x} + v\frac{\partial w}{\partial y} + w\frac{\partial w}{\partial z} = Z - \frac{1}{\rho}\frac{\partial p}{\partial z} + \nu\left(\frac{\partial^2 w}{\partial x^2} + \frac{\partial^2 w}{\partial y^2} + \frac{\partial^2 w}{\partial z^2}\right) \tag{2.1c}$$

in which u, v, w represent the velocity components in the x, y, z directions respectively; X, Y, Z, the components of the body force per unit mass in the corresponding directions; p, pressure; and ν, the kinematic viscosity of the fluid.

The terms like Du/Dt denote the substantive acceleration. They are also known as the Eulerian derivative, material derivative, or co-moving derivative of velocity. The substantive acceleration consists of a local acceleration (in unsteady flow) due to the change of velocity at a given point with time and a convective acceleration due to translation. The operator D/Dt may be applied to density, temperature, etc., to determine their respective Eulerian derivatives.

The derivation of the foregoing equations, known as the Navier-Stokes equations for an incompressible Newtonian fluid with constant viscosity, may be found in many basic reference texts (see e.g., Schlichting 1968) and will not be repeated here.

When gravity is the only body force exerted, a body-force potential may be defined such that $\Omega = -gh$ and

$$X = \frac{\partial \Omega}{\partial x}, \quad Y = \frac{\partial \Omega}{\partial y}, \quad Z = \frac{\partial \Omega}{\partial z}$$

where h is height above a horizontal datum. Then Eqs. (2.1) reduce to

$$\frac{Du}{Dt} = -\frac{1}{\rho}\frac{\partial(p + \rho gh)}{\partial x} + \nu\nabla^2 u \tag{2.2a}$$

$$\frac{Dv}{Dt} = -\frac{1}{\rho}\frac{\partial(p + \rho gh)}{\partial y} + \nu\nabla^2 v \tag{2.2b}$$

$$\frac{Dw}{Dt} = -\frac{1}{\rho}\frac{\partial(p + \rho gh)}{\partial z} + \nu\nabla^2 w \tag{2.2c}$$

or in more convenient vector notation, we have

$$\frac{\partial \mathbf{q}}{\partial t} + (\mathbf{q}\ \text{grad})\ \mathbf{q} = -\frac{1}{\rho}\nabla(p + \rho gh) + \nu\nabla^2 \mathbf{q} \tag{2.3}$$

where $\mathbf{q}$ is the velocity vector and may be written as $\mathbf{q} = iu + jv + kw$.

The Navier-Stokes equations evolved over a period of 18 years starting in 1827 with Navier and culminating with Stokes in 1845. It took another 6 years to firmly set the hypothesis that there is no slip on a boundary or more precisely, the fluid immediately adjacent to the boundary acquires the velocity of the boundary.*

If L is a characteristic length scale over which the velocity varies in magnitude by U and Eqs. (2.1) are expressed in dimensionless form by resort to L and U, then it is seen that the ratio UL/ν, which is a Reynolds number, represents the ratio of the inertial to viscous forces. In a wide class of flows the

---

*"I shall assume, therefore, as the conditions to be satisfied at the boundaries of the fluid, that the velocity of a fluid particle shall be the same, both in magnitude and direction, as that of the solid particle with which it is in contact," (Stokes 1851).

Reynolds number is very large and the viscous terms in the above equations are much smaller than the remaining inertial terms over most of the flow field. A notable exception is in a boundary layer where the velocity gradients are steep and viscous stresses are significant.

2.1.1 Equation of Continuity

The conservation of mass for an incompressible fluid requires that the volumetric dilatation be zero, i.e.,

$$\nabla \cdot q = \frac{\partial u}{\partial x} + \frac{\partial v}{\partial y} + \frac{\partial w}{\partial z} = 0 \tag{2.4}$$

Equation (2.4) is independent of the choice of coordinates, i.e., it is an invariant.

The solution of the four unknowns (u, v, w, and p) from Eqs. (2.3) and (2.4) becomes fully determined when the initial and boundary conditions are specified. The kinematic boundary condition for a nonporous wall is that the velocity of the solid boundary at an arbitrary point must be impressed on the fluid particle immediately adjacent to it. In other words, the normal and tangential components of the velocity relative to the boundary must be zero. The first condition requires that the co-moving derivative of the function $F(x, y, z, t) = 0$, defining the boundary, must be zero,* i.e.,

$$\frac{DF}{Dt} = \frac{\partial F}{\partial t} + u\frac{\partial F}{\partial x} + v\frac{\partial F}{\partial y} + w\frac{\partial F}{\partial z} = 0 \tag{2.5}$$

The second condition (no slip) is based on heuristic reasoning, mathematical simplicity, and experimental justification. The no-slip condition does not have to be satisfied for an ideal fluid. Partly because of this and partly because of the fact that viscosity modifies radically the flow of a real fluid, the inviscid flow assumption cannot serve even as a first approximation to the actual flow in the vicinity of the body.

The boundary conditions at a free surface will be discussed following the introduction of the velocity potential.

---

*The function F for a sphere of radius c moving along the x axis with a velocity U is given by

$$F = (x - Ut)^2 + y^2 + z^2 - c^2 = 0$$

Then Eq. (2.5) yields

$$(u - U)(x - Ut) + vy + wz = 0$$

as the boundary condition which must be satisfied at every point on the sphere.

2.2 ROTATIONAL AND IRROTATIONAL FLOWS

The rates of rotation of a fluid particle about the x, y, z axes are given by (see, e.g., Schlichting 1968)

$$\omega_x = \frac{1}{2}\left(\frac{\partial w}{\partial y} - \frac{\partial v}{\partial z}\right) \tag{2.6a}$$

$$\omega_y = \frac{1}{2}\left(\frac{\partial u}{\partial z} - \frac{\partial w}{\partial x}\right) \tag{2.6b}$$

$$\omega_z = \frac{1}{2}\left(\frac{\partial v}{\partial x} - \frac{\partial u}{\partial y}\right) \tag{2.6c}$$

They are components of the rotation vector $\omega = 1/2 \text{ curl } \mathbf{q}$. The flows for which curl $\mathbf{q} \neq 0$ are said to be rotational because each fluid particle undergoes a rotation as specified by Eqs. (2.6), in addition to translations and pure straining motions. The absence of rotation, i.e., $\omega_x = \omega_y = \omega_z = 0$, does not, however, require that the fluid be inviscid. In other words, in the regions of flow where curl $q = 0$, a real fluid exhibits an irrotational or inviscid-fluid like behavior since the shear stress vanishes.

Rotation is related to two other fundamental concepts, namely, circulation and vorticity. Circulation, Γ, is defined as the line integral of the velocity vector taken around a closed curve, enclosing a surface S within the region of fluid considered. Thus, we have

$$\Gamma = \oint \mathbf{q} \cdot d\mathbf{s} = \oint (u\,dx + v\,dy + w\,dz) \tag{2.7}$$

According to the Stokes theorem,

$$\Gamma = \oint \mathbf{q} \cdot d\mathbf{s} = \int_S \text{curl } \mathbf{q} \cdot d\mathbf{S} = 2\int_S \omega \cdot \mathbf{n}\,dS \tag{2.8}$$

and therefore Eq. (2.7) may be written as

$$\Gamma = \iint 2\omega_x\,dy\,dz + \iint 2\omega_y\,dz\,dx + \iint 2\omega_z\,dx\,dy \tag{2.9}$$

in which twice the components of rotation vector appear. They are said to be the components of the vorticity vector ζ such that $\zeta_x = 2\omega_x$, $\zeta_y = 2\omega_y$, and

$\zeta_z = 2\omega_z$. Thus, it follows from Eq. (2.8) that $\Delta\Gamma = \zeta_n \Delta S$ where ζ_n is the component of the vorticity vector normal to the surface element ΔS. In other words, the flux of vorticity through the surface is equal to the circulation along the curve enclosing the surface.

For reference purposes only, we will note that in a frictionless fluid a fluid element cannot acquire or lose rotation (there are no shear forces to induce such a motion); a vortex tube always consists of the same fluid particles, regardless of its motion; and the circulation remains constant with time. These are the fundamental theorems of vorticity and have been enunciated by Helmholtz and Kelvin. For a detailed discussion of these theorems the reader is referred to classic reference texts such as Batchelor (1967), Milne-Thomson (1960) or Lamb (1932).

In real fluids, vorticity may be generated, redistributed, diffused, and destroyed since frictional forces are not conservative. In other words, vorticity is ultimately dissipated by viscosity to which it owes its generation. For example, the vorticity found in a vortex about four diameters downstream from a circular cylinder is about 70 percent of the vorticity produced at the separation point (Bloor and Gerrard 1966).* The remainder is partly diffused and partly cancelled by the ingestion of fluid bearing oppositely-signed vorticity. One should also bear in mind that the experiments yield only the normal component of vorticity. Thus, the consequences of the stretching and twisting of vortex filaments as a consequence of three dimensionality in the wake and hence the redistribution of vorticity into directions other than the normal one are not accounted for. This relatively simple example points out not only some of the difficulties associated with the use of the inviscid-fluid assumption in attempting to model the motion of real fluids but also the fact that the most important region for a bluff body is not the far wake (commonly associated with the Karman vortex street) but rather that enclosing the body and the near wake where the vorticity is generated, diffused, and dissipated (Sarpkaya and Shoaff 1979).

2.3 VELOCITY POTENTIAL

Irrotational motion exists only when *all* components of the rotation vector are zero, i.e.,

$$\frac{\partial w}{\partial y} - \frac{\partial v}{\partial z} = 0, \quad \frac{\partial u}{\partial z} - \frac{\partial w}{\partial x} = 0, \quad \frac{\partial v}{\partial x} - \frac{\partial u}{\partial y} = 0 \tag{2.10}$$

---

*Schmidt and Tilmann (1972) find a 50% reduction in circulation as the vortices move from a 5-D to 12-D downstream position. They have used a novel experimental technique (1970).

It is then possible to devise a continuous, differentiable, scalar function $\phi = \phi(x, y, z, t)$ such that its gradients satisfy Eqs. (2.10) automatically. In fact, curl $\mathbf{q} = 0$ is the necessary and sufficient condition for the existence of such a function. The difference of potential $\delta\phi$ along a portion of a streamline of length δs is $\delta\phi = v_s \delta s$ or $v_s = \partial\phi/\partial s$. Thus, the velocity is given by the gradient of the potential known as the velocity potential. The velocity potential is analogous to the force potential in a gravitational field, steady electric potential in a homogeneous conductor, and to the steady temperature-distribution function in a homogeneous thermal conductor.*

In Cartesian and cylindrical polar coordinates, the velocity components are thus given by

$$u = \frac{\partial\phi}{\partial x}, \quad v = \frac{\partial\phi}{\partial y}, \quad w = \frac{\partial\phi}{\partial z}, \quad \text{i.e., } \mathbf{q} = \text{grad } \phi \tag{2.11}$$

and

$$v_r = \frac{\partial\phi}{\partial r}, \quad v_\theta = \frac{\partial\phi}{r\partial\theta}, \quad w = \frac{\partial\phi}{\partial z} \tag{2.12}$$

respectively. Evidently, Eqs. (2.11) and (2.12) satisfy Eqs. (2.10) automatically, i.e., the potential flow is irrotational. It is also true that a potential exists only for an irrotational flow.

The introduction of ϕ into the continuity equation (2.4) results in a second order linear differential equation, known as the Laplace equation

$$\nabla^2\phi = \frac{\partial^2\phi}{\partial x^2} + \frac{\partial^2\phi}{\partial y^2} + \frac{\partial^2\phi}{\partial z^2} = 0 \tag{2.13}$$

This equation was first introduced by Laplace in his book on Celestial Mechanics and is of fundamental importance in many branches of physics, mechanics, and mathematics. The solutions of Laplace's equation are known as harmonic functions. The real and imaginary parts of a complex function F(z) are such harmonic functions. Their linear combinations are also solutions of the Laplace equation.

The direct determination of a harmonic function which satisfies the given boundary conditions is often a difficult problem. Intuition, heuristic reasoning, experience, and numerous methods (e.g., finite-difference, finite-element, and

*Current flow = $I = k_1 \partial E/\partial s$, heat flow = $h = k_2 \partial\theta/\partial s$ where k_1 and k_2 are the electrical and thermal conductivity, respectively; E, the steady electrical potential at a point; and θ, the steady temperature-distribution function.

relaxation methods) must be called upon not only to obtain a solution but also to ascertain that the solution based on the inviscid-flow assumption is a reasonable approximation to the actual behavior of the fluid. No general guidelines can be given to determine when and when not an idealized solution is a reasonable approximation. In general, flows or regions of flow which are rapidly accelerating or dominated by inertial forces may be treated by potential flow methods. Flow about a sphere at Reynolds numbers in the order of unity cannot be treated by inviscid flow methods since it is dominated by viscous forces even though the flow is essentially unseparated. On the other hand, efflux of jets, jet impingement, jet deflection, flow with relatively small amplitudes of oscillation about a large cylinder or streamlined body may be treated with potential flow methods with due regard to diffraction and conditions of radiation.

Often the inviscid-fluid assumption for the unseparated flow of an actual fluid may be used as a first approximation to the outer flow (e.g., flow about a foil at moderate angles of attack). However, it must be emphasized that the potential flow solution (in whole or in part, e.g., about an entire circular cylinder or only in the forebody of the cylinder) does not constitute even a first order approximation. The state of the art is such that analytical results, in particular those based on the zero vorticity assumption, must be compared with those obtained experimentally.

2.4 EULER'S EQUATIONS AND THEIR INTEGRATION

The assumption of zero shear enables one to reduce Eqs. (2.1) to

$$\frac{Du}{Dt} = X - \frac{\partial p}{\rho \partial x}, \quad \frac{Dv}{Dt} = Y - \frac{\partial p}{\rho \partial y}, \quad \frac{Dw}{Dt} = Z - \frac{\partial p}{\rho \partial z} \qquad (2.14a, b, c)$$

These are the celebrated Euler equations and have been obtained by Leonhard Euler about 100 years before the evolution of the Navier-Stokes equations and about 60 years after Newton discovered a major part of mechanics. Their beauty and originality lie in Euler's recognition that Newton's laws apply to every part of every system, whether discrete or continuous. The inviscid fluid assumption also implies that the pressure of the fluid is normal to the surface on which it acts.

The use of the conditions of irrotationality [Eqs. (2.6)] and the force potential enable one to reduce the three Euler equations into one equation

$$\frac{1}{2} q^2 + \frac{\partial \phi}{\partial t} - \Omega + \frac{p}{\rho} = F(t) \qquad (2.15)$$

where $q^2 = u^2 + v^2 + w^2$ and $F(t)$ is an arbitrary function of time only. The important point is that, at any instant, the left-hand side of Eq. (2.15) has the same value at all points of the region of irrotational motion, and not merely about a streamline. Frequently, F(t) is absorbed into ϕ since this does not affect the physical quantities of interest. However, in some nonlinear problems this may cause difficulties [e.g., Whitham (1974)] and must remain isolated.

For steady flow $\partial\phi/\partial t = 0$, then

$$\frac{1}{2}q^2 + \frac{p}{\rho} + gh = \text{Constant} \tag{2.16}$$

This is of course the familiar Bernoulli equation and enables one to determine the pressure distribution once the potential function and hence the velocity distribution are obtained from the solution of Laplace's equation. It must be remembered that Eq. (2.16) is valid for steady flows with the provisions that either the flow is irrotational or the calculation is confined to one streamline or the velocity vector and the rotation vector coincide.

An essential feature of irrotational flow is that it represents the instantaneous response of the fluid, uninfluenced by its previous history, to the immediately prevailing boundary conditions, i.e., the properly posed boundary conditions determine it uniquely. For the case of flow generated by a solid body moving through homogeneous fluid, the appropriate boundary condition is that the normal components of velocity of the solid boundary must be impressed upon the fluid adjacent to it. In a general form this is expressed as

$$\frac{\partial\phi}{\partial n} = V_n \quad \text{on the surface} \tag{2.17}$$

where n is the direction normal to the surface and V_n is the velocity of the surface normal to itself. If the boundary is rigid and fixed, such as seabed, then V_n is zero and we have simply

$$\frac{\partial\phi}{\partial n} = 0 \tag{2.18}$$

In the case of a free surface, as on a water wave, a kinematic and a dynamic boundary condition are needed. The kinematic condition states that any particle which lies at the free surface at any instant will never leave it. This leads to (see e.g., Lamb 1932)

$$\frac{\partial\phi}{\partial z} = \frac{\partial\eta}{\partial t} + u\frac{\partial\eta}{\partial x} + v\frac{\partial\eta}{\partial y} \quad \text{at } z = \eta \tag{2.19}$$

where $z = \eta(x, y, t)$ represents the free surface, (for additional discussion of the surface conditions see Chapter IV and also Lighthill 1979).

The dynamic free surface condition requires that the pressure difference across the interface results in a force normal to the boundary which is due wholly to surface tension. This condition takes the form

$$p = p_a + \sigma\left(\frac{1}{R_1} + \frac{1}{R_2}\right) \tag{2.20}$$

where σ is the surface tension, R_1 and R_2 are the radii of curvature of the free surface in any two orthogonal directions and $p - p_a$ is the pressure difference across the interface.

When the free surface is uncontaminated so that σ is taken as zero, when the flow is irrotational such that Eq. (2.15) describes the pressure p within the fluid, and when the (atmospheric) pressure just outside the liquid is constant, the free surface condition reduces to

$$\frac{\partial \phi}{\partial t} + \frac{1}{2} q^2 + g\eta = F(t) \quad \text{at } z = \eta \tag{2.21}$$

2.5 KINETIC ENERGY

The kinetic energy of a fluid region is given by

$$T = \frac{1}{2}\rho \iiint (u^2 + v^2 + w^2)\, dx\, dy\, dz \tag{2.22}$$

which, for an ideal-fluid flow with a single-valued potential, may be written as

$$T = \frac{1}{2}\rho \int \phi \frac{\partial \phi}{\partial n}\, dS \tag{2.23a}$$

where n is the outward normal; and dS, the elemental area. The integral is evaluated over the boundary. For a two-dimensional inviscid-fluid flow Eq. (2.23a) reduces to [see e.g., Milne-Thomson (1960)]

$$T = i\frac{\rho}{4} \oint_C F(z)\, \overline{dF(z)} \tag{2.23b}$$

where F(z) is the complex velocity potential and $z = x + iy$. Equations (2.23a) and (2.23b) are often used to determine the added mass of a body or the force which must be added to the force necessary to accelerate the body in vacuo. This subject will be discussed later.

2.6 STREAM FUNCTION

Lagrange's stream function (first introduced by d'Alembert) is a scalar quantity which describes not only the geometry of a *two-dimensional flow* but also the components of the velocity vector at any point and the flow rate between any two streamlines.

For the type of flow and fluid under consideration, the flow rate between the two streamlines is independent of the path of integration. Thus, for a flow from left to right (see Fig. 2.1), one has

$$u = \frac{\partial \psi}{\partial y} \quad \text{and} \quad v = -\frac{\partial \psi}{\partial x} \tag{2.24}$$

Thus, the partial derivatives of ψ with respect to any direction give the velocity components in the direction 90 degree clockwise to that direction. The definition of stream function does not require that the motion be irrotational. In other words, ψ exists irrespective of whether the flow is rotational or irrotational, as long as it is continuous.

For an irrotational two-dimensional flow, the condition of zero rotation yields

$$\omega_z = \frac{\partial z}{\partial x} - \frac{\partial u}{\partial y} = \frac{\partial^2 \psi}{\partial x^2} + \frac{\partial^2 \psi}{\partial y^2} = \nabla^2 \psi = 0 \tag{2.25}$$

Lagrange's stream function for an irrotational two-dimensional flow is orthogonal to the potential function and as such may be regarded as the imaginary part of the complex function $F(z) = \phi + i\psi$. Also, ϕ and ψ in $F(z)$ may be interchanged since both satisfy Laplace's equation. Furthermore, each has to satisfy the condition leading to the existence of the other in order to represent a continuous, two-dimensional, irrotational flow.

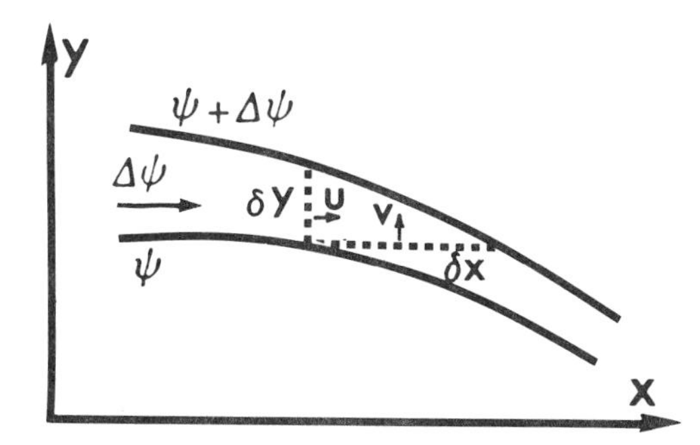

Fig. 2.1. Relation between ψ and the velocity components u and v.

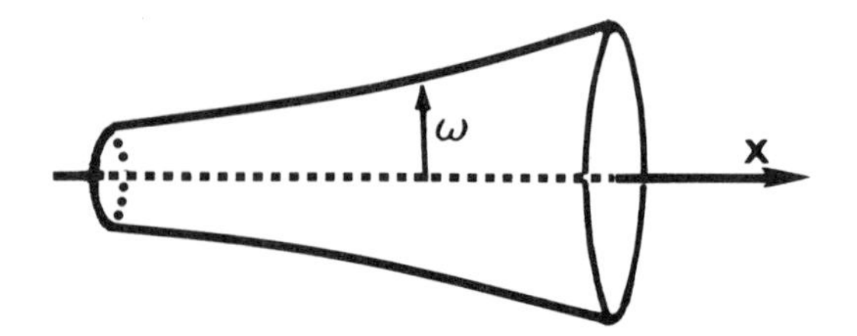

Fig. 2.2. Axisymmetric stream tube.

In polar coordinates the velocity components become

$$v_r = \frac{\partial \psi}{r\partial \theta} = \frac{\partial \phi}{\partial r} \quad \text{and} \quad v_\theta = -\frac{\partial \psi}{\partial r} = \frac{\partial \phi}{r\partial \theta} \tag{2.26}$$

Stokes' stream function which is applicable only to *axisymmetric* flows is based on the same concept, i.e., definition of a scalar function which yields the velocity components at any point, gives a measure of the flow rate, and thus automatically satisfies the equation of continuity. In cylindrical coordinates, the velocity components are given by (see Fig. 2.2)

$$u = \frac{1}{\overline{\omega}} \frac{\partial \psi}{\partial \overline{\omega}} = \frac{\partial \phi}{\partial x}$$

$$v_{\overline{\omega}} = -\frac{1}{\overline{\omega}} \frac{\partial \psi}{\partial x} = \frac{\partial \phi}{\partial \overline{\omega}} \tag{2.27}$$

which satisfy the equation of continuity given by

$$\frac{\partial(\overline{\omega} v_{\overline{\omega}})}{\partial \omega} + \frac{\partial(\overline{\omega} u)}{\partial x} = 0 \tag{2.28}$$

If the motion is assumed to be irrotational, then

$$\frac{\partial v_{\overline{\omega}}}{\partial x} - \frac{\partial u}{\partial \overline{\omega}} = \frac{\partial^2 \psi}{\partial x^2} + \frac{\partial^2 \psi}{\partial \overline{\omega}^2} - \frac{1}{\overline{\omega}} \frac{\partial \psi}{\partial \overline{\omega}} = 0 \tag{2.29}$$

In terms of ϕ, Eq. (2.29) is given by

$$\frac{\partial^2 \phi}{\partial x^2} + \frac{\partial^2 \phi}{\partial \overline{\omega}^2} + \frac{1}{\overline{\omega}} \frac{\partial \phi}{\partial \overline{\omega}} = 0 \tag{2.30}$$

The sign difference in the last terms of Eqs. (2.29) and (2.30) should be noted.

It is possible to represent the velocity as a function of dual stream functions of three-dimensional-flow fields which include, as special cases, both Lagrangian and Stokesian stream functions (see, e.g., Robertson 1965).

2.7 BASIC FLOW PATTERNS

Even the drastic assumption of an inviscid flow does not lead to a direct solution of the equation of Laplace since the boundary conditions are often quite difficult to satisfy. Most of the practically useful solutions are indirect and require the combination of simpler known solutions. When the superposition of basic flow patterns is not sufficient to obtain a solution one must resort to other exact or approximate methods such as conformal transformations, Fourier series, eigen-functions, spherical or ellipsoidal harmonics, methods of relaxation, finite difference, finite elements, etc. The application of any one of these methods requires considerable work and ingenuity depending on the complexity of the problem and the accuracy desired. In any case, one must at the onset ascertain that the time and effort required to attain a solution is commensurate with the need for a solution and with the idealizations imposed on the real flow. For example, the use of the distributed sources and source panels to analyze the diffraction effects on a large circular cylinder is quite reasonable when separation is not likely to occur. On the other hand, the application of the same method to large bodies with sharp corners may be questionable since separation occurs at all relative displacements regardless of the magnitude of displacement.

Tables 2.1 and 2.2 give the characteristics of basic two- and three-dimensional flow patterns. Plots of corresponding potential and stream functions may be found in many reference texts (see e.g., Milne-Thomson 1960, Robertson 1965).

2.8 FORCE ON A CYLINDER

Because of its central importance in fluid mechanics, the flow about a cylinder and the resulting forces will be described in some detail.

The proper combination of uniform flow with a doublet represents the inviscid flow about a circular cylinder at rest. We have (see Table 2.1)

$$F(z) = U\left(z + \frac{c^2}{z}\right) \quad \text{and} \quad \psi = -\mu \frac{\sin\theta}{r} + Ur\sin\theta, \quad r \geqslant c \tag{2.31}$$

where $\mu = Uc^2$ and c is the radius. The potential function becomes

$$\phi = U(r + c^2/r)\cos\theta \tag{2.32}$$

Table 2.1 Potential and Stream Functions for Two-Dimensional Flows.

Uniform flow	$\phi = Ux,\ \psi = Uy,\ F(z) = Uz$
Point source	$\phi = (Q/2\pi)\,\mathrm{Ln}r,\ \psi = (Q/2\pi)\theta,\ F = (Q/2\pi)\,\mathrm{Ln}z$
Velocities	$u = \partial\phi/\partial x = \partial\psi/\partial y,\ v = \partial\phi/\partial y = -\partial\psi/\partial x,$ $\dfrac{dF(z)}{dz} = u - iv,\ v_r = \dfrac{\partial\psi}{r\partial\theta} = \dfrac{\partial\phi}{\partial r},\ v_\theta = -\dfrac{\partial\psi}{\partial r} = \dfrac{\partial\theta}{r\partial\theta}$
Doublet	$\phi = \mu x/r^2,\ \psi = \mu y/r^2,\ F = \mu/z$
Vortex	$\phi = (\Gamma/2\pi)\theta,\ \psi = (\Gamma/2\pi)\,\mathrm{Ln}r,\ F = -(i\Gamma/2\pi)\,\mathrm{Ln}z$
Uniform flow at angle α	$\phi = U(x\cos\alpha + y\sin\alpha)$ $\psi = U(y\cos\alpha - x\sin\alpha)$ $F = Uze^{-i\alpha}$
Flow about a circular cylinder	$\phi = Ur\cos\theta + (Uc^2/r)\cos\theta$ $\psi = Ur\sin\theta - (Uc^2/r)\sin\theta$ $F = Uz + Uc^2/z$
Flow about a cylinder with circulation	$F = U(z + c^2/z) - (i\Gamma/2\pi)\,\mathrm{Ln}z$
Line source of uniform strength over length L	$\phi = (Q/4\pi L)\displaystyle\int_0^L \mathrm{Ln}[(x-\xi)^2 + y^2]\,d\xi$ $\psi = (Q/4\pi L)\displaystyle\int_0^L [\tan^{-1}\{y/(x-\xi)\}]\,d\xi$

Table 2.2 Potential and Stream Functions for Axisymmetric Flows.

Uniform flow	$\phi = Ux = Ur\cos\theta \quad \psi_s = 0.5Ur^2\sin^2\theta = 0.5U\omega^2$ $u = \frac{\partial\phi}{\partial x}, \quad v_\omega = \frac{\partial\phi}{\partial\omega}, \quad \Delta Q = 2\pi\Delta\psi_s$ $u = \frac{1}{\omega}\frac{\partial\Psi_s}{\partial\omega}, \quad v_\omega = -\frac{1}{\omega}\frac{\partial\psi_s}{\partial x}$
Source (3-D)	$\phi = -Q/(4\pi r), \quad \psi_s = -(Q/4\pi)\cos\theta$
Sink (3-D)	$\phi = +Q/(4\pi r), \quad \psi_s = +(Q/4\pi)\cos\theta$
Doublet (3-D)	$\phi = \mu\cos\theta/r^2, \ \psi_s = -\mu\sin^2\theta/r$
Flow about a sphere of radius c	$\phi = Ur[1 + c^3/(2r^3)]\cos\theta$ $\phi = Ur[1 + G(r)]\cos\theta$ $\psi = 0.5Ur^2(1 - c^3/r^3)$
Translation of a sphere in a fluid	$\phi = [Uc^3/(2r^2)]\cos\theta \quad \psi_s = -[Uc^3/(2r)]\sin^2\theta$ (direction of motion is in the x-direction)
Axisymmetric line source of uniform distribution	$\phi = [-Q/(4\pi L)]\,Ln[(r_2 - x + L)/(r_1 - x)]$ $\psi = [-Q/(4\pi L)](r_1 - r_2)$ Q = total strength of the line source

The velocity at the surface of the circular cylinder, necessarily tangent to the cylinder, is

$$v_\theta = 2U\sin\theta \tag{2.33}$$

Then the pressure on the cylinder reduces to

$$p = 0.5\,\rho U^2(1 - 4\sin^2\theta) \tag{2.34}$$

The symmetry of the pressure distribution with respect to x- and y-axes and hence the absence of both the drag and lift forces are well-known since d'Alembert.

If the ambient flow is a function of time, i.e., $U = U(t)$, then the pressure is given by Bernoulli's equation as

$$p = -0.5\,\rho q^2 - \rho\frac{\partial\phi}{\partial t} \tag{2.35}$$

Then the force exerted on the cylinder by the fluid in the x-direction becomes

$$F = -\int_C p\,ds\cos\theta = \int_C 0.5\rho q^2\cos\theta\,ds + \int_C \rho\frac{\partial\phi}{\partial t}\cos\theta\,ds \tag{2.36}$$

Note that θ is measured in the counterclockwise direction. The second integral is clearly zero because of symmetry, as seen above. Thus, we have

$$F = \rho \int_C \frac{\partial \phi}{\partial t} \cos \theta \, ds \tag{2.37}$$

Since

$$\left.\frac{\partial \phi}{\partial t}\right|_{r=c} = \frac{dU}{dt} (r + c^2/r)\big|_{r=c} \cos \theta = 2c \frac{dU}{dt} \cos \theta \tag{2.38}$$

one has

$$F = 2c^2 \rho \frac{dU}{dt} \int_0^{2\pi} \cos^2 \theta \, d\theta = 2\pi \rho c^2 \frac{dU}{dt} \tag{2.39}$$

The fact that one half of this force is due to the pressure gradient to accelerate the flow and one half due to the singularity (doublet) producing the cylinder may be made clearer by considering a more general case.

The potential function for a two-dimensional cylinder of any form may be written as (no additional singularities outside the cylinder)

$$\phi = Ur \cos \theta + Ur\, G(r) \cos \theta \tag{2.40}$$

in which the last term represents the disturbance made to the potential field of the stream by the presence of the cylinder. $G(r)$ is of order $0(r^{-n})$, $(n \geqslant 2)$, and must necessarily tend to zero at infinity.

With reference to Fig. 2.3, one has

$$F = -\int_C p \, ds \sin \alpha \tag{2.41}$$

or

$$F = 0.5\rho \int_C q^2 \sin \alpha \, ds + \frac{dU}{dt} \int_C r_c \cos \theta \sin \alpha \, ds + \rho \frac{dU}{dt} \int_C r_c G(r_c) \cos \theta \sin \alpha \, ds \tag{2.42}$$

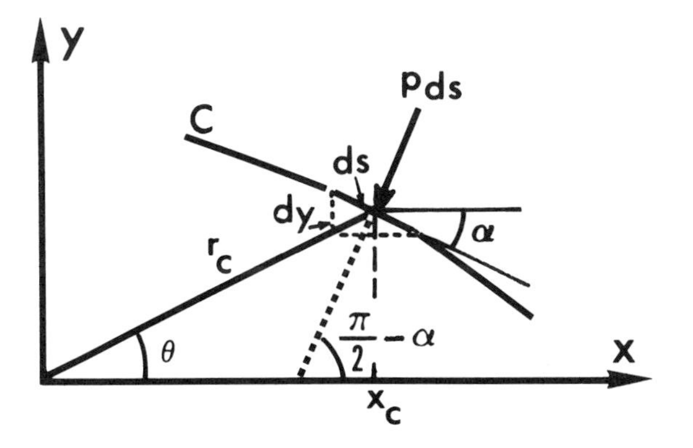

Fig. 2.3. Force on a Cylinder C.

In the absence of circulation about the cylinder, the first integral is clearly zero due to the Kutta-Joukowsky theorem. The second integral reduces to

$$F_1 = \rho \frac{dU}{dt} \int_C x_c \, dy = \rho S_0 \frac{dU}{dt} \tag{2.43}$$

where S_0 represents the cross-sectional area or the volume of the cylinder per unit length. The third integral may be evaluated only if the equation of the curve representing the cylinder, i.e., $r_c = G(\theta)$ is known. In any case, it may be written as

$$F_2 = \rho S_0 \frac{dU}{dt} \int_C \frac{G(r_c)}{S_0} r_c \cos\theta \sin\alpha \, ds \tag{2.44}$$

or

$$F_2 = \rho S_0 \frac{dU}{dt} \int_C G(r_c) \frac{dS}{S_0}, \quad dS = x_c \, dy = 0.5 r_c^2 \, d\theta \tag{2.45}$$

Introducing an added mass M_a, and an added-mass coefficient defined by $C_a = M_a/\rho S_0$, Eq. (2.45) reduces to

$$C_a = \int_C G(r_c) \, dS/S_0 \tag{2.46}$$

For the circular cylinder $G(r_c) = 1$ and thus $C_a = 1$.

In general, for unseparated unsteady flow about a cylinder, one has

$$F = F_1 + F_2 = \rho S_0 \underline{(1 + C_a)}\, dU/dt = \rho S_0 \underline{C_m}\, dU/dt \tag{2.47}$$

The purpose of the foregoing was to illustrate the relationship between C_a and C_m and not to offer a method of evaluation for the added mass. As will be seen later, there are easier and more general methods to evaluate the 21 components of the added mass coefficients.

Let us now consider the unsteady motion of a cylinder normal to its axis in a fluid otherwise at rest. The potential function may be represented by

$$\phi = Ur\, G(r) \cos \theta \tag{2.48}$$

where G(r) is of order r^{-n}, $(n = 2, 3, \text{etc.})$. Then the kinetic energy of the fluid becomes (see Eq. 2.23 and Fig. 2.3),

$$T = 0.5\rho \int \phi \frac{\partial \phi}{\partial n}\, ds = 0.5\rho \int U\phi \sin \alpha\, ds \tag{2.49}$$

Inserting Eq. (2.48) in (2.49) and simplifying, one has

$$T = 0.5\, \rho S_0 U^2 \int_C G(r_c)\, dS/s_0 \tag{2.50}$$

where dS is an elemental area of the cross section and the integral is evaluated over the cylinder surface. The work done per unit time on the fluid must equal the time rate of increase of the kinetic energy of the fluid. We, therefore, have

$$FU = dT/dt = M_a U\, dU/dt = \rho S_0 U\, dU/dt \int_C G(r_c)\, dS/S_0 \tag{2.51}$$

or

$$C_a = M_a/\rho S_0 = \int_C G(r_c)\, dS/S_0 \tag{2.52}$$

which is identical with that given by Eq. (2.46).

In general, the body has a mass M_b of its own and experiences a resistance (in real fluids) primarily due to separation of the flow or the resulting pressure forces. This is called the *form drag.* Assuming that the instantaneous values of the drag and inertial forces can be added, the total force acting on a cylinder moving unidirectionally with a time-dependent velocity U(t) becomes

$$F = 0.5\, C_d(t)\, \rho A_p |U| U + [M_b + \rho \forall C_a(t)]\ dU/dt \tag{2.53}$$

For a body at rest in a unidirectional time-dependent flow, the force exerted on the body becomes

$$F = 0.5\, C_d \rho A_p |U| U + \rho(1 + C_a) \forall\ dU/dt \tag{2.54}$$

in which A_p is the projected area of the body on a plane normal to the flow; and $\forall$, the volume of the body. In general, C_a and C_d depend on time, Reynolds number, relative displacement of the fluid, and on the parameters characterizing the history of the motion. Consequently, each time-dependent motion must be considered as unique and the most appropriate means of analysis and evaluation of the force-transfer coefficients must be discovered through intuition, heuristic reasoning, and above all through experimental observations and measurements. As will be noted later, the separation of the flow gives rise to a time-dependent transverse force or lift force regardless of whether the ambient flow is steady or time-dependent. If the relative displacement of fluid in one direction is not large enough, separation does not necessarily lead to an alternating vortex shedding.

The discussion of the fluid forces acting on a cylinder will not be complete without the consideration of the relative motion of the body in a fluid stream whose velocity may vary with both time and distance in the direction of cylinder motion, i.e., $U = U(x, t)$. Let us assume that the body is subjected to a displacement x, velocity $\dot{x}$, and acceleration $\ddot{x}$ in the direction of the incident stream and the diffraction effects are negligible. Then the force acting on the cylinder becomes (see e.g., Newman 1977)

$$F = 0.5 \rho C_d A_p |U - \dot{x}| (U - \dot{x}) + \rho \forall (1 + C_a) \left[\frac{\partial U}{\partial t} + (U - \dot{x}) \frac{\partial U}{\partial x} \right] - \rho \forall C_a \ddot{x} \tag{2.55}$$

The first term on the right-hand side of Eq. (2.55) represents the form drag; the second term, the inertial force due to the local and convective accelerations of the fluid about the body; and the third term, the inertial force due to the motion of the body, as it would be in a fluid otherwise at rest. The negative sign in front of the last term is due to the fact that the inertial fluid force acting on the accelerating body is in the opposite direction to the drag and inertial forces resulting from the motion of the fluid about the body.

Equation (2.55) may also be written as*

$$F = 0.5\rho A_p C_d |U - \dot{x}|(U - \dot{x}) + \rho \forall C_m \left[\frac{\partial U}{\partial t} - \ddot{x}\frac{C_a}{C_m} + (U - \dot{x})\frac{\partial U}{\partial x} \right] \tag{2.56}$$

The simplicity of this equation should not obscure the following facts. Firstly, the drag and inertia coefficients are not time-invariant and depend on the Reynolds number, relative motion of the fluid, history of the motion, relative roughness, etc. The variation of the velocity at a given point with time and at a given time with space, the omnidirectionality of the ambient flow, motion of the body in directions other than those of the ambient flow, interference with the wakes of other neighboring bodies, wall- and free-surface proximity effects, just to mention a few of the practically important conditions, play significant roles in the determination of the drag and inertia coefficients. Secondly, the form drag is not always necessarily expressible in terms of the square of the relative velocity. In fact, it may be necessary to devise more suitable expressions for the form drag depending on the relative magnitudes of the time-dependent velocities $U(t)$ and $\dot{x}(t)$.

2.9 GENERALIZATION OF THE CONCEPT AND EVALUATION OF ADDED MASS

In order that Newton's second law ($F = ma$) accurately predict any one of the three quantities (F or m or a), the remaining two quantities must be known accurately. In vacuum, this is a relatively simple matter as far as the mass of the accelerating body is concerned. In a fluid medium of a given density, the determination of the effective mass of the body is not as easy and requires the consideration of the motion of the fluid as well as that of the body.

About 180 years ago Chevalier DuBuat, experimenting with spheres oscillating in water, and later Bessel in 1828, experimenting with spherical pendulums in air and water, found that it is necessary to attribute to the sphere a virtual mass greater than the ordinary mass of the sphere. Bessel represented the increase of inertia by a mass equal to k times the mass of the fluid displaced by the body. His k values were 0.6 and 0.5 for air and water, respectively.

The mathematical study of the phenomenon was commenced by Poisson who disregarded the viscosity and obtained $k = 0.5$ for the sphere, a result

---

*The generalization of Eq. (2.55) to nonlinear flows and the discussion of the use of DU/Dt versus $\partial U/\partial t$ in Eq. (2.54) are taken up in Section 5.3.1.

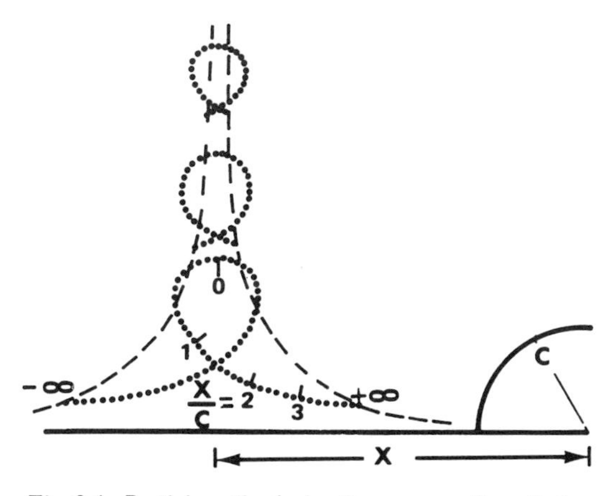

Fig. 2.4. Particle paths during the passage of a cylinder.

which was subsequently confirmed by Green, Plana, Stokes, and Lamb, all of whom used different methods (Dryden, Murnaghan, Bateman 1956).

Although considerable progress has been made on the theoretical and experimental techniques, and the added mass and the added-mass moment of inertia have been determined for a wide range of shapes of bodies, no new physical meaning have been attached to them. The basic definition of added mass remained as 'the quotient of the additional force required to produce the accelerations throughout the fluid divided by the acceleration of the body.'

Surprisingly enough, as early as 1888, Riecke recognized the essential fact that for a sphere moving with constant velocity throughout an infinite inviscid fluid, the individual fluid particles which are pushed aside by the sphere in its forward motion do not return to their former positions. The paths of the individual particles are not closed curves but of the shape similar to that shown in Fig. 2.4, in which c represents the radius of the cylinder. The broken lines show the initial and final positions of the particles before and after the passage of the cylinder. Hence, besides pushing the particles aside temporarily in passing, the cylinder also displaces the fluid particles permanently in the direction of its motion. However, the importance of this displacement and its relation to added mass was not recognized by Riecke. Sir Charles Darwin (1953) has shown that this permanently displaced mass of the fluid enclosed between the initial and final positions of fluid particles, is in fact the added mass itself. It can be shown in general that the motion of a body through an inviscid fluid media is always accompanied by a fluid-mass transport and that this mass is the added mass which unveils itself only if the body is accelerated.

The preceding explanation, although adding considerably to our understanding of the added mass, does not take into account the effect of the wake or the cavity-induced mass. Furthermore, for an accelerating or decelerating body (or fluid about a body), the induced mass varies with the instantaneous shape and volume of the wake or cavity as well as with their rates of change. The instantaneous magnitude of the added mass and added-mass moment of inertia in transient conditions depend on the time history of the motion. It is clear from the preceding discussion that the variation of the wake geometry with time causes marked changes not only in C_d but also in C_a.

Added mass may be expressed in various ways. The use of each analytical expression depends on the simplicity of its evaluation. Let us examine in detail the various forms for two- and three-dimensional bodies.

The total kinetic energy of a body of mass m moving in a fluid otherwise at rest may be written as

$$T = T_b + T_f = 0.5U^2 \left[m + \rho \iiint (q/U)^2 \, d\forall \right] \tag{2.57}$$

The quantity

$$M_a = \rho \iiint (q/U)^2 \, d\forall = 2T_f/U^2 \tag{2.58}$$

is the added mass and T_f represents the kinetic energy of the fluid. For the unseparated inviscid flow, q/U is independent of U, i.e., it depends only on the flow pattern. Hence, the above integral is independent of time even though U may vary with time. In other words, M_a is a constant associated with the body, the type of motion which specifies q(x, y, z, t), and the density of fluid. For the separated motion of a real fluid, q/U is not independent of time even if U is constant. Thus, M_a varies with time and depends on the basic parameters characterizing the flow, e.g., the history of the fluid or the body motion, etc. It is also evident that the force acting on the body varies with time not only because of the variation of the velocity of the ambient flow but also because of the variation of the fluid velocity with space and time in the wake of the body. Thus, in general dM_a/dt is not zero and one has

$$\frac{dT}{dt} = (m + M_a) U \frac{\partial U}{\partial t} + \frac{1}{2} U^2 \frac{\partial M_a}{\partial t} \tag{2.59}$$

Now let us consider an inviscid fluid about a cylinder and let $w_x(x)$ and $w_y(z)$ denote the complex velocity potentials for the unit velocities in the x- and y-directions respectively. The velocity-square dependent component of the force is zero in the absence of circulation. The time-dependent or inertial component of the force acting on the cylinder due to $w_x(z)$ is given by

Blasius' theorem (see e.g., Milne-Thomson 1960) as

$$Y - iX = \rho \frac{\partial U}{\partial t} \oint w_x(z)\, dz \tag{2.60}$$

in which X and Y are the components of the force in the x- and y-directions. Since the X component of the force is given, as seen before, by [see Eq. (2.47)]

$$X = (\rho S_0 + M_a) \frac{\partial U}{\partial t}$$

one has

$$\frac{1}{\partial U/\partial t}\left[Y - i(\rho S_0 + M_a) \frac{\partial U}{\partial t}\right] = \rho \oint w_x(z)\, dz \tag{2.61}$$

Defining

$$Y = A_{xy} \frac{\partial U}{\partial t} \quad \text{and} \quad M_a = A_{xx}$$

one has

$$A_{xy} - i(\rho S_0 + A_{xx}) = \rho \oint w_x(z)\, dz \tag{2.62}$$

which may be written as

$$C_{xy}^a - i\left(\frac{S_0}{S_r} + C_{xx}^a\right) = \frac{1}{S_r} \oint w_x(z)\, dz \tag{2.63}$$

with

$$C_{xy}^a = A_{xy}/\rho S_r \quad \text{and} \quad C_{xx}^a = A_{xx}/\rho S_r$$

in which S_0 is the actual cross-sectional area of the cylinder and S_r is a suitable reference area. For a circular cylinder one may write $S_0 = S_r = \pi c^2$. For a thin plate of width 2a moving broad-side on, $S_0 = 0$ and one ordinarily uses $S_r = \pi a^2$.

A_{xx} and A_{xy} are the components of the added-mass tensor and C_{xx}^a and C_{xy}^a are the added-mass coefficients. The first index (here x) represents the direction of fluid motion; and the second index (here y), the transverse direction.

The time-dependent force acting on the cylinder due to $w_y(z)$, the potential for the unit fluid velocity in the y-direction, is given by

$$(\rho S_0 + A_{yy}) - iA_{yx} = \rho \oint w_y(z)\, dz \tag{2.64}$$

or

$$\left(\frac{S_0}{S_r} + C^a_{yy}\right) - iC^a_{yx} = \frac{1}{S_r} \oint w_y(z)\, dz \tag{2.65}$$

Once again, C^a_{yy} and C^a_{yx} represent the added-mass coefficients for the directions of fluid motion indicated by the first index. Suffice it to note that $C^a_{xy} = C^a_{yx}$ or in general $C^a_{ij} = C^a_{ji}$ because of Green's reflection principle. Also, it is easy to show that the superposition principle in inviscid two-dimensional flows would have been violated had C_{ij} been different from C_{ji}. In practical terms, the foregoing simply states that the inertial fluid force acting on the body in the y-direction due to the fluid motion in the x-direction is equal to that acting in the x-direction due to the fluid motion in the y-direction, provided that the fluid has identical accelerations in both directions. Also, for the type of fluid motion under consideration, it does not matter how the fluid has reached that particular acceleration and what the magnitude of the velocity was at that time.

Before generalizing the foregoing, one should note that the fluid motion in the x- and y-directions resulted in three added-mass coefficients, namely C^a_{xx}, C^a_{yy}, and C^a_{xy}. In general, a body may undergo translation in and rotation about all three axes. Then it is more advantageous to use the kinetic energy expression given by

$$T_f = \frac{1}{2} \rho \int \phi \frac{\partial \phi}{\partial n}\, dS \tag{2.23 repeated}$$

Letting

$$\phi = U_i \phi_i = U_1 \phi_1 + U_2 \phi_2 + \cdots + U_6 \phi_6 \tag{2.66}$$

in which U_1, U_2, and U_3 represent the translational components of the velocity of the body; ϕ_1, ϕ_2, and ϕ_3, the corresponding potentials for *unit velocity*; U_4, U_5, and U_6, the angular velocities; and ϕ_4, ϕ_5, and ϕ_6, the corresponding potentials for unit angular velocity (see Fig. 2.5). In other words, each ϕ represents the velocity potential due to a body motion with unit velocity in the mode indicated by the index i.

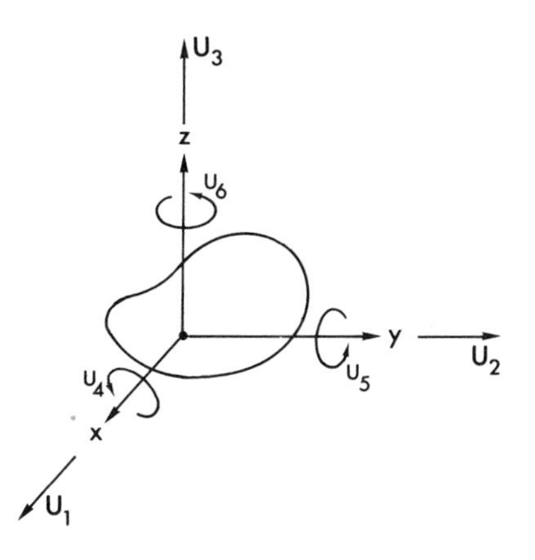

Fig. 2.5. Rotational and translational motions of a body.

Substituting Eq. (2.66) in Eq. (2.23), one has

$$\frac{2T_f}{U_i U_j} = A_{ij} = \rho \int \phi_i \frac{\partial \phi_j}{\partial n} dS \tag{2.67}$$

or

$$C_{ij}^a = \frac{A_{ij}}{\rho \forall_r} = \frac{1}{\forall_r} \int \phi_i \frac{\partial \phi_j}{\partial n} dS \tag{2.68}$$

where $\forall_r$ represents a suitable reference volume and dS, the elemental surface area.

Of the 36 added-mass coefficients, 9 are for translation, 9 for rotation, and 18 result from the interaction between translation and rotation. Because of the fact that $C_{ij}^a = C_{ji}^a$ for inviscid fluids, one has 6, 6, and 9 components, respectively, or 21 in all.

The coefficients C_{ij}^a are independent of time and the translational and rotational velocities and accelerations. They depend only on the shape of the body, the proximity of other bodies or free surface, and the choice of the coordinates. Only the sum of the six elements represented by C_{ii}^a is independent of the choice of coordinates.

The foregoing could have been presented with greater degree of mathematical sophistication (see e.g., Yih 1969). However, the type of flows considered herein and the motion of real fluids in general are such that the fundamental assumptions of inviscid fluid and no separation leading to Eq. (2.67) are rarely satisfied.

The special cases for which the added-mass coefficients obtained through the use of the velocity potential may be used are:

a. initial instants of the motion when the body is set in motion impulsively from rest;

b. when the body is subjected to relatively small amplitude oscillations (so as to preclude the effects of separation) at a relatively high frequency;

c. initial instants of the change of velocity of the body when the steady velocity of the body is changed impulsively to a new steady velocity.

The technique of vibrating the body in water is commonly used [Yu 1945, Stelson and Mavis 1955, Sarpkaya 1960, Chandrasekaran and Salni 1972, Fickel 1973, Ceisluk and Colonell 1974, Brooks 1976] to determine the added-mass coefficient and to compare it with that obtained from the potential theory.

Added-mass coefficients for various shapes of bodies may be found in [J. L. Taylor 1930, McLachlan 1932, Wendel 1950, Weinblum 1952, Bryson 1954, Ackermann and Arbhabhirama 1964, Kochin et al. 1964, McConnell and Young 1965, Patton 1965, Sedov 1965, Landweber 1967, Goldschmidt and Protos 1968, Myers et al. 1969, Waugh and Ellis 1969, Meyerhoff 1970, Pope and Leibowitz 1974, Moretti and Lowery 1975, Chen et al. 1976, Au-Yang 1977, Gibson 1977, and Newman 1977). The added mass for a few and relatively simple bodies are given in Table 2.3.*

For two-dimensional bodies the added-mass matrix with respect to another Cartesian coordinate system (r, s) is given by (Sedov 1965)

$$A_{rr} = A_{xx} \cos^2 \alpha + A_{yy} \sin^2 \alpha + A_{xy} \sin 2\alpha$$

$$A_{ss} = A_{xx} \sin^2 \alpha + A_{yy} \cos^2 \alpha - A_{xy} \sin 2\alpha$$

$$A_{rs} = \tfrac{1}{2} (A_{yy} - A_{xx}) \sin 2\alpha + A_{xy} \cos 2\alpha$$

$$A_{r\beta} = (A_{xx} \eta - A_{xy} \xi + A_{x\theta}) \cos \alpha + (A_{xy} \eta - A_{yy} \xi + A_{y\theta}) \sin \alpha$$

$$A_{s\beta} = -(A_{xx} \eta - A_{xy} \xi + A_{x\theta}) \sin \alpha + (A_{xy} \eta + A_{yy} \xi + A_{y\theta}) \cos \alpha$$

$$A_{\beta\beta} = A_{xx} \eta^2 + A_{yy} \xi^2 - 2A_{xy} \xi\eta + 2(A_{x\theta} \eta - A_{y\theta} \xi) + A_{\theta\theta} \qquad (2.69)$$

in which α represents the angle through which the r–s axes are rotated with respect to the x–y axes; ξ and η, the origin of the r–s axes with respect to x–y axes; θ, the rotation about an axis through the origin of the x–y coordinates; and β, the rotation about the origin of the r–s axes (see Fig. 2.6). Evidently, the added-mass expressions with respect to the r–s system are considerably simpli-

*See Appendix A *at the end of this chapter.*

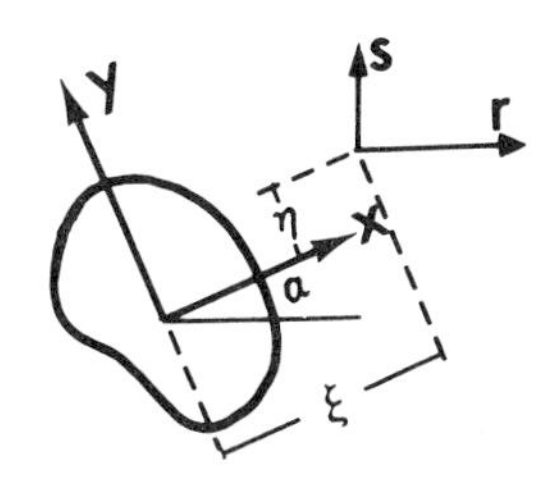

Fig. 2.6. Coordinates (x, y) and (r, s).

fied if the x–y axes are axes of symmetry so that

$$A_{xy} = A_{x\theta} = A_{y\theta} = 0 \quad \text{or} \quad A_{12} = A_{16} = A_{26} = 0$$

In general, the analytical or numerical evaluation of the added-mass coefficients for bodies with relatively complex geometries is somewhat difficult and time consuming. Under such circumstances it is advantageous to use either experimental methods or relatively simple and approximate numerical methods such as strip theory. This method consists of the summing of the added masses of individual two-dimensional strips, provided that (i) the flow over a narrow slice or strip of the body is essentially two-dimensional, (ii) the interaction between the adjacent strips is negligible, (iii) the end effects are relatively small, and/or (iv) the body is sufficiently slender. There are no simple rules to determine as to when and to what degree of approximation will these conditions be satisfied. Evidently, the more slender the body (the flow is assumed to be normal to the long axis of the body), the more suitable is the strip theory. For example, a submarine, a ship hull, or a long cigar-shaped body are more suitable for the strip theory than a cube or a sphere. Table 2.4 shows a comparison of the added-mass coefficients obtained through the use of the strip theory with those obtained experimentally and theoretically. It should be noted that the strip theory cannot take into account the three-dimensional flow effects even if the flow is assumed to be unseparated. Consequently, the predictions of the strip theory differ significantly from those obtained experimentally and theoretically as the body becomes increasingly stubby.

In unseparated flow, the added mass always decreases the natural frequency of the body from that which would be measured in a vacuum. The relationship between the two frequencies may be written as

$$f(\text{in a fluid})/f(\text{in vacuum}) = (1 + A/m)^{-1/2} \tag{2.70}$$

where A is the added mass and m, the mass of the body. For a pipe of mass m per unit length (in air) the above equation may be written as

$$f(\text{in water})/f(\text{in vacuum}) = [1 + (A + m_w)/(m + m_w)]^{-1/2} \tag{2.71}$$

Table 2.4 Added-Mass Coefficients for Finite Cylinders and Plates (Blevins 1979).

Cylinder (length = L, radius = c)			Plate (length = L, width = w)		
	Added Mass/($\pi\rho Lc^2$)			Added Mass/($\pi\rho Lw^2/4$)	
L/(2c)	Strip Theory	Experiment	L/w	Strip Theory	Exact Solution
1.2	1.0	0.62	1.0	1.0	0.5790
2.5	1.0	0.78	2.0	1.0	0.7568
5.0	1.0	0.90	4.0	1.0	0.8718
9.0	1.0	0.96	10.0	1.0	0.9469

in which A represents the added mass (as given in tables) and m_w, the mass of water inside the pipe, all for unit length. Apparently, both the added mass and the fluid inside the pipe play significant roles in determining the vibrational characteristics of structures in the ocean environment. Furthermore, water surrounding the structure increases the so-called fluid damping of the vibrations. This effect is discussed in detail in Chapter VIII.

For the separated motion of real fluids, the added-mass coefficients depend, in general, on the parameters characterizing the history of the motion, time, Reynolds number (suitably defined) in addition to the parameters such as the type and direction of motion, proximity effects, etc. Furthermore, C_{ij}^a is not in general equal to C_{ji}^a. It is also evident that there is often no single mathematical or experimental procedure to separate the inertial component of the force from that attributable to the form drag. This brings us to the fundamental question of 'how does one express the time-dependent force acting on a body undergoing an arbitrary time-dependent motion?' For inviscid fluids, this is a relatively simple matter and the appropriate force and moment expressions have been given by Taylor (1928), Birkhoff (1950), Landweber (1956), Landweber and Yih (1956), Cummins (1957), Milne-Thomson (1960), and Sarpkaya (1963). For real fluids, however, the above question is not answerable. In fact, the determination of resistance in time-dependent flows presents an enormously complex texture of conditions and threatens to remain a perpetual problem. Only in the case of a few and relatively manageable time-dependent motions that one can devise approximate force equations. Even then one must be aware of the fact that the approximate equation may be valid only for a limited range of the governing parameters. Ordinarily, one performs a dimensional analysis of the pertinent parameters (assuming that one knows before hand what the most pertinent parameters are), carries out a series of experiments (with due consideration to scale effects, transition to turbulence, etc.) and attempts to determine the range of validity of the proposed equation. In doing so, the coefficients obtained with steady flows or from potential theory serve as limiting values or simple bench marks. In all practical applications, the awareness of the limitations of the equations and coefficients used is just as important as the equations and coefficients themselves.

In the foregoing, we have not touched upon the added mass of bodies moving at or near the liquid/air interface (see Chapter VI), added mass of fluids in closed systems (see e.g., Sarpkaya 1962), added mass of bodies in an unbounded fluid due to the expansion or collapse of nearby cavities and so on.

For a body undergoing small harmonic oscillations on or near a free surface (see e.g., Frank 1967), the added-mass coefficients have frequency-dependent in-phase and out-of-phase components. For low frequency oscillations in the vertical direction at the liquid/air interface, one may assume $\partial\phi/\partial n = 0$ at the free surface. For large frequency oscillations, however, the free surface condition is given by $\phi = 0$. Extensive research has been devoted to the calculation of the in-phase and out-of-phase components of the resulting force for both low and high-frequency oscillations of various shapes of bodies in surge, sway, yaw, heave, roll, and pitch. For a detailed discussion of the subject the reader is referred to Havelock (1963), Ogilvie (1964), Kim (1966), Vugts (1968), Flagg and Newman (1971), Wehausen (1971), Bai (1977), and Takaki (1977).

2.10 AN EXAMPLE

This example concerns the prediction of the time-dependent inertial force exerted on a finite cylinder by an oscillating air bubble.

Genesis of the Problem

Safety relief valves are used to discharge steam from the boiling water reactors to the suppression pool after a loss-of-coolant accident (LOCA), turbine trip, etc. Ordinarily, the discharge piping of a safety relief line contains air and a column of water determined by the submergence of the safety-relief line in the suppression pool. Following the valve actuation, first the water column and then air and steam are forced into the pool in the form of a high-temperature and high-pressure bubble. The bubble expands and contracts with a characteristic frequency until it breaks through the free surface due to buoyant force. The bubble fluctuation gives rise to time-dependent loads on the walls of the suppression pool and on the structures submerged in the pool. These loads are primarily inertial and the velocity-dependent drag forces are negligible. In other words, bubble-induced fluid displacements are quite small relative to the size of the submerged body.

Solution of the Problem

Consider a *spherical bubble* [modeled as a source of strength $m(t)$] at a distance f away from a finite circular cylinder (Fig. 2.7). The x-component

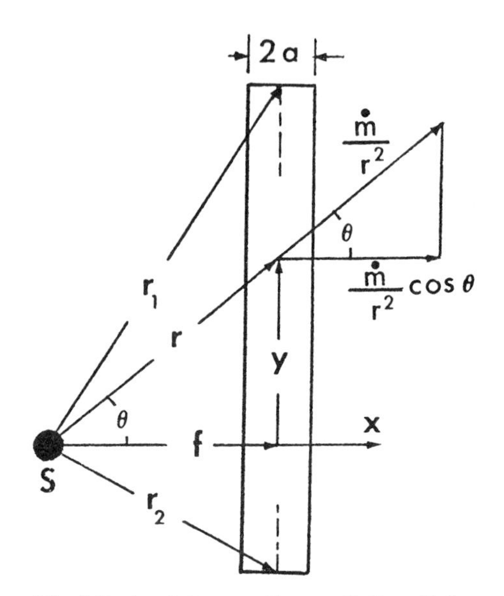

Fig. 2.7. A point source S near a finite cylinder.

of the inertial force on an element dy is

$$dF_x = \rho \forall (1 + C_a) \dot{U}_x \, dy \tag{2.72}$$

where $\forall$ represents the volume per unit length; C_a, the added-mass coefficient; and $\dot{U}_x$, the x-component of the acceleration. For the case shown in Fig. 2.7, one has

$$\forall = \pi a^2, \qquad \dot{U}_x = \frac{\dot{m}}{r^2} \cos\theta = \frac{\dot{m} f}{r^3}$$

$$C_a = 1.0, \qquad y = (r^2 - f^2)^{1/2}$$

Thus, one has

$$F = 2\pi a^2 \rho \dot{m} f \left[\int_f^{r_1} \frac{dr}{r^2 \sqrt{r^2 - f^2}} + \int_f^{r_2} \frac{dr}{r^2 \sqrt{r^2 - f^2}} \right] \tag{2.73}$$

or

$$F = 2\pi \rho f \dot{U}_\infty(t) [(1 - f^2/r_1^2)^{1/2} + (1 - f^2/r_2^2)^{1/2}] \tag{2.74}$$

in which

$$\dot{U}_\infty(t) = \dot{U}(t)\big|_{r=f} = \dot{m}/f^2$$

For an infinitely long circular cylinder, $r_1 = r_2 = \infty$ and

$$F = 4\pi\rho a^2 f\dot{U}_\infty(t) \tag{2.75}$$

which is identical to that obtained by Tung (1978) through the use of Green's function (see Chapter VI). Equation (2.74) becomes increasingly more accurate as the length-to-diameter ratio for the finite cylinder increases, (see Table 2.4).

For a sphere $\forall = 4\pi a^3/3$ and $C_a = 1/2$. Then, Eq. (2.72) reduces to

$$F = \frac{4}{3}\,\pi\rho a^3(1 + 0.5)\,\dot{m}(t)/f^2 = 2\pi\rho a^3\dot{m}(t)/f^2 \tag{2.76}$$

Apparently, the acceleration forces exerted by an oscillating bubble on structures of simple geometry, such as cylinders of finite length (circular cylinders, I-beams, etc.) may be calculated in a straightforward manner and with greater generality without the use of Green's function, as long as the diffraction effects are negligible and the fluid motion is in the inertia-dominated regime, i.e., not affected by separation. However, one must bear in mind that the bubble does not remain spherical or symmetrical. The determination of its instantaneous shape is an exceedingly complex two-phase flow problem. In fact, the analysis of LOCA-induced loads on submerged structures due to the formation of a mushroom-like vortex ring (prior to the vent clearing) and the growth and motion of bubbles (following the vent clearing) is in need of considerable theoretical and experimental work. The foregoing relatively simply example shows once again that hydrodynamics can provide only a rational description of certain flows, even if the details are not analyzed completely and the most practical problems require a strong interaction between theory and laboratory and numerical experiments. As Sir Geoffrey Ingram Taylor (1974) put it "Though the fundamental laws of the mechanics of the simplest fluids, which possess Newtonian viscosity, are known and understood, to apply them to give a complete description of any industrially significant process is often far beyond our power."

2.11 REFERENCES

Ackerman, N. L. and Arbhabhirama, A. 1964. Viscous and Boundary Effects on Virtual Mass. *Jour. Engineering Mechanics Div. ASCE*, Vol. 90, EM4, pp. 123–130.

Au-Yang, M. K. 1977. Generalized Hydrodynamic Mass for Beam Mode Vibration of Cylinders Coupled by Fluid Gap. *Jour. Applied Mechanics, ASME*, Vol. 44, pp. 172–174.

Bai, K. J. 1977. The Added-Mass of Two-Dimensional Cylinders Heaving in Water of Finite Depth. *Journal of Fluid Mechanics*, Vol. 81, pp. 85–105.

Basset, A. B. 1888. On the Motion of a Sphere in a Viscous Liquid. *Phil. Trans. Royal Soc.* London, Vol. 179, pp. 43–63. (See also *A Treatise on Hydrodynamics*, Vol. 2, (Chp. 21), 1888, Cambridge: Deighton, Bell and Co.), (Also, New York, Dover Publications, Inc., 1961).

Batchelor, G. K. 1967. *An Introduction to Fluid Dynamics.* Cambridge Univ. Press, Bentley House, London.

Birkhoff, G. 1950. Hydrodynamics–*A Study of Logic, Fact and Similitude.* Princeton Univ. Press, Princeton, N.J.

Blevins, R. D. 1977. *Flow-Induced Vibration.* Van Nostrand Reinhold, New York.

Bloor, M. S. and Gerrard, J. H. 1966. Measurements on Turbulent Vortices in a Cylinder Wake. *Proc. Royal Soc. London*, Vol. 294A, pp. 319–342.

Brooks, J. E. 1976. Added Mass of Marine Propellers in Axial Translation. David Taylor NSRDC, Report No. 76-0079.

Bryson, A. E. 1954. Evaluation of the Inertia Coefficients of the Cross-Section of a Slender Body. *Jour. Aeronautical Sci.*, Vol. 21, pp. 424–427.

Chandrasekaran, A. R., Salni, S. S. and Malhotra, M. M. 1972. Virtual Mass of Submerged Structures. *Jour. Hydraulics Div. ASCE*, Paper No. 8923.

Chen, S. S., Wambsganss, M. W. and Jendrzejczyk, J. A. 1976. Added Mass and Damping of a Vibrating Rod in Confined Viscous Fluid. *Jour. Applied Mechanics, ASME*, Vol. 98, pp. 325–329.

Ciesluk, A. J. and Colonell, J. M. 1974. Experimental Determination of Hydrodynamic Mass Effects. AD-781 910.

Cummins, W. E. 1957. The Force and Moment on a Body in a Time-Varying Potential Flow. *Jour. Ship Research*, Vol. 1, pp. 7–18.

Darwin, Sir Charles. 1953. Notes on Hydrodynamics. *Proc. Cambridge Phil. Soc.*, Vol. 49, pp. 342–354.

Dryden, H. L., Murnaghan, F. D. and Bateman, H. 1956. *Hydrodynamics.* Dover Publications, Inc., New York.

Fickel, M. G. 1973. Bottom and Surface Proximity Effects on the Added Mass of Rankine Ovoids. AD-775 022. (Thesis submitted to the Naval Postgraduate School, Monterey, Calif.)

Flagg, C. N. and Newman, J. N. 1971. Sway Added-Mass Coefficients for Rectangular Profiles in Shallow Water. *Jour. Ship Research*, Vol. 15, pp. 257–265.

Frank, W. 1967. Oscillation of Cylinder in or Below the Free Surface of Deep Fluids. NSRDC Report 2357.

Gibson, R. J. and Wang, H. 1977. Added Mass of Pile Groups. *Jour. Waterways etc., Div. ASCE*, WW2, pp. 215–223.

Goldschmidt, V. W. and Protos, A. 1968. Added Mass of Equilateral Triangular Cylinders. *Jour. Eng. Mechanics Div. ASCE*, EM6, pp. 1539–1545.

Hamilton, W. S. 1972. Fluid Force on Accelerating Bodies. *Proc. of the 13th Coastal Engineering Conf., ASCE*, New York, Vol. III, pp. 1767–1782.

Havelock, T. H. 1963. *Collected Papers.* ONR/ACE-103, U.S. Government Printing Office, Washington, D.C.

Kim, W. D. 1966. On a Freely Floating Ship in Waves. *Jour. Ship Research*, Vol. 10, pp. 182–191 + 200.

Kochin, N. E., Kibel, I. A. and Roze, N. V. 1964. *Theoretical Hydromechanics.* English translation of the 5th ed., Wiley and Sons, New York.

Lamb, Sir Horace. 1932. *Hydrodynamics.* (6th ed.), Dover Publications, New York.

Landweber, L. 1956. On a Generalization of Taylor's Virtual Mass Relation for Rankine Bodies. *Quarterly Jour. Appl. Math.*, Vol. 14, pp. 51-62.

Landweber, L. 1967. Vibration of a Flexible Cylinder in a Fluid. *Jour. Ship Research*, Vol. 11, pp. 143-150.

Landweber, L. and Yih, C.-S. 1956. Forces, Moments, and Added Masses for Rankine Bodies. *Jour. of Fluid Mechanics*, Vol. 1, pp. 319-336.

Lighthill, M. J. (Sir), 1979. Waves and Hydrodynamic Loading. *BOSS'* 79, London, pp. 1-40.

McConnell, K. G. and Young, D. F. 1965. Added Mass of a Sphere in a Bounded Fluid. *Jour. Eng. Mechs. Div. ASCE*, Vol. 91, EM4, pp. 147-164.

McLachlan, N. W. 1932. The Accession to Inertia of Flexible Discs Vibrating in a Fluid. *Proc. Phys. Soc. London*, Vol. 44, pp. 546-555.

Meyerhoff, W. K. 1970. Added Masses of Thin Rectangular Plates Calculated from Potential Theory. *Jour. Ship Research*, Vol. 14, pp. 100-111.

Milne-Thomson, L. M. 1960. *Theoretical Hydrodynamics.* (4th ed.), The MacMillan Co., New York.

Moretti, P. M. and Lowery, R. L. 1975. Hydrodynamic Inertia Coefficients for a Tube Surrounded by a Rigid Tube. *Jour. Pressure Vessel Tech.*, Vol. 97, pp. 345-351.

Myers, J., Holm, C. H. and McAllister, R. F. 1969. *Handbook of Ocean Engineering.* McGraw-Hill, New York.

Newman, J. N. 1977. *Marine Hydrodynamics.* MIT Press, Camb. Mass.

Nielsen, J. N. 1960. *Missile Aerodynamics.* McGraw-Hill, New York.

Odar, F. and Hamilton, W. S. 1964. Forces on a Sphere Accelerating in a Viscous Fluid. *Jour. Fluid Mechanics*, Vol. 18, pp. 302-314.

Ogilvie, T. F. 1964. Recent Progress Toward the Understanding and Prediction of Ship Motions. In *5th Symposium on Naval Hydrodynamics*, U.S. Government Printing Office, Washington, D.C., pp. 3-128.

Patton, K. T. 1965. Tables of Hydrodynamic Mass Factors for Translational Motion. ASME Paper No. 65-WA-UNT.

Pope, L. D. and Leibovitz, R. C. 1974. Intermodal Coupling Coefficients for a Fluid-Loaded Rectangular Plate. *Jour. Acoustical Soc. America*, Vol. 56, pp. 408-415.

Riecke, E. 1888. Notes on Hydrodynamics. (in German), *Nachr. Ges. Wiss. Göttingen, Math-Phys.* Klasse, pp. 347-350.

Robertson, J. M. 1965. *Hydrodynamics in Theory and Application.* Prentice Hall, Inc., Englewood Cliffs, N.J.

Sarpkaya, T. 1960. Added Mass of Lenses and Parallel Plates. *Jour. Eng. Mechs. Div. ASCE*, Vol. 86, EM3, pp. 141-152.

Sarpkaya, T. 1962. Unsteady Flow of Fluids in Closed Systems. *Jour. Eng. Mechs. Div. ASCE*, Vol. 88, pp. 1-5.

Sarpkaya, T. 1963. Lift, Drag, and Added-Mass Coefficients for a Circular Cylinder Immersed in a Time-Dependent Flow. *Jour. Applied Mechs., Trans. ASME*, Vol. 85, pp. 13-15.

Sarpkaya, T. and Shoaff, R. L. 1979. Inviscid Model of Two-Dimensional Vortex Shedding by a Circular Cylinder. *AIAA Journal*, Vol. 17, pp. 1193-1200.

Saunders, H. E. 1964. *Hydrodynamics in Ship Design.* Society of Naval Architects and Marine Engineers.

Sedov, L. I. 1965. *Two-Dimensional Problems in Hydrodynamics and Aerodynamics.* Translated from Russian and edited by C. K. Chu et al., Interscience Publishers, New York.

Schlichting, H. 1968. *Boundary-Layer Theory.* McGraw-Hill, New York.

Schmidt, D. W. von, and Tilmann, P. M. 1970. Experimental Study of Sound-Wave Phase Fluctuations Caused by Turbulent Wakes. *Jour. Acoust. Soc. Amer.* Vol. 47, pp. 1310–1319.

Schmidt, D. W. von, and Tilmann, P. M. 1972. Uber die Zirkulationsentwicklung in Nachlaufen von Rundstaben. *Acoustica*, Vol. 27, pp. 14–22.

Stelson, T. E. and Mavis, F. T. 1955. Virtual Mass and Acceleration in Fluids. *Proc. ASCE*, Vol. 81, Separate No. 670, pp. 1–9.

Stokes, Sir George Gabriel. 1851. One the Effect of the Internal Friction on Fluids on the Motion of Pendulums. *Cambridge Phil Trans.* IX, (also *Mathematical and Physical Papers*, Camb. Univ. Press, Vol. 3).

Takaki, M. 1977. On the Hydrodynamic Forces and Moments Acting on the Two-Dimensional Bodies Oscillating in Shallow Water. Rep. Res. Inst. Appl. Mech., Kyushu Univ., Vol. 25, pp. 1–64.

Taylor, Sir Geoffrey I. 1928. The Energy of a Body Moving in an Infinite Fluid, with Application to Airships. *Proc. Roy. Soc. London*, Vol. 120A, pp. 13–21.

Taylor, G. I. (Sir). 1974. The Interaction Between Experiment and Theory in Fluid Mechanics. *Annual Reviews of Fluid Mechanics*, Vol. 6, pp. 1–16.

Taylor, J. Lockwood. 1930. Some Hydrodynamical Inertia Coefficients. *Phil Magazine*. Vol. 9, pp. 161–183.

Tung, C. 1978. Acceleration Forces Generated by an Oscillating Air Bubble. *Trans. Amer. Nuclear Soc.*, Vol. 28, pp. 418–419.

Vugts, J. H. 1968. The Hydrodynamic Coefficients for Swaying, Heaving and Rolling Cylinders in a Free Surface. *International Shipbuilding Progress*, Vol. 15, pp. 251–276.

Waugh, J. G. and Ellis, A. T. 1969. Fluid-Free-Surface Proximity Effect on a Sphere Vertically Accelerated from Rest. *Jour. Hydronautics*, Vol. 3, pp. 175–179.

Wehausen, J. V. 1971. The Motion of Floating Bodies. *Ann. Reviews of Fluid Mechanics*, Vol. 3, pp. 237–268.

Weinblum, G. 1952. On Hydrodynamic Masses. DTMB Report 809.

Wendel, K. 1950. Hydrodynamische Masses und Hydrodynamische Masses-tragheits-Momente. *Jahrbuch der Schiffbautechnischen Gesellschaft*, Vol. 44, Translation no. 260, DTNSRDC, 1956.

Whitham, G. B. 1974. *Linear and Nonlinear Waves.* Wiley Interscience, New York.

Yih, C.-S. 1969. *Fluid Mechanics–A Concise Introduction to the Theory.* McGraw-Hill, New York.

Yu, Y. T. 1945. Virtual Masses of Rectangular Plates and Parallel Pipes in Water. *Jour. Appl. Mechs.*, Vol. 16, pp. 724–729.

APPENDIX A

Table 2.3 Added Masses of Various Bodies

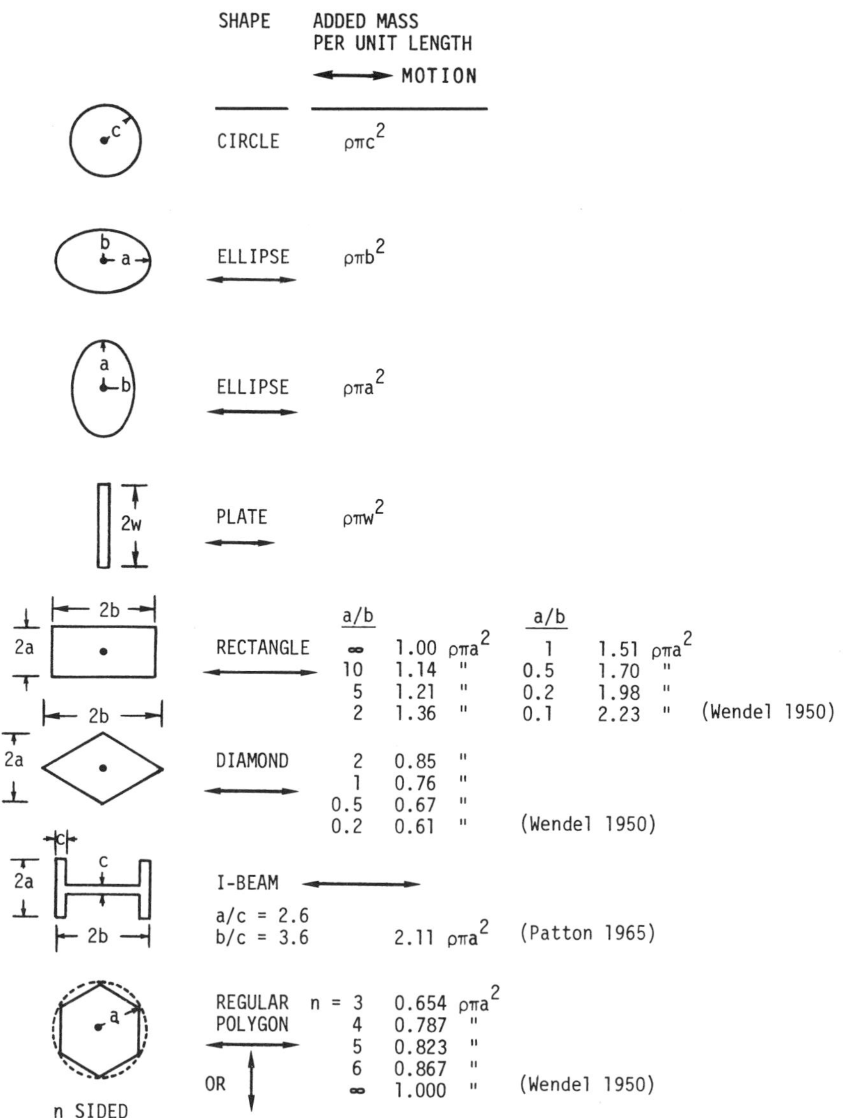

SHAPE	ADDED MASS PER UNIT LENGTH ⟷ MOTION
CIRCLE	$\rho\pi c^2$
ELLIPSE ⟷	$\rho\pi b^2$
ELLIPSE ⟷	$\rho\pi a^2$
PLATE ⟷	$\rho\pi w^2$
RECTANGLE ⟷	a/b = ∞: 1.00 $\rho\pi a^2$; 10: 1.14 "; 5: 1.21 "; 2: 1.36 "; 1: 1.51 $\rho\pi a^2$; 0.5: 1.70 "; 0.2: 1.98 "; 0.1: 2.23 " (Wendel 1950)
DIAMOND ⟷	a/b = 2: 0.85 "; 1: 0.76 "; 0.5: 0.67 "; 0.2: 0.61 " (Wendel 1950)
I-BEAM ⟷ a/c = 2.6, b/c = 3.6	2.11 $\rho\pi a^2$ (Patton 1965)
REGULAR POLYGON ⟷ OR ↕ (n SIDED)	n = 3: 0.654 $\rho\pi a^2$; 4: 0.787 "; 5: 0.823 "; 6: 0.867 "; ∞: 1.000 " (Wendel 1950)

Table 2.3 (Continued)

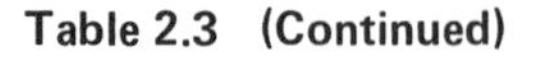

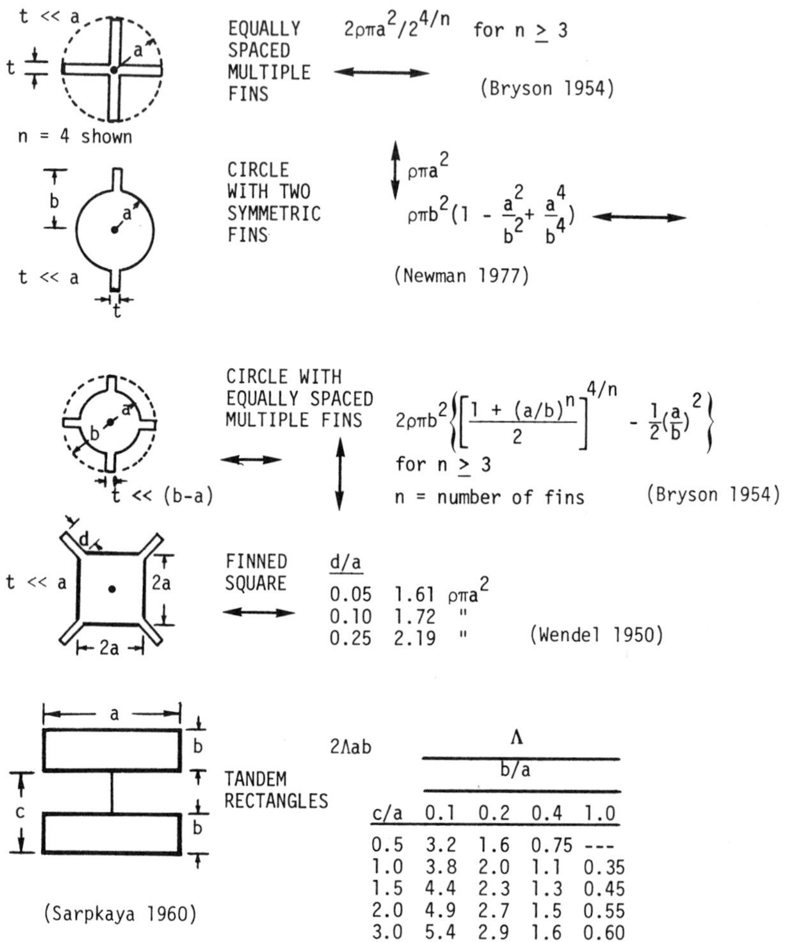

$2\Lambda ab$

	Λ			
	b/a			
c/a	0.1	0.2	0.4	1.0
0.5	3.2	1.6	0.75	---
1.0	3.8	2.0	1.1	0.35
1.5	4.4	2.3	1.3	0.45
2.0	4.9	2.7	1.5	0.55
3.0	5.4	2.9	1.6	0.60
4.0	5.6	3.0	1.6	0.70

Table 2.3 (Continued)

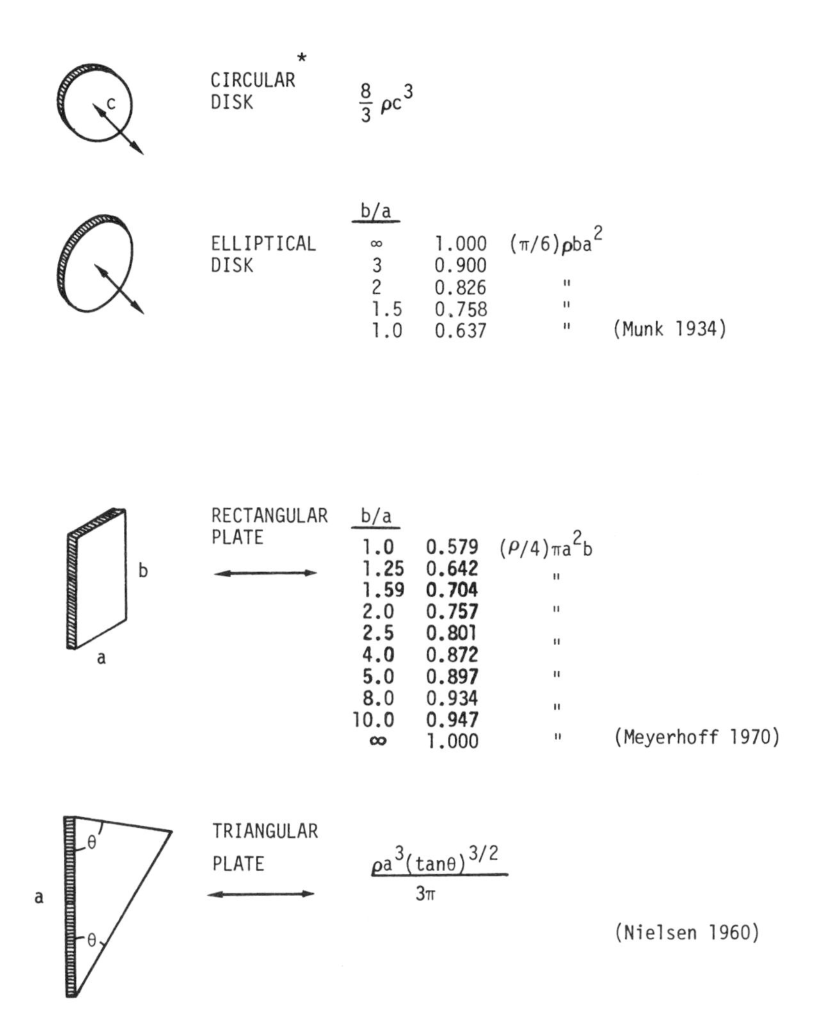

Body		Added mass		
CIRCULAR DISK*		$\frac{8}{3}\rho c^3$		
ELLIPTICAL DISK	b/a			
	∞	1.000	$(\pi/6)\rho ba^2$	
	3	0.900		
	2	0.826	"	
	1.5	0.758	"	
	1.0	0.637	"	(Munk 1934)
RECTANGULAR PLATE	b/a			
	1.0	0.579	$(\rho/4)\pi a^2 b$	
	1.25	0.642	"	
	1.59	0.704		
	2.0	0.757	"	
	2.5	0.801	"	
	4.0	0.872		
	5.0	0.897	"	
	8.0	0.934	"	
	10.0	0.947		
	∞	1.000	"	(Meyerhoff 1970)
TRIANGULAR PLATE		$\frac{\rho a^3 (\tan\theta)^{3/2}}{3\pi}$		(Nielsen 1960)

* All bodies shown above are assumed to be very thin.

Table 2.3 (Continued)

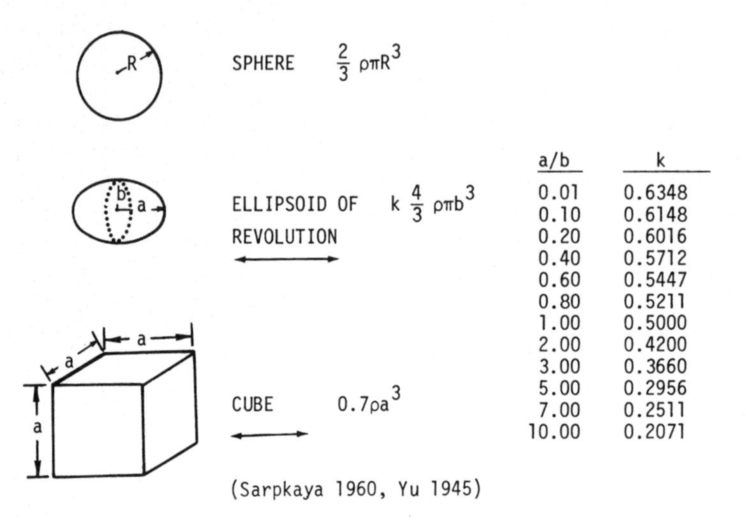

SPHERE $\frac{2}{3}\rho\pi R^3$

ELLIPSOID OF REVOLUTION $k\,\frac{4}{3}\rho\pi b^3$

a/b	k
0.01	0.6348
0.10	0.6148
0.20	0.6016
0.40	0.5712
0.60	0.5447
0.80	0.5211
1.00	0.5000
2.00	0.4200
3.00	0.3660
5.00	0.2956
7.00	0.2511
10.00	0.2071

CUBE $0.7\rho a^3$

(Sarpkaya 1960, Yu 1945)

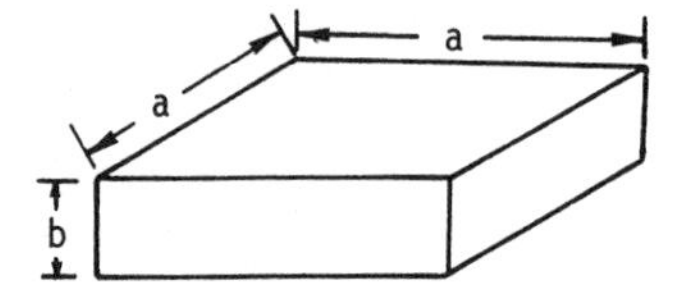

RECTANGULAR BLOCK $k\,\rho a^2 b$

b/a	k
0.5	1.32
0.6	1.15
0.8	0.86
1.0	0.70
1.2	0.57
1.6	0.45
2.0	0.35
2.4	0.30
2.8	0.26
3.6	0.22

(Sarpkaya 1960)

Table 2.3 (Continued)

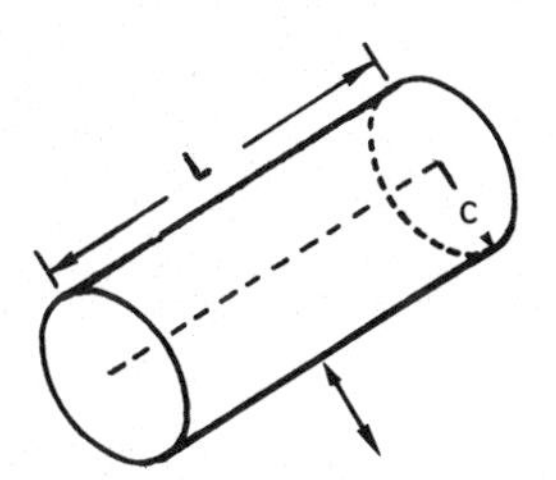

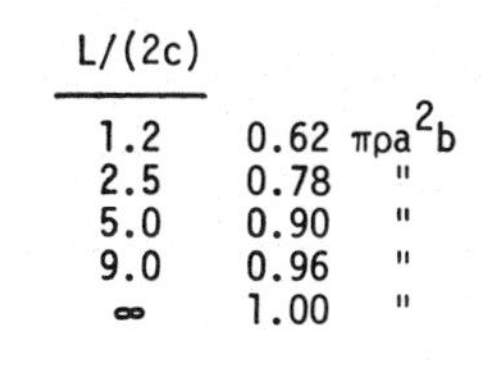

L/(2c)	
1.2	0.62 $\pi\rho a^2 b$
2.5	0.78 "
5.0	0.90 "
9.0	0.96 "
∞	1.00 "

(Wendel 1950)

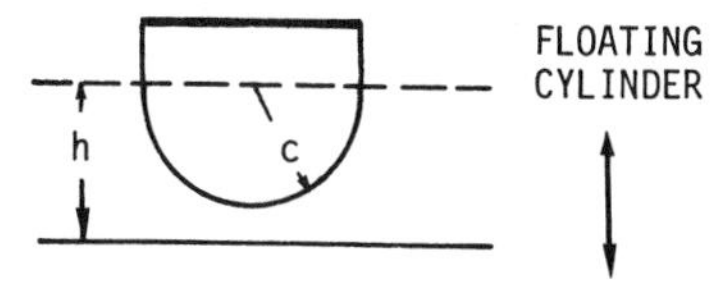

FLOATING CYLINDER

$k \frac{\pi}{2} c^2$

h/c	k
1.2	1.83
1.5	1.45
2.0	1.22
3.0	1.09
5.0	1.03
∞	1.00

(Bai 1977)

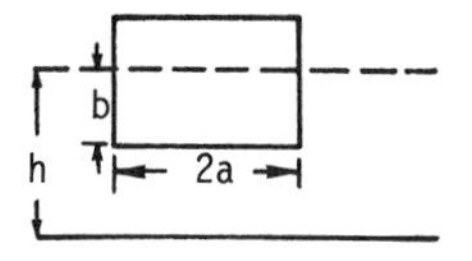

FLOATING RECTANGLE

$2k_1\rho ab$

	k_1			
	a/b			
b/h	0.2	0.5	1.0	2.0
0.0	4.75	2.14	1.18	0.67
0.2	4.92	2.25	1.29	0.78
0.4	5.43	2.63	1.66	1.16
0.6	6.63	3.56	2.53	2.02
0.8	10.15	6.46	5.23	4.62

$2k_2\rho ab$

h/b	k_2
1.1	5.52
1.2	3.49
1.5	2.11
3.0	1.35
8.0	1.21
∞	1.19

(Bai 1977, Flagg and Newman 1971)

3
Flow Separation and Time-Dependent Flows

3.1 INTRODUCTION

Separation is the key phenomenon for most of the flows dealt with in this chapter. Real fluids cannot move relative to a solid boundary and separate under the effect of adverse pressure gradient as momentum is consumed by both wall shear and pressure gradient. An essential feature of such flows is that there is mutual interaction between the viscous layers and the outer flow. The pressure gradient, and hence the production of vorticity at the body surface, and the position of the breakaway streamline cannot be determined independently. Each determines the other.

The importance of separation is best understood in terms of its consequences. Historically, it was the occurrence and perplexing consequences of separation that led Prandtl in 1904 to his boundary-layer or 'transition-layer' concept. It is rather ironic that even though the boundary-layer has revolutionized fluid dynamics and brought hydraulics and hydrodynamics closer, the phenomenon which gave impetus to its inception still remains enigmatic.

3.2 CONSEQUENCES OF SEPARATION

Observations as well as numerical experiments show that the wake of a bluff body is comprised of an alternating vortex street. The character of the vortices immediately behind the cylinder and in the wake further downstream depend, for a steady ambient flow about a smooth cylinder, on the Reynolds number and the intensity and length-scale of the turbulence present in the ambient flow. For a time-dependent flow, the instantaneous state as well as the past history of the flow play significant roles and it is not possible to give a general set of normalized parameters. Often heuristic reasoning, laboratory and numerical experiments will have to supplement the information gathered fróm relatively idealized solutions in order to correctly identify the most important governing parameters.

The separation point may be mobile as on a circular cylinder or fixed at an edge, e.g., an edge of an inclined flat plate or of the base of a triangular section. When the separation point is fixed, the drag and pressure coefficients vary considerably less than those for the circular cylinder. However, they are not completely independent of the Reynolds number effect, especially in the lower ranges where various transitions are occurring in the wake (Roshko and Fiszdon 1969). It is not correct to assume that the problems associated with the downstream of fixed separation points are less complex than those associated with the downstream of mobile separation points. For example, for flow over symmetrical wedges the base pressure appears to be insensitive to the nose angle (Roshko 1955, 1970; Tanner 1964). The reason for this experimental fact is not clear and contrary to intuitive expectations.

When the separation point is mobile, its motion is coupled with the shedding of vortices. The vorticity feeding the shear layer fluctuates with time since the separation point moves into higher or lower velocity regions. The separation occurs beyond the point of maximum velocity (about 10 degrees downstream of this), i.e., the pressure gradient is positive at the mobile separation point whereas it is negative up to the fixed separation point. In addition and to further complicate the matters, the flow upstream of the separation points is affected by the occurrence and motion of the separation points or by the circumstances leading to the formation, growth, and motion of the wake.

The front stagnation point oscillates about its mean position in such a manner that its motion is 180 degrees out of phase (Dwyer and McCroskey 1973) with that of the separation point. The streamlines emanating from the separation points may be convex (at a fixed separation point or at a mobile separation point at the front of a cylinder) or concave (when separation occurs on the back of the cylinder).

It is evident from the foregoing that there is a fundamental interaction between the body and the separated flow particularly in the region enclosing the body and its near wake. The dynamics of this interaction is of major importance in determining the time-dependent fluid resistance and the characteristics of flow-induced vibrations. Unfortunately, it is not yet possible to predict theoretically the behavior of the flow when separation leads to a large scale wake comprised of alternating vortices. This is particularly true for a time-dependent flow about a cylinder (e.g., wave motion, two-dimensional harmonic flow with or without a mean flow, etc.). In such cases, the separation points undergo large excursions and the roles played by the forebody (part of the body upstream of the separation points) and the afterbody (downstream of the separation points) are periodically interchanged. Consequently, various models devised for finite wakes (for an excellent review see Wu 1972) and cavities are not of much assistance. Furthermore, the boundary-layer equations (if they can ever be integrated through the separation points interacting with the wake) can deal only with flows with relatively small Reynolds numbers.

When the vortices, vortex feeding layers, and even the boundary layer over the forebody become turbulent, significant changes take place in the forces acting on the body. This is vividly illustrated through measurements and motion pictures. For example, the expository experimental study by Roshko and Fiszdon (1969) have shown that when the Reynolds number lies between about 1 and 50, the entire flow is steady and laminar. In the range of Reynolds numbers from about 50 to 200, the flow retains its laminar character but the near wake becomes unstable and oscillates periodically. At still higher Reynolds numbers but below about 1500, turbulence sets in and spreads downstream. In the region between 1500 and 2×10^5, the transition and turbulence gradually move upstream along the free shear layers and the far wake becomes increasingly irregular. When the transition coincides with the separation point at a Reynolds number of about 5×10^5 (depending on the intensity of the free-stream turbulence and the peculiarities of the test apparatus), the flow undergoes first a laminar separation, followed by a reattachment to the cylinder, and then a turbulent separation to form a narrower wake. This results in a large fall in both the lift and the drag coefficient. Because of the more diverse interest in the steady drag force, the phenomenon leading to the sharp decrease in drag became known as the drag crisis. The fact that the lift force as well as other characteristics of the flow also undergo dramatic changes in the same Reynolds number range suggests that the phenomenon be called *resistance crisis*. It is very important to remember that the Reynolds number at which the resistance crisis occurs depends very much on the character of the flow. It is not correct to assume that the resistance crisis will occur at the same Reynolds number for a time-dependent flow. For a uniformly accelerating flow about a cylinder, the resistance crisis may occur at higher Reynolds numbers. On the other hand, for a decelerating flow or for a periodic flow about a cylinder, the resistance crisis may occur at smaller Reynolds numbers. In general, it is worth remembering that when the flow is sensitive to the variations in a particular parameter, it becomes equally sensitive to others whose influence may have been otherwise negligible, (e.g., roughness, free stream turbulence, vibrations, end conditions, uniformity of the flow, etc.). It is unfortunate that often the critical values of the parameters corresponding to idealized two-dimensional experiments are remembered and the experiments are designed for other flow situations as if the critical conditions were to occur at the corresponding steady-state values.

In the absence of roughness and excessive free-stream turbulence, the transition in drag coefficient between Re = 5×10^5 and Re = 7×10^5 is currently interpreted (e.g., Roshko 1961; Tani 1964; Batchelor 1967 p. 342; Jones, Cincotta and Walker 1969) to be due to transition of the separated boundary layer to a turbulent state, the formation of a separation bubble, reattachment of the rapidly spreading turbulent free-shear layer, and finally, separation of the turbulent boundary layer at a position further downstream from the first point of laminar separation. The reduction of the wake size as a consequence of the retreat of the separation points then results in a smaller form drag (see Fig. 3.1 where these different ranges are distinguished).

The subsequent increase in the drag coefficient between Re = 10^6 and Re = 10^7 is then interpreted to be a consequence of the transition to a turbulent state of the attached portion of the boundary layer. At higher Reynolds numbers the trend is not clear and additional experiments, however difficult to perform, are needed. The best conjecture is that dramatic changes are not likely to occur in the boundary layers at Reynolds numbers several orders of magnitude larger than 10^7.* The diffusion and cancellation of vorticity may increase. These in turn lead to higher Strouhal numbers and lower drag coefficients. Be that as it may, there is, at present, a large Reynolds-number gap, between the highest Reynolds numbers attained in the experiments and the Reynolds numbers to be encountered by the designers of marine structures.

The experimental observations cited above do not consider the physical movement of the body, free-stream turbulence, and the roughness effects. Experience has shown that elastic structures near linear-resonance conditions can develop flow-induced oscillations by extracting energy from the flow about them. The oscillations of the structure modify the flow and give rise to nonlinear interaction. This is in addition to any nonlinearity which can arise from the restoring force (variable support stiffness) and/or from response-dependent structural damping. The understanding of these nonlinear interactions is of paramount importance, (see Chapter VIII). The free-stream turbulence and roughness effects are discussed in Section 3.5.

For a more detailed discussion of separation and flow about cylinders the

*There is not yet two universally accepted names for the flow regimes beyond critical. Roshko (1961) classified the flow regimes as subcritical, critical, transcritical, and supercritical. Szechenyi (1975) adopted the classification of subcritical, critical, supercritical, and transcritical. Others (Miller 1977) preferred to call the last flow regime postcritical. The authors feel that none of these descriptions is without some ambiguity and prefer to identify the flow regimes as subcritical, critical, supercritical, and post-supercritical. This definition is used throughout the text, (see Fig. 3.1).

	A Subcritical	B Critical	C Supercritical	D Post-supercritical
Boundary layer	laminar	transition	turbulent	turbulent
Separation	about 82 deg.	transition	120 - 130 deg.	about 120 deg.
Shear layer near separation	laminar		laminar separation, bubble turbulent reattachment	turbulent
Strouhal number	$S = 0.212 - \frac{2.7}{Re}$	transition	0.35 - 0.45	about 0.29
Wake	Re<60 laminar; 60<Re<5000 vortex street Re > 5000 turbulent	not periodic		
Approximate Re range	$< 2x10^5$	$2x10^5$ to $5x10^5$	$5x10^5 - 3x10^6$	$> 3x10^6$

Fig. 3.1a. Incompressible flow regimes and their consequences.

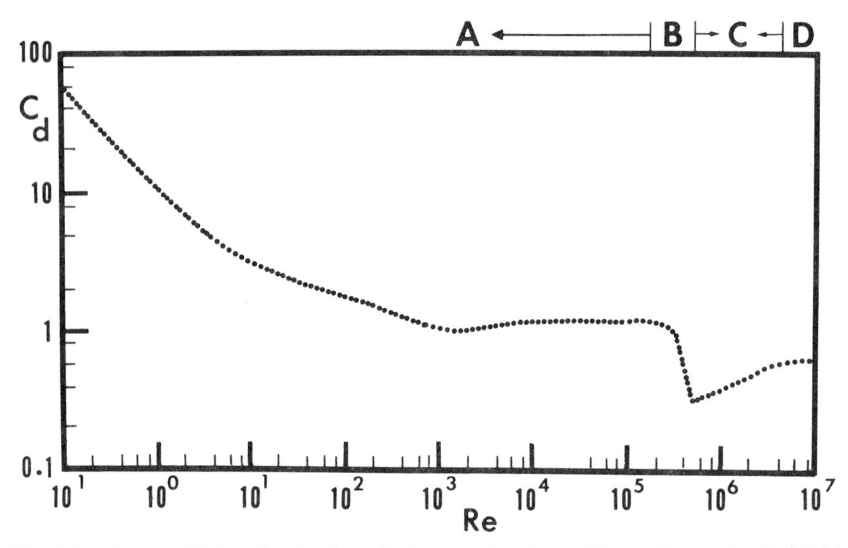

Fig. 3.1b. Drag coefficient for circular cylinders as a function of Reynolds number (Schlichting 1968).

reader is referred to Marris (1964), Morkovin (1964), Lienhard (1966), Chang (1970), Mair and Maull (1971), Berger and Wille (1972), Wille (1974), Sarpkaya (1979), and Sarpkaya and Shoaff (1979a,b). A comprehensive list of references may be found in Nelligan (1974).

3.3 STROUHAL NUMBER

In 1878, Strouhal discovered a relationship between the vortex-shedding frequency, cylinder diameter, and the velocity of the ambient flow in connection with his work on a special method of creation of sound. This relationship, denoted by $S_0 = f_0 D/V$, is known as the Strouhal number. The net result of the alternate vortex shedding is an oscillating side thrust, upon a cylinder of suitable form, in a direction away from the last detached vortex. This side thrust or lift force (sometimes referred to as lateral force or transverse force or out-of-plane force) exists practically at all Reynolds numbers regardless of whether the body is allowed to respond dynamically or not. In other words, the alternate shedding of vortices gives rise to a transverse pressure gradient. Unsteady hydrodynamic loads arising from this pressure gradient acting on the afterbody can excite dynamic response. Body natural frequencies near the exciting frequency raise the specter of load and response enhancement. Thus, the spectral content of the forcing functions are important to dynamic structural response analysis.

The shedding process may be random (broadband) over a portion of the Reynolds number range for which a statistical response analysis is required based on the spectral content. Also required is a measure of the correlation length along the cylinder. If the correlation lengths are present, the structural analysis must be three-dimensional, and the total load on the body is reduced. Evidently, the characterization of the vortex shedding process by a simple frequency is a practical simplification. Power spectral density analyses of unsteady cylinder loads reveal that in certain Reynolds number regimes the shedding process is practically periodic and can be characterized by a single Strouhal frequency.

At subcritical Reynolds numbers, the energy containing frequencies are confined to a narrow band, and the Strouhal number is about 0.2 for smooth cylinders (see Fig. 3.2). It must be emphasized that only an average Strouhal number may be defined for Reynolds numbers larger than about 20,000. In the critical Reynolds number regime, a broad band power spectral density is usually observed for a rigidly held cylinder. At higher Reynolds numbers, the Strouhal frequency rises to about 0.3 (Roshko 1961) and the shedding process is quasi-periodic. The spectral content of the exciting forces is particularly important for bodies which may undergo in-line and/or transverse oscillations since the vortex shedding frequency locks on to the frequency of the transverse oscillations of the cylinder when the vortex shedding frequency is in the neighborhood of the natural frequency of the cylinder. This often raises questions as to what happens when the flow is in the critical regime. The question is a complex one since the so-called critical regime does not necessarily occur at Reynolds numbers corresponding to that for a stationary cylinder. The motion of the cylinder changes the boundary-layer characteristics, the occurrence of the laminar separation bubble (assumed to be responsible for the transition in the critical and supercritical regions), and

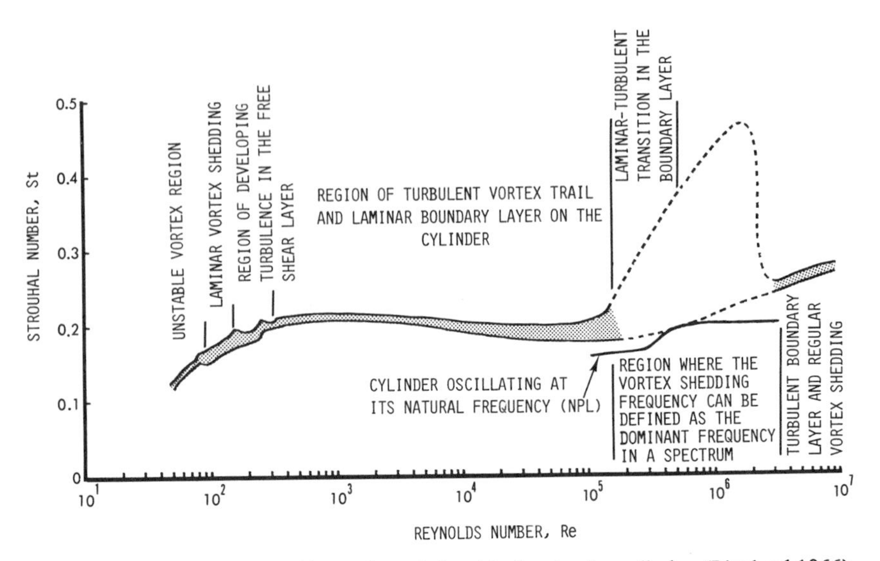

Fig. 3.2. The Strouhal-Reynolds number relationship for circular cylinders (Lienhard 1966).

the Strouhal freuqency. Thus, calculations based on stationary cylinder values in the critical regime may prove to be false. Experiments conducted at the National Physical Laboratory (1969), (see Fig. 3.2), have shown that when the cylinder is free to oscillate the sharp rise in the Strouhal number does not occur and remains at a value nearly equal to that found at subcritical Reynolds numbers.

Various attempts have been made to devise a universal Strouhal number which would remain constant for differently shaped two-dimensional and axisymmetric bodies (Roshko 1954, 1955; Bearman 1967; Calvert 1967; Simmons 1977). The universal Strouhal numbers introduced by Roshko and Bearman employ the wake width as the characteristic length and velocity outside the shear layers as the characteristic velocity. Accordingly, Roshko defined S* as

$$S^* = f_0 h'/U_b \tag{3.1}$$

in which h' is the distance between the shear layers and is obtained from the notched-hodograph theory of Roshko (1954) and U_b, the velocity at the edge of the shear layer, is related to the base pressure C_{pb} and the velocity of the ambient flow V by

$$U_b = V(1 - C_{pb})^{1/2} \tag{3.2}$$

For a circular cylinder, flat plate, and 90-degree wedge, Roshko found a fairly constant value of $S^* = 0.163 \pm 0.01$ over most of the Reynolds number range examined.

Bearman (1967) defined S^* as

$$S^* = f_0 b/V \tag{3.3}$$

where b is the lateral spacing between the vortices. The value of b was obtained both from Karman's stability analysis and Kronauer's minimum drag criteria (Bearman 1967). Bearman found that when S^* is plotted against the base pressure parameter $K = (1 - C_{pb})^{1/2}$, $S^* = 0.181$ over a wide range of K for a variety of bluff-body shapes.† Even though this finding appears to justify the use of the Kronauer's stability criteria, one must bear in mind several important facts. Firstly, S^* is linearly dependent on b, whereas the street drag coefficient is very insensitive to changes in b/ℓ_v (ℓ_v is the longitudinal spacing of vortices of one row). Secondly, there is no experimental evidence that the vortices (in the range of Reynolds numbers where they may be observed) arrange themselves so that the drag will be minimum (Kronauer's stability criteria).

The definition of the Strouhal number by Calvert (1967) and Simmons (1977) is given by

$$S^* = f_0 \ell_w / KV \tag{3.4}$$

in which ℓ_w represents measured wake width (the distance between the u_{rms} peaks across the wake at $x = \ell_f$, the formation length); u_{rms}, the root-mean-square of the fluctuating component of velocity and ℓ_f, the streamwise distance from separation edges to C_p (minimum). Calvert found a value of $S^* = 0.19$ for axisymmetric bodies and Simmons obtained $S^* = 0.163$ for various two-dimensional models.

The foregoing definitions suffer from the obvious drawback that they require the solution of the wake formation problem first, or the measurement of one or more flow characteristics. Thus their value is not so much in their ability to predict but rather to uncover the intricate relationship, say, between the flow velocity, vortex-shedding frequency, wake width, base pressure, formation length, time-mean of the rate of shedding of vorticity, etc. It appears from the foregoing that isolating a small number of simple scales to describe the wake-development may not be possible.

Theoretical or semi-empirical predictions of the wake characteristics have been attempted by various researchers. Birkhoff (1953) demonstrated that the longitudinal spacing ℓ_v is trivially invariant since the longitudinal velocity of vortices is uniformly bounded. He has shown further that in an inviscid fluid

---

†It is easy to show that $K^2 = 0.5 |d\Gamma/dt|/V^2$ where $|d\Gamma/dt|$ is the absolute value of the rate of change of circulation at the separation point.

the lateral spacing also remains constant at a value equal to the initial spacing of the shear layers. These led Birkhoff to the conclusion that b/ℓ_v is determined by its initial value and not by von Karman's stability criteria of $b/\ell_v = 0.281$. The use of a wake oscillator model ('the wake swings from side to side, somewhat like the tail of a swimming fish') and some experimentally determined values led Birkhoff to $S_0 = 0.2$. One must keep in mind that vorticity in real vortices is not concentrated in points, the vortices are non-circular and distort and rotate as they move downstream (Davies 1976), vorticity diffuses and is swept across the wake (Zdravkovich 1969), and dissipated by turbulence (Berger and Wille 1972). Vortices are subjected to strain fields imposed by nearby vortices. The resulting patterns are ever changing vortex shapes encompassing elliptic and pearlike geometries. The complexity of the interaction between strained distorting vortices is further exacerbated by the addition of turbulence to the wake, as this is likely to produce a more diffusive vorticity distribution and thus an additional shear field (Davies 1976). Thus, the near constancy of the wake characteristics in the range of Reynolds numbers where a vortex street might be observed is primarily due to the slow variation of ℓ_v, b, and S_0 with respect to the strength and deformation of vortices. Nevertheless, theoretical idealizations of the wake give some clues about the asymptotic behavior of the wake. In particular, one obtains from $f_0 \ell_v = V - u_s$, where $V - u_s$ is the velocity relative to the body, and from the use of von Karman's stability criteria that

$$S_0 = \frac{1}{\ell_v/D} - \frac{\Gamma/VD}{2\sqrt{2}\,(\ell_v/D)^2} \tag{3.5}$$

where Γ represents the strength of a vortex. Assuming that ℓ_v/D remains nearly constant ($\ell_v/D \simeq 5$), one concludes that the smaller the strength of the vortices, the larger is the Strouhal number (smaller drag and narrower wake) and vice versa. The vortices which are subjected to smaller dissipation linger a little while longer in the near wake relative to those which are dissipated more, i.e., strong vortices get stronger. However, the dependence of S_0 on Γ/VD is rather weak. This explains in part the reason for the success of the discrete vortex models in predicting the Strouhal number fairly accurately (Clements 1977, Sarpkaya 1975) in spite of the fact that the calculated vortex strengths are about 35 percent larger than those estimated experimentally. However weak, the dependence of S_0 on Γ/VD becomes a primary factor in the locking-on of the vortex shedding to the natural frequency of the body. The vortices which are stronger continue to be fed by their shear layers on a longer time period, thus further reducing the Strouhal number. In this process the mobility of the separation points is important but not necessary.

Sacksteder (1978) pursued a theoretical approach to determine the Strouhal number at large Reynolds numbers by perturbing the d'Alembert flow (see Eq. 2.31) and obtained $S_0 = 0.2028$. His analysis does not consider the wake formation and thus it is not expected that flows around objects that induce large wakes could be treated with a simple perturbation of the unseparated flow.

The constancy of the Strouhal number over a broad range of Reynolds numbers does not imply that the base pressure remains constant and that a single two-dimensional vortex emanates from a separation line. In reality, there is not only a phase shift between various sections along the vortex, separated by a correlation length (the equivalent length over which the velocity fluctuations at similar points in the wake may be described as perfectly correlated), but also variations in both the intensity and the frequency of vortex segments.

The variation of the base pressure with Reynolds number, in the range where the Strouhal number remains practically constant (see Fig. 3.3), may be related to the variation of the mean vorticity flux or to the variation of the correlation length with the Reynolds number, turbulence, length-to-diameter ratio, and surface roughness (Phillips 1965; Etkin et al., 1958; Prendergast 1958; El Baroudi 1960; Humphreys 1960). Table 3.1, as compiled by King (1977), gives an approximate idea about the typical values of the correlation length. The net effect of the spanwise variations of the vortex tube is that the transverse force (lift) coefficient obtained from a pressure integration is not necessarily identical with that obtained from the direct measurements of the lift force. Partial spanwise correlation leads to variations in both the frequency and the amplitude of the lift force, the variation of the latter being more pronounced than that of the former.

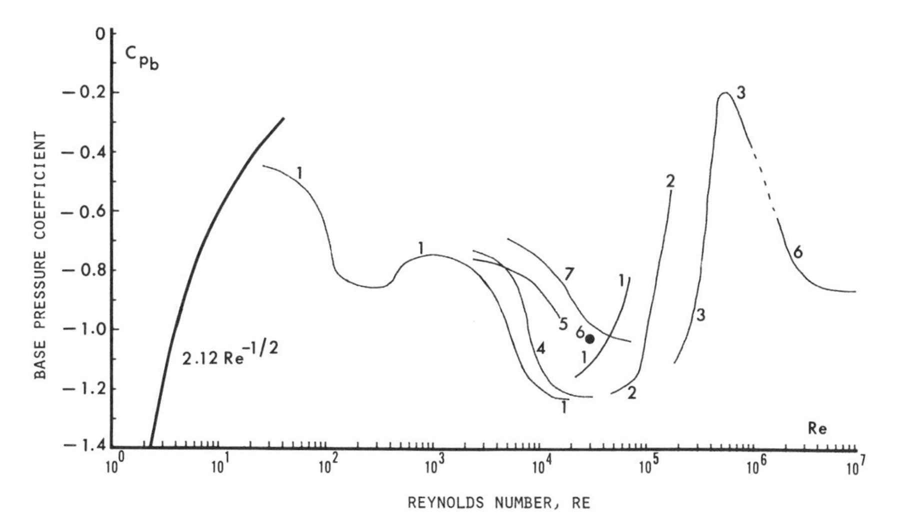

Fig. 3.3. Variation of the base-pressure coefficient with the Reynolds number for circular cylinders: 1, Thom (1928); 2, Fage and Falkner (1931); 3, Flachsbart (1932); 4, Schiller and Linke (1933); 5, Roshko (1953); 6, Fage (1931); 7, Gerrard (1965).

Table 3.1 Correlation Lengths.

Reynolds No. Range	Correlation Length	Source
$40 < Re < 150$	15D–20D	Gerlach (1970)
$150 < Re < 10^5$	2D–3D	Gerlach (1970)
$10^4 < Re < 4.5 \times 10^4$	3D–6D	El Baroudi (1960)
$Re < 10^5$	0.05D	Gerlach (1970)
$Re = 2 \times 10^5$	1.56D	Humphreys (1960)

The lack of correlation exists not only spanwise but also chordwise (Gerrard 1965; Vickery 1966; Chaplin et al., 1971; Wilkinson et al., 1974) and the chordwise correlation is related to the spanwise correlation. Comparison between various results suggests that with increasing Reynolds number over the range 10^4 to 10^5, the chordwise correlation for square and circular cross-section cylinders is maintained or improved. This leads to an increase in fluctuating lift coefficient. The reasons for these variations are not quite clear. The end effects (Humphreys 1960; Stansby 1974; Gowda 1975; Etzold and Fiedler 1976), wall boundary layers, freestream turbulence (McGregor 1957; Berger 1964; Surry 1969), non-uniformity of the flow are mentioned often as possible reasons. The complexity of the three-dimensional nature of the flow about a cylinder is clearly demonstrated with measurements by Tournier and Py (1978).

It would not be correct to assume that the mobility of the separation points is primarily responsible for the imperfect coherence. Even bodies such as 90-degree wedges, square cylinders, with fixed separation lines, do not exhibit perfect correlation. However, the variation of the base-pressure coefficient for bodies with mobile separation lines is greater than that for bodies with fixed separation lines (Roshko 1970). In fact, for flow over symmetrical wedges the base pressure appears to be insensitive to the nose angle (Roshko 1970).

3.4 NEAR WAKE AND PRINCIPAL DIFFICULTIES OF ANALYSIS

As noted earlier, the strength of the vortices plays an important role, particularly in the near wake. Laboratory and numerical experiments have shown that (Fage & Johansen 1928, Roshko 1955, Abernathy & Kronauer 1962, Bloor & Gerrard 1966, Mair & Maull 1971, Schmidt and Tilmann 1972) the net circulation of a rolled-up vortex of the street is 40 to 60 percent smaller than that generated in the boundary layer during a shedding cycle. Prandtl determined that the initial vorticity decreases to about half where the first vortex centers appear. Vorticity is ultimately dissipated by viscosity to which it owes its generation. Nevertheless, one may think of loss of circulation through cancellation of oppositely-signed vorticity. Primarily, there are three mechanisms whereby oppositely-signed vorticity are brought close together: vorticity generated on the forebody is carried by the shear layers near that generated on the afterbody; vorticity of the deformed and cut sheet is carried across the near wake by the entrainment of the irrotational fluid; and finally, vorticity is swept across the entire wake (Zdravkovich 1969). The percentage quoted in the literature (Berger and Wille 1972) for the total loss of vorticity often imply that the vortices, once having acquired a certain circulation, retain that circulation throughout the rest of their motion. The fact that circulation decreases continuously with time or distance is demonstrated clearly by the experiments of Schmidt and Tilmann (1972) and Bloor and Gerrard (1966). The amount of vorticity generated in the boundary layers and the amount dissipated are of prime importance not only for the flow past stationary bluff bodies but also for those undergoing resonant oscillations. In fact, the entire bluff body problem may be reduced to the determination of the vorticity distribution throughout the flow field. This is not yet possible for Reynolds numbers larger than about 100. The determination of the vortex strengths is difficult and sensitive to the theoretical and experimental means employed.

It is evident from the foregoing and from a more detailed study of the references cited that the description of the near wake of a bluff body is in a primitive state. Much of what is known about the consequences of separation has come from laboratory experiments. It has not yet been possible to develop a numerical model with which experiments may be conducted to explain the observed or inferred relationships between various parameters and to guide and complement the laboratory experiments. The principal difficulties are as follows:

(1) Separation Points. They represent a mobile boundary between two regions of vastly different scales. This in turn leads to complex physical nonuniformities in relatively narrow regions which cannot be handled within the framework of the boundary-layer theory (Williams 1977). Finite difference and Marker and Cell (MAC) techniques require in such regions very small time increments. The discretization of the continuous process of vorticity generation by line vortices in the vicinity of a mobile or fixed singular point (discrete vortex model) strongly affects the existing nonuniformities and promotes earlier separation. Attempts to preserve the prevailing flow conditions, say by limiting the influence of the nascent vortices, while satisfying a relatively simple separation criteria lead to hydrodynamical inconsistencies and non-disposable parameters (Sarpkaya and Shoaff 1979a,b).

(2) Reynolds Number. Finite difference schemes for bluff-body flows are limited to relatively small Reynolds numbers whether the scale of the flow is assumed to be governed by a constant viscosity or by a constant eddy viscosity. The large recirculation region of the flow is often comprised of turbulent vortices even when the boundary layer is laminar. The transition to turbulence moves upstream in the shear layers as the Reynolds number is increased from about 10^3 to 5×10^4. At $Re = 5 \times 10^4$, it reaches the shoulder of the cylinder (Bloor

1964). It does not move appreciably further upstream before the critical Reynolds number is reached. Thus, the distribution, turbulent diffusion, and decay of vorticity and the interaction between the wake and the boundary layers cannot be subjected to numerical simulation without recourse to some heuristic turbulence models and inspired insight.

The representation of the wake by clouds of point vortices or discretized spiralling sheets (see, e.g., Abernathy and Kronauer 1962, Fink and Soh 1974, Clements 1977, Sarpkaya and Shoaff 1979a) is not immune to scaling problems. In fact, not a particular Reynolds number but only a particular flow regime may be specified, depending on the separation criteria used.

(3) Three-Dimensionality. As noted earlier, even a uniform flow about a stationary cylinder exhibits chordwise and spanwise variations. These three-dimensional effects may play a major role in the stretching of vortex filaments and in the redistribution of vorticity in all directions. The numerical models are not in a position to account for such complex effects. One may hope to assess the effects of three dimensionality by means of two-dimensional numerical experiments.

3.5 LIFT OR TRANSVERSE FORCE

The lift force data are far less plentiful and far less consistent than the drag force data (see Fig. 3.4). The scatter is attributed to various causes. Humphreys (1960) has experimentally demonstrated the effect of the end gaps. He obtained relatively larger lift coefficients when the cylinder ends were sealed at the wall. In general, many researchers noted the sensitivity of the lift force to the stream turbulence. The degree of rigidity of the mounting of the cylinders may also play an important role. As noted earlier, even the smallest transverse oscillations of a cylinder (say, amplitude of oscillation/diameter = 0.05) distinctly regularises the vortex shedding and considerably increases the spanwise correlation (see, e.g., Blevins 1977).

The shedding of the vortices gives rise also to oscillations in the drag force at a frequency twice the vortex-shedding frequency. The magnitude of such oscillations are relatively small (Drescher 1956, McGregor 1957, Fung 1960). In spite of that, however, violent in-line oscillations at a frequency twice the vortex-shedding frequency may develop (Wootton et al. 1974) and even alter the character of vortex formation in the near wake.

3.6 FREE STREAM TURBULENCE AND ROUGHNESS EFFECTS

An extensive review of the effects of roughness and other factors on the flow past circular cylinders has been presented by Engineering Sciences Data Unit (ESDU) (1970). This work shows both the importance and lack of complete understanding of these effects (see also Dalton 1977).

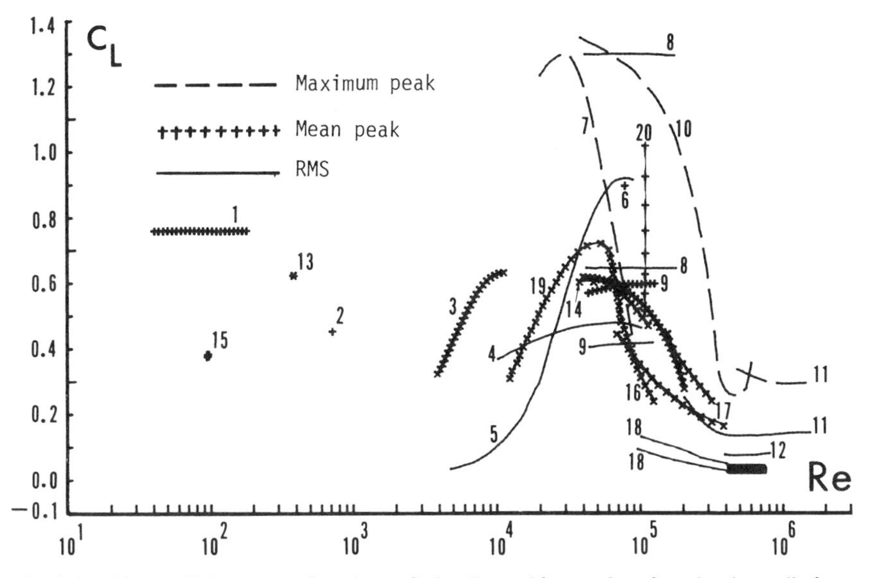

Fig. 3.4. Lift coefficient as a function of the Reynolds number for circular cylinders: 1, Phillips (1956); 2, Schwabe (1935); 3, Bishop and Hassan (1963); 4, Keefe (1962); 5, Gerrard (1961); 6, Bingham et al. (1952); 7, Macovsky (1958); 8, Vickery and Watkins (1962); 9, McGregor (1957); 10, Humphreys (1960); 11, Fung (1960); 12, Schmidt (1965); 13, Jordan and Fromm (1972); 14, Macovsky (1958); 15, Dawson and Marcus (1970); 16, Weaver (1971); 17, Goldman (1958); 18, Bublitz (1971); 19, Warren (1962); 20, Schmidt (1965).

Fage and Warsap (1929) were the first to investigate the effects of grid-generated turbulence, tripping wires, and surface roughness on the flow past circular cylinders (see also Goldstein 1938, and Schlichting 1968, for partial accounts of this study). Fage and Warsap increased surface roughness both by covering their cylinders with different grades of abrasive paper and by installing tripping wires in the boundary layer. They changed the free-stream turbulence level by placing a coarse rope mesh at different distances upstream from the cylinders. In all cases ($10^4 < \mathrm{Re} < 2.5 \times 10^5$), the transitions shifted strongly to lower Re with increasing disturbance. While a variation in upstream turbulence did not appear to alter the severity of the dip in the drag coefficients, increases in cylinder roughness very definitely did. In fact, the dip in C_d almost vanished on an extremely rough cylinder. Fage and Warsap attributed the increase in the drag coefficient with roughness in the supercritical Reynolds number range to the retardation of the boundary-layer flow by roughness and, hence, earlier separation. They have also noted that 'when the surface is very rough the flow around the relatively large excrescences and so around the cylinder, is unaffected by a change in a large value of the Reynolds number.'

Achenbach (1968) made measurements of pressure and skin-friction over a

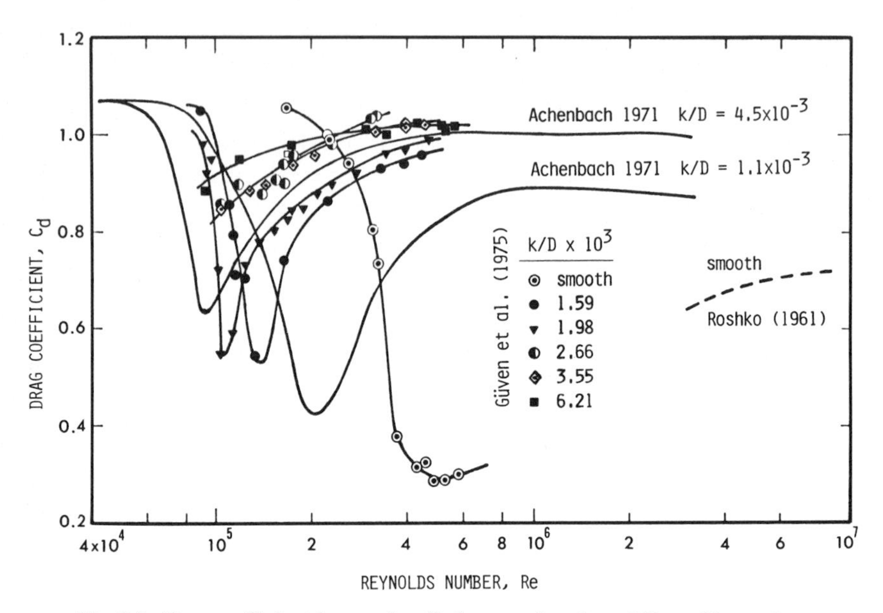

Fig. 3.5. Drag coefficient for rough cylinders as a function of Reynolds number.

Reynolds number up to Re = 3×10^6. His measurements showed, among other things, that beyond some large value of Re the drag coefficient becomes independent of Re, in conformity with Fage and Warsap's suggestion, and that, in the range of Reynolds-number independence, the drag coefficient shows a definite dependence on the relative roughness k/D (see Fig. 3.5). Achenbach also found that the larger roughness results in higher skin-friction and hence in greater retardation of the boundary layer. These in turn result in earlier separation and larger absolute values of the base pressure coefficient. One should not infer from this that flows over rough cylinders with identical separation points will result in identical pressure distributions and drag coefficients. Such a correspondence between separation position and drag coefficient does not exist because of the fact that the pressure distribution is affected not only by the location of separation but also by the boundary layer ahead of separation. For example, Achenbach obtained an identical separation angle of $\theta_s = 110$ degrees for k/D = 1.1×10^{-3} at Re = 4.3×10^5 and for k/D = 4.5×10^{-3} at Re = 3×10^6, but the pressure distribution and the drag coefficients for the two cases were considerably different. The foregoing is an over-simplification of the complex interaction between the wake, separation points, and the flow over the forebody of the cylinder. One cannot intelligently isolate the flow over a given region or at a singular point so as to explain the observations or measurements over the remainder without getting involved in a circular argument.

There are considerable differences between the results of Miller, Guven, Achenbach, and Fage and Warsap. This discrepancy is attributed (see e.g., Guven 1975) to the effect of the end gaps left between the test cylinder and the extension pieces attached to the tunnel walls as noted by Fage and Warsap. With such gaps, the wake of the cylinder is supplied with high pressure fluid from the front and as a result smaller values of C_d are expected since the base pressure is increased over the value it would attain otherwise. In addition, the spanwise correlation may have been different in the two studies so far cited not only because of the three-dimensional flow through the gaps but also because of the difference between the length-to-diameter ratios used.

Batham (1973) has reported experiments on the effects of surface roughness of the continuously-distributed type ($k/D = 2.17 \times 10^{-3}$) and of freestream turbulence on the mean and fluctuating pressure distributions on circular cylinders at $Re = 1.11 \times 10^5$ and $Re = 2.35 \times 10^5$. Szechenyi (1975) carried out an extensive investigation in which he measured steady drag coefficients and unsteady lift coefficients of rough-walled cylinders over a range of Reynolds numbers up to $Re = 6.5 \times 10^6$. Both Batham (1973) and Szechenyi (1975) were interested in simulating the pressure distributions at high Reynolds numbers through the use of roughness. Szechenyi suggested that the drag coefficient in the supercritical regime is a function of the roughness Reynolds number Vk/ν *only* for $k/D = 1.16 \times 10^{-4}$ to 2×10^{-3} (see also Armitt 1968). Guven, Patel and Farell (1975) argue that this observation is at variance with their own measurements as well as with previous findings.

Guven (1975) has undertaken a comprehensive investigation of the effect of surface roughness on cylinders and found that (a) roughness has a strong influence on circular cylinders even at very large Reynolds numbers; (b) beyond some large values of Re which depend on the surface roughness, the pressure distribution becomes independent of Re and is determined by the characteristics of the surface roughness; (c) freestream turbulence appears to have no influence on the pressure distribution at such large Reynolds numbers; (d) the effects of external ribs are generally similar to those of distributed roughness even though there are strong local influences of ribs; (e) at large Re, a cylinder with $k/D = 10^{-5}$ behaves essentially as a smooth cylinder even though the boundary layer may be fully rough, because the additional boundary resistance due to such small surface roughness is very small; (f) for distributed roughness, the major effects of roughness on the pressure distribution are observed in the range of k/D up to 2.5×10^{-3}. For large values of k/D, additional effects of the pressure distribution are relatively small.

Guven (1975) proposed an analytical model, with several assumptions and approximations, which seems to yield good agreement in general between experiment and analysis.

For a highly irregular and relatively rough surface subjected to marine environment, the Reynolds number and the roughness Reynolds number are quite large. Under these circumstances, one finds, as will be seen later, that the drag and inertia coefficients approach nearly constant values, independent of Re and $Re_k = Vk/\nu$. This still does not eliminate the need for the apparent diameter of the cylinder in calculating the projected area and the effective volume of the structures (for use in calculating the inertial force). If a marine-roughened pipe is available, an average diameter may be directly measured including the average height of the protrusions. If, on the other hand, a structure is to be built in a location for which no prior information exists regarding the growth of marine roughness, it is advisable either to place pipes at various depths in location to gather such pertinent information as a function of time or to take strong anti-fouling measures.

In the foregoing we have dealt only with the drag coefficient. As noted earlier, vortex shedding gives rise to a transverse force. Szechenyi (1975) found that the lift coefficient C_L is a function of the roughness Reynolds number. For $Re_k > 10^3$, he found C_L (rms) = 0.25 to 0.30 at a Strouhal number of $f_0D/V = 0.25$ to 0.27. This Strouhal number is quite close to that obtained by Roshko (1961) at post-supercritical Reynolds numbers with smooth cylinders. Roshko attributed the re-establishment of the regular vortex shedding to a return to the generation of two-dimensional vortices. As will be discussed later, similar results are obtained with sand-roughened cylinders in two-dimensional harmonic flows. The importance of this observation is that roughness may enhance the possibility of the occurrence of hydroelastic oscillations partly by regularizing the vortex shedding, partly by increasing the coherence length (Shih and Hove 1977), and partly by increasing the total lift force. Finally, there does not seem to be much information on the effect of roughness on the lift and drag coefficients for yawed cylinders in uniform flow (Glenny 1966). Such information will be useful not only for marine engineering but also in aerodynamics in connection with the determination of in-plane and out-of-plane forces acting on artificially roughened slender bodies at large angles of attack.

For additional information on the effects of roughness in boundary layers the reader is referred to papers by Betterman (1966), Dvorak (1969), Simpson (1973), Roberson et al. (1974), Furuya et al., (1976), Granville (1978), and Knight and McDonald (1979). The effect of roughness on resistance in wavy and harmonic flows will be treated separately.

Miller (1977) investigated the influence of marine roughness upon the drag coefficients in steady flow in a wind tunnel. He found that marine growths and sand roughness of similar size have comparable effects as far as the drag coefficient is concerned. Miller suggested that the drag coefficients obtained from tests in steady flow over rough cylinders may be applicable to wave flows at least when the loading is predominantly drag. This opinion is not commonly shared, not at least for the reasons suggested by Miller [see discussions to Miller's paper (1977)].

Guven (1975) used the average size of particles on commercially available sandpapers and the smooth cylinder diameter to define k/D. Miller (1977) used for k the reciprocal of the grit number for the sand paper and the physical size of the particles (pearl barley) he glued on the smooth cylinders. He inferred, from a comparison of the drag coefficient versus Re curves for the marine-roughened and barley roughened cylinders, that the representative k/D value for the marine roughened cylinders ranged from 1.5×10^{-2} to 4.4×10^{-2}.

Not all of the differences between the drag-coefficient versus Re curves for rough cylinders are attributable to the test conditions (e.g., intensity and scale of turbulence in the ambient flow, nonuniformity of the velocity distribution, end-gap effects, length-to-diameter ratio, blockage effect, method of measurement, flexibility of the mounting of the cylinder, care taken in gluing the edges joining the sand paper, etc.). The difficulty of uniquely specifying 'roughness' is partly responsible for the differences between the steady-flow data reported by various workers, particularly in the drag-crisis region.

It has been known for quite some time (see Schlichting 1968) that the effective surface roughness may be larger or smaller than the nominal relative roughness based on the geometric size of the roughness element depending on the shape and arrangement of the roughness elements. In other words, the shape, size and physical distribution, and packing of the roughness elements, in a given flow demand some justification for the one-parameter characterization of the effects of roughness in terms of k/D. There exists very little systematic investigation of the influence of roughness density for three-dimensional roughnesses. Dvorak (1969) collated the data on roughness density and presented it as a function of a parameter λ_r where

$$\lambda_r = (\text{total surface area})/(\text{plain roughness area})$$

He concluded that the peak skin-friction drag is associated with values of λ_r near 5.

It would be desirable to express the roughness in terms of an equivalent sand roughness, determined in a manner suggested by Schlichting (1968). He used a special experimental channel of rectangular cross-section with three side-walls and one long, interchangeable side-wall whose roughness was varied to suit the experiment. By measuring the velocity distribution in the central section, he determined the shearing stress and hence B in the universal equation

$$\frac{u}{v^*} = 5.75 \log y/k + B$$

where $B = 8.5$ for $k = k_s$ in the completely rough regime. Then the equivalent sand roughness k_s, was obtained from

$$5.75 \log k_s/k = 8.5 - B$$

The use of k_s obtained in this manner may not be satisfactory for time-dependent flows nor the time and effort required be commensurate with the need.

Experiments with relatively small rigid roughness elements glued on a large smooth cylinder do not reflect all the complexities encountered in expressing the roughness of a pipe subjected to marine environment. A structure in the marine environment may be covered with rigid (scale, barnacles, mussels, etc.) and soft (seaweeds, anemones) excrescences. The thickness of the accumulated growth may considerably increase the effective diameter or the characteristic size of the structure. Furthermore, the size of the accumulated growth may change with time and along the structure depending on the prevailing temperature, currents, ecological effects of the structure on the existing marine life, etc.* Thus, it is not a simple matter to decide what fraction of the protrusions constitutes an average roughness or equivalent sand roughness and what fraction constitutes an increase in the radius of a circular cylinder.

3.7 SPECIAL TIME-DEPENDENT FLOWS

3.7.1 Introduction

Unsteady motion is of great interest in the solution of many applied technical problems in fluid mechanics, such as the motion of bodies through fluids, fluid motion in or about bodies, free-surface flow phenomena, and also in the motion of explosion products.

The improvement of experimental techniques and the development of high speed computers, which made possible the numerical solution of the governing nonlinear differential equations, have stimulated research on time-dependent flows. Some of this research produced results which are oblivious of physical needs. Some of the findings are, out of necessity, applicable only to very special situations. It is as well to face from the outset the fact that current theoretical research is in a position to offer very little quantitative help to the engineer concerned with unsteady flows. Observation is still the key word.

As in many other fields, the literature on time-dependent flows is growing so rapidly that it is all but impossible for any one person to follow it in detail.

*In the North Sea, for example, the thickness of the accumulated roughness reached about 8 inches in two years with a rate of growth of about one inch per month. As a consequence, the diameter of certain members increased 4 to 12 inches.

Thus, even after the most careful attempts at simplification and uniformity of exposition, the material treated here still involves a heterogeneous combination of mathematical predictions, experimental facts and empirical equations that may at first appear unpalatable to many readers. Here we have attempted to present not a survey but a sampling of incompressible unsteady flows.

There are many outstanding lectures, survey articles, and critically composed chapters in various books on time-dependent flows. To be sure, some are restricted to aeronautical engineering applications and some to the exposition of only classical laminar flow problems. Temple (1953) reviewed the progress that had been made until 1953 in the theory of unsteady flows about wings and slender bodies. Jones (1962), in his Minta Martin Lecture in 1962, described the aerodynamic aspects of separated unsteady boundary layers. The unsteady aerodynamics of potential flows has been reviewed by Garrick (1966). The phenomenon of periodic vortex shedding from a symmetrical bluff body, an unsteadiness which is caused by hydrodynamic instability rather than (but not necessarily) by a well-defined time-dependence of the ambient flow, has been the subject of extensive review (Rosenhead 1953, Wille 1960, Marris 1964, Morkovin 1964, Wille 1966, Mair and Maull 1971, Berger and Wille 1972, McCroskey 1977), (see also papers in ASME Symposium on Unsteady Flow 1968). The study of the theory of separation-free time-dependent laminar flows has enjoyed particular attention, partly due to its practical significance and partly due to its relative mathematical simplicity. Surveys of Stewartson (1960), Stuart (1963), and Rott (1964) are outstanding reflections of the current state of the art. It is well worthwhile to study these contributions prior to embarking on research on unsteady flows and transient motions.

The hydrodynamics of unsteady flows, in particular those set in motion impulsively from rest, is most aptly described by Sedov (1965). The unsteady motion of continuous media, with particular attention to the unsteady motion of gases, is the subject of a large volume of work by Stanyukovich (1960). The foregoing list of references is by no means complete and there are, to be sure, a great many other survey articles dealing directly or peripherally with the unsteady motion of viscous or inviscid, separated or unseparated, and cavitating flows.

The results of nearly a hundred years of ingenious research on the motion of jets, wakes, and cavities (essentially treated as steady and inviscid) were brought together by Birkhoff and Zarantonello (1957). Since then, the revolutionary changes in computers and numerical methods have generated numerous new results and concepts which are scattered throughout the literature.

The flow of fluid involving separation is beyond the reach of rigorous calculation, even if the flow field is assumed to be the same as for an inviscid fluid. In reality, the flow behind a bluff body, moving steadily through a fluid, is accompanied by large-scale unsteadiness. Thus, any type of unsteadiness of the ambient flow and/or of the motion of the body introduces additional changes in the characteristics of the flow.

The formation of a wake gives rise not only to a form drag, as would be the case if the motion were steady, but also to significant changes in the inertial forces. The velocity-dependent form drag is not the same as that for the steady flow of a viscous fluid, and the acceleration-dependent inertial resistance is not the same as that for an unseparated unsteady flow of an inviscid fluid. In other words, the drag and the inertial forces are interdependent as well as time-dependent. Although indirect, the role of viscosity is paramount in that its consequences are separation, vortex formation and shedding, and resultant alterations in the virtual mass. The specification of these various aspects provides a basis for the correlation of theoretically predicted and observed forces. Evidently, the coefficients obtained for unseparated unsteady flows are not applicable to occurrences in which the duration of flow in one direction is long enough and the body form blunt enough for separation to occur. As noted previously, it is thus clear that it is necessary to determine the relationships between various resistance components in terms of the unsteadiness of the ambient flow, the geometry of the body, the degree of the upstream turbulence, the roughness of the object, and the past history of the flow.

As in a great many steady flow problems, where the potential flow analysis provides many approximate solutions and a strong impetus for the need for the exact or approximate solution of the Navier-Stokes equations, it is especially challenging in unsteady flows to try to approximate real wake behavior using models involving potential theory (see e.g., Sarpkaya and Shoaff 1979). One may expect that some of these approximations will provide realistic solutions (in the sense of matching the observed characteristics) and will help to make it possible to introduce, eventually, the effects of roughness, viscosity, gravity, turbulence, past history, etc. The comparison between the theoretical results and the experimental observations may not be all 'sweetness and light,' but the objective is not always the achievement of a favorable comparison but rather the careful selection of those found to be reasonably accurate from among many cases tried.

Any agreement between the experimental results and the approximate theoretical predictions should be regarded only as a first step towards further exploration of the potential flow methods. Impulsive flow experiments necessarily involve a period of initial acceleration from rest to a uniform velocity, whereas the numerical methods do not. Thus, the effects of history of the fluid motion, among other things, are not apparent in the comparison and they may have significant effects on both the instantaneous value of the resistance as well as on the time to reach that instantaneous value of the resistance.

3.7.2 Impulsively Started Flows

Such flows are common examples of non-steady boundary layers and have some practical importance particularly when an accident sets the fluid impulsively in motion about the bodies immersed in it (e.g., a loss-of-coolant accident in boiling-water nuclear reactors).

Impulsively started flow is one of those unsteady flow situations for which analytical and numerical solutions exist at least for small times and relatively low Reynolds numbers.

At the early stages of the motion the vorticity does not have enough time to diffuse. Thus, the boundary layers are very thin and the flow is essentially irrotational. The fluid force acting on the body is primarily inertial and the inertia coefficient is $C_m = 1 + C_a$, C_a being the added mass coefficient obtained from the potential flow theory. For bodies without sharp corners (e.g., a circular cylinder), the separation does not occur immediately. Furthermore, it does not necessarily initiate at the downstream stagnation point (as in the case of an elliptic cylinder).

For two-dimensional cylinders, it can be shown that the separation begins after a time t_s at a place where the absolute value of dU/dx is largest. The relationship between t_s and dU/dt is (see Schlichting 1968)

$$1 + \left(1 + \frac{4}{3\pi}\right)\frac{dU}{dt}\, t_s = 0 \tag{3.6}$$

For a circular cylinder started impulsively from rest to a constant velocity, the distance covered until separation begins is $s = 0.351c$, c being the radius of the cylinder. The separation begins at the rear stagnation point. For a uniformly accelerating circular cylinder the same distance is $s = 0.52c$ and obviously greater than that for the case of impulsive motion.

For axisymmetric bodies t_s is given by the expression (Schlichting 1968)

$$1 + t_s\left[\frac{dU}{dx}\left(1 + \frac{4}{3\pi}\right) + 0.15\frac{U}{r}\frac{dr}{dx}\right] = 0 \tag{3.7}$$

For a sphere impulsively set in motion $s = 0.392c$. The distance covered by the sphere until separation begins is larger, as in the case of the cylinder, when the sphere is accelerated uniformly from rest. Evidently, the rate of acceleration as well as the history of acceleration is important in the calculation of the relative distance covered prior to the occurrence of separation.

For a circular cylinder undergoing harmonic oscillations, the flow may be assumed to accelerate uniformly, at least during the early stages of acceleration.

Then the relative distance or preferably π times that distance, denoted by K, is $K = 2\pi s/2c = 0.52\pi = 1.63$. For bodies with sharp corners separation starts immediately and the added mass coefficient is not necessarily equal to that given by the potential theory for the unseparated flow.

The role played by separation on the added mass is an intriguing one and must be understood clearly. Consider an impulsive change superimposed on an already established flow pattern. Just prior to the impulsive change, the drag coefficient is given by its steady state value at the corresponding Reynolds number. Sears, as reported by Rott (1964), has shown that 'the initial motion following the impulsive change of the conditions consists of the superposition of the velocity pattern existing just before the change and the inviscid flow velocity pattern due to the impulsive boundary values.' In other words, at the initial instants of the impulsive change the drag coefficient is equal to its steady state value and $C_m = 1 + C_a$, (C_a being equal to that given by the potential theory, e.g, $C_m = 2$ for a circular cylinder). As time progresses, neither C_d nor C_a remains the same and changes with the evolution of the flow, ever dominated by the past history and ever affected by the gross features of the current state. Thus, it may be said that the changes in the added mass coefficient come about not because of separation but rather by the changes in the state of separated flow, e.g., additional motion of separation points, the increase or decrease of the rate of circulation, etc.

Theoretical investigations of the impulsively started motion of a circular cylinder in a fluid otherwise at rest are confined mostly to early times and very small Reynolds numbers. Such a motion was first considered by Blasius (1908) and his work was later extended by Goldstein and Rosenhead (1936), Görtler (1944, 1948), Schuh (1953), Watson (1955), and Wundt (1955). It was found, as noted earlier, that, after a certain lapse of time, the boundary layer separates from the surface of the cylinder; the time and location of separation depending on the Reynolds number and the bluffness of the body. The separation points then move rapidly around the cylinder until at large times they coincide with the average positions of the points of laminar separation for steady flow.

Finite difference techniques have been employed by several investigators (Payne 1958, Hirota and Miyakoda 1965, Kawaguti and Jain 1966, Wang 1967, Jain and Rao 1969, Rimon 1969, Son and Hanratty 1969, Thoman and Szewczyk 1969, Honji 1972, Mehta and Lavan 1972, Collins and Dennis 1973a and 1973b, Wu and Thompson 1973, Lugt and Haussling 1974, Telionis and Tsahalis 1974, Bar-lev and Yang 1975, Panniker and Lavan 1975, Cebeci 1979). Among others, Tuann and Olson (1976) employed the finite-element method.

Experiments at relatively low Reynolds numbers have been reported by Schwabe (1935), Taneda and Honji (1971), Taneda (1972), and Coutanceau and Bouard (1977).

There are very few experimental data for impulsively started flow at sufficiently high Reynolds numbers, i.e., for Reynolds numbers in the supercritical and post-supercritical regimes. This is partly because of the experimental difficulties encountered in establishing a vibration-free impulsively-started steady flow and partly because of the instrumentation required to measure the transient quantities involved. In fact, the force that acts on the cylinder in impulsive flow at relatively high Reynolds numbers has been measured directly only by Sarpkaya (1966, 1978).

Schwabe (1935) used a circular cylinder with a radius of $c = 1.7$ inch. Experiments were conducted in an open channel with water. The velocity of the cylinder was $V = 0.328$ inch/sec. The Reynolds number was about 600. Velocities were determined from the path lengths of the particles suspended on the water surface. Other quantities such as the time rate of change of the difference of the square of velocities, rate of change of circulation, and the radius of curvature of the streamlines needed to calculate the drag coefficient from the equation of Bernoulli were all determined by graphical methods. A careful examination of Schwabe's work tends to indicate that a considerable amount of experimental error may have existed in the evaluation of the pressures and hence in the resulting drag coefficient because of the very small velocities involved in the calculations (Schwabe's drag coefficient is about twice the steady-state value, and still increasing, when the cylinder has moved about 9 body radii).

Bingham et al. (1952) carried out a number of experiments in a shock tube to observe the influence of Reynolds and Mach number on the impulsive loading of a 0.5 inch cylinder for the Reynolds number range of 3.1×10^4 and 7.7×10^4 and the Mach number range of 0.15 to 0.4. The pressures were determined from the density fields and the drag coefficients were calculated by an integration of the pressures. The variation of the drag coefficient with Vt/c, for a given set of Re and M values and from one set of M, Re to another, is quite significant. As pointed out by Bingham et al., this is partly because of the growth and shedding of vortices and partly because of the different rates of decay of the initial high pressure on the rear of the cylinder due to passage of the two branches of the original shock. It appears that it is rather difficult to extract accurate lift- and drag-force information from the shock-tube experiments through the use of interferograms.

Friberg (1965) 'quickly' immersed a circular cylinder into the steady and uniform free-stream flow field of a water table. Asher and Dosanjh (1968) conducted experiments in a shock tube in a manner similar to that done by Bingham et al., and determined the characteristics (e.g., position of vortices, their relative velocities, Strouhal number, etc.) of the wake. Contrary to their assertions, their experiments suffer from the same drawbacks as those of Bingham et al. Suffice it to note that, other than numerical experiments, there is no mechanical or traveling-shock system which is capable of generating a truly impulsive flow. Efforts to generate impulsive or uniformly-accelerated flow at high Reynolds

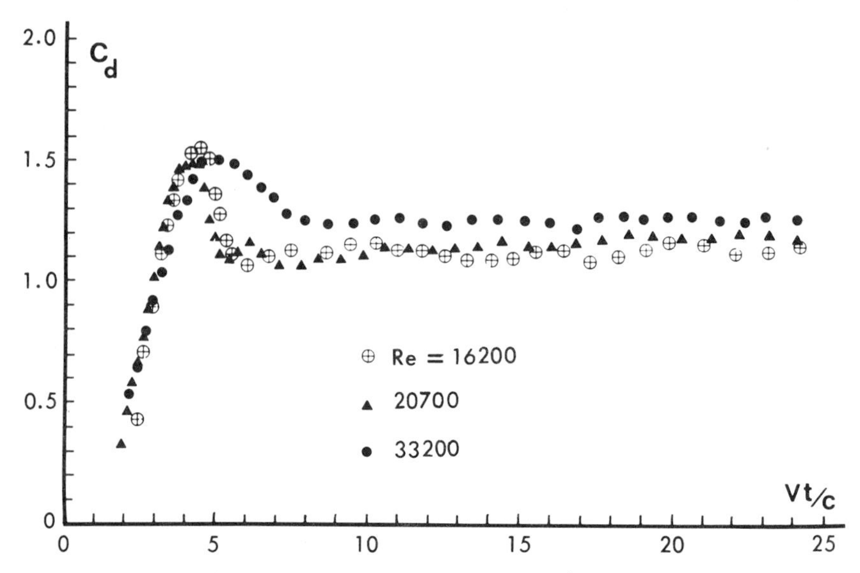

Fig. 3.6a. Measured drag coefficient as a function of Vt/c for impulsively-started flow (Sarpkaya 1978).

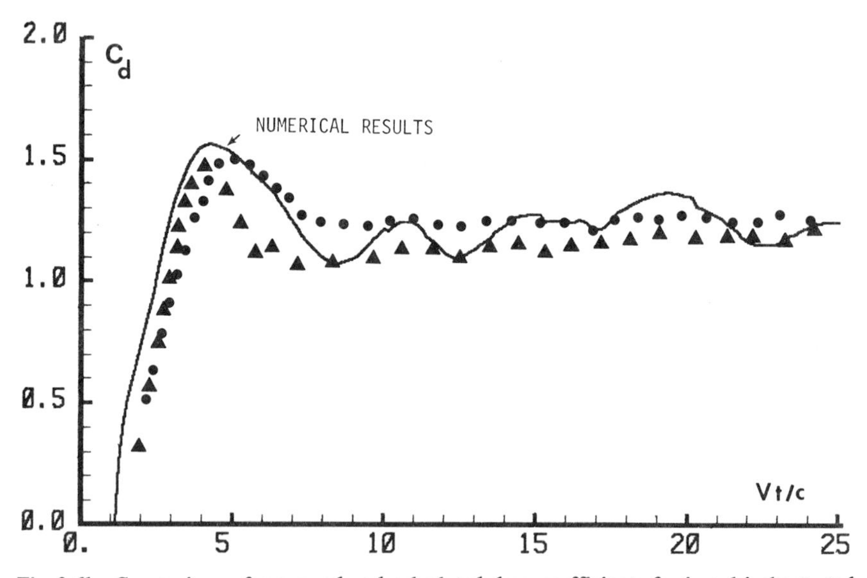

Fig. 3.6b. Comparison of measured and calculated drag coefficients for impulsively-started flow (Sarpkaya and Shoaff 1979a).

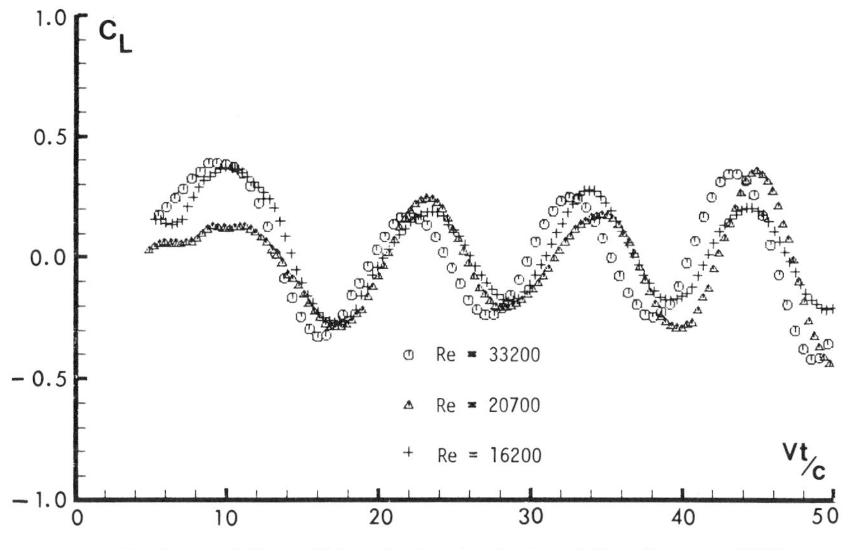

Fig. 3.7. Measured lift coefficient for impulsively-started flow (Sarpkaya 1978).

numbers or accelerations in a liquid medium may be hampered by the generation of compression and rarefaction waves and regions of intense cavitation. These are some of the difficulties of the experiments with impulsive flows.

Sarpkaya (1978) has recently repeated his previous work in a larger vertical water tunnel with considerably more sophisticated instrumentation and measured simultaneously both the in-line and transverse forces. The representative data are shown in Figs. 3.6 and 3.7. The numerical simulation through the use of the discrete vortex model (Sarpkaya and Shoaff 1979a and 1979b) of the evolution of the early stages of the motion is shown in Fig. 3.8 for successive values of Vt/c. The growth and motion of the vortices in the later stages of the fully established flow are shown in Fig. 3.9.

It is seen from Fig. 3.6 that the drag coefficient in the initial stages ($Vt/c \simeq 4$) of an impulsively started flow can exceed its steady value by as much as 30 percent. In the early periods of the flow, vorticity is slow to diffuse and therefore accumulates rapidly in the close vicinity of the cylinder. Although the growing vortex soon reaches unstable proportions and separates from its shear layer, the growth of the vortices are so rapid that the vortices become much larger than their quasi-steady-state size (at about the same positions) before they separate from their shear layers. This leads to the observed large drag coefficient, [see also Roos and Willmarth (1971) for a similar observation with spheres]. Shortly after the onset of asymmetry, the drag coefficient decreases sharply and

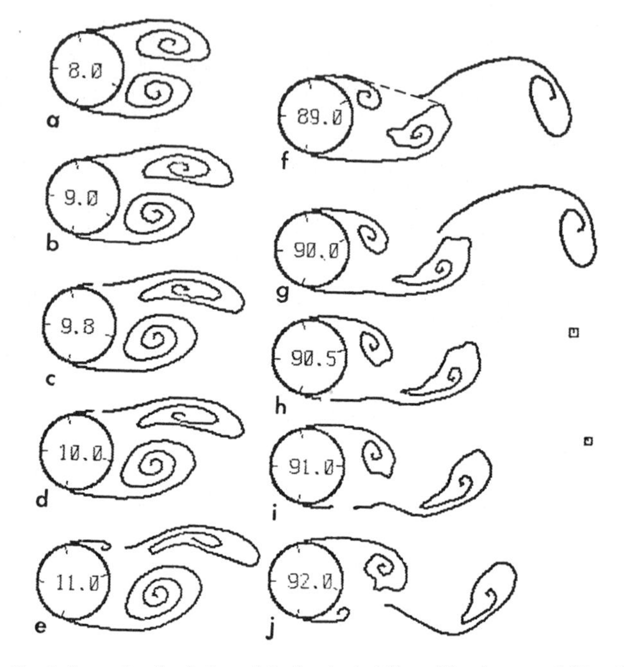

Fig. 3.8. Evolution of wake in impulsively-started flow (Sarpkaya and Shoaff 1979a).

the lift coefficient begins to increase. Subsequently, it oscillates with the frequency f_0 of the shedding of vortices, from the same side of the cylinder.

The impulsive flow has long been regarded as analogous to the evolution of separated flow about slender bodies moving at high angles of attack in the subsonic to moderately supersonic-velocity range. The approximate flow similarity between the development of the crossflow with distance along an inclined body of uniform diameter and the development with time of the flow on a cylinder in impulsive flow is known as the 'cross-flow analogy.' This analogy was first suggested by Allen and Perkins (1951) and subsequently used by many other researchers to calculate the in-plane normal force and the out-of-plane force (side force normal to the plane of flight). A detailed discusion of the analogy, extensive measurements for various nose-shapes and body combinations, and the most pertinent references may be found in (Thomson and Morrison 1969, 1971; Thomson 1972; Bostock 1972; Lamont 1973; Lamont and Hunt 1976; Wardlaw 1974; and Ericsson and Reding 1979). It must be noted that the analogy is far from perfect. A blunt-nosed cylinder at high angles of attack generates a stationary asymmetric vortex array which is similar to the Karman vortex street. Thus, the approximate space-time equivalence is possible because the vortices

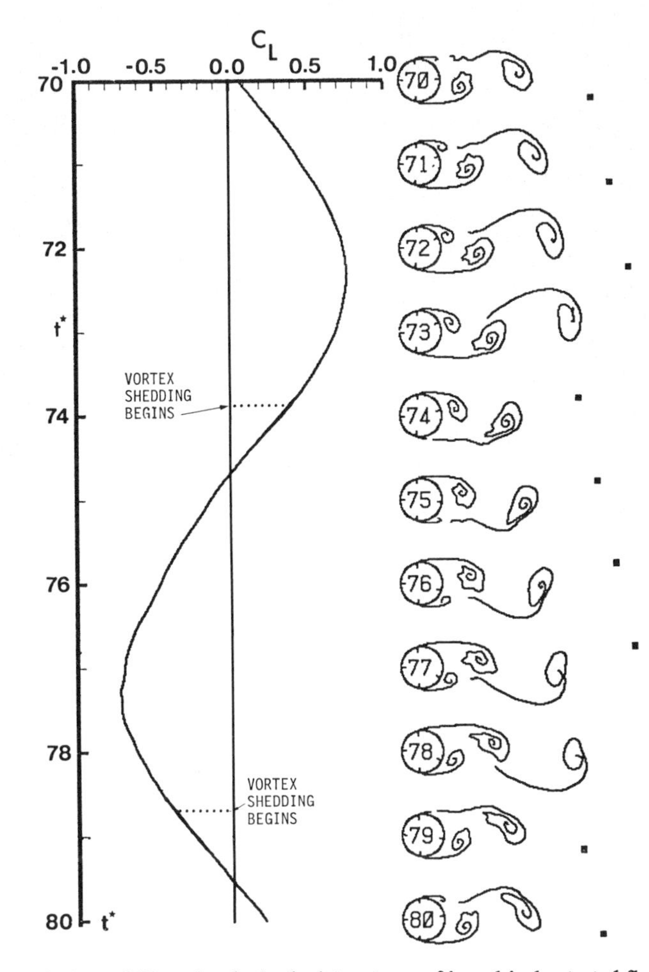

Fig. 3.9. Evolution of lift and wake in the later stages of impulsively-started flow (Sarpkaya and Shoaff 1979a).

have an axial degree of freedom for their lift-off to make up the asymmetric array. On a pointed slender body the first asymmetric vortex pair originates at the apex and the above analogy no longer applies. Furthermore, asymmetric vortices give rise to a coupling between longitudinal and lateral degrees of freedom, often discontinuous and associated with hysteresis effects. Consequently, the phenomenon ceases to be a simple two-dimensional space-time equivalence and becomes a complex three-dimensional fluid-structure interaction problem (for additional references see Ludwieg 1979).

3.7.3 Uniformly Accelerating Flows

Additional insight into the role played by time dependence may be gained by considering a relatively more manageable case; namely, the unidirectional flow with constant acceleration and the unidirectional acceleration of a body in a stationary fluid.* The immediate consequence of the use of constant acceleration is that equations describing the motion may be reduced to a workable form dependent only on the relative displacement of the fluid and the Reynolds number.

Unidirectional acceleration of a body in a stationary fluid has been investigated by Iversen and Balent (1951); Stelson and Mavis (1955); Bugliarello (1956); Keim (1956); Laird and Johnson (1956); Laird, Johnson, and Walker (1959); Sarpkaya and Garrison (1963); Odar and Hamilton (1964); Mellsen, Ellis, and Waugh (1966); Hjelmfelt and Mockros (1967); Mavis (1970); Hamilton and Lindell (1971); and Hamilton (1972).

Keim (1956) conducted experiments with cylinders and one disk accelerated vertically from rest in a water tank. The driving force was held constant and acceleration and velocity were measured as functions of time. A single force coefficient combining the effects of drag and inertia was plotted as a function of $(dV/dt)D/V^2$, acceleration modulus. This correlated the data fairly well, although Reynolds number effect was noted, ($300 < Re < 4{,}750$). In the experiments by Laird and his associates, the results were again expressed in terms of total resistance coefficient in which the effects attributable both to velocity and to acceleration were necessarily superposed. For a study of the flow patterns about a circular cylinder and a flat plate normal to a uniformly decelerated flow the reader is referred to Tatsuno and Taneda (1971). The measured added masses compared well to the spherical body of shape equal to the sphere plus the boundary-layer displacement.

The studies by Mavis (1970) and Lai (1973) dealt respectively with the determination of added mass of disks and displacement of spheroids.

Sarpkaya and Garrison (1963) have shown, through the use of the generalized Lagally's theorem (see Sarpkaya 1963) that the total resistance for a circular cylinder in a flow with unidirectional constant acceleration may be written as

$$F = \tfrac{1}{2}\rho V^2 D\, G(s/D) + \tfrac{1}{4}\rho\pi D^2 (dV/dt)\, H(s/D) \tag{3.8}$$

*Mathematically, there is a direct relation between the two flow situations. As shown by Batchelor (1967), "The equation of motion of a fluid in the moving frame is therefore identical in form with that in an absolute frame provided that the fictitious body force $-F_0$ (assuming a non-rotating frame) per unit mass acts upon the fluids in addition to the real body and surface forces $\cdots$ F_0 is simply the apparent body force that compensates for the translation acceleration."

where G(s/D) and H(s/D) represent two functions dependent on the relative displacement s/D; and s, the displacement of the ambient flow; namely, $s = 0.5Vt = 0.5(dV/dt)t^2$. Equation (3.8) suggests that the total force for at least this particular flow may be written as

$$F = C_d \rho D V^2/2 + C_m \rho \frac{\pi D^2}{4} (dV/dt) \quad (3.9a)$$

or

$$F/(0.25\pi\rho D^2 \; dV/dt) = C = C_m + \frac{4}{\pi}\frac{s}{D} C_d \quad (3.9b)$$

in which C_d is commonly known as the drag coefficient and C_m, the inertia coefficient. It is evident from the foregoing that a unique relationship should exist between C_d and C_m since both depend on s/D. The dependence of C_d and C_m on the Reynolds number does not appear in Eqs. (3.8) and (3.9) because these equations have been developed through the use of the inviscid flow assumption.

The results obtained by Sarpkaya and Garrison are shown in Figs. 3.10 through 3.12. The maximum value of the Reynolds number was 5.2×10^5. Evidently, C_d reaches a maximum value at $s/D = 2.5$. This value compares well with those obtained with an impulsively started flow for which $s/D = 2$. As s/D increases, C_d and C_m fluctuate between certain limits with ever-diminishing amplitudes. For $s/D > 20$, $dC/d(s/D) = 1.53$, $C_m = 1.3$ and $C_d = 1.2$. The experiments should be repeated at higher Reynolds numbers to determine the effect of transition in the boundary layer.

There is always the temptation to use $C_m = 2$ and $C_d = 1.2$ in Eq. (3.9a) to calculate the total force acting on a cylinder in a flow with constant acceleration assuming that the boundary layer over the forebody of the cylinder remains laminar. Acceleration is expected to stabilize the flow and increase the critical Reynolds number over that commonly accepted for the steady flow over a cylinder. There is, at present, no experimental data regarding the occurrence of the 'drag crisis' in accelerating flows.

The comparison of the results obtained with $C_m = 2$ and $C_d = 1.2$ with those calculated through the use of Fig. 3.10 shows that the use of $C_m = 2$ and $C_d = 1.2$ overpredicts the resistance early in the transient and then tends to underpredict it later in the transient. In general, the use of the ideal value of the inertia coefficient and the steady-state value of the drag coefficient cannot be recommended in calculating the total resistance through the use of Eq. (3.9a).

Odar and Hamilton (1964) and Hamilton (1972) proposed the inclusion of

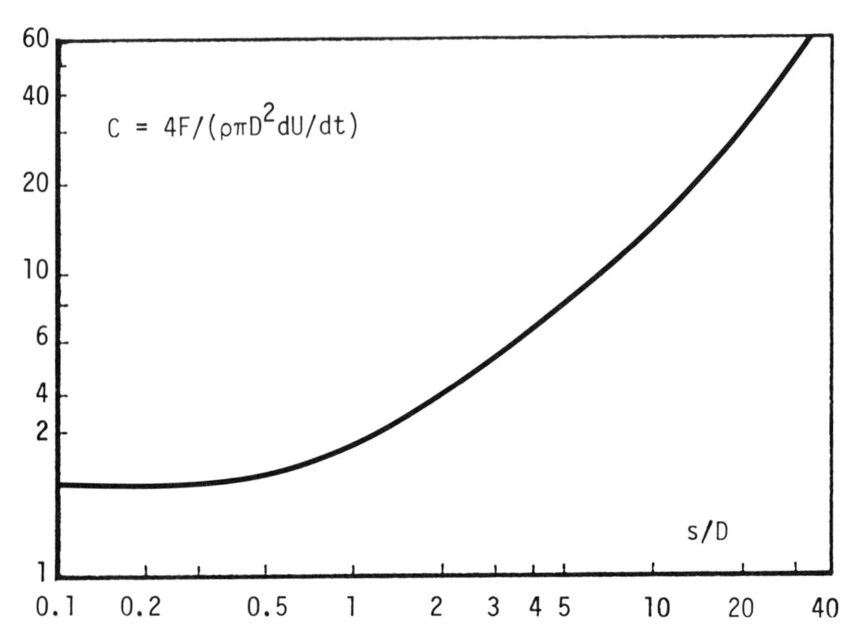

Fig. 3.10. Variation of the total force coefficient with relative displacement in uniformly-accelerated flow.

a 'history term' or 'Basset term'* in Eq. (3.9a). This resulted in an equation with three force-transfer coefficients. Their calculations have been based on very low Reynolds numbers (about 60) and it is not yet clear under which circumstances the history term is significant.

It is apparent from the foregoing that the more general the flow the more complex the problem. In fact, only for relatively simple cases (e.g., uniformly accelerating flow about a cylinder) the total resistance can be expressed as a sum of the drag and inertia forces and the dependence of C_d and C_m on the relative displacement and the Reynolds number can be demonstrated through the use of dimensional analysis.

In general the velocity may vary arbitrarily or as a function of the resistance

---

*Basset (1888) was the first to introduce the history term in connection with the calculation of force acting on a sphere of radius R as

$$F = 6\pi\mu RV + \frac{1}{2}\left(\frac{4}{3}\pi R^3\right)\rho\frac{dV}{dt} + 6R^2(\pi\mu\rho)^{1/2}\int_0^t \frac{(dV/dt')\,dt'}{t-t'}$$

where the last term represents the history effect.

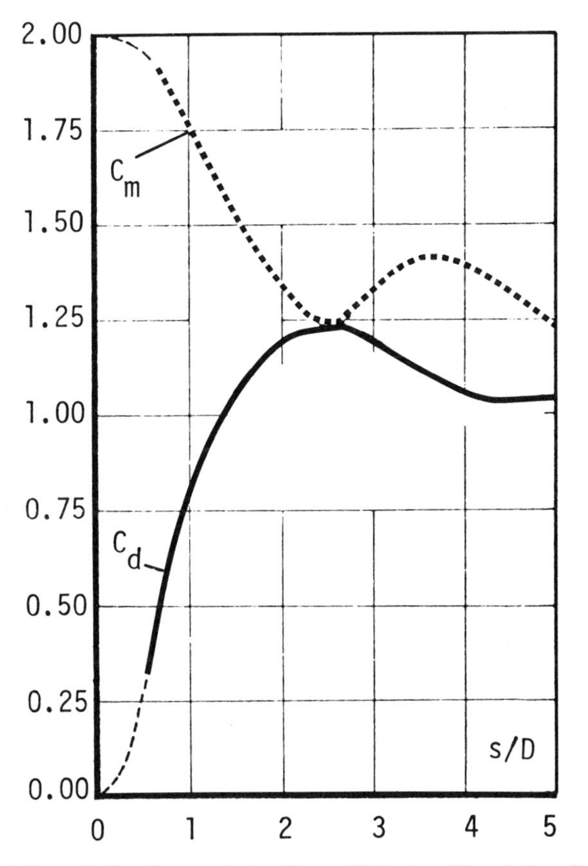

Fig. 3.11. Variation of the drag and inertia coefficients with relative displacement in uniformly-accelerated flow.

experienced by the body (consider the case of a buoyant sphere or surface-piercing object released from rest or set in motion impulsively at some depth below the ocean surface). Under these circumstances one does not know in advance either the fluid resistance or the trajectory of the motion of the body which depends on all the forces (buoyancy, weight, fluid resistance) acting on the body, (see e.g., Clark and Robertson 1961, Viets 1971, Tyler and Salt 1978).

A priori it is evident that both the in-line and transverse forces depend on the instantaneous velocity V of the ambient flow; first, second, and higher order instantaneous accelerations of flow; diameter D of the body; density ρ and the kinematic viscosity ν of the fluid; and finally, on time t. Cavitation, compressibility, yaw, roughness, and proximity effects are not considered below for

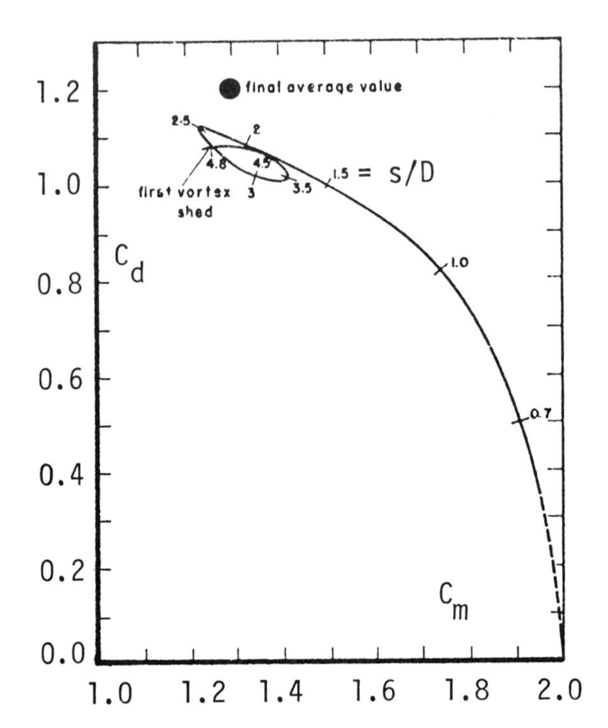

Fig. 3.12. The interrelationship between the drag and inertia coefficients in uniformly-accelerated flow.

sake of simplicity. Then a simple dimensional analysis yields

$$\frac{F}{\rho \frac{\pi D^2}{4} \frac{\partial V}{\partial t}} = f\left\{\frac{VD}{\nu}, \frac{Vt}{D}, \frac{D}{V^2}\frac{\partial V}{\partial t}, \ldots, \frac{D^n}{V^{n+1}}\frac{\partial^n V}{\partial t^n}\right\} \tag{3.11}$$

It is apparent that, when the acceleration is held constant, Eq. (3.11) reduces to

$$4F/(\pi\rho D^2\ dV/dt) = f(Re, s/D) \tag{3.12}$$

Even then the decomposition of F into a drag and inertial force requires some special analysis [such as the use of the discrete vortex analysis (Sarpkaya 1963, 1968; Sarpkaya and Shoaff 1979) or the free streamline theory (McNown 1957) based on inviscid flow assumptions] and relatively difficult experiments and measurements.

It has been noted in connection with the discussion of impulsively-started flows that a drag overshoot occurs at about $s/D \simeq 2$ because of the rapid accumulation of vorticity in the two symmetrically growing vortices (see Fig. 3.6).

In uniformly-accelerating flows, a milder drag overshoot is found (see Fig. 3.11) at about $s/D \simeq 2.5$. Evidently, this is because the rate of accumulation of vorticity is not as large. For all other flows whose velocity is an arbitrary function of time, the drag overshoot may range from about 10 percent (uniformly accelerating flow) to about 30 percent (impulsively-started flow).

3.8 HARMONICALLY OSCILLATING (BODIES) FLOWS

3.8.1 Introduction

Small amplitude harmonic oscillations have been used extensively to determine the added mass of bodies of various shapes (see e.g., Stelson and Mavis 1955, Sarpkaya 1960, Chandrasekaran et al., 1972, Fickel 1973, Ciesluk and Colonell 1974, Skop et al., 1976, Brooks 1976). We will be concerned here primarily with the relatively large amplitude oscillations of flow around the body where the separation of flow plays a very important role. The wavy flow about cylinders will be discussed separately.

There are several fundamental differences between the unidirectional flow and the harmonically oscillating flow and between the harmonically oscillating flow and the wavy flow over a cylinder. When a cylinder is subjected to a harmonic flow normal to its axis, the flow does not only accelerate from and decelerate to zero but changes direction as well during each cycle. This produces a reversal of the wake from the downstream to the upstream side whenever the velocity changes sign. The separation points undergo large excursions. The boundary layer over the cylinder may change from fully laminar to partially or fully turbulent states and the Reynolds number may range from subcritical to post-supercritical over a given cycle. The vortices which have been or are being formed or shed during the first half of the flow period are also reversed around the cylinder during the wake reversal giving rise to a transverse force (as a consequence of their mere convection) with or without additional vortex shedding. This is particularly pronounced for amplitudes of flow oscillation for which the number of newly formed vortices during the period of flow reversal is not much greater than that of the vortices which have survived dissipation (laminar and/or turbulent) and convected around the cylinder during the wake reversal. Particularly significant are the changes in the lift, drag, and inertial forces when the reversely-convected vortices are not symmetric.

The wavy flows are of course relatively more complex (Chapters IV and V). Aside from the effects of the free surface, the orbital motion of the fluid particles give rise to three-dimensional flow effects. The rotation of the wake about a horizontal cylinder and the exponential decay of the representative wave velocity along a vertical cylinder further complicate the matters. In fact, it is because of these reasons that a number of investigators preferred to separate the additional effects brought about by the waviness of the flow from those resulting from the periodic reversal of the flow in a simply harmonic rectilinear motion.

Harmonic flow about cylinders have been investigated by a number of researchers (Keulegan and Carpenter 1958; Heinzer and Dalton 1969; Rance 1969; Thirriot, Longree and Barthet 1971; Hamman and Dalton 1971; Mercier 1973; Sarpkaya 1975a, 1976a-e; Dalton et al., 1976; Yamamoto and Nath 1976; Sarpkaya 1977a; Maull and Milliner 1978; Bearman and Graham 1979; Bearman, Graham, and Singh 1979; Maull and Milliner 1979; Singh 1979).*

As in the case of other time-dependent flows, the most serious difficulty with harmonic flows lies in the description of the time-dependent force itself. Some insight may be gleaned into the nature and decomposition of this force from a remarkable paper by Stokes (1851) on the motion of pendulums. Stokes has shown that the force acting on a sphere oscillating in a liquid with the velocity $U = -A\omega \cos \omega t$ is given by

$$F(t) = \frac{\rho \pi D^3}{6}\left(\frac{1}{2} + \frac{9}{2}\sqrt{\frac{2\mu}{\rho\omega D^2}}\right)\frac{dU}{dt} + 3\pi\mu DU\left(1 + \frac{1}{2}\sqrt{\frac{\rho\omega D^2}{2\mu}}\right) \tag{3.13}$$

This force is composed of two parts: an inertial force and a drag force, linearly-dependent on acceleration and velocity, respectively. Evidently, the fluid motion is assumed to be unseparated. Both components of the force depend on viscosity.

The decomposition of the time-dependent force into the two said components is somewhat arbitrary. The same force may be decomposed into three or four parts and each part may be given a separate meaning. For example, we may write

$$F(t) = \frac{1}{2}\left(\frac{\rho\pi D^3}{6}\right)\frac{dU}{dt} + 3\pi\mu DU + \frac{9}{2}\left(\frac{\rho\pi D^3}{6}\right)\sqrt{\frac{2\mu}{\rho\omega D^2}}\,\frac{dU}{dt} + \frac{3}{2}\pi\mu DU\sqrt{\frac{\rho\omega D^2}{2\mu}} \tag{3.14}$$

in which the first term on the right-hand side represents the added-mass (its ideal value) times acceleration; second term, the linear viscous resistance to the steady motion of a sphere at very small Reynolds numbers (say $Re < 1$); the third term, either the effect of history or the motion on the inertial force or simply the

*Even though the Keulegan-Carpenter work has been carried out in a standing-wave vessel, it has been included among the works cited above, which dealt with strictly harmonic motion, primarily because the vertical component of velocity in the experiments of Keulegan and Carpenter never exceeded 5 percent of the maximum horizontal velocity.

viscous effects in harmonic motion on the acceleration-dependent forces; and the last term, the history effect on the linear drag. Also, one may combine the last two terms and regard them as history-dependent modifications to the ideal values of the inertia and drag forces.

As yet a theoretical analysis of the problem for separated flow is difficult and much of the desired information must be obtained both numerically and experimentally. In this respect, the experimental studies of Morison and his co-workers (1950) on the forces on piles due to the action of progressive waves have provided a useful and somewhat heuristic approximation. The forces are divided into two parts, one due to the drag, as in the case of flow of constant velocity, and the other due to acceleration or deceleration of the fluid. This concept necessitates the introduction of a drag coefficient C_d and an inertia coefficient C_m in the expression for force. In particular, if F is the force per unit length experienced by a cylinder, then

$$F = \frac{1}{2} C_d \rho D |U| U + C_m \rho \frac{\pi D^2}{4} \frac{dU}{dt} \tag{3.15}$$

where U and dU/dt represent respectively the undisturbed velocity and the acceleration of the fluid. It is assumed that the wave slope and the associated pressure gradient are roughly constant across the diameter of the body and the wave scattering is negligible.

In Stokes sphere problem where the Reynolds number is very small, drag is proportional to the first power of velocity. In Morison's equation, drag is proportional to the square of the velocity since the flow is separated and the drag is primarily due to pressure rather than the skin friction. It is evident that Morison's equation is an heuristic extension to separated time-dependent flows of the solution obtained by Stokes. It is also evident that the validity of the equation and the limits of its application will have to be determined experimentally.

The fact that the drag and inertia coefficients in the Morison form of the resistance equation depend on both the Reynolds number ($Re = U_m D/\nu$ where $U_m = A\omega$) and the relative amplitude A/D or $K = 2\pi A/D$ may be demonstrated by writing Morison's force and Stoke's force for a harmonically oscillating flow ($U = -U_m \cos \omega t$) about a sphere at rest as

$$F = -\frac{1}{2} \rho \frac{\pi D^2}{4} C_d U_m^2 |\cos \omega t| \cos \omega t + \rho \frac{\pi D^3}{6} C_m U_m \omega \sin \omega t$$

or

$$\frac{F}{\frac{1}{2} \rho \frac{\pi D^2}{4} U_m^2} = -C_d |\cos \omega t| \cos \omega t + \frac{8\pi}{3} \frac{1}{K} C_m \sin \omega t \tag{3.16}$$

and, from Eq. (3.13),

$$\frac{F}{\frac{1}{2}\rho\frac{\pi D^2}{4}U_m^2} = -\frac{24}{Re}\left(1+\frac{1}{2}\sqrt{\pi\frac{Re}{K}}\right)\cos\omega t + \frac{8\pi}{3K}\left(\frac{3}{2}^* + \frac{9}{2}\sqrt{\frac{K}{Re}}\right)\sin\omega t \tag{3.17}$$

Equation (3.17) yields

$$C_d = \frac{24}{Re}\left(1+\frac{1}{2}\sqrt{\pi\frac{Re}{K}}\right) = \frac{24}{Re}\left(1+\frac{1}{2}\sqrt{\pi\beta}\right) \tag{3.18}$$

and

$$C_m = \frac{3}{2} + \frac{9}{2}\sqrt{\frac{1}{\pi}\frac{K}{Re}} = \frac{3}{2} + \frac{9}{2}\sqrt{\frac{1}{\pi\beta}} \tag{3.19}$$

where $\beta = Re/K = D^2/\nu T$ and $T = 2\pi/\omega$.

C_d and C_m for the Stokes force depend on both K and Re. However, there is a unique relationship between C_d and C_m, dependent only on Re, i.e.,

$$\left(\frac{C_d}{24/Re} - 1\right)\left(C_m - \frac{3}{2}\right) = \frac{9}{4} \tag{3.19a}$$

24/Re is the steady-flow drag coefficient for a sphere in the Stokes regime and the constant $\frac{3}{2}$ is the ideal value of C_m for a sphere. Thus, in unseparated Stokes flow the oscillations increase both the drag and the inertia coefficient above their corresponding steady-state values. The fact that this is not always so for separated flows will become apparent later. Experiments show that only for small values of K and β, C_m exceeds its ideal potential-flow value.

The foregoing is of little importance to the designer of offshore structures but it is very instructive as are all basic investigations, however limited their scope or range of applicability may be.

On the basis of irrotational flow around a cylinder, C_m should be equal to 2 (cylinder at rest, the fluid accelerating; otherwise $C_m = 1$), and one may suppose that the value of C_d should be identical with that applicable to a constant velocity. However, numerous experiments have shown that this is not the case and

*The pressure gradient to accelerate the flow about a sphere increases the constant $\frac{1}{2}$ in Eq. (3.13) to $\frac{3}{2}$, [see Eq. (2.54)].

that C_d and C_m show considerable variations from those just cited above. Even though no one has suggested a better alternative, the use of the Morison's equation gave rise to a great deal of discussion on what values of the two coefficients should be used. Furthermore, the importance of the effect of viscosity, roughness, rotation of the velocity vector, upstream turbulence, spanwise coherence, free surface, yaw, and the effect of neighboring elements remained in doubt since experimental evidence published over the past twenty five years has been quite inconclusive. The problem has further been compounded by the difficulty of accurately measuring the velocity and accelerations to be used in Morison equation. In general, the nature of the equation rather than the lack of precision of measurements or the difficulty of calculating the kinematic characteristics of the flow from the existing wave theories has been criticized.

The drag and inertia coefficients obtained from a large number of field tests, as compiled by Wiegel (1964) show extensive scatter whether they are plotted as a function of the Reynolds number or the so-called Keulegan-Carpenter number K, ($K = U_m T/D$). The reasons for the observed scatter of the coefficients remained largely unknown. The scatter was attributed to several reasons or combinations thereof such as the irregularity of the ocean waves, free-surface effects, three-dimensional nature of the flow, inadequacy of the *averaged* resistance coefficients to represent the actual variation of the nonlinear force (particularly when there are only one or two vortices in the wake), omission of some other important parameter which has not been incorporated into the analysis, the effect of ocean currents on separation, vortex formation, and hence on the forces acting on the cylinders, etc.

3.8.2 Fourier-Averaged Drag and Inertia Coefficients

The first systematic evaluation of the Fourier-averaged drag and inertia coefficients has been made by Keulegan and Carpenter (1958) at relatively low Reynolds numbers through measurements on submerged horizontal cylinders and plates placed in the node of a standing wave, applying theoretically derived rather than measured values of the velocities and accelerations.

Keulegan and Carpenter expressed the force in terms of a Fourier series assuming the force to be an odd-harmonic function of $\theta = 2\pi t/T$, i.e., $F(\theta) = -F(\theta + \pi)$, as

$$2F/(\rho D U_m^2) = 2\,[A_1 \sin\theta + A_3 \sin 3\theta + A_5 \sin 5\theta + \cdots$$
$$B_1 \cos\theta + B_3 \cos 3\theta + B_5 \cos 5\theta + \cdots] \qquad (3.20)$$

Keulegan and Carpenter were able to reconcile Eq. (3.20) with the equation proposed by Morison by writing Eq. (3.20) in the following form

$$2F/(\rho D U_m^2) = \frac{\pi^2}{K} C_m \sin\theta + 2[A_3 \sin 3\theta + A_5 \sin 5\theta + \cdots]$$
$$-C_d |\cos\theta| \cos\theta + 2[B_3' \cos 3\theta + B_5' \cos 5\theta + \cdots] \tag{3.21}$$

in which U is assumed to be given by $U = -U_m \cos\theta$. Evidently, Eq. (3.21) reduces to the equation proposed by Morison, i.e., to

$$2F/(\rho D U_m^2) = \frac{\pi^2}{K} C_m \sin\theta - C_d |\cos\theta| \cos\theta \tag{3.22}$$

provided that the coefficients C_m and C_d are independent of θ, i.e., each term has the same constant value (dependent on K and Re) and A_n and B_n are zero for n equal to or greater than 3.

The Fourier averages of C_d and C_m are obtained by multiplying both sides of Eq. (3.22) once with $\cos\theta$ and once with $\sin\theta$ and integrating between the limits $\theta = 0$ and $\theta = 2\pi$. This procedure yields

$$C_d = -\frac{3}{4}\int_0^{2\pi} \frac{F\cos\theta}{\rho D U_m^2}\, d\theta \tag{3.23}$$

$$C_m = \frac{2U_m T}{\pi^3 D}\int_0^{2\pi} \frac{F\sin\theta}{\rho D U_m^2}\, d\theta \tag{3.24}$$

The drag and inertia coefficients may also be evaluated through the use of the method of least squares. This method consists of the minimization of the error between the measured and calculated forces. Letting F_m represent the instantaneous measured force and F_c the force calculated through the use of Eq. (3.22), and writing

$$E^2 = (F_m - F_c)^2 \tag{3.25}$$

and $dE^2/dC_m = 0$ and $dE^2/dC_d = 0$, one obtains

$$C_{dls} = -\frac{8}{3\pi}\int_0^{2\pi} \frac{F_m |\cos\theta| \cos\theta}{\rho D U_m^2}\, d\theta \tag{3.26}$$

and $C_{mls} = C_m$ as given in Eq. (3.24). Apparently, the Fourier analysis and the method of least squares yield identical C_m values and the C_d values differ only slightly.

The difference between the measured and calculated forces, particularly in the neighborhood of the maximum forces, may be further minimized by choosing the square of the measured force as the weighting factor in the least-squares analysis. Thus writing,

$$E^2 = F_m^2 (F_m - F_c)^2 \tag{3.27}$$

and $dE^2/dC_d = 0$ and $dE^2/dC_m = 0$, one has

$$C_d = \frac{2}{L\rho D U_m^2} \frac{f_5 f_3 - f_4 f_2}{f_4 f_1 - f_3 f_3} \tag{3.28}$$

and

$$C_m = \frac{T^3}{L\rho A D^2 \pi^3} \frac{f_5 f_1 - f_3 f_2}{f_4 f_1 - f_3 f_3} \tag{3.29}$$

in which A and T represent, respectively, the amplitude and period of the oscillations and L, the length of the cylinder. The functions f_i are given by

$$f_1 = \int_0^{2\pi} F^2 \cos^4 \theta \, d\theta, \qquad f_2 = \int_0^{2\pi} F^3 |\cos \theta| \cos \theta \, d\theta$$

$$f_3 = \int_0^{2\pi} F^2 \sin \theta \cos \theta \, |\cos \theta| \, d\theta, \qquad f_4 = \int_0^{2\pi} F^2 \sin^2 \theta \, d\theta$$

$$f_5 = \int_0^{2\pi} F^3 \sin \theta \, d\theta \tag{3.30}$$

Equations (3.28) and (3.29) may be shown to reduce to Eqs. (3.23) and (3.24) by replacing F^n in Eqs. (3.30) by F^{n-2} and carrying out the necessary integrations in which F does not appear.

One can show through the use of Eq. (3.22) that the rate of change of force with time is zero at the time of maximum acceleration and is proportional to C_m/KT at the time of maximum velocity. Thus, the determination of C_m, in particular through the use of force at the time of maximum acceleration, depends on the particular values of C_m, K and T, and may not be quite accurate. In general, it is recommended that either the Fourier-averaged or the least-squares averaged force-transfer coefficients be used.

Equation (3.22) also shows that the maximum in-line force does not occur at the time of maximum velocity, but rather it leads the maximum velocity. The maximum force coefficient C_F (spp) may be calculated from Eq. (3.22) to yield

$$C_F(spp) = C_d + \frac{\pi^4 C_m^2}{4 C_d K^2} \quad \text{for} \quad K > \frac{\pi^2 C_m}{2 C_d} \tag{3.31}$$

and

$$C_F(spp) = \frac{\pi^2 C_m}{K} \quad \text{for} \quad K < \frac{\pi^2 C_m}{2 C_d}$$

In general, the dividing value of K may be assumed to be about $K = 10$. For relatively small values of K, the force is said to be inertia dominated. For relatively large values of K, the force is drag dominated. The ratio of the maximum inertial force to maximum drag force is

$$R = \frac{\pi^2 C_m}{K C_d} \tag{3.32}$$

For purposes of orientation, it will be useful to indicate the range of K values over which the various components of force become predominant. These are

K smaller than about 10 . . . inertia increasingly important
K larger than about 15 . . . drag increasingly important
K larger than about 5 . . . lift force important

A more detailed discussion of the various loading regimes will be taken up later.

3.8.3 Experimental Studies on C_d and C_m

It is recognized that the coefficients cited above are not constant throughout the cycle and are either time-invariant averages or peak values at a particular moment in the cycle. A simple dimensional analysis of the flow under consideration shows that the time-dependent coefficients for a uniformly-roughened cylinder may be written as

$$\frac{2F}{\rho L D U_m^2} = f\left[\frac{U_m T}{D}, \frac{U_m D}{\nu}, \frac{k}{D}, \frac{t}{T}\right] \tag{3.33}$$

in which F represents the in-line or the transverse force. Equation (3.33) combined with Eq. (3.22), taking for now the latter to be valid, yields

$$C_d = f_1(K, Re, k/D, t/T) \tag{3.34a}$$

$$C_m = f_2(K, Re, k/D, t/T) \tag{3.34b}$$

There is no simple way to deal with Eqs. (3.34) even for the most manageable time-dependent flows. Another and perhaps the only other alternative is to eliminate time as an independent variable and consider suitable time-invariant averages as given by Eqs. (3.23) and (3.24). Thus, one has

$$[C_d, C_m, C_L, \cdots] = f_i(K, Re, k/D) \tag{3.35}$$

For periodically oscillating flows the Reynolds number is not necessarily the most suitable parameter. The primary reason for this is that U_m appears in both K and Re. Thus, replacing Re by $Re/K = D^2/\nu T$ in Eq. (3.35), on has

$$C_i \text{ (a coefficient)} = f_i(K, \beta, k/D) \tag{3.36}$$

in which $\beta = D^2/\nu T$ and is called the 'frequency parameter' by Sarpkaya (1976a). The dependence of C_d and C_m on β has already been noted in connection with the discussion of the Stokes sphere problem.

From the standpoint of dimensional analysis, either the Reynolds number or β could be used as an independent variable. β is constant for a series of experiments conducted with a cylinder of diameter D in water of uniform and constant temperature if T is kept constant. Then the variation of a force coefficient with K may be plotted for constant values of β. Subsequently, one can easily recover the Reynolds number from $Re = \beta K$ and connect the points, on each β = constant curve, representing a given Reynolds number.

From the standpoint of laminar boundary layer theory, β represents the ratio of the rate of diffusion through a distance δ (i.e., ν/δ^2 where δ is the boundary-layer thickness) to the rate of diffusion through a distance D (i.e., ν/D^2). This ratio is also equal to $(D/\delta)^2$ and, when it is large, gradients of velocity in the direction of flow are small compared with the gradients normal to the boundary, a situation to which the boundary-layer theory is applicable (Rosenhead 1963).*

Let us now re-examine the Keulegan-Carpenter data (1958) partly to illustrate the use and significance of β as one of the governing parameters and partly to take up the question of the effect of Reynolds number on the force coefficients.

The data given by Keulegan and Carpenter may be represented by 12 different values of β. The drag and inertia coefficients are plotted in Figs. 3.13 and

*Note that the boundary-layer theory is not *directly* applicable, even at large values of β, to calculate the forces acting on an oscillating body without taking into consideration the modifications of the surface pressure due to boundary layer and separation.

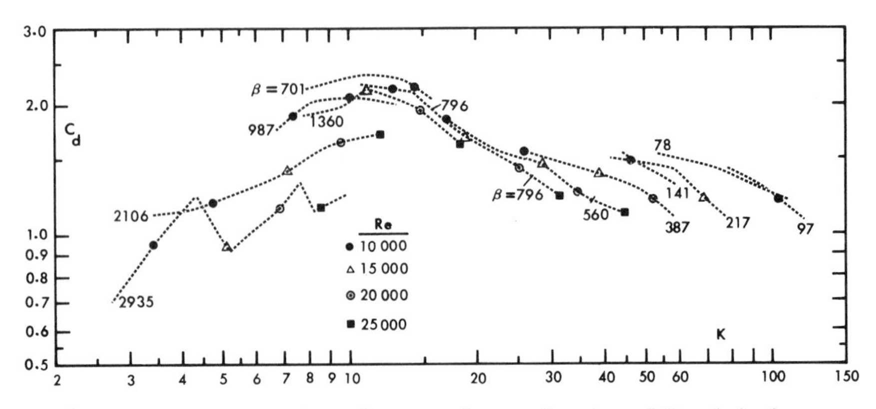

Fig. 3.13. Replot of the Keulegan-Carpenter C_d as a function of K and the frequency parameter (Sarpkaya 1976a).

3.14 and connected with straightline segments. The identification of the individual data points in terms of the cylinder diameter, as was done by Keulegan and Carpenter and also by Sarpkaya (1975a) irrespective of the β values, gives the impression by a scatter in the data and invites one to draw a mean drag curve through all data points. Such a temptation is further increased by the fact that the data for each β span over only a small range of K values. Thus, the drawing of such a mean curve obscures the dependence of C_d and/or C_m on β and hence on Re.

Also shown in Figs. 3.13 and 3.14 are points representing four selected Reynolds numbers. The corresponding K values for each Re and β were calculated from $K = Re/\beta$. The points corresponding to the selected Reynolds numbers are reproduced in Figs. 3.15 and 3.16. These figures show, within the range

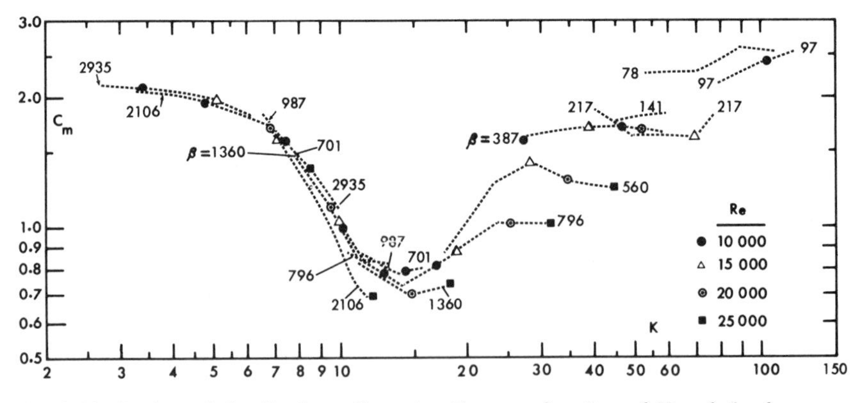

Fig. 3.14. Replot of the Keulegan-Carpenter C_m as a function of K and the frequency parameter (Sarpkaya 1976a).

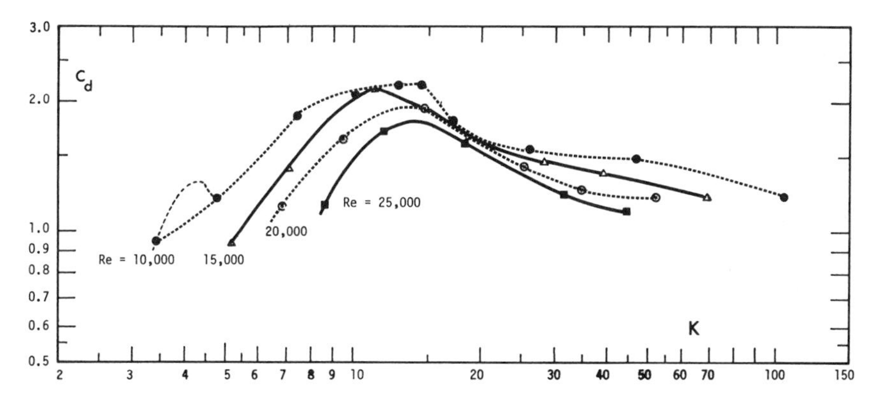

Fig. 3.15. C_d versus K. Replot of the Keulegan-Carpenter data for various values of the Reynolds number (Sarpkaya 1976a).

of Re and K values encountered in Keulegan-Carpenter data, that (a) C_d depends on both K and Re and decreases with increasing Re for a given K; and that (b) C_m depends on both K and Re for K larger than approximately 15 and *decreases* with increasing Re. A similar analysis of Sarpkaya's data (1975a) also shows that C_d and C_m depend on both K and Re and that C_m *increases* with increasing Re. Notwithstanding this difference in the variation of C_m between the two sets of data, Figs. 3.15 and 3.16 put to rest the long-standing controversy regarding the dependence or lack of dependence of C_d and C_m on Re and show the importance of β as one of the controlling parameters in interpreting the data, in interpolating the K values for a given Re, and in providing guidelines for further experiments as far as the range of K and β are concerned.

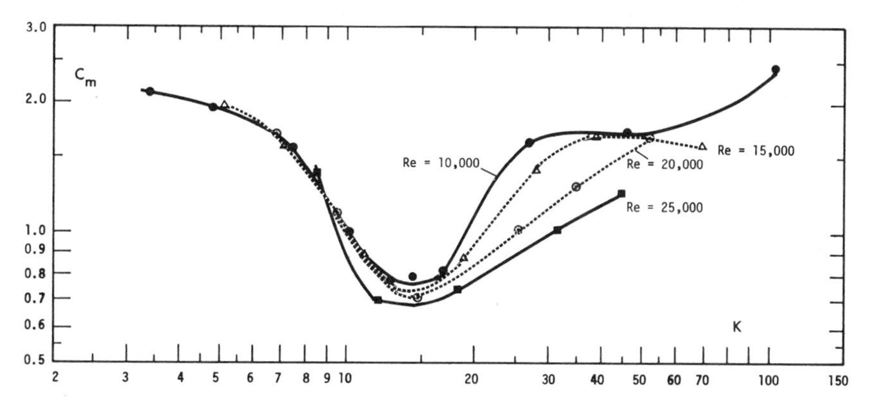

Fig. 3.16. C_m versus K. Replot of the Keulegan-Carpenter data for various values of the Reynolds number (Sarpkaya 1976a).

As noted earlier, there has been growing awareness of the fact that the coefficients obtained at relatively low Reynolds numbers may not be applicable at higher Reynolds numbers, that the transverse forces acting on the elements of offshore structures may be as much or more important (e.g., for the dynamic response of the structural elements) than the in-line forces given by the Morison formula, and that the initial or growing marine roughness may significantly alter the forces acting on the structure. In view of the foregoing considerations, Sarpkaya (1976a, 1976b) conducted a series of experiments with smooth and sand-roughened cylinders in a novel, U-shaped vertical water tunnel. In these experiments the drag and inertia coefficients have been evaluated through the use of Eqs. (3.23) and (3.24), i.e., through the use of the Fourier analysis.

Figures 3.17 and 3.18 show C_d versus K and C_m versus K for five representative values of β. There is very little scatter in the data even though the figures represent the results of four independent runs. Mean lines drawn through the data shown in Figs. 3.17 and 3.18 are presented in Figs. 3.19 and 3.20 together with the constant Reynolds-number lines obtained through the use of $K = Re/\beta$. Evidently, there is a remarkable correlation between the force coefficients, Reynolds number, and the Keulegan-Carpenter number.

Figures 3.19 and 3.20 show that C_d and C_m do not vary appreciably with Re for Re smaller than about 20,000 and help to explain the conclusions previously reached by Keulegan and Carpenter (1958) and by Sarpkaya (1975a). The entire data are shown as a function of Re for constant values of K in Figs. 3.21 and 3.22. These figures clearly show that C_d decreases with increasing Re to a value of about 0.5 (depending on K) and then gradually rises to a constant value (post-supercritical value) within the range of Reynolds numbers encountered.

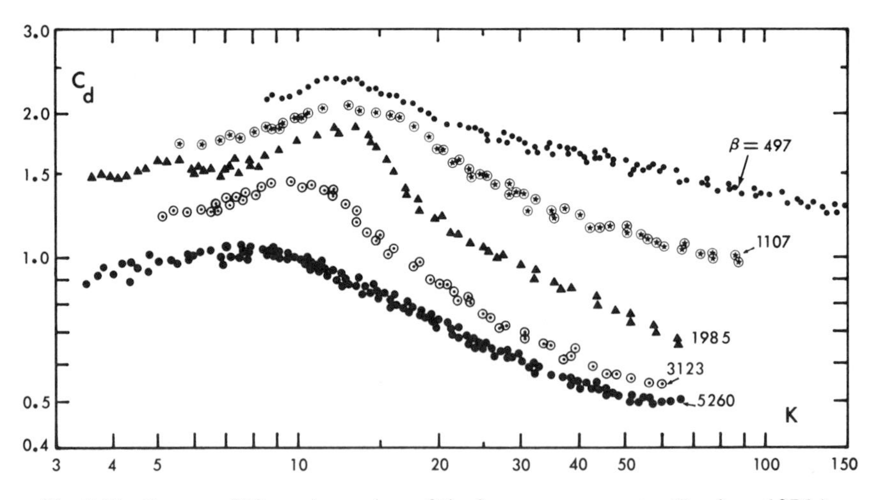

Fig. 3.17. C_d versus K for various values of the frequency parameter (Sarpkaya 1976a).

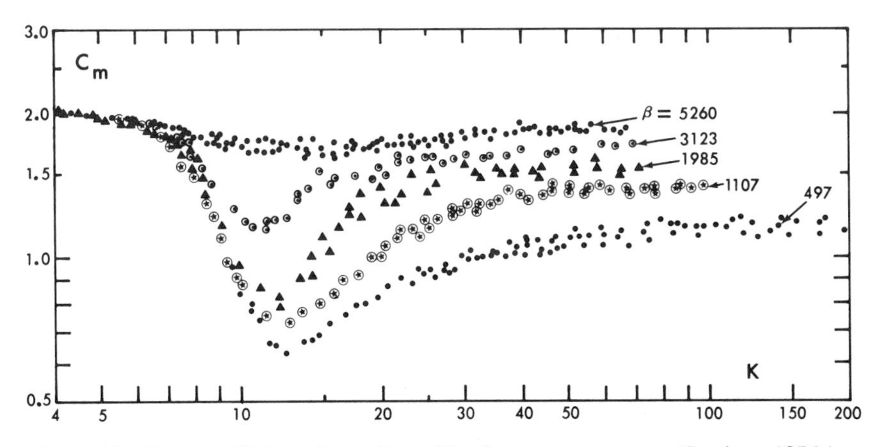

Fig. 3.18. C_m versus K for various values of the frequency parameter (Sarpkaya 1976a).

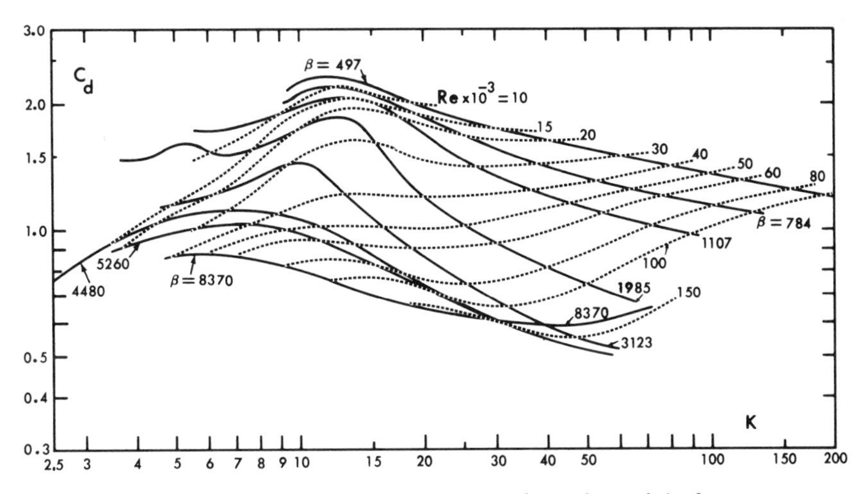

Fig. 3.19. C_d versus K for various values of the Reynolds number and the frequency parameter (Sarpkaya 1976a).

The inertia coefficient increases with increasing Re, reaches a maximum, and then gradually approaches a value of about 1.85. It will be recalled that the Keulegan-Carpenter data indicated an opposite trend. It is now believed that the Keulegan-Carpenter data for C_m are not quite reliable for K values larger than about 15.

Figure 3.21 shows that the drag coefficient for a cylinder in harmonically oscillating flow is not always larger than that for steady flow at the same Reynolds number. As noted earlier, Stokes analysis for the unseparated flow about an oscillating sphere yielded a drag coefficient which is always larger than its

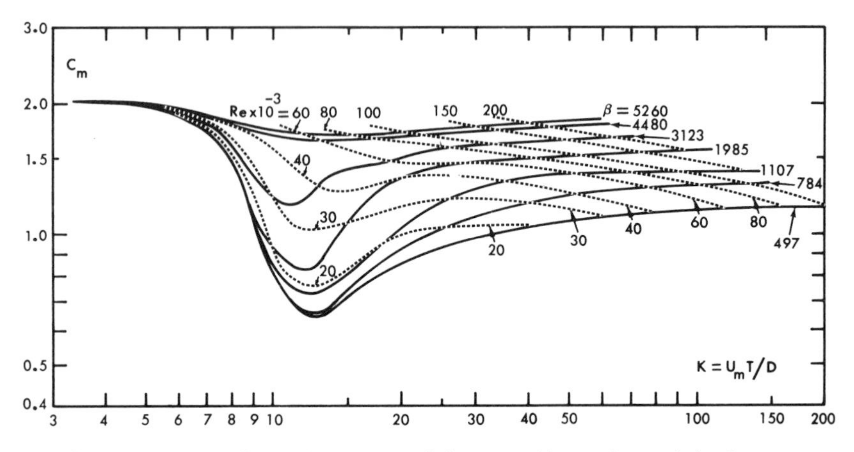

Fig. 3.20. C_m versus K for various values of the Reynolds number and the frequency parameter (Sarpkaya 1976a).

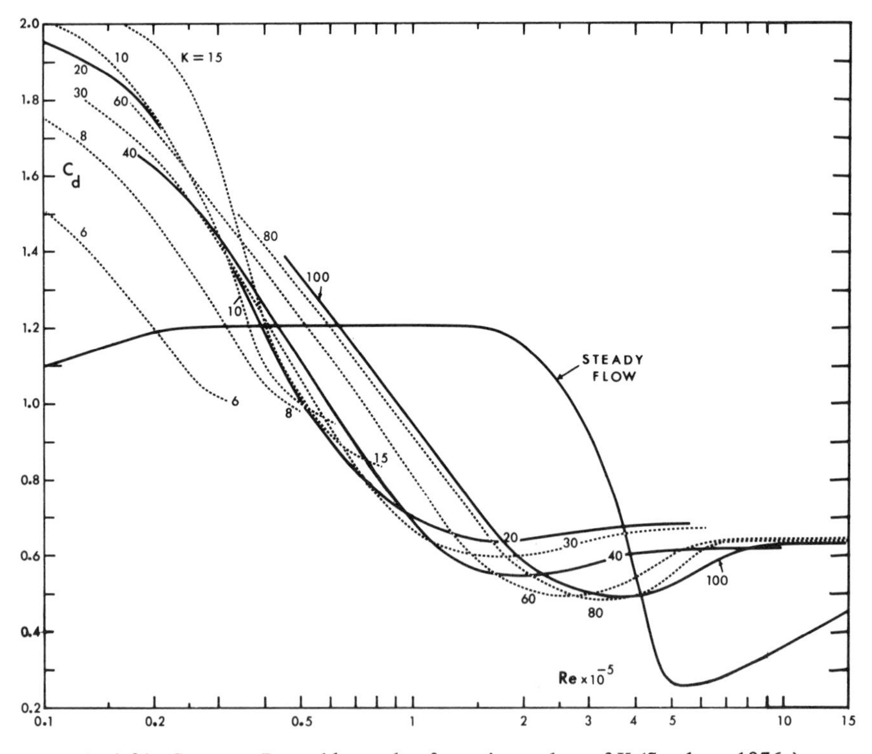

Fig. 3.21. C_d versus Reynolds number for various values of K (Sarpkaya 1976a).

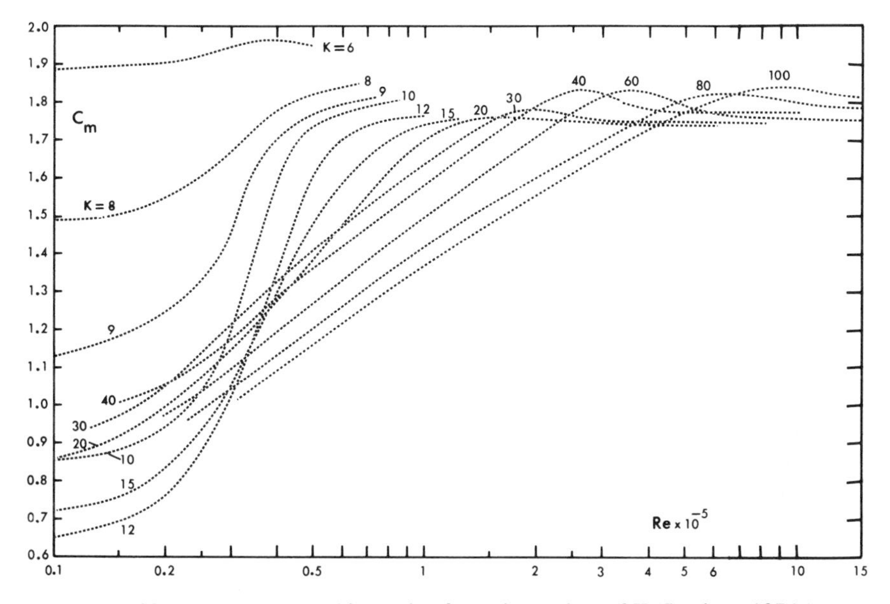

Fig. 3.22. C_m versus Reynolds number for various values of K (Sarpkaya 1976a).

corresponding steady-state value. According to Fig. 3.21, for K = 100, for example, C_d for the oscillating flow is larger than its steady-state value for Re smaller than about 60,000 and larger than about 400,000. In the range of Reynolds numbers between the two cited above, the drag coefficient for the oscillating flow is considerably lower than that for steady flow. The reason for this is surely the earlier transition in the boundary layers. This and other characteristics of the force coefficients raise several questions which may be explained only partially on the basis of the observations with steady flows. Some of these questions are: why does the transition begin sooner and span over a larger range of Reynolds numbers; why does the added-mass coefficient C_a, ($C_a = C_m - 1$), become negative for certain values of K and Re (see Fig. 3.22); is there a unique relationship between C_m and C_d; etc.

It is a well known fact that the occurrence of drag crisis in steady cross-flow about a cylinder depends on the length and scale of turbulence in the ambient flow, blockage and length-to-diameter ratio of the cylinder, the vibration amplitude and frequency of the test body, surface roughness, and on other peculiarities of the wind- or water-tunnel in which the experiments are performed. In fact, it is for this reason that the minimum value of the drag coefficient and the base pressure in steady flow are widely scattered. Since the formation of the separation bubbles are largely responsible for the low values of C_d, one could state that the formation and the extent of the separation bubbles are very sensitive to the factors cited above. *A priori,* one would expect the same to occur in harmonic flow about a cylinder. During a given cycle, the flow at both sides of the cylinder contains a number of vortices and large scale turbulence. Thus, it is natural to assume that they would give rise to earlier transition.

The effect of the growth and motion of vortices on the increase of the drag coefficient relative to that for the steady flow needs further discussion. It has been shown by Sarpkaya (1963) that the variation of the characteristics of vortices in the neighborhood of a cylinder strongly affects the lift, drag, and the inertia coefficients. It has also been shown (Sarpkaya 1966) that the drag in the initial stages of an impulsively-started flow about a circular cylinder can exceed its steady value by as much as 30 percent (see also Sarpkaya and Shoaff 1979a and 1979b). These findings are relevant to the present study in a qualitative sense. During the periods of high acceleration in harmonic flow, vorticity is slow to diffuse and therefore accumulates rapidly in the close vicinity of the cylinder. Although the growing vortex soon reaches unstable proportions and separates from its shear layer, the growth of the vortices are so rapid that the vortices become much larger than their quasi-steady-state-state size before they separate from their shear layers. This would ordinarily lead to a larger drag force. However, the maximum drag does not occur at the time of maximum acceleration or maximum velocity. Evidently, at the initial stages of acceleration, the vortices in the downstream side of the cylinder are not yet fully grown. Furthermore, the convection of the vortices shed in the previous cycle towards the cylinder help to reduce the pressure on the upstream side of the cylinder and prevent the drag force from reaching large values. As the velocity increases, the vortices on the upstream side move towards the top and bottom of the cylinder and loose their influence on the pressure distribution on the upstream face of the cylinder. The vortices on the downstream side of the cylinder, now fully grown and ready to shed, give rise to a large drag force. By the time the velocity reaches its maximum, the vortices which are now shed from the downstream side are further convected downstream partly by the action of the outer vortices (and their images) existing in the flow and partly by the base flow itself. Thus, the drag force begins to decrease by the time the velocity reaches its maximum.

The role played by the vortices becomes most pronounced if the duration of flow in one direction is not too long (e.g., the amplitude-to-diameter ratio of about two). This is partly because of the influence of the vortices in the immediate vicinity of the cylinder is not tempered by other vortices further upstream and downstream and partly because of the incomplete or fractional shedding of vortices. In fact, for very small Keulegan-Carpenter numbers (K from 5 to 15), a vortex may shed only from one side of the cylinder which in turn leads to the formation of a vortex pair on only one side of the cylinder (Zdravkovich and Namork 1977). Such phenomena give rise to dramatic changes in the drag force and significantly increase the transverse force.

The number of vortices shed in each cycle and the intensity of turbulence depend on the relative motion of the fluid and the Reynolds number. In a given cycle the flow may start in one direction as a subcritical flow with boundary layers separating in a laminar state. As the flow speed or the instantaneous Reynolds number increases, the flow may enter a critical state. The background turbulence and vortices may easily and often locally disrupt the separation lines. When this happens, the flow may be affected over a considerable length of the cylinder and the base pressure along the span is no longer uniform. Such phenomena have been carefully noted by Bearman (1969) in connection with steady flows. If the flow continues in the same direction with ever increasing instantaneous Reynolds numbers, the separation bubbles may completely disappear and part of the boundary layer may become turbulent.

The foregoing discussion shows that the time-averaged force coefficients reflect only in a very crude way the state of an extremely complex time-dependent flow. The key to the understanding of the instantaneous behavior of the lift, drag, and inertia forces is the understanding of the formation, growth, and motion of vortices. At present, this is possible only experimentally and in a qualitative sense. Numerical attempts based on the discrete vortex model (see e.g., Stansby 1977, 1979; Graham 1979; Sawaragi and Nakamura 1979) require several assumptions and some fine tuning. Furthermore, the neglect of the rate of change of circulation of the vortices with time inevitably leads to an inertia coefficient of $C_m = 2$. It has been shown by Sarpkaya (1963) that the time-dependence of the circulation of the vortices causes profound changes in the force coefficients and the deviation of C_m from its ideal value of 2 is directly related to the rate of change of circulation with time (this is partially accounted for in Sawaragi and Nakamura's work).

Maull and Milliner (1978) examined the interdependence of the in-line and transverse forces and the production and motion of vortices in a sinusoidally-oscillating flow in a small U-tube at Reynolds numbers smaller than about 4,000. They proposed that the variation of the drag coefficient during a cycle may be considered as the addition of two terms, the inertia term with $C_m = 2$, and a further term which is a function of the movement of the vortices produced. As noted above, the use of $C_m = 2$ is questionable since there is no resemblance between the ideal unseparated potential flow and the periodic separated flow under consideration. This will become more evident as we later examine the instantaneous values of the drag, inertia, and the added mass coefficients.

Bearman and Graham (1979) measured the in-line and transverse force on several cylindrical bodies in plane oscillatory flow in a small U-tube over a range of Keulegan-Carpenter numbers from 3 to 70 at relatively small Reynolds numbers. They have noted large cycle to cycle variations in computed values of C_d and C_m, even though the bulk flow in the U-tube was closely repetitive. The length-to-height ratio of the test section of their U-tube was about half that of Sarpkaya (1976a) and the corners of their U-tube were not streamlined.

Furthermore, they had to filter the electrical signals to remove small amplitude vibrations superimposed on them. Bearman and Graham concluded that different modes of vortex shedding occur over different ranges of K (as previously shown by others, e.g., Keulegan and Carpenter 1958) but although the strength of the vortices shed varies from body to body increasing with acuteness of the edges, there is a marked similarity between the pattern of shedding from the different bodies (flat plates, and diamond and circular cylinders). Their measurement of the in-line force showed that Morison's equation is rather inaccurate for sharp-edged bodies at low K. The transverse forces exhibited rather large and slowly varying modulation of the amplitude over a large number of cycles. Bearman and Graham suggested, as did Maull and Milliner (1978), that the r.m.s. amplitude of the in-line force may be more useful than the usual division into drag and inertia components (i.e., the linear-quadratic combination of the forces based on the instantaneous values of the acceleration and velocity).

In a related study Maull and Milliner (1979) studied the forces on a circular cylinder having a complex periodic motion. They have found that not only are second harmonic components of the drag but also other harmonics are altered. The lift force had components at the harmonics, at frequencies half way between the harmonics and at other unrelated frequencies.

Studies on the motion of the separation points and the distribution of the instantaneous pressure around the cylinder may help to clarify the relationships between the vortex shedding, transverse force, and the drag and inertia components of the in-line force (see e.g., Grass and Kemp 1979; and Matten, Hogben, and Ashley 1979 for studies along these lines).

3.8.4 Relationship Between Averaged C_d and C_m Values

The relationship between C_d and C_m has been of special concern and will be re-examined here. A plot of C_m versus C_d shows that there is not a unique relationship between them, independent of K and Re (see Fig. 3.23). A similar plot may be prepared by maintaining Re constant at suitably selected values of Re and plotting C_m versus C_d corresponding to the same value of K. Hogben (1976) suggested a conceptual modelling of the interaction leading to an explicit formula for the interdependence of C_d and C_m. It is based on a highly simplified and somewhat intuitive reasoning. It appears that more sophisticated models are needed to express the relationship between the two coefficients.

3.8.5 Transverse Force and the Strouhal Number

Vortex shedding and the resulting alternating force in steady flow have been studied extensively and the existing data on lift coefficients have been presented in Fig. 3.4. In spite of the considerable interest, however, the transverse force in

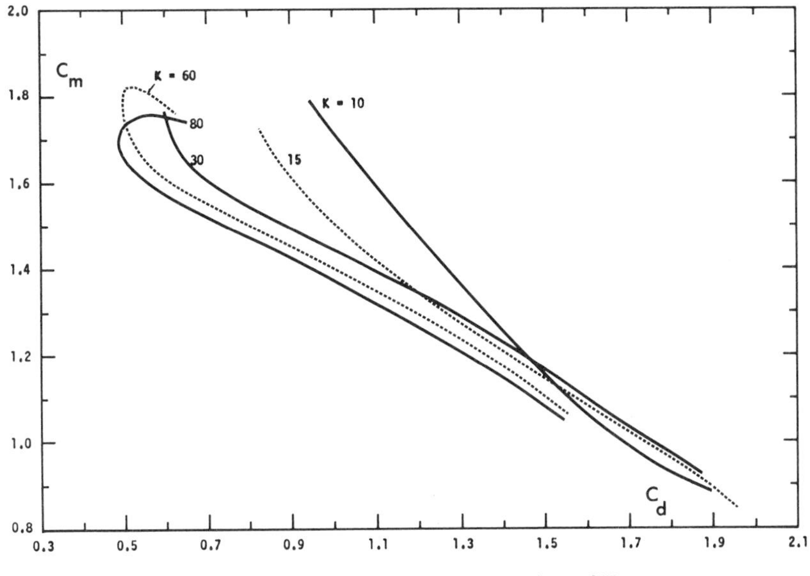

Fig. 3.23. C_m versus C_d for various values of K.

harmonic flows received very little attention. Recently, it became clear from observations of the oscillations of long piles and strumming of cables that the lift forces are important not only because of their magnitude but also because of their alternating nature, which under certain circumstances may lead to the phenomenon known as the lock-in or vortex synchronization. This phenomenon may cause failure due to fatigue and increased in-line force. Obviously, the total instantaneous force acting on the structure is increased by the lift force and modified by the oscillations of the body, mode shape of the oscillation, strum-suppression devices, the ovalling of the cross-section, etc.

The transverse force exerted on a cylinder in harmonically oscillating flows (or bodies) has been measured by Mercier (1973), Sarpkaya (1975a, 1976a–c), and Maull and Milliner (1978, 1979).* Mercier oscillated a small cylinder at relatively low Reynolds numbers (about 12,000) and observed peak lift coefficients for K values between 10 and 20.

Sarpkaya measured the transverse force acting on smooth and sand-roughened cylinders for a wide range of Reynolds numbers, Keulegan-Carpenter numbers, and relative roughnesses. Figures 3.24 and 3.25 show the lift coefficient [de-

*In wavy flows and at very small Reynolds numbers Chang (1964), Bidde (1971), Wiegel and Delmonte (1972), Isaacson (1974), and Isaacson and Maull (1976) have measured the transverse force acting on a vertical cylinder. For a critical discussion of these works see Sarpkaya (1976a) and Isaacson (1974).

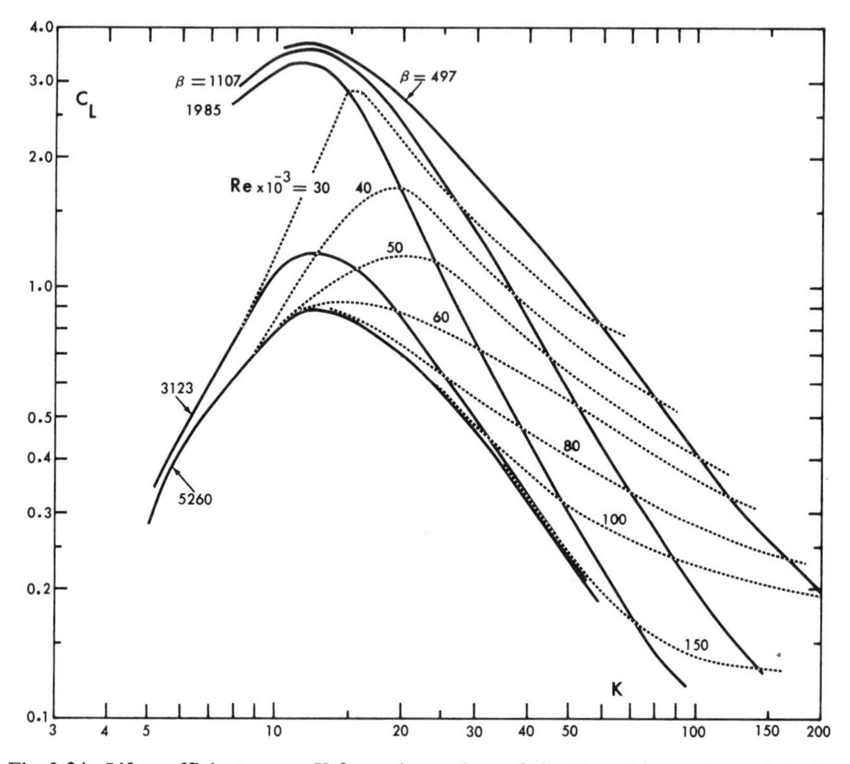

Fig. 3.24. Lift coefficient versus K for various values of the Reynolds number and the frequency parameter.

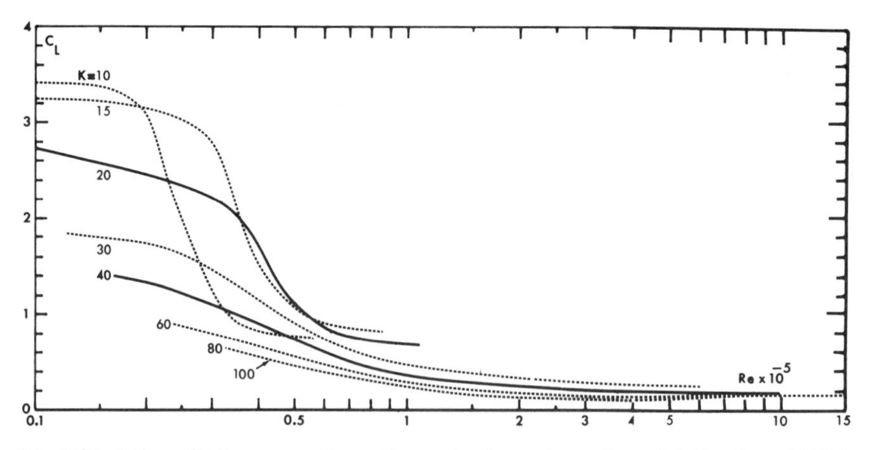

Fig. 3.25. Lift coefficient versus Reynolds number for various values of K (Sarpkaya 1976a).

fined by C_L = (lift force)/[$0.5\rho LDU_m^2$)] as a function of K (for various values of β) and as a function of Re (for various values of K). Evidently, C_L (max) reaches its maximum value in the neighborhood of K = 12 and decreases sharply with increasing K.

As noted earlier, the lift force is a consequence of the pressure gradient across the wake. The alternating pressure gradient increases with increasing asymmetry of the strength and position of the vortices. Thus, one may conclude that the flow for K values in the neighborhood of 12 must exhibit maximum asymmetry. Part of this asymmetry is due to the shedding of one or two new vortices and partly due to the convection of the vortices shed in the previous half cycle back towards the cylinder. Neither the convected vortices nor the ones shed newly are necessarily symmetrical and result in a maximum wake asymmetry (Zdravkovich and Namork 1977). The interaction of the convected vortices with the flow about the cylinder is such as to enhance the asymmetry by forcing the shedding of a new vortex from the side of the cylinder nearest to the convected vortex. This is easily understood if one considers the high velocities induced on the cylinder by the convected vortex and its image inside the cylinder.

Figure 3.25 shows that for Re smaller than about 20,000, C_L depends primarily on K. In the Reynolds number range from about 20,000 to 100,000, C_L depends, to varying degrees, both on Re and K. Above a Reynolds number of about 100,000, the dependence of C_L on Re and K is quite negligible and certainly obscured by the scatter in the data. However, the magnitude of the lift force relative to the in-line force is not negligible.

The minimum value of K at which lift or the asymmetry in the vortices develop is, by the very nature of the vortices, extremely sensitive to the experimental conditions. Experiments by Sarpkaya (1976a) have shown that there is a 90 percent chance that the asymmetry will appear at K = 5. At K = 4, there is only a 5 percent chance that the asymmetry will appear for very short periods of time. The determination of the onset of asymmetry is of special importance not only in connection with the ocean structures but also with bodies of revolution flying at high angles of attack (see e.g., Lamont and Hunt 1976). It has previously been noted in connection with the discussion of impulsively-started flows and uniformly-accelerating flows (see sections 3.7.2 and 3.7.3) that the asymmetry sets in when s/D is equal to about 3 for the impulsive motion and to about 4.8 for the uniformly accelerating motion. Assuming K = 5 for the onset of asymmetry in harmonic motion, one finds that $s/D = A/D = 5/(2\pi) = 0.8$. Strictly speaking, a direct comparison of the harmonically oscillating flow with unidirectional flows is not justified. Furthermore, the effect of flow deceleration is very profound in promoting instability and earlier separation. Nevertheless, a simple-minded comparison of the relative distances covered by the fluid in various flow situations prior to the onset of asymmetry shows that asymmetry occurs much sooner in oscillating flows as would be expected.

Aside from its magnitude, the most important feature of the transverse force is its frequency of oscillation. This frequency varies with K, Re, and time in a given cycle and also from cycle to cycle. This leads to multiple lift coefficients at a given K and Re (Sarpkaya 1976a and Maull and Milliner 1978). As noted by Maull and Milliner, this multiplicity raises doubts about any elementary analysis of the lift signal since it must be non-stationary in a statistical sense and therefore the r.m.s. value, for instance, will be a function of the length of the record taken.

Experiments show that (see e.g., Maull and Milliner 1978) the occurrence of peaks in the r.m.s. lift is closely associated with the progressive occurrence of higher order components in the frequency domain. With increasing K the r.m.s. force reaches its first maximum at approximately $K = 13$ where the dominant frequency is twice the particle frequency. The next maximum occurs at $K = 18$ and is associated with a dominant frequency component at three times the particle frequency. At higher K, the lift coefficient decreases and the said frequency component is associated with higher and higher multiples of the particle frequency. This is evidenced by Fig. 3.26 where the frequency ratio $f_r = f_v/f_w$ (the ratio of the maximum frequency in a cycle, defined here as the reciprocal of the shortest interval between two maxima, to f_w) is plotted for a smooth cylinder. A point on each line represents the maximum value of K for a given Re and f_r. In other words, a line such as $f_r < 4$ means that the alternating force does not contain frequencies larger than $f_r = 4$ for K and Re values in the region to the left of the line. Intermediate values of f_r such as $f_r = 3, 5$, etc. are not shown to keep the figure relatively simple.

Several facts are of special importance and will be discussed in detail. Firstly, Fig. 3.26 begins with $K = 5$. As noted earlier, there is occasional vortex shedding for K values between 4 and 5. Secondly, each $f_r = N$ line does not represent an absolute line of demarkation between the frequencies $N - 1$ and $N + 1$. Occasionally, a frequency of $N + 1$ will occur on the $N - 1$ side of the N line, and vice versa. Thirdly, the frequency of vortex shedding is not a pure multiple of the flow oscillation frequency. At first this would appear anomalous but a closer examination of the behavior of the vortices shows that a fractional value of f_r is perfectly understandable. Evidently, f_r, as an integer, is a measure of the number of vortices actually shed during a cycle. However, those vortices which do not break away from their shear layers before the flow is reversed are partially developed and result in incomplete shedding. Thus, the fractional part of f_r indicates an incomplete shedding. This is particularly true for f_r in the neighborhood of 2 or 3. Finally, it should be noted that the frequencies cited above correspond to the vortices shed at or near the maximum velocity in the cycle. As noted earlier, the vortex shedding frequency is not constant throughout the cycle. Furthermore, the variation of the shedding frequency is not a simple harmonic function as it would be if the vortex shedding process directly responded to the instantaneous velocity.

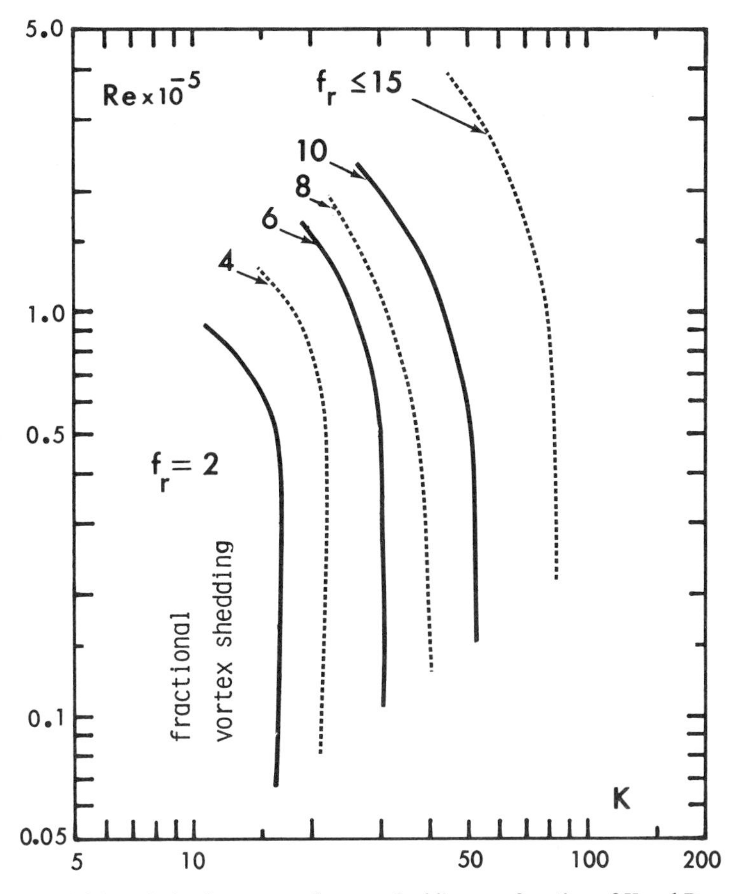

Fig. 3.26. Relative frequency of vortex shedding as a function of K and Re.

The fractional shedding of vortices for K values in the region of 5 to 15 cause incalculable changes in the flow pattern, and in the in-line and transverse forces. In fact, no separated, time-dependent flow is more complex than the one in which there are only one or two vortices. For small values of K, two vortices begin to develop at the start of the cycle in one direction but the vortices do not acquire identical strengths due to various reasons. As the flow reverses, the larger of the vortices is swept past the cylinder but the weaker one dissipates partly due to turbulent diffusion. The consequences of this single shedding are that the in-line force becomes asymmetrical and the vortex which is swept away plays an important role in the formation of the new vortices when the flow reverses its direction. The dominant vortex establishes, by its sense of rotation (and the opposite sense of rotation of its image inside the cylinder), a preferred location for the generation of a new dominant vortex. The new vortex and the one con-

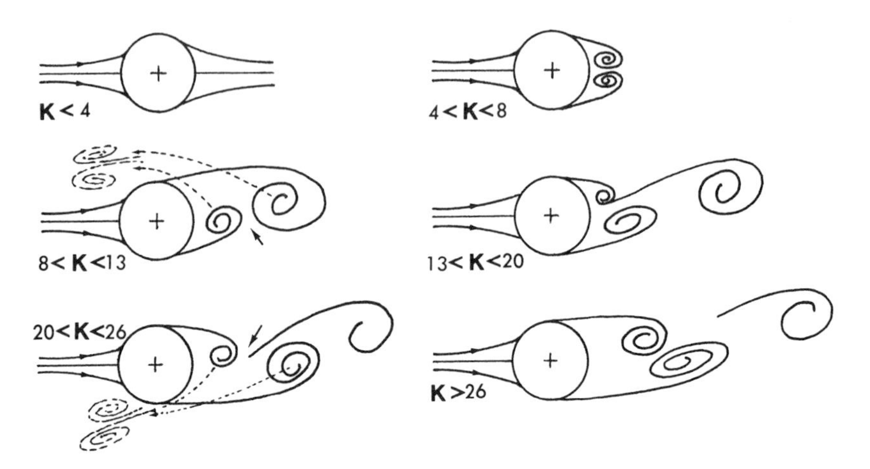

Fig. 3.27. Schematic of the evolution of vortices in various ranges of the Keulegan-Carpenter number.

vected downstream may form a pair (Zdravkovich and Namork 1977), increase the transverse pressure gradient and thus give rise to significant lift forces as noted earlier. For certain values of K, the convected vortex during the flow reversal may be rapidly thrown out of the flow field due to the large local transport velocities induced at it by the newly formed vortices (see Fig. 3.27). The foregoing points out the complexity of the growth and motion of vortices in harmonic flow and the reasons for the difficulty in devising a universal expression for the time-dependent force (such as Morison's equation) which would be equally applicable to harmonic flows at all Re and K values. This matter will be taken up later in discussing the merits and shortcomings of the Morison equation.

As noted in connection with steady flows, it is customary to express the vortex shedding frequency in terms of the Strouhal number. In harmonic flow there is not a unique Strouhal number. For sake of simplicity, one may define a Strouhal number in terms of the maximum velocity of the flow as $St = f_v D/U_m = f_r/K$. Sarpkaya's experiments (1976a) have shown that St so defined remains reasonably constant at 0.22 for f_r larger than about 3. Figure 3.26 shows that St depends on both Re and K. For very large values of Re (i.e., in the post-supercritical region) St rises to about 0.3. The average value of St based on all vortices shed during a given cycle was found to be between 0.14 and 0.16.

3.8.6 Roughness Effects on C_d, C_m, C_L, and St in Harmonic Flow

Of the scores of papers dealing with fluid loading on offshore structures only few have treated the effect of roughness on the force-transfer coefficients. Yet it is a fact that the structures in the marine environment become gradually covered with rigid as well as soft excrescences. Thus, the fluid loading and the structural response due to identical ambient flow conditions may be significantly different from that experienced when the structure was clean partly because of the 'roughness effect' of the excrescences on the flow (the boundary layer) and partly because of the increase of the 'effective diameter' and the effective mass (natural frequency and damping) of the elements of the structure.

In the absence of any data appropriate to harmonic or wavy flows, it has been assumed that the drag coefficients obtained from tests in steady flow over artificially- or marine-roughened cylinders are applicable to wave force calculations at least when the loading is predominantly drag (Miller 1977).

It is not generally appreciated that the consequences of all nearly steady flows are not always identical to those of steady flows. The case in point is the harmonic flow under consideration. Even for large amplitudes of oscillation, there is only a finite vortex street comprised of vortices of nearly equal strength due to the nearly steady nature of the flow. As the flow reverses, the situation is not the same as that of a uniform flow (with or without free stream turbulence) approaching a roughened cylinder but rather that of a finite vortex street approaching a rough-walled cylinder. Such a flow cannot be regarded identical to steady flow with some turbulence of fairly uniform intensity and scale as the data presented herein show.

The salient features of the influence of roughness on the cross-flow around a cylinder in *steady flow* have been discussed previously. It has been pointed out that the supercritical value of the drag coefficient depends on both the character of the flow and the surface condition of the cylinder; the drag coefficient in the post-supercritical region returns more or less to its steady subcritical value; the larger the effective roughness, the larger is the retardation of the boundary layer; and that the disturbances generated by the roughness elements cause an incalculable change in the critical region of the flow. It appears from the foregoing that roughness should play an equally significant role on the characteristics of a sinusoidally oscillating flow about a circular cylinder.

Sarpkaya (1976b, 1977a) carried out a series of experiments with sand-roughened cylinders in harmonic flow. Part of his results are shown in Figs. 3.28 and 3.29 for two values of K as a function of Re. Each curve on each plot corresponds to a particular relative roughness. Also shown on each figure is the corresponding drag and inertia coefficient for the smooth cylinder at the corresponding K value.

The k/D = constant curves on each plot undergo changes similar to those found for steady flow about rough cylinders (see Fig. 3.5). For a given relative roughness, the drag coefficient does not significantly differ from smooth-cylinder value at very low Reynolds numbers. As the Reynolds number increases, C_d for the rough cylinder decreases rapidly, goes through the region of drag crisis at a Reynolds number considerably lower than that for the smooth-cylinder

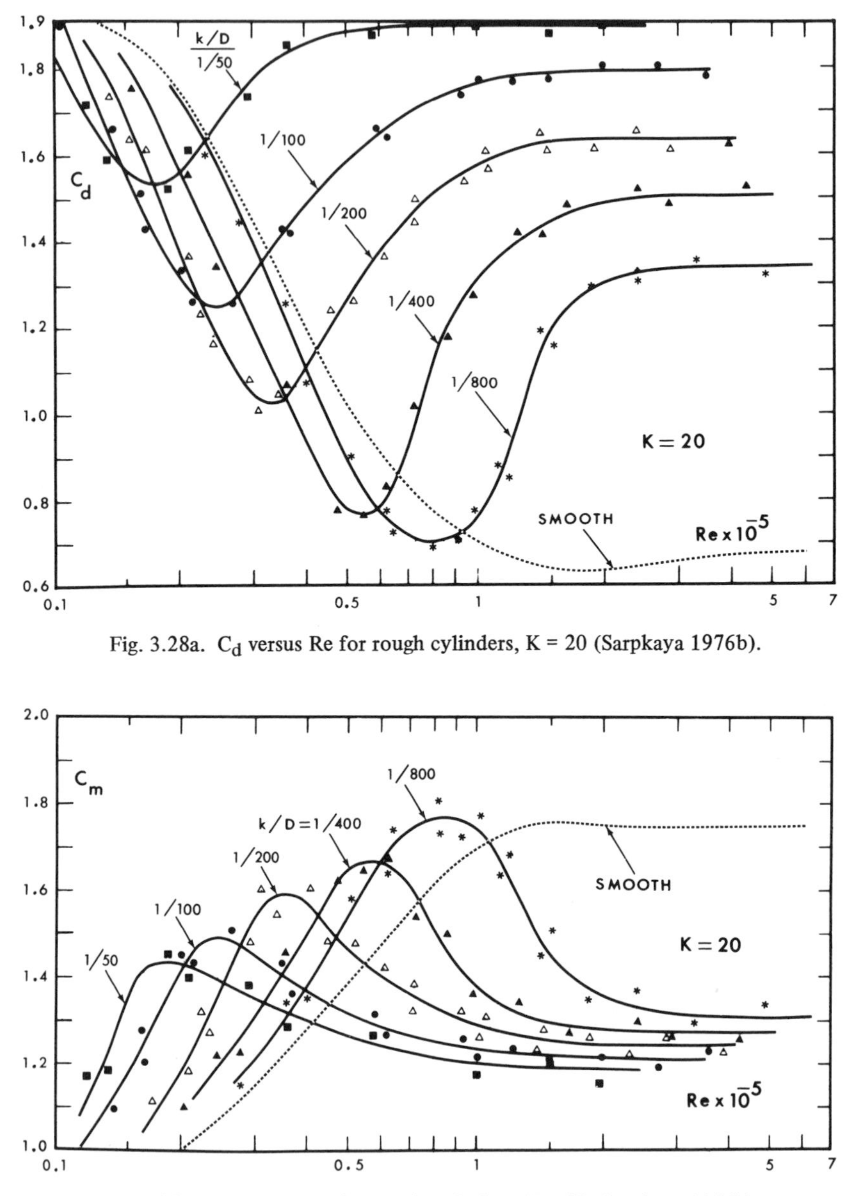

Fig. 3.28a. C_d versus Re for rough cylinders, K = 20 (Sarpkaya 1976b).

Fig. 3.28b. C_m versus Re for rough cylinders, K = 20 (Sarpkaya 1976b).

and then rises sharply to a nearly constant post-supercritical value. The larger the relative roughness the larger is the magnitude of the minimum C_d and the smaller is the Reynolds number at which that minimum occurs. However, there appears to be a minimum Reynolds number below which the results for rough

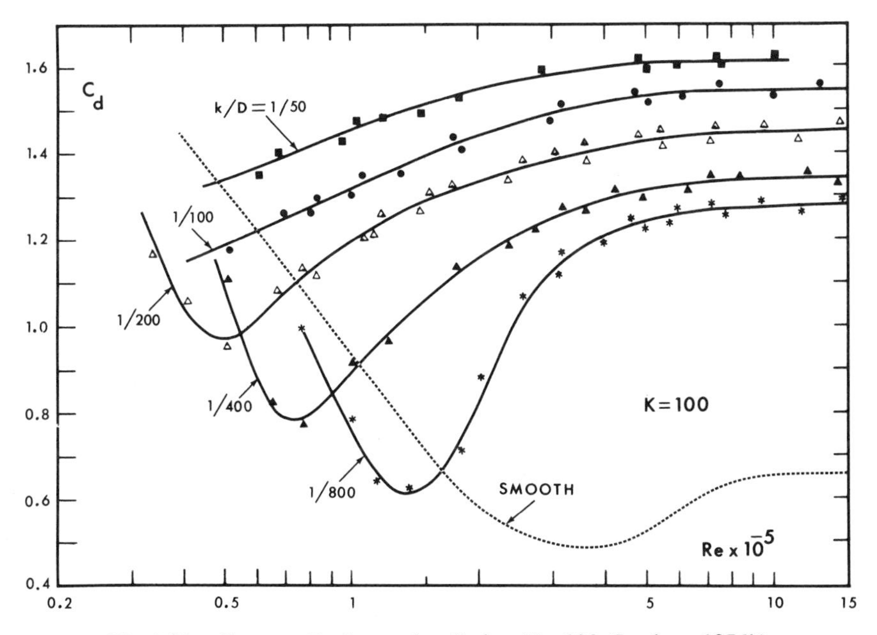

Fig. 3.29a. C_d versus Re for rough cylinders, K = 100 (Sarpkaya 1976b).

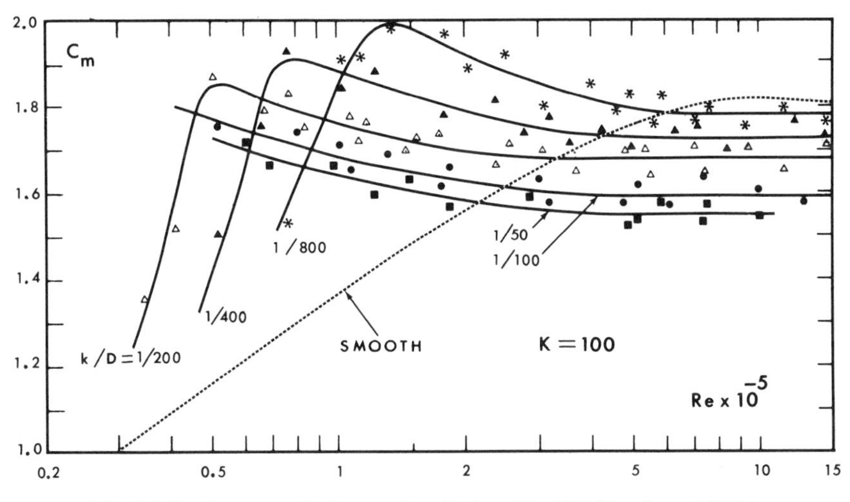

Fig. 3.29b. C_m versus Re for rough cylinders, K = 100 (Sarpkaya 1976b).

cylinders do not significantly differ from those corresponding to smooth cylinders. In other words, the Reynolds number must be sufficiently high for the roughness to play a role on the drag and flow characteristics of the cylinder.

The figures for the drag coefficient also exhibit a few other interesting features. First, even a relative roughness as small as 1/800 can give rise to post-super-critical drag coefficients which are considerably higher than those for the smooth cylinder. Further increases in roughness have a smaller effect than the initial change from a smooth to the first rough cylinder. Secondly, the asymptotic values of the drag coefficient (within the range of K and Re values encountered) for roughened cylinders can reach values which are considerably larger than those obtained with steady flows over cylinders of similar roughness ratio. In other words, it is not safe to assume that the post-supercritical drag coefficient in harmonic flows will be identical to those found in steady flows and will not exceed a value of about unity.

In steady flow about a cylinder, roughness precipitates the occurrence of drag crisis and gives rise to a minimum drag coefficient which is larger than that obtained with a smooth cylinder. This is partly because of the transition to turbulence of the free shear layers at relatively lower Reynolds numbers (due to disturbances brought about by the roughness elements) and partly because of the retardation of the boundary-layer flow by roughness (higher skin friction) and, hence, earlier separation. In harmonic flow about a cylinder, roughness appears to play an even more complex role because of the time-dependence of the boundary layer and the position of the separation points. In particular, the magnitude of the transverse force (to be discussed later) strongly suggests that the combined effect of uniformly-distributed roughness and time-dependence (even in the drag-dominated region of K values) is *to increase the strength of the vortices and the spanwise coherence* relative to that in steady flow at the same Reynolds number about the same cylinder.

The flow visualization studies of Grass and Kemp (1979) of the oscillatory flow past smooth and rough cylinders have shown that the angle of separation follows a well defined time-varying path through each half cycle of oscillation (see Fig. 3.30). In the case of rough cylinders, the separation angle initially reduces much more rapidly and remains considerably smaller than that for the smooth cylinder. As noted by Grass and Kemp, this observation is consistent with the larger drag coefficient measured by Sarpkaya (1976b) under closely similar flow and relative roughness conditions.

The Reynolds number at which the drag crisis occurs gives rise to an 'inertia crisis.' For a given relative roughness, C_m rises rapidly to a maximum at a Reynolds number which corresponds to that at which C_d drops to a minimum. At relatively higher Reynolds numbers, C_m decreases somewhat and then attains nearly constant values which are lower than those corresponding to the smooth cylinders. It is also apparent from the inertia coefficient curves that the smaller the relative roughnesses the larger is the maximum inertia coefficient. For relatively smaller roughness such as $k/D = 1/800$, the terminal value of C_m is not entirely unexpected. It has been previously noted that whenever there is a rise in the drag coefficient, there also is a decrease in the inertia coefficient. It is

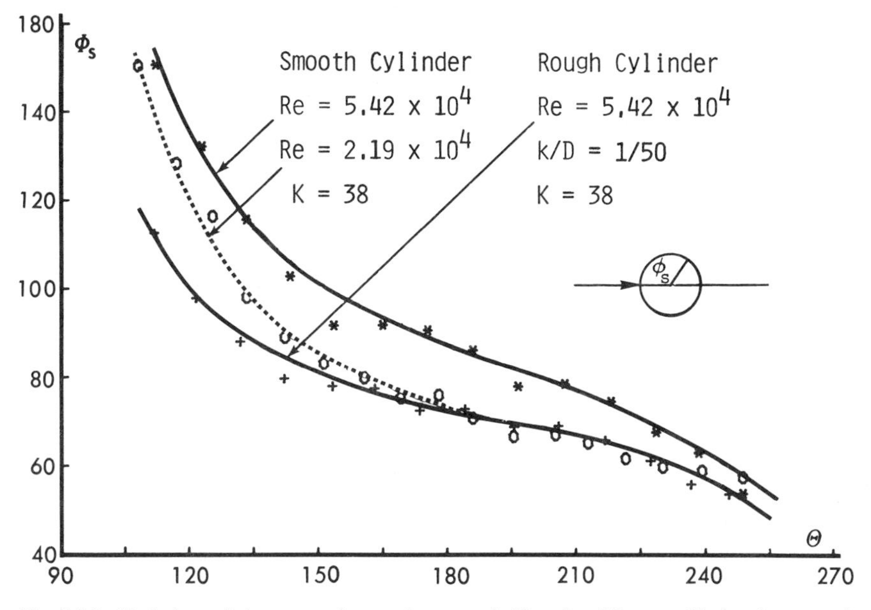

Fig. 3.30. Variation of the separation angle over a half cycle of flow oscillation for smooth and rough cylinders (Grass and Kemp 1979).

apparent from Eqs. (3.23) and (3.24) that C_d increases and C_m decreases as the phase difference between the maximum force and the maximum velocity decreases. Thus, the noted counter-variation of C_d and C_m is a consequence of the use of Morison's equation with time-invariant coefficients and not a consequence of a fluid mechanical phenomenon.

Sarpkaya's rough cylinder data for K = 100 are plotted in Fig. 3.31 as a function of the *roughness Reynolds number*, defined by $Re_k = U_m k/\nu$, for all values

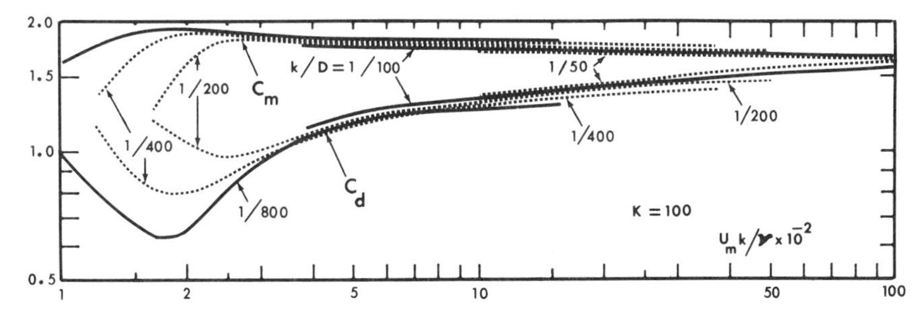

Fig. 3.31. Drag and inertia coefficients as a function of the roughness Reynolds number (Sarpkaya 1976b).

of k/D. Similar plots may be prepared for other values of K through the use of the data given by Sarpkaya (1976b).

It is rather remarkable that C_d and C_m become practically independent of k/D for Re_k larger than about 300. In other words, for sufficiently large values of the roughness Reynolds number, the drag and inertia coefficients for a uniformly-roughened cylinder in a given harmonic flow are determined by the height of the excrescences (above the mean diameter) rather than by the diameter of the cylinder (fully rough regime). The importance and the consequences of this result for post-supercritical Reynolds number simulation for steady flow over roughened cylinders have been discussed by Szechenyi (1975). It must be emphasized that the designer needs to know the apparent diameter of the pipe in order to predict accurately the forces acting on the roughened pipe (drag increases with D and inertia, with D^2). Furthermore, increased marine growth results in higher values of measured response at low frequencies. Consequently, there is a strong need for the acquisition of marine-fouling data for various oceans and construction sites. Only through such information can one determine the effective diameter of a pipe as a function of water depth at the construction site.

The maximum lift coefficient for the roughened cylinders is presented in Fig. 3.32 as a function of K for various values of β and for representative values of the relative roughness k/D. The lift coefficient for rough cylinders does not vary appreciably with either Re or β. It is rather surprising that the smooth cylinder data at relatively low values of β form more or less the upper limit of the rough cylinder data.

The Strouhal number St for rough cylinders remains essentially constant at a value of about 0.22. To be sure, there are variations from one cylinder to another and from a given combination of Re and K to another. Nevertheless, the Strouhal number is fairly constant for all roughnesses, relative amplitudes, and Reynolds numbers (larger than about 20,000). This fact is of special importance in determining the in-line and transverse vibrational response of the elements of a structure to wave-induced forces. One must, however, bear in mind the fact that the spanwise coherence along a vertical cylinder in the ocean environment may be reduced by the variation of the velocity vector with time and depth and that the lift coefficients cited above represent the maximum possible values.

3.8.7 Instantaneous Values of C_d, C_m, C_L, and the Negative Added Mass

For the past thirty years workers on resistance in wavy flows dealt primarily with the problem of determining the time-invariant averages of the drag and inertia coefficients. It is clear now that this is but one, and sometimes secondary, aspect of the problem. There is often no difficulty in finding suitable values of

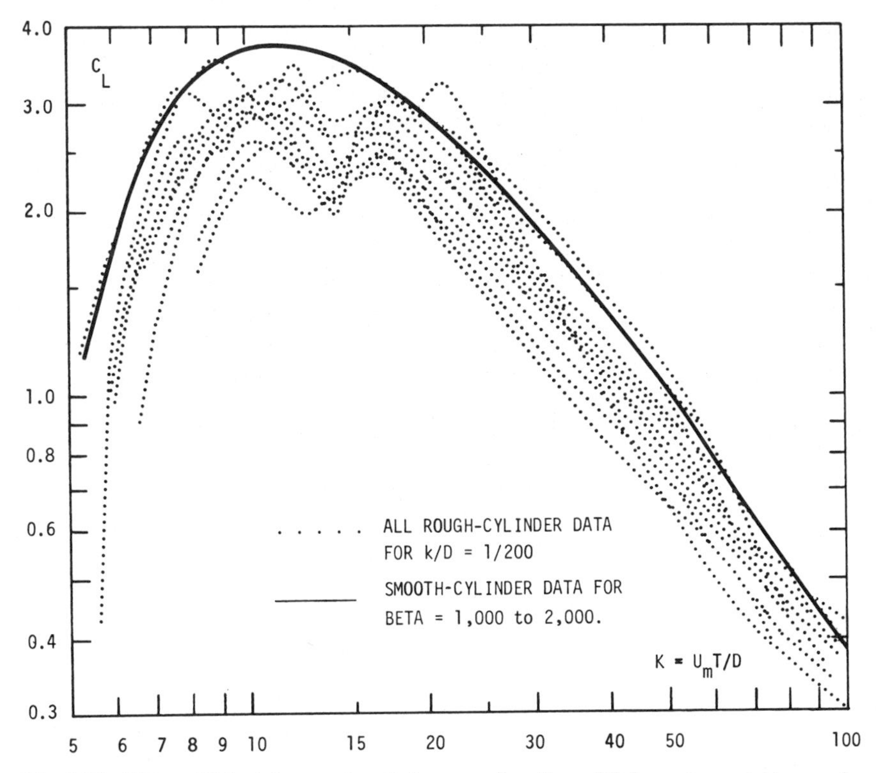

Fig. 3.32. Lift coefficient for rough cylinders as a function of K for various relative roughnesses (Sarpkaya 1976b).

C_d and C_m which will lead to the calculation of the in-line force to some accuracy dependent on K, Re, k/D, and the kinematics of the ambient flow, but the remaining problem is far more complex. The main stream of modern work in steady and time-dependent flows is concerned with the phenomena in the near wake.

In steady flow, the near wake is dominated by and large by the newest generation of vortices. In periodic flow, where the sweeping back of the vortices is an inevitable feature of the flow, the body is surrounded by a cluster of young and old generations of vortices. The formation, growth, and motion of the new vortices strongly affect and are affected by the older vortices which have survived diffusion and dissipation. The congregation of the old and new generation of vortices, with their differing memories residing in their vorticity, gives rise to numerous interesting phenomena. Some vortices never grow to full strength, separation points undergo large excursions, and some vortex couples of opposite sign quickly remove themselves away from the main cluster of vortices. The transverse force is induced partly by the shedding of the newest vortices, as in steady flow, and partly by the older vortices returning to their generator (the body). These phenomena lead to incalculable changes in the pressure distribution and point out once again the fact that even at relatively high Keulegan-Carpenter numbers there must be significant differences between the steady uniform flow and the oscillating flow about a bluff body. The assumption that at high Keulegan-Carpenter numbers (say $K = 100$) the oscillatory flow must approximate a steady flow is quite false.

Clearly, the time-invariant drag and inertia coefficients have very little physical significance. They only lead to a rough estimate of the in-line force dependent on K, Re, k/D, and the wave theory with respect to which they are calibrated. The use of the averaged coefficients cannot, therefore, be expected to represent the time-dependent force with equal degree of accuracy for all combinations of K and Re. As will be discussed in greater detail in Section 3.8.8, one can, in Morison's equation, disregard the drag term for K less than about 8; the inertia term, for K greater than about 25; and the linear-quadratic sum concept, with constant coefficients, for the intermediate values of K. This fact has been recognized by Keulegan and Carpenter (1956) and led them to the calculation of the instantaneous values of C_d and C_m and to the evaluation of a 'remainder force function' ΔR, to account for the differences between the measured force and the one predicted through the use of the averaged C_d and C_m in the linear-quadratic sum.

Keulegan and Carpenter determined the local values of C_d and C_m from the observed values of force using Morison's equation. Two sets of evaluations were made. It was assumed in the first evaluation that for $\theta_1 = \pi/2 + \alpha$ and $\theta_2 = \pi/2 - \alpha$, where α is an angle less than $\pi/2$, the coefficients C_d and C_m each have equal values, since these are the phases where the accelerations dU/dt are equal and the velocities U are equal in absolute value although of opposite signs. This is true also for $\theta_1 = 3\pi/2 + \alpha$ and $\theta_2 = 3\pi/2 - \alpha$. In the second evaluation it was again assumed that for $\theta = \pi + \beta$ and $\theta = \pi - \beta$, where β is an angle less than π, the coefficients C_d and C_m each have equal values, since these are the phases where the velocities are equal and the accelerations are equal in absolute value although of opposing signs.

Keulegan and Carpenter's method of the evaluation of the instantaneous values of C_d and C_m is not acceptable since it assumes that there is no distinction between the accelerating and decelerating flows as long as the absolute values of the corresponding velocities and accelerations are equal. The correct method to determine the instantaneous values of $C_d(\theta)$ and $C_m(\theta)$ is to solve them from a set of Morison-type equations written at $\theta = \theta_n$ and $\theta = \theta_n + \Delta\theta$, assuming that $C_d(\theta)$ and $C_m(\theta)$ remain constant in the interval $\Delta\theta$, (about 3 degrees). Such a procedure has been applied by Sarpkaya (1980) to the mea-

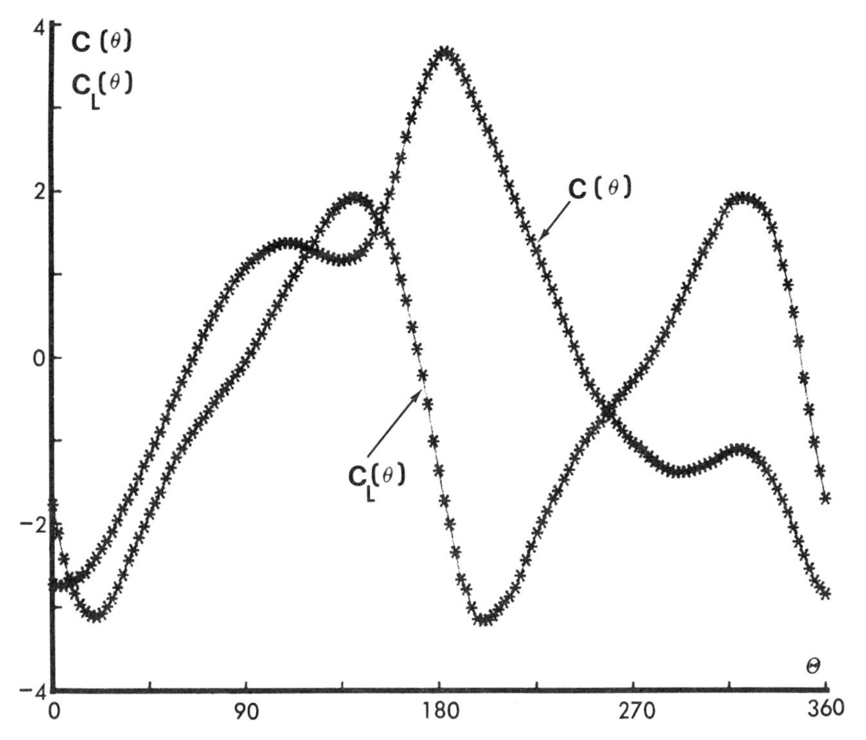

Fig. 3.33. Normalized in-line and transverse forces as a function of time, K = 12 and Re = 12,540 (Sarpkaya 1980).

sured in-line force for a wide range of K and Re values. Figure 3.33 shows $C(\theta) = 2F/\rho DU_m^2$ and $C_L(\theta)$ as a function of θ for K = 12 and Re = 12,540. The Fourier-averaged drag and inertia coefficients for this particular data are $C_d = 2.13$ and $C_m = 0.70$. Figure 3.34 shows the instantaneous values of $C_d(\theta)$ and $C_m(\theta)$. Several observations can be made. Firstly, the averaged C_m is less than unity and the added mass coefficient C_a is negative according to $C_a = C_m - 1$. Secondly, neither $C_d(\theta)$ nor $C_m(\theta)$ is symmetrical with respect to $\theta = \pi/2$ or π or $3\pi/2$, as they would have been had one used the Keulegan-Carpenter method in determining the instantaneous values of C_d and C_m. In other words, the accelerating and decelerating flows with identical absolute values of the corresponding velocities and accelerations do not exert identical forces on the cylinder. Thirdly, both $C_d(\theta)$ and $C_m(\theta)$ exhibit large variations during a given cycle (C_m shows large negative values), particularly for the intermediate values of K. The large increase in $C_d(\theta)$ at $\theta = \pi/2$ and $3\pi/2$ is due to the fact that at these angles the velocity vanishes ($U = -U_m \cos\theta$). It is easy to

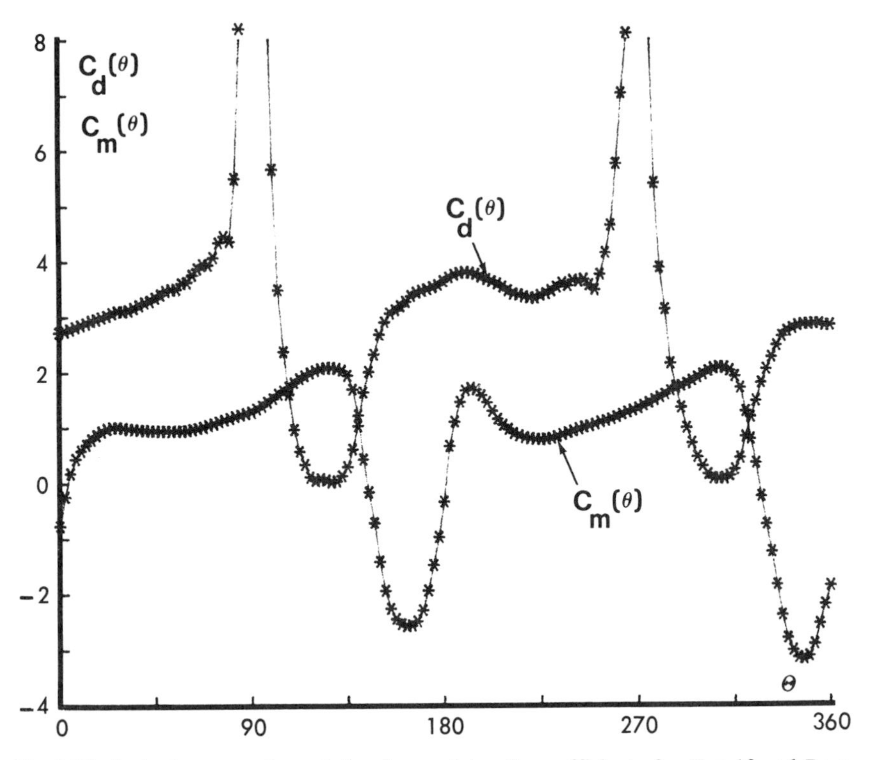

Fig. 3.34. Instantaneous values of the drag and inertia coefficients for K = 12 and Re = 12,540 (Sarpkaya 1980).

verify that the use of $C_d(\theta)$ and $C_m(\theta)$ in Morison's equation yields an in-line force which is identical to that measured. Such a verification has been done for all the examples worked out even though it has been expected as a natural outcome of the method used.

Figure 3.34 and others for the intermediate range of K values (for β less than about 2500) show that the large variations in $C_d(\theta)$ and $C_m(\theta)$ cannot be approximated by two averaged coefficients. This unhappy result leaves one with very few choices: (a) Morison's equation is not applicable in the so-called drag-inertia dominated regime; (b) one should use the instantaneous values of C_d and C_m, which is a nearly impossible task; and (c) Morison's equation should be modified with the addition of a remainder term or terms which will account for and minimize the difference between the measured and calculated forces. Suffice it to note that for K smaller than about 8 and larger than about 20, for a cylinder, the difference between C_d and $C_d(\theta)$ and C_m and $C_m(\theta)$ rapidly decreases and the instantaneous force is accurately predicted by the linear-quadratic sum with constant coefficients.

The next question to be dealt with is the negative added mass and inertia coefficient. So far no satisfactory explanation has been offered for this intuitively anomalous result. Generally, the motion of the vortices is suggested as the reason. This is too broad a generalization which may be invoked as an explanation for all the features of time-dependent separated flows. It will now be shown that the negative mass is a consequence of the averaging process and the instantaneous value of C_a never becomes negative.

The inertial part of the in-line force may be written, as first suggested by McNown and Keulegan (1959), as

$$F_{in} = \rho \frac{\pi D^2}{4} C_m(\theta) \frac{dU}{dt} = \frac{d}{dt}\left[C_a(\theta)\rho \frac{\pi D^2}{4} U\right] + \rho \frac{\pi D^2}{4} \frac{dU}{dt} \tag{3.37}$$

where $\theta = 2\pi t/T$ and $C_a(\theta)$ is the instantaneous value of the added-mass coefficient. The second term represents the rate of change of momentum and the last term, the usual buoyant force due to the pressure gradient. With $U = -U_m \cos\theta$, Eq. (3.37) simplifies to

$$\frac{d}{d\theta}[C_a(\theta)\cos\theta] = [1 - C_m(\theta)]\sin\theta \tag{3.38}$$

Integrating, one has

$$C_a(\theta) = -1 - \frac{1}{\cos\theta}\int_{\pi/2}^{\theta} C_m(\theta)\sin\theta\, d\theta \tag{3.39}$$

Equation (3.39) may be evaluated through the use of $C_m(\theta)$. Sarpkaya (1980) has performed such calculations for various values of K, particularly in the intermediate region. Figure 3.35 shows the variation of $C_a(\theta)$ for the $C_m(\theta)$ values shown in Fig. 3.34. Note that the plot shows $5C_a(\theta)$ in order to accentuate the variation of $C_a(\theta)$ with θ. Also shown in Fig. 3.35 is $C_L(\theta)$ for the same data (see Fig. 3.33). Figures 3.36 through 3.38 show plots similar to Figs. 3.33 through 3.35 for K = 15 and Re = 18,390 (for which $C_d = 1.82$ and $C_m = 0.85$).

Numerous calculations have shown that $C_a(\theta)$ varies with time, *never becomes negative*, and oscillates between zero and one. Furthermore, there is a remarkable correlation between $C_a(\theta)$ and $C_L(\theta)$, i.e., the shedding of each vortex causes similar changes in $C_L(\theta)$ and $C_a(\theta)$.

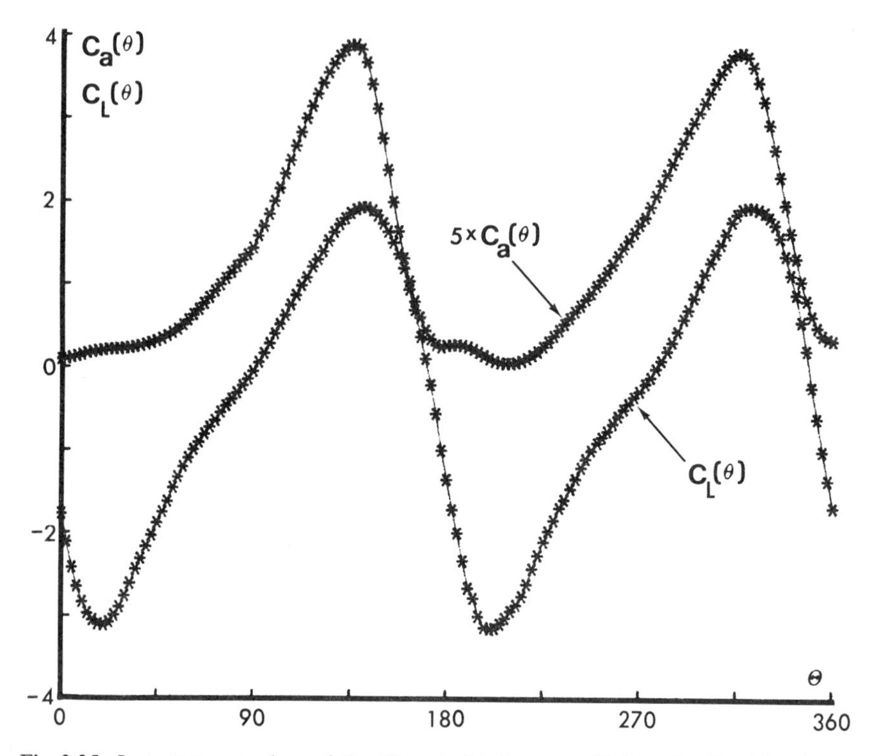

Fig. 3.35. Instantaneous values of the lift and added mass coefficients for K = 12 and Re = 12,540 (Sarpkaya 1980).

The first term on the right hand side of Eq. (3.37), which may be written as,

$$\rho \frac{\pi D^2}{4} \left[C_a(\theta) \frac{dU}{dt} + U \frac{dC_a(\theta)}{dt} \right] \tag{3.40}$$

shows that when the variable mass is growing $[dC_a(\theta)/dt > 0]$, the average C_a is much larger than that required to define the quantity of fluid being accelerated. The shedding of a vortex brings about an abrupt reduction in the fluid mass (negative dC_a/dt) which can and does reduce the average added mass coefficient even to negative values (for β less than about 2,500, in the drag-inertia dominated regime). Thus, the averaged negative added mass is a consequence of the averaging process and does not actually contradict reality. The fact that C_a becomes negative only for certain values of K and for β less than about 2,500 shows that not only the number of vortices but also the viscous effects play an important role in yielding average negative mass. This process may also be strongly affected by the correlation or lack thereof along the cylinder.

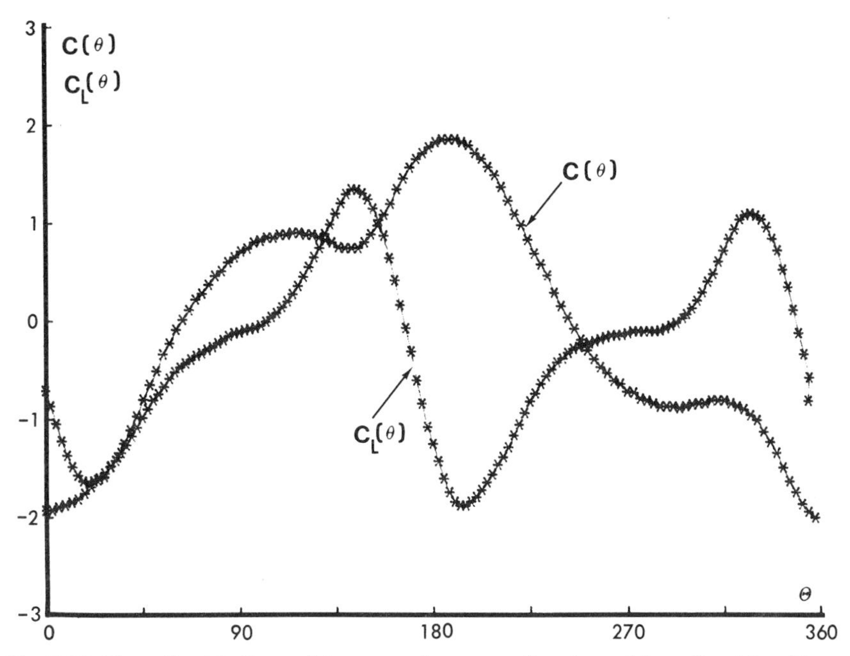

Fig. 3.36. Normalized in-line and transverse forces as a function of time, K = 15 and Re = 18,390 (Sarpkaya 1980).

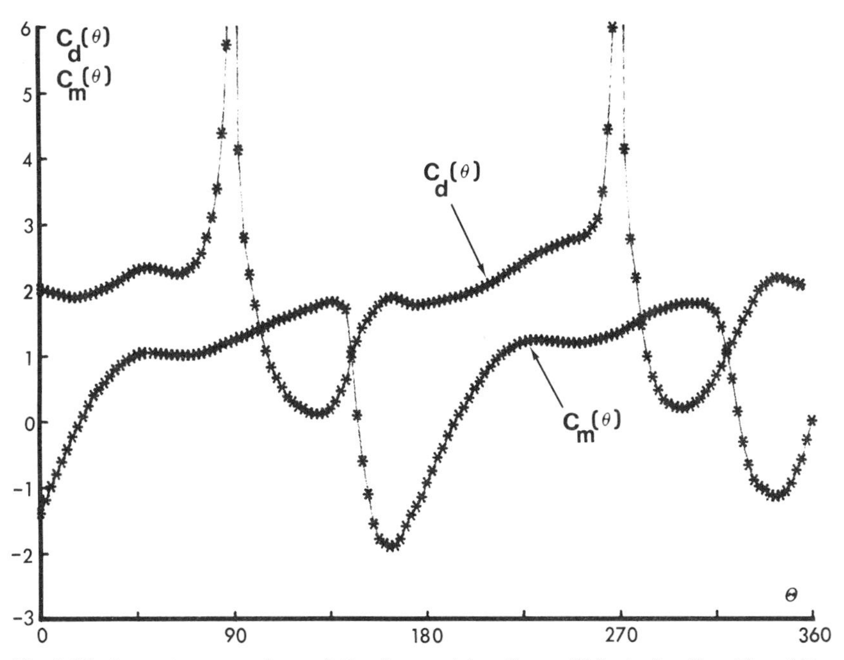

Fig. 3.37. Instantaneous values of the drag and inertia coefficients for K = 15 and Re = 18,390 (Sarpkaya 1980).

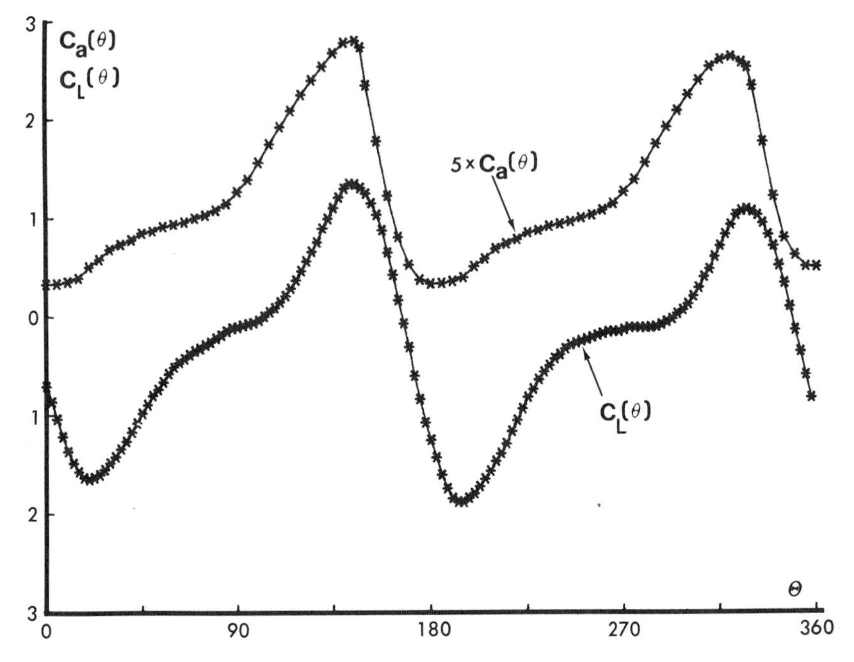

Fig. 3.38. Instantaneous values of the lift and added mass coefficients for K = 15 and Re = 12,540 (Sarpkaya 1980).

The variations of $C(\theta)$, $C_a(\theta)$, $C_m(\theta)$, $C_d(\theta)$, and $C_L(\theta)$ with respect to each other exhibit extremely interesting features in terms of the growth and motion of vortices. These will not be shown here because of space limitations.

3.8.8 A Critical Assessment of Morison's Equation

Morison's equation was proposed as an approximate solution to a complex problem. Its critical assessment means different things to different people. To an offshore engineer it usually connotes the examination of the applicability of the said equation to the prediction of the wave and current induced forces acting on offshore structures within acceptable and reliably definable error limits. This, in turn, means a closer look at the accuracy of the kinematic inputs, the method of evaluation of the averaged coefficients, the quality of the data used in their evaluation, sensitivity and the degree of dependence of the predictions of Morison's equation on Reynolds number, Keulegan-Carpenter number, relative roughness, body orientation, proximity effects, parameters specifying the characteristics of the motion of fluid particles, etc. Such an assessment does not allow one to draw conclusions regarding the intrinsic nature of the equation.

Both the form of the equation and the uncertainties that go into the characterization of the ocean environment are jointly responsible for the differences between the measured and calculated forces. In other words, it is not a meaningful exercise to relegate the errors only to one or the other. Thus, as far as the ocean environment is concerned, Morison's equation, *with calibrated coefficients*, is tolerated in light of all other uncertainties and hidden and intentional safety factors that go into the design.

To a fluid dynamicist the critical assessment of Morison's equation means the understanding of why, how, and when a linear-quadratic sum represents the force acting on a body immersed in a given time-dependent separated flow. Clearly, a wholesale assessment of Morison's equation is not very meaningful. Equally important is the distinction between the desire to improve it by the addition of one or more semi-empirically chosen terms and the desire to understand the instantaneous behavior of the resistance in terms of the formation, growth, and motion of vortices. The last objective is the most difficult to fulfill in view of the fact that even in simple steady, smooth, uniform, two-dimensional, separated flow past a smooth circular cylinder one does not have a theoretical or numerical solution which explains all the characteristics of the flow as a function of the Reynolds number. Existing numerical models require some fine tuning to match the predictions to the experimental results.

It is because of the foregoing reasons that this discussion is restricted mainly to in-line forces in harmonically-oscillating planar motion about a cylinder. Clearly, harmonic flow is the simplest of all oscillating flows and the determination of its kinematics does not require the use of intermediate theories whose applicability is subject to separate assessment. We do not, however, conclude that the applicability or lack of applicability of Morison's equation to one particular flow situation precludes its use in other flow situations.

Morison's equation yields no information about the transverse force and seems to be adapted best to a range of Keulegan-Carpenter numbers smaller than about 8 and larger than about 25, where the complex problems associated with the motion of a few vortices are not as much pronounced.

It has already been shown in Section 3.8.7 that the instantaneous values of C_d and C_m differ significantly from their Fourier-averaged values. This difference is reflected by the mismatch between the measured and calculated forces. Figure 3.39 is one of many such examples where the residue is a significant fraction of the maximum force.

Numerous attempts have been made either to improve Morison's equation or to devise new equations (see e.g., Barnouin et al., 1979). So far no satisfactory results have been obtained. It appears that it would be rather difficult to abandon the linear-quadratic sum since it works rather well outside a narrow range of Keulegan-Carpenter numbers. Thus, it appears to be preferable to attempt to improve the equation rather than to devise a new one. In fact, this was the

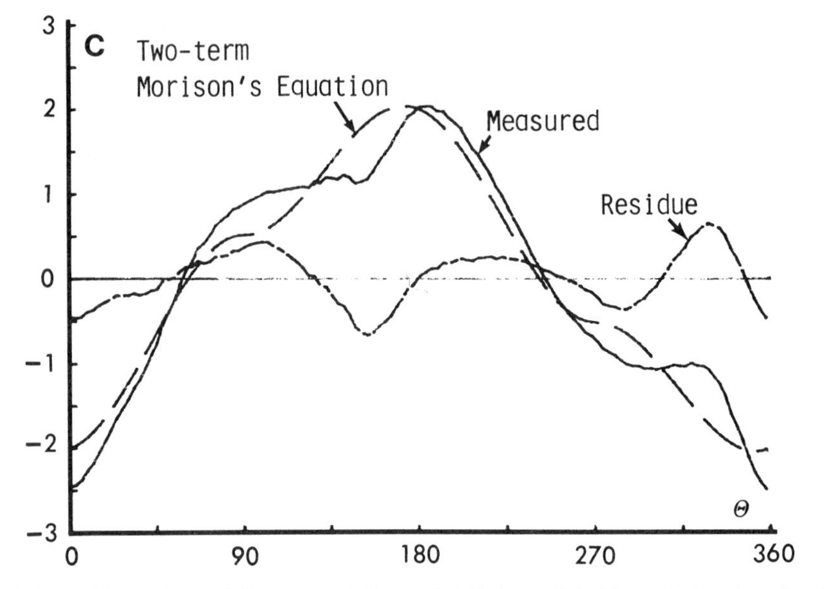

Fig. 3.39. Comparison of the measured force with that calculated from the two-term Morison equation, K = 14 and Re = 27,800.

original proposal of Keulegan and Carpenter who expressed the force as (it is assumed that the diffraction effects are negligible)

$$\frac{2F}{\rho D U_m^2} = \frac{\pi^2}{K} C_m \sin\theta - C_d \cos\theta\,|\cos\theta| + \Delta R \tag{3.41}$$

where ΔR represents the residue given by [see Eq. (3.21)]

$$\Delta R = 2\,[A_3 \sin 3\theta + A_5 \sin 5\theta + \cdots] + 2\,[B_3' \cos 3\theta + B_5' \cos 5\theta + \cdots] \tag{3.42}$$

in the range of K and Re values where Morison's equation fails to represent the measured force with sufficient accuracy, (large ΔR), Eq. (3.42) may be used to evaluate the coefficients A_n and B_n' through the Fourier analysis. Keulegan and Carpenter approximated ΔR with two terms involving only A_3 and B_3'. This procedure showed considerable improvement over the two-term Morison equation in the range of $10 < K < 25$. The obvious disadvantage of this expanded form of the equation is that it now requires the evaluation of four coefficients, namely, C_d, C_m, A_3, and B_3'. Even then the calculated and measured forces do not always correspond partly due to the existence of other harmonics and partly due to the pronounced effect of the randomness of the shedding, coherence, and motion of a few vortices, vice large number of vortices. This, in turn, requires the addition of two more terms involving A_5 and B'_5. Clearly, the determination of the dependence of six coefficients on the parameters characterizing the phenomenon is a nearly impossible task and is not very practicable for the design of offshore structures. It is partly because of this reason and partly because of the uncertainties of the input parameters that the two-term Morison equation has been used over the past thirty years in spite of its known limitations. The inaccuracies resulting from the use of the said equation have been taken care of partly by nature through the mitigating effects of the ocean environment (e.g., reduced span-wise coherence, omnidirectionality of the waves and currents, etc.) and partly by the designer through the use of hidden and intentional safety factors.

Ideally, one would like to revise Morison's equation with the following constraints: (a) the revision should be fluid-mechanically satisfying; (b) the revised form of the equation should not contain more than the two coefficients already in use, namely, C_d and C_m; (c) the coefficients of the additional terms should be related to C_d and/or C_m through a careful spectral and Fourier analysis of ΔR; and (d) the revised equation should reduce to Morison's equation in the drag and in the inertia dominated regimes of the flow.

An extensive study of the measured and calculated forces and the residues by Sarpkaya (1980) of the data presented previously (Sarpkaya 1976a) has shown that although all harmonics play some role, the third harmonic of the residue is most important. Then, a three-term Morison equation may be written as

$$C = \frac{2F}{\rho D U_m^2} = \frac{\pi^2}{K} C_m \sin\theta - C_d |\cos\theta| \cos\theta + C_3 \cos(3\theta - \phi_3) \quad (3.43)$$

The analysis of the data has shown that both C_3 and ϕ_3 are strong functions of the deviation of C_m from its ideal value of 2 (for a circular cylinder). By properly scaling C_3 and ϕ_3 in terms of $(2 - C_m)$, K and Re, Eq. (3.43) has been reduced to

$$C = \frac{\pi^2}{K} C_m \sin\theta - C_d |\cos\theta| \cos\theta + \beta^{3/4} [(2 - C_m)/(100\,K)]$$
$$\cdot \cos[3\theta + (K - 4)(2 - C_m)\pi/2K] \quad (3.44)$$

Equation (3.44) is empirical in nature but it does satisfy practically all the constraints imposed on its evolution. The results have shown that Eq. (3.44) reduces the r.m.s. value of the residue obtained from the Morison equation by

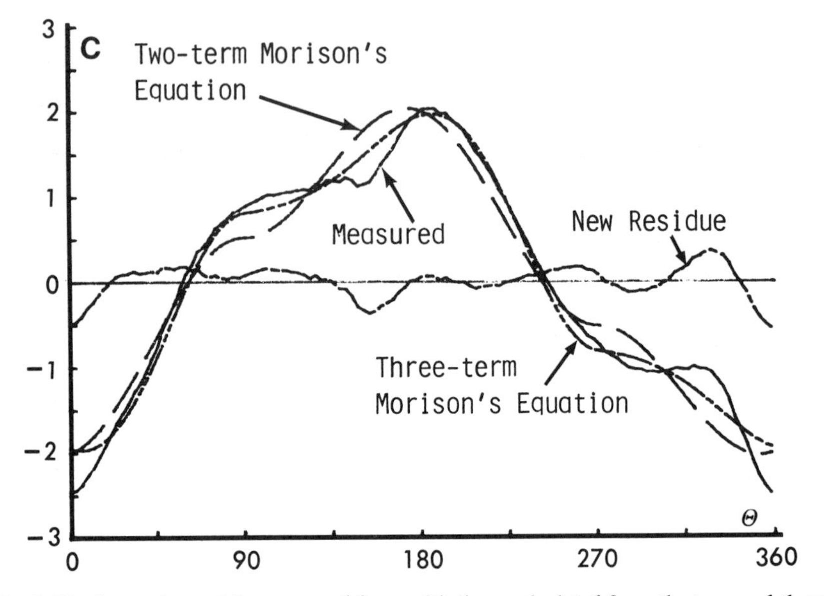

Fig. 3.40. Comparison of the measured force with those calculated from the two- and three-term Morison equation, K = 14 and Re = 27,800.

about 60%, particularly in the drag/inertia dominated regime, and provides a much better fit between the measured and calculated forces (see Fig. 3.40). Additional work along these lines may lead to the substantial improvement of Morison's equation.

In summary, it is suggested that Morison's equation is quite satisfactory in the drag and in the inertia dominated regimes, it is unlikely that an entirely new equation will ever replace it, and that the addition of one more term (with no new coefficients), based on the analysis of the residues, may result in a practically more suitable expression, if not fluid-mechanically more satisfactory explanation, of the resistance in complex time-dependent flows.

3.9 SUPERPOSED MEAN AND OSCILLATORY FLOW ABOUT A CYLINDER OR THE IN-LINE HARMONIC OSCILLATIONS OF A CYLINDER IN STEADY FLOW

This section will be devoted strictly to the discussion of two-dimensional harmonic motion in uniform flow. The effects of currents superimposed on the waves on the loading of offshore structures (platform legs, cross-bracing, drilling risers, mooring lines, etc.) will be taken up in Chapter 5.

The in-line oscillations of a cylinder in uniform flow of velocity $\overline{V}$ has not been studied extensively. Chen and Ballengee (1971) examined the vortex shedding from circular cylinders in an oscillating freestream of 3 Hz with A/D from 15 to 1,000, $D/\overline{V}T = 0.003$, and the Reynolds numbers up to 40,000. They have found that the vortex shedding from a circular cylinder responds instantaneously to the freestream oscillations and that 'the instantaneous Strouhal number stays sensibly constant at 0.2 ± 0.01.' This is rather expected since the amplitude of oscillations is many times that of the cylinder diameter and the flow in the absence of significant accelerations exhibits a quasi-steady behavior.

Hatfield and Morkovin (1972) studied the effect of an oscillating freestream on the unsteady pressure on a circular cylinder ($D/\overline{V}T$ from 0.15 to 0.25, A/D from 0.05 to 0.087, and Re = 50,000). They have found that there is no significant coupling between the small-amplitude freestream oscillations and the vortex shedding. Their results would suggest that the drag coefficient associated with the mean flow would essentially remain constant at its steady state value. This study, unlike the previous one, concentrated on the other extreme of the A/D values. In this case, the flow fluctuations in the neighborhood of the cylinder are essentially uncorrelated and therefore relatively ineffective. Evidently, for values of $D/\overline{V}T$ from about 0.1 to 0.5 and A/D from about 0.2 to 1.0 the interaction of oscillations with the body becomes significant and rather complex. The study of the two extremes of $D/\overline{V}T$ and A/D does not shed much light on the understanding of oscillating flow about bluff bodies.

Mercier (1973) who subjected cylinders to large streamwise oscillations found that the average drag coefficient significantly increases with $D/\overline{V}T$ and that the rate of increase depends on the amplitude-to-diameter ratio in the range of A/D from 0.2 to 3, $D/\overline{V}T$ from 0.1 to infinity, and Re from 4,000 to 16,000. Mercier had considerable difficulty in separating the fluid force from the inertial force acting on the oscillating cylinder and consequently his data exhibited significant scatter.

Davenport (1961) subjected bluff bodies (flat plate, circular and triangular cylinders, and a lattice truss) to small amplitude oscillations in a water flume and evaluated the drag and inertia coefficients through the measurement of the rate of damping of the amplitude of oscillations of the bodies, i.e., without directly measuring the forces acting on the bodies. The experiments were carried out in the range of $D/\overline{V}T$ from 0.02 to 2. The Reynolds number ranged from 1,700 to 8,000. Davenport evaluated C_m and C_d assuming the mean value of the drag coefficient to be equal to that for the corresponding steady flow at the velocity $\overline{V}$ and ignoring the term involving U_m^2. In general the mean value of the drag coefficient does not remain constant and U_m^2 cannot be ignored for $U_m/\overline{V}$ larger than about 0.1.

In a related study, Tseng (1972) conducted experiments with flat plates normal to the stream undergoing freely decaying oscillations. His results have shown that the presence of the mean flow significantly increases the damping force and that the rate of extinction of the oscillations increases monotonically with speed.

Tanida, Okajima, and Watanabe (1973) found that for a circular cylinder oscillating parallel to the flow ($A/D = 0.14$, $D/\overline{V}T$ from zero to 0.5, and $Re = 80$ and 4,000), the vortex synchronization can be observed in a range around double the Strouhal frequency, where vortices are shed with a frequency half the imposed one. The results of Tanida and his associates showed that the mean drag reaches its maximum in the middle of the synchronization range, i.e., $D/\overline{V}T$ between 0.2 and 0.4, and that the sign of C_d is such that no energy can be extracted from the fluid to render the oscillations unstable.* In other words, in-line oscillations are stable for the two Reynolds numbers at $A/D = 0.14$. Tanida conjectured that instability is likely to occur at much higher Reynolds numbers. Unfortunately, their C_d values cannot be relied upon since the inertial force was subtracted only approximately.

Goddard (1972) carried out a numerical analysis of the drag response of a cylinder to streamwise fluctuations through the use of the Navier-Stokes equations for $Re = 40$ and $D/\overline{V}T = 0.019$, 0.12, and 3.18, and for $Re = 200$ and $D/\overline{V}T = 0.149$. He found that for very low frequencies the instantaneous values of the drag correspond very nearly to the quasi-steady solution and that for higher frequencies the drag anticipates the freestream velocity maximum. This work cannot be generalized to higher Reynolds numbers since it is based on the Navier-Stokes equations and since the diffusion of vorticity in the concentrated vortices for Re larger than about 200 is primarily turbulent. It might be possible to use an eddy-viscosity model in additional numerical work for some qualitative results.

From a more practical point of view, in-line oscillations in uniform flow attracted attention partly because of the superposition of waves with currents and partly because of the damaging vortex-induced oscillations encountered in tidal waters. Such oscillations were observed at an oil terminal on the Humber estuary in England during the later 1960s. A brief description of the problems encountered has been given by Sainsbury and King (1971). Subsequently, Wootton et al., (1974) conducted full scale experiments to ascertain the causes of in-line vibrations of the pilings. They have found that the Strouhal number of the far wake remains constant at about 0.23. They have further found that two distinct flow patterns exist in the immediate wake of the cylinder, depending on the particular value of $f_n D/\overline{V}$, where f_n is the natural frequency of the piling. The inverse of this frequency parameter is known as the reduced velocity. One would ordinarily expect that the in-line oscillations will occur when the reduced

*Synchronization or lock-in described in greater detail in Chapter 8.

velocity $U_r = \overline{V}/f_n D$ is equal to about 2.5 since the frequency of the drag oscillations is twice the Strouhal frequency. Wootton et al. observed that the in-line oscillations occur over a range of $U_r = 1$ to 3 and in two distinct resonant response regimes separated by $U_r = 2$. At $U_r = 1.7$ a symmetric vortex shedding was observed. At $U_r = 2$, the vortex shedding changed to the commonly observed form of alternate shedding, indicating a radical change in the cause of excitation and dynamic response.

King (1974) and King et al. (1973) observed that the in-line oscillations occur for a range of reduced velocity from 1.5 to 4. They too have noted two distinct regimes of vortex shedding separated by $U_r = 2.5$.

Tatsuno (1972) towed through still water a small cylinder which was vibrated in-line with the tow direction. His experiments conducted at a Reynolds number of 100 were primarily concerned with the frequency and spacing of vortices. He did not measure any forces.

Crandall et al. (1975) investigated the destructive vibration of trashracks due to fluid-structure interaction and found that the excitation mechanism involved synchronization between the fluctuating drag and the in-line motion of the cylinders. The vibrations occurred at the 4th and 5th mode. The reduced velocity ranged from 1.12 to 1.81. They have noted a drastic change at $U_r = 1.47$ which corresponded to a change in the mode of vibration from the 4th to the 5th mode. Crandall et al. did not measure the vortex shedding frequencies but advanced very interesting concepts regarding the cause of synchronization.

Griffin and Ramberg (1976) conducted similar experiments in air at a Reynolds number of 190. Their results have essentially substantiated the previous observations of Wootton and King. No information was obtained regarding the forces acting on the cylinder.

Evidently, the determination of the forces acting on a cylinder undergoing harmonic in-line oscillations is just as important as the understanding of its kinematics. Since the phenomenon is identical to that where the cylinder is subjected to a time-dependent flow characterized by $U = \overline{V} - U_m \cos\theta$, the evaluation of the force might shed some light on the combined effect of waves and currents on the members of offshore structures.

It is ordinarily assumed (as recommended by the American Petroleum Institute, API, 1977)* that Morison's equation applies equally well to periodic flow with a mean velocity and that C_d and C_m have current-invariant, Fourier- or least-squared averages equal to those applicable to rigid, stationary cylinders in wavy flow. This, in turn, implies that C_d and C_m are independent of the biassed

*API RP 2A, 8th edition, April 1977 states that "Due consideration should be given to the possible superposition of current and waves. In those cases where this superposition is deemed necessary, the current velocity should be added vectorially to the wave velocity before the total force is computed."

convection of vortices and its attendant consequences. The fact that this is not necessarily so is clearly evidenced by the measurements of Mercier (1973), Sarpkaya (1977b), and Verley and Moe (1979). Thus, the effect of the current-harmonic flow combination on the motion of vortices and on the force-transfer coefficients must be carefully examined in light of available data and the limits of application of Morison's equation to such flows must be assessed anew. The latter is particularly important in view of the fact that the drag and inertia coefficients in ocean tests (where there are always some currents and body motion) are evaluated through the use of Morison's equation (note that the values of C_d and C_m may vary considerably from one half wave cycle to another because of the current-induced biassing of the wake and vortex formation and that neither set of coefficients may be identical with those obtained without current).

It has been customary to express the in-line force either as

$$F = 0.5\rho C_{dc} D(\overline{V} - U_m \cos\theta)|\overline{V} - U_m \cos\theta| + (\pi\rho C_{mc} D^2/4)\, dU/dt \tag{3.45}$$

or as (Mercier 1973, Sarpkaya 1977b, Verley and Moe 1979)

$$F/(0.5\rho D\overline{V}^2) = \overline{C}_d - C_{dh}(U_m T/D)^2 (D/\overline{V}T)^2\, |\cos\theta| \cos\theta + C_{mh}\pi^2 (U_m T/D)(D/\overline{V}T)^2 \sin\theta \tag{3.46}$$

The coefficients C_{dc}, C_{mc}, C_{dh}, and C_{mh} are given by their Fourier averages and, in general, $C_{dc} \neq C_{dh} \neq C_d$ and $C_{mc} \neq C_{mh} \neq C_m$ where C_d and C_m are the in-line force coefficients for the harmonic flow alone. All of the coefficients appearing in Eqs. (3.45) and (3.46), including $\overline{C}_d$, are functions of $\overline{V}T/D$, $U_m T/D$ (or A/D), Reynolds number, and the relative roughness (there are no data for the discussion of the cylinder yaw, body-proximity, etc. in wave-current flow).

The said coefficients may be determined only experimentally either in a wave channel with a mean current in which case the additional complexities brought about by the three-dimensionality of the flow tend to obscure the systematic evaluation of the said parameters, or in a uniform flow channel by oscillating the body in the in-line direction, or by moving the body at a constant speed in a harmonically oscillating flow. The results obtained by Mercier (1973), Sarpkaya (1977b), and Verley and Moe (1979), through the use of one or the other of the above methods, have shown that the force coefficients undergo dramatic changes with increasing A/D in the range of $D/\overline{V}T$ values from 0.1 to 0.5.

The question of whether a single drag coefficient C_{dc}, as in Morison's equation for strictly harmonic flow, could suffice to calculate the drag force accurately for a cylinder oscillating in uniform flow may be examined through the use of the data presented by Mercier (1973) and Sarpkaya (1977b). For this purpose assume that the drag portion of the in-line force may be written as

$$F_d/(0.5\rho D\overline{V}^2) = C_{dc}(U_m/\overline{V})^2\,|\overline{V}/U_m - \cos\theta|\,(\overline{V}/U_m - \cos\theta) \quad (3.47)$$

The mean value of C_{dc} is then given by

$$C_{dc}(1)\,(U_m/\overline{V})^2\,(1/2\pi)\int_0^{2\pi} |\overline{V}/U_m - \cos\theta|\,(\overline{V}/U_m - \cos\theta)\,d\theta = \overline{C}_d \quad (3.48)$$

or by

$$C_{dc}(1) = H(1)\overline{C}_d \quad (3.49)$$

where

$$H(1) = (\overline{V}/U_m)^2/G(1) \quad (3.50)$$

and

$$G(1) = \frac{1}{2\pi}\int_0^{2\pi} |\overline{V}/U_m - \cos\theta|\,(\overline{V}/U_m - \cos\theta)\,d\theta \quad (3.51)$$

The functions G(1) and hence H(1) are dependent only on $(\overline{V}/U_m)$ and may be calculated once and plotted as shown in Fig. 3.41.

Similarly, one can calculate the Fourier average of C_{dc} as

$$C_{dc}(2)\,(U_m/\overline{V})^2\int_0^{2\pi} |\overline{V}/U_m - \cos\theta|\,(\overline{V}/U_m - \cos\theta)\cos\theta\,d\theta$$

$$= -C_{dh}\,(U_m/\overline{V})^2\int_0^{2\pi} |\cos\theta|\cos^2\theta\,d\theta$$

$$= -C_{dh}\,(U_m/\overline{V})^2\,(8/3) \quad (3.52)$$

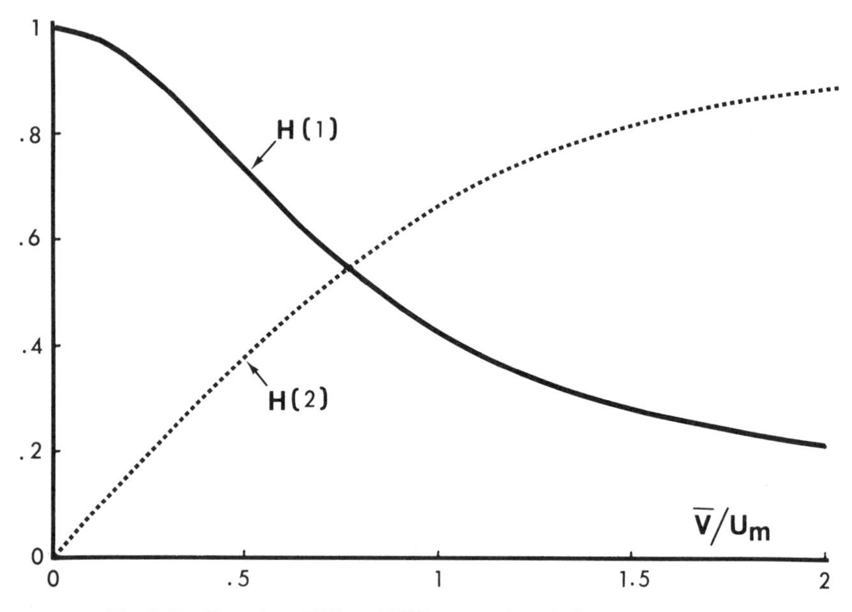

Fig. 3.41. Functions H(1) and H(2) versus the relative current velocity.

which may be reduced to

$$C_{dc}(2)G(2) = -C_{dh} \quad \text{or} \quad C_{dc}(2) = H(2)C_{dh} \tag{3.53}$$

where H(2) is given by

$$H(2) = -(8/3)/\left[\int_0^{2\pi} |\overline{V}/U_m - \cos\theta|\,(\overline{V}/U_m - \cos\theta)\cos\theta\, d\theta\right] \tag{3.54}$$

The Eqs. (3.48) and (3.52) are identical to those obtained by Matten (1976) and Verley and Moe (1979) through the use of slightly different methods. Once again G(2) and H(2) are functions of $\overline{V}/U_m$ as shown in Fig. 3.41.

Sarpkaya's (1977b) results of $\overline{C}_d$ and C_{dh} have been used together with Fig. 3.41 to calculate $C_{dc}(1)$ and $C_{dc}(2)$ through the use of Eqs. (3.49) and (3.53). These coefficients are shown in Figs. 3.42a and 3.42b as a function of $\overline{V}T/D$ for various values of A/D. Evidently, had it been possible to use a single drag coefficient for the calculation of the drag component of the in-line force through the use of Morison's equation [Eq. (3.47)], the coefficients $C_{dc}(1)$ and $C_{dc}(2)$ would have been identical. The fact that this is not quite so is clear from a comparison of Figs. 3.42a and 3.42b.

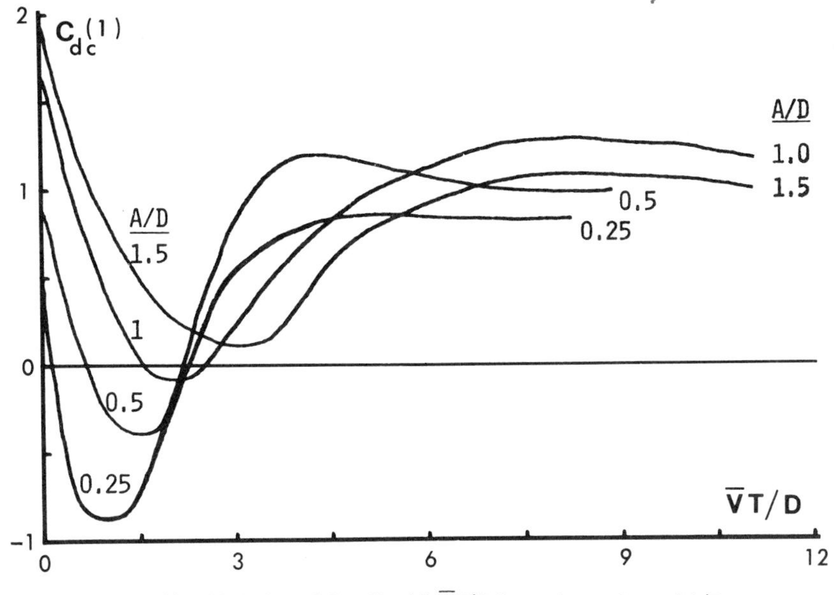

Fig. 3.42a. Variation of $C_{dc}(1)$ with $\overline{V}T/D$ for various values of A/D.

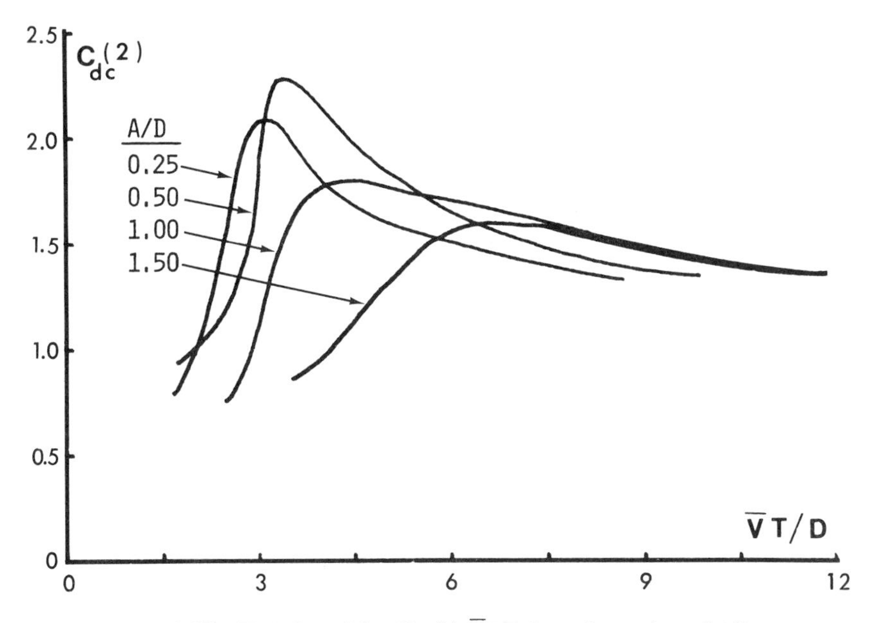

Fig. 3.42b. Variation of $C_{dc}(2)$ with $\overline{V}T/D$ for various values of A/D.

There are several interesting and practically important features of Figs. 3.42a and 3.42b which concern the in-line oscillations of flexible cylinders. It is clear from these figures that $C_{dc}(2)$ is negative for $\overline{V}T/D$ smaller than about 3 and A/D smaller than about 0.3. This means that in the range of $\overline{V}T/D$ values cited, relatively small currents can give rise not only to a dramatic reduction in C_{dc} but even to an energy transfer from the fluid to the cylinder. With negative $C_{dc}(2)$, the cylinder may undergo sustained in-line oscillations with an amplitude dependent on the mass ratio and the material damping of the cylinder and its supports. Experiments have shown that (King 1977) in-line oscillations occur within two adjacent regions. The first is in the range of $1.25 < U_r < 2.5$, maximum amplitudes (about 0.2D) occurring at $U_r = 2.1$. The second region extends from $U_r = 2.7$ to $U_r = 3.8$ with maximum amplitudes at $U_r = 3.2$ [see $C_{dc}(1)$ in Fig. 3.42a]. The first instability region is accompanied by symmetric vortex shedding (as if the flow started impulsively from rest at each cycle) and the second by alternate vortex shedding, as noted earlier.

The reason for the symmetric shedding of vortices in the first instability region may be explained as follows. Consider a relative amplitude of A/D = 0.25. It is easy to show that $\overline{V} = U_m$ for $\overline{V}T/D = 1.57$.* At mid-position (t/T = 0.5), the fluid velocity relative to the cylinder is zero since $\overline{V} = U_m$. As the cylinder moves in the direction of flow, the relative velocity about the cylinder increases from zero to $\overline{V}$. During that period only two small vortices develop symmetrically and remain attached since the relative displacement of the cylinder is very small. It has previously been shown in connection with the discussion of impulsively started or uniformly accelerating flows that the relative displacement of the fluid should be sufficiently large for the vortices to grow and become unstable. A relative displacement of 0.25 or so is not sufficient for the full development of the vortices let alone that of asymmetry. As the cylinder reverses its direction, the relative velocity increases and reaches a value of $2\overline{V}$ at the mid-position of oscillation. In other words, the vortices continue to grow symmetrically and give rise to a drag which is significantly larger than that obtained with a corresponding steady flow. The motion of the cylinder further upstream results in a decelerating flow relative to the cylinder. The vortices, having now fully grown, move away from the cylinder under the influence of the uniform flow and other vortices and their images. During a given cycle, there is not sufficient time for the onset of instability for the vortices to develop an asymmetric configuration. A similar phenomenon is observed in impulsively started flows. The onset of initial asymmetry and the shedding of the first vortex depend very much on the magnitude of the disturbances and the time allowed for the growth of the asymmetry.

---

*$U_m = 2\pi A/T$, $\overline{V}/U_m = (\overline{V}T/D)[D/(2\pi A)]$, $x = -A \sin 2\pi t/T$, $\dot{x} = -U_m \cos 2\pi t/T$.

It is apparent from the foregoing that the symmetric vortex shedding owes its existence to a proper range of $\overline{V}/U_m$ and to relatively small values of A/D. In the region of $\overline{V}T/D$ values for which $\overline{V} > U_m$, the relative velocity about the cylinder is always in the downstream direction. This leads to an alternate vortex shedding as in the case of steady flows, for sufficiently large values of $(\overline{V} - U_m)$. Additional details of the in-line oscillations in the first and second regions of instability are discussed in Chapter 8.

So far attention has been concentrated on C_{dc} and nothing has been said about C_{mc} and its dependence on $\overline{V}T/D$, U_mT/D, Reynolds number, and the relative roughness. It has been shown in Section 3.8.7 that there is a strong relationship between the vortex shedding and the instantaneous value of C_a. Thus, one would anticipate that the biassing of the shedding of vortices by the current would cause profound changes in C_{mc} relative to its no-current value of C_m. This is clearly evident from the data of Mercier (1973), Sarpkaya (1977b), and Verley and Moe (1979). Figures 3.43 and 3.44 show C_{dc} and C_{mc} as a function of $K = U_mT/D$ for various values of $\overline{V}T/D$. This data have been obtained by Verley and Moe (1979) for $\beta = D^2/\nu T \simeq 300$, primarily for the purpose of determining the so-called 'fluid damping' for cylinders oscillating in a current. It must be emphasized that the Reynolds number $Re = U_mD/\nu$ increases with K along each curve, reaching a maximum value of about 12,000 at K = 40. Thus, the use of this data at large Reynolds numbers is not warranted. Nevertheless, Figs. 3.43 and 3.44 show that the current causes profound changes not only in C_{dc} but also in C_{mc} relative to the no-current case, *at the corresponding K and Re values*. For example, at K = 25 (Re = 7,500), C_m = 1.1

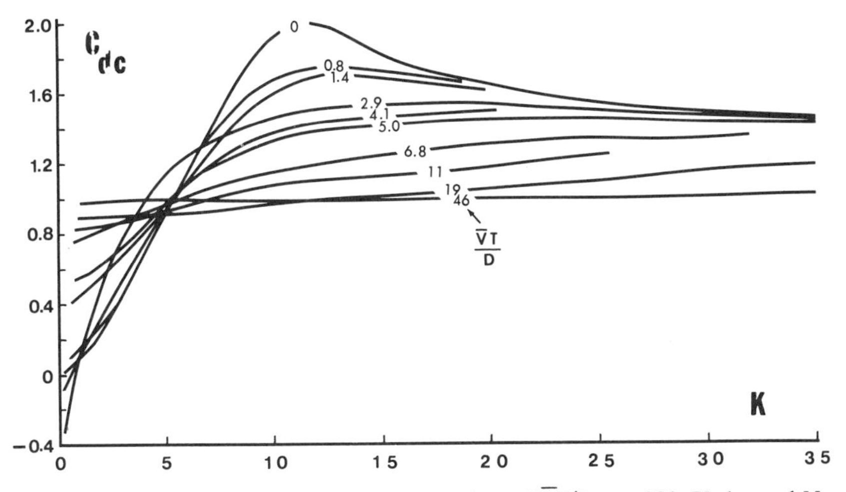

Fig. 3.43. Variation of C_{dc} with K for various values of $\overline{V}T/D$, $\beta \simeq 300$ (Verley and Moe 1979).

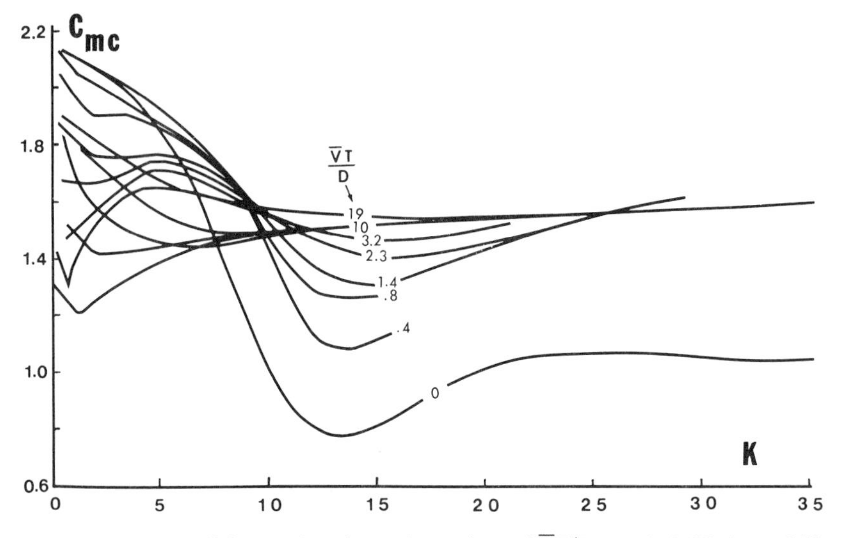

Fig. 3.44. Variation of C_{mc} with K for various values of $\overline{V}T/D$, $\beta \simeq 300$ (Verley and Moe 1979).

(no-current) and $C_{mc} = 1.55$ for $1.4 < \overline{V}T/D < 30$. A value of $\overline{V}T/D = 2.5$ at $K = 25$ corresponds to $\overline{V}/U_m = 0.1$, a situation which is not unusual for the ocean environment except that the Reynolds numbers are considerably larger. It is known that C_m for a smooth cylinder in harmonic flow increases to about 1.8 at large K and Re. Thus, it is possible that the effect of current on C_{mc} at high K and Re numbers may be negligible and C_{mc} may differ little from C_m. Equally important is the fact that as K increases (say K larger than about 25) the inertial component of the in-line force becomes negligible. In the intermediate region of K values, the effect of current is to increase C_{mc} and reduce C_{dc} relative to their no-current values. This may in part account for both the scatter and the relatively smaller value of the drag coefficient obtained from field data through the use of the Morison equation. Clearly, the use of one half wave cycle to calculate C_d and C_m in wave-current flow is not warranted and does not account for the actual behavior of the biassed wake.

A simple calculation shows that the semi-peak-to-peak value of the maximum in-line force for $\overline{V}/U_m < 0.1$ is about 10 percent over estimated by using the no-current values of C_d and C_m relative to that where the actual values of C_{mc} and C_{dc} are used. For only practical purposes, it is tentatively concluded that the use of the modified Morison equation [Eq. (3.45)] for $\overline{V}T/D < 0.1K$, i.e., $\overline{V}/U_m < 0.1$, with C_d and C_m values obtained with pure wave or harmonic motion may lead to reasonably conservative in-line forces.

Evidently, much research is needed on this subject to delineate the effect of current on waves (wave-current interaction) and on wave forces (both in-line and transverse) as a function of K, Re, $\overline{V}T/D$, relative roughness, yaw, and body proximity, particularly at the intermediate and high K and Re values. Additional research is needed on the effect of time-and-depth dependent current (e.g., $V = V_m \sin \omega t$) on all aspects of the phenomenon. It appears that much of this information will have to be obtained experimentally. Flow visualization and the use of refined discrete vortex models might provide some useful clues.

3.10 REFERENCES

Abernathy, F. H. and Kronauer, R. E. 1962. The formation of Vortex Streets. *JFM*, Vol. 13, pp. 1-20.

Achenbach, E. 1968. Distribution of Local Pressure and Skin Friction around a Circular Cylinder. *JFM*, Vol. 46, pp. 321-335.

Achenbach, E. 1971. Influence of Surface Roughness on the Cross-flow Around a Circular Cylinder. *JFM*, Vol. 46, pp. 321-335.

Allen, H. J., and Perkins, E. W. 1951. A Study of Effects of Viscosity on Flow Over Slender Inclined Bodies of Revolution. NACA T. R. No. 1048.

American Petroleum Institute (API). 1977. API Recommended Practice for Planning, Designing, and Constructing Fixed Offshore Platforms. API RP 2A, *American Petroleum Institute Production Department*, 300 Corrigan Tower Bldg., Dallas, Texas.

Armitt, J. 1968. The Effect of Surface Roughness and Free Stream Turbulence on the Flow Around a Model Cooling Tower at Critical Reynolds Numbers. *Proceedings of a Symposium on Wind Effects on Buildings and Structures*, Laughborough University of Technology, England.

Asher, J. A. and Dosanjh, D. S. 1968. An Experimental Investigation of the Formation and Flow Characteristics of an Impulsively Generated Vortex Street. *Journal of Basic Engineering*, ASME. Vol. 90, pp. 596-606.

ASME. 1968. Symposium on Unsteady Flow. (M. Sevik and G. F. Wislicinus, eds.), United Engineering Center, 345 East, 47th St., New York, N.Y., 10017.

Bar-lev, M. and Yang, H. T. 1975. Initial Flow Field over an Impulsively Started Circular Cylinder. *J. Fluid Mech.*, Vol. 72, pp. 625-647.

Barnouin, B., Mattout, M. R. and Sagner, M. M. 1979. Experimental Study of the Validity Domain of Some Formulae for Hydrodynamic Forces for Regular and Irregular Flows. in *Mechanics of Wave-Induced Forces on Cylinders*, (ed. T. L. Shaw), Pitman Advanced Publishing Program, London, pp. 393-405.

Basset, A. B. 1888. On the Motion of a Sphere in a Viscous Liquid. *Phil. Trans., Royal Society of London*, Vol. 179, pp. 43-63. (See also A Treatise on Hydrodynamics, Vol. 2, ch. 21, 1888, Cambridge: Deighton, Bell and Co.), (also New York Dover Publications, Inc., 1961).

Batchelor, G. K. 1967. *An Introduction to Fluid Dynamics*. Cambridge Univ. Press, Bentley House, London.

Batham, J. P. 1973. Pressure Distributions on Circular Cylinders at Critical Reynolds Numbers. *JFM*, Vol. 57, pp. 209-229.

Bearman, P. W. 1967. On Vortex Street Wakes. *JFM*, Vol. 28, pp. 625-641.

Bearman, P. W. 1969. On Vortex Shedding from a Circular Cylinder in the Critical Reynolds Number Regime. *JFM*, Vol. 37, pt. 3, pp. 577-585.

Bearman, P. W. and Graham, J. M. R. 1979. Hydrodynamic Forces on Cylindrical Bodies in Oscillatory Flow. *Proceedings of the Second International Conference on Behaviour of Offshore Structures*, BOSS'79, London, Paper No. 24, pp. 309-322.

Bearman, P. W., Graham, J. M. R. and Singh, S. 1979. Forces on Cylinders in Harmonically Oscillating Flows. in *Mechanics of Wave-Induced Forces on Cylinders*, (ed. T. L. Shaw), Pitman Advanced Publishing Program, London, pp. 437-449.

Berger, E. 1964. Unterdrueckung der Laminaren Wirbelstroemung und des Turbulenzeinsatzes der Karmanschen Wirbelstrasse im Nachlauf eines Schwingenden Zylinders bei kleinen Reynolds-Zahlen. *Jahrbuch der WGLR*, pp. 164-172.

Berger, E. and Wille, R. 1972. Periodic Flow Phenomena. *Ann. Rev. of Fluid Mechanics*, Vol. 4, Annual Reviews Inc., Palo Alto, Calif., pp. 313-340.

Betterman, D. 1966. Contribution on a L'etude de la Convection Forcee Turbulente le Long Plagues Ruguresses. *International Journal of Heat and Mass Transfer*, Vol. 9.

Bidde, D. D. 1971. Laboratory Study of Lift Forces on Circular Piles. *J. Waterways, Harbors and Coastal Engineering Div., ASCE*, Vol. 97, No. WW4, pp. 595-614.

Bingham, H. H., Weimer, D. K. and Griffith, W. 1952. The Cylinder and Semicylinder in Subsonic Flow. Princeton Univ., Department of Physics, Technical Report 11-13.

Birkhoff, G. 1950. *Hydrodynamics–A Study in Logic, Fact, and Similitude*. Princeton Univ. Press, Princeton, N.J.

Birkhoff, G. 1953. Formation of Vortex Streets. *Journal of Applied Physics*, Vol. 24, pp. 98-103.

Birkhoff, G. and Zarantonello, E. H. 1957. *Jets, Wakes and Cavities*. Academic Press, N.Y.

Bishop, R. E. D. and Hassan, A. Y. 1963. The Lift and Drag Forces on a Circular Cylinder in a Flowing Fluid. *Proc. Royal Soc., London*, Vol. 277A, pp. 32-50.

Blasius, H. 1908. Grenzchichten in Flussigkeiten mit Kleiner Reibung. *Z. Math. u. Phys.*, Vol. 56, p. 1-6.

Blevins, R. D. 1977. *Flow-Induced Vibrations*. Van Nostrand Reinhold Co., N.Y.

Bloor, M. S. 1964. The Transition to Turbulence in the Wake of a Circular Cylinder. *JFM*, Vol. 19, pp. 290-304.

Bloor, M. S. and Gerrard, J. H. 1966. Measurements on Turbulent Vortices in a Cylinder Wake. *Proceedings of the Royal Society, London*, Vol. 294A, pp. 319-342.

Bostock, B. R. 1972. Slender Bodies of Revolution at Incidence. Ph.D. dissertation submitted to the Department of Engineering, Univ. of Cambridge.

Brooks, J. E. 1976. Added Mass of Marine Propellers in Axial Translation. David W. Taylor NSRDC, Rept. No. 76-0079.

Bugliarello, G. 1956. The Resistance to Accelerated Motion of Spheres in Water. (in Italian) *La Ricera Scientifica*, Vol. 26, No. 2, pp. 437-442.

Bublitz, P. 1971. Messung der Drücke und Kräfte am ebenen, querangeströmten Kreiszylinder, Teil I: Untersuchungen am ruhenden Kreiszylinder. AVA-Bericht 71 J 11. See also AVA-Bericht 71 J 20, (1971).

Calvert, J. R. 1967. Experiments on the Low-speed Flow Past Cones. *JFM*, Vol. 27, pp. 273-289.

Cebeci, T. 1979. The Laminar Boundary Layer on a Circular Cylinder Started Impulsively from Rest. *J. Computational Physics*, Vol. 31, pp. 153-172.

Chandrasekaran, A. R., Salni, S. S., and Malhotra, M. M. 1972. Virtual Mass of Submerged Structures. *Proc. ASCE., Jour. Hyd. Div.*, paper No. 8923.

Chang, K. S. 1964. Transverse Forces on Cylinders due to Vortex Shedding in Waves, M.S. Thesis, MIT.

Chang, P. K. 1970. *Separation of Flow*. Pergamon Press, New York.

Chaplin, J. R. and Shaw, T. L. 1971. On the Mechanics of Flow-Induced Periodic Forces on Structures. in *Dynamic Waves in Civil Engineering*, (eds. D. A. Howells, I. P. Haigh, and C. Taylor), Wiley-Interscience, London, pp. 73–94.

Chen, C. F. and Ballengee, D. B. 1971. Vortex Shedding from Circular Cylinders in an Oscillating Stream. *AIAA Jour.*, Vol. 9, No. 2, pp. 340-362.

Ciesluk, A. J. and Colonell, J. M. 1974. Experimental Determination of Hydrodynamic Mass Effects. AS-781 910.

Clark, M. E. and Robertson, J. M. 1961. The Kinematics of Buoyant-Body Water Exit. in *Developments in Mechanics*, (ed. Lay & Malvern), Plenum Press, New York, Vol. 1, pp. 430–441.

Clements, R. R. 1977. Flow Representation, Including Separated Regions, Using Discrete Vortices. AGARD Lecture series No. 86.

Collins, W. M. and Dennis, S. C. R. 1973a. The Initial Flow Past an Impulsively Started Circular Cylinder. *Quart. J. Mech. Appl. Math.*, Vol. 26, pp. 53–75.

Collins, W. M. and Dennis, S. C. R. 1973b. Flow Past an Impulsively Started Circular Cylinder. *JFM*, Vol. 60, pp. 105–110.

Coutanceau, M. and Bouard, R. 1977. Experimental Determination of the Main Features of the Viscous Flow in the Wake of a Circular Cylinder in Uniform Translation. Part-1: Steady Flow; Part-2: Unsteady Flow. *JFM*, Vol. 79, pt. 2, pp. 231–256 and 257–272.

Crandall, S. H., Vigander, S., and March, P. A. 1975. Destructive Vibration of Trashracks due to Fluid-Structure Interaction. ASME Paper No. 75-DET-63.

Dalton, C., Hunt, J. P., and Hussain, A. K. M. F. 1976. The Forces on A Cylinder Oscillating Sinusoidally in Water: II, Further Experiments. *Proceedings of the Offshore Technology Conference*, Vol. 11, (OTC-2538), pp. 159–168.

Dalton, W. L. 1977. A Survey of Available Data on the Normal Drag Coefficient of Cables Subject to Cross-Flow. MAR, Inc. Report No. N62583-77-M-R443, (CR 78.001 of the Civil Engineering Lab., Port Hueneme, Calif.)

Davenport, A. G. 1961. The Application of Statistical Concepts to the Wind Loading of Structures. *The Institution of Civil Engineers*, Vol. 19, pp. 449–472.

Davies, M. E. 1976. A Comparison of the Wake Structure of a Stationary and Oscillating Bluff Body, Using a Conditional Averaging Technique. *JFM*, Vol. 75, pp. 209–231.

Delany, N. K. and Sorensen, N. E. 1953. Low Speed Drag of Various Shapes. N.A.C.A. Technical Note 3038.

Dawson, C. and Marcus, M. 1970. DMC–A Computer Code to Simulate Viscous Flow about Arbitrarily Shaped Bodies. *Proc. 1970 Heat Transfer and Fluid Mechanics Institute*, (T. Sarpkaya, Ed.), Stanford Univ. Press, Stanford, California, pp. 323–338.

Drescher, H. 1956. Messung der auf querangestroemte Zylinder Ausgeubten zeitlich veranderten Drucke. *Z. Flugwiss.*, Vol. 4, (1/2), pp. 17–21.

Dvorak, R. A. 1969. Calculation of Turbulent Boundary Layers on Rough Surfaces in Pressure Gradient. *AIAA Journal*, Vol. 7, No. 9, pp. 1752–1759.

Dwyer, H. A. and McCroskey, W. J. 1973. Oscillating Flow Over a Cylinder at Large Reynolds Number. *JFM*, Vol. 61, pt. 4, pp. 753–767.

El Baroudi, M. Y. 1960. Measurements of Two-Point Correlations of Velocity Near A Circular Cylinder Shedding a Karman Vortex Street. Univ. of Toronto, UTIA T.N. 31.

Engineering Sciences Data Unit, (E.S.D.U.). 1970. Fluid Forces Acting on Circular Cylinders for Application in General Engineering. Part-I: Long Cylinders in Two-Dimensional Flow. Item No. 70013, Part-II: Finite Length Cylinders. Item No. 70014, E.S.D.U., 251-259 Regent Street, London, WIR 7 AD, England.

Ericsson, L. E. and Reding, J. P. 1979. Vortex-Induced Asymmetric Loads on Slender Vehicles. Lockheed Missiles and Space Company, Inc. Report No. LMSC-D630807. (See also AIAA Paper No. AIAA-80-0181, January 14–16, 1980).

Etkin, B., Korbacher, G. K. and Keefe, R. T. 1958. Acoustic Radiation from a Stationary Cylinder in a Fluid Stream (Aeolian Tones). U.T.I.A., Rev. No. 13.

Etzold, F. and Fiedler, H. 1976. The Near-Wake Structures of a Cantilevered Cylinder in a Cross Flow. *Zeitschrift fur Flugwissenschaften*, Vol. 24, pp. 77-82.

Fage, A. 1931. Further Experiments on the Flow Around a Circular Cylinder. ARC R & M, No. 1369, Vol. 1, pp. 186-195.

Fage, A. 1937. On the Reynolds Number of Transition. ARC R & M, No. 1765, pp. 69-74.

Fage, A. 1932. The Turbulence in the Wake of a Body (Three-Dimensional Wake of a Long Cylinder) in Water Re to 11×10^2. ARC R & M, No. 1510, pp. 116-127.

Fage, A. and Falkner, V. M. 1931. Further Experiments on the Flow Around a Circular Cylinder. ARC R & M, No. 1369.

Fage, A. and Johansen, R. C. 1928. The Structure of Vortex Sheets. *Aeronautical Research Council*, R and M 1143.

Fage, A. and Warsap, J. H. 1930. The Effects of Turbulence and Surface Roughness on the Drag of a Circular Cylinder, *Aero. Res. Comm.*, London, Reports and Memoranda No. 1283.

Fickel, M. G. 1973. Bottom and Surface Proximity Effects on the Added Mass of Rankine Ovoids. AD-775 022. M.S. Thesis presented to the Naval Postgraduate School, Monterey, CA.

Fink, P. T. and Soh, W. K. 1974. Calculation of Vortex Sheets in Unsteady Flow and Applications in Ship Hydrodynamics. 10th Symposium of Naval Hydrodynamics, Cambridge, Mass. (Superintendent of Documents, U.S. Government Printing Service, Washington, D.C.).

Flachsbart, O. 1932. Winddruck auf Gasbehälter. Reports of the AVA in Göttingen, IVth Series, pp. 134-138.

Frank, W. 1967. Oscillations of Cylinder in or below the Free Surface of Deep Fluids. NSRDC Report No. 2357.

Friberg, E. G. 1965. Measurements of Vortex Separation, Part I: Two-Dimensional Circular and Elliptic Bodies. MIT Aerophysics Lab. Tech. Report 114.

Fung, Y. C. 1960. Fluctuating Lift and Drag Acting on a Cylinder in a Flow at Supercritical Reynolds Numbers. *J. Aerospace Sciences*, Vol. 27, No. 11.

Furuya, Y., Miyata, N., and Fujita, H. 1976. Turbulent Boundary Layer and Flow Resistance on Plates Roughened by Wires. ASME paper No. 76-FE-6.

Garrick, I. E., 1966. Unsteady Aerodynamics of Potential Flows, *Applied Mechanics Surveys*. Spartan Books, Washington, D.C., pp. 965-970.

Gerlach, C. R. and Dodge, F. T. 1970. An Engineering Approach to Tube Flow-Induced Vibrations. *Proceedings of the Conference on Flow-Induced Vibrations in Reactor System Components*, Argonne National Laboratory, pp. 205-225.

Gerrard, J. H. 1961. An Experimental Investigation of the Oscillating Lift and Drag of a Circular Cylinder Shedding Turbulent Vortices. *JFM*, Vol. 11, pt. 2, pp. 215-227.

Gerrard, J. H. 1963. The Calculation of the Fluctuating Lift on a Circular Cylinder and its Application to the Determination of Aeolian Tone Intensity. AGARD Report No. 463.

Gerrard, J. H. 1965. A Disturbance-Sensitive Reynolds Number Range of the Flow Past a Circular Cylinder. *JFM*, Vol. 22, pt. 1, pp. 187-196.

Gerrard, J. H. 1966. The Mechanics of the Formation Region of Vortices Behind Bluff Bodies. *JFM*, Vol. 25, pt. 2, pp. 401-413.

Gerrard, J. H. 1966. The Three-Dimensional Structure of the Wake of a Circular Cylinder. *JFM*, Vol. 25, pt, 1, pp. 143-164.

Gibson, R. J. and Wang, H. 1977. Added Mass of Pile Group. *Journal of the Waterways, Port, Coastal and Ocean Div., ASCE*, WW2, Paper no. 12932, pp. 215–223.

Glenny, D. E. 1966. A Review of Flow Around Circular Cylinders Stranded Cylinders and Struts Inclined to the Flow Direction. Australian Defense Sci. Service Aero. Res. Labs. Mech. Eng. Note 284.

Goddard, V. P. 1972. Numerical Solutions of the Drag Response of a Circular Cylinder to Stream-wise Velocity Fluctuations. Ph.D. Thesis, Univ. of Notre Dame. (See also *Rozprawy Inzynierskie*, Vol. 33, No. 3, 1974, pp. 487–508.)

Goldman, R. L. 1958. Karman Vortex Forces on the Vanguard Rocket. *The Shock and Vibration Bulletin*, No. 26, pt. 2, pp. 171–179.

Goldstein, S. 1938. *Modern Developments in Fluid Mechanics*, Oxford, Clarendon Press.

Goldstein, S. and Rosenhead, L. 1936. Boundary Layer Growth. *Proc. Cambr. Phil Soc.*, Vol. 32, pp. 392–401.

Görtler, H. 1944. Verdrängungswirkung der Laminaren Grenzschicht und Druckwiderstand. *Ing.-Arch.*, Vol. 14, pp. 286–305.

Görtler, H. 1948. Grenzschichtentstehung an Zylindern bei Anfahrt aus der Ruhe. *Arch. d. Math.*, Vol. 1, pp. 138–147.

Gowda, B. H. L. 1975. Some Measurements of the Phenomenon of Vortex Shedding and Induced Vibrations of Circular Cylinders. Deutsche Luft und Raumfahrt Forschungsbericht No. 75-01.

Graham, J. M. R. 1979. Analytical Methods of Representing Wave-Induced Forces on Cylinders. in *Mechanics of Wave-Induced Forces on Cylinders*, (ed. T. L. Shaw), Pitman Advanced Publishing Program, London, England, pp. 133–151.

Granville, P. S. 1978. Similarity-Law Characterization Methods for Arbitrary Hydrodynamic Roughnesses. DWTNSRDC Report No. 78-SPD-815-01.

Grass, A. J. and Kemp, P. H. 1979. Flow Visualisation Studies of Oscillatory Flow Past Smooth and Rough Circular Cylinders. in *Mechanics of Wave-Induced Forces on Cylinders*, (ed. T. L. Shaw), Pitman, London, pp. 406–420.

Griffin, O. M., and Ramberg, S. E. 1976. Vortex Shedding From a Cylinder Vibrating in Line With An Incident Uniform Flow. *JFM*, Vol. 75, pt. 2, pp. 257–271.

Güven, O. 1975. An Experimental and Analytical Study of Surface-Roughness Effects on the Mean Flow Past Circular Cylinders. P.D. Thesis submitted to the Univ. of Iowa, Iowa City, IA.

Güven, O. O., Patel, V. C., and Farrell, C. 1975. Surface Roughness Effects on the Mean Flow Past Circular Cylinders. Iowa Institute of Hydraulic Research Report No. 175, Iowa City, IA.

Hamilton, W. S. 1972. Fluid Force on Accelerating Bodies, *Proceedings of the 13th Coastal Engineering Conference*. ASCE, New York, Vol. III, pp. 1767–1782.

Hamilton, W. S. and Lindell, J. E. 1971. Fluid Force Analysis and Accelerating Sphere Tests. *Jour. of the Hydraulics Div.*, ASCE, HY6, Vol. 97, pp. 805–817.

Hamman, F. H. and Dalton, C. 1971. The Forces on a Cylinder Oscillating Sinusoidally in Water. Trans. ASME, Vol. 93, Series B, No. 4.

Hatfield, H. M. and Morkovin, N. V. 1972. Effect of an Oscillating Free-Stream on the Unsteady Pressure on a Circular Cylinder. ASME paper No. 72-WA/FE-12.

Heinzer, A. and Dalton, C. 1969. Wake Observations for Oscillating Cylinders. *Journal of Basic Engineering*, ASME, Vol. 91, pp. 850–852.

Hirota, I. and Miyakoda, K. 1965. Numerical Solution of Karman Vortex Street Behind a Circular Cylinder. *J. Met. Soc. Japan*, Vol. 43, p. 30.

Hjelmfelt, A. T. and Mockros, L. F. 1967. Stokes Flow Behavior of an Accelerating Sphere. *Jour. of Engineering Mechanics Div.*, ASCE, Vol. 93, No. EM6, p. 87.

Hogben, N. 1976. Wave Loads on Structures. in the *Proceedings of the First International Conference on the Behaviour of Offshore Structures*, BOSS-76, Vol. 1, pp. 187–219.

Honji, H. 1972. Starting Flows Past Spheres and Elliptic Cylinders. Rep. Res. Inst. Appl. Mech., Kyushu Univ., Vol. 19, p. 271.

Humphreys, J. S. 1960. On a Circular Cylinder in a Steady Wind at Transition Reynolds Numbers. *J. Fluid Mech.*, Vol. 9, pp. 603–612.

Isaacson, M. de St. Q. 1974. The Forces on Circular Cylinders in Waves. Ph.D. Thesis submitted to the Dept. of Engineering, Univ. of Cambridge.

Isaacson, M. de St. Q. and Maull, D. J. 1976. Transverse Forces on Vertical Cylinders in Waves. ASCE, Vol. 102, WW1, Paper no. 11934, pp. 49–60.

Iversen, H. W. and Balent, R. 1951. A Correlating Modulus for the Fluid Resistance in Accelerated Motion. *J. Applied Physics*, Vol. 22, No. 3, pp. 324–328.

Iwagaki, Y. and Ishida, H. 1976. Flow Separation, Wake, Vortices, and Pressure Distribution Around a Circular Cylinder under Oscillatory Waves. Proc. of the 15th International Conf. on Coastal Engineering, Honolulu, Hawaii.

Jain, P. C. and Rao, K. S. 1969. Numerical Solution of Unsteady Viscous Incompressible Fluid Flow Past A Circular Cylinder. *Phys. of Fluids*, Supplement, Vol. 12, p. II-57.

Jones, G. W., Jr., Cincotta, J. J., and Walker, R. W. 1969. Aerodynamic Forces on a Stationary and Oscillating Circular Cylinder at High Reynolds Number, NASA Technical Report TR R-300.

Jones, W. P. 1962. Research on Unsteady Flow. *Jour. Aerospace Sciences*, Vol. 29, No. 3, pp. 249–263.

Jordan, S. K. and Fromm, J. E. 1972. Oscillatory Drag, Lift, and Torque on a Circular Cylinder in a Uniform Flow. *Physics of Fluids*, Vol. 15, No. 3, pp. 371–376. (See also AIAA Paper No. 72-111, pp. 1–9).

Kawaguti, M. and Jain, P. C. 1966. Numerical Study of a Viscous Fluid Flow Past a Circular Cylinder. *J. Physical Soc. Japan*, vol. 21, No. 10, p. 2055.

Keefe, R. T. 1962. An Investigation of the Fluctuating Forces Acting on a Stationary Cylinder in a Subsonic Stream, and of the Associated Sound Field. UTIAS Report 76. (See also Jour. Acous. Soc. Amer., Vol. 34, No. 11, 1962, pp. 1711–1719).

Keim, S. R. 1956. Fluid Resistance to Cylinders in Accelerated Motion. *Journal of the Hydraulics Div., ASCE*, Vol. 83, No. HY6.

Keulegan, G. H. and Carpenter, L. H. 1956. Forces on Cylinders and Plates in an Oscillating Fluid. National Bureau of Standards Report No. 4821.

Keulegan, G. H. and Carpenter, L. H. 1958. Forces on Cylinders and Plates in an Oscillating Fluid. *Journal of Research of the National Bureau of Standards*, Vol. 60, No. 5, pp. 423–440.

King, R. 1974. Vortex Excited Structural Oscillations of a Circular Cylinder in Steady Currents. OTC-1948, Offshore Technology Conf., Houston, TX. (See also Ph.D. Thesis by R. King, Loughborough Univ.).

King, R. 1977. A Review of Vortex Shedding Research and its Applications. *Ocean Engineering*, Vol. 4, pp. 141–171.

King. R., Prosser, M. J. and Johns, D. J. 1973. On Vortex Excitation of Model Piles in Water. *Jour. of Sound and Vibration*, Vol. 29, No. 2, pp. 169–188.

Knight, D. W. and McDonald, J. A. 1979. Hydraulic Resistance of Artificial Strip Roughness. *Jour. Hydraulics Div., ASCE*, HY6, pp. 675–689.

Lai, R. S. 1973. Accelerated Motion of a Spheroid in Viscous Fluid. *Proc. ASCE, Jour. Hydraulics Div.*, Paper No. 9809. (See also *JFM*, Vol. 52, pt. 1, 1972, p. 1).

Laird, A. D. K. and Johnson, C. A. 1956. Drag Forces on Accelerated Cylinder. *Jour. of Petroleum Tech.*, Vol. 8, pp. 65–67.

Laird, A. D. K., Johnson, C. A. and Walker, R. W. 1959. Water Forces on Accelerated Cylinders. *Jour. of Waterways and Harbor Div., ASCE*, WW1, pp. 99-119.

Lamont, P. J. 1973. The Out-of-Plane Force on an Ogive Nosed Cylinder at Large Angles of Inclination to a Uniform Stream. Ph.D. Thesis, Univ. of Bristol, England.

Lamont, P. H. and Hunt, B. L. 1976. Pressure and Force Distributions on a Sharp-Nosed Circular Cylinder at Large Angles of Inclination to a Uniform Subsonic Steam. *JFM*, Vol. 76, pt. 3, pp. 519-559.

Lienhard, J. H. 1966. Synopsis of Lift, Drag, and Vortex Frequency Data for Rigid Circular Cylinders. Washington State Univ., College of Engineering, Research Division Bulletin 300.

Ludwieg, H. 1979. Längswirbel in Strömungen. *Zeit. Flugwiss. u. Weltraum Forschung*, Vol. 5, pp. 273-283.

Lugt, H. J. and Haussling, H. J. 1974. Laminar Flow Past and Abruptly Accelerated Elliptic Cylinder at 45° Incidence. *JFM*, Vol. 65, pt. 4, pp. 711-734.

Macovsky, M. S. 1958. Vortex-Induced Vibration Studies. DTMB Report No. 1190.

Mair, W. A. and Maull, D. J. 1971. Bluff Bodies and Vortex Shedding–A Report on Euromech 17. *JFM*, Vol. 45, pt. 2, pp. 209-224.

Marris, A. W. 1964. A Review on Vortex Streets, Periodic Wakes, and Induced Vibration Phenomena. *Jour. of Basic Engineering, Trans. ASME*, Series D, Vol. 86, No. 2, pp. 185-196.

Matten, R. B. 1976. Calculation of Drag Forces on a Circular Cylinder in a Combined Plane Oscillatory and Uniform Flow Field Using Morison's Equation. (Unpublished Communication to the National Maritime Institute, England).

Matten, R. B., Hogben, N., and Ashley, R. M. 1979. A Circular Cylinder Oscillating in Still Water, in Waves and in Currents. in *Mechanics of Wave-Induced Forces on Cylinders*, (ed. T. L. Shaw), Pitman, London, pp. 475-489.

Maull, D. J. and Milliner, M. G. 1978. Sinusoidal Flow Past a Circular Cylinder. *Coastal Engineering*, Vol. 2, pp. 149-168.

Maull, D. J. and Milliner, M. G. 1979. The Forces on a Circular Cylinder having a Complex Periodic Motion. in *Mechanics of Wave-Induced Forces on Cylinders*, (ed. T. L. Shaw), Pitman, London, pp. 490-502.

Mavis, F. T. 1970. Virtual Mass of Plates and Discs in Water. *Proc. ASCE, Jour. Hydraulics Div.*, Paper No. 7593.

McCroskey, W. J. 1977. Some Current Research in Unsteady Fluid Dynamics. *ASME Journal of Fluids Engineering*, Vol. 99, pp. 8-39.

McGregor, D. M. 1957. An Experimental Investigation of the Oscillating Pressures on a Circular Cylinder in a Fluid Stream. Univ. of Toronto, UTIA TN No. 14.

McNown, J. S. 1957. Drag in Unsteady Flow. *Proceedings of the 9th International Congress of Applied Mechanics*, Brussels, pp. 124-132.

McNown, J. S. and Wolf, L. W. 1956. Resistance to Unsteady Flow I–Analysis of Tests with Flat Plates. University of Michigan, Engineering Research Institute Report.

McNown, J. S. and Keulegan, G. H. 1959. Vortex Formation and Resistance in Periodic Motion. *Proc. ASCE*, EM1, Paper No. 1894, pp. 1-6.

McNown, J. S. and Sarpkaya, T. 1960. Vortex Formation and Resistance in Unsteady Flow. Internal Report to Division 5112, Sandia Corporation.

Mehta, U. B. and Lavan, Z. 1972. Starting Vortex, Separation Bubbles and Stall–A Numerical Study of Laminar Unsteady Flow Around an Airfoil. AFOSR Tech. Rep. AFOSR-TR-73-0640. (See also Themis Rep. R-72-11, I.I.T.).

Mellsen, S. B., Ellis, A. T., and Waugh, J. G. 1966. On the Added Mass of a Sphere in a

Circular Cylinder Considering Real Fluid Effects. Ad-636438. (See also ASME paper 66 WA-UNT-6, Nov. 1966).

Mercier, J. A. 1973. Large Amplitude Oscillations of a Circular Cylinder in a Low-Speed Stream. Ph.D. Dissertation, Stevens Institute of Technology.

Miller, B. L. 1977. The Hydrodynamic Drag of Roughened Circular Cylinders. *The Naval Architect, Journal of the Royal Institution of Naval Architects*, No. 2, pp. 55-70.

Morison, J. R., O'Brien, M. P., Johnson, J. W., and Schaaf, S. A. 1950. The Forces Exerted by Surface Waves on Piles. *Petroleum Trans.*, AIME, Vol. 189, pp. 149-157.

Morkovin, M. V. 1964. Flow Around Circular Cylinder–A Kaleidoscope of Challenging Fluid Phenomena. Symposium on Fully Separated Flows, ASME, New York, pp. 102-118.

Nelligan, J. J. 1974. A Survey of the Experimental Methods in Vortex Shedding from Cables and Cylinders. MAR Incorporated, 1335 Rockville Pike, Rockville, MD., Technical Report No. 137.

NPL (National Physical Laboratory). 1968. Strouhal Number of Model Stacks Free to Oscillate. NPL Aero Report No. 1257.

Odar, F. and Hamilton, W. S. 1964. Forces on a Sphere Accelerating in a Viscous Fluid. *JFM*, Vol. 18, pt. 2, pp. 302-314.

Panniker, P. K. G. and Lavan, Z. 1975. Flow Past Impulsively Started Bodies Using Green's Functions. *J. Comp. Phys.*, Vol. 18, pp. 46-52.

Phillips, O. M. 1956. The Intensity of Aeolian Tones. *JFM*, Vol. 1, pp. 607-624.

Prandtl, L. 1904. Uber Flüssigkeitsbewegung bei sehr kleiner Reibung. *Proc. 3rd International Math. Congr.*, Heidelberg.

Prendergast, V. 1958. Measurements of Two-Point Correlations of the Surface Pressure on a Circular Cylinder. Univ. of Toronto, UTIA T.N. 14.

Payne, R. B. 1958. Calculations of Unsteady Viscous Flow Past a Circular Cylinder. *JFM*, Vol. 4, pp. 81-87.

Rance, P. J. 1969. Wave Forces on Cylindrical Members of Structures. Hydraulic Research Station Annual Report.

Relf, E. F. and Simmons, L. F. G. 1925. On the Frequency of Eddies Generated by the Motion of Circular Cylinders through a Fluid. *Phil. Mag.*, 6th series, Vol. 49, pp. 509-515.

Rimon, Y. 1969. Numerical Solution of the Incompressible Time-Dependent Viscous Flow Past a Thin Oblate Spheroid. *Phys. Fluids Supplement*, Vol. 12, p. II-65.

Roberson, J. A., Bajwa, M., and Wright, S. J. 1974. A General Theory for Flow in Rough Conduits. *Journal of Hydraulic Research*, Vol. 12, pp. 223-240.

Roos, F. W. and Willmarth, W. W. 1971. Some Experimental Results on Sphere and Disk Drag. *AIAA Journal*, Vol. 9, No. 2, pp. 285-291.

Rosenhead, L. 1953. Vortex Systems in Wakes. in *Advances in Applied Mechanics*, Academic Press, N.Y., Vol. III, pp. 185-193.

Rosenhead, L. (Ed.). 1963. *Laminar Boundary Layers*. Oxford Univ. Press, Oxford.

Roshko, A. 1953. On the Development of Turbulent Wakes from Vortex Streets. NACA Technical Note 2913.

Roshko, A. 1954. On the Wake and Drag of Bluff Bodies. *J. Aeronautical Sciences*, Vol. 22, pp. 124-132.

Roshko, A. 1955. On the Wake and Drag of Bluff Bodies. *Journal of Aerospace Sciences*, Vol. 22, No. 2, pp. 124-132.

Roshko, A. 1961. Experiments on the Flow Past a Circular Cylinder at very High Reynolds Number. *JFM*, Vol. 10, pp. 345-356.

Roshko, A. 1967. A Review of Concepts in Separated Flow. *Proc. Canadian Congress of Applied Mechanics*, Quebec.

Roshko, A. 1967. Transition in Incompressible Near-wakes. *Physics of Fluids*, Vol. 10, pp. 181–183.

Roshko, A. 1970. On the Aerodynamic Drag of Cylinders at High Reynolds Number. U.S.-Japan Research Seminar on Wind Loads on Structures, Honolulu.

Roshko, A. and Fiszdon, W. 1969. On the persistence of Transition in the Near Wake. *Problems of Hydrodynamics and Continuum Mechanics, Society of Industrial and Applied Mathematics*, Philadelphia, pp. 606–616.

Rott, N. 1964. Theory of Time-Dependent Laminar Flows. in *Theory of Laminar Flows*, F. K. Moore (ed.), Princeton Univ. Press, Princeton, N.J., pp. 395–438.

Sacksteder, R. 1978. On Oscillatory Flows. *The Mathematical Intelligencer*, Vol. 1, pp. 45–51.

Sainsbury, R. N. and King, D. 1971. The Flow-Induced Oscillations of Marine Structures. *Proc. Inst. Civil Engineers*, Vol. 49, pp. 269–302.

Sarpkaya, T. 1960. Added Mass of Lenses and Parallel Plates. *Journal of Engineering Mechanics Div., ASCE*, Vol. 86, No. EM3, pp. 141–152.

Sarpkaya, T. 1962. Unsteady Flow of Fluids in Closed Systems. *Journal of the Engineering Mechanics Div., ASCE*, Vol. 88, pp. 1–5.

Sarpkaya, T. 1963. Lift, Drag, and Added-Mass Coefficients for a Circular Cylinder in a Time-Dependent Flow. *Journal of Applied Mechanics*, Vol. 30, No. 1, *Trans. ASME*, Vol. 85, Series E, pp. 13–15.

Sarpkaya, T. 1966. Separated Flow About Lifting Bodies and Impulsive Flow About Cylinders. *AIAA Jour.*, Vol. 4, pp. 414–420.

Sarpkaya, T. 1968. An Analytical Study of Separated Flow About Circular Cylinders. *Journal of Basic Engineering*, Vol. 90, Series D, No. 4, pp. 511–520.

Sarpkaya, T. 1975a. Forces on Cylinders and Spheres in an Oscillating Fluid. *Journal of Applied Mechanics*, ASME, Vol. 42, pp. 32–37.

Sarpkaya, T. 1975b. An Inviscid Model of Two-Dimensional Vortex Shedding for Transient and Asymptotically Steady Separated Flow over an Inclined Plate. *JFM*, Vol. 68, pt. 1, pp. 109–128.

Sarpkaya, T. 1976a. Vortex Shedding and Resistance in Harmonic Flow About Smooth and Rough Circular Cylinders at High Reynolds Numbers. Report No. NPS-59SL76021, Naval Postgraduate School, Monterey, CA.

Sarpkaya, T. 1976b. In-Line and Transverse Forces on Smooth and Sand-Roughened Cylinders in Oscillatory Flow at High Reynolds Numbers. Report No. NPS-69SL76062, Naval Postgraduate School, Monterey, CA.

Sarpkaya, T. 1976c. Vortex Shedding and Resistance in Harmonic Flow About Smooth and Rough Circular Cylinders. *Proceedings of the International Conference on the Behavior of Offshore Structures* (BOSS'76), Vol. 1, pp. 220–235.

Sarpkaya, T. 1976d. In-Line and Transverse Forces on Cylinders in Oscillatory Flow at High Reynolds Numbers. *Proceedings of the Offshore Technology Conference*, Vol. II, pp. 95–108, (OTC-2533).

Sarpkaya, T. 1976e. Forces on Cylinders Near a Plane Boundary in a Sinusoidally Oscillating Fluid. *Trans. ASME, Jour. Fluids Engineering*, pp. 499–505.

Sarpkaya, T. 1977a. In-Line and Transverse Forces on Cylinders in Oscillatory Flow at High Reynolds Numbers. *Journal of Ship Research*, Vol. 21, No. 4, pp. 200–216.

Sarpkaya, T. 1977b. Unidirectional Periodic Flow About Bluff Bodies. Final Technical Report to NSF, Naval Postgraduate School Report No. NPS-69SL77051, Monterey, CA.

Sarpkaya, T. 1978. Impulsive Flow About a Circular Cylinder. Naval Postgraduate School Technical Report No. NPS-69SL78-008, Monterey, CA.

Sarpkaya, T. 1979. Vortex-Induced Oscillations—A Selective Review. *Journal of Applied Mechanics, Trans. ASME*, Vol. 46, pp. 241–258.

Sarpkaya, T. 1980. A Critical Assessment of Morison's Equation. (to be published).
Sarpkaya, T. and Garrison, C. J. 1963. Vortex Formation and Resistance in Unsteady Flow. *Journal of Applied Mechanics*, Vol. 30, Series E, No. 1, pp. 16-24.
Sarpkaya, T. and Shoaff, R. L. 1979a. Inviscid Model of Two-Dimensional Vortex Shedding by a Circular Cylinder. *AIAA Journal*, Vol. 17, No. 11, pp. 1193-1200.
Sarpkaya, T. and Shoaff, R. L. 1979b. A Discrete-Vortex Analysis of Flow About Stationary and Transversely Oscillating Circular Cylinders. Naval Postgraduate School Technical Report No. NPS-69SL79011, Monterey, CA.
Sawaragi, T. and Nakamura, T. 1979. An Analytical Study of Wave Force on a Cylinder in Oscillatory Flow. Proceedings of the Specialty Conference on Coastal Structures 79, ASCE.
Sedov, L. I. 1965. Two-Dimensional Problems in Hydrodynamics and Aerodynamics. Translated from Russian and edited by C. K. Chu et al., *Interscience Publishers*, New York.
Schiller, L. and Linke, W. 1933. Druck- und Reibungswiderstand des Zylinders bei Reynoldsschen Zahlen 5000 bis 40,000. N.A.C.A. Tech. Memo No. 715.
Schlichting, H. 1968. Boundary-Layer Theory. McGraw-Hill Book Co., New York, 6th ed.
Schmidt, L. V. 1965. Measurements of Fluctuating Air Loads on a Circular Cylinder. *Journal of Aircraft*, Vol. 2, No. 1, pp. 49-55.
Schmidt, D. V. von, and Tilmann, P. M. 1972. Uber die Zirkulationsentwicklung in Nachlaufen von Rundstaben. *Acustica*, Vol. 27, pp. 14-22.
Schuh, H. 1953. Calculation of Unsteady Boundary Layers in Two-Dimensional Laminar Flow. *ZFW*, Vol. 1, pp. 122-131.
Schwabe, M. 1935. Uber Drückermittlung in der Instationaren ebenen Strömung. *Ing.-Arch.*, Vol. 6, pp. 34-50.
Shih, W. C. L. and Hove, D. T. 1977. High Reynolds Number Flow Considerations for the OTEC Cold Water Pipe. Science Applications, Inc., Report No. SAI-78-607-LA, El Segundo, Calif.
Simmons, J. E. L. 1977. Similarities Between Two-Dimensional and Axisymmetric Vortex Wakes. *The Aeronautical Quarterly*, Vol. 28, pt. 1, pp. 15-20.
Simpson, R. L. 1973. A Generalized Correlation of Roughness Density Effects on the Turbulent Boundary Layer. *AIAA Journal*, Vol. 11, pp. 242-244.
Singh, S. 1979. Forces on Bodies in Oscillatory Flow. Ph.D. Thesis, University of London.
Skop, R. A., Ramberg, S. E., and Ferer, K. M. 1976. Added Mass and Damping Forces on Circular Cylinders. ASME Paper No. 76-Pet-3.
Son, J. S. and Hanratty, T. H. 1969. Numerical Solution for the Flow Around a Cylinder at Reynolds Numbers of 40, 200, and 500. *JFM*, Vol. 35, pp. 369-375.
Stansby, P. K. 1974. The Effect of End Plates on the Base Pressure Coefficient of a Circular Cylinder. *Aeronautical Journal*, Vol. 78, No. 757, pp. 36-37.
Stansby, P. K. 1977. An Inviscid Model of Vortex Shedding from a Circular Cylinder in Steady and Oscillatory far Flows. *Proc. Institution of Civil Engineers*, Vol. 63, pt. 2, pp. 865-880.
Stansby, P. K. 1979. Mathematical Modeling of Vortex Shedding from Circular Cylinders in Planar Oscillatory Flows, Including Effects of Harmonics. in *Mechanics of Wave-Induced Forces on Cylinders*, (ed. T. L. Shaw), Pitman, London, pp. 450-460.
Stanyukovich, K. P. 1960. Unsteady Motion of Continuous Media. Pergamon Press, London, (translated by J. G. Adashko and M. Holt).
Stelson, T. E. and Mavis, F. T. 1965. Virtual Mass and Acceleration in Fluids. *Proceedings ASCE*, Vol. 81, Separate No. 670, pp. 670-1 to 670-9.

Stewartson, K. 1960. The Theory of Unsteady Laminar Boundary Layers. in *Advances in Applied Mechanics*, edited by H. L. Dryden et alli, Academic Press, N.Y., Vol. VI, pp. 1–37.

Stokes, G. G. (Sir). 1845. On the Theories of the Internal Friction of Fluids in Motion, and of the Equilibrium and Motion of Elastic Bodies. *Cambridge Phil. Trans.*, Vol. VIII, p. 287.

Stokes, G. G. (Sir). 1851. On the Effect of the International Friction of Fluids on the Motion of Pendulums. *Cambridge Phil. Trans.*, Vol. IX, pt. 2, p. 8. (Also *Mathematical and Physical Papers*, Camb. Univ. Press, Vol. 3).

Strouhal, V. 1878. Über eine besondere Art der Tonerregung. *Ann. Phys. und Chemie*, New Series Vol. 5, p. 216–251.

Stuart, J. T. 1963. Unsteady Boundary Layers. in *Laminar Boundary Layers*, Edited by L. Rosenhead, Oxford Univ. Press, Oxford, pp. 349–406.

Surry, D. 1969. The Effect of High Intensity Turbulence on the Aerodynamics of a Rigid Circular Cylinder at Subcritical Reynolds Numbers. UTIAS Report 142.

Szechenyi, E. 1975. Supercritical Reynolds Number Simulation for Two-Dimensional Flow over Circular Cylinders. *JFM*, Vol. 70, pt. 3, pp. 529–542. (See also *La Recherche Aerospatiale*, May-June 1974).

Taneda, S. 1972. The Development of the Lift of an Impulsively Started Elliptic Cylinder at Incidence. *Jour. Physical Soc. Japan*, Vol. 33, No. 6, pp. 1706–1711.

Taneda, S. and Honji, H. 1971. Unsteady Flow Past a Flat Plate Normal to the Direction of Motion. *Jour. Physical Soc. Japan*, Vol. 30, No. 1, pp. 262–272.

Tani, I. 1964. Low-Speed Flows Involving Bubble Separations. *Progress in Aeronautical Sciences*, Vol. 5, Pergamon Press, pp. 70–103.

Tanida, Y., Okajima, A., and Watanabe, Y. 1973. Stability of a Circular Cylinder Oscillating in Uniform Flow or in a Wake. *JFM*, Vol. 61, pt. 4, pp. 769–784.

Tanner, M. 1964. Totwasser beeinflussung bei Keilstromunger, *Deutsche Luft-und Raumfahrt Forschungsbericht* 64–39.

Tatsuno, M. 1972. Vortex Wakes Behind a Circular Cylinder Oscillating in the Flow Direction. *Bulletin Research Inst.*, Applied Mechs., Kyushu Univ., No. 36.

Tatsuno, M. and Taneda, S. 1971. Flow Patterns Around a Circular Cylinder and a Flat Plate Normal to the Uniformly Decelerated Flow. *Jour. Physical Soc. Japan*, Vol. 31, pp. 1266–1274.

Telionis, D. P. and Tsahalis, D. T. 1974. Unsteady Laminar Separation over Cylinder Started Impulsively from Rest, *Acta Astronautica*, Vol. 1, p. 1487.

Temple, G. 1953. Unsteady Motion. in *Modern Developments in Fluid Dynamics*, Oxford Univ. Press, London, Vol. 1, pp. 325–371.

Thirriot, C., Longree, W. D., and Barthet, H. 1971. Sur La Perte de charge due a un Obstacle en mouvement periodique. Proceedings of the 14th Congress IAHR.

Thom, A. 1928. The Boundary Layer of the Front Portion of a Cylinder. ARC R & M No. 1176.

Thoman, D. C. and Szewczyk, A. A. 1969. Time-Dependent Viscous Flow Over a Circular Cylinder. *Physics of Fluids Supplement*, Vol. 12, p. II-76-II-80; also Tech. Rept. 66-14, Univ. of Notre Dame, Ind., 1966.

Thomson, K. D. and Morrison, D. G. 1971. The Spacing, Position and Strength of Vortices in the Wake of Slender Cylindrical Bodies at Large Incidence. *JFM*, Vol. 50, pt. 4, pp. 751–783.

Thomson, K. D. 1972. The Estimation of Viscous Normal Force, Pitching Moment Side Force and Yawing Moment on Bodies of Revolution of Incidences Up to 90°. WRE-Rep-782 (WR & D), Australian Defense Scientific Service, Melbourne, Aus.

Thomson, K. D. and Morrison, D. F. 1969. The Spacing, Position and Strength of Vortices in the Wake of Slender Cylindrical Bodies at Large Incidence. Australian Dept. of Supply, W.R.E. HSA 25.

Tournier, C. and Py, B. 1978. The Behavior of Naturally Oscillating Three-Dimensional Flow Around a Cylinder. *JFM*, Vol. 85, pp. 161–186.

Tseng, M. 1972. Drag of an Oscillating Plate in a Stream. *Schiffstechnik*, Vol. 19, No. 96, pp. 28–34.

Tuann, S.-Y. and Olson, M. D. 1976. Numerical Studies of the Flow Around a Circular Cylinder by a Finite Element Method. Structural Research Series Report No. 16, ISSN 0318-3378, Univ. of British Columbia, Vancouver, B.C., Canada.

Tyler, A. L. and Salt, D. L. 1978. Periodic Discontinuities in the Acceleration of Spheres in Free Flight. *Journal of Fluids Engineering*, ASME, Vol. 100, No. 1, pp. 17–21.

Verley, R. L. P. and Moe, G. 1979. The Forces on a Cylinder Oscillating in a Current. River and Harbour Laboratory, The Norwegian Institute of Technology, Report No. STF60 A79061.

Vickery, B. J. 1966. Fluctuating Lift and Drag on a Long Cylinder of Square Cross-Section in a Smooth and in a Turbulent Stream. *JFM*, Vol. 25, pt. 3, pp. 481–494.

Vickery, B. J. and Watkins, R. D. 1962. Flow-Induced Vibrations of Cylindrical Structures. *Proc. 1st Australiasian Conf.*, pp. 213–241.

Viets, H. 1971. Motion of Freely Falling Spheres at Moderate Reynolds Numbers. *AIAA Jour.*, Vol. 9, No. 10, pp. 2038–2089.

Wang, C. Y. 1967. The Flow Past a Circular Cylinder which is Started Impulsively From Rest. *Jour. Math. Phys.*, Vol. 46, p. 195.

Ward, E. G. and Dalton, C. 1969. Strictly Sinusoidal Flow Around a Stationary Cylinder. *Journal of Basic Engineering*, Trans. ASME, pp. 707–713.

Wardlaw, A. D., Jr. 1974. Prediction of Yawing Force at High Angle of Attack. *AIAA Jour.*, Vol. 12, No. 8, pp. 1142–1144, (See also NOLTR 73-209, Oct. 1973).

Warren, W. F. 1962. An Experimental Investigation of Fluid Forces on an Oscillating Cylinder. Ph.D. Thesis, Univ. of Maryland, College Park, Md.

Watson, E. J. 1955. Boundary Layer Growth. *Proc. Roy. Soc. A.*, Vol. 231, pp. 104–116.

Weaver, W., Jr. 1961. Wind-Induced Vibrations in Antenna Members. *Proc. ASCE*, Vol. 87, EM1, pp. 141–149.

Wiegel, R. 1964. *Oceanographical Engineering*. Englewood Cliffs, N.J., Prentice-Hall.

Wiegel, R. L. and Delmonte, R. C. 1972. Wave-Induced Eddies and 'Lift' Forces on Circular Cylinders. Univ. of Calif., Berkeley, Tech. Report No. HEL 9-19.

Wilkinson, R. H., Chaplin, J. R., and Shaw, T. L. 1974. On the Correlation of Dynamic Pressure on the Surface of a Prismatic Bluff Body. in *Flow-Induced Structural Vibrations*, (ed. E. Naudascher), Springer-Verlag, Berlin, pp. 471–487.

Wille, R. 1960. Karman Vortex Streets. in *Advances in Applied Mechanics*, Academic Press, N.Y., Vol. VI, pp. 273–293.

Wille, R. 1966. On Unsteady Flows and Transient Motions. in *Progress in Aeronautical Sciences*, Pergamon Press, London, Vol. 7, pp. 195–207.

Wille, R. 1974. Generation of Oscillatory Flows. in *Flow-Induced Structural Vibrations*, (ed. E. Naudascher), Springer-Verlag, Berlin, pp. 1–16.

Williams, J. C. 1977. Incompressible Boundary-Layer Separation. *Annual Reviews of Fluid Mechanics*, Vol. 9, pp. 113–114.

Wootton, L. R., Warner, M. H., and Cooper, D. H. 1974. Some Aspects of the Oscillations of Full-Scale Piles. IUTAM-IAHR Symposium on Flow-Induced Structural Vibrations, Naudascher, E., ed., Springer-Verlag, Berlin, pp. 587–061. (See also 1972 CIRIA Report 40.)

Wu, Th. Y. T. 1972. Cavity and Wake Flows. *Ann. Review of Fluid Mechanics*, Vol. 4, pp. 243–284.

Wu, J. C. and Thompson, J. F. 1973. Numerical Solution of Time-Dependent Incompressible Navier-Stokes Equations Using an Integro-Differential Formulation. *Computers and Fluids*, Vol. 1, p. 197.

Wundt, H. 1955. Wachstum der Laminaren Grenzschicht an schräg angeströmten Zylindern bei Anfahrt aus der Ruhe. *Ing-Arch.*, Vol. 23, pp. 212–230.

Yamamoto, T. and Nath, J. H. 1976. High Reynolds Number Oscillating Flow by Cylinders. *Proceedings of the 15th Coastal Engineering Conference*, Vol. III, pp. 2321–2340.

Zdravkovich, M. M. 1969. Smoke Observations of the Formation of a Karman Vortex Street. *JFM*, Vol. 37, pp. 491–499.

Zdravkovich, M. M. and Namork, J. E. 1977. Formation and Reversal of Vortices around Circular Cylinders Subjected to Water Waves. *ASCE, WW3*, paper No. 13102, pp. 378–383.

4 Wave Theories

4.1 GOVERNING EQUATIONS

The equations of fluid motion, together with the boundary conditions that they are required to meet, have been outlined in Chapter 2, and we are now ready to take up the problem of describing wave motion. Some texts which outline the development and results of wave theories include those by Lamb (1945), Stoker (1957), Wehausen and Laitone (1960), Wiegel (1964), Ippen (1966), Milne-Thompson (1968), Silvester (1974), Whitham (1974), Le Méhauté (1976), Phillips (1977), Horikawa (1978), LeBlond and Mysak (1978) and Sorensen (1978).

In the present context we define a Cartesian coordinate system (x, y, z) with x measured in the direction of wave propagation, z measured upwards from the still water level and y orthogonal to x and z. It is assumed that the waves are two-dimensional in the x-z plane, that they are progressive in the positive x direction and that they propagate over a smooth horizontal bed in water of constant undisturbed depth d. We further assume that the wave maintains a permanent form, that there is no underlying current and that the free surface is uncontaminated. The fluid (water) is taken to be incompressible and inviscid and the flow to be irrotational (but see Section 4.5.2). Figure 4.1 indicates the general form of a wave train conforming to these assumptions. Here the wave height H is the vertical distance from trough to crest, the wave length L is the distance between successive crests, the wave period T is the time interval between successive crests passing a particular point and the wave speed or celerity c is the speed of the wave travelling through the fluid ($c = L/T$). It is often convenient to work

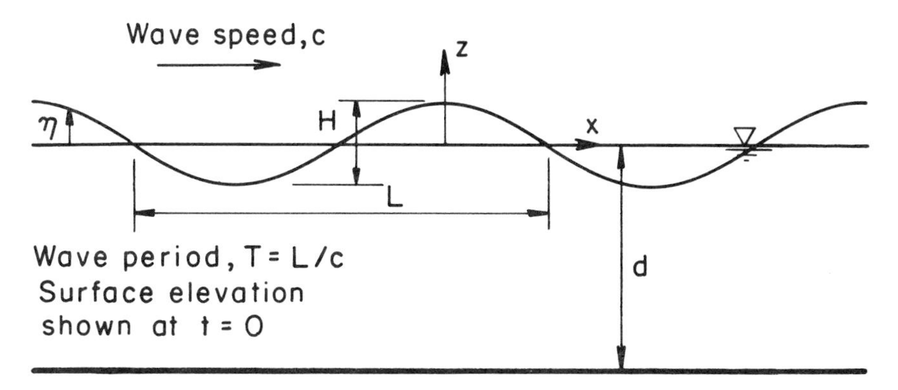

Fig. 4.1. Definition sketch for a progressive wave train.

with the wave angular frequency $\omega = 2\pi/T$ and the wave number $k = 2\pi/L$ (and thus $c = \omega/k$ also).

Any particular wave train is generally specified by the quantities H, L and d or by H, T and d, and the objective of any wave theory is to determine c (and therefore T or L as appropriate) and a description of the water particle motions throughout the flow. Dimensionless parameters are frequently used to characterize a wave train. The wave height is often expressed in terms of H/gT^2, the wave steepness H/L or the relative height H/d. And the water depth is often expressed in terms of the depth parameters d/gT^2 or kd or the relative depth d/L. For steeper waves in shallow water the Ursell number $U = HL^2/d^3$ is often used. Thus a design wave specified by H, T and d may conveniently be characterized, for example, by the parameters H/gT^2 and d/gT^2.

We require to determine the velocity potential ϕ pertaining to the fluid region. This satisfies the Laplace equation

$$\frac{\partial^2\phi}{\partial x^2} + \frac{\partial^2\phi}{\partial z^2} = 0 \tag{4.1}$$

and will be subject to the boundary conditions

$$\frac{\partial\phi}{\partial z} = 0 \qquad \text{at } z = -d \tag{4.2}$$

$$\frac{\partial\eta}{\partial t} + \frac{\partial\phi}{\partial x}\frac{\partial\eta}{\partial x} - \frac{\partial\phi}{\partial z} = 0 \qquad \text{at } z = \eta \tag{4.3}$$

$$\frac{\partial\phi}{\partial t} + \frac{1}{2}\left[\left(\frac{\partial\phi}{\partial x}\right)^2 + \left(\frac{\partial\phi}{\partial z}\right)^2\right] + g\eta = f(t) \quad \text{at } z = \eta \tag{4.4}$$

$$\phi(x, z, t) = \phi(x - ct, z) \tag{4.5}$$

where $\eta(x, t)$ is the free surface elevation measured above the still water level $z = 0$.

The existence of the velocity potential ϕ and the validity of the Laplace equation follow from the assumptions of an irrotational flow and an incompressible fluid. Equation (4.2) corresponds to the boundary condition at the seabed which imposes a zero vertical component on the fluid particle velocity at the seabed. Equations (4.3) and (4.4) represent the kinematic and dynamic free surface boundary conditions respectively. The former describes the condition that the fluid particle velocity normal to the free surface is equal to the velocity of the free surface itself in that direction, while the dynamic condition states that the pressure at the free surface, expressed in terms of the Bernoulli equation, is constant. This latter requirement follows from the assumptions that the atmospheric pressure (immediately above the fluid) is itself constant and that the free surface is uncontaminated (corresponding to a surface tension that may be taken as zero). Equation (4.5) describes the periodic nature of the wave train. In the absence of an underlying current the waves are progressive with a celerity c and are of permanent form: the dependence of variables of interest upon x and t may consequently be written in terms of dependence upon a single variable $(x - ct)$.

Of course some of the assumptions made in order to establish Eqs. (4.1) through (4.5) are seldom justified. Perhaps the most severe are the assumptions that there is no underlying current, that the depth is constant and that the wave train is two-dimensional and of permanent form. However these assumptions are retained for the purpose of developing various wave theories and the influence of currents, variable depth and an irregular wave profile are separately examined later. On the other hand, the irrotationality assumption is generally found to be reasonable outside the (thin) boundary layers at the seabed and free surface. For the present, then, we continue to examine the formulation of wave theories on the basis of all the aforementioned assumptions.

In some cases, where it is convenient to specify that the incident wave direction makes an angle α with the positive x axis, we have merely to replace x by $(x \cos\alpha + y \sin\alpha)$ in any results that are obtained.

An alternative method of developing a wave theory for waves of permanent form involves the choice of a coordinate system travelling with the speed of the wave such that the wave train then appears stationary. This new coordinate system may clearly be chosen as (x', y, z), where $x' = x - ct$. Since any transformation into the new system involves only a steady change in origin, the equations descriptive of the wave motion remain valid—though the variables now become independent of time. Alternatively, the same transformation may be considered as the superposition upon the original system of a uniform current of velocity $-c$ in the x direction. This would render the wave profile stationary and the flow steady in the fixed reference frame (x, y, z).

In either case, it is convenient to retain the notation ϕ and $u = \partial\phi/\partial x$ for the velocity potential and horizontal particle velocity associated with the wave motion only and excluding the steady current c. The horizontal velocity of the fluid motion would then be $u' = u - c$, and the corresponding velocity potential would be $\phi' = \phi - cx$. The velocity potential satisfies the Laplace equation in the (x', z) frame

$$\frac{\partial^2\phi'}{\partial x'^2} + \frac{\partial^2\phi'}{\partial z^2} = 0 \tag{4.6}$$

The boundary conditions applicable to the (x', z) system are obtained by omitting the time dependent terms from Eqs. (4.2), (4.3) and (4.4) and by writing $u' = u - c$ for the horizontal velocity. Thus we have

$$\frac{\partial\phi'}{\partial z} = 0 \qquad \text{at } z = -d \tag{4.7}$$

$$u' \frac{\partial\eta}{\partial x'} - \frac{\partial\phi'}{\partial z} = 0 \qquad \text{at } z = \eta \tag{4.8}$$

$$\frac{1}{2g}(u'^2 + w^2) + \eta = Q \qquad \text{at } z = \eta \tag{4.9}$$

where $w = \partial\phi/\partial z$ is the vertical particle velocity and Q is a constant. Once the velocity potential of the wave motion has been obtained, the solution may be expressed in the original reference frame by taking $\phi(x - ct, z) = \phi'(x', z) + cx$. The results of the steady flow problem are unique on the basis of a particular method of solution and degree of approximation (wave theory). However, in the general case the evaluation of the wave celerity c requires an additional assumption conerning the absence of an underlying current and this aspect is taken up in Section 4.3.2. To a first (linear) approximation alternative possible assumptions need not be distinguished and the wave celerity obtained is unambiguous.

4.2 SMALL AMPLITUDE WAVE THEORY

Two serious difficulties arise in the attempt to obtain an exact solution for a two-dimensional wave train. The first is that the free-surface boundary conditions are nonlinear, and the second is that these conditions are prescribed at the free surface $z = \eta$ which is initially unknown. The simplest and most fundamental approach is to seek a linear solution of the problem by taking the wave height H to be very much smaller than both the wave length L and the still water depth d: that is $H \ll L, d$. The wave theory which results from this additional assumption is referred to alternatively as small amplitude wave theory, linear wave theory, sinusoidal wave theory or as Airy theory. Because of the assumption that $H \ll L, d$, the nonlinear terms in Eqs. (4.3) and (4.4), which involve products of terms of order of the wave height (expressed in a suitably dimensionless form), are then negligible in comparison with the remaining linear terms which are themselves of the order of the wave height. Furthermore, the free-surface boundary conditions may now be applied directly at the still water level $z = 0$. These simplifications, arising from the assumption of small wave height and which are made here on the basis of physical reasoning, will later be shown to represent the first approximation deriving from a formal perturbation procedure.

4.2.1 Theoretical Development

For small amplitude waves, the free-surface boundary conditions as expressed in Eqs. (4.3) and (4.4) reduce to

$$\frac{\partial \phi}{\partial z} - \frac{\partial \eta}{\partial t} = 0 \quad \text{at } z = 0 \tag{4.10}$$

$$\frac{\partial \phi}{\partial t} + g\eta = 0 \quad \text{at } z = 0 \tag{4.11}$$

which may be combined to give

$$\frac{\partial^2 \phi}{\partial t^2} + g \frac{\partial \phi}{\partial z} = 0 \quad \text{at } z = 0 \tag{4.12}$$

$$\eta = -\frac{1}{g}\left(\frac{\partial \phi}{\partial t}\right)_{z=0} \tag{4.13}$$

Bearing in mind the periodicity condition given by Eq. (4.5), the solution to the problem may be obtained by a separation of variables technique in which ϕ is written in the form

$$\phi = Z(z)\Phi(x - ct) \tag{4.14}$$

Substituting this expression for ϕ into the Laplace equation, Eq. (4.1), we obtain two ordinary differential equations

$$\frac{\partial^2 Z}{\partial z^2} - k^2 Z = 0 \tag{4.15}$$

$$\frac{\partial^2 \Phi}{\partial x^2} + k^2 \Phi = 0 \tag{4.16}$$

where the sign of the constant k^2 has been chosen to provide a periodic rather than hyperbolic solution in $(x - ct)$. The general solutions of Z and Φ are

$$Z = A_1 \cosh(kz) + A_2 \sinh(kz) \tag{4.17}$$

$$\Phi = A_3 \cos[k(x - ct)] + A_4 \sin[k(x - ct)] \tag{4.18}$$

where A_1, A_2, A_3 and A_4 are constants to be determined by the boundary conditions.

Time t is defined to be zero when a wave crest crosses the plane $x = 0$. Bearing in mind Eq. (4.13), we must have that $A_3 = 0$. The boundary condition at the seabed Eq. (4.2) implies that $A_2 = A_1 \tanh(kd)$. We now substitute into Eq. (4.14) the expressions we have obtained for Z and Φ to find that the velocity potential may be written in the form

$$\phi = A \frac{\cosh(k(z + d))}{\cosh(kd)} \sin[k(x - ct)] \tag{4.19}$$

where the new constant $A = A_1 A_4$. The constant A may be determined by using Eq. (4.13) to provide an expression for η and relating the range in η to the wave height H. We thus obtain $A = gH/2kc$. Equation (4.19) shows that the velocity potential is periodic in the x direction with a wave length $L = 2\pi/k$ and in time with a wave period $T = 2\pi/kc$. Thus the constant k is recognized as the wave number introduced earlier; while $kc = \omega$, the wave angular frequency. That is $c = L/T = \omega/k$.

The remaining boundary condition, Eq (4.12), may now be used to obtain an expression for c or ω in terms of k, for upon substitution of Eq. (4.19) into Eq. (4.12) we obtain the *linear dispersion relation*

$$\omega^2 = gk \tanh(kd) \tag{4.20a}$$

or

$$c^2 = \frac{g}{k} \tanh(kd) \tag{4.20b}$$

This describes how the wave speed increases with wave length. More generally, the *dispersion relation* for a finite amplitude wave train, expressing c in terms of k, involves also the wave height H and may be developed according to any particular wave theory.

In view of Eqs. (4.20), the expression for the velocity potential may finally be written in the alternative forms

$$\phi = \frac{gH}{2\omega} \frac{\cosh(k(z + d))}{\cosh(kd)} \sin\theta \tag{4.21a}$$

$$\phi = \frac{\pi H}{kT} \frac{\cosh(k(z + d))}{\sinh(kd)} \sin\theta \tag{4.21b}$$

where $\theta = k(x - ct) = kx - \omega t$ is the wave phase angle.

4.2.2 Description of Results

Now that a complete solution has been obtained for ϕ, the other variables of interest may immediately be determined. In particular, an expression for η derives from Eq. (4.13), the velocity components u and w from the appropriate derivatives of ϕ, the particle accelerations from the temporal derivatives of u and w, and the horizontal and vertical particle displacements from integrals of u and w taken over time,

$$\xi = \int_0^t u\, dt = -\frac{H}{2} \frac{\cosh(k(z + d))}{\sinh(kd)} \sin\theta \tag{4.22a}$$

$$\zeta = \int_0^t w\, dt = \frac{H}{2} \frac{\sinh(k(z + d))}{\sinh(kd)} \cos\theta \tag{4.22b}$$

The pressure p is given by the linearized form of the unsteady Bernoulli equation, Eq. (2.15), in which the nonlinear velocity squared terms are omitted in accordance with the present linear approximation,

$$p = -\rho g z - \rho \frac{\partial \phi}{\partial t} \tag{4.23}$$

Expressions for all these quantities are assembled in Table 4.1. In presenting these, it is convenient to use a vertical coordinate $s = z + d$ measured upwards from the seabed. Thus $s = 0$ represents the seabed and $s = d$ represents the still water level. Included in the table are some further quantities of interest: E, P,

Table 4.1 Results of Linear Wave Theory.

Velocity potential	$\phi = \frac{\pi H}{kT} \frac{\cosh(ks)}{\sinh(kd)} \sin\theta$ $= \frac{gH}{2\omega} \frac{\cosh(ks)}{\cosh(kd)} \sin\theta$
Dispersion relation	$c^2 = \frac{\omega^2}{k^2} = \frac{g}{k} \tanh(kd)$
Surface elevation	$\eta = \frac{H}{2} \cos\theta$
Horizontal particle displacement	$\xi = -\frac{H}{2} \frac{\cosh(ks)}{\sinh(kd)} \sin\theta$
Vertical particle displacement	$\zeta = \frac{H}{2} \frac{\sinh(ks)}{\sinh(kd)} \cos\theta$
Horizontal particle velocity	$u = \frac{\pi H}{T} \frac{\cosh(ks)}{\sinh(kd)} \cos\theta$
Vertical particle velocity	$w = \frac{\pi H}{T} \frac{\sinh(ks)}{\sinh(kd)} \sin\theta$
Horizontal particle acceleration	$\frac{\partial u}{\partial t} = \frac{2\pi^2 H}{T^2} \frac{\cosh(ks)}{\sinh(kd)} \sin\theta$
Vertical particle acceleration	$\frac{\partial w}{\partial t} = -\frac{2\pi^2 H}{T^2} \frac{\sinh(ks)}{\sinh(kd)} \cos\theta$
Pressure	$p = -\rho g z + \frac{1}{2} \rho g H \frac{\cosh(ks)}{\cosh(kd)} \cos\theta$
Group velocity	$c_G = \frac{1}{2} \left[1 + \frac{2kd}{\sinh(2kd)} \right] c$
Average energy density	$E = \frac{1}{8} \rho g H^2$
Energy flux	$P = E c_G$
Radiation stress	$S_{xx} = \left[\frac{1}{2} + \frac{2kd}{\sinh(2kd)} \right] E$ $S_{xy} = S_{yx} = 0$ $S_{yy} = \left[\frac{kd}{\sinh(2kd)} \right] E$

c_G and S, but a description of these will be deferred until we have considered a few of the more fundamental features of the solution just obtained.

A distinction is first made between ranges of kd over which certain approximations are useful. The shallow and deep water ranges correspond to $kd < \pi/10$ and $kd > \pi$ respectively, and over these ranges approximate expressions may be substituted for the hyperbolic functions that have been encountered

$$\left.\begin{aligned} \sinh(kd) \simeq \tanh(kd) &\simeq kd \\ \cosh(kd) &\simeq 1 \end{aligned}\right\} \text{for } kd < \pi/10 \tag{4.24a}$$

$$\left.\begin{aligned} \sinh(kd) \simeq \cosh(kd) &\simeq \tfrac{1}{2}\, e^{kd} \\ \tanh(kd) &\simeq 1 \end{aligned}\right\} \text{for } kd > \pi \tag{4.24b}$$

Substituting these into the results of Table 4.1, we obtain the simplified expressions that are summarized in Table 4.2. The complete range of water depths, then, is conveniently divided into the shallow water, intermediate depth and deep water ranges as follows:

$$\text{shallow water waves:} \quad \frac{1}{20} > \frac{d}{L}; \qquad 0.0025 > \frac{d}{gT^2};$$

$$\text{intermediate depth waves:} \quad \frac{1}{20} < \frac{d}{L} < \frac{1}{2}; \qquad 0.0025 < \frac{d}{gT^2} < 0.08;$$

$$\text{deep water waves:} \quad \frac{d}{L} > \frac{1}{2}; \qquad \frac{d}{gT^2} > 0.08.$$

These limits in terms of d/gT^2 are equivalent to those given in terms of d/L by the application of the dispersion relation.

Bearing these approximations in mind, we see that the expressions for the displacements ξ and ζ indicate that the particles travel in closed elliptic orbits as sketched in Fig. 4.2. The amplitude of horizontal velocity (and displacement) decreases with depth according to $\cosh(k(z + d))$, while the amplitude of vertical velocity (and displacement) decreases according to $\sinh(k(z + d))$. Typical profiles relating to the shallow, intermediate and deep water ranges are sketched in Fig. 4.2. Note that at intermediate depths the orbits diminish in amplitude with depth and also become flatter until the vertical component vanishes at the seabed in accordance with the seabed boundary condition; and also that the velocity and acceleration vectors at a given point and time are not colinear. In the shallow water range the elliptic orbits are relatively flat at all depths and diminish in amplitude only gradually with depth. In deep water the particle motions are circular, the amplitude of motion decreasing exponentially with depth, until at

Table 4.2 Shallow and deep water approximations to linear wave theory.

	Shallow Water	Deep Water
Range of validity	$kd < \frac{\pi}{10}$	$kd > \pi$
	$\frac{d}{L} < \frac{1}{20}$	$\frac{d}{L} > \frac{1}{2}$
	$\frac{d}{gT^2} < 0.0025$	$\frac{d}{gT^2} > 0.08$
Velocity potential	$\phi = \frac{\pi H}{k^2 Td} \sin\theta$	$\phi = \frac{\pi H}{kT} e^{kz} \sin\theta$
	$= \frac{gH}{2\omega} \sin\theta$	$= \frac{gH}{2\omega} e^{kz} \sin\theta$
Dispersion relation	$c^2 = \frac{\omega^2}{k^2} = gd$	$c^2 = c_0^2 = \frac{\omega^2}{k^2} = \frac{g}{k}$
Wave length	$L = T\sqrt{gd}$	$L = L_0 = gT^2/2\pi$
Surface elevation	$\eta = \frac{H}{2} \cos\theta$	$\eta = \frac{H}{2} \cos\theta$
Horizontal particle displacement	$\xi = -\frac{H}{2kd} \sin\theta$	$\xi = -\frac{H}{2} e^{kz} \sin\theta$
Vertical particle displacement	$\zeta = \frac{H}{2}\left(1 + \frac{z}{d}\right) \cos\theta$	$\zeta = \frac{H}{2} e^{kz} \cos\theta$
Horizontal particle velocity	$u = \frac{\pi H}{T(kd)} \cos\theta$	$u = \frac{\pi H}{T} e^{kz} \cos\theta$
Vertical particle velocity	$w = \frac{\pi H}{T}\left(1 + \frac{z}{d}\right) \sin\theta$	$w = \frac{\pi H}{T} e^{kz} \sin\theta$
Horizontal particle acceleration	$\frac{\partial u}{\partial t} = \frac{2\pi^2 H}{T^2 (kd)} \sin\theta$	$\frac{\partial u}{\partial t} = \frac{2\pi^2 H}{T^2} e^{kz} \sin\theta$
Vertical particle acceleration	$\frac{\partial w}{\partial t} = -\frac{2\pi^2 H}{T^2}\left(1 + \frac{z}{d}\right) \cos\theta$	$\frac{\partial w}{\partial t} = -\frac{2\pi^2 H}{T^2} e^{kz} \cos\theta$
Pressure	$p = -\rho gz + \frac{1}{2} \rho gH \cos\theta$	$p = -\rho gz + \frac{1}{2} \rho gH e^{kz} \cos\theta$
Group velocity	$c_G = c$	$c_G = \frac{1}{2} c$
Average energy density	$E = \frac{1}{8} \rho gH^2$	$E = \frac{1}{8} \rho gH^2$
Energy flux	$P = Ec$	$P = \frac{1}{2} Ec$
Radiation stress	$S_{xx} = \frac{3}{2} E$	$S_{xx} = \frac{1}{2} E$
	$S_{xy} = S_{yx} = 0$	$S_{xy} = S_{yx} = 0$
	$S_{yy} = \frac{1}{2} E$	$S_{yy} = 0$

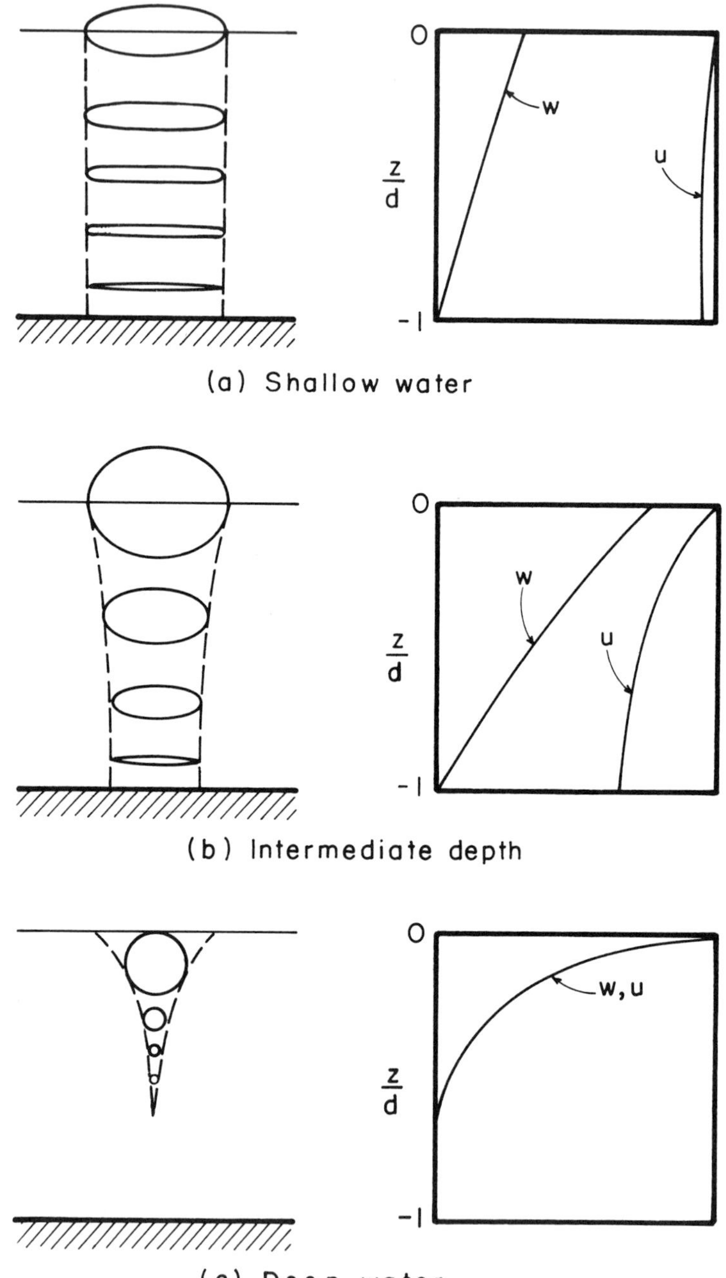

Fig. 4.2. Particle orbits and variation of particle velocity amplitudes with depth.

$z = -L/2$ this amplitude is only 4% [i.e. exp $(-\pi)$] of its value at the still water level. Thus the wave-induced motion may conveniently be considered to penetrate up to a depth of half a wave length below the free surface. The amplitude of the pressure fluctuations within the fluid also attenuates with depth according to the factor

$$K = \frac{\cosh(k(z + d))}{\cosh(kd)} \tag{4.25}$$

K is known as the pressure response factor and may be employed when estimating the wave height from measurements with a pressure sensor at or above the seabed. (The variation of K with depth corresponds to that of u in Fig. 4.2.)

In engineering practice a design wave is usually specified in terms of H, T and d and indeed these are the parameters defining a wave train that are usually the easiest to measure or estimate from observation. In order to calculate the wave number k, it is necessary to solve the dispersion relation implicitly,

$$kd \tanh(kd) = 4\pi^2 \frac{d}{gT^2} \tag{4.26}$$

This may be done by trial and error, by an iterative procedure with a computer, or by resort to a graph or tables. Fig. 4.3 presents the relation between kd and d/gT^2 and may be used to obtain an estimate of kd for a given design wave. The approximations for shallow and deep water waves are also included in the figure. Reasonably accurate solutions for kd involving power series of the right-hand side of Eq. (4.26), taken to different numbers of terms, have been given by Hunt (1979).

The results outlined so far are usually sufficient to deal with a variety of applications. For example, in wave force calculations it is usually the fluid velocity and acceleration components in prescribed directions that are required. Given a design wave specified by H, T and d, the usual procedure would be first to evaluate d/gT^2 and then obtain the corresponding value of kd from a table based on Eq. (4.26) (or from Fig. 4.3). The formulae in Tables 4.1 or 4.2 may then be used directly as necessary.

4.2.3 Some Integral Properties of Waves

Most of the variables already described vary with location and/or time and we now turn to a consideration of various integral properties which take on specific values for a given wave train and which are related mostly to wave energy and momentum. The discussion is generalized so as to apply to a periodic wave train of finite amplitude. Particular results pertaining to linear wave theory may

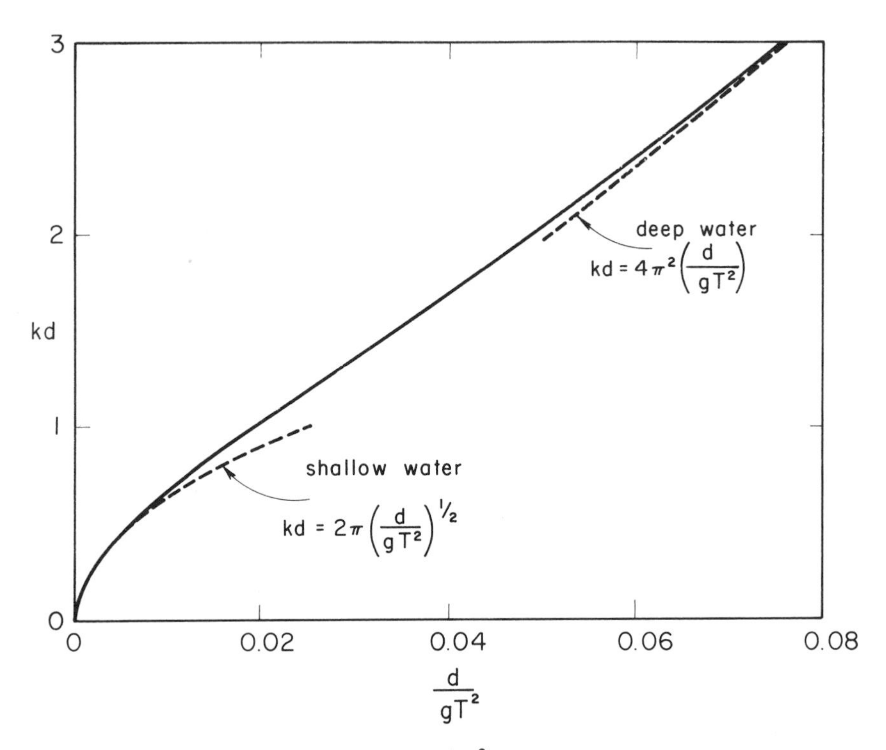

Fig. 4.3. Variation of kd with d/gT^2 for small amplitude waves.

readily be calculated from the general definitions given and many of these are included in Tables 4.1 and 4.2. A useful outline of various integral properties and the relationships between them is given by Longuet-Higgins (1975).

The average kinetic energy of the wave motion per unit horizontal area, denoted K and often termed simply the kinetic energy, may be developed by considering the kinetic energy of a fluid element of height dz, length dx and unit width, and carrying out appropriate integrations. In this way K is given as

$$K = \overline{\int_{-d}^{\eta} \tfrac{1}{2}\, \rho (u^2 + w^2)\, dz} \tag{4.27}$$

where the overbar denotes an average over a wave length or wave period, that is

$$\overline{f} = \frac{1}{L} \int_0^L f\, dx \tag{4.28}$$

(In the case of progressive waves a spatial average must be independent of time.)

The average potential energy of the waves per unit horizontal area, denoted V and termed the potential energy, may be considered in a similar way except that the potential energy of the undisturbed fluid is nonzero (when the datum is taken at $z = -d$) and must be subtracted from the total potential energy in the presence of waves. Thus

$$V = \overline{\int_{-d}^{\eta} \rho g z \, dz} - \int_{-d}^{0} \rho g z \, dz$$

$$= \overline{\int_{0}^{\eta} \rho g z \, dz} = \tfrac{1}{2} \rho g \overline{\eta^2} \tag{4.29}$$

The *average energy density* E, that is the average energy (kinetic and potential) per unit horizontal area of the wave train, is simply the sum of the two expressions in Eqs. (4.27) and (4.29). That is

$$E = K + V \tag{4.30}$$

In the case of small amplitude waves, appropriate substitutions from results already obtained yield

$$K = V = \tfrac{1}{2} E; \qquad E = \tfrac{1}{8} \rho g H^2 \tag{4.31}$$

In order to obtain a consistent order of approximation it will be seen that the upper limit of integration $z = \eta$ may be replaced by $z = 0$ with regard to the kinetic but not the potential energy term. The *wave action denisty* defined as E/ω is also used, such as in applications involving slowly varying wave fields.

The *energy flux* P is the average rate of transfer of energy per unit width across a plane of constant x. An expression for P may be derived by considering the instantaneous rate at which work is done and kinetic and potential energy is transferred across an elemental plane area of unit width, constant x and height dz, integrating this over depth and then taking the time average of the result. P is thereby given as

$$P = \overline{\int_{-d}^{\eta} [p + \tfrac{1}{2} \rho(u^2 + w^2) + \rho g z] \, u \, dz} \tag{4.32}$$

Substituting the unsteady Bernoulli equation, Eq. (2.15), into this equation and noting that $\partial\phi/\partial t = -cu$ for progressive waves, we find that the expression for

P reduces to

$$P = \rho c \overline{\int_{-d}^{\eta} u^2 \, dz}, \tag{4.33}$$

where once again the limit $z = \eta$ may be replaced by $z = 0$ in the case of small amplitude waves. This concept of energy flux is found to have a variety of useful applications when dealing with a number of wave phenomena that involve energy transfer and which include, for example, wave shoaling and wave attenuation to be discussed in Section 4.9.

It is convenient at this point to introduce the *group velocity* c_G which is associated with the energy characteristics of a wave train and which may be defined as

$$c_G = P/E \tag{4.34}$$

The average energy to be supplied by a wave group advancing across still water in order to set up the wave motion in a length ℓ of initially undisturbed water, is simply $E\ell$ per unit width. The average rate of energy transfer in the wave group in the wave propagation direction is P and, in consequence, the quantity of energy $E\ell$ is supplied in time $E\ell/P$. It follows that the group velocity, expressed as the speed of energy transfer associated with the wave group, is P/E.

In physical terms the group velocity is the speed of propagation of a finite group of waves as distinct from the speed of the individual waves themselves. Outside the shallow water range the group velocity is appreciably less than the individual wave speed such that, in a group of finite length, individual waves advance forward within the group and eventually diminish and lose their identity at the leading edge of the group, while new waves are formed at the rear of the group to advance forward within the group in their turn.

A simple method of deriving an expression for the group velocity of small amplitude waves is to consider a wave group arising from the superposition of two wave trains of slightly different wave length and frequency as sketched in Fig. 4.4. Using the notation of that figure, the surface elevation of the resulting group is given as

$$\begin{aligned} \eta_G &= \frac{H}{2} \cos\left[\left(k + \frac{\Delta k}{2}\right) x - \left(\omega + \frac{\Delta\omega}{2}\right) t\right] \\ &\quad + \frac{H}{2} \cos\left[\left(k - \frac{\Delta k}{2}\right) x - \left(\omega - \frac{\Delta\omega}{2}\right) t\right] \\ &= H \cos(kx - \omega t) \cos\left(\frac{\Delta k}{2} x - \frac{\Delta\omega}{2} t\right) \end{aligned} \tag{4.35}$$

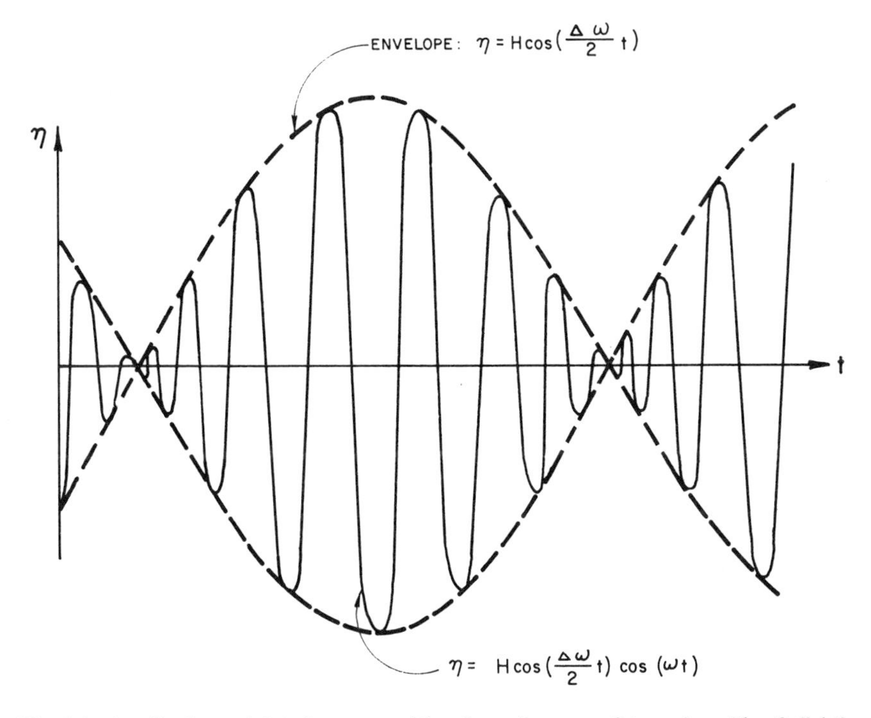

Fig. 4.4. Amplitude modulated wave resulting from the sum of two sinusoids of slightly different frequency.

This represents an amplitude-modulated wave with a length $2\pi/k$ and period $2\pi/\omega$. The envelope's wave length L_G is $2\pi/(\Delta k/2)$ and its period T_G is $2\pi/(\Delta\omega/2)$. The speed of the envelope, given as L_G/T_G, becomes the group velocity c_G in the limit as $\Delta \to 0$. We thus have

$$c_G = \underset{\Delta \to 0}{\text{Limit}} \left(\frac{L_G}{T_G}\right) = \frac{d\omega}{dk}, \tag{4.36}$$

where ω and k are related by the dispersion relation. After performing the necessary algebra, the expression for c_G given in Table 4.1 is readily obtained.

We now consider some integral properties relating to the momentum of the waves. The average wave momentum per unit horizontal area, or impulse I, is simply

$$I = \overline{\int_{-d}^{\eta} \rho u \, dz} \tag{4.37}$$

The *radiation stress* was introduced by Longuet-Higgins and Stewart (1960) and has been later discussed by them (Longuet-Higgins and Stewart 1964) in a simplified manner. It may be defined as the excess flux of momentum due to the presence of the waves and is a directional quantity so that, for example, the subscripts appearing in S_{xy} refer to the flow of x-ward momentum which is being transported in the y direction. The instantaneous flux of horizontal x-momentum across a plane x = constant is given as

$$\int_{-d}^{\eta} (p + \rho u^2)\, dz$$

and the corresponding component of the radiation stress may then be expressed as

$$S_{xx} = \overline{\int_{-d}^{\eta} (p + \rho u^2)\, dz} - \int_{-d}^{0} \rho g z\, dz \tag{4.38}$$

Here the last term represents the corresponding flux of the undisturbed fluid which is subtracted from the total flux due to the presence of the waves.

The instantaneous flux of x-momentum across a plane y = constant contains no contribution from the mean pressure and the corresponding component of radiation stress is therefore given as

$$S_{xy} = \int_{-d}^{\eta} \rho uv\, dz = 0 \tag{4.39}$$

Similarly S_{yx} is zero. An expression for S_{yy} may be developed in a similar manner to that for S_{xx} and we thus have

$$S_{yy} = \overline{\int_{-d}^{\eta} (p + \rho v^2)\, dz} - \int_{-d}^{0} \rho g z\, dz \tag{4.40}$$

The radiation stress has applications to a variety of wave phenomena including wave and current interactions (Longuet-Higgins and Stewart 1961), surf beats (Longuet-Higgins and Stewart 1962), wave set-up (Longuet-Higgins and Stewart 1963) and the like.

More explicit formulae for the various integral properties of small amplitude waves may easily be obtained by appropriate substitution of linear wave theory results already known, and expressions for most of the quantities so far described are given in Tables 4.1 and 4.2.

Additional integral properties are related to the steady flow obtained by choosing a reference frame (x′, z) which moves with the wave train. The horizontal particle velocity is now $u' = u - c$ while the vertical particle velocity w is unaltered. Specifically, in finite amplitude waves c is the wave speed at which $\overline{u} = 0$ (see Section 4.3.2). The mass flux per unit width is denoted $-\mathbf{Q}$ (so that $\mathbf{Q}$ is positive) and is given as

$$\mathbf{Q} = -\int_{-d}^{\eta} \rho u' \, dz \tag{4.41}$$

(An overbar is unnecessary here since the flow is steady and the mass flow rate is independent of time and horizontal distance.) The total head or Bernoulli constant for the steady flow is given as

$$\mathbf{R} = p + \tfrac{1}{2} \rho(u'^2 + w^2) + \rho g(z + d) \tag{4.42}$$

And finally the momentum flux per unit width for the steady flow is

$$\mathbf{S} = \int_{-d}^{\eta} (p + \rho u'^2) \, dz \tag{4.43}$$

These three quantities **Q**, **R**, and **S** were used by Benjamin and Lighthill (1954) to categorize finite amplitude waves, in a paper which emphasized the description of cnoidal waves.

Longuet-Higgins (1975) has described various relationships (many derived for the first time) between different integral properties which include those already outlined and also the mean square velocity at the seabed $\overline{u_b^2} = \overline{u^2}$ at $z = -d$. Some of these are

$$\mathbf{Q} = \rho c d - I \tag{4.44a}$$

$$2K = cI \tag{4.44b}$$

$$S_{xx} = 4K - 3V + \rho \overline{u_b^2}\, d \tag{4.44c}$$

$$P = c(3K - 2V) + \tfrac{1}{2} \overline{u_b^2} (I + \rho c d) \tag{4.44d}$$

$$\mathbf{R} = \tfrac{1}{2} \rho \overline{u_b^2} + \tfrac{1}{2} \rho c^2 + \rho g d \tag{4.44e}$$

$$\mathbf{S} = S_{xx} - 2cI + \rho d \left(c^2 + \frac{gd}{2} \right) \tag{4.44f}$$

The linear theory outlined in this section forms the fundamental tool of coastal and ocean engineering practice for many reasons. It is particularly simple to apply and in many applications does not require sophisticated computer programming as do many other wave theories. It may be applied in the shallow water, intermediate depth and deep water ranges, whereas other wave theories are often applicable over a more restricted range of depths. The linear wave theory may be extended to deal with a variety of situations not readily susceptible to interpretation in terms of nonlinear wave theories. This is particularly the case for investigations of wave diffraction effects and for the statistical treatment of random waves, both these topics being of considerable practical importance. Furthermore, the linear approach has been found to be fairly realistic even when there are quite major departures from the small wave height assumption. A comparison of the predictions of small amplitude wave theory with those of other theories, as well as with experiment, will be made in Section 4.6. In the meantime, we now consider alternative wave theories that may be used to describe a wave train.

4.3 STOKES FINITE AMPLITUDE WAVE THEORY

The linear or small amplitude wave theory described in the preceding section provides a first approximation to the wave motion. In order to approach the complete solution more closely, we may consider a perturbation procedure in which successive approximations are developed. Such a method was used by Stokes (1847, 1880) and more recent contributions to the description of these Stokes waves include those of De (1955), Borgman and Chappelear (1958), Bretschneider (1960), Skjelbreia and Hendrickson (1960) and Tsuchiya and Yamaguchi (1972). The computer extension of this procedure to a high order of approximation is a more recent advance and is mentioned in Section 4.5.4.

4.3.1 Formulation of Stokes Wave Theories

We introduce a perturbation assumption in which the variables describing the flow are developed as power series in terms of a perturbation parameter which is small. It is expected that the various series will converge as the number of terms considered increases. In point of fact, however, convergence towards a complete solution does not occur for steeper waves, unless a different perturbation parameter from that of Stokes is chosen. This represents a complication which has recently received some attention (e.g., Schwartz 1974, Cokelet 1977a), but this should not detract from the practical advantages of the original expansion procedure when applied up to a moderate (e.g. fifth) order of approximation.

The Stokes wave expansion method is formally valid under the conditions $H/d \ll (kd)^2$ for $kd < 1$, and $H/L \ll 1$ (Peregrine 1972). This may be demonstrated by comparing second and first order terms in the expression for velocity potential to be obtained, and requiring that the former is an order of magnitude smaller than the latter. The above conditions place a severe wave height restriction in shallow water and a separate shallow wave expansion procedure (Section 4.4.1) may then be used.

It is assumed that ϕ and associated variables (η, u, w, . . .) may be written in the form

$$\phi = \epsilon\phi_1 + \epsilon^2\phi_2 + \ldots \tag{4.45}$$

in which ϵ is the perturbation parameter. Each subscripted variable appearing in different series is taken as having the same order of magnitude and thus each additional term in the series represents a quantity smaller than the preceding one by a factor of order ϵ. By substituting Eq. (4.45) and corresponding expressions into the governing equations developed earlier, it becomes possible to obtain progressively higher order solutions, each expressed in terms of preceding ones. Thus by substituting Eq. (4.45) into the Laplace equation and the boundary condition at the seabed, and by collecting terms of order $\epsilon, \epsilon^2, \epsilon^3, \ldots$ we obtain respectively

$$\frac{\partial^2 \phi_n}{\partial x^2} + \frac{\partial^2 \phi_n}{\partial z^2} = 0 \qquad \text{for } n = 1, 2, \ldots \tag{4.46}$$

$$\frac{\partial \phi_n}{\partial z} = 0 \qquad \text{at } z = -d \qquad \text{for } n = 1, 2, \ldots \tag{4.47}$$

This is straightforward. Difficulties arise, however, in the free surface boundary conditions which contain nonlinear terms consisting of products, and which are applied at the unknown $z = \eta$ rather than $z = 0$. The free surface conditions give rise to a pair of equations expressing ϕ_n and η_n in terms of $\phi_{n-1}, \eta_{n-1}, \ldots, \phi_1, \eta_1$. In the attempt to make a scheme of this sort more tractable it may be convenient to separate the equations giving ϕ_n and η_n and to rewrite the free surface boundary conditions (e.g. Phillips 1977) as

$$\frac{\partial^2 \phi}{\partial t^2} + g\frac{\partial \phi}{\partial z} + \left[\frac{\partial}{\partial t} + \frac{1}{2}\frac{\partial \phi}{\partial x}\frac{\partial}{\partial x} + \frac{1}{2}\frac{\partial \phi}{\partial z}\frac{\partial}{\partial z}\right]\left[\left(\frac{\partial \phi}{\partial x}\right)^2 + \left(\frac{\partial \phi}{\partial z}\right)^2\right] = 0 \qquad \text{at } z = \eta \tag{4.48}$$

$$\eta = -\frac{1}{g}\left[\frac{\partial \phi}{\partial t} + \frac{1}{2}\left(\frac{\partial \phi}{\partial x}\right)^2 + \frac{1}{2}\left(\frac{\partial \phi}{\partial z}\right)^2\right] \qquad \text{at } z = \eta \tag{4.49}$$

The boundary conditions, when represented in the form

$$f(z) = 0 \qquad \text{at } z = \eta \tag{4.50}$$

may be expanded as a Taylor series in f in order to reduce them to apply directly at z = 0. The Taylor series expansion of Eq. (4.50) is

$$f(z) + \eta\frac{\partial f}{\partial z} + \frac{1}{2!}\eta^2\frac{\partial^2 f}{\partial z^2} + \cdots = 0 \qquad \text{at } z = 0 \tag{4.51}$$

By expressing Eqs. (4.48) and (4.49) in the same form as Eq. (4.50) and by using the power series expressions as are indicated in Eq. (4.45), we obtain successive approximations as already described. The algebra clearly becomes very tedious and it is awkward to obtain a solution beyond the first few approximations (but see Section 4.5.4). For illustrative purposes, however, we now obtain expressions corresponding to the first and second approximations.

The terms of order ϵ appearing in the free surface boundary conditions, Eqs. (4.48) and (4.49), are given by

$$\frac{\partial^2 \phi_1}{\partial t^2} + g\frac{\partial \phi_1}{\partial z} = 0 \qquad \text{at } z = 0 \tag{4.52}$$

$$\eta_1 = -\frac{1}{g}\left(\frac{\partial \phi_1}{\partial t}\right)_{z=0} \tag{4.53}$$

and thus, bearing in mind Eqs. (4.46) and (4.47), the governing equations of the first (linear) approximation are precisely those that were obtained previously and the results of linear wave theory follow.

The terms of order ϵ^2 in Eqs. (4.48) and (4.49) are given respectively as

$$\frac{\partial^2 \phi_2}{\partial t^2} + g\frac{\partial \phi_2}{\partial z} = -\eta_1\frac{\partial}{\partial z}\left[\frac{\partial^2 \phi_1}{\partial t^2} + g\frac{\partial \phi_1}{\partial z}\right] - \frac{\partial}{\partial t}\left[\left(\frac{\partial \phi_1}{\partial x}\right)^2 + \left(\frac{\partial \phi_1}{\partial z}\right)^2\right] \qquad \text{at } z = 0 \tag{4.54}$$

and

$$\eta_2 = -\frac{1}{g}\left[\frac{\partial \phi_2}{\partial t} + \eta_1\frac{\partial^2 \phi_1}{\partial z \partial t} + \frac{1}{2}\left(\frac{\partial \phi_1}{\partial x}\right)^2 + \frac{1}{2}\left(\frac{\partial \phi_1}{\partial z}\right)^2\right] \qquad \text{at } z = 0 \tag{4.55}$$

Equation (4.54) expresses ϕ_2 directly in terms of ϕ_1 and η_1, enabling its solution to be obtained. Similarly an expression for η_2 may be obtained from Eq.

(4.55) once ϕ_2 is known. Along these lines the complete solution to the second order problem may be obtained and the principal results of the second order solution are summarized in Table 4.3.

As an alternative approach, we may of course prefer to work with the coordinate system (x′, z) in which the origin travels with the waves. This clearly gives rise to an equivalent set of equations in which all quantities are independent of time. It is pointed out that in this case it is necessary to express both the celerity c and the Bernoulli constant Q (appearing in Eq. (4.9)) as power series in the form of Eq. (4.45). We have furthermore to take the terms of order ϵ^0 in each of these series as the lowest terms appearing. The reason for this will be appreciated by noting that the first order result for u is an order smaller than that for c. Ippen (1966) provides details of the derivation of the second order solution using this approach.

4.3.2 Definitions of Celerity

It should be made clear at this point that the value of c is somewhat arbitrary and is customarily determined by the requirement that there is no underlying current. The first order solution already obtained was oscillatory with the average horizontal particle velocity equal to zero, and the effect of a uniform current superposed on a first order wave motion will be considered in Section 4.9.3. At higher orders of approximation, however, the requirement that there be no underlying current has to be specified rather more precisely and an additional assumption concerning the wave celerity becomes necessary in order to obtain a unique solution to the problem.

A particular wave theory should provide a unique solution for the steady flow relative to the reference frame (x′, z) which moves with the waves. That is, the corresponding horizontal velocity $u' = u - c$ should be established uniquely. Here u and c are respectively the horizontal particle velocity and the wave celerity relative to the fixed reference frame (x, z) and have yet to be determined on the basis of an additional assumption. The two most usual assumptions were originally proposed by Stokes (1847). The first is that the average horizontal particle velocity u is zero: $\overline{u} = 0$, where the overbar may represent a spatial average. This directly provides a definition for c in terms of u′ as required

$$c = -\overline{u'} \tag{4.56}$$

(This equation applies equally at any elevation within the fluid since the vertical gradient of the average horizontal velocity is zero on account of the irrotationality condition.)

The second assumption is that the average horizontal momentum or impulse I is zero. That is

$$\overline{\int_{-d}^{\eta} u\, dz} = 0 \tag{4.57}$$

Putting $u = c^* + u'$ and recalling that $\overline{\eta} = 0$ by definition, we have a second definition of the wave celerity in terms of u'

$$c^* = -\frac{1}{d}\int_{-d}^{\eta} u'\, dz \tag{4.58}$$

where c^* is used to denote the use of the second assumption in contrast to the first. Note that the overbar has been dropped since the right-hand side of Eq. (4.58) is constant for any wave train, being directly related to the mass flux $-\mathbf{Q}$ in Eq. (4.41),

$$c^* = \frac{\mathbf{Q}}{\rho d} \tag{4.59}$$

It is sometimes convenient (see Section 4.5.4) to define a depth d' such that an analogous equation might apply also to c (first definition)

$$c = \frac{\mathbf{Q}}{\rho d'} \tag{4.60}$$

Thus d' may be interpreted as the depth of a uniform stream moving with velocity c and with the same mass flow rate $-\mathbf{Q}$.

The above two definitions of wave celerity may be incorporated into any chosen wave theory and may be seen to be related by

$$c^* = c - \frac{1}{d}\overline{\int_{0}^{\eta} u'\, dz} \tag{4.61}$$

It may thereby readily be confirmed that the two definitions coincide, $c^* = c$, on the basis of linear wave theory and Stokes second order wave theory, and for deep water and solitary waves of any amplitude.

4.3.3 Results of Stokes Second Order Theory

We continue now with an examination of the results of Stokes second order wave theory as given in Table 4.3. The table shows that most time varying quantities contain second order components at twice the wave frequency superposed on the fundamental components predicted by linear wave theory. This gives rise to wave crests which are steeper and troughs which are flatter than those of a sinusoidal profile, a feature which is generally observed for steeper waves. For higher amplitude waves, the theory predicts a secondary crest within the wave trough provided that

$$\frac{H}{L} \geqslant \frac{1}{3\pi}\left[\frac{\sinh^2(kd)\tanh(kd)}{2+\cosh(2kd)}\right] \tag{4.62}$$

but for such conditions the theory's applicability becomes questionable. Another feature of the second order theory is that the particle paths are no longer closed orbits and there is, in consequence, a gradual (second order) drift of fluid particles involved in the wave motion. This drift velocity, termed the mass-transport velocity, is considered in some detail in Section 4.8.

One important result of the second order theory is that the linear dispersion relation continues to hold. This considerably simplifies the application of the theory since the wave celerity and wave length remain independent of the wave height. At higher orders, however, the celerity does indeed depend on the wave height, and the corresponding theories are consequently more difficult to apply without resort to a computer. In particular, the celerity given to the third order and on the basis of a zero mean Eulerian velocity (first definition) is

$$c^2 = \frac{g}{k}\tanh(kd)\left\{1+\left(\frac{\pi H}{L}\right)^2\left[\frac{9-8\cosh^2(kd)+8\cosh^4(kd)}{8\sinh^4(kd)}\right]\right\} \tag{4.63}$$

Finally, the severe limitations on second order theory imposed by shallow water are again emphasized. It has already been indicated that the condition $H/d \ll (kd)^2$ for $kd < 1$ is applicable, and that otherwise a nonlinear shallow wave theory may be more appropriate. In order to demonstrate the range of validity of Stokes second order theory, Fig. 4.5 shows the ratio $(\epsilon^2 u_2)_{max}/(\epsilon u_1)_{max}$ of the second order to first order maximum horizontal velocity at $z = 0$ as a function of the depth parameter d/L for various values of H/d, as well as for waves of maximum steepness taken to be given by $H/L = 0.142 \tanh(kd)$. This ratio should be small $(0[0.1])$ for the theory to be valid. The curves thus demonstrate how the second order terms becomes large and the theory consequently becomes invalid for steeper waves in shallow water.

4.3.4 Stokes Fifth Order Theory

Skjelbreia and Hendrickson (1960) have presented Stokes wave theory to the fifth order and their approach has found widespread usage in engineering prac-

Table 4.3 Results of Stokes Second Order Wave Theory.

Velocity potential

$$\phi = \frac{\pi H}{kT} \frac{\cosh(ks)}{\sinh(kd)} \sin\theta + \frac{3}{8}\frac{\pi H}{kT}\left(\frac{\pi H}{L}\right)\frac{\cosh(2ks)}{\sinh^4(kd)} \sin 2\theta$$

Dispersion relation

$$c^2 = \frac{\omega^2}{k^2} = \frac{g}{k}\tanh(kd)$$

Surface elevation

$$\eta = \frac{H}{2}\cos\theta + \frac{H}{8}\left(\frac{\pi H}{L}\right)\frac{\cosh(kd)}{\sinh^3(kd)}[2 + \cosh(2kd)]\cos 2\theta$$

Horizontal particle displacement

$$\xi = -\frac{H}{2}\frac{\cosh(ks)}{\sinh(kd)}\sin\theta + \frac{H}{8}\left(\frac{\pi H}{L}\right)\frac{1}{\sinh^2(kd)}\left[1 - \frac{3\cosh(2ks)}{2\sinh^2(kd)}\right]\sin 2\theta + \frac{H}{4}\left(\frac{\pi H}{L}\right)\frac{\cosh(2ks)}{\sinh^2(kd)}(\omega t)$$

Vertical particle displacement

$$\zeta = \frac{H}{2}\frac{\sinh(ks)}{\sinh(kd)}\cos\theta + \frac{3H}{16}\left(\frac{\pi H}{L}\right)\frac{\sinh(2ks)}{\sinh^4(kd)}\cos 2\theta$$

Horizontal particle velocity

$$u = \frac{\pi H}{T}\frac{\cosh(ks)}{\sinh(kd)}\cos\theta + \frac{3}{4}\frac{\pi H}{T}\left(\frac{\pi H}{L}\right)\frac{\cosh(2ks)}{\sinh^4(kd)}\cos 2\theta$$

Vertical particle velocity

$$w = \frac{\pi H}{T}\frac{\sinh(ks)}{\sinh(kd)}\sin\theta + \frac{3}{4}\frac{\pi H}{T}\left(\frac{\pi H}{L}\right)\frac{\sinh(2ks)}{\sinh^4(kd)}\sin 2\theta$$

Horizontal particle acceleration

$$\frac{\partial u}{\partial t} = \frac{2\pi^2 H}{T^2}\frac{\cosh(ks)}{\sinh(kd)}\sin\theta + \frac{3\pi^2 H}{T^2}\left(\frac{\pi H}{L}\right)\frac{\cosh(2ks)}{\sinh^4(kd)}\sin 2\theta$$

Table 4.3 *(Continued)*.

Vertical particle acceleration	$\frac{\partial w}{\partial t} = -\frac{2\pi^2 H}{T^2} \frac{\sinh(ks)}{\sinh(kd)} \cos\theta - \frac{3\pi^2 H}{T^2}\left(\frac{\pi H}{L}\right)\frac{\sinh(2ks)}{\sinh^4(kd)} \cos 2\theta$
Pressure	$p = -\rho g z + \frac{1}{2}\rho g H \frac{\cosh(ks)}{\cosh(kd)} \cos\theta + \frac{3}{4}\rho g H\left(\frac{\pi H}{L}\right)\frac{1}{\sinh(2kd)}\left[\frac{\cosh(2ks)}{\sinh^2(kd)} - \frac{1}{3}\right]\cos 2\theta - \frac{1}{4}\rho g H\left(\frac{\pi H}{L}\right)\frac{1}{\sinh(2kd)}[\cosh(2ks) - 1]$
Average energy density	$E = \frac{1}{8}\rho g H^2 \; (+0[\epsilon^4])$
Energy flux	$P = \frac{1}{16}\rho g H^2 c\left(1 + \frac{2kd}{\sinh(2kd)}\right) \; (+0[\epsilon^4])]$

tice. At this level of approximation the velocity potential may be expressed as the sum of five terms,

$$\frac{k\phi}{c} = \sum_{n=1}^{5} \phi_n' \cosh(nks) \sin(n\theta) \tag{4.64}$$

and the free surface elevation is expressed as

$$k\eta = \sum_{n=1}^{5} \eta_n' \cos(n\theta). \tag{4.65}$$

The coefficients ϕ_n' and η_n' are defined in Table 4.4 in terms of a parameter λ and a series of coefficients A and B. The wave celerity c is given as

$$c^2 = C_0^2(1 + \lambda^2 C_1 + \lambda^4 C_2) \tag{4.66}$$

where C_0 is the celerity given by linear wave theory for the same depth and wave number: $C_0^2 = (g/k)\tanh(kd)$. The various coefficients A, B and C are functions of kd only and are obtained by substituting the expressions for ϕ and η, Eqs. (4.64) and (4.65) respectively, into the governing equations of the problem. Explicit expressions for all these coefficients are given by Skjelbreia and Hendrickson.

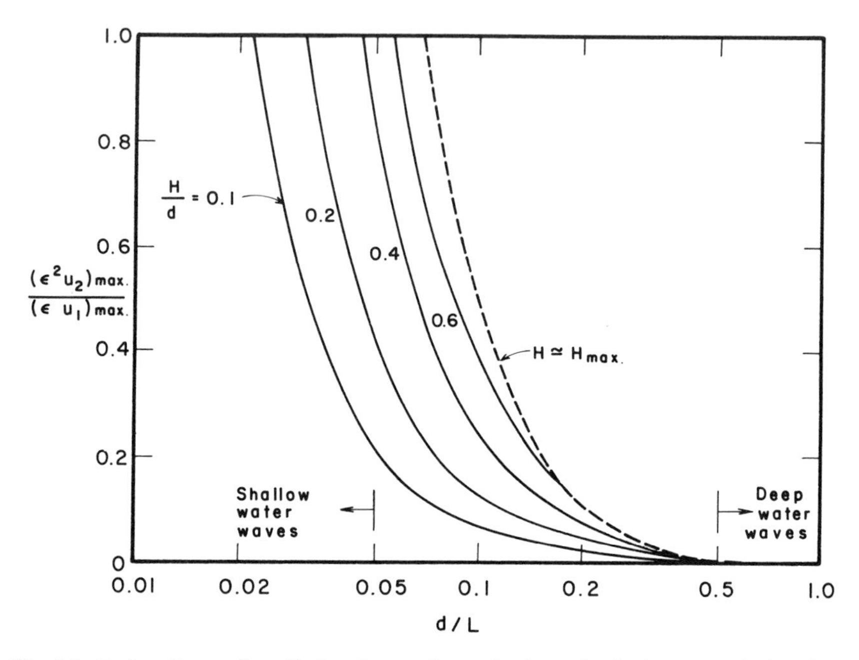

Fig. 4.5. Ratio of second to first order maximum horizontal velocity at $z = 0$, based on Stokes second-order wave theory and shown as a function of d/L for H/d = 0.1, 0.2, 0.4, 0.6 and for $H \simeq H_{max}$ ($H/L = 0.142 \tanh(kd)$).

For a given design wave, λ and k (or in dimensionless form kd) are to be determined before the preceding results can be applied. These two quantities may be obtained by virtue of the following pair of equations.

$$\frac{1}{kd}\left[\lambda + B_{33}\lambda^3 + (B_{35} + B_{55})\lambda^5\right] = \frac{H}{2d} \tag{4.67}$$

$$kd \tanh(kd)\left[1 + C_1\lambda^2 + C_2\lambda^4\right] = 4\pi^2 \frac{d}{gT^2} \tag{4.68}$$

The values of λ and kd corresponding to given values of H/d and d/gT^2 may be determined by an iterative procedure, bearing in mind that the coefficients A, B and C are known functions of kd. The solution will then be complete, and with ϕ_n' and η_n' now known the remaining variables of interest may readily be evaluated. In particular, the horizontal and vertical components of fluid velocities and accelerations, as well as the fluid pressure, are generally required and the expressions for these, together with those for ϕ, η and c already given, are listed in Table 4.4.

It is seen, then, that the development of a computer program based on Stokes

Table 4.4 Results of Stokes Fifth Order Wave Theory.

Velocity potential, ϕ	$\frac{k\phi}{c} = \sum_{n=1}^{5} \phi'_n \cosh(nks) \sin(n\theta)$
Wave celerity, c	$\frac{c^2}{gd} = \frac{\tanh(kd)}{kd} [1 + \lambda^2 C_1 + \lambda^4 C_2]$
Surface elevation, η	$k\eta = \sum_{n=1}^{5} \eta'_n \cos(n\theta)$
Horizontal particle velocity, u	$\frac{u}{c} = \sum_{n=1}^{5} n\,\phi'_n \cosh(nks) \cos(n\theta)$
Vertical particle velocity, w	$\frac{w}{c} = \sum_{n=1}^{5} n\,\phi'_n \sinh(nks) \sin(n\theta)$
Horizontal particle acceleration, $\partial u/\partial t$	$\frac{\partial u/\partial t}{\omega c} = \sum_{n=1}^{5} n^2\,\phi'_n \cosh(nks) \sin(n\theta)$
Vertical particle acceleration, $\partial w/\partial t$	$\frac{\partial w/\partial t}{\omega c} = -\sum_{n=1}^{5} n^2\,\phi'_n \sinh(nks) \cos(n\theta)$
Temporal derivative of ϕ	$\frac{\partial \phi/\partial t}{c^2} = -\sum_{n=1}^{5} n\,\phi'_n \cosh(nks) \cos(n\theta)$
Pressure, p	$\frac{p}{\rho g d} = 1 - \frac{s}{d} - \frac{c^2}{gd}\left\{\frac{\partial \phi/\partial t}{c^2} + \frac{1}{2}\left[\left(\frac{u}{c}\right)^2 + \left(\frac{w}{c}\right)^2\right]\right\}$

where

$\phi'_1 = \lambda A_{11} + \lambda^3 A_{13} + \lambda^5 A_{15}$, $\phi'_2 = \lambda^2 A_{22} + \lambda^4 A_{24}$,
$\phi'_3 = \lambda^3 A_{33} + \lambda^5 A_{35}$, $\phi'_4 = \lambda^4 A_{44}$, $\phi'_5 = \lambda^5 A_{55}$,
$\eta'_1 = \lambda$, $\eta'_2 = \lambda^2 B_{22} + \lambda^4 B_{24}$, $\eta'_3 = \lambda^3 B_{33} + \lambda^5 B_{35}$,
$\eta'_4 = \lambda^4 B_{44}$, $\eta'_5 = \lambda^5 B_{55}$.

The coefficients A, B, C are known functions of kd only, given by Skjelbreia and Hendrickson (1960). The parameters λ and kd are obtained from Eqs. (4.67) and (4.68).

fifth order wave theory is not unduly difficult: for a given design wave, specified in terms of H, T and d, the values of λ and kd may be calculated numerically on the basis of Eqs. (4.67) and (4.68). Hence the values of ϕ_n' and η_n' can be determined, and consequently the velocity and acceleration components and the hydrodynamic pressure at any desired location and instant may be determined on the basis of the expressions in Table 4.4. Note that it is convenient to use the expression given for pressure directly, rather than one obtained by substituting for u and w and then omitting terms above the fifth order: either alternative is given to the same degree of accuracy.

The above approach is based on the first definition of wave celerity. Tsuchiya and Yamaguchi (1972) have emphasized the differences between the two definitions of celerity and have derived results corresponding to those of Skjelbreia and Hendrickson, but taken to the fourth order and based now on the second definition of celerity.

4.4 NONLINEAR SHALLOW WAVE THEORIES

4.4.1 Formulation of Shallow Wave Theories

The low order Stokes finite amplitude wave theories just described are generally inadequate in the shallow water range since many coeffficients of the higher order terms then "blow up": that is they become excessive relative to the lowest order terms. Laitone (1962) has investigated the range of validity of Stokes third order wave theory on a theoretical basis, and suggests that it is most suitable for wave lengths less than about 8 times the depth ($kd > 0.78$). For longer waves a different procedure is appropriate if the effects of finite wave height are to be investigated, and to this end nonlinear periodic wave theories suitable for shallow water have been developed since the last century.

The fundamental theory, termed the cnoidal wave theory, was first developed on an intuitive basis by Korteweg and de Vries (1895). According to this theory the wave characteristics are expressed in terms of the Jacobian elliptic function cn and hence the terminology "cnoidal wave theory" is used. A tpyical cnoidal wave profile is sketched in Fig. 4.6a. One limiting case of this, in which the wave length becomes infinite, corresponds to the solitary wave, whose profile is sketched in Fig. 4.6b. Another limiting case corresponds to shallow water sinusoidal wave theory, with a wave profile as sketched in Fig. 4.6c. Other presentations of a first approximation to this theory have been made by Keulegan and Patterson (1940), Keller (1948) and Laitone (1961). Wiegel (1960, 1964) has given a summary of the first approximation that is directed towards engineering applications. Laitone (1961) and Chappelear (1962) have developed respectively second and third approximations to cnoidal wave theory, the latter involving a numerical procedure rather than explicit formulae. More recently, Fenton (1979) has presented a cnoidal wave theory which is capable of extension to any

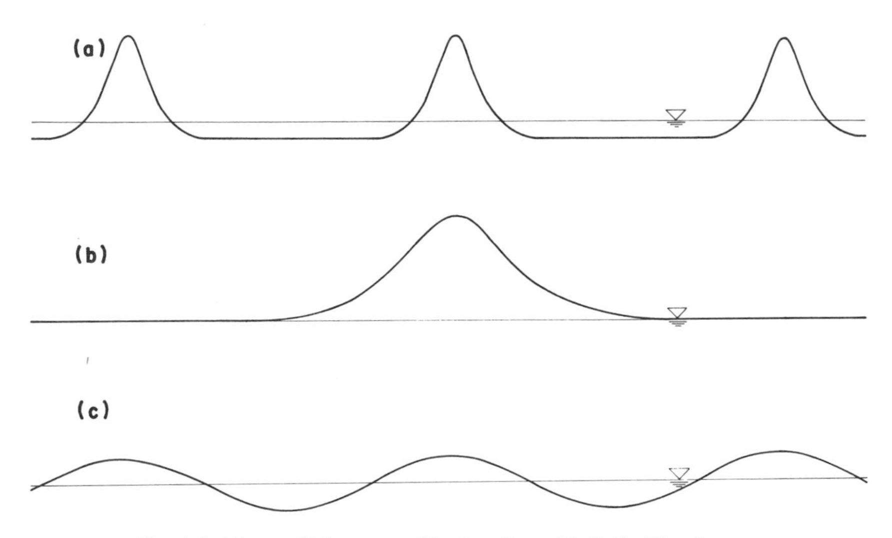

Fig. 4.6. The cnoidal wave profile, together with its limiting forms.

desired order, and which is readily suited to engineering application. Details of the theory to the fifth order have been given by him.

A formal procedure for systematically developing shallow wave theories was introduced by Friedrichs (1948) and has also been outlined by Stoker (1957). In this shallow wave expansion procedure, the variables associated with vertical and horizontal length scales are considered to be given by different orders of magnitude and consequently the vertical ordinate must be "stretched" as a preliminary to the analysis. This procedure is carried out to ensure that power series representations of the relevant variables do in fact contain coefficients of like order. We choose a reference frame (x', s) with origin on the seabed and moving beneath a wave crest in phase with the wave (that is $s = z + d$, and $x' = x - ct$ as before). Suppose now that ℓ is a typical length in the x direction and this is of the same order as the wave length. Then if d is the still water depth as before, the distortion to be applied to obtain a set of dimensionless variables will be as follows:

$$\left.\begin{aligned} &X = \frac{x'}{\ell}, \qquad S = \frac{s}{d}, \qquad U = \frac{u}{\sqrt{gd}}, \\ &W = \frac{w}{\sqrt{gd}}\frac{d}{\ell}, \qquad P = \frac{\Delta p}{\rho g d}, \qquad N = \frac{\eta}{d}. \\ &\left[\text{And } T = \frac{\sqrt{gd}}{\ell}t, \qquad \Phi = \frac{\phi}{\ell\sqrt{gd}}.\right] \end{aligned}\right\} \qquad (4.69)$$

The dimensionless variables obtained in this way are then substituted into the governing equations to yield a set of equations involving the parameter σ which is assumed small

$$\sigma = \left(\frac{d}{\ell}\right)^2 \ll 1 \tag{4.70}$$

Following the approach used by Laitone, the equations of motion expressed in terms of U, W and P become after transformation

$$\frac{\partial W}{\partial X} - \frac{\partial U}{\partial S} = 0 \tag{4.71a}$$

$$\sigma \frac{\partial U}{\partial X} + \frac{\partial W}{\partial S} = 0 \tag{4.71b}$$

$$\sigma\left(U \frac{\partial U}{\partial X} + \frac{\partial P}{\partial X}\right) + W \frac{\partial U}{\partial S} = 0 \tag{4.71c}$$

$$\sigma\left(U \frac{\partial W}{\partial X} + \frac{\partial P}{\partial S} + 1\right) + W \frac{\partial W}{\partial S} = 0 \tag{4.71d}$$

Equations (4.71) correspond in turn to the irrotationality condition, the continuity equation and the momentum equations in the x' and s directions. They are subject to the following boundary conditions

$$P = 0 \qquad \text{at } S = 1 + N \tag{4.72}$$

$$W = \sigma U \frac{\partial N}{\partial X} \qquad \text{at } S = 1 + N \tag{4.73}$$

$$W = 0 \qquad \text{at } S = 0 \tag{4.74}$$

At this stage a perturbation procedure is introduced by expressing the flow variables as power series in σ, in which the coefficient of each power of σ is assumed to be of like order. Substituting these series into the equations defining the problem and equating terms of like powers of σ as described previously for Stokes waves, successive approximations to the solution may be developed.

It is noted in particular that in the expansion of N, the lowest order term N_0 is a constant, and consequently the relative wave height H/d, which is associated with the next term N_1, is therefore of order σ. That is, the shallow wave expansion procedure is in effect based on the assumption

$$\frac{H}{d} = 0\,[(kd)^2] \ll 1 \tag{4.75}$$

This may be contrasted to the conditions assumed for Stokes waves (Section 4.3).

The above procedure is wholly equivalent to the more usual approach involving the velocity potential ϕ or stream function ψ. The Laplace equation for ϕ, expressed in terms of the dimensionless variables used here, becomes

$$\sigma \frac{\partial^2 \Phi}{\partial X^2} + \frac{\partial^2 \Phi}{\partial S^2} = 0 \tag{4.76}$$

A general power series solution to Φ which satisfies this equation as well as the bottom boundary condition, Eq. (4.74), is

$$\begin{aligned} \Phi &= \cos\,(\sqrt{\sigma}\;SD) f(X) \\ &= \left(1 - \sigma \frac{S^2 D^2}{2!} + \sigma^2 \frac{S^4 D^4}{4!} - \cdots\right) f(X) \end{aligned} \tag{4.77}$$

where D is the differential operator d/dX, and f(X) is a function of X only which must be determined by the free-surface conditions. This kind of approach has been used by Fenton (1979) and illustrates how depth variations are expressed as power series in S.

The shallow wave equations to the lowest order are identically those of linearized long wave theory and are obtained by a simplified argument in Section 4.5.1. The next approximation is given as a solution to the ordinary differential equation

$$\frac{\partial^3 F}{\partial X^3} + \alpha F \frac{\partial F}{\partial X} - \beta \frac{\partial F}{\partial X} = 0 \tag{4.78}$$

where α and β are constants and F is a function only of X. This equation has a periodic solution in terms of the Jacobian elliptic function $\mathrm{cn}(q, \kappa)$, where q is the argument and κ the modulus of the function.

More generally, for an arbitrary (unsteady) wave motion corresponding to this procedure, the governing equation to this order may be written in terms of the free surface elevation as (Hammack and Segur 1974)

$$\frac{1}{\sqrt{gd}} \frac{\partial \eta}{\partial t} + \frac{\partial \eta}{\partial x} + \frac{3}{2} \eta \frac{\partial \eta}{\partial x} + \frac{d^2}{6} \frac{\partial^3 \eta}{\partial x^3} = 0 \tag{4.79}$$

By a suitable choice of variables this can be simplified to

$$\frac{\partial f}{\partial \tau} + 6f\frac{\partial f}{\partial \chi} + \frac{\partial^3 f}{\partial \chi^3} = 0 \tag{4.80}$$

where $\chi = (x - \sqrt{gd}\,t)/d$, $\tau = \sqrt{g/d}\,t/6$ and $f(\chi, \tau) = 3\eta/2d$. The above equations, of which Eq. (4.78) is a special case, are alternative forms of the *Korteweg-de Vries equation.* This is a governing equation for shallow wave motions which include both nonlinearity and dispersion effects. These two effects are associated respectively with the terms $f\,\partial f/\partial\chi$ and $\partial^3 f/\partial\chi^3$ in Eq. (4.80). Simplifications to the equation have been made by neglecting either or both of these effects under different circumstances (Hammack and Segur 1978b). Mention is made here of the Ursell number $U = HL^2/d^3$ which has already been introduced and which is useful in assessing the relative importance of nonlinear shallow wave theories. This takes large values for high waves in shallow water, and the application of the complete Korteweg-de Vries equation is appropriate when the Ursell number is of order unity or is moderately large.

4.4.2 Results of Cnoidal Wave Theory

By following the approach already described, Laitone (1961) developed a second approximation to the cnoidal wave theory, and Chappelear (1962) obtained a third approximation which involves a numerical procedure suitable for the application of a computer.

Laitone's results were presented in terms of the trough depth h and the reference frame (x', s) moving with the wave train. If this theory is to be employed for engineering application, it is convenient first to express the results directly in terms of the still water depth d and the fixed reference frame (x, s), with time t taken as zero as a wave crest passes $x = 0$. The fundamental equations describing the solution and incorporating this transformation are presented in Table 4.5. Here the first approximation is obtained merely by omitting the higher order term in each of the expressions given. In this table and in the following discussion the symbol ϵ is used to denote H/d. It should be pointed out that Le Méhauté (1968) detected an error in Laitone's original expression for $(d - h)/h$ and consequently the expressions given in Table 4.5 differ from those that Laitone (1965) subsequently provided, although they do agree with those given later by Yamaguchi and Tsuchiya (1974).

The Jacobian elliptic function modulus κ forms the fundamental parameter in terms of which the solution is expressed, and this ranges from zero to unity, although it is usually values approaching unity that are of greater interest in the present context. It is emphasized that published tables (e.g. Abromowitz and Stegun 1965) often refer to κ^2 (usually denoted by m) rather than to κ itself.

Functions of κ which are encountered here are as follows: K is the complete elliptic integral of the first kind, E is the complete elliptic integral of the second kind, γ is the ratio E/K and $\kappa'^2 = 1 - \kappa^2$. The variations of K, γ and $U_1 = 16\kappa^2 K^2/3$ with κ are shown in Fig. 4.7. The variations with distance x and time t of the different variables of interest are realized through the Jacobian elliptic function argument $q = K(kx - \omega t)/\pi$. It may be noticed that the expression for d/gT^2 is given to a higher order in ϵ than appears consistent. The purpose of this is simply to ensure numerical equality in the relation $c = L/T$. As previously mentioned, the wave celerity is not uniquely determined but depends upon an additional assumption. The celerity, horizontal particle velocity and wave period are affected by the assumption that is made, and the expressions listed in the table without an asterisk are based on Stokes' first definition (zero mean horizontal velocity), whereas the quantities with asterisks are based on the second definition (zero mean horizontal momentum).

The fundamental characteristics of cnoidal waves may be found by examining the first approximation contained in Table 4.5, where it is seen that there is no variation of horizontal velocity amplitude with depth, the vertical velocity varies linearly with depth and the pressure is hydrostatic.

Figure 4.8 shows the behavior of (a) $cn^2 q$ and (b) cn q dn q sn q as functions of phase angle θ for various values of κ. These indicate, again to the first approximation, temporal variations of (a) surface elevation, horizontal particle velocity and pressure, and (b) vertical particle velocity and horizontal particle acceleration. These variables are given to the first approximation by the expressions

$$\frac{\eta}{d} = \frac{u}{\sqrt{gd}} = \frac{p}{\rho g d} - \left(1 - \frac{s}{d}\right) = \epsilon\left(cn^2\, q - \frac{\gamma - \kappa'^2}{\kappa^2}\right) \tag{4.81}$$

$$\frac{w}{\sqrt{gd}} = \frac{1}{g}\frac{\partial u}{\partial t} = \frac{\epsilon\sqrt{3\epsilon}}{\kappa K}\left(\frac{s}{d}\right) \text{cn q dn q sn q} \tag{4.82}$$

To a first approximation the Ursell number is given as

$$U = U_1 = 16\kappa^2 K^2/3 \tag{4.83}$$

so that κ is uniquely related to U_1 as shown in Fig. 4.7, and may be estimated from it. To higher approximations, κ depends on H/d as well and would thus be estimated by an iterative procedure. More usually, though, it is the wave period rather than the wave length that is specified for a design wave, and κ would be estimated instead from the corresponding expression for d/gT^2 or d/gT^{*2} (depending on one's choice of celerity definition). Thus, on the basis of the second

Table 4.5 Results of Cnoidal Wave Theory—Second Approximation.

Trough depth, h	$\frac{h}{d} = 1 - \epsilon h_1 - \epsilon^2 h_2 + 0[\epsilon^3]$
Surface elevation, η	$\frac{\eta}{d} = \epsilon(\mathrm{cn}^2 q - h_1) - \epsilon^2\left[\frac{3}{4}\mathrm{cn}^2 q\,(1 - \mathrm{cn}^2 q) + h_2\right] + 0[\epsilon^3]$
Wave celerity, c	$\frac{c}{\sqrt{gd}} = 1 + \epsilon c_1 + \epsilon^2 c_2 + 0[\epsilon^3]$
	$\frac{c^*}{\sqrt{gd}} = 1 + \epsilon c_1 + \epsilon^2 c_2^* + 0[\epsilon^3]$
Wave length, L	$\frac{L}{d} = \frac{4\kappa K}{\sqrt{3\epsilon}}\{1 - \epsilon \ell_1 + 0[\epsilon^2]\}$
Wave period, T	$\frac{d}{gT^2} = \frac{3\epsilon}{16\kappa^2 K^2}\left\{\left(\frac{1 + \epsilon c_1 + \epsilon^2 c_2}{1 - \epsilon \ell_1}\right)^2 + 0[\epsilon^3]\right\}$
	$\frac{d}{gT^{*2}} = \frac{3\epsilon}{16\kappa^2 K^2}\left\{\left(\frac{1 + \epsilon c_1 + \epsilon^2 c_2^*}{1 - \epsilon \ell_1}\right)^2 + 0[\epsilon^3]\right\}$
Horizontal particle velocity, u	$\frac{u}{\sqrt{gd}} = \epsilon(\mathrm{cn}^2 q - h_1) + \epsilon^2\left\{(f_1 + f_2\,\mathrm{cn}^2 q - \mathrm{cn}^4 q) - \frac{3}{4\kappa^2}\left(\frac{s}{d}\right)^2[\kappa'^2 + 2(2\kappa^2 - 1)\,\mathrm{cn}^2 q - 3\kappa^2\,\mathrm{cn}^4 q]\right\} + 0[\epsilon^3]$
	$\frac{u^*}{\sqrt{gd}} = \epsilon(\mathrm{cn}^2 q - h_1) + \epsilon^2\left\{(f_1^* + f_2\,\mathrm{cn}^2 q - \mathrm{cn}^4 q) - \frac{3}{4\kappa^2}\left(\frac{s}{d}\right)^2[\kappa'^2 + 2(2\kappa^2 - 1)\,\mathrm{cn}^2 q - 3\kappa^2\,\mathrm{cn}^4 q]\right\} + 0[\epsilon^3]$
Vertical particle velocity, w	$\frac{w}{\sqrt{gd}} = \frac{\epsilon\sqrt{3\epsilon}}{\kappa}\left(\frac{s}{d}\right)\mathrm{cn}\,q\,\mathrm{dn}\,q\,\mathrm{sn}\,q\left\{1 + \epsilon\left[f_3 - 2\,\mathrm{cn}^2 q - \left(\frac{s}{d}\right)^2\left(\frac{2\kappa^2 - 1}{2\kappa^2} - \frac{3}{2}\mathrm{cn}^2 q\right)\right] + 0[\epsilon^2]\right\}$
Horizonal particle acceleration, $\partial u/\partial t$	$\frac{1}{g}\frac{\partial u}{\partial t} = \frac{\epsilon\sqrt{3\epsilon}}{\kappa}\mathrm{cn}\,q\,\mathrm{dn}\,q\,\mathrm{sn}\,q\left\{1 + \epsilon\left[f_4 - 2\,\mathrm{cn}^2 q - \left(\frac{s}{d}\right)^2\left(\frac{3(2\kappa^2 - 1)}{2\kappa^2} + \frac{9}{2}\mathrm{cn}^2 q\right)\right] + 0[\epsilon^2]\right\}$

Table 4.5 *(Continued).*

Vertical particle acceleration, $\partial w/\partial t$	$\frac{1}{g}\frac{\partial w}{\partial t} = \frac{3\epsilon^2}{2\kappa^2}\left(\frac{s}{d}\right)(\kappa'^2 + 2(2\kappa^2 - 1)\,\mathrm{cn}^2 q - 3\kappa^2\,\mathrm{cn}^4 q) + 0[\epsilon^3]$
Pressure, p	$\frac{p}{\rho g d} = \frac{\eta}{d} + 1 - \frac{s}{d} - \epsilon^2\left(\frac{3}{4\kappa^2}\right)\left[\left(\frac{s}{d}\right)^2 - 1\right]$ $[\kappa'^2 + 2(2\kappa^2 - 1)\,\mathrm{cn}^2 q - 3\kappa^2\,\mathrm{cn}^4 q] + 0[\epsilon^3]$
Average energy density, E	$\frac{E}{\rho g d^2} = \frac{1}{2}\epsilon^2\{e_1 + \epsilon e_2 + 0[\epsilon^2]\}$
Energy flux, P	$\frac{P}{\rho g d^2\sqrt{gd}} = \epsilon^2\{p_1 + \epsilon p_2 + 0[\epsilon^2]\}$

where

$$q = \frac{K\theta}{\pi} = \frac{K}{\pi}(kx - \omega t)$$
$$h_1 = \{\gamma - \kappa'^2\}/\kappa^2$$
$$h_2 = \{\gamma(\kappa^2 - 2) + 2\kappa'^2\}/4\kappa^4$$
$$c_1 = \{2 - \kappa^2 - 3\gamma\}/2\kappa^2$$
$$c_2 = \{-5\gamma(15\gamma + 19\kappa^2 - 38) - 18\kappa^4 - 88\kappa'^2\}/120\kappa^4$$
$$c_2^* = \{5\gamma(3\gamma - \kappa^2 + 2) - 6\kappa^4 - 16\kappa'^2\}/40\kappa^4$$
$$\ell_1 = \{12\gamma + 5\kappa^2 - 10\}/8\kappa^2$$
$$f_1 = \{-\gamma(6\gamma + 11\kappa^2 - 16) + \kappa'^2(9\kappa^2 - 10)\}/12\kappa^4$$
$$f_1^* = \{\gamma(2\gamma - \kappa^2) + \kappa'^2(3\kappa^2 - 2)\}/4\kappa^4$$
$$f_2 = \{2\gamma + 7\kappa^2 - 6\}/4\kappa^2$$
$$f_3 = \{16\gamma + 19\kappa^2 - 22\}/8\kappa^2$$
$$f_4 = \{4\gamma + 15\kappa^2 - 14\}/8\kappa^2$$
$$e_1 = 2\{-\gamma(3\gamma + 2\kappa^2 - 4) - \kappa'^2\}/3\kappa^4$$
$$e_2 = \{\gamma(15\gamma^2 - 2\kappa^4 - 17\kappa'^2) - \kappa'^2(\kappa^2 - 2)\}/15\kappa^6$$
$$p_1 = \{-\gamma(3\gamma + 2\kappa^2 - 4) - \kappa'^2\}/3\kappa^4$$
$$p_2 = \{\gamma(75\gamma^2 + 60\gamma(\kappa^2 - 2) + 8\kappa^4 + 53\kappa'^2) + 4\kappa'^2(\kappa^2 - 2)\}/30\kappa^6$$

approximation, and adopting the first definition of celerity, Fig. 4.9 shows the variation of $1 - \kappa^2$ with HgT^2/d^2 for different values of H/d, and may be used to estimate κ for prescribed values of H, T and d. (The parameter HgT^2/d^2 is used in preference to d/gT^2 since the influence of H/d is of second order.) Once κ is known, the variables of interest may be determined explicitly by resort to Table 4.5.

As an example of how the second approximation to cnoidal wave theory

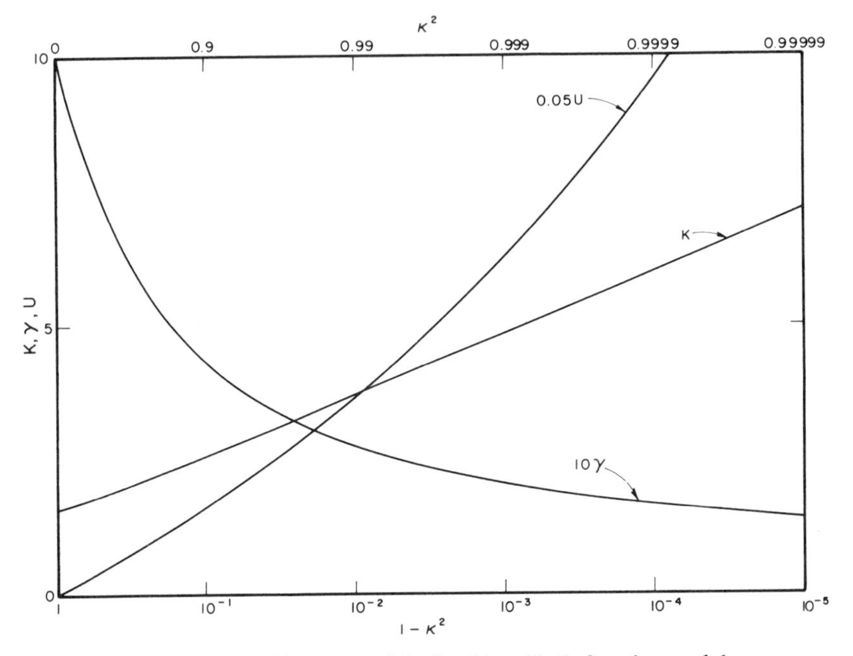

Fig. 4.7. Variations of functions of the Jacobian elliptic function modulus κ.

might be applied, we consider a typical calculation for a design wave specified as $H = 8$ ft, $T = 15$ sec, $d = 25$ ft. This corresponds to $H/d = 0.32$, $d/gT^2 = 0.00345$, $HgT^2/d^2 = 92.7$. Assuming Stokes' first definition of wave celerity we may use an iterative procedure based on the formula for d/gT^2 given in Table 4.5. This gives $\kappa = 0.9974$. (Fig. 4.9 provides an approximation to this result.) Now that κ is known, any variable of interest may be determined. Thus the maximum forward particle velocity at the seabed is given from Table 4.5 by the formula

$$\frac{u_{max\ (s=0)}}{\sqrt{gd}} = \epsilon(1 - h_1) + \epsilon^2(f_1 + f_2 - 1) \tag{4.84}$$

where the functions h_1, f_1 and f_2 of κ are defined in the table. Consequently, for the case being considered we readily obtain that $u_{max\ (s=0)} = 5.22$ ft/s.

The third approximation developed by Chappelear (1962) involves solving three simultaneous equations numerically to obtain three parameters which may essentially be taken as the modulus κ and two other parameters $L_0(=L_1 - L_3$ in Chappelear's notation) and L_3. Once these are known all the variables of interest may be determined directly. By expanding L_0 and L_3 as power series in H/d, Yamaguchi and Tsuchiya (1974) have confirmed that terms of Chappelear's

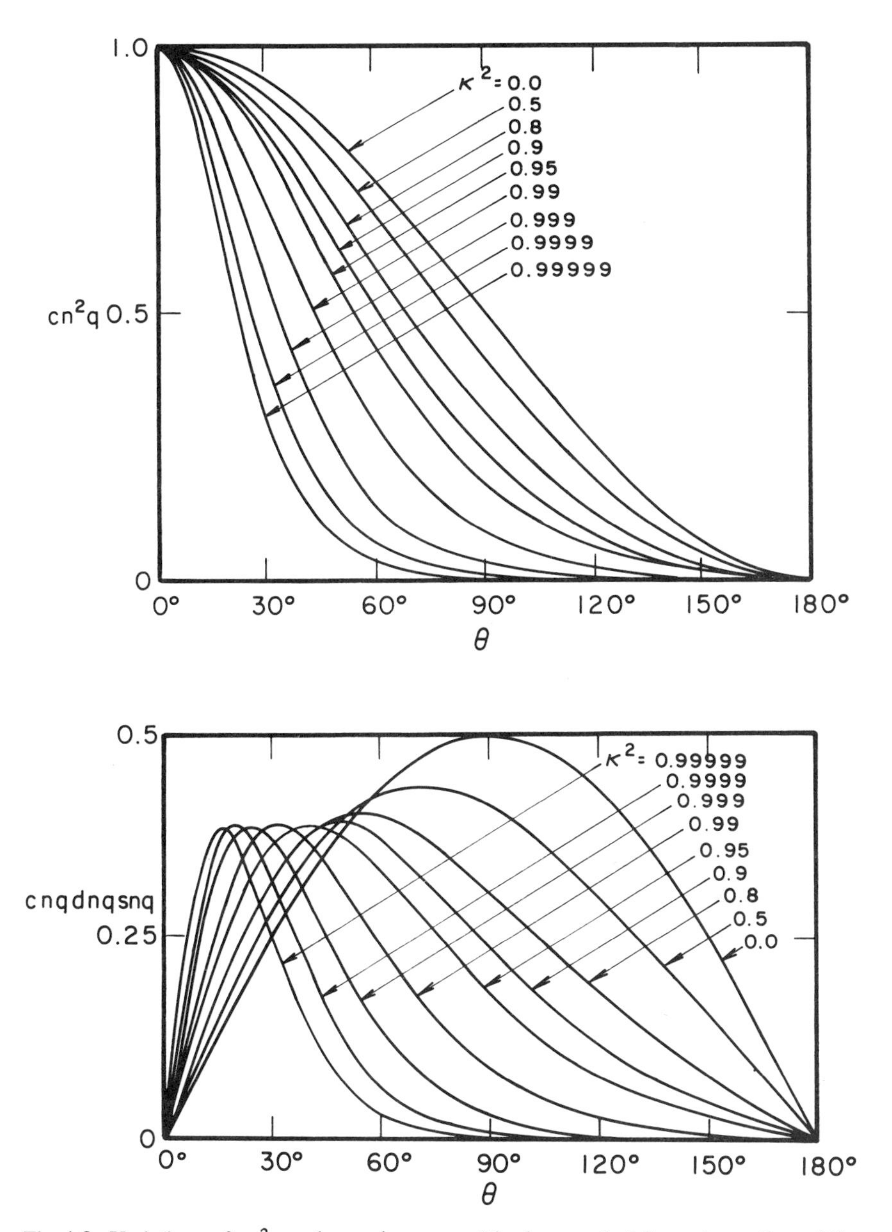

Fig. 4.8. Variations of cn^2q and cn q dn q sn q with phase angle θ for various values of the modulus κ.

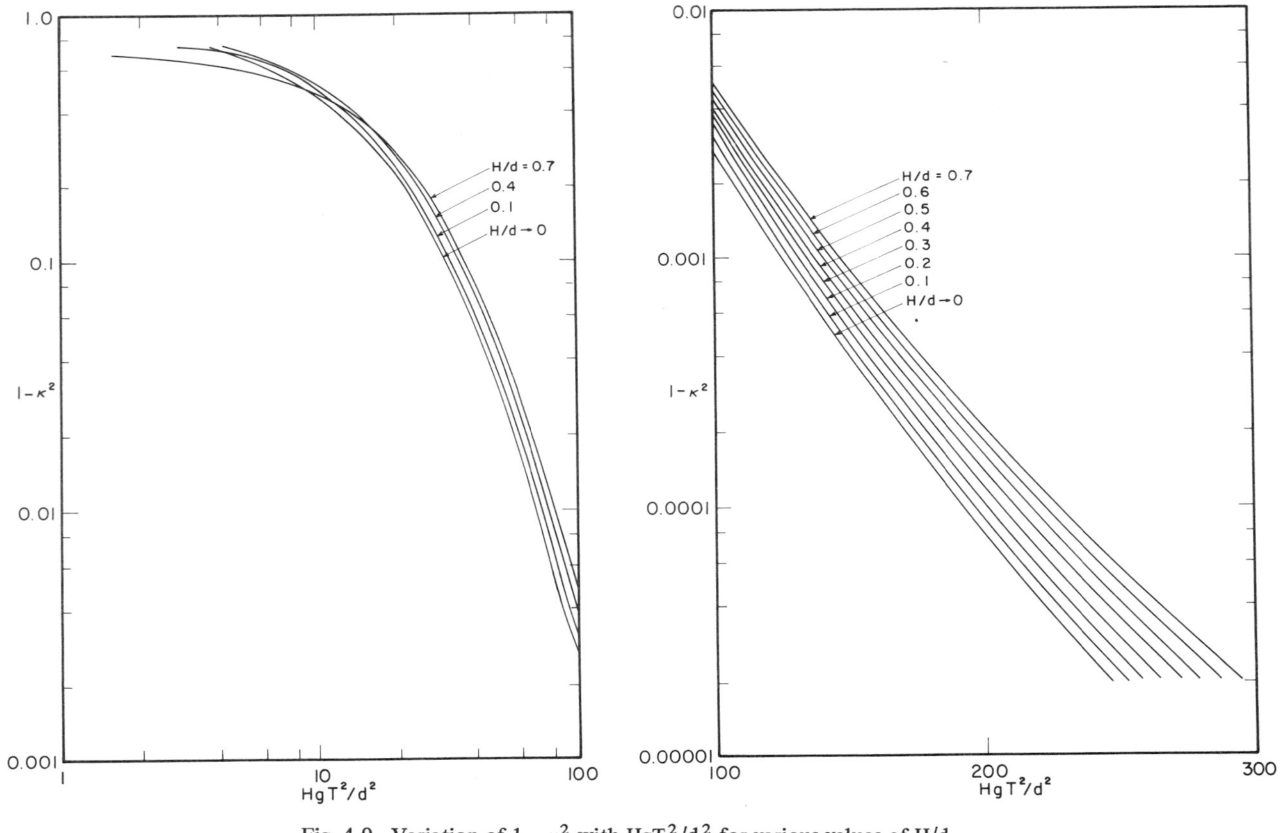

Fig. 4.9. Variation of $1 - \kappa^2$ with HgT^2/d^2 for various values of H/d.

solution up to the second approximation correspond exactly to those of Laitone's solution, when the same definition of wave celerity is used.

Fenton (1979) has recently presented a cnoidal wave theory capable of extension to any desired order. He outlines how κ^2 and H/h, where h is the trough depth, can be evaluated for given design wave parameters H, T and d, or H, L and d. An equation for either wave length or period analogous to those in Table 4.5 and involving H/d and κ^2 must first be solved iteratively for κ^2. The ratio h/d can then be obtained in terms of H/d and κ^2. The alternative height parameter $\epsilon' = H/h$ is then known, and is used in preference to H/d since fewer numerical coefficients are needed to describe the detailed solution. The usual quantities of interest, nondimensionalized in terms of h, g and ρ, are given in terms of κ^2, ϵ' and numerical coefficients which Fenton has tabulated. For example, the wave speed c and the free surface elevation above the seabed η' are expressed as the series

$$\frac{c}{\sqrt{gh}} = 1 + \sum_{i=1}^{5} \left(\frac{\epsilon'}{m}\right)^i \sum_{j=0}^{i} (A_{ij} + \gamma B_{ij}) m^j \tag{4.85}$$

$$\frac{\eta'}{h} = 1 + \sum_{i=1}^{5} \left(\frac{\epsilon'}{m}\right)^i \sum_{j=1}^{i} m^j \sum_{k=1}^{j} C_{ijk}\, \mathrm{cn}^{2k}(q) \tag{4.86}$$

where $m = \kappa^2$, q is the argument of the elliptic functions as before, and A, B and C are numerical coefficients which are provided. Analogous formulae are given for all the usual quantities of interest, including both definitions of the wave celerity, particle velocities and accelerations, pressure and also integral properties such as the average energy density and the energy flux. Fenton's solution is relatively simple to use since all quantities of interest may be obtained explicitly once the derived parameters ϵ' and m have been obtained by iteration.

Two extreme cases of cnoidal wave theory warrant particular attention. Firstly, when κ is much less than unity the theory is not suitable unless the wave height is extremely small in which case the much simpler linear wave theory could instead be employed. In particular as $\kappa \to 0$ and $\epsilon \to 0$ (while ϵ/κ^2 remains finite) we eventually duplicate the results of the shallow water range of sinusoidal wave theory. This limit is obtained quite simply by noting that

$$\left.\begin{aligned} K &\to \pi/2 \\ \gamma &\to 1 \\ \kappa'^2 &\to 1 \\ \mathrm{cn}\, q &\to \cos q \\ \mathrm{dn}\, q &\to 1 \\ \mathrm{sn}\, q &\to \sin q \\ q &\to \theta/2 \end{aligned}\right\} \text{ as } \kappa \to 0 \tag{4.87}$$

On the other hand, as $\kappa \to 1$ we have $K \to \infty$, and the wave length L becomes infinite (while L/K remains finite). This special case corresponds to the solitary wave to be considered in Section 4.4.4.

Finally, it is found convenient in some applications to express the temporal variations of certain quantities as Fourier series in θ. In particular,

$$\mathrm{cn}^2\, q = \sum_{n=0}^{\infty} A_n \cos(n\theta) \tag{4.88}$$

$$\mathrm{cn}\, q\, \mathrm{dn}\, q\, \mathrm{sn}\, q = \frac{\pi}{K} \sum_{n=1}^{\infty} nA_n \sin(n\theta) \tag{4.89}$$

where the Fourier coefficients A_n are functions only of κ, and are given as

$$A_n = \begin{cases} \dfrac{2\pi^2}{\kappa^2 K^2}\left(\dfrac{nr^n}{1 - r^{2n}}\right) & \text{for } n \geqslant 1 \\ \dfrac{\gamma - \kappa'^2}{\kappa^2} & \text{for } n = 0 \end{cases} \tag{4.90}$$

in which

$$r = \exp\left[-\pi K(\kappa')/K(\kappa)\right] \tag{4.91}$$

4.4.3 Hyperbolic Wave Theory

The variation of the complete elliptic integral K with κ was set out in Fig. 4.7. When κ is close but not equal to unity such that K remains finite, an approximate solution may be obtained by taking $\kappa = 1$ and using instead of κ the finite value of K as the fundamental parameter characterizing the waves. This is equivalent to making the approximation

$$\begin{aligned} \kappa^2 &= 1 \\ \kappa'^2 &= 0 \\ \gamma &= 1/K \\ \mathrm{cn}\, q &= \mathrm{sech}\, q \end{aligned} \tag{4.92}$$

and the simplified results that follow from this procedure are indicated in Table 4.6. This method of approximation was introduced by Iwagaki (1968) and leads to the hyperbolic wave theory, so-called because temporal variations are expressed in terms of hyperbolic functions rather than the more awkward Jacobian elliptic functions. Iwagaki suggests that this approximation is reasonable for $K > 3$ which corresponds to $\kappa > 0.98$, a range which may be expressed in terms of the Ursell number as

$$U > 48\left(1 + \frac{\epsilon}{8}\right)^2 \tag{4.93}$$

The approximation may accordingly be used whenever U is greater than about 58.

The expressions given in Table 4.6 are slightly different from those first proposed by Iwagaki since his expressions were based on the original results of Laitone (1961) which, as already mentioned, were subsequently found to contain an error (Le Méhauté 1968). It will also be noted that the solution is no longer strictly periodic since the periodic cn^2q function is replaced by the non-periodic $sech^2q$ function. It follows that the hyperbolic wave formulae only embrace variations of θ in the range $-\pi < \theta < \pi$. There will, furthermore, be a discontinuity in, for example, wave slope when the attempt is made to match conditions at $\theta = \pi$ with those at $\theta = -\pi$ for the neighbouring wave. Thus the first approximation predicts the wave slope at $\theta = \pi$ to be

$$\frac{1}{H/L}\frac{\partial \eta}{\partial x} = -4K\,\mathrm{sech}^2\,K \tanh K \tag{4.94}$$

If we take $K = 3$, which may be considered the lower limit of the approximation, the wave slope at $\theta = \pi$ is then about 12% of the wave steepness instead of zero. However, the wave crests are relatively far apart and this anomaly at the wave troughs is generally unimportant.

The value of K may be determined for a given wave from the expression for d/gT^2 contained in Table 4.6. To the first approximation we have simply

$$K = \left(\frac{3HgT^2}{16d^2}\right)^{1/2} \tag{4.95}$$

Thus in the example treated earlier by cnoidal wave theory, the hyperbolic theory expression for d/gT^2 in Table 4.6 now gives $K = 4.20$. (This compares with $K = 4.17$ from Eq. (4.95).) When the value of K is applied to the expression for horizontal particle velocity, we now have $u_{max\ (s=0)}$ predicted to be essentially the same as previously calculated using the cnoidal wave theory.

4.4.4 Solitary Wave Theory

The solitary wave is a wave of translation in which the surface lies wholly above the mean water level as was indicated in Fig. 4.6b. It was first reported by

Table 4.6 Results of Hyperbolic Wave Theory—Second Approximation.

Trough depth, h

$$\frac{h}{d} = 1 - \frac{\epsilon}{K} + \frac{\epsilon^2}{4K} + 0[\epsilon^3]$$

Surface elevation, η

$$\frac{\eta}{d} = \epsilon\left(\operatorname{sech}^2 q - \frac{1}{K}\right) - \epsilon^2\left[\frac{3}{4}\operatorname{sech}^2 q\,(1 - \operatorname{sech}^2 q) - \frac{1}{4K}\right] + 0[\epsilon^3]$$

Wave celerity, c

$$\frac{c}{\sqrt{gd}} = 1 + \epsilon c_1 + \epsilon^2 c_2 + 0[\epsilon^3]$$

$$\frac{c^*}{\sqrt{gd}} = 1 + \epsilon c_1 + \epsilon^2 c_2^* + 0[\epsilon^3]$$

Wave length, L

$$\frac{L}{d} = \frac{4K}{\sqrt{3\epsilon}}\{1 - \epsilon\ell_1 + 0[\epsilon^2]\}$$

Wave period, T

$$\frac{d}{gT^2} = \frac{3\epsilon}{16K^2}\left\{\left(\frac{1 + \epsilon c_1 + \epsilon^2 c_2}{1 - \epsilon\ell_1}\right)^2 + 0[\epsilon^2]\right\}$$

$$\frac{d}{gT^{*2}} = \frac{3\epsilon}{16K^2}\left\{\left(\frac{1 + \epsilon c_1 + \epsilon^2 c_2}{1 - \epsilon\ell_1}\right)^2 + 0[\epsilon^2]\right\}$$

Horizontal particle velocity, u

$$\frac{u}{\sqrt{gd}} = \epsilon\left[\operatorname{sech}^2 q - \frac{1}{K}\right] + \epsilon^2\left\{[f_1 + f_2 \operatorname{sech}^2 q - \operatorname{sech}^4 q] - \frac{3}{4}\left(\frac{s}{d}\right)^2 \operatorname{sech}^2 q\,(2 - 3\operatorname{sech}^2 q)\right\} + 0[\epsilon^3]$$

$$\frac{u^*}{\sqrt{gd}} = \epsilon\left[\operatorname{sech}^2 q - \frac{1}{K}\right] + \epsilon^2\left\{[f_1^* + f_2 \operatorname{sech}^2 q - \operatorname{sech}^4 q] - \frac{3}{4}\left(\frac{s}{d}\right)^2 \operatorname{sech}^2 q\,(2 - 3\operatorname{sech}^2 q)\right\} + 0[\epsilon^3]$$

Vertical particle velocity, w

$$\frac{w}{\sqrt{gd}} = \epsilon\sqrt{3\epsilon}\left(\frac{s}{d}\right)\operatorname{sech}^2 q \tanh q\left\{1 + \epsilon\left[\frac{1}{8}\left(\frac{16}{K} - 3\right) - 2\operatorname{sech}^2 q - \frac{1}{2}\left(\frac{s}{d}\right)^2 (1 - 3\operatorname{sech}^2 q)\right] + 0[\epsilon^2]\right\}$$

Horizontal particle acceleration, $\partial u/\partial t$

$$\frac{1}{g}\frac{\partial u}{\partial t} = \epsilon\sqrt{3\epsilon}\operatorname{sech}^2 q \tanh q\left\{1 + \epsilon\left[\frac{1}{8}\left(\frac{4}{K} + 1\right) - 2\operatorname{sech}^2 q - \frac{3}{2}\left(\frac{s}{d}\right)^2 (1 + 3\operatorname{sech}^2 q)\right] + 0[\epsilon^2]\right\}$$

Table 4.6 (*Continued*).

Vertical particle acceleration, $\partial w/\partial t$	$\frac{1}{g}\frac{\partial w}{\partial t} = \frac{3\epsilon^2}{2}\left(\frac{s}{d}\right)\text{sech}^2\, q\,(2 - 3\,\text{sech}^2\, q) + 0\,[\epsilon^3]$
Pressure, p	$\frac{p}{\rho g d} = \frac{\eta}{d} + 1 - \frac{s}{d} - \frac{3}{4}\epsilon^2\,\text{sech}^2\, q\left[\left(\frac{s}{d}\right)^2 - 1\right] \cdot [2 - 3\,\text{sech}^2\, q] + 0\,[\epsilon^3]$
Average energy density, E	$\frac{E}{\rho g d^2} = \frac{1}{2}\epsilon^2\left\{-\frac{2}{3K}\left(\frac{3}{K} - 2\right) + \frac{\epsilon}{15K}\left(\frac{15}{K^2} - 2\right) + 0\,[\epsilon^2]\right\}$
Energy flux, P	$\frac{P}{\rho g d^2\sqrt{gd}} = \epsilon^2\left\{-\frac{1}{3K}\left(\frac{3}{K} - 2\right) + \frac{\epsilon}{30K}\left(\frac{75}{K^2} - \frac{60}{K} + 8\right) + 0\,[\epsilon^2]\right\}$

where

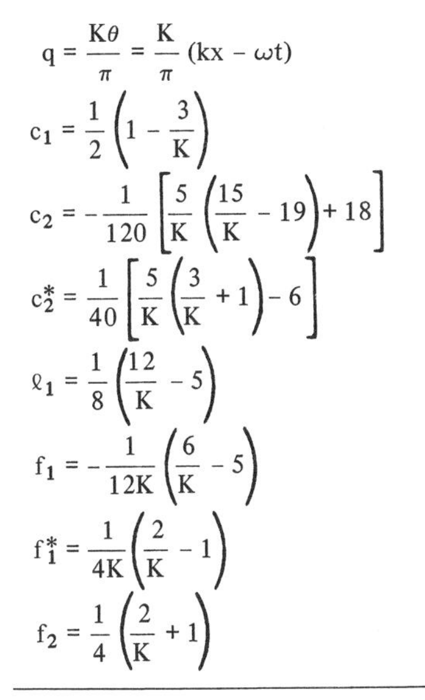

$$q = \frac{K\theta}{\pi} = \frac{K}{\pi}(kx - \omega t)$$

$$c_1 = \frac{1}{2}\left(1 - \frac{3}{K}\right)$$

$$c_2 = -\frac{1}{120}\left[\frac{5}{K}\left(\frac{15}{K} - 19\right) + 18\right]$$

$$c_2^* = \frac{1}{40}\left[\frac{5}{K}\left(\frac{3}{K} + 1\right) - 6\right]$$

$$\ell_1 = \frac{1}{8}\left(\frac{12}{K} - 5\right)$$

$$f_1 = -\frac{1}{12K}\left(\frac{6}{K} - 5\right)$$

$$f_1^* = \frac{1}{4K}\left(\frac{2}{K} - 1\right)$$

$$f_2 = \frac{1}{4}\left(\frac{2}{K} + 1\right)$$

Russell (1844) on the basis of experimental observation, and was first analysed by Boussinesq (1871) and again by Rayleigh (1876)–prior to the discovery of cnoidal waves in 1895. Keller (1948) obtained the first approximation to solitary waves by the expansion method of Friedrichs (1948) and a description of solitary wave theory which is directed towards engineering use has been presented by Munk (1949).

Second and third approximations were later developed by Laitone (1961) and Grimshaw (1971) respectively. More recently highly accurate descriptions of solitary waves have been provided by Fenton (1972), Longuet-Higgins (1974),

Longuet-Higgins and Fenton (1974) and Byatt-Smith and Longuet-Higgins (1976), with due attention given to the solitary wave of maximum height. Fenton (1972) presented a ninth order solution in which Shanks transforms were employed to improve series convergence. Longuet-Higgins and Fenton (1974) carried out computations to a high order, employing power series in a perturbation parameter $\epsilon = 1 - q^2/gd$, where q is the particle velocity at the crest relative to the crest. The series were summed using Padé approximants (see Section 4.5.4). Byatt-Smith and Longuet-Higgins (1976) employed an integral equation method and confirmed many of the features described by Longuet-Higgins and Fenton.

The following discussion is restricted to some of the more fundamental aspects of solitary waves as described by the first or second approximations. The solitary wave may conveniently be considered as a special case of the cnoidal wave in which the wave length and period become infinite as a consequence of taking the limit $\kappa \to 1$. We then have

$$\left.\begin{array}{r}\gamma \to 0 \\ K \to \infty \\ \mathrm{cn}\, q \to \mathrm{sech}\, q\end{array}\right\} \text{as } \kappa \to 1 \tag{4.96}$$

and by making the necessary substitutions in the results already obtained for cnoidal wave theory (Table 4.5) the corresponding expressions relating to the second approximation of solitary wave theory may be deduced and are assembled in Table 4.7.

Because the wave length and period are infinite, only one parameter $\epsilon = H/d$ is needed to specify the solitary wave (in place of two, say H/d and d/gT^2, for periodic waves). K and the wave length L both tend to infinity when $\kappa \to 1$ such that the ratio K/L remains finite and the argument q, which reflects x-ward and temporal variations, may to a first approximation be expressed as

$$q = \underset{\kappa \to 1}{\mathrm{Limit}} \left[\frac{2K}{L}(x - ct)\right] \simeq \frac{\sqrt{3\epsilon}}{2d}(x - ct) \tag{4.97}$$

To lowest order the solitary wave profile varies as $\mathrm{sech}^2 q$, and the horizontal particle velocity and the pressure within the fluid show similar kinds of variation. That is, to a first approximation,

$$\frac{\eta}{H} = \frac{u}{\sqrt{gd}\,(H/d)} = \frac{\Delta p}{\rho g H} = \mathrm{sech}^2\, q, \tag{4.98}$$

where Δp is the pressure at a point due to the presence of the wave and taken

Table 4.7 Results of Solitary Wave Theory—Second Approximation.

Surface elevation, η	$\frac{\eta}{d} = \epsilon \operatorname{sech}^2 q - \frac{3}{4}\epsilon^2 \operatorname{sech}^2 q \tanh q + 0[\epsilon^3]$
Wave celerity, c	$\frac{c}{\sqrt{gd}} = \frac{c^*}{\sqrt{gd}} = 1 + \frac{1}{2}\epsilon - \frac{3}{20}\epsilon^2 + 0[\epsilon^3]$
Horizontal particle velocity, u	$\frac{u}{\sqrt{gd}} = \frac{u^*}{\sqrt{gd}} = \epsilon \operatorname{sech}^2 q + \epsilon^2 \operatorname{sech}^2 q \left\{\frac{1}{4} - \operatorname{sech}^2 q - \frac{3}{4}\left(\frac{s}{d}\right)^2 (2 - 3\operatorname{sech}^2 q)\right\} + 0[\epsilon^3]$
Vertical particle velocity, w	$\frac{w}{\sqrt{gd}} = \epsilon\sqrt{3\epsilon}\left(\frac{s}{d}\right)\operatorname{sech}^2 q \tanh q \left\{1 - \epsilon\left[\frac{3}{8} + 2\operatorname{sech}^2 q + \frac{1}{2}\left(\frac{s}{d}\right)^2 (1 - 3\operatorname{sech}^2 q)\right] + 0[\epsilon^2]\right\}$
Horizontal particle acceleration, $\partial u/\partial t$	$\frac{1}{g}\frac{\partial u}{\partial t} = \epsilon\sqrt{3\epsilon}\operatorname{sech}^2 q \tanh q \left\{1 + \epsilon\left[\frac{1}{8} - 2\operatorname{sech}^2 q - \frac{3}{2}\left(\frac{s}{d}\right)^2 (1 + 3\operatorname{sech}^2 q)\right] + 0[\epsilon^2]\right\}$
Vertical particle acceleration, $\partial w/\partial t$	$\frac{1}{g}\frac{\partial w}{\partial t} = \frac{3\epsilon^2}{2}\left(\frac{s}{d}\right)\operatorname{sech}^2 q (2 - 3\operatorname{sech}^2 q) + 0[\epsilon^3]$
Pressure, p	$\frac{p}{\rho g d} = \frac{\eta}{d} + 1 - \frac{s}{d} - \frac{3}{4}\epsilon^2 \operatorname{sech}^2 q\left[\left(\frac{s}{d}\right)^2 - 1\right](2 - 3\operatorname{sech}^2 q) + 0[\epsilon^3]$

where

$$q = \frac{\sqrt{3\epsilon}}{2d}\left(1 - \frac{5}{8}\epsilon\right)(x - ct)$$

relative to that in the undisturbed fluid. The solitary wave profile is given to the second approximation in Fig. 4.10 for various values of $\epsilon = H/d$.

The total pressure range Δp occuring at a point in the fluid is given to the second approximation as

$$\frac{\Delta p}{\rho g H} = 1 - \frac{3}{4}\epsilon\left[1 - \left(\frac{s}{d}\right)^2\right] \tag{4.99}$$

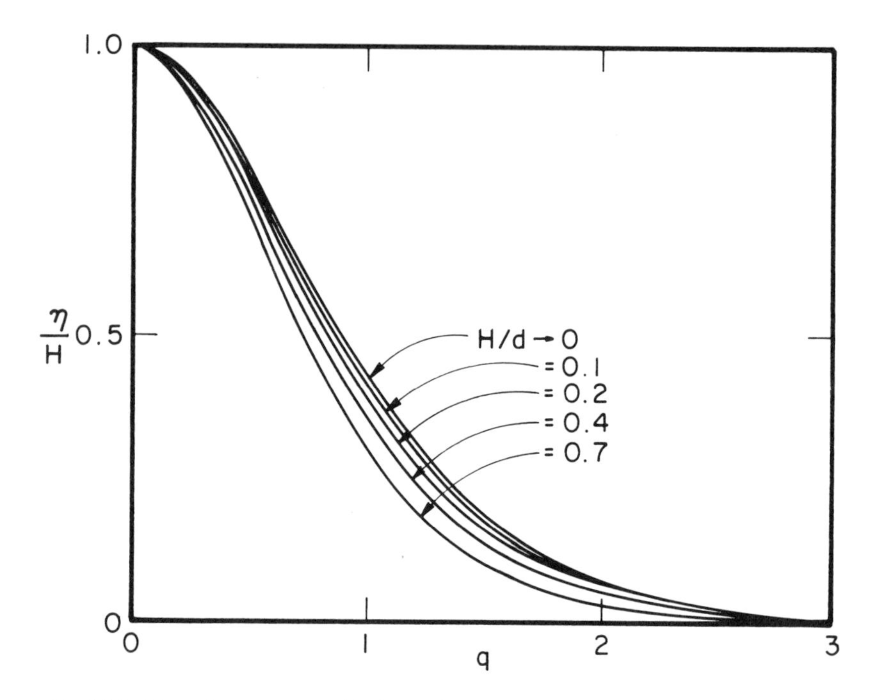

Fig. 4.10. The solitary wave profile (second approximation) for various values of H/d.

Thus the wave height required to produce a given pressure variation Δp on the seabed can be estimated from the simple formula

$$\frac{\Delta p}{\rho g H} = \frac{1}{2}\left[1 + \left(1 - \frac{3\Delta p}{\rho g d}\right)^{1/2}\right] \tag{4.100}$$

The horizontal particle acceleration is obtained simply by taking the temporal derivative of u and is given to the first approximation as

$$\frac{1}{g}\frac{\partial u}{\partial t} = \epsilon\sqrt{3\epsilon}\ \mathrm{sech}^2\, q \tanh q \tag{4.101}$$

Consequently the fluid acceleration is a maximum in front of the crest and occurs at a point given by $q = 0.66$ and has a magnitude $(\partial u/\partial t)_{max} = 0.385g\ \epsilon\sqrt{3\epsilon}$.

Finally, it is emphasized that because the solitary wave is a wave of translation and never exhibits horizontal particle velocities in the opposite direction to that of wave advance, there is a net displacement of fluid in the direction of wave advance. This net displacement is given to the second order as

$$\frac{\xi}{d} = 4\left(\frac{\epsilon}{3}\right)^{1/2}\left\{1 + \frac{3}{8}\epsilon + 0[\epsilon^2]\right\} \tag{4.102}$$

This is the same at all depths and the total volume per unit width of fluid displaced by the passage of the wave is therefore given directly as ξd. To a higher approximation ξ increases towards the free surface.

4.5 OTHER WAVE THEORIES

Additional wave theories which are now considered include the linearized long wave theory (e.g. Stoker 1957), Gerstner's (1802) trochoidal wave theory, Dean's (1965) stream function theory and various other modern theories based on computer applications (e.g. Monkmeyer 1970, Schwartz 1974, Cokelet 1977a and Bloor 1978).

The long wave theory has application to tsunami propagation, tidal motion, storm surge, flood waves and the like, but it is not of primary interest in the present context and only a brief development of the governing equations is therefore given. Gerstner's (1802) trochoidal theory is largely of historic interest even though it has been applied to engineering problems to a limited extent. It differs from most other wave theories in that it involves a rotational fluid motion. The remaining theories mentioned above have been developed relatively recently and depend on numerical schemes requiring the use of a computer.

4.5.1 Linearized Long Wave Theory

The linearized long wave theory is developed from the basic assumptions that, firstly, the wave height is small so that all nonlinear terms in the governing equations may be neglected, and, secondly, that the wave length is much larger than the water depth so that the vertical particle acceleration may also be neglected. It follows from these two requirements that the horizontal particle velocity is invarient with depth and the pressure is hydrostatic. These simplifications prove most useful in obtaining solutions by numerical methods for unsteady flows and/or flows with complex boundaries. More detailed outlines are given by Stoker (1957) and by Le Méhauté (1976).

We restrict our discussion here to a two-dimensional motion in which the still water depth $d(x)$ is not a constant but is a function of x. The equations of motion, after omitting the nonlinear terms, are

$$\frac{\partial u}{\partial t} = -\frac{1}{\rho}\frac{\partial p}{\partial x} \tag{4.103}$$

$$\frac{\partial w}{\partial t} = -\frac{1}{\rho}\frac{\partial p}{\partial z} - g \tag{4.104}$$

$$\frac{\partial u}{\partial x} + \frac{\partial w}{\partial z} = 0 \tag{4.105}$$

These correspond respectively to the momentum equations in the x and z directions and to the continuity equation. Here it may be necessary under certain conditions to include a body force term on the right-hand side of Eq. (4.103). The boundary conditions are

$$w = \frac{\partial \eta}{\partial t} \qquad \text{at } z = \eta \tag{4.106}$$

$$p = 0 \qquad \text{at } z = \eta \tag{4.107}$$

$$w = u\frac{\partial d}{\partial x} \qquad \text{at } z = -d \tag{4.108}$$

These are respectively the kinematic and dynamic free surface boundary conditions and the kinematic boundary condition at the seabed. In view of the further assumption that the vertical acceleration is negligible, the integration of Eq. (4.104) and the application of Eq. (4.107) yield

$$p = \rho g(\eta - z) \tag{4.109}$$

That is, the pressure within the fluid is hydrostatic. Substitution of Eq. (4.109) into Eq. (4.103) gives

$$\frac{\partial u}{\partial t} + g\frac{\partial \eta}{\partial x} = 0 \tag{4.110}$$

which implies that the horizontal particle velocity is uniform over depth, $u = u(x, t)$. We now integrate the equation of continuity, Eq. (4.105), from $z = -d$ to $z = \eta$, make use of the boundary conditions given in Eqs. (4.106) and (4.108), and take $\eta \ll d$ to obtain eventually

$$\frac{\partial \eta}{\partial t} + \frac{\partial (ud)}{\partial x} = 0 \tag{4.111}$$

Equations (4.110) and (4.111) may finally be combined to obtain the one-dimensional wave equation in the alternative forms

$$\frac{\partial^2 \eta}{\partial t^2} - c^2 \frac{\partial^2 \eta}{\partial x^2} = 0 \tag{4.112}$$

and

$$\frac{\partial^2 (ud)}{\partial t^2} - c^2 \frac{\partial^2 (ud)}{\partial x^2} = 0 \tag{4.113}$$

where $c = \sqrt{gd}$ is the wave speed.

Under conditions where d is constant, the compound variable ud in Eq. (4.113) may be replaced by u and in this case c will also be a constant. The general solution of Eq. (4.112) may now be written as

$$\eta = f_1(x - ct) + f_2(x + ct) \tag{4.114}$$

If η is known for all x at an initial time t_0, or for all values of t at a location x_0, then the subsequent motion can be determined by a finite difference scheme or by the method of characteristics. The adoption of this technique leads to a fundamental approach for treating problems on the basis of the linearized long wave theory.

It is fairly routine to extend this development to a two-dimensional wave motion and also to obtain governing equations for problems which include various additional effects. For example, it is necessary in some cases to include a body force term in the momentum equations in order to take into account Coriolis effects, bottom friction, wind stresses and so on. Another modification necessary in the context of tsunami generation is to consider the depth d to be time dependent and in this case the bottom boundary condition becomes

$$w = u \frac{\partial d}{\partial x} - \frac{\partial d}{\partial t} \tag{4.115}$$

and consequently Eq. (4.111) must now be replaced by

$$\frac{\partial (d + \eta)}{\partial t} + \frac{\partial (ud)}{\partial x} = 0 \tag{4.116}$$

while Eq. (4.110) remains unaltered.

The flow in an estuary or inlet of variable width may be accounted for in terms of the sectional area A and the flow velocity U averaged across A. By considering a control volume of length dx in the flow direction, the continuity equation may be written here as

$$UA - \left(U + \frac{\partial U}{\partial x}\,dx\right)\left(A + \frac{\partial A}{\partial x}\,dx\right) = \frac{\partial \eta}{\partial t}\,\ell(x)\,dx \tag{4.117}$$

where $\ell(x)$ is the inlet width at the free surface. This equation reduces to

$$\ell\frac{\partial \eta}{\partial t} + \frac{\partial (UA)}{\partial x} = 0 \tag{4.118}$$

and this may be used together with Eq. (4.110) in which u is replaced by the averaged velocity U.

It is seen from the above discussion how alternative governing equations may be derived for different kinds of shallow wave problems. A discussion of other such problems and their solutions is outside the scope of the present text and is not pursued further.

4.5.2 Trochoidal Wave Theory

The trochoidal wave theory was introduced by Gerstner (1802) and has been adequately described by Milne-Thompson (1968). This theory differs from most other finite amplitude wave theories in that it depends on a rotational fluid motion and also in that the solution is an exact one under the assumptions made. The wave motion may be described in terms of circular particle orbits whose radii decrease exponentially with depth. The instantaneous position (ξ, ζ) of a given particle is

$$\xi = x_0 + \frac{1}{k}\exp(kz_0)\sin(kx_0 - \omega t) \tag{4.119}$$

$$\zeta = z_0 - \frac{1}{k}\exp(kz_0)\cos(kx_0 - \omega t) \tag{4.120}$$

where x_0 and z_0 are the coordinates of the center of that particle's orbit. In this notation the origin of z is not at the still water level and remains to be determined. It may readily be confirmed that the motion described does indeed satisfy the continuity equation.

Surfaces of constant pressure, which include the free surface, are given in parametric form in terms of an angle α as

$$\begin{aligned} kz &= kz_0 - \exp(kz_0)\cos\alpha \\ kx - \omega t &= \alpha + \exp(kz_0)\sin\alpha \end{aligned} \tag{4.121}$$

These curves are trochoids: they are generated by a point fixed on a disc of radius $1/k$ at a distance of $\exp(kz_0)/k$ from its center when the disc rolls in the positive x direction under the line $z = z_0 + 1/k$. These locii are indicated in Fig. 4.11. The limiting case with $z_0 = 0$ corresponds to the point lying on the outer edge of the disc and the resultant locus becomes a cycloid. On the other hand, as $z_0 \to -\infty$, the generating point (P in Fig. 4.11) becomes arbitrarily close to the disc's center and then traces a horizontal line.

The pressure along a trochoidal surface is given as

$$p = \rho g(z_s - z_0) + \tfrac{1}{2}\rho c^2 \left[\exp(2kz_0) - \exp(2kz_s)\right] \tag{4.122}$$

where z_s is the value of z_0 for particles forming the free surface, and the reference pressure at the free surface is taken as zero. The wave celerity c is given by

$$c = \left(\frac{g}{k}\right)^{1/2} \tag{4.123}$$

which agrees with the result of linear wave theory for deep water waves (as are being considered here). The still water level is located at the coordinate

$$z = z_s - \frac{1}{2k}\exp(2kz_s) \tag{4.124}$$

and the wave height, deriving from Eq. (4.121), is given as

$$H = \frac{2}{k}\exp(kz_s) \tag{4.125}$$

It follows that the wave steepness is

$$\frac{H}{L} = \frac{1}{\pi}\exp(kz_s) \tag{4.126}$$

which has a maximum value of $1/\pi = 0.3183$ when $z_s = 0$, and the surface profile then assumes the limiting case of a cycloid (see the uppermost profile in Fig. 4.11b).

The average energy density E is given by

$$E = \frac{1}{8}\rho g H^2 \left[1 - \frac{1}{2}\left(\frac{\pi H}{L}\right)^2\right] \tag{4.127}$$

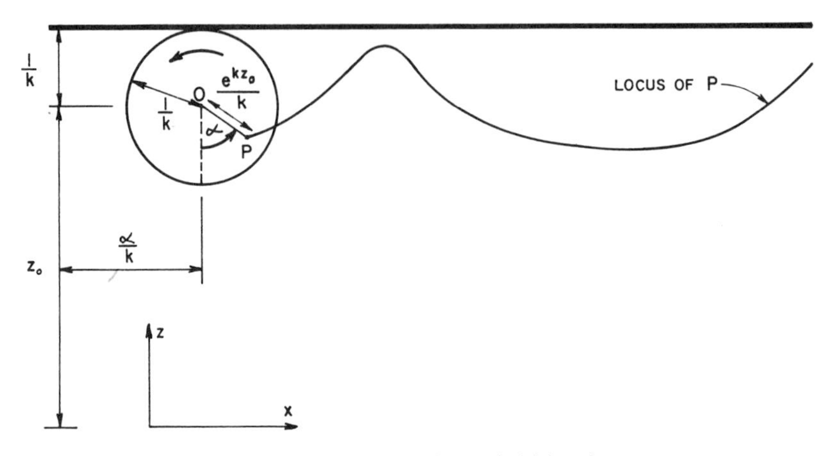

Fig. 4.11a. Generation of a trochoidal surface.

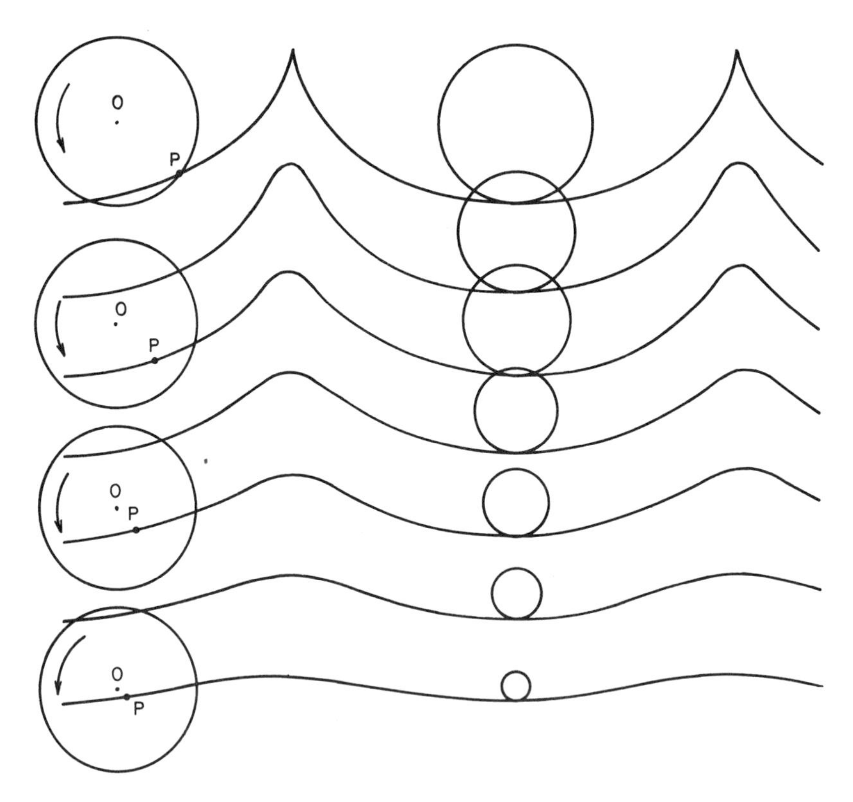

GENERATING CIRCLES PARTICLE ORBITS

Fig. 4.11b. Particle orbits in a trochoidal wave train.

which when $H/L \ll 1$ corresponds to the result of small amplitude wave theory. Unlike most other wave theories the fluid motion is not irrotational. The vorticity in the fluid is

$$\xi = -2kc\left[\frac{\exp(2kz_0)}{1 - \exp(2kz_0)}\right] \tag{4.128}$$

Here the negative sign indicates that the vorticity is in the opposite sense to the particle orbital motion indicated in Fig. 4.11. It follows that the physical realization of such waves seldom occurs, except under a few possible circumstances as when waves are progressing against a wind which induces a vorticity within the fluid in the opposite sense to the particle motions.

Trochoidal wave theory is not generally used in engineering design, in part because of the questionable assumption made concerning the rotationality of the fluid, and in part because of the more attractive alternatives which are available and which provide either greater simplicity or greater accuracy or both.

4.5.3 Stream Function Theory

Dean (1965) introduced a numerical method for predicting two-dimensional wave characteristics which is based on a stream function representation of the flow and which has since attained fairly widespread application. This approach somewhat supercedes a not dissimilar technique proposed earlier by Chappelear (1961) and which was based instead on a velocity potential representation.

Dean applied himself to the solution of two separate problems:

(a) how to obtain numerically a stream function representation of a wave with a prescribed profile, and
(b) how to obtain numerically a stream function representation of periodic waves of permanent form with prescribed period, height and still water depth.

The approach adopted by Dean is capable of generalization and affords a solution where a free surface pressure distribution and uniform current are also prescribed. Dean's second problem, (b) above, is the one tackled by other wave theories. An outline of its solution will be given here in the special case of constant free surface pressure (as is usually assumed in wave theories) and in the absence of an underlying current. The effect of a uniform current can be accounted for directly in terms of the solution obtained (Section 4.9.3).

We choose a coordinate system (x, z) moving with the waves, thus ensuring that the problem is reduced to one of steady flow. This corresponds to the system (x', z) previously referred to, but for simplicity the primes are now omitted.

Because the flow is two-dimensional, a stream function ψ exists, and the irrotationality condition implies that this must satisfy the Laplace equation within the fluid,

$$\frac{\partial^2 \psi}{\partial x^2} + \frac{\partial^2 \psi}{\partial z^2} = 0. \tag{4.129}$$

This is subject to the seabed and the two free surface boundary conditions,

$$\frac{\partial \psi}{\partial x} = 0 \qquad \text{at } z = -d \tag{4.130a}$$

$$w = u \frac{\partial \eta}{\partial x} \qquad \text{at } z = \eta \tag{4.130b}$$

$$\frac{1}{2g}(u^2 + w^2) + \eta = Q \qquad \text{at } z = \eta \tag{4.130c}$$

in which Q is a Bernoulli constant as appearing in Eq. (4.9). The stream function $\psi = \psi(x, z)$ is assumed to be of the form

$$\psi(x, z) = cz + \sum_{n=1}^{N} X_n \sinh(nk(z + d)) \cos(nkx) \tag{4.131}$$

This is an even function of kx corresponding to a symmetrical wave profile. (Should the surface pressure not be constant, then terms in sin (nkx) will be necessary to describe the resulting asymmetrical profile.) Now Eq. (4.131) satisfies the Laplace equation, the bottom boundary condition and the kinematic free surface condition. The latter provides that the surface value of the stream function $\psi(x, \eta)$ is a constant, denoted ψ_η, and therefore Eq. (4.131) applied at $z = \eta$ gives a relation between η and ψ_η as

$$\psi_\eta = c\eta + \sum_{n=1}^{N} X_n \sinh(nk(\eta + d)) \cos(nkx) \tag{4.132}$$

The coefficients X_n, the wave number k and the surface value of the stream function have now to be determined by the dynamic free surface boundary condition. For convenience the unknown wave number k may be denoted X_{N+1} and an initial estimate of the unknowns X_n may be made, say, by using linear wave theory: thus the initial values of all X_n except X_1 and X_{N+1} are zero. Successive approximations may then be obtained by any suitable numerical procedure. Dean's approach was to minimize the error in the fit with the dynamic free surface boundary condition, Eq. (4.130c). This error E may be defined as

$$E = \frac{1}{I} \sum_{i=1}^{I} (Q_i - \overline{Q})^2 \tag{4.133}$$

where i is an index which ranges from 1 to I as x takes successive values spanning one complete wave length, Q_i represents a value of the Bernoulli constant computed according to Eq. (4.130c) at the i-th position of x, and $\overline{Q}$ represents the actual or average value of the Bernoulli constant. Thus if the dynamic free surface boundary condition were satisfied exactly $Q_i = \overline{Q}$ for all i and there would be no error. $\overline{Q}$, which describes the average value of Q_i, may be a prescribed constant, or may be progressively adjusted according to the definition

$$\overline{Q} = \frac{1}{I} \sum_{i=1}^{I} Q_i \tag{4.134}$$

It is convenient to use the superscript (j) to denote values pertaining to the j-th cycle of the iteration process. Q_i depends on the assigned values of X_n and a subsequent approximation to $Q^{(j)}$ is written as

$$Q_i^{(j+1)} = Q_i^{(j)} + \sum_{n=1}^{N+1} \frac{\partial Q_i^{(j)}}{\partial X_n} X'_n \tag{4.135}$$

where X'_n represents the correction in the value of each X_n as the iteration process is advanced through one cycle,

$$X_n^{(j+1)} = X_n^{(j)} + X'_n \tag{4.136}$$

The error $E^{(j+1)}$ can then be written in terms of X'_n by substituting Eq. (4.135) into Eq. (4.133). In the case where $\overline{Q}$ is a specified constant we then have

$$E^{(j+1)} = \frac{1}{I} \sum_{i=1}^{I} \left\{ \left(Q_i^{(j)} + \sum_{n=1}^{N+1} \frac{\partial Q_i^{(j)}}{\partial X_n} X'_n \right) - \overline{Q} \right\}^2 \tag{4.137}$$

The values of X'_n may now be determined according to the least squares criterion which requires that E is a minimum. Adopting this procedure we put $\partial E/\partial X'_n = 0$ to obtain a set of equations which yields the required value of each X'_n. The subsequent approximation to $X_n^{(j)}$, that is $X_n^{(j+1)}$, is then known, Eq. (4.136).

The free surface elevation $\eta^{(j+1)}$ may next be estimated by resort to Eq. (4.131) applied at the free surface $z = \eta$, and using $\psi_\eta^{(j)}$ and the coefficients $X_n^{(j+1)}$, which are the most recently refined values available. That is Eq. (4.132) is written in the form

$$\eta^{(j+1)} = \frac{1}{c}\left\{\psi_\eta^{(j)} - \sum_{n=1}^{N} X_n^{(j+1)} \sinh\left(nk(\eta^{(j)} + d)\right) \cos(nkx)\right\} \quad (4.138)$$

The improved value of the surface streamline $\psi_\eta^{(j+1)}$ may now be determined by ensuring that the mean value of $\eta^{(j+1)}$ is zero. Thus

$$\psi_\eta^{(j+1)} = \frac{1}{L}\int_0^L \left[\sum_{n=1}^{N} X_n^{(j+1)} \sinh\left(nk(\eta^{(j+1)} + d)\right) \cos(nkx)\right] dx \quad (4.139)$$

Corresponding values of the velocity components u and w at the free surface may also be obtained, and a new set of values of Q_i, that is $Q_i^{(j+1)}$, may thus be calculated using Eq. (4.130c). All quantities corresponding to the $(j + 1)$-th iteration are now known and thus the whole iteration process can now be repeated.

This completes the outline of the iterative procedure. When the wave height is used to define a particular problem then additional steps in the iterative procedure are necessary, but in practice it is simplest to repeat the procedure until the wave height obtained is sufficiently close to the specified value.

Tables are available for the practical application of the stream function theory (Dean 1974). These list various quantities of engineering interest such as the particle velocities and accelerations and the pressure at specified locations and instants, as well as various integral quantities related to the energy and momentum of the wave train. These are tabulated for a range of values of $d/L_0 = 2\pi(d/gT^2)$ and for wave heights given by $H/H_b = 0.25, 0.50, 0.75$ and 1.0, where H_b is the maximum (breaking) wave height at a given d/L_0. As an example of what may be obtained by the stream function approach, Fig. 4.12 reproduces Dean's (1965) result for a near-breaking wave in fairly shallow water with $H/d = 0.7984$, $L/d = 17.373$ and $d/gT^2 = 0.00466$.

Dalrymple (1974a,b) has extended the stream function theory to permit a non-uniform (shear) current to be specified and this treatment is referred to in Section 4.9.3. Mention should also be made of the somewhat related extended velocity potential method (sometimes termed the EXVP method) described by Lambrakos and Brannon (1974). The method was developed to enable treatment of arbitrary wave profiles which may vary in shape during propagation, or which may have a separately specified crest elevation to wave height ratio. This

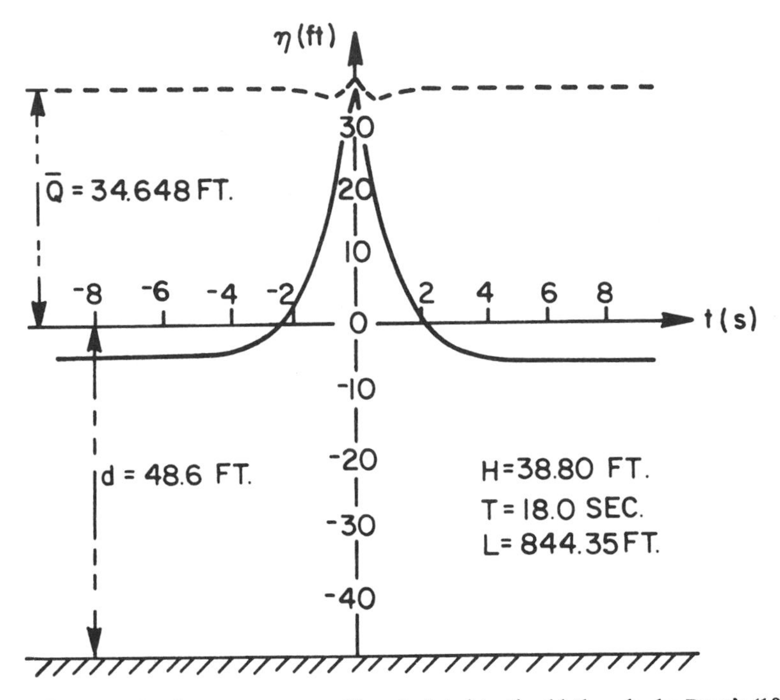

Fig. 4.12. Example of a steep wave profile calculated to the 11-th order by Dean's (1965) stream function theory.

procedure involves a double Fourier series expansion of the velocity potential as

$$\phi(x, z, t) = \sum_{n,m}^{N,M} \cosh(k_n(z + d)) [A_{nm} \cos(k_n x - \omega_m t) + B_{nm} \sin(k_n x - \omega_m t)] \tag{4.140}$$

The unknown Fourier coefficients are determined by a least squares minimization technique applied to the free surface boundary conditions and carried out over time and the x direction. The technique is useful to deal with, for example, a series of consecutive waves with independent characteristics.

4.5.4 Modern Computational Methods

The requirement of an accurate description of steep waves over a complete range of depths is not adequately fulfilled by the Stokes and cnoidal wave theories in the forms described in Sections 4.3 and 4.4. Several authors have attempted to obtain solutions to a high degree of accuracy by resort to digital computer, and in this regard we have already encountered Dean's (1965) stream function theory which involves an iterative scheme requiring a computer. Other contributions to the accurate prediction of the behavior of steep waves include those of Monkmeyer (1970) (see also Monkmeyer and Kutzbach (1965) for the deep water case), Longuet-Higgins (1973, 1975), Sasaki and Murakami (1973), Schwartz (1974), Byatt-Smith and Longuet-Higgins (1976), Cokelet (1977a), Longuet-Higgins and Fox (1977, 1978), Bloor (1978), Fenton (1979) and Vanden-Broeck and Schwartz (1979).

Some of this work has revealed certain drawbacks in Stokes classical expansion procedure. In particular, Schwartz (1974) discovered that the Stokes small amplitude expansion method does not converge for waves above a certain steepness less than the maximum. Also Longuet-Higgins and Fenton (1974) and Longuet-Higgins (1975) have shown, for solitary and deep water waves respectively, that many characteristics, such as the wave speed and wave energy, are not monotonic functions of wave height, but in fact eventually decrease with increasing height as the maximum wave height is reached for a given wave length.

The approaches adopted are generally expressed in terms of the complex potential: under the conditions of incompressibility and irrotationality, a two-dimensional flow may be treated by the theory of complex variables and the complex potential $w(z)$ may then be defined in the usual way as $w(z) = \phi + i\psi$, where w is a function of the complex variable $z = x + iy$. Here a two-dimensional coordinate system is used with x measured against the direction of wave propagation (for the present) and y vertically upwards from a distance d' above the seabed, and the flow is reduced to a steady one by taking the origin to move with the wave: the fluid moves from left to right. At the free surface $\psi = 0$, while at the seabed $\psi = -cd'$. The depth d' was referred to in Section 4.3.2 as the depth of a uniform flow with velocity c and with the same flow rate as in the wave motion. This depth is related to the still water depth d by the relation involving the two definitions of wave celerity: $d' = d(c^*/c)$.

Some of the procedures may conveniently be considered in terms of a conformal transformation. In this, the physical z plane is transformed to a new ζ plane such that the fluid boundaries of the wave motion are now mapped into some known configuration and the fluid motion in the ζ plane is easily defined. In the following outline it is convenient to define the unit of length by taking $k = 1$.

Prior to the discovery of the steep wave difficulties mentioned already, Monkmeyer (1970) employed a conformal transformation between the z and ζ planes which is given as

$$z = i\left\{\ln(\zeta) + \frac{\pi}{2K}\sum_{j=1}^{\infty}\frac{a_j}{j}\exp\left[i\,\mathrm{am}\left(-\frac{2jK}{\pi}\ln(\zeta)\right)\right]\right\} \qquad (4.141)$$

where K is the complete elliptic integral of the first kind and am( ) is the amplitude of the elliptic integral of the first kind (e.g. Abramowitz and Stegun 1965). Milne-Thompson (1968) provides a useful background to the general procedure involved. The coordinates describing ζ are conveniently given in radial form in terms of a radius r and an angle χ as $\zeta = re^{i\chi}$. As a result of this transformation, Eq. (4.141), the fluid region in the physical z plane is mapped into an annulus in the ζ plane occupied between concentric circles of radii $r = 1$ and $r = r_0 < 1$. The free surface over one wave length is then transformed into the outer circle and the seabed into the inner circle. This mapping of the boundaries is indicated in Fig. 4.13. Here the boundary DEAB transforms independently of the choice of the coefficients a_j. In fact the transformation of the seabed portion of this boundary, EA, indicates that the radius r_0 is directly related to the relative depth, for by applying Eq. (4.141) at $z = -id'$ we have $r_0 = \exp(-d')$. On the other hand, the complete transformation of the free surface BCD depends on appropriate values being assigned to the coefficients a_j. The boundary conditions in the z plane are applicable in the ζ plane as soon as they have been appropriately transformed, and the actual flow in the ζ plane which satisfies these boundary conditions is simply that of a potential vortex with a clockwise flow,

$$w(\zeta) = ic \ln(\zeta) \tag{4.142}$$

The problem therefore reduces to the determination of sufficient values of a_j to ensure that the results have the desired accuracy. These are obtained by making use of the dynamic free surface boundary condition.

The limiting cases of the transformation are obtained by letting the Jacobian elliptic function modulus κ tend to zero or unity. As $\kappa \to 0$ we have $K \to \pi/2$, $r_0 \to 0$, $d' \to \infty$ and the deep water limit is reached; and as $\kappa \to 1$ we have $K \to \infty$, $r_0 \to 1$, $d' \to 0$ and the solitary wave limit is reached. Monkmeyer did not go on to present detailed results of the theory and specifically did not investigate maximum wave heights, although in an earlier paper Monkmeyer and Kutzbach (1965) obtained a maximum deep water wave steepness $H/L = 0.1442$ by using the theory to the fifteenth order for the deep water case.

In passing it is mentioned that Longuet-Higgins (1973) has described a very simple approximation to the form of steep deep water waves by using the conformal transformation

$$\zeta = \zeta_0 \exp(-iz) \tag{4.143}$$

where ζ_0 is a constant (cf. Eq. (4.141)). He considered the fluid region occupied by a regular hexagon in the ζ plane as a transformation of six consecutive progressive waves in the physical z plane. The angle at the wave crest is then 120° in agreement with Stokes' (1880) limit for the steepest wave, and the free sur-

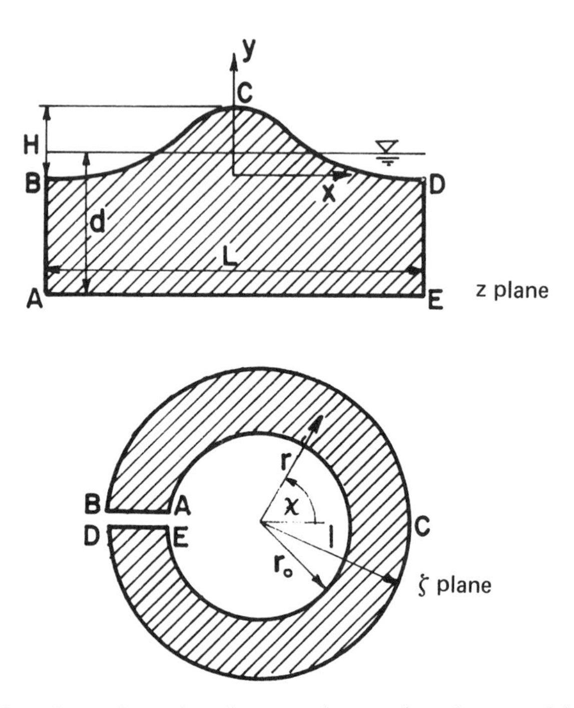

Fig. 4.13. Conformal transformations between the z and ζ planes used by Monkmeyer (1970) and Schwartz (1974).

face profile is then given by

$$k\eta = 6 \ln\left(\sec\frac{\theta}{6}\right) \qquad \text{for } -\pi < \theta < \pi \tag{4.144}$$

where η is here the free surface elevation measured above the wave trough and θ is the phase angle also measured from the wave trough. This formula predicts the wave steepness to be $H/L = 0.1374$ which differs from Schwartz's (1974) more accurate figure of 0.1412 for the maximum wave steepness by less than 3%.

Schwartz (1974) made considerable progress in carrying out wave calculations to high order. He employed a similar transformation formula to that used by Monkmeyer and given by

$$z = i\left\{\ln(\zeta) + \sum_{j=1}^{\infty} \frac{a_j}{j}\left(\zeta^j - \frac{r_0^{2j}}{\zeta^j}\right)\right\} \tag{4.145}$$

Once again the wave is mapped into the ζ plane as indicated in Fig. 4.13 and as before $r_0 = \exp(-d')$, with the limits $r_0 \to 0$ and $r_0 \to 1$ corresponding to the two extreme cases of deep water and the solitary wave respectively. The complex potential in the ζ plane representing a potential vortex is identical to before, Eq. (4.142), and the new set of coefficients a_j must again be determined numerically by considerations based on the dynamic free surface boundary condition. This condition may be written in terms of the complex velocity $q = dw/dz = u - iv$, and by virtue of Eqs. (4.142) and (4.145) this eventually provides a series of equations for the coefficients a_j and the celerity c, and thus also involves the depth parameter r_0 and the Bernoulli constant. Schwartz expanded the coefficients a_j and associated variables as series in a perturbation parameter ϵ and for the first time was able to carry out the calculations to a very high order with a computer. The series are analytically extended and summed using Padé approximants. These express a power series as a rational fraction which duplicates the original series to the known order, but which contains an infinite number of terms. The Padé approximant provides a much better approximation to the infinite sum than that obtained by simply summing the known terms. Their use provides for greatly improved convergence and high accuracy.

The choice of ϵ influences the nature of the results obtained. Stokes had in effect adopted the first coefficient a_1 as the perturbation parameter and this is indeed found entirely satisfactory for describing lower amplitude waves, for example by using the third and fifth order solutions. However, Schwartz investigated the use of a_1 in a high-order solution and discovered that as the wave height approaches the maximum value, a_1 no longer increases monotonically with height. Thus for steeper waves a high order solution based on a_1 is not convergent, and it is possible for two different heights to correspond to a given value of a_1. To avoid this difficulty Schwartz adopted instead the perturbation parameter $\epsilon = \frac{1}{2}kH$ and in the manner already outlined obtained results to order ϵ^{117} for deep water and to order ϵ^{48} for other depths. Deep water wave profiles based on this work and relating to various wave steepnesses H/L are reproduced in Fig. 4.14. The steepest wave in deep water is found to be given as H/L = 0.1412.

Cokelet (1977a) has employed a similar technique to that of Schwartz (1974) but used instead a perturbation parameter defined as

$$\epsilon = \left(1 - \frac{q_c^2 q_t^2}{c^4}\right)^{1/2} \tag{4.146}$$

where c is the wave speed, q_c and q_t are the fluid speeds at the crest and trough respectively in the (x, y) reference frame which moves with the waves. Unlike the parameter used by Schwartz, it is known *a priori* that ϵ ranges from 0 to 1. Cokelet pointed out that the conformal transformation procedure is not essential to the method, but that it is wholly equivalent (after Stokes 1880) to expressing z as a series in w. Cokelet worked with a flow moving from right to left (deriv-

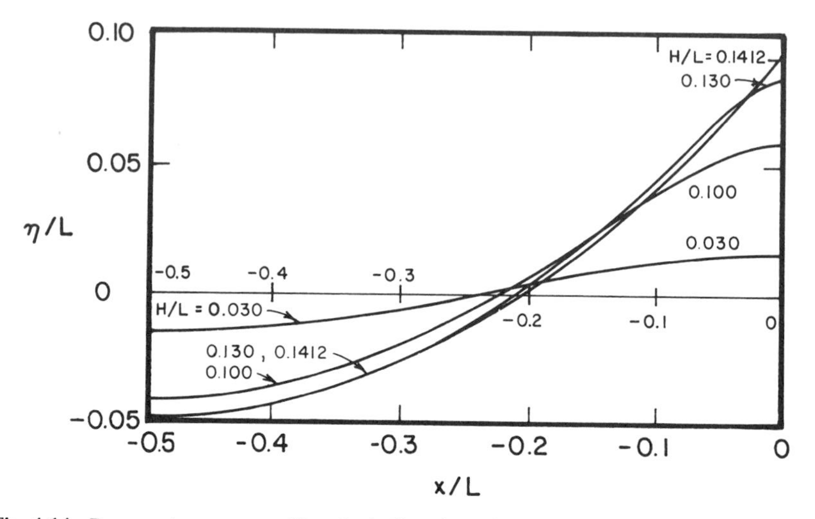

Fig. 4.14. Deep water wave profiles, including that of the steepest wave, as calculated by Schwartz (1974).

ing from x measured in the direction of wave propagation) and this relationship between z and w is

$$z(w) = -\frac{w}{c} + i \sum_{j=1}^{\infty} \frac{a_j}{j} \left(e^{ijw/c} - e^{-2jd} e^{-ijw/c}\right) \tag{4.147}$$

(The equivalent result with x defined in the opposite direction is obtained by substituting Eq. (4.142) and $r_0 = \exp(-d')$ into Eq. (4.145).)

Cokelet also summed the series using Padé approximants and was then able to carry out reliable calculations for waves of all heights up to the highest. He presented results for exp $(-d')$ ranging from 0 to 0.9 in steps of 0.1, this corresponding to kd varying from deep water conditions to as low as about 0.1. The results include the wave profile, wave celerity (both definitions) and various integral properties of the wave train. Cokelet's graphical results of the wave celerity (first definition) are reproduced in Fig. 4.15.

Bloor (1978) employed an alternative conformal transformation scheme which maps the region occupied by the fluid into the lower half of the ζ plane. The transformation of the fluid boundaries is given by the formula

$$\frac{dz}{d\zeta} = \frac{C}{\zeta} \exp\left\{\int_0^{\infty} \frac{1}{\pi} \frac{\alpha(t)}{\zeta - t} \, dt\right\} \tag{4.148}$$

Here α is the angle of the fluid boundary in the (physical) z plane to the x axis,

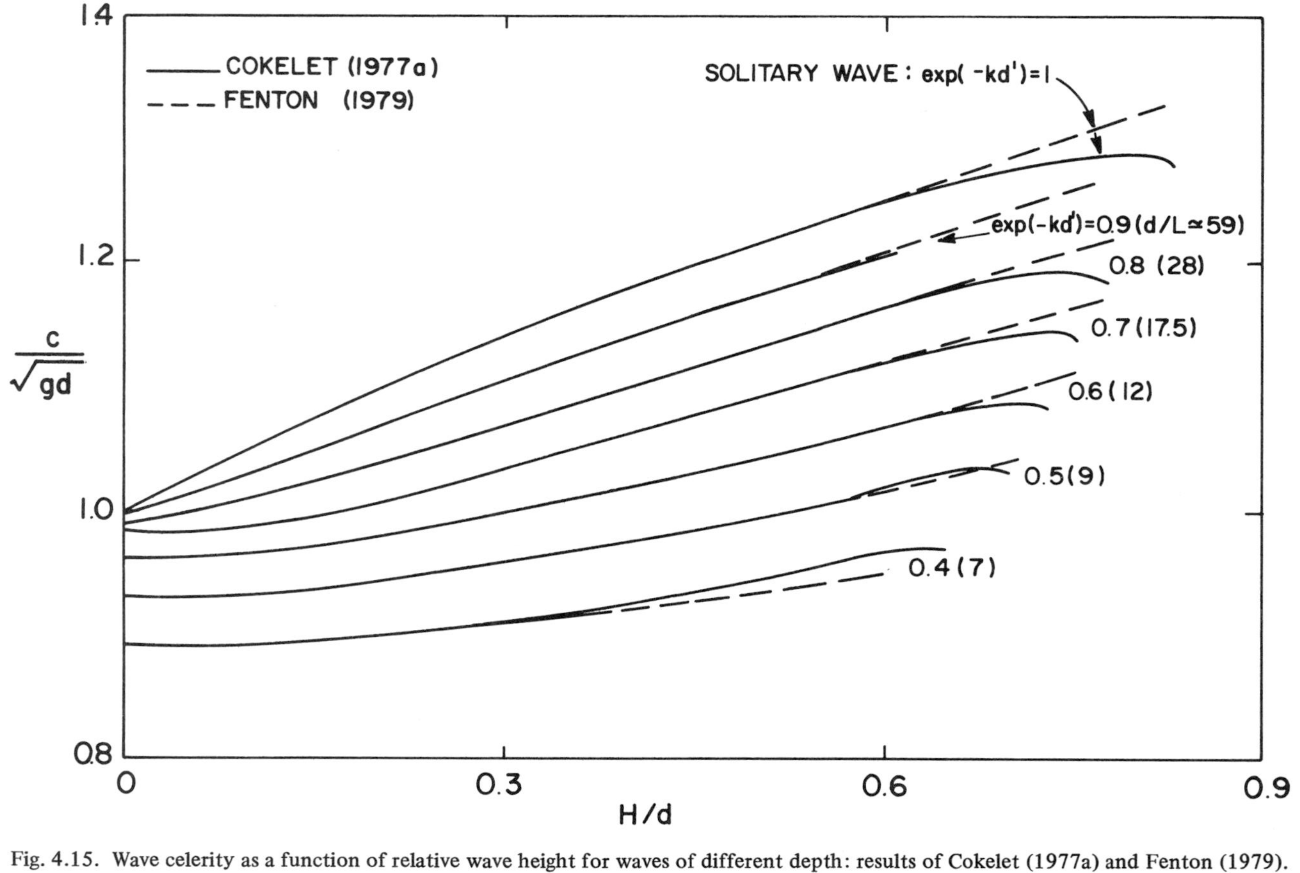

Fig. 4.15. Wave celerity as a function of relative wave height for waves of different depth: results of Cokelet (1977a) and Fenton (1979).

and C is a constant. The complex potential in the ζ plane corresponds simply to a source at the origin and so is known,

$$w = \frac{Q}{\pi} \ln(\zeta) \tag{4.149}$$

where Q/π is the source strength. Thus the velocity components in the z plane may be expressed as

$$u - iv = \frac{dw}{dz} = \frac{dw}{d\zeta}\bigg/\frac{dz}{d\zeta} = \frac{Q}{\pi C} \exp\left\{\int_0^{\infty} -\frac{1}{\pi}\frac{\alpha(t)}{\zeta - t}\,dt\right\} \tag{4.150}$$

By applying the free surface condition, the problem is reduced to obtaining the solution of an equation for α. This is done by expanding α in a suitable manner. Bloor shows that when α is small the linear wave solution is retrieved, while for steep waves the method provides good agreement with previous calculations (Chappelear 1961, Schwartz 1974). Bloor also applies the method to capilliary waves and indicates that it may be extended to wave motions about bodies.

Finally, mention is made of recent contributions by Schwartz and Vanden-Broeck (1979) and Vanden-Broeck and Schwartz (1979), in which accurate solutions have been obtained by a somewhat different transformation procedure. The former paper treats the deep water case but includes the effect of surface tension, while the latter paper extends the analysis (excluding surface tension effects) to finite depths and provides accurate solutions for very shallow waves (d/L as low as 1/120). The fluid region in the physical z plane is once again transformed to an annulus in the ζ plane as indicated in Fig. 4.13, but the transformation now involves a boundary-integral formulation over the free-surface in the ζ plane ($r = 1$). Results are compared with those of Cokelet (1977a) and the method apparently provides better convergence in shallow water.

The different approaches described here have been developed quite recently and have not yet found widespread engineering application as have, say, the Stokes fifth order and the stream function theories. The calculations of Schwartz (1974) and Cokelet (1977a) appear to be the most accurate ones carried out for a symmetrical wave train in water of uniform depth under a wide range of conditions. It is emphasized that their calculations are carried out for specified values of d' and ϵ, and the wave height H (in the case of Cokelet) and the still water depth d for a given wave length are essentially derived parameters. This differs from an explicit calculation for a given design wave specified in terms of H, T and d or H, L and d. However, this presents no real difficulty and the method should soon be used to provide a comprehensive assessment of the reliability of other wave theories and may also be applied directly to engineering problems.

4.6 COMPARISON OF WAVE THEORIES

The problem of selecting the most suitable wave theory for a particular application invariably arises in engineering situations. This is difficult to resolve since for specified values of H, T and d different wave theories might better reproduce different characteristics of interest and there can be no unique answer. It is emphasized then that any comparison must be considered only in relation to a particular characteristic and no inferences may necessarily be made regarding other characteristics of the wave motion.

One further aspect to be considered when selecting a wave theory concerns the increased effort necessary to utilize a more sophisticated theory. For example, when making preliminary or tentative calculations for a particular problem it may be best to employ linear wave theory and avoid recourse to a computer. Or again, Schwartz's (1974) approach may be found to be superior to Stokes fifth order theory for certain conditions, but the availability and simplicity of the latter may be the governing factor in a choice between the two. Comparisons between various theories may be made on both theoretical as well as experimental grounds and these are now reviewed in turn.

4.6.1 Comparisons Based on Theory

Laitone (1962) has compared the second approximation to cnoidal waves with the Stokes third order wave theory by examining expressions of wave speed predicted by the two theories. Laitone suggests that the cnoidal theory should not be applied when $L/d < 5$ ($kd > 1.26$) but suggests that it is better than Stokes theory for either $L/d > 8$ ($kd < 0.79$) or $U = HL^2/d^3 > 48.3$.

Dean (1970) has compared several wave theories on a theoretical basis. The criterion he used was the closeness of fit of the predicted motion to the complete problem formulation. Since the Laplace equation and bottom boundary condition are exactly satisfied in all the theories considered, the error of fit to the two nonlinear free surface boundary conditions was used as the criterion of validity. The wave theories examined included linear wave theory, Stokes third and fifth order theories, cnoidal (first and second approximations), solitary (first and second approximations) and the stream function theories. Dean found that the first order cnoidal, the linear, the Stokes fifth order and the stream function theories were generally the most suitable over the ranges indicated in Fig. 4.16. He emphasizes, however, that the method used to assess the theories does not necessarily imply the best overall theory, and he also suggests that this kind of comparison may be biased in favor of the lower order theories.

Le Méhauté (1976) has presented a convenient plot showing the approximate limits of validity of various wave theories as shown in Fig. 4.17. It is stressed however that Le Méhauté indicates that this plot is not based on any quantita-

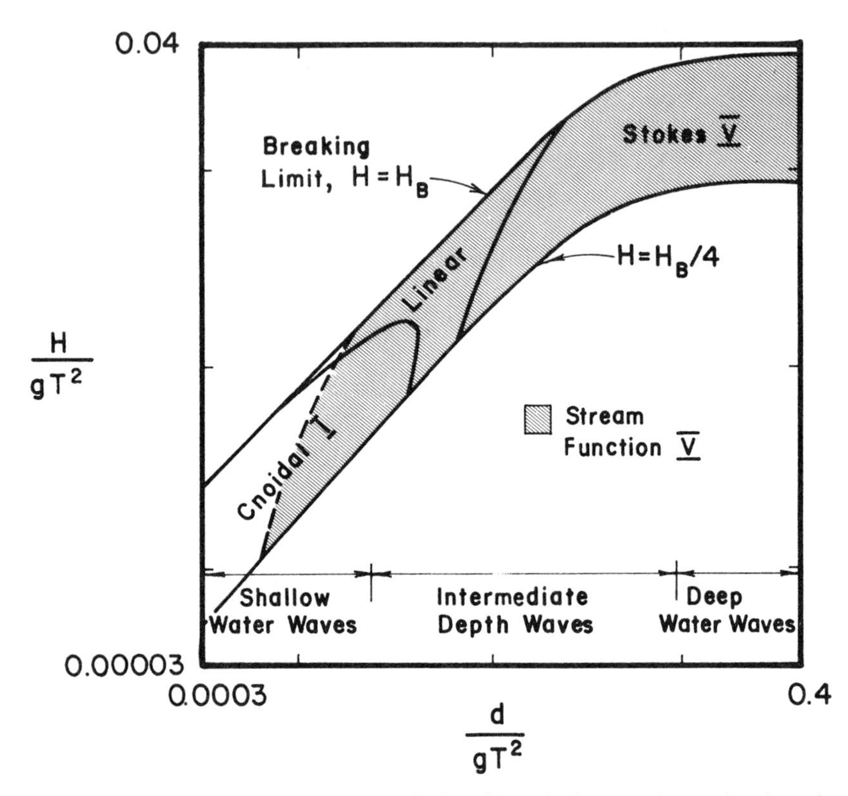

Fig. 4.16. Ranges of wave theories giving the best fit to the dynamic free surface boundary condition (Dean 1970).

tive investigation and so is somewhat arbitrary. Even so, there is some agreement with Dean's comparison in that for higher waves cnoidal wave theory is recommended for the shallow water range and Stokes high order theory for the deep water range.

A comparison between the second approximation to cnoidal wave theory and Stokes second order theory has been made by Isaacson (1978a) on the basis of the mass-transport velocity near the seabed. His curves, given in Fig. 4.18, provide a convenient means of delineating the suitability of either theory and suggest that the cnoidal wave theory is superior when approximately

$$\frac{H}{d} > 350 \left(\frac{d}{gT^2}\right)^{3/2} \tag{4.151}$$

(H/d is limited by wave breaking.) This of course relates to the particular characteristic mentioned, but the mass-transport velocity depends largely on the mean-square of the orbital velocity at the seabed and so provides a comparison which

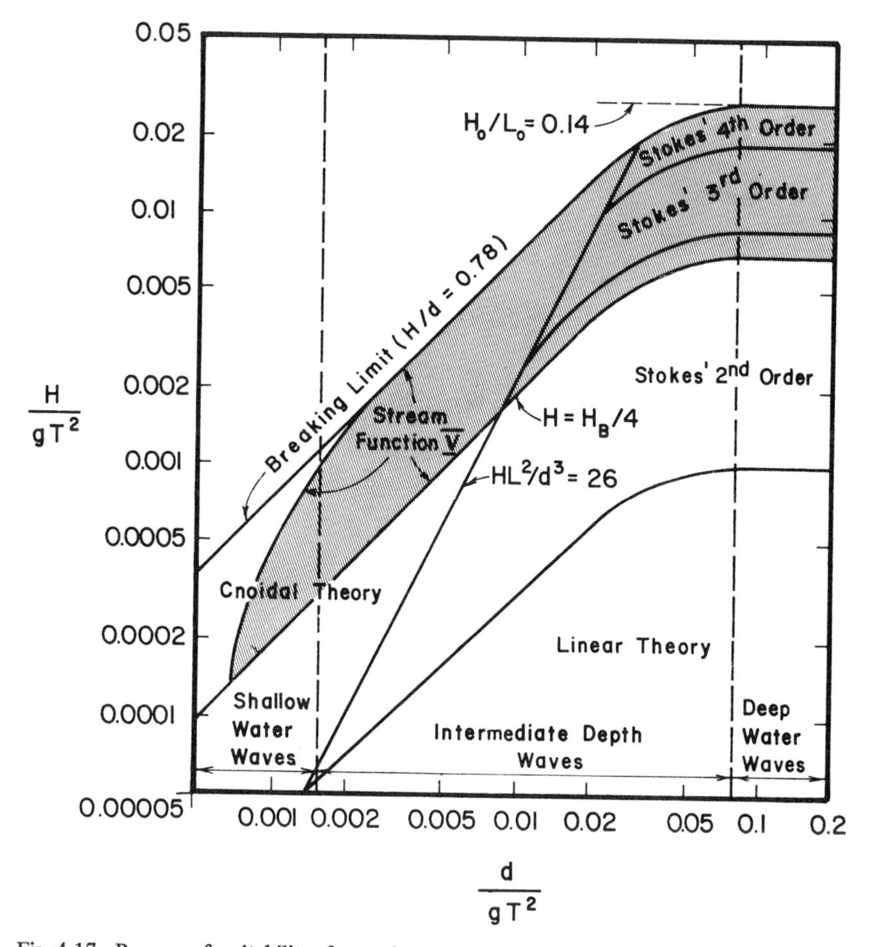

Fig. 4.17. Ranges of suitability for various wave theories as suggested by Le Méhauté (1976).

depends on temporal variations throughout a cycle rather than upon an instantaneous value of some quantity such as the maximum orbital velocity at the seabed.

Cokelet (1977a) has compared various results obtained by his approach with those of Dean (1965, 1974), Von Schwind and Reid (1972), Schwartz (1974), Longuet-Higgins (1975) and others. He reports that agreement with Dean's stream function theory for deep water waves is good for waves up to moderate wave steepness, but that Dean's values for the maximum deep water wave steepness and corresponding celerity show some differences from his own results. Dean obtained $H/L = 0.1394$ and $kc^2/g = 1.2221$ which compare with Cokelet's results, $H/L = 0.1411$ and $kc^2/g = 1.1928$ (the maximum value of kc^2/g is about

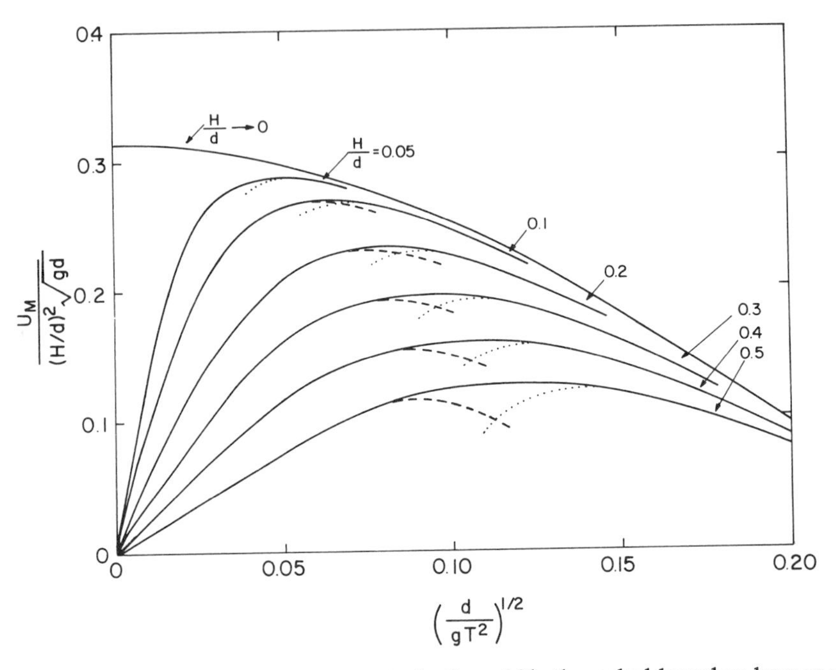

Fig. 4.18. Non-dimensional mass-transport velocity outside the seabed boundary layer as a function of $(d/gT^2)^{1/2}$ for various values of H/d (Isaacson 1978a). ---, · · · · , departure of cnoidal and Stokes wave theory predictions from the proposed curves. The curve $H/d \rightarrow 0$ corresponds to sinusoidal wave theory.

1.1945). These figures show a difference of about 2% in the maximum wave steepness and for typical design waves which have a smaller height the differences are not likely to be significant for engineering purposes. It would seem, however, that Cokelet's results remain accurate in relatively shallow water: he presents results for kd as low as about 0.1 ($d/gT^2 \simeq 0.0003$). Finally, it must be borne in mind that particle velocities and accelerations are perhaps the quantities of greatest importance in wave force calculations. Although the methods of Cokelet and Schwartz are capable of generating these quantities, at the present time they have not yet found widespread usage in this regard.

Fenton (1979) has compared wave celerity predictions of the fifth order cnoidal wave theory which he has developed with the accurate results of Cokelet (1977a). The comparison is shown in Fig. 4.15 and indicates that the cnoidal theory predictions are reliable for kd up to about 1.0 even for quite steep waves.

Stokes fifth order theory and stream function theory are both used frequently in wave force calculations, and it is instructive to compare the predictions of either theory. In deeper water corresponding predictions tend to overlap, whereas at intermediate depths the temporal and spatial variations predicted by the fifth-order Stokes theory based on Skjelbreia and Hendrickson (1960) may contain unrealistic fluctuations or "bumps" not reproduced by the stream function theory. In even shallower water the wave profiles given by Stokes theory may be triple-crested (i.e. they contain two secondary crests in addition to the primary crest over a wave length), or alternatively the pertinent equations may not be susceptible to solution. Ebbesmeyer (1974) has proposed that the presence of these bumps is an indication of the validity of Skjelbreia and Hendrickson's theory, and he presented the range of conditions applicable to each of the situations mentioned as reproduced in Fig. 4.19. This figure indicates the regions in which (a) wave profiles are smooth, (b) wave profiles contain bumps, and (c) wave profiles are triple-crested or no solution exists. A comparison between Fig. 4.16 and 4.19 will make clear the conditions in which stream function theory shows itself as particularly superior.

As an example of these differences in prediction, Fig. 4.20 compares the variation of the free surface elevation as calculated by the Stokes fifth order theories based on Skjelbreia and Hendrickson (1960) and by De (1955) and by the stream function theory taken to fifth order, for two waves with correspond-

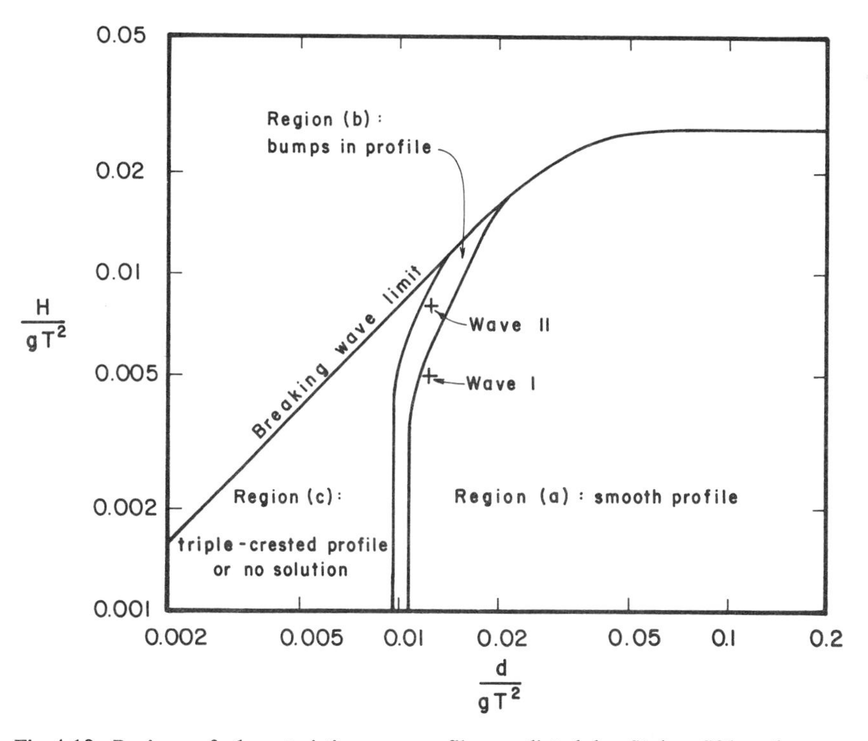

Fig. 4.19. Regions of characteristic wave profiles predicted by Stokes fifth order wave theory (Ebbesmeyer 1974). Waves I and II refer to those shown in Fig. 4.20.

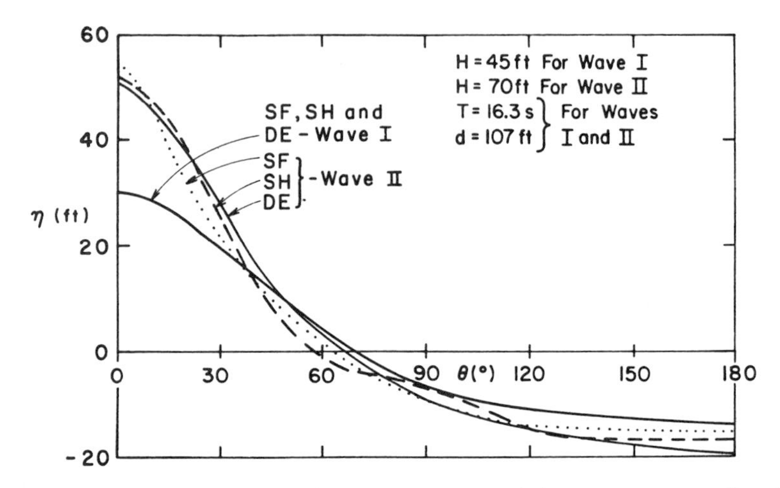

Fig. 4.20. Comparison of wave profiles for two specified design waves, as predicted by Stokes fifth order wave theory based on Skjelbreia and Hendrickson (1960)(SH), and De (1955)(DE), and by fifth order stream function theory (SF). Calculated wave lengths are $L_{SH} = L_{DE} = 948.1$ ft., $L_{SF} = 934.9$ ft. for Wave I; and $L_{SH} = 1080$ ft., $L_{DE} = 981.2$ ft., $L_{SF} = 997.4$ ft. for Wave II.

ing depth and period, d = 107 ft, T = 16.3 sec, and with heights of 45 ft and 75 ft in turn (Dalton, personal communication). These two design waves correspond to the two points I and II in Fig. 4.19. The corresponding distinction between the smooth profiles and the profile based on Skjelbreia and Hendrickson (1960) which contains bumps is clearly illustrated in Fig. 4.20. The different wave lengths predicted by the three wave theories are given in the figure caption.

In summary, it is apparent that modern wave theories (e.g. Schwartz 1974, Cokelet 1977a) are able to predict accurately the flow behavior over the complete range of design wave parameters H, L and d. The available accuracy is in excess of typical engineering requirements, when compared to the somewhat arbitrary choice of celerity definition used, and to the unreliabilities in design wave selection or in other steps of the entire design process. Availability or the ability to apply a wave theory fairly easily are also of major concern. From this viewpoint, it would appear that the Stokes and cnoidal fifth order theories are relatively simple to use, and yet provide sufficient accuracy for most engineering purposes. Fenton (1979) recommends that the cnoidal theory may be used for $L/d < 8$ and the Stokes theory otherwise. At lower wave heights, the ranges of validity of these two theories overlap to a greater extent.

4.6.2 Comparisons Based on Experiment

The suitability of one theory over another from a theoretical viewpoint is not necessarily reflected in better agreement with experimental data gathered either in the laboratory or in the field. Once again the suitability of a particular theory depends upon which characteristic is being compared. Le Méhauté, Divoky and Lin (1968) have reported on a series of experiments used to assess the validity of various wave theories for relatively shallow conditions. This was based on measurements of the maximum particle velocity and indicated that no single theory could be considered outstanding. Goda's (1964) empirical correction to linear wave theory, linear theory itself and stream function theory (compared in a subsequent study, Dean and Le Méhauté 1970) were however found to provide better agreement than did the remaining ones.

Tsuchiya and Yamaguchi (1972) have compared theoretical predictions with measurements of wave celerity, horizontal and vertical particle velocities at various depths in phase with a wave crest and trough, and temporal variations of the free surface elevation and the horizontal and vertical velocity components. In this comparison they used their own experimental data as well as earlier data of Goda (1964), Le Méhauté, Divoky and Lin (1968), Iwagaki and Sakai (1969) and Iwagaki and Yamaguchi (1968); and the theories considered were linear wave theory, Stokes fourth order theory (both definitions of celerity), Stokes fifth order theory (De's (1955) version, first definition of celerity), and the second and third approximations to cnoidal wave theory (both definitions of celerity).

A few of their comparisons are reproduced in Fig. 4.21 which shows vertical distributions of horizontal particle velocity in phase with a wave crest. Tsuchiya and Yamaguchi found that finite amplitude wave theories predict particle velocities well, but they did not recommend any one theory as being the most suitable over any particular range. They suggest, however, that the poor agreement obtained by Le Méhauté, Divoky and Lin (1968) may have been due to the apparatus used, and that the second definition of celerity gives better agreement with measured wave speeds than does the first definition.

Fenton (1979) has compared the measurements of horizontal velocities under a wave crest obtained by Le Méhauté, Divoky and Lin (1968) and by Iwagaki and Sakai (1969) with the predictions of fifth order cnoidal wave theory. He finds that, over its range of validity, the cnoidal theory provides better agreement with the measurements than do the Stokes or stream function theories.

Grace (1976) has measured wave pressures together with particle velocities and accelerations in the ocean environment, and found a favorable comparison between the observed maximum horizontal velocity and the predictions of linear wave theory. He emphasizes that different waves possessing identical values of H, T and d will differ in their detailed behavior. As an example, the temporal variations of the free surface elevation and near-bottom particle velocity and acceleration for three waves with the same values of H, T and d are set out in Fig. 4.22.

Ohmart and Gratz (1978) have compared water surface profiles and horizontal particle velocities and accelerations measured at a site in the Gulf of

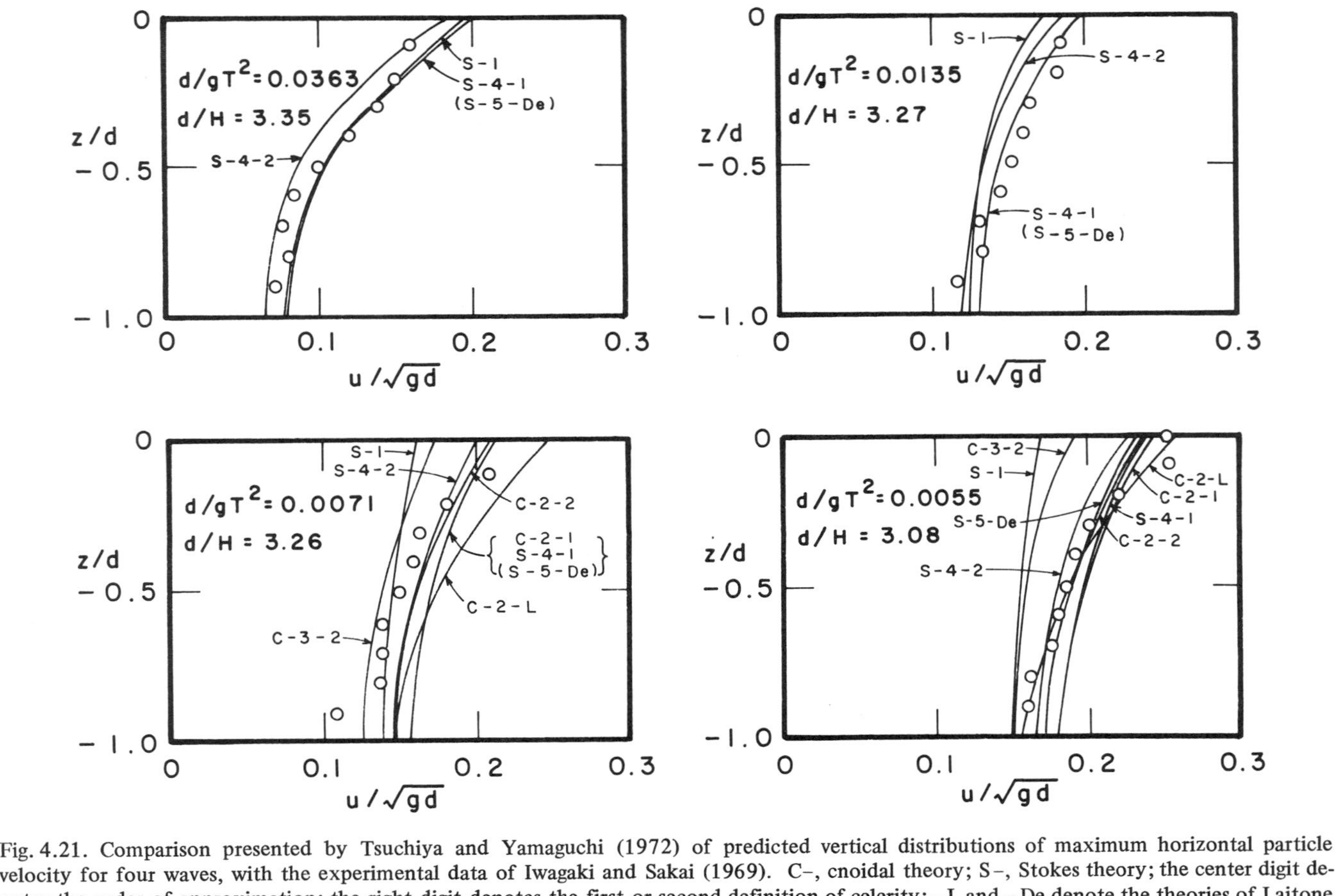

Fig. 4.21. Comparison presented by Tsuchiya and Yamaguchi (1972) of predicted vertical distributions of maximum horizontal particle velocity for four waves, with the experimental data of Iwagaki and Sakai (1969). C–, cnoidal theory; S–, Stokes theory; the center digit denotes the order of approximation; the right digit denotes the first or second definition of celerity; –L and –De denote the theories of Laitone (1961) and De (1955) respectively.

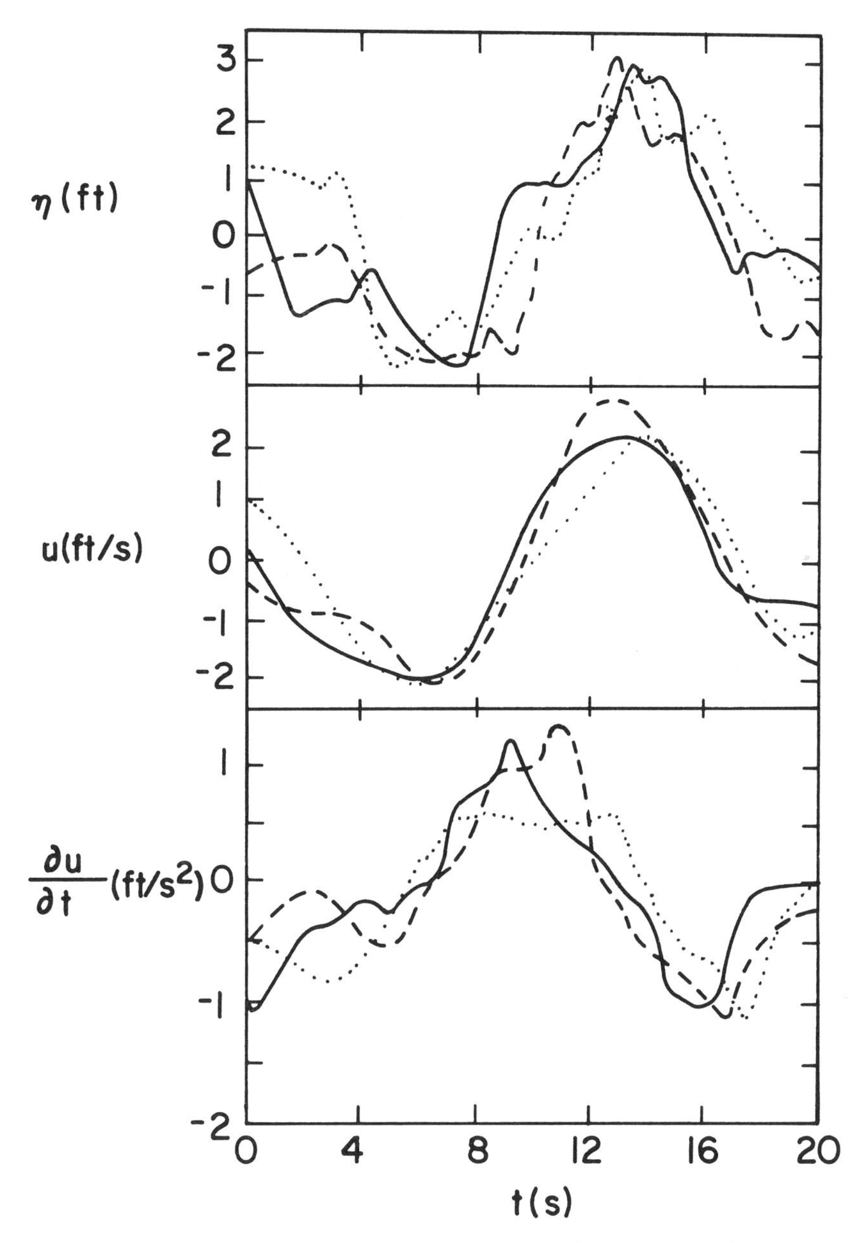

Fig. 4.22. Comparison of measured temporal variations of surface elevation and near-bottom particle velocity and acceleration for three waves of identical gross characteristics (Grace 1976).

Mexico with the predictions of linear, Stokes fifth order and stream function (both regular and irregular) wave theories. As might be expected, they found that the irregular form of the stream function theory provided the best fit to the data. Otherwise, they found that for the conditions examined the predictions of both the Stokes fifth order theory and the regular stream function theory were indistinguishable, and that except for the higher waves even the linear theory predictions were close to those of the Stokes fifth order theory.

As a final comment, it is of practical importance to have an accurate knowledge of particle velocities and accelerations in steep near-breaking waves, especially in the context of wave force calculations. However, it appears that comprehensive comparisons with experiment under such more extreme conditions are relatively unavailable and thus, in spite of the sophistication of wave theories that may be employed, uncertainties remain in the prediction of particle kinematics for very steep waves.

4.7 SIMILARITIES WITH HARMONIC FLOW

A major simplification arises when it is possible to represent the wave-induced flow at a given depth by a horizontal two-dimensional simple harmonic flow. Considerable experimental work has been carried out with this harmonic flow used to simulate the wave motion (see Chapters 3 and 8). In order to fix ideas, we may consider the wave motion around a section of a vertical pile, while bearing in mind that the Morison equation, already developed in Chapter 3 and discussed further in Chapter 5 with respect to wave motion, expresses the sectional force on the pile in terms of the horizontal fluid velocity and acceleration at that section.

A two-dimensional harmonic flow can be reproduced more conveniently in the laboratory than can the corresponding wave motion, and in particular it is possible to attain conditions at a relatively high Reynolds number. From a research viewpoint, the use of a two-dimensional flow serves to isolate some of the more fundamental aspects of a wave/structure interaction problem. Indeed, reliable laboratory measurements of wave force coefficients have been made for such a flow and the relationship of the flow to the initial wave motion forms an essential step in the solution of many wave force problems. It is necessary, then, to consider carefully the nature of the similarities and dissimilarities between these two flows.

A fundamental result of linear wave theory is that the horizontal velocities, displacements and accelerations are simple harmonic with the amplitude of motion dependent on the section depth. This flow is to be compared with one represented simply by $u = U_m \cos(\omega t)$, where U_m is independent of location. We see from this that in order to simulate more completely the flow field generated by the wave motion certain additional effects will have to be included.

These are

(a) the cyclical variation over a wave length of horizontal velocity, together with other kinematics,
(b) vertical velocities,
(c) vertical gradient of horizontal velocity amplitude,
(d) harmonics of the horizontal velocity not predicted by linear wave theory, and
(e) free surface effects.

When investigating a problem confined within a range of x which is small compared with the wave length (such as the wave interaction with a slender structural element), the first effect (a) above should be unimportant. This is in contrast to the case of wave interaction with a large structure extending over an appreciable fraction of the wave length where the procedure described here is not too useful, and the approaches of Chapter 6 then apply.

When a "strip" approach is used such that the flow (and force) at any section (elevation) is assumed independent of the flow at neighboring sections, the effects (b) and (c) are considered unimportant. But in certain situations, such as where the force on a pile is influenced by the extent of vertical vortex correlation, the effects (b) and (c) will clearly assume a particular significance. The relative importance of these two effects may be characterized by certain simple formulae. Vertical velocities may be examined by reference to the ratio of vertical to horizontal velocity amplitudes given by linear wave theory which is simply $\tanh(k(z + d))$. Similarly, the vertical gradient of horizontal particle displacement, which is a suitable dimensionless measure of effect (c), is given as $\frac{1}{2}kH \sinh(k(z + d))/\sinh(kd)$. The maximum values of these two ratios occur at the free surface and are, respectively,

$$\left(\frac{w_{max}}{u_{max}}\right)_{z=0} = \tanh(kd) \tag{4.152a}$$

$$\left(\frac{1}{\omega}\frac{\partial u_{max}}{\partial z}\right)_{z=0} = \frac{1}{2}kH \tag{4.152b}$$

and it follows that both the effects are relatively unimportant in small amplitude shallow water waves and in any case they become insignificant for conditions near the seabed.

The horizontal velocity at a given location in a wave flow may generally be represented as a Fourier series

$$u = \sum_{n=0}^{\infty} A_n \cos(n\omega t) \tag{4.153}$$

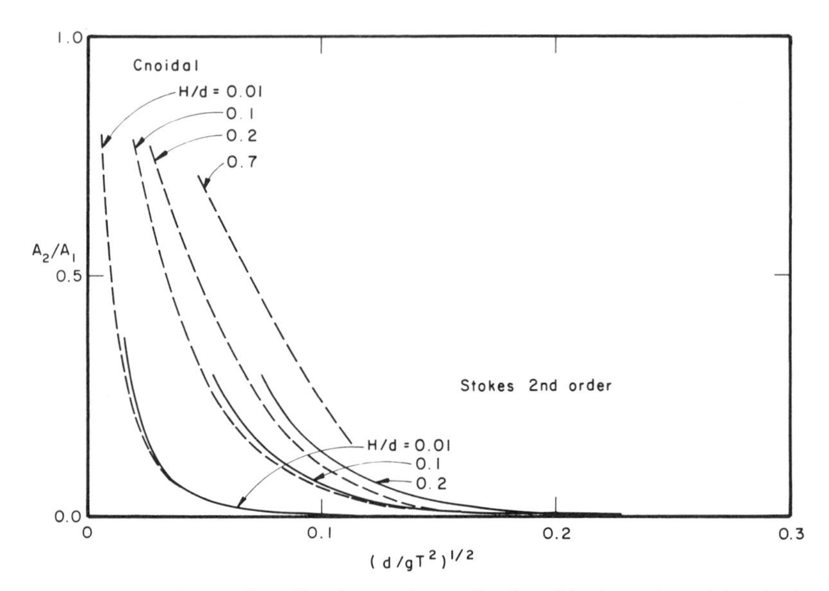

Fig. 4.23. Ratio of second to first harmonic amplitudes of horizontal particle velocity at $z = 0$ as a function of $(d/gT^2)^{1/2}$ for various values of H/d. ——, Stokes second order theory, – –, second approximation to cnoidal theory.

so that one possible indication of the extent of harmonics present is given simply by the ratio A_2/A_1, which may be established on the basis of any chosen wave theory appropriate to the given conditions. This ratio at $z = -d$, given both by Stokes second order wave theory (which gives $A_n = 0$ for $n > 2$) and the second approximation to cnoidal wave theory, is presented in Fig. 4.23 as a function of $(d/gT^2)^{1/2}$ for various values of H/d. It is clearly seen that for shallower waves, say $(d/gT^2)^{1/2} < 0.1$, the ratio is significant even for waves of modest height.

Another feature which is significant in the comparison of a uniform harmonic flow and a wave flow concerns the existence of convective accelerations. These are absent in the (uniform) harmonic flow, whereas in a wave flow they are present to the second order on account of the spatial variations of velocity. Thus the distinction between local and total (i.e. local plus convective) accelerations may be important in nonlinear wave flows. For example, the inertia force as applied to the Morison equation is usually taken as proportional to one or the other, even though these may differ noticeably in nonlinear waves. The consequences of this are discussed elsewhere (Sections 5.3 and 6.8.4), but it is appropriate here to illustrate these differences for typical nonlinear waves. Figure 4.24 shows the variations of local horizontal acceleration $\partial u/\partial t$ and total horizontal acceleration Du/Dt $(=\partial u/\partial t + u\, \partial u/\partial x + w\, \partial u/\partial z)$ over a wave cycle on the basis of the stream function theory for two design waves of the same

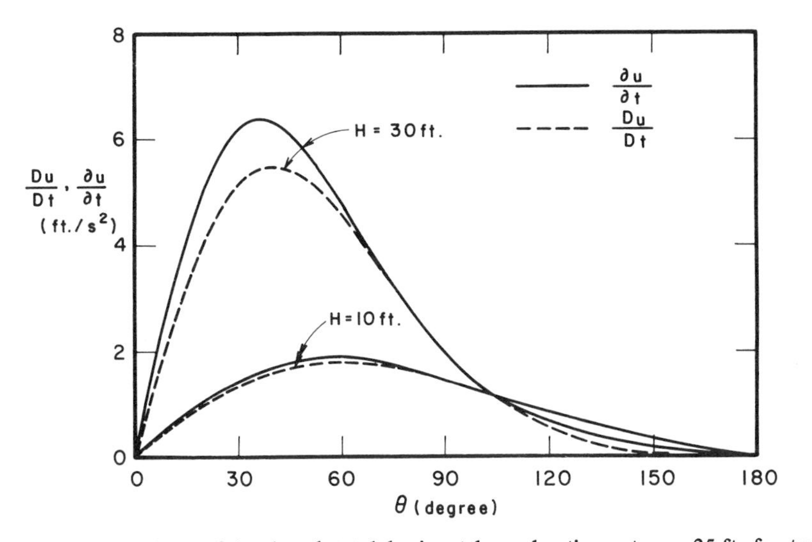

Fig. 4.24. Comparison of local and total horizontal accelerations at z = −25 ft. for two waves of height H = 10 ft. and 30 ft., and both with T = 14 sec, d = 50 ft. The curves are based on stream function theory.

depth and period, as indicated in the figure caption. The kind of differences observed, particularly for the steeper wave, must be borne in mind when considering applications involving inertia coefficients with nonlinear waves (Isaacson 1979), (see Section 5.3.1).

When conditions near the water surface are to be investigated the use of a two-dimensional oscillatory flow is not really appropriate at relatively small Keulegan-Carpenter numbers. This subject is discussed in some detail in Section 5.3.4.

A related reference flow that may be used in turn to simulate the two-dimensional harmonic flow about a body is one in which the identical body is oscillated in otherwise still water. The pair of two-dimensional flows that have then to be compared are as follows:

(a) the fluid remote from a stationary body has a uniform sinusoidal velocity, and
(b) the identical body has an equal velocity in the opposite direction while the fluid is now otherwise stationary.

A formal examination of this change in reference frame may be made (e.g. Batchelor 1967), and it may readily be shown that

(i) the fluid velocities relative to the body and hence the "flow patterns" in both cases are identical,

(ii) the flow (b) with the oscillating body has a pressure gradient acting in the direction of the fluid acceleration relative to the body which is an amount

$$\frac{\partial p}{\partial x} = \rho \frac{\partial u}{\partial t} \tag{4.154}$$

in addition to that of flow (a),

(iii) the effect of this pressure gradient on the longitudinal force is that the force on the body in the flow (b) is smaller than that on the body in the flow (a) by an amount $\rho A \partial u/\partial t$, where A is the sectional area of the body,

(iv) the transverse force on the body is identical in both flows.

It is emphasized that the comparison applies to flows of a real fluid and no assumption is made regarding the absence of viscous effects.

It is thus clear that the two flow situations are identical with the exception of the inertial force due to the pressure gradient necessary to accelerate the fluid. There may, however, be significant differences in the performance of experiments with oscillating flows and bodies. In general, the oscillation of a body in a fluid otherwise at rest offers severe vibration problems, free surface disturbances, and requires that the inertial force due to the accelerating mass of the body be subtracted from the total measured force. Finally, it is quite difficult to measure both the in-line and transverse components of the fluid force acting on an oscillating cylinder. For an additional discussion of the foregoing see Chapter 3.

The above outline has been made mainly in the context of wave motion past a vertical cylinder, but the remarks made can be extended to the flow in the neighbourhood of the seabed which is also of great practical importance in connection with submarine pipelines, sediment problems and so on.

4.8 MASS TRANSPORT IN WAVES

As already mentioned in Section 4.3, the fluid particle motions in a two-dimensional wave train include a steady drift, the speed of which is termed the *mass-transport velocity*. This was first predicted theoretically by Stokes (1880) for irrotational waves with a sinusoidal first order motion. However, the magnitude of the predicted mass-transport velocity is somewhat arbitrary, depending on the definition of wave celerity adopted.

An expression for the horizontal mass-transport velocity may be developed by representing the (Lagrangian) horizontal velocity of a given particle at the instantaneous position (ξ, ζ) as a Taylor series of the (Eulerian) horizontal velocity at a given point (0, 0). This is

$$u(\xi, \zeta, t) = u(0, 0, t) + \xi \frac{\partial u(0, 0, t)}{\partial x} + \zeta \frac{\partial u(0, 0, t)}{\partial z} + \ldots \tag{4.155}$$

The usual perturbation procedure is now used in which velocity components are represented in the form

$$f = \sum_{n=1}^{\infty} \epsilon^n f_n \tag{4.156}$$

where ϵ is a perturbation parameter. Bearing in mind that the time-averaged first order velocity is taken as zero (i.e. there is no underlying current in the absence of the wave motion), the substitution of Eq. (4.156) into the Taylor series Eq. (4.155) shows that the horizontal mass-transport velocity is a second order quantity which is given to the first approximation as

$$\begin{aligned} U_M &= \epsilon^2 U_{M2} = \epsilon^2 \overline{u_2(\xi, \zeta, t)} \\ &= \epsilon^2 \left\{ \overline{u_2} + \overline{\frac{\partial u_1}{\partial x} \int^t u_1 \, dt'} + \overline{\frac{\partial u_1}{\partial z} \int^t w_1 \, dt'} \right\} \end{aligned} \tag{4.157}$$

where an overbar denotes a temporal mean and the velocity components shown without arguments are taken to represent the usual Eulerian components calculated by wave theory. Thus even when the mean Eulerian velocity vanishes (implying that $\overline{u_2} = 0$) in accordance with the first definition of wave celerity, the mass-transport velocity itself is generally nonzero by virtue of the remaining terms in Eq. (4.157) above. When the first order expressions of linear wave theory are substituted into Eq. (4.157) and when $\overline{u_2}$ is assigned a constant value (implying the superposition of a uniform second order current with a related value of c), U_M is eventually given as

$$U_M = \frac{H^2 \omega k}{8 \sinh^2 (kd)} \cosh [2k(z + d)] + C \tag{4.158}$$

where C is an arbitrary constant (which is of order the first term on the right-hand side of Eq. (4.158)). If C is chosen such that the total horizontal mass transport averaged over depth is zero (as might be expected in a wave channel), we then have

$$U_M = \frac{H^2 \omega k}{8 \sinh^2 (kd)} \left[\cosh [2k(z + d)] - \frac{\sinh (2kd)}{2kd} \right] \tag{4.159}$$

Some profiles with depth of this mass-transport velocity, expressed in a suitably dimensionless form, are presented in Fig. 4.25a for different values of kd. One particular consequence of this approach is that the predicted mass-transport velocity at the seabed is then in the opposite direction to that of wave propagation which is in contrast to observation.

In 1953, Longuet-Higgins (1953) presented a general theory of mass transport in which the effects of fluid viscosity were included and which predicted for the first time the observed forward mass transport near the seabed. Longuet-Higgins found that both the mass-transport velocity just outside the laminar boundary layer at the seabed, as well as the vertical gradient of the mass-transport velocity just below the free surface boundary layer are established uniquely and are independent of the coefficient of viscosity. These are given respectively as

$$U_M = \frac{5 H^2 \omega k}{16 \sinh^2 (kd)} \qquad \text{at } z = -d \tag{4.160a}$$

$$\frac{\partial U_M}{\partial z} = H^2 \omega k^2 \coth (kd) \qquad \text{at } z = 0 \tag{4.160b}$$

where the elevations shown refer to the corresponding interior boundary layer edges.

Longuet-Higgins obtained two equations, termed the *conduction* and *convection equations*, which describe the mass transport in the fluid interior under different circumstances. When the wave amplitude a is much smaller than the boundary layer thickness $\delta = (2\nu/\omega)^{1/2}$ (which is of course generally untrue in practice), the mean second order motion is governed by the "conduction" equation

$$\nabla^4 \overline{\psi_2} = 0 \tag{4.161}$$

where $\overline{\psi_2}$ is the stream function corresponding to $(\overline{u_2}, \overline{w_2})$. This equation may be solved by applying the boundary conditions Eqs. (4.160) and making the

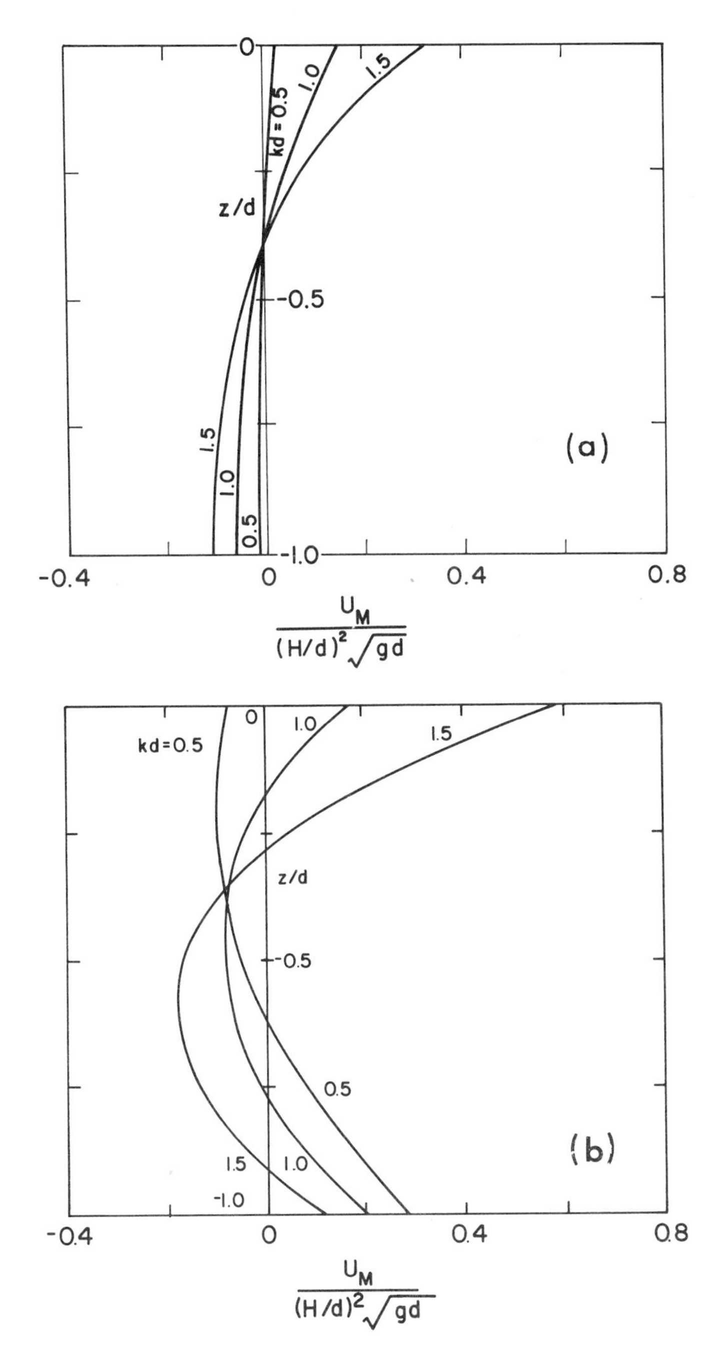

Fig. 4.25. Non-dimensional mass-transport velocity profiles with depth for various values of the depth parameter kd. (a) Stokes (1880) solution for irrotational flow. (b) Longuet-Higgins (1953) "conduction" solution.

additional assumption that the total horizontal mass transport averaged over depth is zero. The resulting solution is

$$U_M = \frac{H^2 \omega k}{16 \sinh^2 (kd)} \left\{ 2 \cosh [2k(z + d)] + 3 \right.$$

$$+ 3 \left[\frac{\sinh (2kd)}{2kd} + \frac{3}{2} \right] \left[\left(\frac{z}{d} \right)^2 - 1 \right]$$

$$\left. + kd \sinh (2kd) \left[3 \left(\frac{z}{d} \right)^2 + 4 \left(\frac{z}{d} \right) + 1 \right] \right\} \tag{4.162}$$

A number of profiles showing the mass-transport velocity variation with depth for different values of kd and deriving from Eq. (4.162) are plotted in Fig. 4.25b. These show notable differences from those in Fig. 4.25a which correspond to the mass transport for a wholly irrotational flow.

The "convection" equation becomes valid when, as is invariably the case in practice, the wave amplitude a is much larger than the boundary layer thickness δ. This equation is given as

$$\left(U_{M2} \frac{\partial}{\partial x} + W_{M2} \frac{\partial}{\partial z} \right) \nabla^2 \overline{\psi_2} = 0 \tag{4.163}$$

where U_{M2}, W_{M2} are the mass-transport velocity components based on Eq. (4.156). Unfortunately the "convection" equation provides results which depend on the specification of additional boundary conditions at the upstream and downstream ends of a channel and so are difficult to interpret.

It appears, then, that neither of the two solutions may be of practical use: the former because the assumption of such a small wave height is unrealistic, and the latter because the solution is not readily obtained. One other difficulty with the "conduction" solution Eq. (4.162) is that it becomes unbounded as the depth d becomes infinite: this can be seen by putting $z = 0$ or $-d$ in Eq. (4.162) and letting kd become large. Furthermore, Sleath (1973) and Liu and Davis (1977) have indicated that the viscous attenuation of the waves (Section 4.9.6) influences the mass-transport velocity distribution over depth. Liu and Davis (1977) develop a modification to Longuet-Higgins' "conduction" solution which, for the case of progressive waves, may be expressed as

$$U_M = \frac{1}{4} H^2 \omega k A^2 \left\{ \frac{\cosh [2k(z+d)]}{2 \sinh^2 (kd)} + \coth (kd) \sin (2kz) + \frac{1}{2} \overline{p}_x \right.$$

$$+ \left[\frac{3}{4} \frac{1}{\sinh^2 (kd)} + \sin (2kd) \coth (kd) \right.$$

$$\left. \left. - \frac{1}{2} \overline{p}_x \right] \frac{\cos (2kz)}{\cos (2kd)} \right\} \tag{4.164}$$

where A is a factor with a value of unity at a reference location and which varies with x, reflecting the slow decay due to viscous attenuation. Also $\overline{p}_x$ is the mean pressure gradient imposed on the flow, which is related to the net drift over depth, and so can be selected (except in deep water) to provide for no net drift as before.

One important feature of this solution is that it is derived without any restriction placed on the wave amplitude to boundary layer thickness ratio a/δ, although Liu and Davis suggest that when a/δ is large the solution might be unstable and other solutions may then exist. Furthermore, unlike Longuet-Higgins' "conduction" solution, Liu and Davis' solution is bounded as $kd \to \infty$, but at certain critical values of kd ($\pi/4, 3\pi/4, 5\pi/4, \ldots$) a quasi-steady state is not reached.

Mass transport measurements have been made by Russell and Orsorio (1958) and these indicate that mass transport profiles with depth are in many cases quite similar to Longuet-Higgins' "conduction" solution, even though the condition $a \ll \delta$ did not hold. Liu and Davis' modified solution, which exhibits profiles similar to those of the original solution, does not require the condition $a \ll \delta$ and so helps explain the observed profiles of Russell and Orsorio.

The mass-transport velocity near the seabed is of particular practical importance since it influences the sediment transport processes there. Longuet-Higgins (1958) showed that his previous result for the mass-transport velocity outside the bottom boundary layer may also be valid when the boundary layer is turbulent and does therefore have application in the field. The mass transport in the bottom laminar boundary layer has since been extended by Sleath (1972) to a second approximation based on Stokes finite amplitude wave theory, and by Isaacson (1976a, c) to both first and second approximations based on cnoidal wave theory. Isaacson (1978a) also compared the mass-transport velocity outside the bottom boundary layer as predicted by the second approximation to cnoidal theory with that predicted by the Stokes second order theory. These results have already been reproduced in Fig. 4.18. In this figure the portions of the curves applicable to shallow water correspond to cnoidal theory, while those for deeper water correspond to the Stokes theory. Departures of the predictions of either theory from the suggested curves are indicated by the broken lines. Although the onset of turbulence limits the range of validity of these curves, it is suggested that they remain valid within the range of accuracy set by the first approximations and so may continue to apply when the boundary layer is turbulent. The question of transition to a turbulent boundary layer is discussed in Section 4.9.6. The results of Isaacson's (1978a) experiments, as well as those of the previous experiments of Brebner and Collins (1961) agree reasonably well with these predictions. Other measurements of mass transport had earlier been made by Russell and Orsorio (1958), as already mentioned, and also by Allen and Gibson (1959).

4.9 TRANSFORMATION OF WAVES

We have so far been concerned with the behavior of a regular wave train of permanent form propagating over a smooth horizontal bed in the absence of an underlying current. We turn now to consider some of the more fundamental changes that a wave train may undergo, such as may be caused by variations in water depth, the presence of currents, or obstacles in the flow. Because the intention here is simply to provide a background to the central topic of wave forces, the presentation is necessarily concise and only some of the more basic features of each aspect are outlined.

4.9.1 Wave Shoaling

As a wave train propagates into shallow water, such characteristics as the wave height and wave length alter. This process is described as wave shoaling. The solution to the complete boundary value problem in which the boundary condition at the seabed takes account of the variation of depth is difficult, but provided the seabed slope is gentle, the shoaling effect may be estimated on the basis of a particular wave theory under the assumptions that

(a) the motion is two-dimensional,
(b) the wave period remains constant,
(c) the average rate of energy transfer in the direction of wave propagation is constant, and
(d) the selected wave theory applies to the local wave characteristics at any given depth.

These assumptions in turn require that the seabed has a gentle (but not necessarily uniform) slope resulting in negligible wave reflection, and that energy is neither supplied (by the wind) nor dissipated (by wave breaking or by friction/

percolation at the seabed). In practice these assumptions often hold reasonably well up to the point of wave breaking. The case of three-dimensional motion is treated in the following section.

Conditions in deep water are chosen as a reference and are denoted by the subscript 0 so that on the basis of linear wave theory

$$c_0 = \frac{gT}{2\pi}, \qquad L_0 = \frac{gT^2}{2\pi}, \qquad k_0 = \frac{4\pi^2}{gT^2} \tag{4.165}$$

Since the wave period is constant we have

$$ck = c_0 k_0 = \omega = \text{constant} \tag{4.166}$$

and applying the dispersion relation at either location

$$gk \tanh(kd) = gk_0 = \omega^2 = \text{constant} \tag{4.167}$$

From Eqs. (4.166) and (4.167) we must therefore have

$$\frac{c}{c_0} = \frac{k_0}{k} = \frac{L}{L_0} = \tanh(kd) \tag{4.168}$$

The dispersion relation is given as

$$kd \tanh(kd) = 4\pi^2 \frac{d}{gT^2} = 2\pi \frac{d}{L_0} \tag{4.169}$$

and indicates that kd is a unique function of d/gT^2. It follows that the ratios in Eq. (4.168) are uniquely determined at any given depth, d/gT^2. Note that it is convenient to use d/gT^2, or equivalently d/L_0, to represent the depth in place of kd since the latter contains the wave number which itself varies with depth.

Furthermore, since the average rate of energy transfer **P** is independent of depth we have

$$\tfrac{1}{8}\rho g H^2 c_G = \tfrac{1}{8}\rho g H_0^2 c_{G0} = \text{constant} \tag{4.170}$$

so that

$$\frac{H}{H_0} = \left(\frac{c_{G0}}{c_G}\right)^{1/2} = \left(\frac{2\cosh^2(kd)}{2kd + \sinh(2kd)}\right)^{1/2} \tag{4.171}$$

Thus H/H_0 is also uniquely related to the depth d/gT^2. This relationship, together with the dependence of $c/c_0(=L/L_0)$, c_G/c and $(H/L)/(H_0/L_0)$ on d/gT^2 and d/L_0 are presented in Fig. 4.26. In the shallow water range, provided the waves have not broken, the usual approximation results in the following simplified relationships,

$$\frac{c}{c_0} = \frac{L}{L_0} = 2\pi\left(\frac{d}{gT^2}\right)^{1/2} = \left(2\pi\ \frac{d}{L_0}\right)^{1/2} \tag{4.172}$$

$$\frac{H}{H_0} = \left(16\pi^2\ \frac{d}{gT^2}\right)^{-1/4} = \left(8\pi\ \frac{d}{L_0}\right)^{-1/4} \tag{4.173}$$

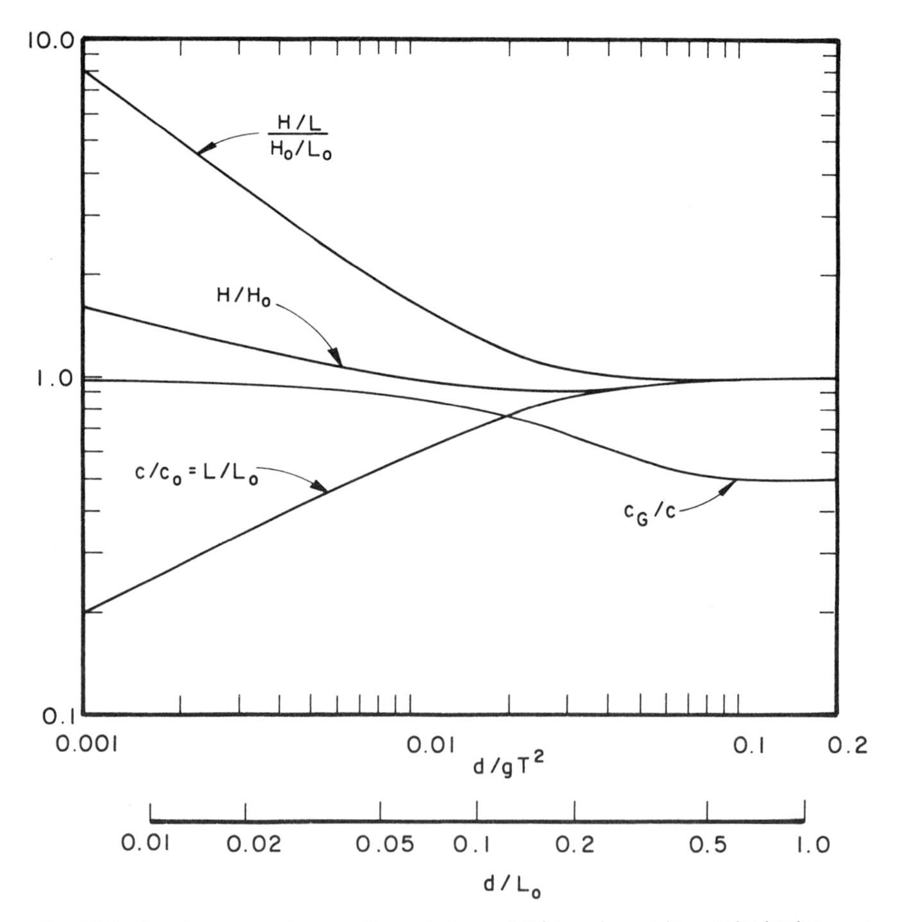

Fig. 4.26. Shoaling curves showing the variations of H/H_0, $c/c_0 = L/L_0$, $(H/L)/(H/L)_0$ and c_G/c with d/gT^2 given by linear wave theory.

Similar relationships to those given above may be derived on the basis of a finite amplitude wave theory, although it must be borne in mind that a particular theory may not apply over the whole range of depths. Koh and Le Méhauté (1966) have obtained shoaling characteristics on the basis of Stokes third and fifth order theories; and Tsuchiya and Yamaguchi (1976) have done so on the basis of the second approximation to cnoidal wave theory. In the latter case the characteristics cannot be extended to deep water and so are matched to those provided by Stokes finite amplitude theory for an intermediate depth. However, it is emphasized (Section 4.4.1) that in general a smooth transition between the results of cnoidal and Stokes theories cannot be expected. In the case of any finite amplitude theory, the various ratios c/c_0, H/H_0 etc. no longer depend solely on the single variable d/gT^2 but also on the deep water wave steepness H_0/L_0. A family of curves is therefore required to describe the shoaling variations. Stiassnie and Peregrine (1980) have recently computed these with considerable accuracy by applying the solutions of Cokelet (1977a) for periodic waves and of Longuet-Higgins and Fenton (1974) for solitary waves.

Solitary wave shoaling has also been studied both theoretically and experimentally by Camfield and Street (1969). The theory has been given by Grimshaw (1970), and extended by him (1971) to include terms up to order $(H/d)^3$. The predicted height variations were then found to compare reasonably with the measurements of Camfield and Street. The behavior of solitary waves due to gradually varying conditions, including shoaling, is reviewed by Miles (1980).

4.9.2 Wave Refraction

When waves approach a bottom slope obliquely, the portion in shallower water travels more slowly than that in deeper water in accordance with the dispersion relation, and consequently the line of the wave crest is bent so as to become more closely aligned with the bottom contours. This is the phenomenon of wave refraction. Figure 4.27a shows this process for a small time interval δt, occuring across a contour on either side of which the depths are taken as constant and to differ by a small amount. The wave crest travels a distance ℓ so that the speeds in the regions 1 and 2 are given as

$$c_1 = \frac{\ell_1}{\delta t} = \frac{s \sin \alpha_1}{\delta t} \tag{4.174a}$$

$$c_2 = \frac{\ell_2}{\delta t} = \frac{s \sin \alpha_2}{\delta t} \tag{4.174b}$$

It follows that the wave speed obeys Snell's law

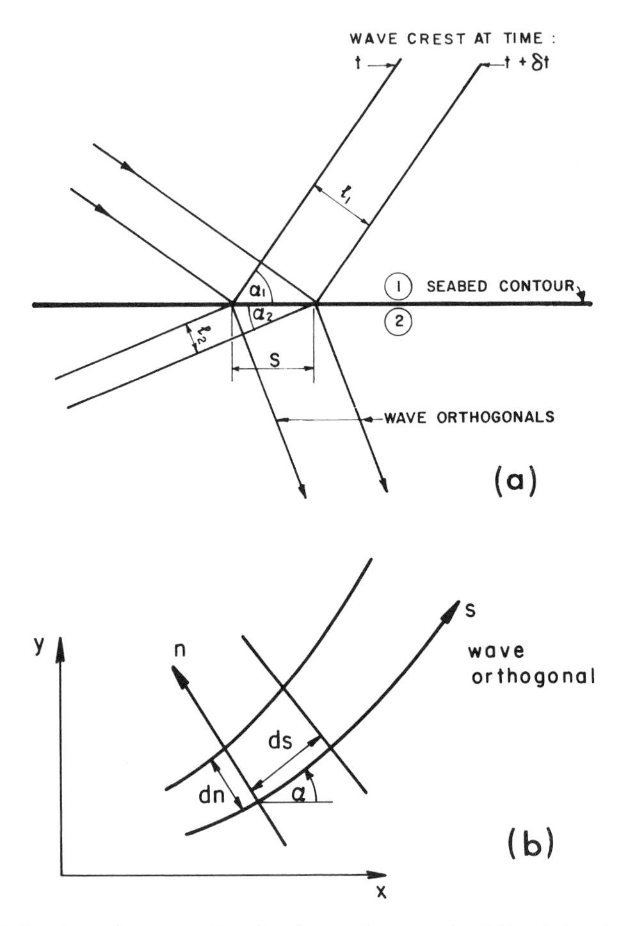

Fig. 4.27. Refraction of wave crests and orthogonals over a short time interval. (a) Relative to a seabed contour, (b) relative to a prescribed coordinate system (x, y).

$$\frac{c_1}{c_2} = \frac{\sin \alpha_1}{\sin \alpha_2} \tag{4.175}$$

in which α is the angle the wave crest makes with the bottom contour and the subscripts denote appropriate regions. This relation may be applied to deeper and deeper contours so that eventually we may use the deep water conditions as a reference, and write for any general depth

$$\frac{c}{c_0} = \frac{\sin \alpha}{\sin \alpha_0} \tag{4.176}$$

This is the basis for the development of various numerical or geometric schemes for tracing paths of wave orthogonals (lines orthogonal to the wave crests) from deep water to shoaling water in accordance with the contours describing a particular region. Such methods are not described here and are quite tedious. Reference may be made to the U.S. Army Shore Protection Manual (1973) for a description of available geometrical procedures and, for example, to Griswold (1963), Battjes (1968), Jen (1969), Keulegan and Harrison (1970) and Skovgaard, Jonsson and Bertelsen (1975) for more detailed discussions and for descriptions of computational approaches used. In computer solutions of wave refraction, it is often convenient to work with a set of ray equations describing the paths of the orthogonals (rays). Consider the directions s and n directed along and perpendicular to a wave orthogonal as indicated in Figure 4.27b. With reference to the differential lengths ds and dn as indicated in the figure, a differential form of Snell's law, Eq. (4.176), is readily obtained as

$$\frac{d\alpha}{ds} = -\frac{1}{c}\frac{dc}{dn} \tag{4.177}$$

(The x-y coordinate system may have any prescribed orientation since $d\alpha$ is independent of that orientation.) By the chain rule, Eq. (4.177) may be expressed as

$$\frac{d\alpha}{ds} = \frac{1}{c}\left(\sin\alpha\frac{dc}{dx} - \cos\alpha\frac{dc}{dy}\right) \tag{4.178}$$

and we have also

$$\frac{dx}{ds} = \cos\alpha \tag{4.179}$$

$$\frac{dy}{ds} = \sin\alpha \tag{4.180}$$

The above ray equations, Eqs. (4.178)-(4.180), may be solved numerically to determine the variation of α and hence the paths of the orthogonals. Some authors write the equations in terms of dt (by putting ds = c dt), and the solution can then be obtained by a time-stepping procedure.

The variation of wave height for refracted waves may be estimated as before by considerations of energy transfer. Once more it is assumed that energy is neither supplied nor dissipated and also that energy is not transferred laterally along a wave crest (Battjes 1968). It follows from these assumptions that the rate of energy transfer between two wave orthogonals may be taken as constant. It must be remembered, however, that the distance between two neighbouring orthogonals b is no longer constant as in the two-dimensional shoaling case, and the equation of energy transfer must be modified from Eq. (4.170) to read

$$\tfrac{1}{8}\,\rho g H^2\, c_G b = \tfrac{1}{8}\,\rho g H_0^2\, c_{G0} b_0 = \text{constant} \tag{4.181}$$

This implies that

$$\frac{H}{H_0} = \left(\frac{b_0}{b}\right)^{1/2} \left(\frac{c_{G0}}{c_G}\right)^{1/2} \tag{4.182}$$

and the ratio $(c_{G0}/c_G)^{1/2}$ is given as before, Eq. (4.171), as a function of depth only

$$\sqrt{\frac{c_{G0}}{c_G}} = \left(\frac{2\cosh^2(kd)}{2kd + \sinh(2kd)}\right)^{1/2} \tag{4.183}$$

This ratio $(c_{G0}/c_G)^{1/2}$ accounts for the wave height increase due to shoaling in the absence of refractive effects and is termed the *shoaling coefficient* K_s. The ratio $(b_0/b)^{1/2}$ is an effect of the refraction process, can thus be obtained from a refraction diagram, and is termed the *refraction coefficient* K_r. Consequently we can write

$$\frac{H}{H_0} = K_r K_s \tag{4.184}$$

Alternatively, the separation of orthogonals may be described by an orthogonal separation factor $\beta = b/b_0 = 1/\sqrt{K_r}$. A differential equation giving β as a function of s may be established (Munk and Arthur 1952) and used in conjunction with Eqs. (4.178)-(4.180).

It is sometimes also convenient to refer to the deep water unrefracted wave height H_0' with respect to any specific point and which is defined by

$$\frac{H}{H_0'} = K_s, \qquad \frac{H_0'}{H_0} = K_r \tag{4.185}$$

That is H_0' is the deep water wave height that would result in the given wave height H at a specific point if only shoaling had occured.

A fundamental example of the process of wave refraction is represented by a wave train obliquely approaching a straight shoreline with a uniform seabed slope as shown in Fig. 4.28. The deep water approach angle (between wave crest

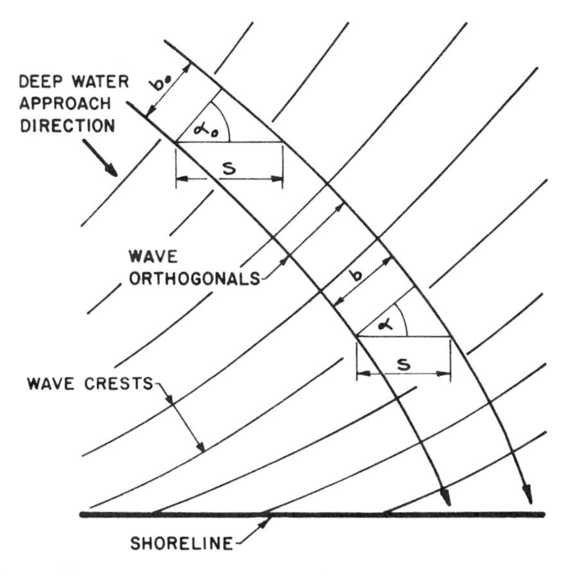

Fig. 4.28. Refraction of waves obliquely approaching a straight shoreline with uniform bed slope.

and bottom contour) is α_0. The ratios c/c_0, and L/L_0 are given as before by Eq. (4.168), while the variation of α is controlled by Snell's law, Eq. (4.176). Therefore,

$$\frac{c}{c_0} = \frac{L}{L_0} = \frac{\sin \alpha}{\sin \alpha_0} = \tanh (kd) \tag{4.186a}$$

with

$$kd \tanh (kd) = 4\pi^2 \frac{d}{gT^2} \tag{4.186b}$$

Referring now to Fig. 4.28, we see that the distance s is clearly independent of location and consequently

$$\frac{b}{\cos \alpha} = \frac{b_0}{\cos \alpha_0} = \text{constant} \tag{4.187}$$

The variation of wave height is therefore given as

$$\frac{H}{H_0} = \left[\frac{1 - \sin^2 \alpha_0 \tanh^2 (kd)}{\cos^2 \alpha_0}\right]^{-1/4} \left[\frac{2 \cosh^2 (kd)}{2kd + \sinh (2kd)}\right]^{1/2} \tag{4.188}$$

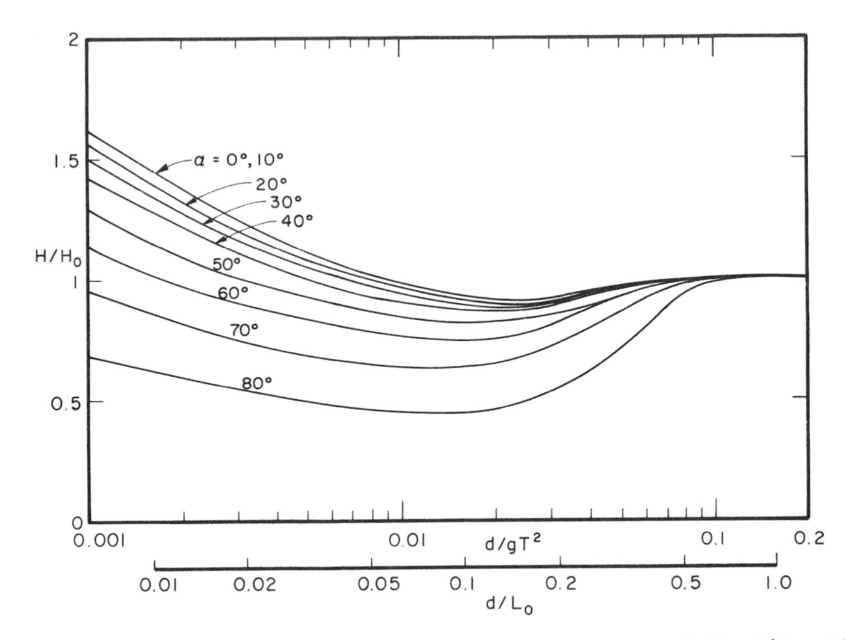

Fig. 4.29. Oblique wave refraction: variation of dimensionless wave height H/H_0 with d/gT^2 for various deep water approach angles α_0.

The variation of H/H_0 with d/gT^2 for any given value of the deep water approach angle α_0 may thereby be determined numerically, and corresponding curves are shown in Fig. 4.29. In shallow water, provided that the waves remain unbroken, the shoaling relations given by Eqs. (4.186) and (4.188) can be simplified and expressed explicitly in terms of d/gT^2 and α_0,

$$\frac{c}{c_0} = \frac{L}{L_0} = 2\pi\left(\frac{d}{gT^2}\right)^{1/2} \tag{4.189}$$

$$\frac{H}{H_0} = \left[\frac{1 - 4\pi^2 (d/gT^2) \sin^2 \alpha_0}{\cos^2 \alpha_0}\right]^{-1/4} \left[16\pi^2 \frac{d}{gT^2}\right]^{-1/4} \tag{4.190}$$

Other fundamental examples of wave refraction include the convergence of wave orthogonals over a submarine ridge, and the divergence of orthogonals over a submarine trough. The wave heights in these regions will be increased and decreased respectively.

4.9.3 Wave Interaction with Currents

A variety of problems is encountered in the interaction of water waves with currents. One fundamental aspect of this interaction concerns the behavior of a regular wave train propagating in the direction of a steady uniform current. Extensions to this case include waves approaching a uniform current obliquely, waves propagating either parallel to or obliquely to a non-uniform current which may vary horizontally with distance either along the stream or across the stream, or which may be uniform in a horizontal plane but may vary with depth. The "current" may be unsteady as in the interaction between long and short waves. A separate aspect concerns currents induced by waves including the mass-transport velocity, which has been discussed in Section 4.8, longshore currents and rip currents. A comprehensive review by Peregrine (1976) on the interaction of waves and currents treats these and other associated problems. The following outline is restricted to some of the more basic aspects of wave-current interactions.

We first treat the description of a wave train in the presence of a *uniform and steady current* U, and will subsequently consider the propagation of waves in a varying current field. It will be recalled from Sections 4.1 and 4.3 that wave theory provides a unique solution for the steady wave flow in the reference frame (x', z) moving with the waves, and it is only by making an assumption concerning the absence of an underlying current that the motion relative to the fixed reference frame (x, z) becomes fully determined. That is, Eqs. (4.6)-(4.9) establish uniquely the horizontal particle velocity u' relative to the reference frame (x', z), and upon writing $u' = u - c$ and taking u, the horizontal particle velocity in the fixed reference frame, to provide for no underlying current (e.g. $\bar{u} = 0$) we may then obtain the wave celerity c.

If we now take account of an underlying uniform, steady and colinear current U, then the horizontal velocity u' is unaltered but it is now conveniently expressed as $u' = u + U - c_c$, where c_c is the wave speed in the presence of the current and relative to the fixed reference frame, and u is the wave-induced oscillatory particle velocity which can now be subjected to the same assumption as before (e.g. $\bar{u} = 0$). We therefore have $c_c = c + U$ with the usual dispersion relation applying to c, which denotes the wave speed in the absence of the current, or alternatively when the current is present it is the wave speed relative to that current. (This argument has not so far been restricted to small amplitude waves, but if such were the case we should have $c = [(g/k) \tanh (kd)]^{1/2}$.)

The above distinction between a subscripted and an unsubscripted symbol is applied to other variables in the following discussion: that is a variable with subscript c is one pertaining to the presence of a current and measured relative to a fixed reference frame. The corresponding unsubscripted variable pertains to the absence of a current, or alternatively in the current's presence is equivalent to that variable measured relative to the current.

The wave length and wave number k are unaltered by a change in reference frame so that the result $c_c = c + U$ may be expressed in terms of the wave frequency or period. That is $\omega_c = \omega + kU$ or $1/T_c = 1/T + kU/2\pi$. (In the case of small amplitude waves $\omega = 2\pi/T = [gk \tanh (kd)]^{1/2}$.) When the current makes an angle α with the direction of wave propagation, then the velocity component along the x axis, U cos α, affects the wave speed and frequency as described above, while the component along the y axis induces a fluid motion along the wave crests but does not affect the wave speed. That is, for example,

$$\omega_c = \omega + kU \cos \alpha \tag{4.191}$$

For a given wave number k and a given current specified by U and α, it is a straightforward matter to calculate the modified wave frequency ω_c since ω may first be calculated in the usual manner by any chosen wave theory. But in practice a design wave in the presence of a current is most conveniently specified in terms of H, T_c, d, U and α (as for example might be obtained from recordings in a wave-current field). In order to apply the results of any wave theory, the problem essentially reduces to a determination of the wave number k, for then the details of the flow become available by the standard application of the wave theory. In principle this provides no real difficulty for ω and k are both initially unknown in Eq. (4.191) but they are related by a dispersion relation so that the problem is soluble.

A graphical method of solution was proposed by Jonsson, Skovgaard and Wang (1970) for linear waves, and Hedges (1978) has emphasized that the method may be combined with any chosen wave theory. Eq. (4.191) may be re-written as

$$\frac{\omega}{\sqrt{g/d}} = \frac{\omega_c}{\sqrt{g/d}} \left\{1 - kd\left(\frac{U \cos \alpha}{\omega_c d}\right)\right\} \tag{4.192}$$

The dispersion relation of the chosen wave theory gives the left hand side as a known (generally implicit) function of kd for a given value of H/d. And the right-hand side is directly given as a known linear function of kd. Both the left and the right-hand sides of Eq. (4.192) are thus plotted as functions of kd and the intersections of the curves provide the possible solutions.

In the case of linear waves the dispersion relation may be substituted directly into the left-hand side of Eq. (4.192) and we have

$$\pm\sqrt{kd \tanh (kd)} = \frac{\omega_c}{\sqrt{g/d}} \left\{1 - kd\left(\frac{U \cos \alpha}{\omega_c d}\right)\right\} \tag{4.193}$$

Note that Jonsson et al. only adopted the positive root on the left-hand side (corresponding to ω being positive), but Peregrine (1976) indicates that a further solution corresponding to the negative root is possible. That is ω is negative for a positive ω_c (or equivalently if ω is considered positive only then negative values of ω_c should now be admitted). This corresponds physically to the current giving rise to a change in direction of wave travel: ω and ω_c are of opposite signs. The graphical procedure applied to linear waves is carried out in Fig. 4.30a and this shows four possible solutions labelled A to D.

A physical interpretation of the different solutions may be made in terms of the directions of wave crest travel and wave energy transfer. Discussions of the propagation of wave energy in the presence of a uniform current have been given, among others, by Longuet-Higgins and Stewart (1960) and Whitham (1962, 1974). The group velocity in the presence of a current and taken relative to a fixed reference frame is written as $c_{Gc} = c_G + U \cos \alpha$. Thus, in Fig. 4.31 we consider a given wave speed c and corresponding group velocity c_G for waves in the absence of a current, and we examine the influence on such waves of a current with different possible values of $U \cos \alpha$. For the whole range of $U \cos \alpha$ values, the figure indicates that four distinct solutions giving rise to differences in the directions of $U \cos \alpha$, c_c and c_{Gc} may be possible. These correspond to the solutions A to D in Fig. 4.30a and may be identified as follows.

Solution A: the current has a component in the direction of wave propagation (when considered both in the presence and in the absence of the current) such that both the wave speed and the group velocity are increased: that is both the wave crests and the wave energy are swept downstream at a faster rate than in the current's absence.

Solution B: the current has a component in the direction opposite that of wave propagation (when considered both in the presence and in the absence of the current) such that both the wave speed and the wave energy are swept upstream but at a slower rate than in the current's absence.

Solution C: the current has a component in the direction opposite that in which the waves would propagate in the current's absence and has a magnitude such that the wave energy (or a wave group) is carried downstream while the individual wave crests propagate upstream. In the case of a finite group of waves, this implies that the group as a whole is swept downstream while the individual crests propagate upstream, eventually to lose their identity at the upstream edge of the group.

Solution D: the current has a component in the direction opposite that in which the waves would propagate in the current's absence, and has a relatively large magnitude such that both the wave crests and the wave energy are carried downstream. Unlike solution A, the waves in the current's absence travel with speed c in the opposite direction to those in the presence of the current and therefore ω and ω_c are of opposite sign.

Of the four solutions described here it is the first two, A and B, that are really of engineering interest. These correspond to currents whose magnitudes in the wave propagation direction are relatively small such that the directions of wave

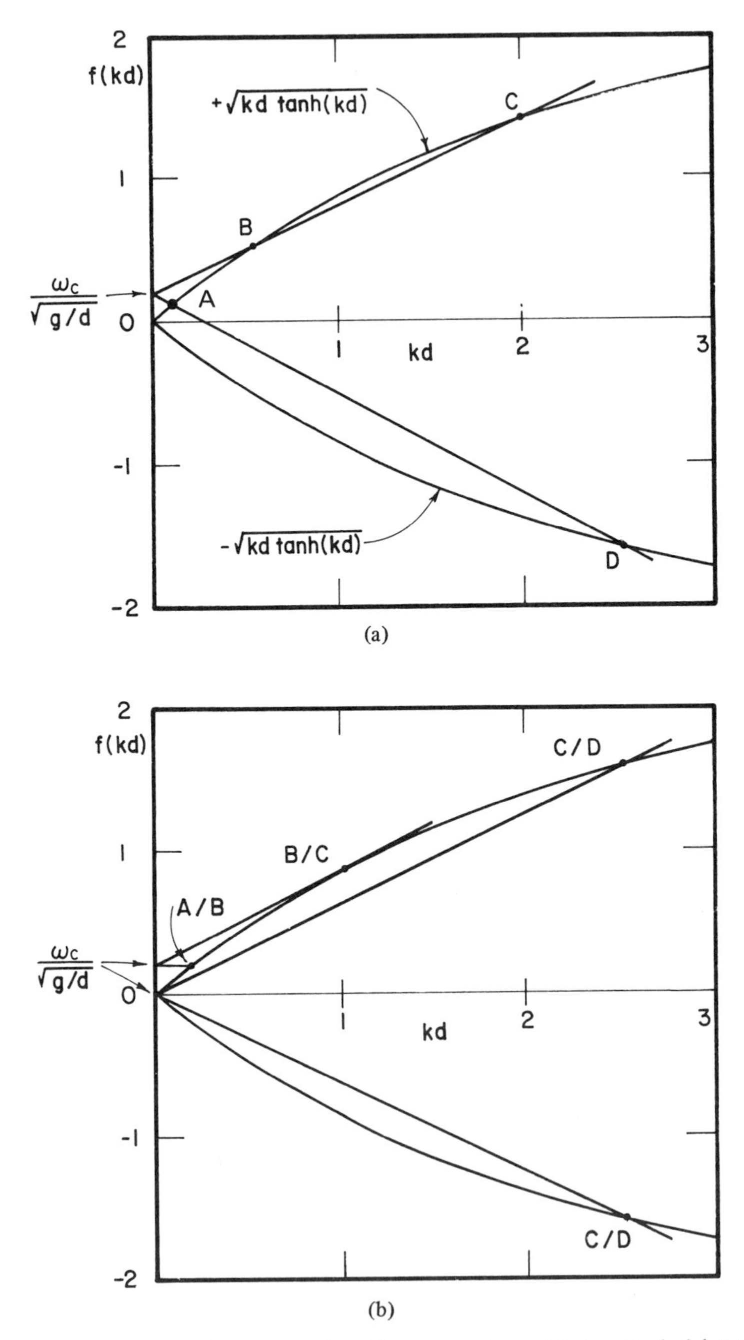

(a)

(b)

Fig. 4.30. Waves propagating on a steady uniform current: graphical method of determining wave length for specified wave frequency and water depth.

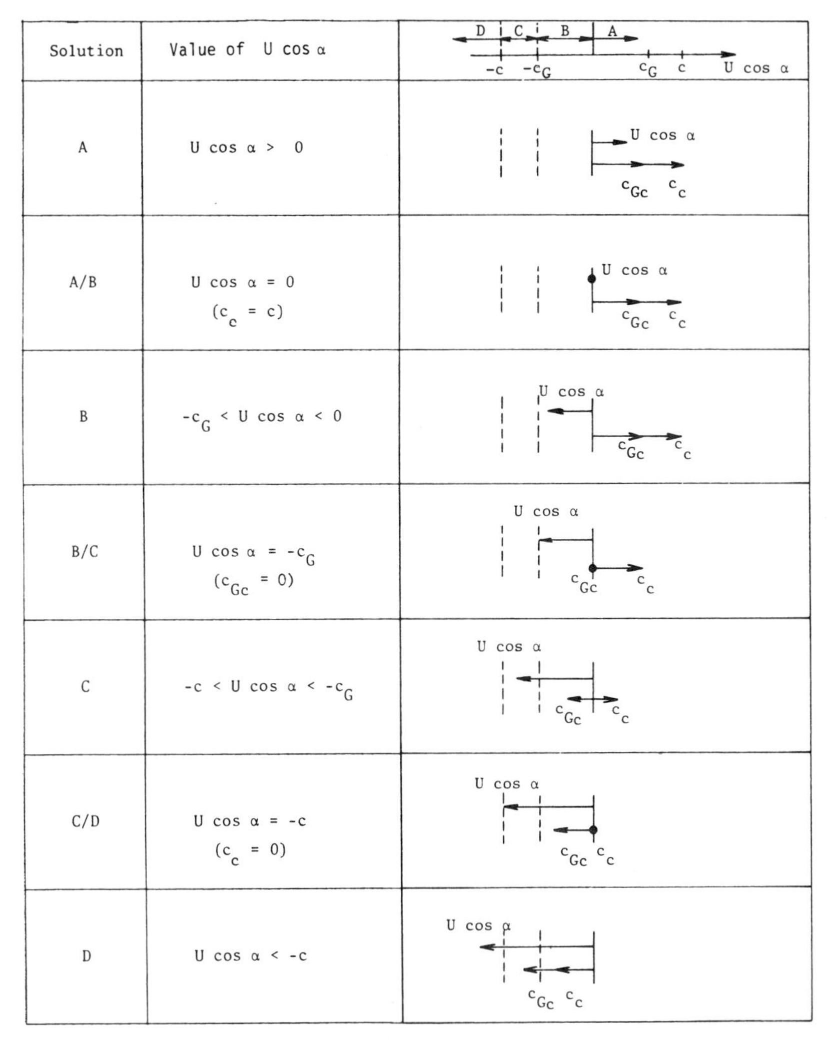

Solution	Value of U cos α	D C B A; −c −c_G c_G c U cos α
A	$U \cos\alpha > 0$	U cos α; c_{Gc} c_c
A/B	$U \cos\alpha = 0$ $(c_c = c)$	U cos α; c_{Gc} c_c
B	$-c_G < U \cos\alpha < 0$	U cos α; c_{Gc} c_c
B/C	$U \cos\alpha = -c_G$ $(c_{Gc} = 0)$	U cos α; c_{Gc} c_c
C	$-c < U \cos\alpha < -c_G$	U cos α; c_{Gc} c_c
C/D	$U \cos\alpha = -c$ $(c_c = 0)$	U cos α; c_{Gc} c_c
D	$U \cos\alpha < -c$	U cos α; c_{Gc} c_c

Fig. 4.31. Relationship of the alternative solutions A to D to the value of U cos α. The sketch indicates how the different solutions correspond to different relative directions of U cos α, c_{Gc} and c_c, as U cos α is varied for a given c (and c_G).

travel and wave energy transfer are no different when considered relative to the current.

The special cases when adjacent solutions coincide are of some interest and we now briefly consider these with the aid of Figs. 4.30b and 4.31.

Solutions A and B coincide when there is no current in the direction of wave propagation, U cos $\alpha = 0$. This case requires no further comment.

Solutions B and C coincide when $c_{Gc} = 0$, that is when $U \cos \alpha = -c_G$. Taking $\alpha = 0°$, the current $U = -c_G$ is now termed a *stopping current* or an *arresting current* since it just stops the wave energy from being transmitted upstream. It is given on the basis of linear wave theory as

$$\frac{U}{v_0} = -\frac{1}{4} \tanh(kd) \left\{1 - \left(\frac{2kd}{\sinh(2kd)}\right)^2\right\} \tag{4.194}$$

where $v_0 = g/\omega_c = gT_c/2\pi$, and we see that in deep water the stopping current is given simply as $U = -\frac{1}{4} v_0$. (In shallow water c_c and ω_c become zero and U is then equal to $-\sqrt{gd}$.) For finite amplitude waves the situation is more difficult to deal with, but for any wave height, the greatest magnitude of the stopping current is found to correspond approximately to $U = -0.3\, v_0$ (e.g. Peregrine 1976). In practice however waves will generally break in the presence of a strong adverse current. Dalrymple and Dean (1975) investigated the stopping current for waves of maximum height on the basis of stream function theory and they presented the corresponding dependence of U/v_0 on $2\pi d/gT_c^2$. They also considered the influence of a current on the breaking wave height and this aspect is touched upon in Section 4.9.8.

Solutions C and D coincide when $U = -c$, that is the waves are stationary. This is often observed to be due to obstacles in a stream. The wave frequency is now zero, $\omega_c = 0$, and in Fig. 4.30b the straight line therefore passes through the origin so that solutions C and D both provide the same result. (ω may be considered to be either positive or negative.)

The case of a wave train propagating over a nonuniform and/or unsteady current is less simple. It is generally assumed that the currents are "large scale" in that the current time and length scales are much larger than the wave period and wave length respectively. More precisely, the assumption is made that $k \gg (\partial U/\partial x)/U$ and $\omega \gg (\partial U/\partial t)/U$. Such problems have been approached in various ways. The most powerful approach is that based on the calculus of variations and employing an averaged Lagrangian, which is a function of the various parameters defining the situation. The method was largely developed by Whitham (1965a,b, 1967a,b) and has been outlined by him in detail (Whitham 1974). A separate approach used by Phillips (1977) involves the separation of variables into steady and unsteady terms and then taking the time average of unsteady depth-integrated terms. Previously, Longuet-Higgins and Stewart (1961) had treated the interaction of waves and nonuniform currents by means of the concept of radiation stress, mentioned in Section 4.2.

Their treatment involves a small amplitude wave train propagating in deep water over a nonuniform current varying in the x direction. When a current of magnitude U is flowing in the direction of wave propagation, the wave speed relative to fixed axes is given as before by $c + U$, where c is the speed in the absence of the current. Taking the wave period to remain constant, we have

$$k(c + U) = k_0(c_0 + U_0) = \omega = \text{constant} \tag{4.195}$$

where the subscript 0 refers to reference values, say at $x = 0$. The deep water dispersion relation gives

$$kc^2 = k_0 c_0^2 = g = \text{constant} \tag{4.196}$$

Combining Eqs. (4.195) and (4.196) we may eventually obtain

$$\frac{1}{c}\frac{dc}{dx} = -\frac{1}{2k}\frac{dk}{dx} = \frac{1}{c + 2U}\frac{dU}{dx} \tag{4.197a}$$

and thus

$$\frac{c}{c_0} = \sqrt{\frac{k_0}{k}} = \frac{1}{2(1 + \gamma)}\left\{1 + \left[1 + 4(1 + \gamma)\frac{U}{c_0}\right]^{1/2}\right\} \tag{4.197b}$$

where $\gamma = U_0/c_0$. This gives the wave speed or wave number at any given location in terms of the local current speed and the reference current and wave speeds.

Longuet-Higgins and Stewart found corresponding relations for the changes in wave amplitude a from considerations of an energy transfer equation involving the radiation stresses in the waves. These results depend on the manner in which the current field is made to compensate for the horizontal gradient of U to ensure that the continuity equation is satisfied. Two cases were considered:

(a) when there is a vertical upwelling from below, $V = 0$, $W \neq 0$, and
(b) when there is a horizontal inflow from the sides, $V \neq 0$, $W = 0$.

The corresponding variations in wave amplitude may be expressed in terms of the reference values a_0, c_0 and U_0 and are given as

$$\frac{a}{a_0} = \left[\frac{c_0(c_0 + 2U_0)}{c(c + 2U)}\right]^{1/2} \quad \text{for case (a)} \tag{4.198a}$$

$$\frac{a}{a_0} = \left[\frac{c(c_0 + 2U_0)}{c_0(c + 2U)}\right]^{1/2} \quad \text{for case (b)} \tag{4.198b}$$

Here the variation of c with U has already been given in Eq. (4.197) and thus the wave amplitude at any prescribed location may also be determined. Figure 4.32 shows the variation of a/a_0, c/c_0 and k/k_0 with U/c_0 for these two cases when the subscript 0 denotes values corresponding to $U = 0$. It is seen that when an adverse current U approaches $-\frac{1}{2}c = -\frac{1}{4}c_0$ the wave height becomes very large and in consequence the waves may be expected to break when propagating against a current of this magnitude. The above relations refer to deep water conditions, as mentioned, but may be generalized to arbitrary depths.

Mention is made in passing of a contribution by Crapper (1972) who employed the averaged Lagrangian method of Whitham (1965a,b) to the case of large amplitude waves and found that the rate of growth is then somewhat less than that predicted by the theory of Longuet-Higgins and Stewart.

Longuet-Higgins and Stewart (1961) also considered the case of a current with a vertical axis of shear and making an oblique angle with the wave direction as depicted in Fig. 4.33. The wave length measured along the y direction, denoted k_y, must be independent of location so that

$$k \sin \theta = k_y = \text{constant} \tag{4.199}$$

The wave frequency also remains constant, implying that

$$k(c + V \sin \theta) = \omega = \text{constant} \tag{4.200}$$

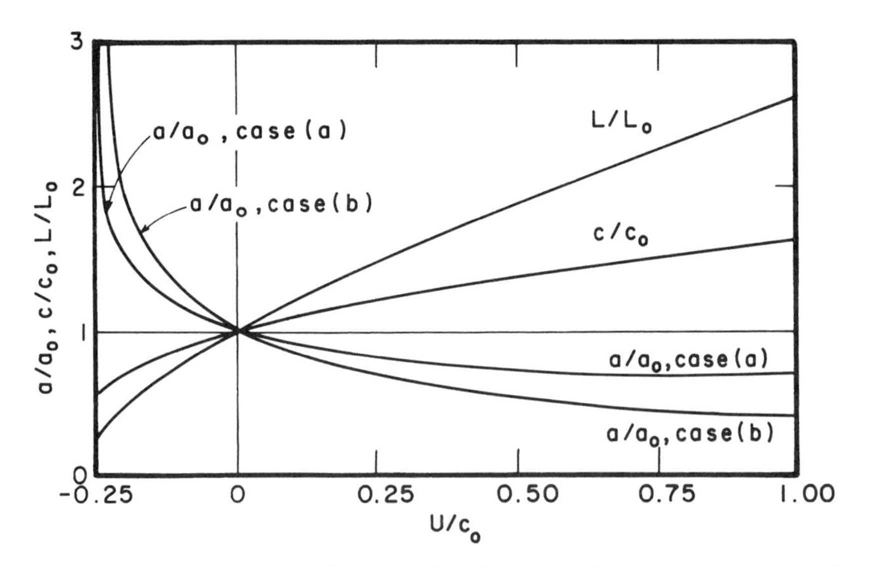

Fig. 4.32. Variation of wave amplitude, length and celerity with colinear current magnitude U. The subscript 0 denotes values at $U = 0$. Case (a): vertical upwelling, $V = 0$, $W \neq 0$; Case (b): horizontal inflow, $V \neq 0$, $W = 0$.

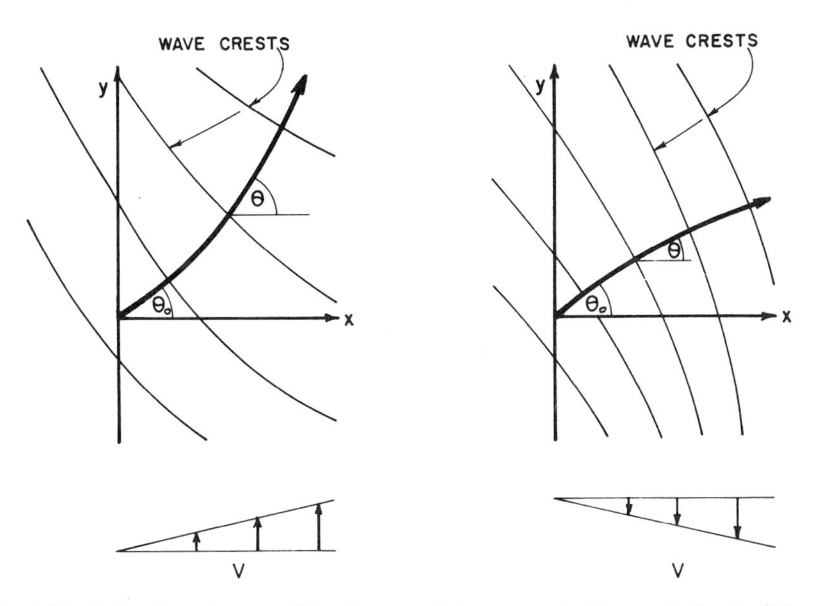

Fig. 4.33. Refraction of waves obliquely approaching a current with a vertical axis of shear.

And, furthermore, the deep water dispersion relation once more applies

$$kc^2 = g = \text{constant} \tag{4.201}$$

Using the subscript 0 to denote conditions when V = 0, the preceding three relations may be used to obtain

$$\frac{c}{c_0} = \left(\frac{k}{k_0}\right)^{-1/2} = \left(\frac{\sin\theta}{\sin\theta_0}\right)^{1/2} = \frac{1}{1 - V\sin\theta_0/c_0}. \tag{4.202}$$

And the variation of wave amplitude is now found to be given as

$$\frac{a}{a_0} = \left(\frac{\sin 2\theta}{\sin 2\theta_0}\right)^{-1/2} \tag{4.203}$$

Consequently, for a given current field showing variations in the horizontal plane, the paths of wave orthogonals and corresponding variations in wave amplitude may be determined. Such a process of wave refraction by a current may be compared to the refraction by depth variations considered in Section 4.9.2. In particular, Skovgaard and Jonsson (1976) have reported on a computational model in which the combined effects of depth refraction and current refraction on the development of Stokes second order waves are calculated.

Another important class of nonuniform currents are those which are uniform in the horizontal plane but which vary with depth (shear flows). The flow is not now irrotational and possesses a vorticity $\xi = dU/dz$. A simple case is a current varying linearly with depth and given as $U = U_s + \xi z$, where U_s is the current velocity at the still water level and ξ is the vorticity which is taken as constant. Biesel (1950) obtained a solution to this problem for small amplitude waves and gave the dispersion relation in terms of the wave frequency ω_c relative to a fixed reference frame as

$$(\omega_c - kU_s)^2 = gk \tanh(kd)\left[1 + \frac{\xi}{g}\left(U_s - \frac{\omega_c}{k}\right)\right] \tag{4.204}$$

This can be used to determine the wave number k for specified values of U_s, ξ, ω_c and d. Sarpkaya (1955) has described experiments involving the interaction of waves with nonuniform currents. And Peregrine (1976) reviews the interaction of small amplitude waves with shear flows of different profiles.

Dalrymple (1974a, b) extended Dean's (1965) stream function theory to include a current which has a linear variation as above, or a bi-linear variation with depth (i.e. a continuous profile made up of two joined linear variations). He found a similar wave length modification due to the vorticity as predicted by the linear wave theory, although the actual wave lengths predicted by the two theories may themselves be quite different. Finally, discussions of wave-current induced forces on small bodies and of the modifications to the Morison equation are presented in Section 5.3.5, while the topic of wave-current induced forces on large bodies is mentioned in Section 6.9.

4.9.4 Wave Reflection

When a wave train of small amplitude encounters a vertical wall located, say, in the plane $x = 0$, the waves are reflected in order to satisfy the boundary condition imposed by the wall, $u = 0$ at $x = 0$, in place of the previous conditions requiring a wave train of permanent form. This reflection condition is simply satisfied in terms of the component incident progressive wave train by superposing a second reflected progressive wave train travelling in the opposite direction. The surface elevations of the component waves are

$$\begin{aligned}\eta_i &= \tfrac{1}{2} H \cos(kx - \omega t)\\ \eta_r &= \tfrac{1}{2} H \cos(kx + \omega t)\end{aligned} \tag{4.205}$$

where the subscripts i and r denote the incident and reflected waves respectively. Each component motion separately satisfies the governing equations and since these are linear for small amplitude waves, they may be superposed so that the combined motion will continue to be a solution. Furthermore, the horizontal particle velocities of the two component wave trains are always equal and opposite at x = 0 and the combined wave motion does therefore satisfy also the new condition u = 0 at x = 0. The free surface elevation of the combined wave form is then given as

$$\eta = \eta_i + \eta_r = H \cos(kx) \cos(\omega t) \tag{4.206}$$

This represents a standing wave in which the trough to crest height is 2H, which is twice that of the incident wave train. The standing wave train profile is sketched in Fig. 4.34. At the nodes given by cos (kx) = 0, η is continuously zero, as are the vertical velocity w and the hydrodynamic component of the pressure, $(p + \rho gz)$; while at the antinodes, given by cos (kx) = ±1 and which include the wall, the horizontal velocity u is zero, while the maximum variations occur in the other quantities η, w and $(p + \rho gz)$. Table 4.8 contains formulae for these and other quantities in a standing wave train.

The characteristics of this standing wave system may be used to describe free oscillations in a two-dimensional basin of constant depth. Such a basin can contain a standing wave system with an antinode at a closed end, at which the boundary condition of zero horizontal velocity is satisfied, and with a node at an end open to an infinite sea, at which the boundary condition of zero hydrodynamic pressure (or free surface elevation) is satisfied. If ℓ is the basin length, free oscillations in a closed basin can accordingly occur at a wave length $L = 2\ell$ for the first (lowest) mode, and more generally at $L = 2\ell/n$ for the n-th mode; while in an open-ended basin, the corresponding formulae are $L = 4\ell$ for the first mode and $L = 4\ell/(2n - 1)$ for the n-th mode. The corresponding wave periods can be found by applying the dispersion relation. In the usual case of shallow water the natural periods of the n-th mode are thus

$$T_n = \begin{cases} \dfrac{2\ell}{n\sqrt{gd}} & \text{for a closed basin} \\[2ex] \dfrac{4}{(2n-1)\sqrt{gd}} & \text{for an open-ended basin} \end{cases} \tag{4.207}$$

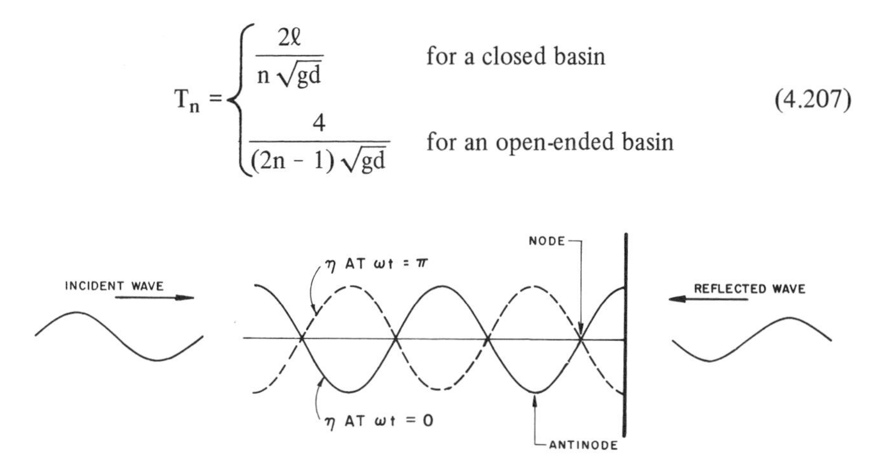

Fig. 4.34. Standing wave profile shown as the sum of incident and reflected wave trains.

Table 4.8 Results For Small Amplitude Standing Waves.

Velocity potential	$\phi = -\dfrac{2\pi H}{kT}\dfrac{\cosh(ks)}{\sinh(kd)}\cos(kx)\sin(\omega t)$
	$= -\dfrac{gH}{\omega}\dfrac{\cosh(ks)}{\cosh(kd)}\cos(kx)\sin(\omega t)$
Dispersion relation	$c^2 = \dfrac{\omega^2}{k^2} = \dfrac{g}{k}\tanh(kd)$
Surface elevation	$\eta = H\cos(kx)\cos(\omega t)$
Horizontal particle displacement	$\xi = -H\dfrac{\cosh(ks)}{\sinh(kd)}\sin(kx)\cos(\omega t)$
Vertical particle displacement	$\zeta = H\dfrac{\sinh(ks)}{\sinh(kd)}\cos(kx)\cos(\omega t)$
Horizontal particle velocity	$u = \dfrac{2\pi H}{T}\dfrac{\cosh(ks)}{\sinh(kd)}\sin(kx)\sin(\omega t)$
Vertical particle velocity	$w = -\dfrac{2\pi H}{T}\dfrac{\sinh(ks)}{\sinh(kd)}\cos(kx)\sin(\omega t)$
Horizontal particle acceleration	$\dfrac{\partial u}{\partial t} = \dfrac{4\pi^2 H}{T^2}\dfrac{\cosh(ks)}{\sinh(kd)}\sin(kx)\cos(\omega t)$
Vertical particle acceleration	$\dfrac{\partial w}{\partial t} = -\dfrac{4\pi^2 H}{T^2}\dfrac{\sinh(ks)}{\sinh(kd)}\cos(kx)\cos(\omega t)$
Pressure	$p = -\rho gz + \rho gH\dfrac{\cosh(ks)}{\cosh(kd)}\cos(kx)\cos(\omega t)$
Average energy density	$E = \frac{1}{4}\rho gH^2$
Radiation stress	$S_{xx} = \left[\dfrac{1}{2} + \dfrac{2kd}{\sinh(2kd)}\right]E$
	$S_{xy} = S_{yx} = 0$
	$\overline{S_{yy}} = \left[\dfrac{kd}{\sinh(2kd)}\right]E$

where

H = incident wave height = $\frac{1}{2}$ standing wave height

We now turn to the case of a wave train approaching a vertical wall obliquely. This oblique reflection produces a three-dimensional wave pattern as sketched in Fig. 4.35. If the incident wave orthogonals make an angle α with the wall, which is taken to lie along the x axis, the incident wave surface elevation is given

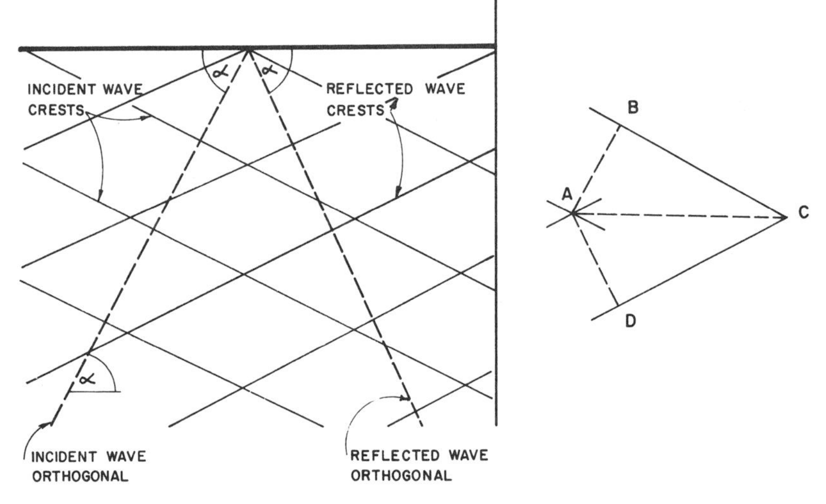

Fig. 4.35. Oblique reflection of a wave train.

as

$$\eta_i = \tfrac{1}{2} H \cos(kx \cos\alpha + ky \sin\alpha - \omega t) \tag{4.208}$$

The reflected wave surface elevation will then be

$$\eta_r = \tfrac{1}{2} H \cos(kx \cos\alpha - ky \sin\alpha - \omega t) \tag{4.209}$$

and it is readily shown that these combine to satisfy the condition of zero velocity normal to the wall. The surface elevation of the combined wave pattern may be written as

$$\eta = H \cos(kx \cos\alpha - \omega t) \cos(ky \sin\alpha) \tag{4.210}$$

from which it may be seen that the lines parallel to the wall, given by $\cos(ky \sin\alpha) = 0$, represent nodes along which the surface elevation is constantly zero, while those lines given by $\cos(ky \sin\alpha) = \pm 1$ (which include the wall) represent antinodes along which η is a maximum or a minimum. These short-crested waves propagate in a direction parallel to the wall at a speed of $c = \omega/(k \cos\alpha)$, a result which may readily be interpreted in physical terms by considering the inset of Fig. 4.35. Here it will be seen that in time δt the crest of the incident wave advances a distance $AB = (\omega/k)\delta t$, while the maximum crest elevation of the combined wave form advances a distance $AC = (\omega/k)\delta t/\cos\alpha$, so that its speed is $\omega/(k \cos\alpha)$ as mentioned. It will also be noticed that for $\alpha = 0°$ the incident and reflected wave trains are coincident so that there is no reflection in the usual sense.

In most cases met with in practice and exemplified by sloping or permeable walls, wave reflection is not complete and the amplitude of the reflected wave train is in consequence less than that of the incident wave. We now treat this case, considering a normal reflection such that the incident waves approach a barrier orthogonally. A *reflection coefficient* K_r is defined as the ratio of reflected wave height to incident wave height, and the total (combined) surface elevation may then be written as

$$\eta = \tfrac{1}{2} H [\cos(kx - \omega t) + K_r \cos(kx + \omega t)] \tag{4.211}$$

where the origin of x is not specified. The wave height of the resulting wave form may be obtained by rearranging this equation in the form

$$\eta = \tfrac{1}{2} H A(kx) \cos[\delta(kx) - \omega t] \tag{4.212}$$

where

$$A^2(kx) = (1 + K_r)^2 \cos^2(kx) + (1 - K_r)^2 \sin^2(kx)$$

$$= 1 + K_r^2 + 2K_r \cos(2kx) \tag{4.213}$$

$$\tan[\delta(kx)] = \left(\frac{1 - K_r}{1 + K_r}\right) \tan(kx) \tag{4.214}$$

Thus at a given value of x the trough to crest height is H A(kx) and the variation of wave height with x for any value of K_r is easily obtained. The maximum and minimum wave heights are thus found to be

$$H_{max} = (1 + K_r) H \tag{4.215a}$$

$$H_{min} = (1 - K_r) H \tag{4.215b}$$

These formulae provide a simple method of measuring the reflection coefficient and incident wave height H in a partially reflected wave system. A wave probe is traversed in the direction of wave propagation in order to measure the maximum and minimum wave heights and the values obtained may then be applied to the formulae

$$K_r = \frac{H_{max} - H_{min}}{H_{max} + H_{min}} \tag{4.216a}$$

$$H = \tfrac{1}{2} (H_{max} + H_{min}) \tag{4.216b}$$

Since the reflection coefficient is generally less than unity, the incident wave energy is not wholly reflected and a part of it will be dissipated and/or transmitted beyond the barrier. In such cases it is convenient to define a *transmission coefficient* K_t as the ratio of transmitted to incident wave heights, and also a coefficient of energy conservation K_c, which is the ratio of the energy conserved in the reflected and transmitted wave motions to the energy content of the incident waves. Since the average energy density is proportional to the square of the wave height, we have

$$K_c = \frac{H_r^2 + H_t^2}{H_i^2} = K_r^2 + K_t^2 \tag{4.217}$$

Thus by measuring the reflection and transmission coefficients the fraction of energy dissipated is readily determined.

The reflection and transmission coefficients are important characteristics in breakwater design and have been predicted and/or measured for a variety of configurations. Theoretical predictions have generally been based on the assumption of zero energy dissipation, corresponding to $K_r = (1 - K_t^2)^{1/2}$. Results obtained with various reference geometries include those by Ursell (1947) for a vertical plate extending down from the free surface; Dean (1945) for a vertical plate extending upwards from deep water; Dean and Ursell (1959) for a semi-immersed circular cylinder; Dean (1948), Ursell (1950) and Ogilvie (1963) for a fully immersed circular cylinder; Black and Mei (1969) for a rectangular section, either semi-immersed or resting on the seabed; and Newman (1974) for a pair of closely spaced vertical plates extending down from the free surface and in deep water. A notable result in the case of a fully immersed circular cylinder is that $K_r = 0$ and $K_t = 1$, although the waves encounter a phase shift in passing over the cylinder.

Permeable breakwaters generally involve significant energy dissipation and have been treated by semi-theoretical approaches involving empirical coefficients (e.g. Hayashi et al. 1966, Marks and Jarlan 1968, Richey and Sollitt 1970, Hattori 1972, and Van Weele and Herbich 1972). These may be comprised of perforated walls, closely spaced piles and so on. A detailed discussion of these different categories of breakwater is not given herein. Suffice it to note that Van Weele and Herbich (1972) conducted experiments in a wave tank with pile configurations consisting of groups of sixteen piles. They have presented data on wave reflection and transmission in terms of wave steepness and the transverse and longitudinal spacing between piles.

In general any porous structure or densely populated rigid or elastic or floating bodies exhibit the same kind of behavior and may serve as a wave reflection and/or energy dissipation barrier. In this regard the use of floating breakwaters has received considerable attention in recent years. Applications are particularly suited to small craft harbors where wave conditions are not too severe such that the breakwater dimensions need not be too large: as a rough guideline, the width of a floating breakwater in the incident wave direction must typically be about L/5 or greater in order to provide a significant reduction in the transmitted wave height. A variety of concepts have been proposed and include moored floating spheres, scrap-tire breakwaters, "A-frame" timber breakwaters, floating pontoons of various configurations, and so on. The Proceedings of the Floating Breakwater Conference (Kowalski 1974) is a useful reference on the subject. Discussion of particular concepts and reviews of the factors involved include papers by Adee and Martin (1974), Richey and Nece (1974), Adee (1976) and Seymour (1976).

4.9.5 Wave Diffraction

When a wave train encounters a vertical barrier the wave motion penetrates into the region of geometric shadow by the process of diffraction. This phenomenon is of considerable practical importance in establishing the wave action behind breakwaters and around small islands or large offshore structures.

General methods of solution have been discussed in detail by Stoker (1957) and only a brief description is given here. In the usual manner the fluid is assumed incompressible and the motion irrotational so that the velocity potential ϕ satisfies the Laplace equation. The treatment is based on linear wave theory in which the wave height is assumed sufficiently small for nonlinear terms to be omitted. The linearization of the motion and the specification of an incident wave permits the velocity potential ϕ to be considered as the sum of components describing the incident (subscript w) and scattered (subscript s) wave motions

$$\phi = \phi_w + \phi_s \tag{4.218}$$

The free surface and bottom boundary conditions apply both to ϕ and to ϕ_w and therefore also to ϕ_s

$$\frac{\partial^2 \phi_s}{\partial t^2} + g \frac{\partial \phi_s}{\partial z} = 0 \qquad \text{at } z = 0 \tag{4.219}$$

$$\eta_s = -\frac{1}{g} \left(\frac{\partial \phi_s}{\partial z} \right)_{z=0} \tag{4.220}$$

$$\frac{\partial \phi_s}{\partial z} = 0 \qquad \text{at } z = -d \tag{4.221}$$

There will, however, be an additional boundary condition due to the presence of the body and relating ϕ_s to ϕ_w in such a manner that the velocity normal to the body surface is zero at that surface. That is

$$\frac{\partial \phi}{\partial n} = \frac{\partial \phi_w}{\partial n} + \frac{\partial \phi_s}{\partial n} = 0 \qquad \text{at the body surface} \tag{4.222}$$

where n is distance normal to the body surface.

The boundary value problem for ϕ_s as described so far may be separately satisfied by scattered waves which are incoming from the far field, or outgoing, or any linear combination of these. In order to ensure that a physically acceptable and unique solution is obtained, one more boundary condition is imposed. This is termed the *radiation condition* in which the solution is restricted to correspond to outgoing waves only. This was introduced and discussed by Sommerfeld (1949) in the context of generalized wave theory, and has also been discussed by Stoker (1957) with reference to surface waves.

If we consider a short length of a vertical circular cylindrical surface lying in the far field and enclosing the region of interest, the scattered waves will be virtually plane across this surface and may generally be represented as $f_1(r - ct) + f_2(r + ct)$, where r is radial distance measured from a point in the region of interest. For outgoing waves only, the latter component must be absent, and the solution near this boundary should therefore satisfy

$$\frac{\partial \phi_s}{\partial r} + \frac{1}{c}\frac{\partial \phi_s}{\partial t} = 0 \tag{4.223}$$

Furthermore, because of the directional spread of wave energy, the amplitude of the scattered waves, or ϕ_s, will decay like $1/\sqrt{r}$. Hence a spreading factor $\sqrt{r}$ should be applied to ensure that Eq. (4.223) remains meaningful. If ϕ_s is expressed in complex form, the radiation condition may thus be written as

$$\underset{r \to \infty}{\text{Limit}} \sqrt{r}\left(\frac{\partial \phi_s}{\partial r} - ik\phi_s\right) = 0 \tag{4.224}$$

This condition is generally satisfied when ϕ_s takes an asymptotic form proportional to $\exp(-ikr)/\sqrt{r}$.

The above description concerns wave scattering in the horizontal plane (two-dimensional case). The radiation condition may be generalized for application to a function of any number of dimensions, but the only other case which is of interest in the present context is the one-dimensional case corresponding to a vertical plane flow. Directional spreading no longer occurs, and so the corresponding radiation condition is simply

$$\underset{r \to \infty}{\text{Limit}} \left(\frac{\partial \phi_s}{\partial r} - ik\phi_s \right) = 0 \tag{4.225}$$

(and $r \equiv |x|$ in this case).

In some cases it may be necessary to include a reflected component of velocity potential, denoted ϕ_r, which does not satisfy the radiation condition (e.g. as in the case of two-dimensional perfect reflection), as well as the diffracted or scattered wave component ϕ_s which does. That is, one takes

$$\phi = \phi_w + \phi_r + \phi_s \tag{4.226}$$

in which both ϕ_w and ϕ_r are known and only ϕ_s is to satisfy the radiation condition. Section 6.1 contains further discussion of the governing equations of a diffraction problem and of the validity of the assumptions made, particularly the assumption that flow separation is absent so that the flow can be taken as irrotational. For the present we continue with a brief review of solutions to the problem for the common case of vertical barriers.

The incident wave potential may be written in complex form as

$$\phi_w = A \frac{\cosh (k(z + d))}{\cosh (kd)} e^{i(kx - \omega t)} \tag{4.227}$$

where $A = -igH/2\omega$. (The real parts of this and subsequent complex expressions are taken to correspond to the physical realization of the quantity described.) Provided that the barrier is vertical and extends from the seabed (or deep water) up to the free surface, the total wave potential may be expressed in a corresponding form

$$\phi = A \frac{\cosh (k(z + d))}{\cosh (kd)} f(x, y) e^{-i\omega t} \tag{4.228}$$

Now for ϕ to satisfy the Laplace equation f must itself satisfy the Helmholtz equation

$$\frac{\partial^2 f}{\partial x^2} + \frac{\partial^2 f}{\partial y^2} + k^2 f = 0 \tag{4.229}$$

Finally, f is also subject both to the radiation condition and to the body surface boundary condition. The latter now reads

$$\frac{\partial f}{\partial n} = 0 \qquad \text{at the body surface} \tag{4.230}$$

Closed-form solutions have been obtained for a few special geometries. One of these which is of considerable practical interest is a straight semi-infinite breakwater as sketched in Fig. 4.36a. The solution was given by Penny and Price (1952) and the results have been described in detail by Wiegel (1962, 1964). The situation is best described in terms of a cylindrical coordinate system as in Fig. 4.36a. The incident wave potential, corresponding to the absence of the breakwater, may be expressed from Eq. (4.227) in the cylindrical coordinate system by writing $x = r \cos \theta$. The solution to the total potential is now written in a corresponding form with $f(x, y)$ being replaced by the function $F(r, \theta)$ which remains to be determined.

The solution for F may be expressed in terms of Fresnel integrals and is given as

$$F(r, \theta) = f(u_1) \exp\left[-ikr \cos(\theta - \theta_0)\right] + f(u_2) \exp\left[-ikr \cos(\theta + \theta_0)\right] \tag{4.231}$$

where θ_0 describes the direction of the incident waves as shown in Fig. 4.36a. Also

$$\begin{aligned} u_1 &= 2\left(\frac{kr}{\pi}\right)^{1/2} \sin\left[\tfrac{1}{2}(\theta - \theta_0)\right] \\ u_2 &= -2\left(\frac{kr}{\pi}\right)^{1/2} \sin\left[\tfrac{1}{2}(\theta + \theta_0)\right] \end{aligned} \tag{4.232}$$

and

$$f(u) = \tfrac{1}{2}\left[(1 + C + S) - i(S - C)\right] \tag{4.233}$$

where C and S are the cosine and sine Fresnel integrals respectively with argument u. (For negative u, one may invoke $C(-u) = -C(u)$ and $S(-u) = -S(u)$.) The wave height at any point (r, θ) is given in terms of the incident wave height H_i as

$$H = 2\eta_{max} = -\frac{2}{g}\left|\frac{\partial \phi}{\partial t}\right|_{z=0} = H_i \, |F(r, \theta)| \tag{4.234}$$

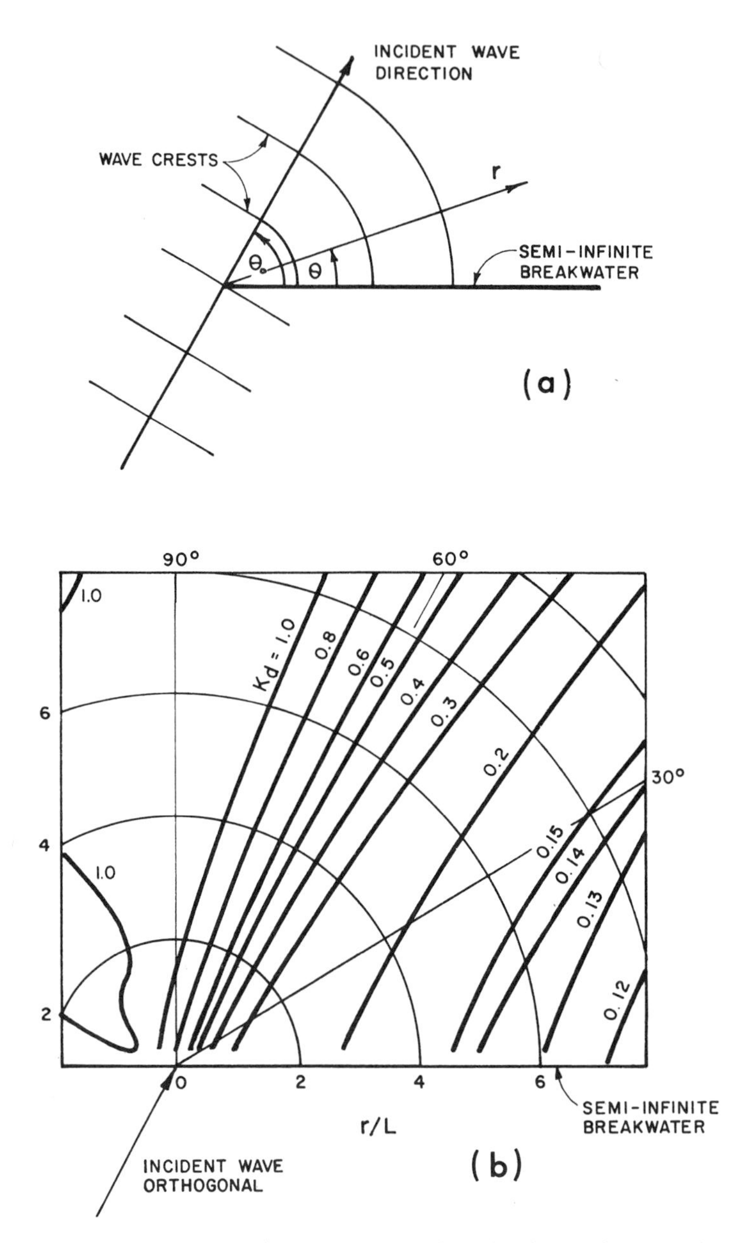

Fig. 4.36. Wave diffraction around a straight semi-infinite breakwater for an incident wave approach angle $\theta_0 = 60°$.

At this point we introduce a *diffraction coefficient* K_d which is defined as the ratio of the wave height at any point in the diffracted wave field to the incident wave height and which is therefore given as

$$K_d = |F(r, \theta)| \tag{4.235}$$

The positions of wave crests calculated at a particular instant with $\theta_0 = 60°$ are shown in Fig. 4.36a and corresponding contours of the diffraction coefficient are set out in Fig. 4.36b. It is not too laborious a task to obtain such contours for any given θ_0 by using a computer. Wiegel (1962) has presented such plots for values of θ_0 in multiples of 15° (see also, U.S. Army Shore Protection Manual 1973). Various approximations have also been suggested (e.g. Larras 1966, Silvester 1974) which permit a desk calculation of the solution.

Other reference geometries involving wave diffraction that have been considered in the context of breakwaters include the diffraction through a gap between two semi-infinite breakwaters (either colinear or inclined to each other) and the diffraction around a straight offshore breakwater. Morse and Rubinstein (1938) considered these cases in the context of electromagnetic waves. For the case of a breakwater gap, Johnson (1951) has presented the solution graphically for practical application. The case of an offshore breakwater has been considered by Stiassnie and Dagan (1972) and by Goda and Yoshimura (1973). The diffraction of waves around a large vertical circular cylinder has also been obtained (MacCamy and Fuchs 1954) and this is of particular practical interest in the context of wave loading on large offshore structures. This latter problem will be considered in detail in Section 6.2.

At this point we mention one important generalization which deals with the combined effects of wave diffraction around vertical barriers, as well as wave refraction due to gradual depth variations. Berkhoff (1972) derives the governing equations for this case by a separation of variables approach, and a power series expansion involving a parameter $\nu = m(d/L)$, where m is the seabed slope assumed to be small. The velocity potential may thereby be expressed in the form given by Eq. (4.228), but now the potential function $f(x, y)$ must satisfy the governing equation

$$\frac{\partial}{\partial x}\left(cc_G \frac{\partial f}{\partial x}\right) + \frac{\partial}{\partial y}\left(cc_G \frac{\partial f}{\partial y}\right) + \omega^2 \frac{c_G}{c} f = 0 \tag{4.236}$$

where c_G is the group velocity, and both c and c_G of course depend on depth and therefore on location (x, y). Equation (4.236) reduces to the Helmholtz equation, Eq. (4.229), for deep water or for uniform depth.

Another aspect of wave diffraction which is of some importance is that of long wave amplification in a harbor. This may be caused, for example, by storm surge or a tsunami and may induce large and damaging horizontal motions within the harbor at certain resonant frequencies. The mathematical treatment of the problem has been reviewed by Miles (1974) and numerical methods of solution by Mei (1978). As with the other cases of wave diffraction considered so far, the incident and reflected wave potentials are specified and the scattered wave potential must be determined so that the boundary condition of zero velocity normal to the harbor boundary, together with a far field radiation condition, must both be satisfied.

One reference case corresponds to a rectangular harbor of constant depth open to the sea through a central gap along one side. This was initially treated by Miles and Munk (1961) and later by Ippen and Goda (1963) who compared theoretical predictions of the wave amplification with experimental measurements. The fundamental resonance occurs at a wave length somewhat greater than $L = 4\ell$, where ℓ is the length of the harbor, and this limit corresponds to the two-dimensional free oscillations in an open-ended basin as considered in Section 4.9.4.

The more general case of harbors of arbitrary shape and either variable or constant depth must be treated by various computational approaches. Hwang and Tuck (1970) presented a numerical procedure in which the harbor boundary is represented as a distribution of "wave sources," with strengths chosen to ensure that the boundary condition of zero normal velocity along the harbor boundary is satisfied. This method is described in the context of wave forces in Section 6.4.3. Other methods have involved dividing the fluid region into two portions: one within the harbor and one exterior to the harbor, and applying matching conditions at the boundary between these regions (Lee 1971); finite difference schemes (Raichlen 1965, Raichlen and Naheer 1976) and finite or hybrid element methods (Sakai and Tsukioka 1975, Chen and Mei 1974 and Bettess and Zienkiewicz 1977). Miles (1971) has emphasized the analogy between harbor oscillations and electric circuits.

4.9.6 Wave Attenuation

It was assumed in Section 4.9.1 that during the propagation of a wave train, energy is neither supplied to the wave motion nor dissipated. Although this assumption is usually reasonable, some energy dissipation does in fact generally occur and may need to be taken into account in certain problems. The energy of laboratory generated waves is dissipated by friction at the bottom and sidewalls of a channel and as a result of surface tension acting at the free surface. In the case of ocean waves at intermediate and shallow depths, energy is dissipated before breaking by friction and percolation at the seabed. This dissipation of wave energy leads to a reduction of wave height termed *wave damping* or *wave attenuation.* (In the open ocean there is a decrease in wave heights largely on account of angular dispersion: waves spread out from the generating area over a range of directions so that their energy is distributed over an increasing area, with the result that the energy density and consequently the wave height reduces.)

In the laboratory the problem may be treated theoretically since (a) percolation does not occur (except when it is planned for), and (b) the boundary layers on the bottom and side-walls are invariably laminar so that laminar boundary layer theory may be applied. In laboratory situations, however, surface tension acting at the free surface also has a significant influence on the degree of damping and introduces uncertainties into the predicted damping.

Two kinds of approximations are generally made in the development of a wave damping theory: one concerns the smallness of the wave height such that higher order terms involving a wave height perturbation parameter are neglected; the other concerns a boundary layer approximation for finite depths in which the energy dissipation occurs principally within the bottom and side-wall boundary layers. The latter may be considered as the first of successive approximations involving powers of $k\delta$ which is a dimensionless boundary layer thickness, $\delta = (2\nu/\omega)^{1/2}$.

The general theoretical method employed is to calculate the rate of energy dissipation within the fluid, which is given for two-dimensional motion in terms of the Rayleigh dissipation function as

$$2F = \mu \int \left[2\left(\frac{\partial U}{\partial x}\right)^2 + 2\left(\frac{\partial W}{\partial z}\right)^2 + \left(\frac{\partial U}{\partial z} + \frac{\partial W}{\partial x}\right)^2 \right] dx\, dz \qquad (4.237)$$

where μ is the dynamic viscosity of the fluid, and the integral is taken over the region occupied by the fluid. The corresponding function for three-dimensional motion is given in Lamb (1945). The average rate of energy dissipation per unit length in the x direction (and per unit width in the two-dimensional case), denoted D, can be expressed in terms of F and is simply $D = 2F/L$. A knowledge of the flow field, particularly in the boundary layers, will therefore enable D to be evaluated.

Alternatively D may be expressed in terms of the shear stresses acting at the fluid boundaries. For a two-dimensional oscillatory motion on the channel bottom, D is given as $D = \overline{\tau U}$, where τ is the shear stress at the bottom, U is the fluid velocity at the outer edge of the boundary layer and the overbar denotes a temporal mean. This representation is particularly useful for the case of a turbulent boundary layer. In the simpler case of a laminar boundary layer induced by a sinusoidal velocity fluctuation of amplitude U_m the result is

$$D = \tfrac{1}{8}\, \rho\nu U_m \qquad (4.238)$$

The rate of energy transfer across a plane x = constant will also be required and may be obtained from the particular wave theory being adopted. Thus in the case of linear wave theory, this is

$$P = \frac{1}{8} \rho g H^2 c_G b \tag{4.239}$$

where b is the appropriate width in the y direction. The difference in average rates of energy flux crossing planes a short distance dx apart is due to the average dissipation between those planes, and an energy balance equation may accordingly be developed. That is $dP/dx = -D$. In this way an expression for the spatial gradient of wave height may be determined.

Making the usual sinusoidal and boundary layer approximations, the wave height is found to decay exponentially and may be written as

$$H = H_0 e^{-\alpha x} \tag{4.240}$$

where H_0 is the height at some reference point and α is an *attenuation factor* which may be defined explicitly as

$$\alpha = -\frac{1}{H}\frac{dH}{dx} \tag{4.241}$$

A dimensionless *attenuation coefficient* may also be introduced in the form

$$a = \alpha d = -\frac{d}{H}\frac{dH}{dx} \tag{4.242}$$

Biesel (1949) investigated the case of two-dimensional sinusoidal waves in water of constant and finite depth and calculated the attenuation coefficient. Hunt (1952) extended this to include the case of waves propagating in a channel of finite width over a gently sloping bed. For this case

$$\alpha = \frac{k\delta}{b}\left[\frac{kb + \sinh(2kd)}{2kd + \sinh(2kd)}\right] \tag{4.243}$$

This factor is made up of components due to bottom (subscript b) friction and side-wall (subscript w) friction, and in terms of the attenuation coefficient a we have

$$a = a_b + \left(\frac{2d}{b}\right) a_w \tag{4.244}$$

with

$$a_b = \frac{k\delta}{2}\left[\frac{2kd}{2kd + \sinh(2kd)}\right] \quad (4.245a)$$

$$a_w = \frac{k\delta}{2}\left[\frac{\sinh(2kd)}{2kd + \sinh(2kd)}\right] \quad (4.245b)$$

For deep water waves there is no bottom boundary layer so that, in the two-dimensional case, the dissipation occurs only in the fluid interior. This problem was treated in the last century by Basset (1888) (see also Lamb 1945). Under these conditions the depth d is of course no longer a relevant variable and it is found that

$$\alpha = 4k^2\nu\left(\frac{k}{g}\right)^{1/2} = 2k(k\delta)^2 \quad (4.246)$$

which, as is to be expected, is an order $k\delta$ higher than that which was given in Eq. (4.243).

The various results quoted so far assume that surface tension effects are absent. Van Dorn (1966) has considered the extreme surface tension effect of an immobile surface and obtains the corresponding additional component a_s of the attenuation coefficient to be

$$a_s = a_b \sinh(kd) \quad (4.247)$$

where a_b is given by Eq. (4.245a). It appears, then, that surface tension effects are expected to become increasingly important for deeper waves, and in the deep water limit we have $\alpha_s = k^2\delta$. The role of surface tension is to permit a non-zero shear stress at the free surface with consequent changes in the boundary condition there and so also in the structure of the free surface boundary layer. Further discussion of the role of surface tension on wave damping is given by Miles (1967) and by Mei and Liu (1973).

The damping of a solitary wave was investigated by Keulegan (1948) who, on the basis of the assumptions already mentioned in the context of periodic waves, found that the wave height attentuation is not exponential but follows an inverse power law

$$\left(\frac{H}{d}\right)^{-1/4} - \left(\frac{H_0}{d}\right)^{-1/4} = \frac{1}{12}\left(1 + \frac{2d}{b}\right)\left(\frac{\nu}{d\sqrt{gd}}\right)^{1/2}\left(\frac{x}{d}\right) \quad (4.248)$$

Consequently the attenuation coefficient is now dependent on the wave height itself and is given according to Keulegan's formula as

$$a = \frac{\nu^{1/2} H^{1/4}}{3g^{1/4} d} \left(1 + \frac{2d}{b}\right) \tag{4.249}$$

Isaacson (1976b) and Miles (1976) have obtained expressions for the damping of cnoidal waves to a first approximation. Isaacson gives

$$a = \frac{\nu}{2\delta\sqrt{gd}} f(\kappa) \tag{4.250}$$

where κ is the modulus of the Jacobian elliptic functions describing the waves and $f(\kappa)$ may be calculated numerically. The limiting cases $\kappa \to 0$ and $\kappa \to 1$ correspond of course to the damping of shallow water sinusoidal and solitary waves respectively. Isaacson (1977) subsequently extended the theory of cnoidal wave damping to a second approximation and presented graphically the variation of the attenuation coefficient with d/gT^2 for various values of H/d. It was found that attenuation coefficients for shallow water are considerably larger than those predicted by sinusoidal wave theory, whereas for intermediate depths they are slightly smaller. In deeper water the cnoidal theory of course breaks down, and furthermore surface tension effects and energy dissipation terms neglected in the boundary layer approximation become increasingly important. It follows that over the whole range of depths we may expect that measured attenuation coefficients will be considerably in excess of predictions based on sinusoidal wave damping theory, provided the boundary layer remains laminar. This feature is generally supported by laboratory measurements (e.g. Eagleson 1962, Iwagaki and Tsuchiya 1966 and Treloar and Brebner 1970).

The preceding discussion concerns laminar boundary layers and is therefore pertinent with regard to laboratory-generated waves. When the boundary layer is turbulent the rate of energy dissipation and therefore the wave height attenuation will differ appreciably from the predictions based on laminar boundary layer theory. Theoretical developments concerning the turbulent boundary layer at the seabed, primarily directed to a determination of the velocity profile within the boundary layer, include those made by Jonsson (1963, 1980), Kajiura (1968), Noda (1971) and Johns (1975).

For a turbulent boundary layer, the friction coefficient $\tau_m / \frac{1}{2} \rho U_m^2$ is independent of Reynolds number but depends on the bed roughness (e.g. Riedel, Kamphuis and Brebner 1972). Here τ_m and U_m are respectively the amplitudes of the shear stress at the bed and the fluid velocity just outside the bottom boundary layer. Assuming that the corresponding instantaneous values remain in phase (Kajiura 1968), the average rate of energy dissipation will be proportional to U_m^3 and therefore to H^3. Also the average energy transfer rate still varies as H^2, Eq. (4.239), and therefore the turbulent damping law is eventually given by dH/dx being proportional to H^2. This was found by Battjes (1965) to be true for the damping due to roughness strips projecting from the sides of a wave channel, and by Isaacson (1978b) for the case of waves propagating along a straight channel lined by sloping crushed-rock banks.

For the smooth bed case considered here, the condition of transition to a turbulent boundary layer has not been conclusively established. The condition may, however, be expressed in terms of a Reynolds number $Re = U_m\delta/\nu$, which is based on the oscillatory boundary layer thickness $\delta = (2\nu/\omega)^{1/2}$ and the velocity amplitude U_m just outside the boundary layer. A relatively low value of 160 suggested by Brebner and Collins (1961) (see also Collins 1963) has been shown to be unfounded (Isaacson 1978a). Values of 566 obtained by Li (1954) and 500 obtained by Sergeer (1966) are probably more reliable indications of a transition Reynolds number, although transition may be expected to take place over a range of Reynolds numbers rather than at one specific value. In particular, Hino, Sawamoto and Takasu (1976) have considered the problem of boundary layer transition in an oscillatory pipe flow, as indicated by experiments employing a hot-wire anemometer to observe velocity fluctuations. The flow is found to be intermittently turbulent over a wide range of conditions, and they distinguish between varying degrees of turbulent flow which they describe as weakly turbulent, conditionally turbulent and fully turbulent. Specifically, conditional turbulence exists when marked turbulence occurs in the decelerating phase of the flow, while a laminar-like flow occurs in the accelerating phase. The critical Reynolds number for the onset of conditional turbulence was found to be independent of pipe size (relative to boundary layer thickness) and to occur at about 550.

But of course the problem is not so simply defined in practice. One important feature to be considered concerns the presence of a pre-existing steady current on which the oscillatory flow is superposed. For discussions of the stability of this pulsating Poiseuille flow, the reader is referred to Sarpkaya (1966) and to Davis (1976). The case of a rough fixed bed is of particular importance, and the corresponding turbulent boundary layer has been investigated by a number of authors, (for a review see Jonsson (1980).) In particular, Jonsson (1980) has proposed a method of predicting velocity profiles for this case. Other difficulties in describing the practical situation include attempting to account for a granular seabed giving rise to bedforms, fluid percolation through the seabed, and so on.

4.9.7 Wave Instability and Evolution

Theories used to describe a wave train of permanent form have been outlined in the earlier part of this chapter, and such waves can generally be realized as states of perfect dynamic equilibrium. However, the stability of such a wave train in the face of a disturbance is a separate and important matter than requires attention. The topic of wave instabilities has been reviewed recently by Yuen and Lake (1980). The wave breaking instability resulting from too large a wave height is described separately in Section 4.9.9.

Benjamin and Feir (1967) and Benjamin (1967) demonstrated that Stokes waves are in fact unstable when subject to certain disturbances, this instability occuring for deeper wave conditions given by $kd > 1.363$. Benjamin and Feir considered a basic nonlinear (Stokes) wave train together with a disturbance comprising of a pair of progressive wave modes at frequencies and wave numbers very slightly different from those of the fundamental. This corresponds to a slow modulation or beating of the basic wave train, as may for example be produced in a wave tank by a paddle motion containing a slight low frequency modulation. The instability was demonstrated by examining the nonlinear free surface boundary conditions corresponding to the nonlinear interaction between the Stokes waves and the disturbance wave modes. The Benjamin-Feir instability was also demonstrated shortly afterwards by Whitham (1967a) using a variational approach.

Bryant (1974) has separately considered the stability of waves of permanent form and of arbitrary depth and so his treatment covers the case of both cnoidal and Stokes waves. The stability was examined with reference to a periodic disturbance with wave length greater than or equal to that of the fundamental and travelling in the same direction. The wave train was found to be generally stable to such disturbances, although the margin of stability may be very slight for a wide range of situations.

An associated problem to that of permanent wave train stability is the investigation of how a wave motion does in fact evolve from a disturbance acting on initially still water, such as may be produced by a wave paddle. The resulting wave motion will not generally be of permanent form. We interest ourselves primarily in a long smooth initial disturbance, that is one corresponding to a moderately large Ursell number. In this case secondary crests appear within a primary wave length and travel at speeds that differ from that of the primary crest, each crest speed depending upon that crest's elevation. An arbitrary disturbance finally evolves into a finite number of positive permanent solitary-like waves called *solitons*, and these are followed by a train of oscillatory waves about the still water level as sketched in Fig. 4.37a. The number of solitons produced depends entirely on the initial disturbance and we note, in particular, that when the water level introduced by the disturbance is everywhere below the still water level then no solitons will emerge. If, however, a net positive volume of fluid is introduced by the disturbance then at least one soliton is formed. The behavior of solitons, including the interaction between them, has been described by Zabusky and Galvin (1971) and by Segur (1973).

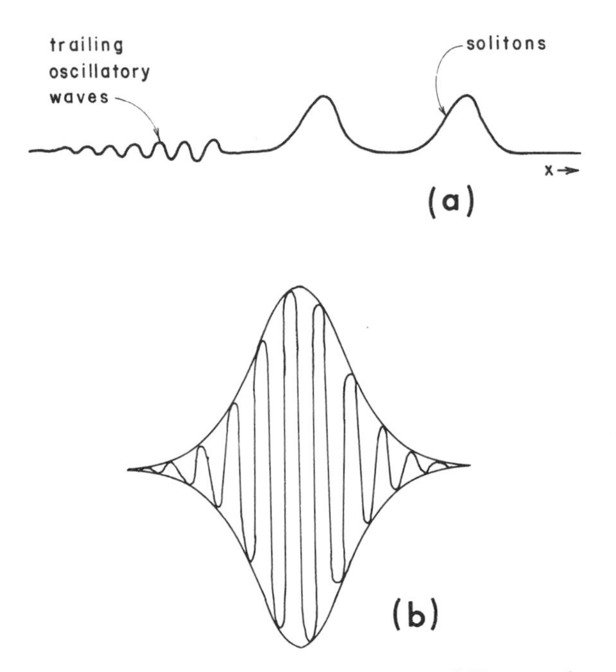

Fig. 4.37. Sketches of (a) solitons with trailing waves, and (b) an envelope soliton.

The oscillatory wave train trailing the solitons is dispersive with the result that longer waves move ahead while shorter waves remain near the rear of the train. This and other developments described here may be predicted by the Korteweg-de Vries equation (Section 4.4.1), which includes within its scope both nonlinear and dispersive effects, but omits the effect of viscous dissipation. The equation may be written in a compact form, Eq. (4.80) as

$$\frac{\partial f}{\partial \tau} + 6f\frac{\partial f}{\partial \chi} + \frac{\partial^3 f}{\partial \chi^3} = 0 \tag{4.251}$$

A linearized approximation to this, obtained by omitting the second (nonlinear) term, is sometimes used as a simpler formulation. The text by Leibovitch and Seebass (1974) contains several useful discussions of this equation and alternatives to it.

Several comparisons of numerical predictions based on the Korteweg-de Vries equation with experimental results have been made. In particular, Madsen, Mei and Savage (1970) have examined the breakdown of smooth shallow waves into forms exhibiting several secondary crests; and Zabusky and Galvin (1971) have compared the numerically predicted and observed numbers and locations of secondary crests and troughs. They found good correspondence, but where crest amplitudes are concerned the comparison was less satisfactory since these are more noticeably influenced by dissipation. Hammack and Segur (1974) have also carried out comparisons of soliton behavior and they later (Hammack and Segur 1978a) investigated separately the properties of the trailing oscillatory wave train.

These oscillatory waves are relatively short and generally satisfy the Benjamin-Feir instability criterion $kd > 1.363$. They consequently exhibit modulational instability and subsequently behave as carrier waves subjected to low frequency modulation changes. Although the Korteweg-de Vries equation predicts the occurence of the oscillatory wave train, this equation is itself unsuitable for describing the development of the train since the waves are short, whereas the equation is valid for long (shallow) waves. The evolution of the oscillatory waves is more appropriately described by the nonlinear Schrödinger equation which was derived in this context by Zakharov (1968) and by Hasimoto and Ono (1972). Yuen and Lake (1975) also derived it using Whitham's variational formulation taken to the correct order. The nonlinear Schrödinger equation may be written in the form used by Yuen and Lake (1975) as

$$i\left[\frac{\partial A}{\partial t} + \frac{1}{2}\left(\frac{\omega_0}{k_0}\right)\frac{\partial A}{\partial x}\right] - \left(\frac{\omega_0}{8k_0}\right)\frac{\partial^2 A}{\partial x^2} - \frac{1}{2}\omega_0 k_0 \, |A|^2 A = 0 \qquad (4.252)$$

where A is the complex wave envelope with the free surface elevation given by $\eta = A \exp\,[i(k_0 x - \omega_0 t)]$. ω_0 and k_0 are constants representing the angular frequency and wave number respectively of the carrier wave. This equation predicts the initial occurence of the Benjamin-Feir instability and also the subsequent emergence in general of a number of nonlinear wave packets termed *envelope solitons.* These are analogous to the solitons described earlier, but the soliton profile here relates to the envelope of the carrier wave as sketched in Fig. 4.37b. The behavior of envelope solitons has been described by Yuen and Lake (1975).

Comparisons of nonlinear wave evolution as predicted by the nonlinear Schrödinger equation and as observed have been presented by Yuen and Lake (1975), Lake et al. (1977) and Hammack and Segur (1978a). In particular, Lake et al. (1977) demonstrate that eventually the modulation periodically increases and decreases with time and the nonlinear resonant interactions do not therefore lead to a disintegration or loss of coherence of the wave form as had been earlier suggested. They describe how dissipation (due to viscosity, capillary waves and wave breaking) may account for differences between their experimental results and the predictions based on the nonlinear Schrödinger equation. Recent contributions to the study of nonlinear wave instabilities have also been made by Fornberg and Whitham (1978) and by Longuet-Higgins (1978).

4.9.8 Waves of Maximum Height

For waves of a given length, in water of uniform and given depth, there exists a maximum wave height beyond which the wave form is unstable. It is of both practical and theoretical interest to establish this maximum height, and considerable progress has now been made in this direction. The behavior of breaking waves is of associated interest and will be considered shortly. Reviews of recent research into waves of maximum wave height and into breaking wave properties have been given by Longuet-Higgins (1976), Cokelet (1977b) and Peregrine (1979).

Stokes (1880) suggested that a suitable criterion for the steepest wave is that the particle velocity at the crest is just equal to the wave speed so that any further increase in particle velocity would cause an instability. Such a wave has a stagnation point at the crest, and in consequence the crest must be sharp and contains an angle of 120°. Using Stokes' criterion, Michell (1893) found that the limiting steepness in deep water is given as $H/L = 0.142 \simeq 1/7$.

As described in Section 4.5.4, Schwartz (1974) found that the perturbation parameter that had been used in the Stokes expansion procedure does not increase with wave height as the waves approach the maximum steepness, and that the Stokes series expansion for obtaining higher order approximations then fails to converge. Schwartz used instead $\frac{1}{2}kH$ as an alternative parameter and was then able to obtain convergence. He thereby confirmed Yamada's (1957a) earlier prediction of 0.1412 for the maximum deep water wave steepness. Cokelet (1977a) has since used a method similar to that of Schwartz but employing a different expansion parameter and he obtained a value of 0.141065.

At the other extreme, the maximum height of a solitary wave is limited by the still water depth, rather than the wave length which is now infinite. McCowan (1891) applied Stokes' criterion to the solitary wave and obtained the maximum height as $H/d = 0.78$. Chappelear (1959) has obtained a value of 0.87, while Yamada (1957b) and Lenau (1966) each obtained 0.83. More recently, Longuet-Higgins and Fenton (1974) used a new expanison parameter and carried out calculations for a solitary wave to a high order, employing Padé approximants for summing series. They obtained the maximum height of the solitary wave as $H/d = 0.827$, and they also found that the speed of this wave is given as $c/\sqrt{gd} = 1.286$. This satisfies the relationship $\frac{1}{2}c^2/gd = H/d$ which must be valid for the limiting wave by a simple application of the Bernoulli equation. Longuet-Higgins and Fenton also found that the maximum wave celerity occurs for a wave of less than the maximum height, $H/d = 0.790$, and is given as $c/\sqrt{gd} = 1.294$. This kind of feature has already been noted for deep water and intermediate depth waves (Section 4.5.4).

Longuet-Higgins (1974) has suggested a simple and accurate approximation for the limiting solitary wave. The profile suggested is given as

$$\frac{\eta}{d} = 1.5389 \exp\left(-1.0495 \frac{x}{d}\right) - 0.7093 \exp\left(-1.4630 \frac{x}{d}\right) \tag{4.253}$$

and this is very close to the predictions of Yamada (1957b), Lenau (1966) and also Longuet-Higgins and Fenton (1974). For example, this predicts the maximum wave height as $H/d = 0.8296$ and the corresponding wave celerity to be $c/\sqrt{gd} = 1.288$. Byatt-Smith and Longuet-Higgins (1976) have also carried out calculations for steep solitary waves using a different method involving the solution of an integral equation and they essentially confirm many of the results given by Longuet-Higgins and Fenton.

For finite and uniform depths, Miche (1944) proposed that the maximum wave steepness is given as

$$\frac{H}{L} = 0.142 \tanh(kd) \tag{4.254}$$

This simple formula serves as a useful engineering approximation over a wide range of depths, but gives a severe overestimate $H/d = 0.89$ for the shallow wave limit. Yamada and Shiotani (1968) have obtained the maximum wave steepness as a function of d/L and Schwartz (1974) has since obtained essentially identical results. Figure 4.38 shows this relationship and includes Miche's result for purposes of comparison. Schwartz also presented the variation of c^2/gd with H/d, where c is the wave celerity based on Stokes first definition and H is the maximum wave height. This information can be used to present the maximum wave height in terms of the water depth and wave period. Thus Fig. 4.39 presents H/gT^2 as a function of d/gT^2 on the basis of Stokes first definition of wave celerity.

It will be noted from the above outline that considerable progress has recently been made on the study of very steep nearbreaking waves. Further recent theoretical contributions have been made by Longuet-Higgins and Fox (1977, 1978) and a useful summary of the recent theoretical work on steep waves has been given by Longuet-Higgins (1976). It must be emphasized once again that the various results mentioned have been obtained on theoretical grounds and that in practice waves will invariably become asymmetric and break before the predicted maximum heights can actually occur. This feature has been discussed in terms of the non-monotonicity of various integral properties with wave height by several of the authors already referred to.

As mentioned in Section 4.9.3, the influence of a uniform current on the maximum wave height has been examined by Dalrymple and Dean (1975). Wave theory gives the maximum wave height H in the form H/L as a function of d/L as in Fig. 4.38. However, in a wave-current field it is generally the wave period T_c relative to a fixed reference frame that is specified and thus it becomes neces-

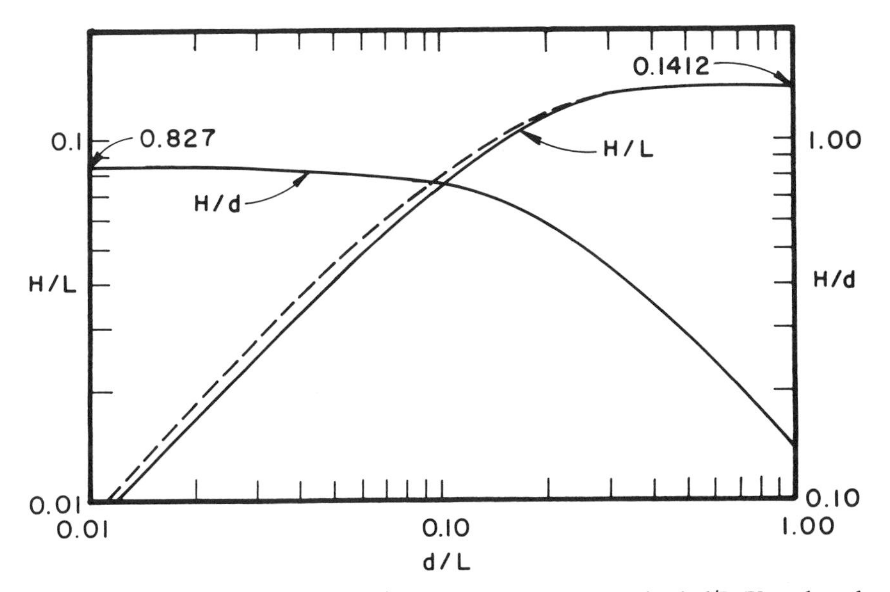

Fig. 4.38. Maximum wave steepness H/L as a function of relative depth d/L (Yamada and Shiotani 1968, and Schwartz 1974). ---, H/L = 0.142 tanh (kd).

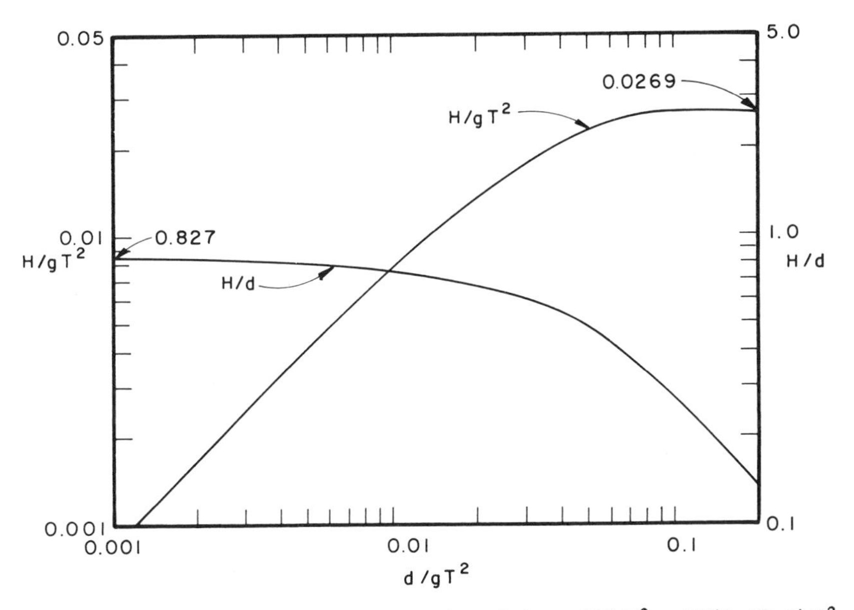

Fig. 4.39. Maximum wave height presented as the variations of H/gT^2 and H/d with d/gT^2, based on the first definition of celerity and derived from the results of Schwartz (1974).

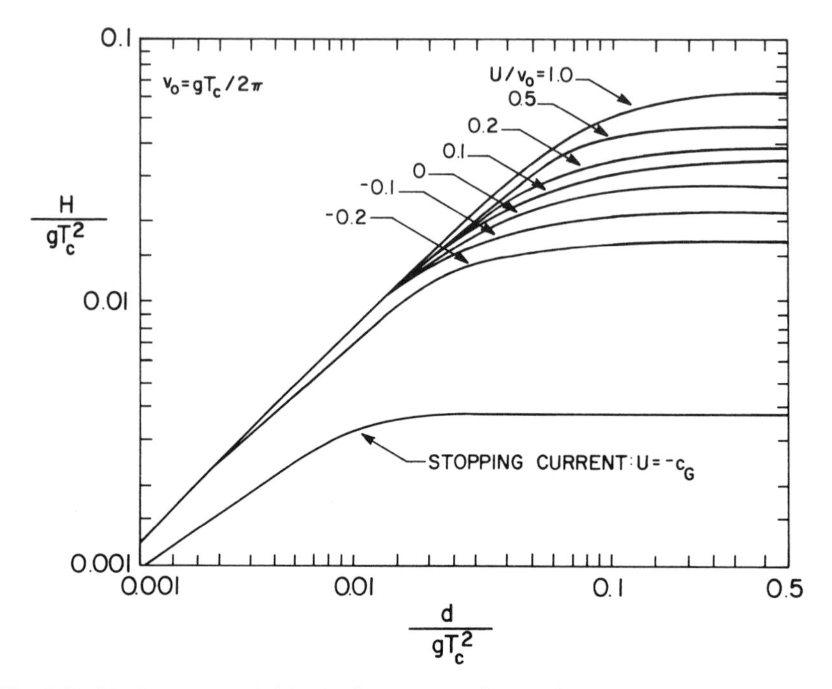

Fig. 4.40. Maximum wave height in the presence of a steady uniform current (Dalrymple and Dean 1975).

sary first to calculate the appropriate value of L. This procedure was indicated in Section 4.9.3 and may be applied to any suitable wave theory. Dalrymple and Dean performed the necessary calculations using the stream function theory and presented the maximum wave height directly in terms of T_c and the current U. Their results are shown in Fig. 4.40.

4.9.9 Breaking Waves

Waves may break as a consequence of various effects, but in coastal engineering one is most often concerned with breaking due to shoaling as waves approach the shore. Reviews of breaking waves include those already mentioned in connection with waves of maximum height (Longuet-Higgins 1976, Cokelet 1977b, and Peregrine 1979). From the viewpoint of engineering design, Tenaud and Graillot (1975) also review breaking waves and their effects on coastal structures.

The depth at breaking d_b, the wave height at breaking H_b, and the form of the breaking wave are usually of primary interest, and under conditions of wave shoaling these depend on the deep water wave steepness H_0/L_0 and the beach slope m.

Although experimental data describing these quantities have shown considerable scatter (Galvin 1969, Goda 1970), empirical relationships have nevertheless been fitted for purposes of design (e.g. Weggel 1972, U.S. Army Shore Protection Manual 1973). These indicate d_b and H_b expressed in suitably dimensionless forms, together with the breaker type, as functions of H_0/L_0 and m. The breaker index d_b/H_b is often used in this context.

Breaking waves are generally classified as *spilling*, *plunging*, *collapsing* or *surging* (Galvin 1968). These different types of breaking waves are sketched in Fig. 4.41, and their descriptions, which may be used as a basis for definition, have been given by Galvin (1968, 1972). With reference to Fig. 4.41, spilling breakers are characterized by foam spilling from the crest down over the forward face of the wave. They occur in deep water or on gentle beach slopes. Plunging breakers occur on moderately steep beach slopes, and are noticeably distinct with a well defined jet of water forming from the crest and falling onto the water surface ahead of the crest. The collapsing wave foams lower down the forward face of the wave and is a transition type between plunging and surging breakers. Surging breakers occur on relatively steep beaches in which there is considerable wave reflection with some foam forming near the beach surface.

Thus the breaker type is generally a continuous variable, proceeding from spilling to plunging to collapsing to surging for increasing values of beach slope, for waves of the same deep water steepness. Galvin (1968) indicates that the formation of a particular breaker type depends on the parameter $\beta = H_0/(L_0 m^2)$ or on $H_b/(gT^2m)$. Thus spilling breakers usually form when $\beta > 5$, plunging when $5 > \beta > 0.1$, collapsing when $\beta \simeq 0.1$, and surging when $\beta < 0.1$. Alternatively, Fig. 4.42 indicates the approximate ranges in parameters H_0/L_0 and m over which various kinds of breakers form on the basis of Goda's (1970) data. However, it must be borne in mind that the distinction in breaker type is not always clear cut, and such parameter ranges are necessarily inexact. Indeed, it appears (Peregrine 1979) that the distinction between breaker types is to a certain extent one of scale and, in particular, a spilling breaker initially forms by the crest curling over and plunging on a much smaller scale relative to the wave size than occurs in a more typical plunging wave. Other features of breaking waves concern the distance a breaker travels (Galvin 1969), and the subsequent flow behavior which in many cases corresponds to the formation of a bore.

Associated theoretical advances into wave breaking are relatively recent. Longuet-Higgins and Cokelet (1976) (see also Longuet-Higgins 1976, Cokelet 1979) have developed a numerical method for calculating the flow in a deforming wave for instants well after the free surface passes through a vertical. The procedure can so provide details of the velocity and pressure variations within the wave immediately prior to it breaking and in the plunging jet of the breaking wave. The flow is assumed irrotational and periodic in space (though not in

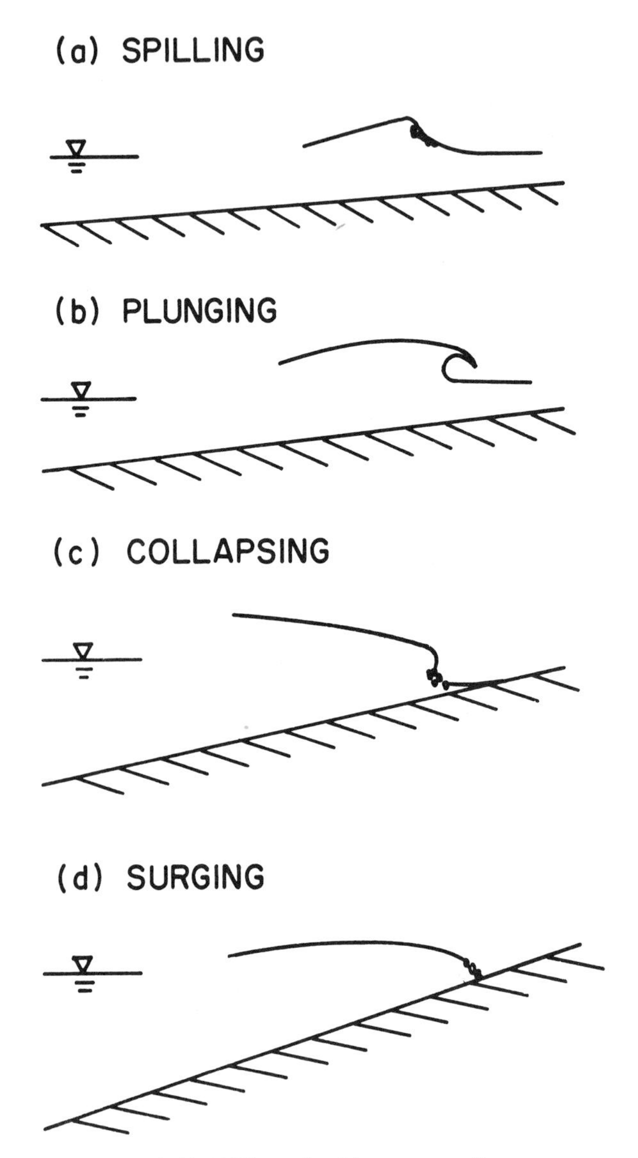

Fig. 4.41. Different breaking wave profiles.

time). In order to trace the progress of the waves by a time-stepping method, the initial free surface profile and initial velocity potential and normal and tangential velocity variations along the free surface are prescribed, and one requires corresponding values a short time interval later. In particular, the calculation of the normal velocity involves a conformal transformation in which the profile

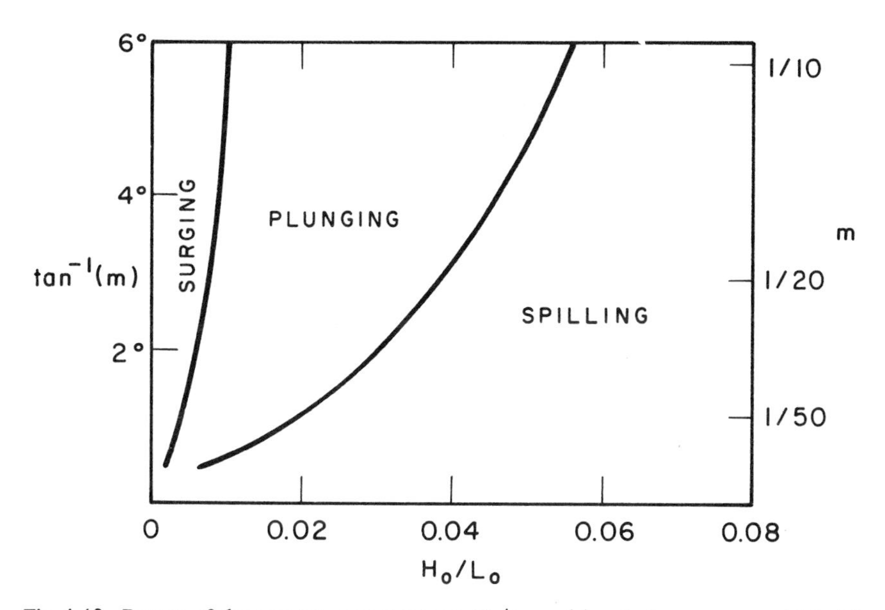

Fig. 4.42. Ranges of deep water wave steepness H_0/L_0 and beach slope m over which different kinds of breakers form.

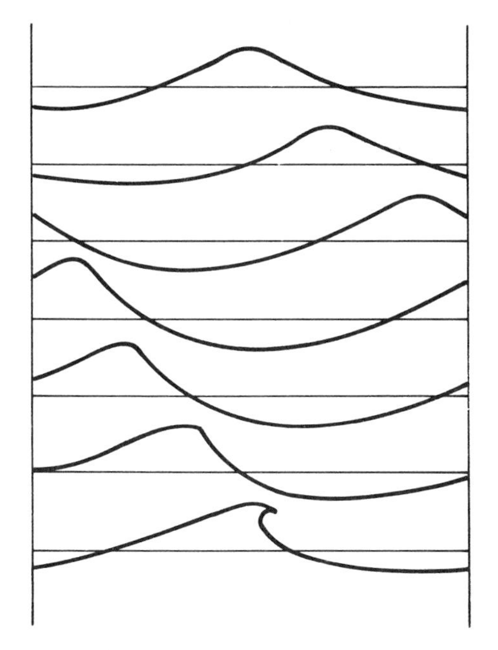

Fig. 4.43. Successive free surface profiles over one wave length of a progressive wave demonstrating its breaking due to an imposed pressure distribution (Longuet-Higgins and Cokelet 1976).

over one wave length is transformed to a closed contour, and then an integral equation around this contour is solved by a discretization procedure.

An example of the method is provided by subjecting a steep symmetric deep-water wave to a prescribed free-surface pressure variation during half a wave cycle. Calculated profiles of the wave train at various times after the pressure variation is initially imposed and extending beyond the time it is removed are reproduced in Fig. 4.43.

4.10 REFERENCES

Abramowitz, M. and Stegun, I. A. 1965. *Handbook of Mathematical Functions*. Dover, New York.

Adee, B. H. 1976. Floating Breakwater Performance. *Proc. 15th Coastal Eng. Conf.*, Honolulu, Vol. III, pp. 2777–2789.

Adee, B. H. and Martin, W. 1974. Analysis of Floating Breakwater Performance. *Proc. Floating Breakwater Conf.*, Univ. of Rhode Island, pp. 21–40.

Allen, J. and Gibson, D. H. 1959. Experiments on the Displacement of Water by Waves of Various Heights and Frequencies. *Proc. Inst. Civil Eng.*, Vol. 13, pp. 363–386.

Basset, A. B. 1888. *A Treatise on Hydrodynamics*, Vols. I and II, Deighton Bell, London. Reprinted 1961 by Dover, New York.

Batchelor, G. K. 1967. *An Introduction to Fluid Dynamics*. Cambridge University Press.

Battjes, J. A. 1965. Wave Attenuation in a Channel with Roughened Sides. *Proc., Coastal Engineering Santa Barbara Specialty Conference, ASCE*, Santa Barbara, pp. 425–460.

Battjes, J. A. 1968. Refraction of Water Waves. *J. Waterways and Harbors Div., ASCE*, Vol. 94, No. WW4, pp. 437–452.

Benjamin, T. B. 1967. Instability of Periodic Wavetrains in Nonlinear Dispersive Systems. *Proc. Roy. Soc.*, Ser. A, Vol. 299, pp. 59–75.

Benjamin, T. B. and Feir, J. E. 1967. The Disintegration of Wave Trains in Deep Water. Part 1. Theory. *JFM*, Vol. 27, pp. 417–430.

Benjamin, T. B. and Lighthill, M. J. 1954. On Cnoidal Waves and Bores. *Proc. Roy. Soc.*, Ser. A, Vol. 224, pp. 448–460.

Berkhoff, J. C. W. 1972. Computation of Combined Refraction-Diffraction. *Proc. 13th Coastal Eng. Conf.*, Vancouver, Vol. I, pp. 471–490.

Bettess, P and Zienkiewicz, O. C. 1977. Diffraction and Refraction of Surface Waves Using Finite and Infinite Elements. *Int. J. for Numerical Methods in Eng.*, Vol. 11, pp. 1271–1290.

Biesel, F. 1949. Calcul de L'atmortissement D'une Houle Dans un Liquide Visqueux de Profondeur Finie. *La Houille Blanche*, Vol. 4, No. 5, pp. 630–634.

Biesel, F. 1950. Etude Théoretique de La Houle en Eau Courante. *La Houille Blanche*, Vol. 5A, pp. 279–285.

Black, J. L. and Mei, C. C. 1969. Scattering of Surface Waves by Rectangular Obstacles in Water of Finite Depth. *JFM*, Vol. 38, pp. 499–511.

Bloor, M. I. G. 1978. Large Amplitude Surface Waves. *JFM*, Vol. 84, pp. 167–179.

Borgman, L. E. and Chappelear, J. E. 1958. The Use of the Stokes-Struik Approximation for Waves of Finite Height. *Proc. 6th Conf. on Coastal Eng.*, Gainsville and Miami Beach, pp. 252–280.

Boussinesq, J. 1871. Théorie de L'intumescence Liquide, Appelée Onde Solitaire ou de Translation se Propageant Dans un Canal Rectangulaire. *Comptes Rendus Acad. Sci.*, Paris, Vol. 72, pp. 755–759.

Brebner, A. and Collins, J. I. 1961. Onset of Turbulence at the Bed Under Periodic Gravity Waves. *Trans. Eng. Inst. Can.*, Vol. 5, pp. 55-62.

Bretschneider, C. L. 1960. A Theory for Waves of Finite Height. *Proc. 7th Conf. on Coastal Eng.*, The Hague, pp. 146-183.

Bryant, P. J. 1974. Stability of Periodic Waves in Shallow Water. *JFM*, Vol. 66, pp. 81-96.

Byatt-Smith, J. G. B. and Longuet-Higgins, M. S. 1976. On the Speed and Profile of Steep Solitary Waves. *Proc. Roy. Soc.*, Ser. A, Vol. 350, pp. 175-189.

Camfield, F. E. and Street, R. L. 1969. Shoaling of Solitary Waves on Small Slopes. *J. Waterways Harbors and Coastal Eng. Div., ASCE*, Vol. 95, No. WW1, pp. 1-22.

Chappelear, J. E. 1959. On the Theory of the Highest Waves. U.S. Army Corps of Engineers, *Beach Erosion Board*, Tech. Memo No. 116.

Chappelear, J. E. 1961. Direct Numerical Calculation of Wave Properties. *J. Geophys. Res.*, Vol. 66, pp. 501-508.

Chappelear, J. E. 1962. Shallow Water Waves. *J. Geophys. Res.*, Vol. 67, pp. 4693-4704.

Chen, H. S. and Mei, C. C. 1974. Oscillations and Wave Forces in a Man-Made Harbor in the Open Sea. *Proc. 10th Symp. Naval Hydrodynamics*, Cambridge, Mass., pp. 573-596.

Cokelet, E. D. 1977a. Steep Gravity Waves in Water of Arbitrary Uniform Depth. *Phil. Trans. Roy. Soc.*, Ser. A, Vol. 286, pp. 183-230.

Cokelet, E. D. 1977b. Breaking Waves. *Nature*, Vol. 267, pp. 769-774.

Cokelet, E. D. 1979. Breaking Waves–The Plunging Jet and the Interior Flow Field. In *Mechanics of Wave-Induced Forces on Cylinders*, ed. T. L. Shaw, Pitman, London, pp. 287-301.

Collins, J. I. 1963. Inception of Turbulence at the Bed Under Periodic Gravity Waves. *J. Geophys. Res.*, Vol. 68, pp. 6007-6014.

Crapper, G. D. 1972. Nonlinear Gravity Waves on Steady Non-Uniform Currents. *JFM*, Vol. 52, pp. 713-724.

Dalrymple, R. A. 1974a. A Finite Amplitude Wave on a Linear Shear Current. *J. Geophys. Res.*, Vol. 79, pp. 4498-4504.

Dalrymple, R. A. 1974b. Models for Nonlinear Water Waves in Shear Currents. *Offshore Tech. Conf.*, Houston, Paper No. OTC 2114, pp. 843-856.

Dalrymple, R. A. and Dean, R. G. 1975. Waves of Maximum Height on Uniform Currents. *J. Waterways Harbors and Coastal Eng. Div., ASCE*, Vol. 101, No. WW3, pp. 259-268.

Davis. S. H. 1976. The Stability of Time-Periodic Flows. *Annual Review of Fluid Mech.*, Vol. 8, pp. 57-74.

De, S. C. 1955. Contributions to the Theory of Stokes Waves. *Proc. Camb. Phil. Soc.*, Vol. 51, pp. 718-736.

Dean, R. G. 1965. Stream Function Representation of Nonlinear Ocean Waves. *J. Geophys. Res.*, Vol. 70, pp. 4561-4572.

Dean, R. G. 1970. Relative Validities of Water Wave Theories. *J. Waterways Harbors and Coastal Eng. Div., ASCE*, Vol. 96, No. WW1, pp. 105-119.

Dean, R. G. 1974. Evolution and Development of Water Wave Theories for Engineering Application, Vols. I and II. U.S. Army, Coastal Eng. Res. Center, *Special Report No. 1*, Fort Belvoir, Virginia.

Dean, R. G. and Le Méhauté, B. 1970. Experimental Validity of Water Wave Theories. *National Structural Eng. Meeting*, ASCE, Portland, Oregon.

Dean, R. G. and Ursell, F. 1959. Interaction of a Fixed Semi-immersed Circular Cylinder with a Train of Surface Waves. *Hydrodynamics Lab.*, M.I.T., Tech. Rept. No. 37, Mass.

Dean, W. R. 1945. On the Reflection of Surface Waves by a Submerged Plane Barrier. *Proc. Camb. Phil. Soc.*, Vol. 41, pp. 231-238.

Dean. W. R. 1948. On the Reflection of Surface Waves by a Submerged Circular Cylinder. *Proc. Camb. Phil. Soc.*, Vol. 44, pp. 483-491.

Eagleson, P. S. 1962. Laminar Damping of Oscillatory Waves. *J. Hyd. Div., ASCE*, Vol. 88, No. HY3, pp. 155-181.

Ebbesmeyer, C. C. 1974. Fifth Order Stokes Wave Profiles. *J. Waterways Harbors and Coastal Eng. Div., ASCE*, Vol. 100, No. WW3, pp. 264-265.

Fenton, J. 1972. A Ninth-Order Solution for the Solitary Wave. *JFM*, Vol. 53, pp. 257-271.

Fenton, J. D. 1979. A High-Order Cnoidal Wave Theory. *JFM*, Vol. 94, pp. 129-161.

Fornberg, B. and Whitham, G. B. 1978. A Numerical and Theoretical Study of Certain Nonlinear Wave Phenomena. *Phil. Trans. Roy. Soc.*, Ser. A, Vol. 289, pp. 373-404.

Friedrichs, K. O. 1948. On the Derivation of the Shallow Water Theory, Appendix to "The Formation of Breakers and Bores" by J. J. Stoker. *Commun. Pure Appl. Math.*, Vol. 1, pp. 81-85.

Galvin, C. J. 1968. Breaker Type Classification on Three Laboratory Beaches. *J. Geophys. Res.*, Vol. 73, pp. 3651-3659.

Galvin, C. J. 1969. Breaker Travel and Choice of Design Wave Height. *J. Waterways Harbors and Coastal Eng. Div., ASCE*, Vol. 95, No. WW2, pp. 175-200.

Galvin, C. J. 1972. Wave Breaking in Shallow Water. In *Waves on Beaches and Resulting Sediment Transport*, Ed. R. E. Meyer, Academic Press, New York, pp. 413-456.

Gerstner, F. 1802. Theorie Der Wellen. *Abhandlungen der Koniglichin Bohimschen Gesellschaft der Wissenschaften*, Prague (see also Gilbert, *Annalen der Physik*, Vol. 32, pp. 412-425, 1809).

Goda, Y. 1964. Wave Forces on a Vertical Circular Cylinder and a Proposed Method of Wave Force Computation. *Report of Port and Harbor Technical Res. Inst.*, No. 8, pp. 1-74.

Goda, Y. 1970. A Synthesis of Breaker Indices. *Trans. Jap. Soc. Civ. Eng.*, Vol. 2, pp. 227-230.

Goda, Y. and Yoshimura, T. 1973. Discussion of "Wave Diffraction by Detached Breakwater," by M. Stiassnie and G. Dagan. *J. Waterways Harbors and Coastal Eng. Div., ASCE*, Vol. 99, No. WW2, pp. 285-288.

Grace, R. A. 1976. Near-Bottom Water Motion Under Ocean Waves. *Proc. 15th Coastal Eng. Conf.*, Honolulu, Vol. III, pp. 2371-2386.

Grimshaw, R. 1970. The Solitary Wave in Water of Variable Depth. *JFM*, Vol. 42, pp. 639-656.

Grimshaw, R. 1971. The Solitary Wave in Water of Variable Depth. Part 2. *JFM*, Vol. 46, pp. 611-622.

Griswold, G. M. 1963. Numerical Calculation of Wave Refraction. *J. Geophys. Res.*, Vol. 68, pp. 1715-1723.

Hammack, J. L. and Segur, H. 1974. The Korteweg-de Vries Equation and Water Waves. Part 2. Comparison with Experiments. *JFM*, Vol. 65, pp. 289-314.

Hammack, J. L. and Segur, H. 1978a. The Korteweg-de Vries Equation and Water Waves. Part 3. Oscillatory Waves. *JFM*, Vol. 84, pp. 337-358.

Hammack, J. L. and Segur, H. 1978b. Modelling Criteria for Long Water Waves. *JFM*, Vol. 84, pp. 359-374.

Hasimoto, H. and Ono, H. 1972. Nonlinear Modulation of Gravity Waves. *J. Phys. Soc. Japan*, Vol. 33, pp. 805-811.

Hattori, M. 1972. Transmission of Water Waves Through Perforated Wall. *Coastal Eng. in Japan*, Vol. 15, pp. 69-79.

Hayashi, T., Hattori, M., Kano, T., and Shirai, M. 1966. Hydraulic Research on the Closely Spaced Pile Breakwater. *Proc. 10th Coastal Eng. Conf.*, Tokyo, pp. 873-884.

Hedges, T. S. 1978. Some Effects of Currents on Measurement and Analysis of Waves. *Proc. Inst. Civil Eng.*, Vol. 65, pp. 685–692.

Hino, M., Sawamoto, M. and Takasu, S. 1976. Experiments on Transition to Turbulence in an Oscillatory Pipe Flow. *JFM*, Vol. 75, pp. 193–208.

Horikawa, K. 1978. *Coastal Engineering, an Introduction to Ocean Engineering*, Wiley, New York.

Hunt, J. N. 1952. Viscous Damping of Waves Over an Inclined Bed in a Channel of Finite Width. *La Houille Blanche*, Vol. 7, No. 6, pp. 836–842.

Hunt, J. N. 1979. Direct Solution of Wave Dispersion Equation. *J. Waterway Port Coastal and Ocean Div., ASCE*, Vol. 105, No. WW4, pp. 457–459.

Hwang, L. S. and Tuck, E. O. 1970. On the Oscillations of Harbours of Arbitrary Shape. *JFM*, Vol. 42, pp. 447–464.

Ippen, A. T., ed. 1966. *Estuary and Coastline Hydrodynamics*. McGraw-Hill, New York.

Ippen, A. T. and Goda, Y. 1963. Wave Induced Oscillations in Harbors: The Solution for a Rectangular Harbor Connected to the Open Sea. *Hydrodynamics Lab.*, M.I.T., Tech. Rept. No. 59, Mass.

Isaacson, M. de St. Q. 1976a. Mass Transport in the Bottom Boundary Layer of Cnoidal Waves. *JFM*, Vol. 74, pp. 401–413.

Isaacson, M. de St. Q. 1976b. The Viscous Damping of Cnoidal Waves. *JFM*, Vol. 75, pp. 449–457.

Isaacson, M. de St. Q. 1976c. The Second Approximation to Mass Transport in Cnoidal Waves. *JFM*, Vol. 78, pp. 445–457.

Isaacson, M. de St. Q. 1977. Second Approximation to Gravity Wave Attenuation. *J. Waterway Port Coastal and Ocean Div., ASCE*, Vol. 103, No. WW1, pp. 43–55.

Isaacson, M. de St. Q. 1978a. Mass Transport in Shallow Water Waves. *J. Waterway Port Coastal and Ocean Div., ASCE*, Vol. 104, No. WW2, pp. 215–225.

Isaacson, M. de St. Q. 1978b. Wave Damping Due to Rubblemound Breakwaters. *J. Waterway Port Coastal and Ocean Div., ASCE*, Vol. 104, No. WW4, pp. 391–405.

Isaacson, M. de St. Q. 1979. Nonlinear Inertia Forces on Bodies. *J. Waterway Port Coastal and Ocean Div., ASCE*, Vol. 105, WW3, pp. 213–227.

Iwagaki, Y. 1968. Hyperbolic Waves and their Shoaling. *Proc. 11th Coastal Eng. Conf.*, London, pp. 125–144.

Iwagaki, Y. and Sakai, T. 1969. Experiment on Horizontal Water Particle Velocity of Finite Amplitudes Waves. (In Japanese.) *Proc. 16th Conf. Coastal Eng. in Japan*, pp. 15–21. Also, Horizontal Water Particle Velocity of Finite Amplitude Waves. *Proc. 12th Coastal Eng. Conf.*, Washington, D.C., pp. 309–326.

Iwagaki, Y. and Tsuchiya, Y. 1966. Laminar Damping of Oscillatory Waves Due to Bottom Friction. *Proc. 10th Coastal Eng. Conf.*, Tokyo, pp. 149–174.

Iwagaki, Y. and Yamaguchi, M. 1968. Studies on Cnoidal Waves (sixth report)–Limiting Condition for Application of Cnoidal Wave Theory. *Annuals, Disaster Prevention Res. Inst.*, Kyoto Univ., No. 11B, pp. 477–502. (In Japanese.)

Jen, Y. 1969. Wave Refraction Near San Pedro Bay, California. *J. Waterways Harbors and Coastal Eng. Div., ASCE*, Vol. 95, No. WW3, pp. 379–393.

Johns, B. 1975. The Form of the Velocity Profile in a Turbulent Shear Wave Boundary Layer. *J. Geophys. Res.*, Vol. 80, pp. 5109–5112.

Johnson, J. W. 1951. Generalized Wave Diffraction Diagrams. *Proc. 2nd Coastal Eng. Conf.*, Houston.

Jonsson, I. G. 1963. Measurements in the Turbulent Wave Boundary Layer. *10th Congress*, IAHR, London, Vol. 1, pp. 85–92.

Jonsson, I. G. 1980. A New Approach to Oscillatory Rough Turbulent Boundary Layers. *Ocean Eng.*, Vol. 7, pp. 109–152.

Jonsson, I. G., Skovgaard, O., and Wang, J. 1970. Interaction Between Waves and Currents. *Proc. 12th Coastal Eng. Conf.*, Washington, D.C., pp. 489–508.

Kajiura, K. 1968. A Model of the Bottom Boundary Layer in Water Waves. *Bull. Earthquake Res. Inst.*, Vol. 46, pp. 75–123.

Keller, J. B. 1948. The Solitary Wave and Periodic Waves in Shallow Water. *Commun. Appl. Math.*, Vol. 1, pp. 323–339.

Keulegan, G. H. 1948. Gradual Damping of Solitary Waves. *J. Res. Nat. Bureau Standards*, Vol. 40, pp. 487–498.

Keulegan, G. H. and Harrison, J. 1970. Tsunami Refraction Diagrams by Digital Computer. *J. Waterways Harbors and Coastal Eng. Div., ASCE*, Vol. 96, No. WW2, pp. 219–233.

Keulegan, G. H. and Patterson, G. W. 1940. Mathematical Theory of Irrotational Translation Waves. *J. Res. Nat. Bureau Standards*, Vol. 24, pp. 47–101.

Koh, C. Y. and Le Méhauté, B. 1966. Wave Shoaling. *J. Geophys. Res.*, Vol. 71, pp. 2005–2112.

Korteweg, D. J. and De Vries, G. 1895. On the Change of Form of Long Waves Advancing in a Rectangular Canal, and on a New Type of Long Stationary Waves. *Phil. Mag.*, 5th Series, Vol. 39, pp. 422–443.

Kowalski, T., ed. 1974. *Proceedings of the Floating Breakwater Conference*, Univ. of Rhode Island, Newport, Rhode Island.

Laitone, E. V. 1961. The Second Approximation to Cnoidal and Solitary Waves. *JFM*, Vol. 9, pp. 430–444.

Laitone, E. V. 1962. Limiting Conditions for Cnoidal and Stokes Waves. *J. Geophys. Res.*, Vol. 67, pp. 1555–1564.

Laitone, E. V. 1965. Series Solutions for Shallow Water Waves. *J. Geophys. Res.*, Vol. 70, pp. 995–998.

Lake, B. M., Yuen, H. C., Rungaldier, H., and Ferguson, W. E. 1977. Nonlinear Deep Water Waves: Theory and Experiment. Part 2, Evolution of a Continuous Wave Train. *JFM*, Vol. 83, pp. 49–74.

Lamb, H. 1945. *Hydrodynamics*. 6th ed., Dover, New York; also Cambridge University Press, 1932.

Lambrakos, K. F. and Brannon, H. R. 1974. Wave Force Calculations for Stokes and Non-Stokes Waves. *Proc. Offshore Technology Conf.*, Houston, Paper No. OTC 2039, pp. 47–60.

Larras, J. 1966. Diffraction de La Houle Par Les Obstacles Rectilinges Semi-Indefinis Sous Incidence Oblique. *Cah. Oceanogr.*, Vol. 18, pp. 661–667.

LeBlond, P. H. and Mysak, L. A. 1978. *Waves in the Ocean*. Elsevier, Amsterdam.

Lee, J-J. 1971. Wave Induced Oscillations in Harbors of Arbitrary Geometry. *JFM*, Vol. 45, pp. 375–394.

Leibovich, S. and Seebass, A. R., eds. 1974. *Nonlinear Waves*. Cornell Univ. Press, Ithaca, New York.

Le Méhauté, B. 1968. Mass Transport in Cnoidal Waves. *J. Geophys. Res.*, Vol. 73, pp. 5973–5979.

Le Méhauté, B. 1976. *An Introduction to Hydrodynamics and Water Waves*. Springer-Verlag, Dusseldorf.

Le Méhauté, B., Divoky, D., and Lin, A. 1968. Shallow Water Waves: A Comparison of Theories and Experiments. *Proc. 11th Coastal Eng. Conf.*, London, pp. 87–107.

Lenau, C. W. 1966. The Solitary Wave of Maximum Amplitude. *JFM*, Vol. 26, pp. 309–320.

Li, H. 1954. Stability of Oscillatory Laminar Flow Along a Wall. U.S. Army Corps of Engineers, *Beach Erosion Board*, Tech. Memo. No. 47.

Liu, A. and Davis, S. H. 1977. Viscous Attenuation of Mean Drift in Water Waves. *JFM*, Vol. 81, pp. 63–84.

Longuet-Higgins, M. S. 1953. Mass Transport in Water Waves. *Phil. Trans. Roy. Soc.*, Ser. A, Vol. 245, pp. 535–581.

Longuet-Higgins, M. S. 1958. The Mechanics of the Boundary Layer Near the Bottom in a Progressive Wave. (Appendix to a paper by R. C. H. Russell and J. D. C. Osorio). *Proc. 6th Coastal Eng. Conf.*, Miami, pp. 184–193.

Longuet-Higgins, M. S. 1973. On the Form of the Highest Progressive and Standing Waves in Deep Water. *Proc. Roy. Soc.*, Ser. A, Vol. 331, pp. 445–456.

Longuet-Higgins, M. S. 1974. On the Mass, Momentum, Energy and Circulation of a Solitary Wave. *Proc. Roy. Soc.*, Ser. A, Vol. 337, pp. 1–13.

Longuet-Higgins, M. S. 1975. Integral Properties of Periodic Gravity Waves of Finite Amplitude. *Proc. Roy. Soc.*, Ser. A, Vol. 342, pp. 157–174.

Longuet-Higgins, M. S. 1976. Recent Developments in the Study of Breaking Waves. *Proc. 15th Coastal Eng. Conf.*, Honolulu, Vol. I, pp. 441–460.

Longuet-Higgins, M. S. 1978. The Instabilities of Gravity Wave of Finite Amplitude in Deep Water. I Superharmonics. II Subharmonics. *Proc. Roy. Soc.*, Ser. A, Vol. 360, pp. 471–488, pp. 489–505.

Longuet-Higgins, M. S. and Cokelet, E. D. 1976. The Deformation of Steep Surface Waves on Water. I. A Numerical Method of Computation. *Proc. Roy. Soc.*, Ser. A, Vol. 350, pp. 1–25.

Longuet-Higgins, M. S. and Fenton, J. D. 1974. On the Mass, Momentum, Energy and Circulation of a Solitary Wave. II. *Proc. Roy. Soc.*, Ser. A, Vol. 340, pp. 471–491.

Longuet-Higgins, M. S. and Fox, M. J. H. 1977. Theory of the Almost Highest Wave: The Inner Solution. *JFM*, Vol. 80, pp. 721–742.

Longuet-Higgins, M. S. and Fox, M. J. H. 1978. Theory of the Almost Highest Wave. Part 2. Matching and Analytic Extension. *JFM*, Vol. 85, pp. 769–786.

Longuet-Higgins, M. S. and Stewart, R. W. 1960. Changes in the Form of Short Gravity Waves on Long Waves and Tidal Currents. *JFM*, Vol. 8, pp. 565–583.

Longuet-Higgins, M. S. and Stewart, R. W. 1961. The Changes in Amplitude of Short Gravity Waves on Steady Non-Uniform Currents. *JFM*, Vol. 10, pp. 529–549.

Longuet-Higgins, M. S. and Stewart, R. W. 1962. Radiation Stress and Mass Transport in Gravity Waves with Application to "Surf-Beats." *JFM*, Vol. 13, pp. 481–504.

Longuet-Higgins, M. S. and Stewart, R. W. 1963. A Note on Wave Set-Up. *J. Marine Res.*, Vol. 21, pp. 4–10.

Longuet-Higgins, M. S. and Stewart, R. W. 1964. Radiation Stress in Water Waves: A Physical Discussion, with Applications. *Deep Sea Res.*, Vol. 11, pp. 529–549.

MacCamy, R. C. and Fuchs, R. A. 1954. Wave Forces on Piles: A Diffraction Theory. U.S. Army Corps of Engineers, *Beach Erosion Board*, Tech. Memo. No. 69.

McCowan, J. 1891. On the Solitary Wave. *Phil. Mag.*, Vol. 32, pp. 45–58.

Madsen, O. S., Mei, C. C., and Savage, R. P. 1970. The Evolution of Time-Periodic Long Waves of Finite Amplitude. *JFM*, Vol. 44, pp. 195–208.

Marks, W. and Jarlan, G. L. E. 1968. Experimental Studies on a Fixed Perforated Breakwater. *Proc. 11th Coastal Eng. Conf.*, London, pp. 1121–1140.

Mei, C. C. 1978. Numerical Methods in Water Wave Diffraction and Radiation. *Ann. Rev. Fluid Mech.*, Vol. 10, pp. 393–416.

Mei, C. C. and Liu, L. F. 1973. The Damping of Surface Gravity Waves in a Bounded Liquid. *JFM*, Vol. 59, pp. 239–256.

Miche, R. 1944. Mouvements Ondulatoires des Mers en Profondeur Constante on Décroissante. *Annales des Ponts et Chaussées*, pp. 25–78, 131–164, 270–292, 369–406.

Michell, J. H. 1893. On the Highest Waves in Water. *Phil. Mag.*, Vol. 36, pp. 430–435.

Miles, J. W. 1967. Surface-Wave Damping in Closed Basins. *Proc. Roy. Soc.*, Ser. A, Vol. 297, pp. 459–475.

Miles, J. W. 1971. Resonant Response of Harbors: An Equivalent Circuit Analysis. *JFM*, Vol. 46, pp. 241–265.

Miles, J. W. 1974. Harbor Seiching. *Ann. Rev. Fluid Mech.*, Vol. 6, pp. 17–35.

Miles, J. W. 1976. Damping of Weakly Nonlinear Shallow-Water Waves. *JFM*, Vol. 76, pp. 251–255.

Miles, J. W. 1980. Solitary Waves. *Annual Review of Fluid Mech.*, Vol. 12, pp. 11–43.

Miles, J. W. and Munk, W. 1961. Harbor Paradox. *J. Waterways Harbors and Coastal Eng. Div., ASCE*, Vol. 87, No. WW3, pp. 111–130.

Milne-Thompson, L. M. 1968. *Theoretical Hydrodynamics*. 5th ed., MacMillan, New York.

Monkmeyer, P. L. 1970. Higher Order Theory for Symmetrical Gravity Waves. *Proc. 12th Coastal Eng. Conf.*, Washington, D.C., pp. 543–562.

Monkmeyer, P. L. and Kutzbach, J. E. 1965. A Higher Order Theory for Deep Water Waves. *Proc., Coastal Eng. Santa Barbara Specialty Conf., ASCE*, Santa Barbara, pp. 301–326.

Morse, P. M. and Rubenstein, P. J. 1938. The Diffraction of Waves by Ribbons and Slits. *Phys. Rev.*, Vol. 54, pp. 895–898.

Munk, W. H. 1949. The Solitary Wave Theory and its Application to Surf Problems. *Annals New York Acad. Sci.*, Vol. 51, pp. 376–423.

Munk, W. H. and Arthur, R. S. 1952. Wave Intensity Along a Refracted Ray. *Gravity Waves, National Bureau of Standards, Circular 521*, U.S. Govt. Printing Office, pp. 95–109.

Newman, J. N. 1974. Interaction of Water Waves with Two Closely Spaced Vertical Obstacles. *JFM*, Vol. 66, pp. 97-106.

Noda, H. 1971. On the Oscillatory Flow in Turbulent Boundary Layers Induced by Water Waves. *Bull. Disaster Prev. Res. Inst.*, Kyoto Univ., Vol. 20, pp. 127–144.

Ogilvie, T. F. 1963. First- and Second-Order Forces on a Cylinder Submerged Under a Free Surface. *JFM*, Vol. 16, pp. 451–472.

Ohmart, R. D. and Gratz, R. L. 1978. A Comparison of Measured and Predicted Ocean Wave Kinematics. *Proc. Offshore Technology Conf.*, Houston, Paper No. OTC 3276, pp. 1947–1957.

Penny, W. G. and Price, A. T. 1952. The Diffraction Theory of Sea Waves by Breakwaters. *Phil. Trans. Roy. Soc.*, Ser. A, Vol. 244, pp. 236–253.

Peregrine, D. H. 1972. Equations for Water Waves and the Approximation Behind Them. In *Waves on Beaches and Resulting Sediment Transport*, ed. R. E. Meyer, Academic Press, New York, pp. 95–121.

Peregrine, D. H. 1976. Interaction of Water Waves and Currents. *Advances in Applied Mechanics*, Academic Press, New York, Vol. 16, pp. 9–117.

Peregrine, D. H. 1979. Mechanics of Breaking Waves–A Review of Euromech 192. In *Mechanics of Wave-Induced Forces on Cylinders*, ed. T. L. Shaw, Pitman, London, pp. 204–214.

Phillips, O. M. 1977. *The Dynamics of the Upper Ocean*, 2nd ed., Cambridge Univ. Press.

Raichlen, F. 1965. Long Period Oscillations in Basins of Arbitrary Shapes. *Proc., Coastal Eng. Santa Barbara Specialty Conf., ASCE*, Santa Barbara, pp. 115–145.

Raichlen, F. and Naheer, E. 1976. Wave Induced Oscillations of Harbors with Variable Depth. *Proc. 15th Coastal Eng. Conf.*, Honolulu, Vol. IV, pp. 3536–3556.

Rayleigh, Lord. 1876. On Waves. *Phil. Mag.*, Vol. 1, pp. 257–279, (see also *Papers*, Vol. 1, Cambridge Univ. Press).

Richey, E. P. and Nece, R. E. 1974. Floating Breakwaters–State of the Art. *Proc. Floating Breakwater Conf.*, Univ. of Rhode Island, pp. 1–19.

Richey, E. P. and Sollitt, C. K. 1970. Wave Attenuation by Porous Walled Breakwater. *J. Waterways Harbors and Coastal Eng. Div., ASCE*, Vol. 96, No. WW3, pp. 643-663.
Riedel, H. P., Kamphuis, J. W. and Brebner, A. 1972. Measurement of Bed Shear Stress Under Waves. *Proc. 13th Coastal Eng. Conf.*, Vancouver, Vol. I, pp. 587-603.
Russell, J. S. 1844. Report on Waves. *14th Meeting Brit. Assoc. Adv. Sci.*, pp. 311-390.
Russell, R. C. H. and Osorio, J. D. C. 1958. An Experimental Investigation of Drift Profiles in a Closed Channel. *Proc. 6th Coastal Eng. Conf.*, Miami, pp. 171-193.
Sakai, F. and Tsukioka, K. 1975. Application of the Finite Element Method to Surface Wave Analysis–The 2nd Report: The Analysis of Harbor Oscillations. *Coastal Eng. in Japan*, Vol. 18, pp. 45-52.
Sasaki, K. and Murakami, T. 1973. Irrotational Progressive, Surface Gravity Waves Near the Limiting Height. *J. Ocean Soc. Japan*, Vol. 29, pp. 94-105.
Sarpkaya, T. 1955. Oscillatory Gravity Waves in Flowing Water. *J. Eng. Mech. Div., ASCE*, Vol. 87, No. 815, pp. 1-33.
Sarpkaya, T. 1966. Experimental Determination of the Critical Reynolds Number for Pulsating Poiseuille Flow. *J. Basic Eng., Trans. ASME*, Ser. D, Vol. 88, pp. 589-598.
Schwartz, L. W. 1974. Computer Extension and Analytic Continuation of Stokes' Expansion for Gravity Waves. *JFM*, Vol. 62, pp. 553-578.
Schwartz, L. W. and Vanden-Broeck, J-M. 1979. Numerical Solution of the Exact Equations for Capillary-Gravity Waves. *JFM*, Vol. 95, No. 1, pp. 119-139.
Segur, H. 1973. The Korteweg-de Vries Equation and Water Waves. Solution of the Equation. Part 1. *JFM*, Vol. 59, pp. 721-736.
Sergeer, S. I. 1966. Fluids Oscillations in Pipes at Moderate Reynolds Numbers. *Soviet Fluid Dynamics*, Vol. 1, pp. 21-22.
Seymour, R. J. 1976. Tethered Float Breakwater: A Temporary Wave Protection System for Open Ocean Construction. *Proc. Offshore Technology Conf.*, Houston, Paper No. OTC 2545, pp. 253-264.
Silvester, R. 1974. *Coastal Engineering, I.* Elsevier, Amsterdam.
Skjelbreia, L. and Hendrickson, J. A. 1960. Fifth Order Gravity Wave Theory. *Proc. 7th Coastal Eng. Conf.*, The Hague, pp. 184-196.
Skovgaard, O. and Jonsson, I. J. 1976. Current Depth Refraction Using Finite Elements. *Proc. 15th Coastal Eng. Conf.*, Honolulu, Vol. I, pp. 721-737.
Skovgaard, O., Jonsson, I. G., and Bertelsen, J. A. 1975. Computation of Wave Heights Due to Refraction and Friction. *J. Waterways Harbors and Coastal Eng. Div., ASCE*, Vol. 101, No. WW1, pp. 15-32.
Sleath, J. F. A. 1972. A Second Approximation to Mass Transport by Water Waves. *J. Mar. Res.*, Vol. 30, pp. 295-304.
Sleath, J. F. A. 1973. Mass-Transport in Water Waves of Very Small Amplitude. *J. Hyd. Res.*, Vol. 11, pp. 369-383.
Sommerfeld, A. 1949. *Partial Differential Equations in Physics.* Academic Press, New York.
Sorensen, R. M. 1978. *Basic Coastal Engineering.* Wiley, New York.
Stiassnie, M. and Dagan, G. 1972. Wave Diffraction by Detached Breakwater. *J. Waterways Harbors and Coastal Eng. Div., ASCE*, Vol. 98, No. WW2, pp. 209-224.
Stiassnie, M. and Peregrine, D. H. 1980. Shoaling of Finite Amplitude Surface Waves on Water of Slowly-Varying Depth. *JFM*, Vol. 97, pp. 783-805.
Stoker, J. J. 1957. *Water Waves.* Interscience, New York.
Stokes, G. G. 1847. On the Theory of Oscillatory Waves. *Trans. Camb. Phil. Soc.*, Vol. 8, pp. 441-455. Also *Math Phys. Papers*, Vol. 1, Camb. Univ. Press, 1880.
Tenaud, R. and Graillot, A. 1975. Ouvrages de Protection dans la Zone de Deferlement des Houles. *La Houille Blanche*, Vol. 7/8, pp. 537-558.

Treloar, P. D. and Brebner, A. 1970. Energy Losses Under Wave Action. *Proc. 12th Coastal Eng. Conf.*, Washington, D.C., pp. 257-267.

Tsuchiya, Y. and Yamaguchi, M. 1972. Some Considerations on Water Particle Velocities of Finite Amplitude Wave Theories. *Coastal Eng. in Japan*, Vol. 15, pp. 43-57.

Tsuchiya, Y. and Yamaguchi, M. 1976. Wave Shoaling of Finite Amplitude Waves. *Proc. 15th Coastal Eng. Conf.*, Honolulu, Vol. I, pp. 497-506.

Ursell, F. 1947. The Effect of a Fixed Vertical Barrier on Surface Waves in Deep Water. *Proc. Camb. Phil. Soc.*, Vol. 43, pp. 374-382.

Ursell, F. 1950. Surface Waves on Deep Water in the Presence of a Submerged Circular Cylinder. I. *Proc. Camb. Phil. Soc.*, Vol. 46, pp. 141-152.

U.S. Army, Coastal Engineering Research Center. 1973. *Shore Protection Manual.*

Van Dorn, W. G. 1966. Boundary Dissipation of Oscillatory Waves. *JFM*, Vol. 24, pp. 769-779.

Van Weele, B. J. and Herbich, J. B. 1972. Wave Reflection and Transmission for Pile Arrays. *Proc. 13th Coastal Eng. Conf.*, Vancouver, Vol. III, pp. 1935-1953.

Vanden-Broeck, J-M. and Schwartz, L. W. 1979. Numerical Computation of Steep Gravity Waves in Shallow Water. *Phys. of Fluids*, Vol. 22, pp. 1868-1871.

Von Schwind, J. J. and Reid, R. O. 1972. Characteristics of Gravity Waves of Permanent Form. *J. Geophys. Res.*, Vol. 77, pp. 420-433.

Weggel, J. R. 1972. Maximum Breaker Height for Design. *Proc. 13th Coastal Eng. Conf.*, Vancouver, Vol. I, pp. 419-432.

Wehausen, J. V. and Laitone, E. V. 1960. Surface Waves. In *Handbuch der Physik*, ed. S. Flugge, Springer-Verlag, Berlin, Vol. IX, pp. 446-778.

Whitham, G. B. 1962. Mass, Momentum and Energy Flux in Water Waves. *JFM*, Vol. 12, pp. 135-147.

Whitham, G. B. 1965a. Nonlinear Dispersive Waves. *Proc. Roy. Soc.*, Ser. A, Vol. 283, pp. 238-261.

Whitham, G. B. 1965b. A General Approach to Linear and Nonlinear Dispersive Waves Using a Lagrangian. *JFM*, Vol. 22, pp. 273-283.

Whitham, G. B. 1967a. Nonlinear Dispersion of Water Waves. *JFM*, Vol. 27, pp. 399-412.

Whitham, G. B. 1967b. Variational Methods and Applications to Water Waves. *Proc. Roy. Soc.*, Ser. A, Vol. 299, pp. 6-25.

Whitham, G. B. 1974. *Linear and Nonlinear Waves*. Wiley, New York.

Wiegel, R. L. 1960. A Presentation of Cnoidal Wave Theory for Practical Application. *JFM*, Vol. 7, pp. 273-286.

Wiegel, R. L. 1962. Diffraction of Waves by a Semi-Infinite Breakwater. *J. Hyd. Div., ASCE*, Vol. 88, No. HY1, pp. 27-44.

Wiegel, R. L. 1964. *Oceanographical Engineering*. Prentice-Hall, Englewood Cliffs, N.J.

Yamada, H. 1957a. Highest Waves of Permanent Type on the Surface of Deep Water. *Rep. Res. Inst. Appl. Mech.*, Kyushu Univ., Vol. 5, pp. 37-52.

Yamada, H. 1957b. On the Highest Solitary Wave. *Rep. Res. Inst. Appl. Mech.*, Kyushu Univ., Vol. 5, pp. 53-67.

Yamada, H. and Shiotani, T. 1968. On the Highest Water Waves of Permanent Type. *Bull. Disaster Prev. Res. Inst.*, Kyoto Univ., Vol. 18, p. 1.

Yamaguchi, M. and Tsuchiya, Y. 1974. Relation Between Wave Characteristics of Cnoidal Wave Theory Derived by Laitone and Chappelear. *Bull. Disaster Prev. Res. Inst.*, Kyoto Univ., Vol. 24, pp. 217-231.

Yuen, H. C. and Lake, B. M. 1975. Nonlinear Deep Water Waves: Theory and Experiment. *Phys. of Fluids*, Vol. 18, pp. 956-960.

Yuen, H. C. and Lake, B. M. 1980. Instabilities of Waves on Deep Water. *Annual Review of Fluid Mech.*, Vol. 12, pp. 303–334.

Zabusky, N. J. and Galvin, C. J. 1971. Shallow-Water Waves, the Korteweg-de Vries Equation and Solitons, *JFM*, Vol. 47, pp. 811–824.

Zakharov, V. E. 1968. Stability of Periodic Waves of Finite Amplitude on the Surface of a Deep Fluid. *Sov. Phys. J. Appl. Mech. Tech. Phys.*, Vol. 9, pp. 86–94.

5
Wave Forces on Small Bodies

5.1 INTRODUCTION

The state of understanding of the assumptions and uncertainties that go into the prediction of fluid loading on offshore structures has been reviewed by a number of people (see e.g., Hogben et al., 1977, Bea and Lai 1978). The emerging fact is that the current body of analytical, experimental, and operational knowledge is still inadequate to describe the complex realities of flüid loading and dynamic response of offshore structures. The primitive state of the description of the wake of bluff bodies is not the total cause of all uncertainties. Winds, waves, currents, ice, earthquake, soil movement, ship collision, and variations in material qualities are random in nature and often difficult to model as inputs. The designer is forced to make idealizations, simplifying assumptions, and extrapolations (with intentional and hidden margins of safety) with the hope that the structure will accommodate with adequate safety the uncertainties associated with its design. Clearly, the choice of a couple of hydrodynamic force coefficients is only a small part of the entire design process. It is also clear that what can be written on wave forces on small bodies* cannot adequately account for

*Throughout this chapter the word 'small' is used with reference to the diameter-wave length ratio. As such, the diffraction effects are negligible and both the drag and inertial forces may be important.

the experience of a seasoned designer. We describe here only the relatively-idealized analysis of various components and recent model and field tests which may be integrated with sufficient experience to describe the actual loading of offshore platforms composed of small diameter members. Such an exposition may also help to increase the communication between researchers and engineers in the offshore industry.

5.2 PRINCIPAL FACTORS OF ANALYSIS AND DESIGN

5.2.1 Analysis and Design

The development of rules and regulations in specifying the nominal and extreme loading conditions, the translation of these conditions into hydrodynamic loadings, and the execution of formal calculations for the structure constitute the essence of analysis and design. These may be performed through the use of two distinct methods: deterministic (for the extreme loading conditions) and stochastic (for the nominal loading conditions). The deterministic method may in turn be pseudo-static or time-dependent.

The extreme conditions are those which give rise to low-cycle fatigue and largest hydrodynamic loadings in general. The nominal conditions are those which give rise to high-cycle fatigue under operational conditions and stem from the waves and currents that constitute the vast majority of the structure's exposure to fluid loading.

The pseudo-deterministic analysis of the extreme loading conditions is based on a design wave (chosen statistically with specified wave height, period, and direction) and a wave theory (such as Stokes 5th) to calculate the fluid velocities and accelerations along the axis of each structural element. The wave is assumed to be long crested and to propagate without change of form. The instantaneous sectional force is calculated [in-line force through the use of Morison's equation together with C_d and C_m values appropriate to that section, i.e., $C_d(K, Re, k/D)$ and $C_m(K, Re, k/D)$, and the transverse force through the use of the appropriate lift coefficient, i.e., $C_L(K, Re, k/D)$]. The instantaneous force on the element is obtained through the integration of the sectional forces. Then the forces are summed vectorially to get the total force and the overturning moment on the whole structure. The method also allows for the determination of the maximum and intermediate values of the total force and moment on the structure as well as on the individual elements. These in turn are used as input for static strength analysis. A quick and very rough estimate of the in-line force and moment acting on a pile may be obtained through the use of the linear wave theory, Morison's equation, and constant drag and inertia coefficients.*

In deeper water, the interaction of time-dependent ocean environment with a dynamically responsive structure leads to complex resonance conditions and gives rise to larger stresses than would be predicted from a pseudo-static analysis. For example, Brannon et al., (1974) have shown that the dynamic response can double the static wave load for a 900-ft platform. Ruhl (1976) reported that the dynamic response of a deep water platform produced by three successive 60-ft, 10 second waves is larger than that due to a 78-ft, 12 second design wave. Thus, dynamic-response analysis is not only desirable but also necessary for structures built in deeper waters.

The analysis begins with the synthesis of a representative wave train with a dominant period and significant wave height. The kinematics of the flow field are then determined by using methods such as the linear filtering technique. Then the wave is propagated in space and the time-dependent hydrodynamic force on all elements is calculated with the linearized Morison equation, assuming the structure to remain stationary. Subsequently, the calculated forces are incorporated into the equation of motion of the structure, together with the appropriate mass, stiffness, and damping characteristics, to predict the time-dependent response. The predictions may be improved by successive iterations to a degree afforded by the accuracy of mass, damping, stiffness, sea spectra, and the force coefficients.

The nominal condition loadings often require the use of stochastic spectral analysis for the determination of the fatigue and dynamic response of the struc-

---

*Morison's equation, integrated to still water level, yields

$$F_x = 0.5\,\rho C_d D d\left(\frac{gH^2}{8d}\right)\left(1 + \frac{2k'd}{\sinh 2k'd}\right)|\cos\omega t|\cos\omega t - C_m\left(\rho\frac{\pi D^2}{4}d\right)\left(\frac{gH}{2d}\right)\tanh k'd\sin\omega t \quad (5.1)$$

and

$$M = 0.5\rho C_d D\left(\frac{HL}{4T}\right)^2\left[\left(\frac{k'd}{\sinh k'd}\right)^2 + \frac{2k'd\cosh k'd}{\sinh k'd} - 1\right]|\cos\omega t|\cos\omega t - C_m\left(\rho\frac{\pi D^2}{4}\right)\left(\frac{HL^2}{2T^2}\right)\left[\frac{k'd\sinh k'd - \cosh k'd + 1}{\sinh k'd}\right]\sin\omega t \quad (5.2)$$

in which D represents the pile diameter; d, depth of water or pile length to the still water level; H, wave height; L, wave length; and $k' = 2\pi/L$. The coefficients C_d and C_m are assumed constant. They may be chosen using the surface values of K and Re. In any case, Eqs. (5.1) and (5.2) yield crude estimates of F_x and M. In general, the variations of C_d, C_m, and C_L with the local values of K, Re, and the relative roughness cannot be ignored. Also, the transverse force must be taken into account.

ture. For this purpose one needs one or more wave spectra (such as Pierson-Moskowitz spectra), (see e.g., Michel 1967, Borgman 1969a, Mobarek 1965), and the force-transfer function for each point and wave direction for the entire frequency range. Then the resulting force-response spectra is calculated (Borgman 1965, Pierson and Holmes 1965, and Wiegel 1969). The method is valid only when the superposition principle is applicable, i.e., if the nonlinear loads such as drag forces are small in comparison with the linear loads such as inertial forces (St. Denis 1973). In this approach, all calculations are performed in the frequency domain instead of the time-domain as in the case of deterministic time-dependent analysis (see section 7.4).

5.2.2 Design Wave and Force Characterization

As noted above, a structure must withstand both the forces exerted by an extreme wave and the fatigue due to the accumulated effect of cyclic loading imposed by nominal waves (see, e.g., Marshall 1976, van Koten 1976).

Extreme wave heights are expressed in terms of wave heights having a low probability of occurrence. The probability P_n that a design wave with a return period of N years will be exceeded in a given duration of n years is given by (see section 7.5.6)

$$P_n = 1 - (1 - 1/N)^n \tag{5.3}$$

For the North Sea, the DnV (Det Norske Veritas 1974) has adopted a 100-year wave (a 100-ft wave with a period of 15 seconds)* as the design criterion, while the DTI (Department of Energy, U.K., 1974) suggests a 50-year return interval. For the Gulf of Mexico area, experience and the majority of operators have indicated that an appropriate return interval for use is about 100 years. The wave period associated with the design-wave height is specified for applications. Both API (American Petroleum Institute, 1977) and DTI have suggestions on the design-wave periods. We will not deal here further with many theories and methods that have been developed for the prediction and hindcasting of wave conditions. These are but briefly discussed in chapter 7.

Design waves are assumed to be uniformly crested and propagate across a hypothetical ocean surface with constant shape and speed. The character of the forces generated by these waves depend on the relative size and shape of the structure. In this chapter it is assumed that in the region near the body the kinematics of the undisturbed flow field do not change in the incident wave direction. Consequently, the force may be 'drag dominated' (this is primarily

*The probability of such a wave actually being exceeded within a period of 100 years is $P_n = 1 - (1-1/100)^{100} = 0.634$.

the case with design waves past space frame structures), 'inertia dominated,' or a combination thereof for intermediate Keulegan-Carpenter numbers. It should be noted in passing that during the period of interaction of a wave with a given element, there is always a time interval during which the inertial forces are predominant and a time interval during which the drag forces are dominant. Evidently, then, the expressions 'drag dominated' or 'inertia dominated' simply refer to the relative magnitudes of the two forces during a given cycle.

An approximate quantitative mapping of the various loading regimes, expressed in terms of wave height, structure diameter and depth below the water surface, may be obtained by assuming some appropriate values for the drag and inertia coefficients, a suitable wave theory, and a wave-length-to-wave-height ratio. Such an exercise has been carried out by Hogben (1976) by assuming a linear water wave for deep water. He assumed $C_d = 0.6$ for the post-supercritical regime and $C_d = 1.2$ for the subcritical regime. The inertia coefficient C_m was assumed to be equal to 2 for both regimes. Figures 5.1a and 5.1b show the various loading regimes for a wave-length-to-wave-height ratio of 15. Evidently, at depths greater than about half the wave length, the amplitudes of wave disturbances and corresponding wave forces are very small. It should be emphasized that Fig. 5.1 represents only ball-park values for a specific set of conditions and should not be regarded as a design criterion. The designer should be aware of the various loading regimes and should prepare his own figures through the use of the parameters and wave conditions most appropriate to his design.

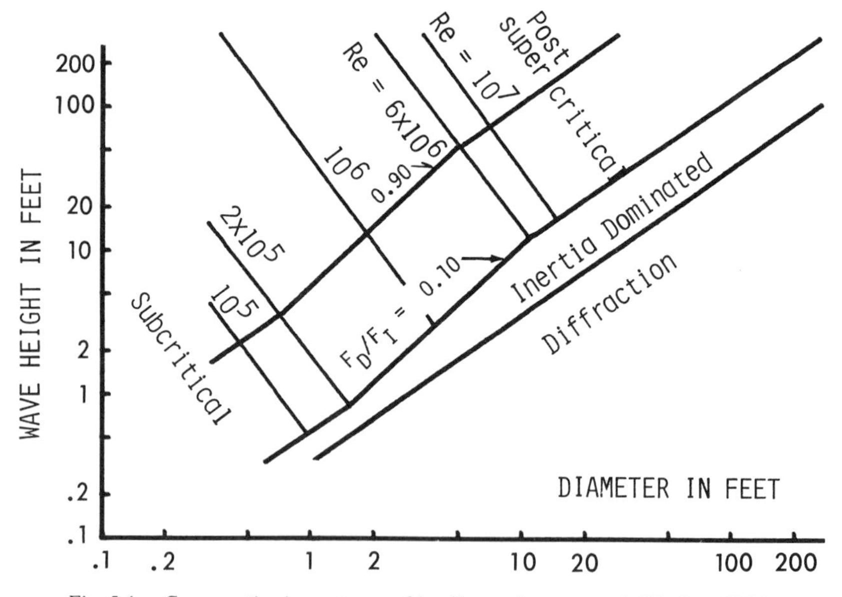

Fig. 5.1a. Comparative importance of loading regimes at z = 0 (Hogben 1976).

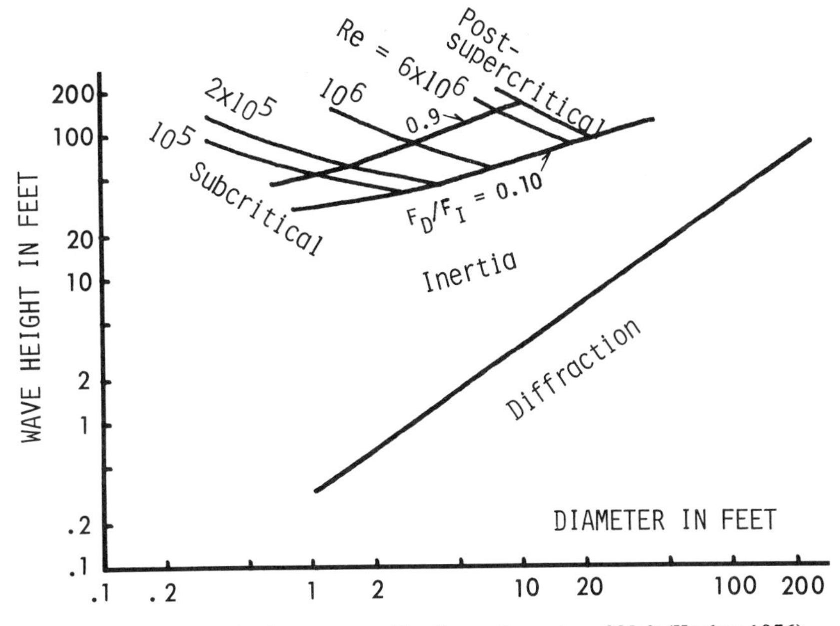

Fig. 5.1b. Comparative importance of loading regimes at z = 330 ft (Hogben 1976).

The inherent instability of the oceans gives rise to complex waves and forces which may be substantially different from those assumed to exist for design purposes. In fact, limited data suggest that real waves exert forces which do not act in uniform directions on members of a structure in space and which change in direction and magnitude as the wave propagates. It appears that the forces predicted through the use of a design wave will be larger than those produced on the structure by a real wave field of comparable characteristics, i.e., the real ocean environment has mitigating effects which can be quantified only approximately.

5.3 FORCE COEFFICIENTS

5.3.1 Sources of Uncertainty, Wave Kinematics, and Methods of Data Evaluation

As in harmonic and uniformly accelerating flows, Morison's equation is used for estimating wave-induced forces on offshore structures. It may be written as

$$F = 0.5\rho C_d A_p |U| U + \rho C_m \forall \frac{dU}{dt} \tag{5.4}$$

in which A_p represents the projected frontal area; $\forall$, the displaced volume of the structure; and U, the velocity of the ambient flow.

In wave force calculations U is taken as the horizontal component of the wave particle velocity and the convective acceleration terms are often ignored, i.e., it is assumed that $dU/dt \simeq \partial U/\partial t$ (but see Section 4.7 and Fig. 4.24 for a comparison of local and total accelerations). Furthermore, it is assumed that in the region near the cylinder the kinematics of the undisturbed flow do not change in the incident wave direction. For cylinder diameters larger than about 20 percent of the wave length the inertia force is no longer in phase with the acceleration and the diffraction effects must be taken into consideration (see chapter 6).

Considerable uncertainty exists regarding the meaning and application of the inertia force for nonlinear flows in which convective accelerations are not negligible. Isaacson (1979) has shown that the complete expression for the inertial force in the x direction is given by

$$F_x = (\rho\forall + m_{11})\left(\frac{\partial u}{\partial t} + u\frac{\partial u}{\partial x}\right) + (\rho\forall + m_{22})\, v\frac{\partial v}{\partial x} + (\rho\forall + m_{33})\, w\frac{\partial w}{\partial x}$$

$$+ m_{12}\left[\frac{\partial v}{\partial t} + \frac{1}{2}\frac{\partial(uv)}{\partial x}\right] + m_{13}\left[\frac{\partial w}{\partial t} + \frac{1}{2}\frac{\partial(uw)}{\partial x}\right] + m_{23}\left[\frac{1}{2}\frac{\partial(vw)}{\partial x}\right] \tag{5.5}$$

in which m_{ij} represent the added-masses. For bodies which are symmetrical about three orthogonal planes, Eq. (5.5) reduces to

$$F_x = \rho\forall\frac{Du}{Dt} + m_{11}\left(\frac{\partial u}{\partial t} + u\frac{\partial u}{\partial x}\right) + m_{22}v\frac{\partial v}{\partial x} + m_{33}w\frac{\partial w}{\partial x} \tag{5.6}$$

Note that in the case of a uniformly accelerated flow one retrieves the well-known result

$$F_x = (\rho\forall + m_{11})\frac{\partial u}{\partial t} = \rho\forall C_m\frac{\partial u}{\partial t} \tag{5.7}$$

For a *body moving* with the velocity $\dot{x}(t)$ in a flow field represented by U(x, t), with the restrictions that $v = w = 0$ and $m_{12} = m_{13} = 0$, Eq. (5.5) may be generalized to

$$F_x = (\rho\forall + m_{11})\left[\frac{\partial U}{\partial t} + (U - \dot{x})\frac{\partial U}{\partial x}\right] - m_{11}\ddot{x} \tag{5.8}$$

which is identical with the inertial component of the force given by Eq. (2.55).

Isaacson (1979) has demonstrated through several examples that the inertia forces calculated in the conventional manner will generally overestimate the actual force. In other words, the use of the local acceleration in lieu of total acceleration does not affect the practical application of the Morison equation which in any case depends on experimental coefficients.

In a more general form Morison's equation for a circular cylinder may be written as (Borgman 1958)

$$F = \begin{Bmatrix} F_x \\ F_y \end{Bmatrix} = 0.5\,\rho C_d D \begin{Bmatrix} V_x \\ V_y \end{Bmatrix} \sqrt{V_x^2 + V_y^2} + 0.25\,\rho\pi D^2 C_m \begin{Bmatrix} a_x \\ a_y \end{Bmatrix} \tag{5.9}$$

For a horizontal cylinder in a uniform flow field with $V_x = -U_m \cos\theta$ and $V_y = -V_m \sin\theta$, Eq. (5.9) reduces to

$$\frac{2F_x}{\rho D U_m^2} = \frac{\pi^2 D}{U_m T} C_m \sin\theta - C_d \sqrt{\cos^2\theta + (V_m/U_m)^2 \sin^2\theta}\,\cos\theta \tag{5.10}$$

and

$$\frac{2F_y}{\rho D V_m^2} = -\frac{\pi^2 D}{V_m T} C_m \sin\theta - C_d \sqrt{(U_m/V_m)^2 \cos^2\theta + \sin^2\theta}\,\sin\theta \tag{5.11}$$

Equations (5.10) and (5.11) state that the instantaneous in-line force in the x- and y-directions is equal to the sum of the projections, on the respective axis, of the instantaneous values of the total-velocity-square dependent drag force and the total-acceleration dependent inertial force. This implies that the flow over a cycle may be regarded as a juxtaposition of planar flows with instantaneous velocities and accelerations given by

$$q = \sqrt{U_m^2 \cos^2\theta + V_m^2 \sin^2\theta}, \qquad a = \sqrt{U_m^2 \sin^2\theta + V_m^2 \cos^2\theta} \tag{5.12}$$

The vortices do not move with the velocity of the ambient flow and the wake does not rotate about the cylinder at the same rate as the ambient velocity vector. In other words, one must be aware of the fact that the writing of Morison's equation in vectorial form does not necessarily imply that the behavior of the wake can be correctly represented by it. One must also note that the Fourier averages of C_d and C_m are no longer given by Eqs. (3.23) and 3.24).

Equations (5.10) and (5.11) cannot be written using only the x- or only the y-component of the velocity in Morison's equation. This will assume that the drag component of the in-line force is proportional to the square of the projected velocity rather than to the square of the instantaneous total velocity (note that U and V are assumed to be normal to the axis of the cylinder). Finally, it is because of the assumptions noted above that the drag components of the forces given by Eqs. (5.10) and (5.11) become linear for $U_m = V_m$, (fluid particles undergoing circular orbits).

The integration of Eq. (5.9) over depth and summation over N vertical piles yield a model equation for total forces as

$$F_T = \sum^{N} \int_{-d}^{0} [0.5\ \rho C_d(K, Re, k/D) D \mathbf{V_n} |V_n| + 0.25\ \pi \rho D^2 C_m(K, Re, k/D) \mathbf{a_n}]\ dz \quad (5.13)$$

where $\mathbf{a_n}$ and $\mathbf{V_n}$ are the accelerations and velocities along the n-th pile.

Morison's equation gave rise to a great deal of discussion on what values of the two coefficients should be used. Furthermore, the importance of roughness, rotation of the velocity vector, orientation of the cylinder, proximity of other members, spanwise coherence, currents, free surface, etc., has remained in doubt since experimental evidence published over the past twenty-five years has been quite inconclusive. The problem has further been compounded by the difficulty of accurately measuring the velocity and acceleration to be used in Morison's equation. In general, the nature of the equation rather than the lack of precision of measurements or the difficulty of calculating the kinematics of the flow from the existing wave theories has been criticized.

The problem is particularly complicated in oceans, because in general, the sea surface is nonlinear and irregular and the waves may originate from a range of directions. Until recently it has not been possible to adequately measure the kinematics of the waves. This shortcoming required that the kinematics used in the correlation of the measured forces with the Morison equation be established through the use of a suitable theory. An additional complication is that the wave theory simply cannot establish the presence of a steady current. Furthermore, it can be demonstrated that the kinematics of the wave are affected by nonlinear currents at other depths, above or below the point of measurement, even if a current does not exist at the point of measurement. Consequently, it is important to distinguish between the force coefficients obtained with strictly harmonic flows, monochromatic waves, irregular two-dimensional waves, and three-dimensional waves. In general, the instantaneous values of wave height and period are insufficient to enable confident prediction of wave forces to be made. A consideration of the previous history is also necessary.

Kinematics of the flow field are often calculated rather than measured. As discussed in detail in chapter 4, there are several wave theories* which might be used in calculating the velocities and accelerations. These theories deal with

---

*Stokes Fifth Order Theory–Skjelbreia and Hendrickson (1961), Solitary Wave Theory–Reid and Bretschneider (1953), Modified Solitary Wave Theory–Munk (1949), Chappelear Theory–Chappelear (1961), Stream Function Theory–Dean (1965), and Extended Velocity Potential Theory–Lambrakos and Brannon (1974).

two-dimensional waves and ignore the effect of the three-dimensionality of the flow.

Stokes 5th order theory is commonly used to describe the characteristics of steep nonlinear waves in relatively deep water. Airy's theory yields fairly similar results even if it is extended beyond the limits of linear assumptions to predict the velocities above the still water level. Consequently, it has been used on occasion to calculate the wave kinematics and the resulting wave forces by performing the integrations up to the actual water surface rather than to the still water level, as it is commonly done in linear analysis. Dean's stream function method (1965), which essentially is a Fourier expansion of the stream function, is used quite often since it provides the best fit to the surface boundary conditions even for irregular waves. In shallower waters, the effects of nonlinearity become significant and more appropriate theories such as cnoidal theory must be used for estimating the velocities and accelerations.

Each theory, be it linear or nonlinear, analytical or numerical, has its own limitations and ranges of applicability. Numerous works have been carried out to develop design guidelines for selecting the appropriate wave theory for specific site conditions. Dean's (1970) analytical validity approach is concerned with the agreement between a particular solution and a set of governing equations which includes the kinematic and dynamic free surface boundary conditions. The experimental validity approach of Le Mehaute, et al. (1968) is concerned with the agreement between a theory and data obtained from field and/or laboratory studies.

It is very costly to make measurements in the ocean environment. Furthermore, it is not easy to interpret the results particularly in separating the effect of currents from the kinematics of waves (Peregrine 1976, Thomas 1979). Thus, it may be generally stated that it is difficult to reconstruct the precise kinematics of the flow at the level of force measurements either through the use of approximate wave theories or through direct measurements even though the latter is clearly more desirable. The designer must be aware of the fact that the interference of the wake with the velocity probe (e.g., a magnetic velocimeter) is a serious problem. If the arm connecting the probe to the pile is too long (so as to avoid interference) then the vibrations of the probe may become important. In either case the wake of the probe itself may be important and must be carefully taken into account through proper calibration under similar flow conditions.

The combined wave and current flow about a structure (say a vertical cylinder extending from the ocean bottom through the free surface) gives rise to an exceedingly complex, separated, time-dependent flow.

The velocity vector rotates 360 degrees and it is not possible to think in terms of simple separation points, laminar separation bubbles, turbulent reattachment, etc. The flow regime may change from subcritical all the way to post-supercritical at a given elevation during the passage of a single wave and at a given time along the pile.

The shear layer is not fed equal amounts of vorticity along the cylinder partly because of the variation of velocity and partly because of the variation of the flow regime. Consequently, the vortices do not detach at the same instant along the cylinder. The spanwise coherence is considerably reduced and the mean coherence length (or correlation length) varies with depth, wave, and flow history. This picture is of course further complicated by the fact that the wake is swept back and forth. In general, the vortices move with a velocity smaller than that of the ambient flow. The currents help them to move further away from the cylinder and the periods of flow deceleration precipitate instability and cause profound changes in the vorticity distribution in the nearwake and in the base pressure.

The currents do not necessarily exist at all elevations at a given time. They are not necessarily uniform or collinear with the waves.

The flow about a horizontal cylinder presents equally complex problems. As noted earlier, even the uniform flow about a cylinder does not result in a perfectly two-dimensional flow. The spanwise coherence depends on the length-to-diameter ratio of the cylinder, upstream disturbances, Reynolds number, etc. In wavy flows, the velocity vector and hence the wake rotates about the cylinder.

So far we have been concerned with the kinematics of the flow both about the cylinder and in the ambient flow. We will now discuss briefly the methods of evaluation of a given set of data. It will become clear that there is not a unique method of evaluation of a given set of kinematic and dynamic data and that there is always some bias in data interpretation.

Some of the most frequently used methods are as follows:

A. Consider only one wave period, e.g., the wave between two crests or two troughs. Then calculate the kinematics of the flow using an appropriate wave theory, if it has not already been measured. The evaluation of the force coefficients and their meaning then depend on the methods of evaluation and on the force measured:

1. Calculate C_d and C_m using either the Fourier-averaging technique or the method of least squares. Evidently, neither of these methods assure that the maximum calculated force be equal to the maximum measured force. Furthermore, the quality of representation of the measured force by the linear-quadratic sum with constant coefficients depends on K, Re, relative roughness, highly variable vortex effects, and the irregularities and nonlinearities of the incident wave train.

2. Calculate C_d and C_m using the force at the points corresponding to maximum velocity and maximum acceleration.* This method may yield reliable results for C_d in the drag dominated region and small differences between the

*Note that for nonlinear waves $U = 0$ does not necessarily correspond to $\dot{U}$(maximum) and vice-versa.

measured and calculated forces (because the phase difference between the maximum velocity and the maximum forces becomes very small in the drag-dominated region). This method also yields accurate results in the inertia dominated region for the C_m values. It is not, however, recommended in a wide range of K and Re values where both drag and inertia may be of equal importance and highly dependent on the history effects. For sake of reference it should be noted that the ratio of the maximum drag force to the maximum inertial force, at a particular depth for a given wave and cylinder or for a given K and Re, is given by

$$F_d(\max)/F_i(\max) = (C_d/C_m)(K/\pi^2) \tag{5.14a}$$

When the above ratio is larger than 0.5, F_d/F_i at *the instant of maximum force* is given by

$$F_d/F_i = 2[F_d(\max)/F_i(\max)]^2 - 0.5 \tag{5.14b}$$

3. Calculate C_d and C_m from Morison's equation by writing it once for the maximum force and once for the zero force together with the corresponding velocities and accelerations. This method may not always yield stable results as far as the correlation of the drag and inertia coefficients with K and Re is concerned. This is primarily due to the fact that small fluctuations in the force maximum or in the velocity at the time of zero force, as a consequence of vortex-induced oscillations, can result in large differences in the inertia coefficient in the drag-dominated region and large differences in the drag coefficient in the inertia-dominated region. However, the method does insure that the measured and calculated maximum forces coincide regardless of what happens during the remainder of the cycle and whether one can establish any correlation or not.

4. Evaluate C_d over short segments of waves in which drag force is dominant and C_m over short segments in which inertia force is dominant (Kim and Hibbard 1975). For this purpose a decision must be made regarding the length of each segment. Then the trial values of C_d and C_m are iterated until the measured force over the wave cycle or half wave cycle is fairly well represented by the linear quadratic sum of forces.

Dean (1976) has presented a detailed discussion of the influence of loading regime on the uncertainties in C_d and C_m which shows that the larger scatter in published coefficients could, at least partially, be attributed to lack of discrimination in the selection of suitable data for each particular coefficient.

The methods cited above give a measure of the drag and inertia coefficients for each wave or half wave cycle. They may also be regarded as time-invariant averages for the particular wave and section of the cylinder on which the forces were measured.

From time to time the force acting on the entire cylinder is measured or the force is deduced from the bending moment acting at the bottom of the cylinder through the use of a suitable wave theory. In this case there are more alternatives and less certainty in the calculation of the drag and inertia coefficients. One can use the surface values of the velocity and acceleration or the velocity and acceleration at a suitable point along the cylinder. In any case, the coefficients calculated represent *time and space averaged* quantities. The hopes of correlating such coefficients with any of the basic parameters are further reduced through this double-averaging process. If this method is to be used, the results should be applied only to situations which are almost exactly the same or do not significantly differ from the original model or prototype conditions from which the data have been deduced.

Another important feature of the methods discussed above is that the drag and inertia coefficients obtained for two identical waves (assuming the use of the same wave theory in calculating the kinematics of the flow field) may not be the same. The forces acting on a segment of the cylinder or on the entire cylinder and the kinematics of the flow field may depend not only on the wave under consideration but also on the waves which have gone by. In other words, a consideration of the previous flow history on the kinematics of the flow and hence on the force coefficients is necessary.

In irregular waves the instantaneous values of wave height, crest height, and wave period are insufficient to enable confident prediction of the wave force to be made. The coefficients obtained from isolated wave forces and kinematics may require parameters such as the ratio of the successive wave heights and periods (H_{i-2}/H_i, H_{i-1}/H_i, H_{i+1}/H_i, H_{i+2}/H_i and T_{i-2}/T_i, . . . T_{i+2}/T_i where i-th wave represents the one used for calculations) in addition to several of the suitabiy defined parameters such as Reynolds number, Keulegan-Carpenter number, Hi/D, $H_i^2/\nu T$, H_i/gT_i^2, z/d, etc., (z = a particular depth, d = depth of ocean, g = gravitational acceleration).

B. Consider a series of waves over a suitable time interval. In this case also one can adopt several methods to deduce the drag and inertia coefficients (see chapter 7 for additional discussion).

1. Use a spectral analysis to achieve the best correspondence between the spectra of the measured and calculated forces. Borgman (1967, 1969), Brown and Borgman (1966), and Wilson (1965) used Airy's linear wave theory and the linearized version of the drag force. The resulting drag and inertia coefficients showed wide scatter possibly as a result of linearizing the drag force term.

2. Use of the probabilistic distribution in time of wave forces. Pierson and Holmes (1965) pursued this particular method and found that Morison's equation is appropriate for small diameter cylinders and the drag and inertia coefficients are not constant along the length of a pile. The method involves some uncertainty due to the non-Gaussian nature of wave-force time records. Bretschneider (1967) derived expressions for the probability of occurrence in a wave record of certain C_d and C_m values through the use of the linear wave theory.

3. Represent the wave train with the stream function and calculate the kinematics of the flow field. Then, using any one of the methods cited in part A (preferably the Fourier averaging or the method of least squares) calculate the drag and inertia coefficients for one particular wave in the middle of a few successive waves. Subsequently, move the window of the wave train to repeat the procedure. Such calculations consume considerable time but it is evident that the effect of flow history is incorporated into the calculations of the force coefficients with as much accuracy as desired.

4. It is possible to represent the wave train with a superposition of monochromatic sinusoidal waves and then calculate the kinematics of the flow field. In general the use of such a method will not be recommended even though it has been used on occasion.

5. Use a mean-square method together with the Morison equation to evaluate C_d and C_m (Bishop 1978). Starting from Morison's equation, Bishop has shown that

$$\overline{F^2} = (0.5\,\rho DLC_d)^2\,\overline{u^4} + (0.25\,\pi\rho LD^2C_m)^2\,\overline{\left[\frac{du}{dt}\right]^2} \tag{5.15}$$

where the bars denote the mean values of the squares of the respective measured quantities integrated over one or more discrete regular cycles or over a long enough interval to include several cycles in an irregular sea. The coefficients C_d and C_m are determined from any one or more pairs of equations set up from different samples of the measurements. This method is intermediate between the wave-by-wave method and the spectral one. As would be expected, the variability in the coefficients due to highly variable effects and irregularities in the incident wave train is averaged out by taking larger sampling times.

The foregoing points out certain simple facts concerning the evaluation of the force-transfer coefficients from laboratory and field tests. Firstly, the collection of full-scale data from structures at sea is difficult and involves considerable uncertainty. Secondly, the interpretation of the results in terms of suitable parameters is subject to ambiguity. In fact, the drag and inertia coefficients obtained through the use of one method should not be compared with those obtained through the use of another one. Thirdly, the flow is definitely three-dimensional and its consequences cannot be evaded by measuring sectional forces rather than the total force on the entire pile. Finally, the experimental conditions in the ocean environment cannot be controlled or repeated. These facts coupled with equally complex human factors entering into the acquisition, evaluation, and the style and degree of completeness of the dissemination of the information generated lead to considerable scatter in the drag, inertia, and lift coefficients. Apparently, the appreciation of the facts leading to the scatter does not necessarily enable one to quantify these factors or remedy the situation but it gives the designer at least a sense of understanding and security within the scope of his overall design philosophy. Additional in situ measurements may help to *calibrate the Morison equation* for application to more or less similar conditions in a given region but they are not likely to help to uncover the degree of importance of the parameters involved. Evidently, the answer lies neither in the use of the steady-flow drag coefficients with an inertia coefficient near its ideal value nor in the use of the coefficients obtained with relatively idealized and controlled experiments without an appreciation of the mitigating effects of the ocean environment. It would be unreasonable to argue on the one hand that steady-flow coefficients should be used at high K and Re values (Miller 1977) and on the other hand to argue that even for relatively high K values, one must expect significant differences between oscillatory and uniform incident flow (Pearcey 1979). It is a well-known fact that the flow conditions in the real environment do not resemble either steady flow or harmonically oscillating flow or two-dimensional wavy flow (orbital motion, sweeping of the wake to and fro over the body, and the spectral nature of waves and currents).

The true purpose of relatively-idealized experiments (e.g., uniform harmonic flow about circular cylinders) is not to provide coefficients for immediate use in the design of offshore structures but rather, and more importantly, to determine whether the linear combination of a linear inertial force with a nonlinear drag force can predict, with sufficient accuracy, the measured time-dependent forces. Should this prove to be the case, one can then determine the role played by each controllable parameter in the variation of the coefficients quantifying the drag and inertial forces. This by no means ensures that the said two-term linear superposition will continue to hold true for more complex flow kinematics and body shapes to the same degree of accuracy as in idealized experiments in the flow regimes defined by K and Re.

It has been shown in section 3.8.8 that Morison's equation does not correctly predict the measured force in the drag-inertia dominated regime even for a harmonically oscillating planar flow. It is not expected that it will hold better for more complex wavy flows. The improvement of the said equation is clearly desirable with the addition of one or more terms. Even then the revised form of the equation may be suitable only for the conditions on which the revision is based. It appears that the comparison of the numerous drag and inertia coefficients, particularly in the range $8 < K < 25$, is not a realistic and fluid-mechanically satisfying exercise. Clearly, *both the form of the Morison equation and the uncertainties that go into the characterization of the ocean environment are jointly responsible for the differences between the measured and calculated forces.* It is not meaningful to relegate the errors only to one or the other. Morison's equation could continue to provide an approximate answer to an approximately-defined problem, in spite of the uncertainties in its form and input, only when it is carefully calibrated and fine-tuned with respect to a wave theory and local conditions and tolerated in light of all other uncertainties and hidden and intentional safety factors that go into the final design of a structure.

As to the future, it is only through the combination of well documented and thought out laboratory and field experiments and sensitivity analyses that the rules for the design of safe marine structures can be continuously refined.

5.3.2 A Brief Summary of the Literature Giving Explicit C_d and C_m Values

A comprehensive summary of the data on force-transfer coefficients has been presented by the British Ship Research Association (1976), (see also Hogben, et al., 1977). Their tables, citing briefly the characteristics of each investigation, will not be reproduced here. The differences in the test conditions, methods of measurement and data evaluation do not permit a critical and comparative assessment of the drag and inertia coefficients obtained in each investigation. Instead, we will describe here only the results of the most important wave projects.

Some of the data came from the measurements carried out in the actual sea conditions and some from laboratory experiments. Reid (1958) measured forces on a section of an 8.625-inch cylinder in water of 30-ft depth. The kinematics of the flow were calculated from the wave profile and the drag and inertia coefficients through the use of the least squares technique. Reid did not take into consideration the effect of the currents and structural vibrations.

Wiegel, Beebe, and Moon (1957) made measurements at the pacific Coast (Davenport, California) on various sections of a 6.625-inch cylinder. The wave kinematics were determined from linear theory. Local forces, calculated using linear theory and average values of C_d and C_m differed from measured forces by up to about 100 percent, (see also Figs. 5.2 and 5.3). Only the average values of C_d and C_m over small increments of K and Re bore some resemblence to those obtained under controlled laboratory conditions (Evans 1969). The large scatter was attributed to roughness, turbulence, wind waves and shallow water effects. The effect of currents was not considered. Structural vibration, which resulted in fatigue failure, was observed. Figures 5.2 and 5.3 show the results obtained by Wiegel et al., and Reid, superimposed on the harmonic flow results obtained by Sarpkaya (1976a). It appears that the re-evaluation of the data of Wiegel, et al., with due consideration to Reynolds number might partially explain the reason for the large scatter. Notwithstanding this observation, it would still be difficult to delineate the effect of currents and roughness. It has previously been shown that roughness causes profound changes in both the drag and inertia coefficients.

Wilson (1965) presented the results of an extremely complex analysis of wave force data from an experiment conducted with a 30-inch diameter pile in confused-sea conditions in the Gulf of Mexico. He developed procedures by which measured and observed raw data may be reduced to a form suitable for computer analysis. These involved the use of numerical filters with which unwanted high frequency effects, such as those due to mechanical vibration, may be removed from force records. The numerical filters were adapted and used in

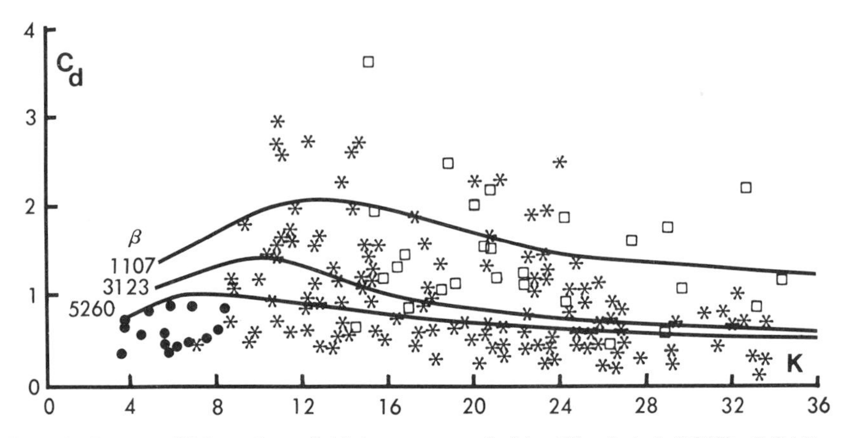

Fig. 5.2. Drag coefficients from field data, as compiled by Wiegel et al. (1957). Solid lines from Sarpkaya (1976a).

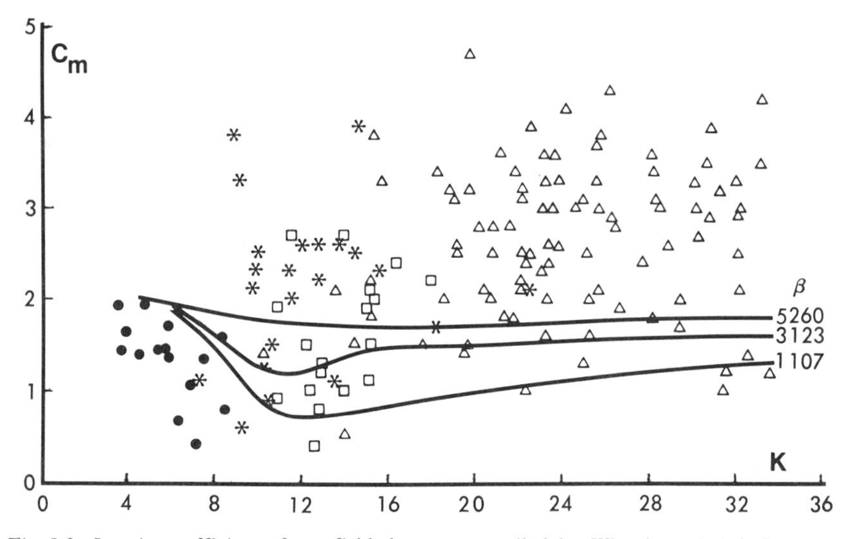

Fig. 5.3. Inertia coefficients from field data, as compiled by Wiegel et al. (1957). Solid lines from Sarpkaya (1976a).

connection with small amplitude wave theory to derive expressions for the water particle velocity and acceleration in terms of wave surface elevation, at any depth. The expressions were then used in the data analysis to obtain C_d and C_m.

Two wave projects were undertaken by* California Research Co., Shell Oil

*The names of the companies are listed here as they were at the time of the execution of the project.

Co., and Humble Oil and Refining Company at the Gulf of Mexico and the results (Thrasher and Aagaard 1969) were evaluated by several researchers (Aagaard and Dean 1969, Evans 1969, Wheeler 1969, Hudspeth et al., 1974). The first project consisted of the measurement of pressure distribution at various elevations on 1 to 4-ft diameter cylinders. The structure was subjected to three hurricanes and two tropical storms during a period from 1954 to 1958. The second project was performed in water of 100-ft depth with a 3.7-ft diameter pile. The pressure distributions on one foot sections allowed the calculation of the forces and their direction. The wave profile was recorded simultaneously with pressures. Wave heights up to 45 ft were recorded during the passage of Hurricane Carla in 1961. None of the evaluators of the data of the two projects considered the effect of currents. The results have demonstrated clearly that the force-transfer coefficients strongly depend on the particular wave theory used in the evaluation of the wave kinematics. In a more constructive sense one may regard the evaluation of this or other ocean test data as a calibration of Morison's equation against a particular wave theory in the range of wave heights and periods and the wave forces that have been measured for certain pile diameters.

The results of the data obtained at the Bass Straits, Australia, were presented by Kim and Hibbard (1975). The test pile was 38 ft long and 12.75 inch in diameter. It was subjected to rather small amplitude waves. Drag and inertia coefficients were calculated for individual waves from measured water-particle velocities and wave forces exerted on an 18-inch long sleeve-type load cell. The overall mean value of the drag coefficient under crests (0.72) was found to be significantly larger than that under the troughs (0.54). With corrections to the measured velocities, the overall average of C_d was found to be 0.61 and $C_m = 1.20$. The agreement between the measured and calculated forces was good in the drag dominated part of the wave cycle, and fair in the inertia dominated region.

A large scale experiment was undertaken by Exxon Production Research Company to evaluate present wave force calculation procedures for fixed, space-frame structures (Geminder and Pomonik 1979; Haring and Spencer 1979; Heideman, Olsen, and Johansson 1979; Dean, Lo, and Johansson 1979; Bendat, Richman, Osborne, and Silbert 1979; Borgman and Yfantis 1979; and Haring, Olsen, and Johansson 1979). This highly instrumented $20 \times 40 \times 120$ ft platform was installed in 66-ft water depth in the Gulf of Mexico. Data obtained include local wave forces on clean and barnacle-covered sensors, local wave kinematics, total base shear and overturning moment on the structure, forces on a simulated group of well conductors, and impact forces on a member above mean water level (discussed in Section 5.5).

Heideman, Olsen, and Johansson (1979) used two methods to evaluate the drag and inertia coefficients. The first was the least-squared-error procedure for each half wave cycle. The instantaneous in-line velocity in Morison's equation included both the wave velocity and the projection of the current velocity. The second method consisted of the evaluation of C_d over short segments of waves in which drag force was dominant and of C_m over short segments in which inertia force was dominant. The in-line force was taken as the projection of the normal force on the velocity vector. The normal force was measured with wave force transducers (WFT) of 16 inch O.D. and 32 inch length, built into the vertical legs at the four corners of the structure, at a depth of −15 ft. The normal water velocity was measured with an electromagnetic current meter (ECM) located 4.67 ft from the WFT axis, i.e., ECM was at 3.5D from the WFT axis. The force coefficients exhibited large scatter particularly for $K < 20$, (see Fig. 5.4). The scatter decreased considerably in the range $20 < K < 45$. It is not clear whether this is a genuine reduction in scatter or whether it is a consequence of the fewer data points in the drag-dominated range. In the drag-inertia dominated regime ($8 < K < 20$), C_d for a 0.5-inch barnacle-encrusted WFT ($k/D \simeq 1/32$, assuming that the entire barnacle height can be attributed to roughness) ranged from 0.6 to about 2. The inertia coefficient for the clean WFT was 1.51 (method 1) and 1.65 (method 2); and for the fouled WFT, 1.25 and 1.43 respectively, for the two methods of evaluation. The apparent C_d seemed to approach an asymptotic limit of about 0.68 for clean WFTs and about 1.0 for fouled WFTs. Heideman et al., attributed the scatter in C_d and C_m to random wake encounters. It is postulated that if the cylinder encounters its wake on the return half cycle

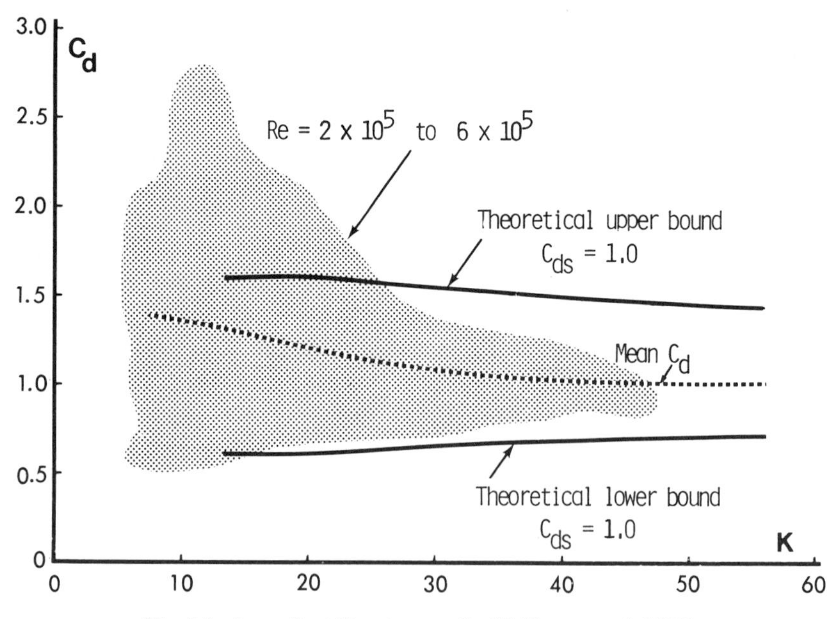

Fig. 5.4. Ocean Test Structure results (Heideman et al. 1979).

but the current meter does not, then the actual incident velocity will be greater than measured and the apparent C_d calculated from the measured force and velocity will be higher than the true C_d. Conversely, if the current meter encounters the wake on the return half cycle but the cylinder does not, then the apparent C_d will be too low. Clearly, the encounter of the wake with the current meter and the biassing of the wake by the current are extremely important. This is evidenced by the fact that the values of C_d and C_m vary considerably from one half wave cycle to another even for the same wave. Thus, it is desirable to evaluate C_d and C_m with due consideration to the effect of current, wave spreading, and the irregularities superimposed on each wave.

Heideman et al., (1979) concluded that (1) Morison's equation with constant coefficients can be made to fit measured local forces and kinematics satisfactorily over individual half wave cycles; (2) most of the scatter in the C_d results can be explained by the random wake encounter concept; (3) local deviations in apparent C_d are not spatially correlated in any given wave; (4) C_d results from Sarpkaya's experiments (1976a) represent an upper band to Cd values that may be expected in random three-dimensional oscillatory flow; (5) for $Re > 2 \times 10^5$, the apparent C_d depends on surface roughness and, for members that are nearly in the orbit plane, on K; (6) asymptotic C_d results from the test data in random three-dimensional oscillatory flow are consistent with steady flow data for the same relative roughness; and (7) C_m is greater for smooth cylinders than for rough cylinders, while the reverse is true for C_d.

Borgman and Yfantis (1979) determined the spectral estimates of C_d and C_m which represent the averages of C_d and C_m simultaneously over the whole structure. They have argued that such 'average' estimates may be more relevant than the local values since the average coefficients take into account directional and random cancellations throughout the structure. This may be true for the determination of the total force and overturning moment acting on the entire structure but not for the determination of the forces acting on the individual members of the structure.

The wave force data from the Christchurch Bay tower (Wheatley 1976) have been analyzed by Bishop (1979), (see also Pearcey and Bishop 1979), through the use of the mean-square method described earlier [see Eq. (5.15)]. The total data were divided into small samples for the derivation of mean square parameters. The sample duration has been varied to study its effect on the force coefficients. Bishop found that C_d and C_m [as defined by Eq. (5.15)] are quite stable for integration intervals larger than about 4 minutes and show increasing variations as the integration interval is reduced. The best fit values of C_d and C_m differed between the two runs evaluated and also for the main column and the wave staff of the test facility. For the wave staff, C_d was 0.73 and $C_m = 1.22$ for one run and 1.66 for the other run. For the main column C_d was forced to 1.0 and C_m was about 1.85. The Reynolds number ranged from 10^5 to 10^6 and the Keulegan-Carpenter number from 2 to 30. There appeared to be significant differences in the force coefficients due to current, for both the wave staff and the main column. The difference was largest on the smaller wave staff.

Ohmart and Gratz (1979) reported the results obtained from wave forces and horizontal particle velocities measured during Hurricane Edith. The determination of the force coefficients was based on the Morison equation using a least-squares approach. The data base included a Reynolds number range of 3×10^5 to 3×10^6. Overall, the use of an inertia coefficient of 1.5 and a constant drag coefficient of 0.7 provided the best fit of measured and predicted forces. For the peak forces, the best fit was obtained using $C_d = 0.7$ and $C_m = 1.7$. These values are in good agreement with those obtained by Sarpkaya (1976a) at Re = 1.5×10^6.

Gaston and Ohmart (1979) measured the total wave force and overturning moment on a smooth and roughened 15-ft long, 1-ft diameter, vertical cylinder under conditions of periodic and random waves in a wave tank. Drag and inertia coefficients have been determined for the smooth and three degrees of roughened cylinders ($k/D = \frac{1}{96}, \frac{1}{32}$, and $\frac{1}{24}$) by the least-squares method, using the measured in-line moment and predicted kinematics from the irregular stream-function theory. Typical Reynolds numbers were 2×10^5 to 3×10^5, based on the r.m.s water-particle velocity data. The results have shown that the drag coefficient is significantly affected by roughness. In fact, the change from the smooth to the rough surface approximately doubled the drag coefficient. Further increases in roughness had a lesser effect than the initial change from a smooth to the first rough cylinder ($k/D = \frac{1}{96}$). Gaston and Ohmart gave the average values of C_d and C_m as: $C_d = 0.77$ and $C_m = 1.81$ for the smooth cylinder; $C_d = 1.34$ and $C_m = 1.87$ for the cylinder with $k/D = \frac{1}{96}$; $C_d = 1.41$ and $C_m = 1.99$ for the cylinder with $k/D = \frac{1}{32}$; and $C_d = 1.42$ and $C_m = 2.01$ for the cylinder with $k/D = \frac{1}{24}$.

Matten (1977) conducted some experiments in small amplitude waves at the National Maritime Institute (England) to determine the effect of roughness by comparing the ratio of the *total force* acting on a rough and a smooth cylinder situated some distance apart. He worked in the range of Reynolds numbers from 20,000 to 60,000 and Keulegan-Carpenter numbers from about zero to 25. This is of course the range of K and Re values where significant transitions occur and it is rather difficult to draw conclusions regarding the effect of roughness at high Reynolds numbers since roughness may in fact decrease the drag due to transition rather than increase it (see Fig. 3.28 and 3.29). Aside from this limitation, Matten had considerable difficulty in obtaining consistent measurements because of the irregularity of the waves and because the forces experienced by the rough and smooth cylinders were related differently not only to the wave but also to the preceding waves. Consequently, Matten was unable to present any drag or inertia coefficient data. His calculations of the ratio of the resistances experienced by the rough and smooth cylinders have shown increases in conformity with those obtained by Sarpkaya at the corresponding K and Re values. The fact that Matten's K and Re values were based on surface kinematics of the waves, that C_d and C_m vary dramatically with decreasing K and Re below the free surface, and that no attempt was made to evaluate systematically C_d and C_m for a given segment of the cylinder make Matten's results practically unusable for any other situation. In spite of that, however, Hogben et al., (1977) used Matten's results as strong supporting evidence for their recommendations regarding the use of rough-cylinder results in steady flow. Sarpkaya's data (1976a) with sand-roughened cylinders have clearly demonstrated that the effect of roughness on resistance is not the same in steady and harmonically oscillating flows even at relatively high K and Re values. Consequently, the conjecture that "when the drag force is much larger than the inertia force in wavy flows the drag coefficients for steady flow will be applicable" is approximately correct only for *smooth* cylinders and quite incorrect for roughened cylinders.*

Heaf (1979) presented a detailed discussion of the effect of marine growth on the performance of fixed offshore platforms in the North Sea. He has pointed out that the marine growth influences the loading of an offshore platform in at least five ways: (1) increased tube diameters, leading to increased projected area and displaced volume and hence to increased hydrodynamic loading; (2) increased drag coefficient, leading to increased hydrodynamic loading; (3) increased mass and hydrodynamic added mass, leading to a reduced natural frequency and hence to an increased dynamic amplification factor; (4) increased structural weight, both in the water and above the water level in air; and (5) effect upon hydrodynamic instabilities, such as vortex shedding. One might also add that roughness decreases the separation angle and increases the correlation length, vortex strength, and the lift coefficient. Equally important is the fact that roughness accumulates most in the uppermost region of the structure where the wave and current induced velocities are largest.

Heaf (1979) reported experiments with cylinders held vertically in a towing tank where the wave loading was due to a highly irregular wave train. This work indicated that in the supercritical Reynolds number region of flow applicable to offshore structures, the surface roughness due to marine growth will increase the local value of C_d by 150 percent or more over the C_d value for a smooth cylin-

*Recently (1980), Sarpkaya enlarged the test section of his U-shaped water tunnel from 3 ft by 3 ft to 3 ft by 4.7 ft and repeated the experiments with sand-roughened cylinders, originally reported in Sarpkaya (1976a). The most recent data did not deviate more than ±2 percent from those obtained previously, showing conclusively that the blockage effects were negligible.

der. Heaf has concluded that the values of C_d and C_m predicted by Sarpkaya's work (1976a) seem to be the most appropriate data for predicting the effect of different heights of surface roughness on C_d and C_m and thereby in the wave loading. The fact that this data was obtained in two-dimensional harmonic flow probably means that such an approach would lead to a conservative design. The reason for this is that there are a number of mitigating effects of the ocean environment. For example, in a real sea-state there are waves of many directions applied simultaneously to make up the design wave height. There is also reduced spacial correlation over the structure as a whole and the design process applies, arbitrarily, the maximum current at the same time and in the same direction as the design wave. Consequently, the similarity between the drag coefficients obtained from the field tests and those obtained with steady uniform flow over similar cylinders (e.g., Miller 1977) under controlled laboratory conditions is rather fortuitous and is a result of the reduced spanwise coherence in confused seas.

The quantification of the mitigating effects is a complex statistical problem. The conventional design codes do not necessarily identify the most likely modes of failure for a large steel jacket. Hence, it is difficult to use the codes to quantify the importance of marine growth in relation to other design variables. A reliability analysis gives a more rigorous measurement of the relative importance of roughness and shows that marine growth (rigid as well as soft excrescences*) is among the most important parameters (Heaf 1979), (See also Houghton 1968, 1970).

5.3.3 Suggested Values for Force-Transfer Coefficients

Considerable attention has been devoted to the differences between the uniform harmonic flow, monochromatic waves, and the waves in an ocean environment. Not only the kinematics of the flow but also the force-transfer coefficients become increasingly uncertain as one comes closer to the conditions of the ocean environment. Consequently, there is substantial latitude in design practices as evidenced by the recommendations of various authoritative sources (see Table 5.1).

The reason for this latitude stems partly from the existence of limited number

*Preliminary tests by Burnett (1979) have shown that a cylinder covered in seaweed experiences significantly larger forces (as much as 220 percent that of the smooth cylinder) than a similar cylinder covered in hard roughness (in Burnett's experiments K was smaller than about 15). The reason for this increase is partly due to the inertial effect of the sweeping back and forth of the seaweed fronds, partly due to the increased form- and skin-friction drag, and partly due to the increase of turbulence introduced into the incident flow. Clearly, additional research is needed for all practically significant values of K and Re in order to quantify the effect of soft excrescences on wave loading on structures.

Table 5.1 Approaches to Design Practice in Static Wave Force Calculation (Space-Frame Structures, Deep Water)

	API RP2A (April 1977)	DNV Rules (July 1974)	U.K. DTI Guidance Notes (March 1974)
Wave kinematics:	"defensible" (e.g., Stokes 5th, Stream Function)	Stokes 5th	"Appropriate to the water depth"
Drag coefficient, C_d:	0.6–1.0 (not smaller than 0.6)	0.5–1.2	"Reliable experimental results"
Inertia coefficient, C_m:	1.5–2.0 (not smaller than 1.5)	2	
	Recognizes that C_d, C_m depend on wave theory.	Other C_d, C_m acceptable with different wave theory. $C_d > 0.7$ at high Reynolds No.	

of measurements in the ocean environment (particularly at high Reynolds numbers), partly from the scatter in the coefficients and the bias in the evaluation of the data, and partly from the desire to apply the force coefficients obtained with three-dimensional waves, currents, and flows to idealized design waves and conditions.

One may justifiably question the dissimilarity between the source of data and the conditions of their application. In using the characteristics of the uniformly crested design waves in Morison's equation together with the coefficients obtained from single-pile force measurements in actual seas one is implying that either there is not much difference between actual seas and design waves and/or the practice of applying the coefficients obtained under non-idealized conditions to idealized design conditions leads to conservative results. Common sense and some limited data suggest that real waves exert forces which do not act in uniform directions on members of a structure in space and which change in direction and magnitude as the wave propagates. This in turn would suggest that the drag coefficients obtained from ocean tests might be smaller than those which might have resulted from monochromatic waves of similar height and period.

It is on the basis of the foregoing that we will make a distinction between the values suggested by the analysis of the measurements made at sea and those for the design of space-frame structures subjected to idealized waves. Hopefully, the analysis will lead to the refinement of the methods of evaluation of the coefficients and to some additional understanding of the effects of three-dimensionality of the flow. This will depend in part on the number of structures built and measurements made at a given site over a number of years.

The major difficulty in suggesting the most appropriate force coefficients consistent with the idealization of the design is that it is impossible to conduct controlled laboratory experiments with high enough waves or at high enough Reynolds and Keulegan-Carpenter numbers. This being the case, one may assume that harmonic flows bear sufficient resemblance to wavy flows and that the effect of the tangential component of wave velocity on a vertical pile is to destabilize the boundary layer and thus lower the critical Reynolds number. Justification for the stated assumption comes partly from the comparison of the measured forces on vertical piles in wavy flows with those predicted through the use of the harmonic-flow results, both at relatively low and comparable K and Re values, and partly from a series of experiments regarding the effect of the orbital motion on resistance in harmonic flow.

Susbielles et al. (1971) used the harmonic flow results of Keulegan and Carpenter to calculate the local wave forces on a vertical pile and obtained agreement with measured forces to within 10 percent. Chakrabarti (1980a) conducted a wave-tank test with fixed vertical tubes (K ranged from zero to 85; and Re, from 2×10^4 to 3×10^4) and found reasonably good agreement between his C_d and C_m values and those obtained by Sarpkaya (1976a) for K smaller than about 40, in the same range of Re values. Wave-tank tests of Susbielles et al., and of Chakrabarti, as well as the field measurements, yield somewhat larger C_m values than those obtained with planar oscillatory flows in the range of K values from about 8 to 20 (interactive or critical or drag-inertia dominated regime). This is also the region where almost all experiments, particularly those conducted in the ocean environment, show the largest scatter. This is partly because of the inherently random nature of the shedding and subsequent mutual interaction of the vortices, partly the fractional shedding of vortices, and partly the reduced spanwise coherence of vortices. In fact, the smaller the spanwise coherence, the larger the C_m in the said region. It is primarily because of these reasons that the development of a fluid-mechanically satisfying unified wave-force equation which would account for the randomness of shedding, memory effects, etc., is rather difficult. It appears that such an equation will have to have both a probabilistic and a deterministic character to it.

The values for the force-transfer coefficients suggested herein are based partly on the foregoing considerations, partly on a careful perusal of the pertinent literature, and partly on the suggestions of the authoritative sources (Table 5.1) and reflect the authors' best current judgment.

For smooth vertical pipes, it is recommended that

(i) a suitable wave theory (e.g., Stokes 5th, stream function) be used to calculate the local K and Re values prevailing at a given depth at the center of a given pile segment;

(ii) Figures 3.21 and 3.22 be used to obtain the local drag and inertia coefficients for the corresponding K and Re values;

(iii) for $Re > 1.5 \times 10^6$ (the upper limit of Re in Figs. 3.21 and 3.22), $C_d = 0.62$ and $C_m = 1.8$ be used;

(iv) the total in-line force acting on the entire pile be calculated by summing the forces acting on all segments [see Eq. (5.13)];

(v) the transverse force be calculated in the same manner using the lift coefficient given in Fig. 3.25. For $Re > 1.5 \times 10^6$, a lift coefficient of 0.20 is recommended.

For marine roughened vertical pipes, it is recommended that

(i) the effective diameter of the pipe (essentially the average diametral distance between the protrusions) be first estimated or determined as accurately as possible on the basis of past experience with structures at the same site; and

(ii) the total force acting on the pile be calculated, as previously described, through the use of a suitable wave theory, Morison's equation, apparent diameter of the pipe, and the drag and inertia coefficients given in Figs. 5.5a and 5.5b as a function of the roughness Reynolds number $Re_k = U_m k/\nu$. The force coefficients may be found by interpolation for values of K intermediate to those shown in Figs. 5.5a and 5.5b.

The examples of a marine roughened pipe are presented in Figs. 5.6a and 5.6b. The first of these figures show randomly-distributed rigid excrescences and the second, a combination of rigid and soft elements. In either case, it is not necessary to know the roughness height to any great precision. The fact that this is so is evidenced by Figs. 3.31, 5.5a and 5.5b where C_d and C_m vary rather gradually with the roughness Reynolds number.

Figures 5.5a and 5.5b are based partly on Sarpkaya's unpublished data with marine roughened cylinders (Figs. 5.6a and 5.6b) and partly on Figs. 3.28 and 3.29. In fact, the differences between the Figs. 3.28 and 3.29 and the Figs. 5.5a and 5.5b stem primarily from the greater emphasis given to the data obtained with marine-roughened cylinders.

It is recommended that the lift force be calculated using Fig. 3.32. For large values of K and Re, C_L should be assumed to be equal to about 0.25 for marine-roughened cylinders.

The Strouhal number, defined by $f_v D/U_m$, may be taken 0.22. Neither the rigid nor the soft excrescences mitigate the effects of vortex shedding in waves. The soft roughness elements (e.g., seaweed, kelp, grass, etc.) do not suppress the vortex shedding or cable strumming. Soft elements do not appear to have the right natural frequency to interfere with the shedding of the vortices as evidenced by flow visualization experiments.

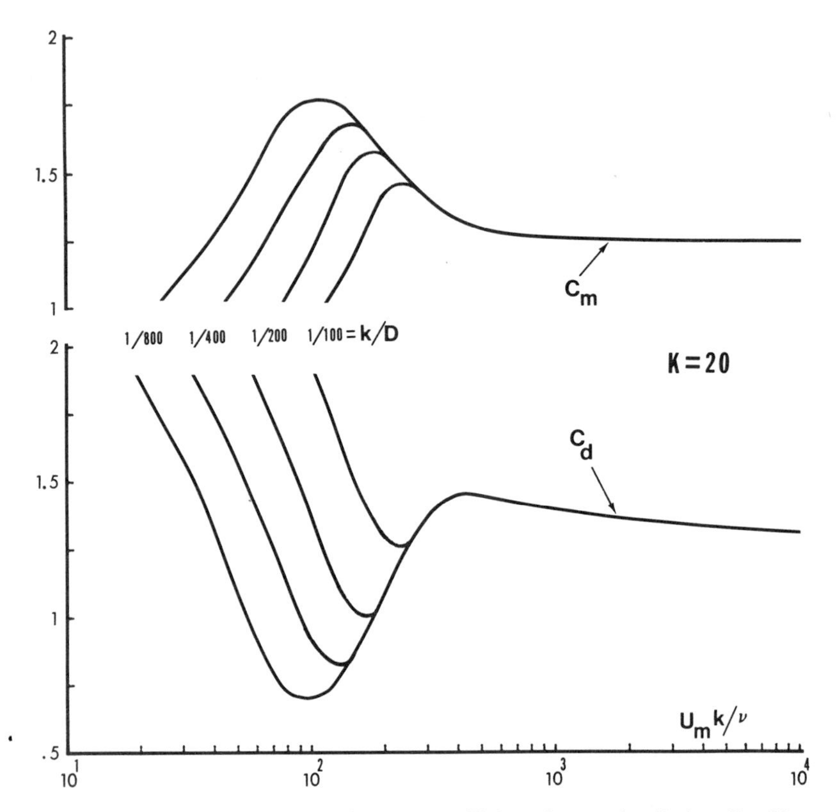

Fig. 5.5a. Recommended drag and inertia coefficients for rough cylinders, K = 20.

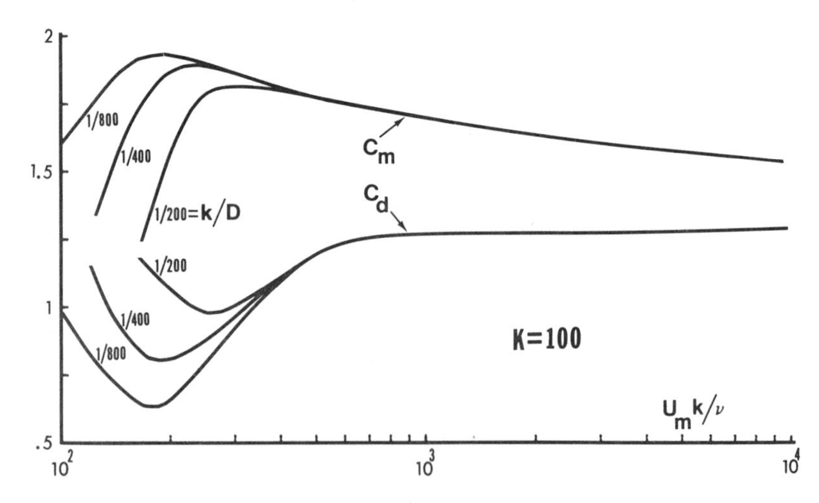

Fig. 5.5b. Recommended drag and inertia coefficients for rough cylinders, K = 100.

Fig. 5.6a. A marine roughened pipe with rigid excrescences.

Fig. 5.6b. A marine roughened pipe with soft excrescences.

5.3.4 Effect of Orbital Motion on Force Coefficients for Horizontal Cylinders

Relatively little attention has been paid to the wave forces (excluding wave slamming) acting on horizontal cylinders. In fact, there is no known full-scale data. Evidently, for laboratory experiments conducted with harmonic flows (e.g., Rance 1969, Sarpkaya 1976a) or with oscillating cylinders (e.g., Yamamoto and Nath 1976) in a fluid otherwise at rest it does not make any difference whether the cylinder is held vertical or horizontal as long as its axis is normal to the flow.

Yamamoto and Nath (1976) conducted both oscillating cylinder and wave flow experiments on cylinders placed at various distances from the bottom of a wave channel. Their force traces contained considerable noise. The drag and inertia coefficients were evaluated by considering the inline force at the instants of maximum velocity and maximum acceleration respectively. The kinematics of the wave flow (presumably at the position of the center of the horizontal cylinder) were determined from the Airy wave theory through the use of the measured wave height and period. Their oscillating-cylinder data and the wave-flow data do not overlap, i.e., they have obtained no wave-flow data in the region of oscillating-cylinder data and vice-versa. Furthermore, whereas the cylinders were oscillated in a plane at a distance of $e = 6D$ from the bottom, the 3-inch cylinder for the wave-flow experiments was placed at $e = 3D$ and about 6 feet below the mean free surface. Yamamoto and Nath's data fall somewhat below those obtained by Sarpkaya (1976a). They have offered no discussion on the effect of orbital motion and stated that "the general trends of the oscillatory cylinder data and the wave force data match continuously" and that "this indicates the physical similarity between the two flow situations." These conclusions appear to be unwarranted and encompass a broader range than Yamamoto and Nath may have intended to. It seems that, for the waves they have generated, the water particles had a very flat elliptical orbit (a nearly linear harmonic motion) at the position of the cylinder. This together with their method of evaluation of the data and the use of calculated rather than measured wave kinematics may explain their conclusions and the implied lack of the effect of orbital motion on the force coefficients.

It is not an easy matter to isolate the effect of orbital motion on cylinder resistance in wavy or harmonic flows, particularly in the range of practical K and Re values. In wave flows, the cylinder has to be placed horizontally at various depths below the free surface and the partical velocity together with the instantaneous forces be measured. In doing so, one has to vary the wave height systematically so as to obtain nearly identical K and Re values. This is a nearly impossible task. Consequently, the measurements made with waves of nearly identical characteristics reflect not only the effect of the orbital motion but also the effects of the variation of K and Re with depth. Finally, one must also bear in mind the three-dimensional nature of the flow about the cylinder due to spanwise and chordwise instabilities.

In harmonic flows, the cylinders will have to be oscillated harmonically in the transverse direction to simulate the orbital motion. In this case also there are limitations as to the ranges of the K and Re values and the ratio of the amplitudes of the vertical and horizontal velocities.

Maull and Norman (1979) carried out experiments with horizontal cylinders in a wave tank at relatively low K and Re values ($K < 24$ and $Re < 4000$, both based on the horizontal velocity which was calculated from the wave profile using second-order wave theory). They have expressed the total in-line force coefficient as $C_F' = F_x'/(0.5\, \rho D U'^2)$ where F_x' is the r.m.s. force in the wave-propagation direction and U' is the r.m.s. of the particle velocity in the horizontal direction at the position of the cylinder center. The lift coefficient C'_L was expressed in a similar manner, i.e., $C'_L = F'_y/(0.5\ \rho D V'^2)$ where F'_y and V' are the r.m.s. values of force and velocity normal to the direction of wave propagation. The results have shown that C'_F and C'_L are only functions of K based on the velocity at the position of the cylinder and the ratio of the vertical to horizontal velocities, V'/U', at that point, ignoring any effects of viscosity. C'_F decreased with increasing V'/U' relative to the case where $V'/U' = 0$, (planar harmonic flow). C'_L exhibited a similar behavior, showing in both cases that the rotation of the wake about the horizontal cylinder due to the orbital motion of the fluid particles is important, at least in the region where $K < 24$ and $Re < 4000$. In other words, Morison's equation with the coefficients C_d and C_m obtained for the $V'/U' = 0$ situation cannot be applied to predict the in-line force on a horizontal cylinder in wavy flows. The force coefficients C_d, C_m, and C_L are functions of K, Re, V/U (also k/D if the cylinder is rough).

Koterayama (1979) measured the wave forces on small horizontal cylinders (0.4 to 3.2 inch diameter) and found that C_d and C_m are smaller than those obtained with planar harmonic motion and with vertical cylinders in waves. His V/U ratios and Reynolds numbers were not specified. Ramberg and Niedzwecki (1979) conducted experiments with small horizontal cylinders in waves ($K < 12$ and $\beta = 430$ and $\beta = 610$) and obtained C_d and C_m values similar to those of Maull and Norman (1979) and Koterayama (1979). Preliminary studies with cylinders oscillating normal to the harmonic flow have shown that (Sarpkaya 1980) C_d and C_m, as given by Eqs. (5.10) and (5.11), for $V_m/U_m = 1$ and $K > 20$ are about one half of those for $V_m/U_m = 0$. Clearly, additional experiments are needed at much higher Reynolds and Keulegan-Carpenter numbers for the determination of wave forces on horizontal cylinders.

5.3.5 Effect of Currents

There are numerous physical circumstances in which interactions between waves, currents and structures occur. The analysis of the interaction of waves with preexisting and/or wind-or wave-generated currents and the interaction of the modified wave-current combination with rigid or elastic structures and their components require different mathematical approaches, relevant observations, and experiments that are applicable to all or some of these physical circumstances.

Basically, the wave-current combination may be treated either as a complex fluid-mechanical phenomenon where the interaction of waves and current is taken into consideration or as a relatively simple phenomenon where the interaction is ignored and the current is simply superimposed on waves. If the current is in the direction of wave propagation, the wave amplitude decreases and its length increases. If the current opposes the wave, the wave becomes steeper and shorter. One must also bear in mind that if there is a current varying with depth it may affect the waves on the water surface since the wave motion also varies with depth.

An elegant mathematical review of the interaction of water waves and currents has been presented by Peregrine (1976) where an extensive list of references may be found (see also Jonsson 1976).

Measurements of wave-current interaction phenomena are scarce. Among the first to perform substantial controlled experiments of this nature was Sarpkaya (1955). He made measurements of wave-phase velocity, amplitude, wave length, and shape of ascending and decending waves in a long wave-current flume and determined the neutral stability conditions for waves propagating with constant amplitude. Additional experiments were conducted by Inman and Bowen (1963) in connection with sand transport by waves and currents.

Dalrymple and Dean (1975), following the simple superposition principle, related waves of maximum height on currents to equivalent waves in still water. Their procedure does not address the problem of wave-current interaction. Their objective can be stated as follows: Given a wave propagating in still water, determine its height and length on a constant current. It is easy to show that the wave celerity C' and the wave length L' (both with the presence of a current $\overline{V}$) are related by

$$(C' - \overline{V})^2 = \frac{gL'}{2\pi} \tanh \frac{2\pi d}{L'} \tag{5.16}$$

where d represents the water depth. Dalrymple and Dean gave a number of examples to illustrate the use of the superposition method to account for the presence of uniform and steady currents (see Fig. 4.40).

In general no wave propagation is possible when the celerity of energy transmission is equal to the mean velocity of flow. For small amplitude waves, this yields (Sarpkaya 1955)

$$\frac{\overline{V}}{\sqrt{gd}} = \frac{1}{2}\left[1 + \frac{4\pi d/L'}{\sinh 4\pi d/L'}\right] [(L'/2\pi d) \tanh (2\pi d/L')]^{1/2} \tag{5.17}$$

which shows that the waves could be stopped against a current of $\overline{V}/C' = \frac{1}{3}$ for large values of d/L'. This conclusion is an approximate one, however, since the waves near breaking become more peaked at the crest and slow down as shown by Longuet-Higgins (1961, 1976), (see also Section 4.9.3).

Little information exists on the effect of current plus wave interaction on hydrodynamic loading of offshore structures. The complexity of the problem stems from several facts. Firstly, an analytical solution of the separated, time-dependent flow is not yet possible even for relatively idealized situations. Secondly, waves and currents are omnidirectional and the directional distribution of energy is anisotropic. Finally, an experimental investigation of the problem in the practically significant range of Reynolds number, Keulegan-Carpenter number, relative current velocity, and suitably defined current gradient is practically impossible. Some tests may be carried out in the ocean by towing cylindrical models in head, following, beam and quartering seas, with various forward speeds for a range of wave heights and periods. In doing so one should measure in-line and transverse forces, in addition to the kinematics of the waves, and make sure that the model does not undergo transverse oscillations.

The results obtained with superposed mean and two-dimensional harmonic flow about a cylinder have been discussed in Section 3.9. It has been shown, for at least relatively small Reynolds numbers, that for sufficiently small values of $\overline{V}/U_m$ the drag coefficient for a current-harmonic flow combination may be considerably smaller than that for a harmonic flow alone. Qualitatively, this corresponds to a region where the effect of the current on the convection of vortices is most important, (i.e., small K).

Tung and Huang (1976) examined the influence of wave-current interactions on some of the statistical properties of fluid force through the use of Morison's equation and the superposition of the wave and current velocities. In this connection one may also wish to study an important contribution by Moe and Crandall (1977).

Tung and Huang assumed that the fluid motion is irrotational; the waves are in deep water; small wave theory is valid; the sea-surface is unidirectional, stationary, and normally distributed; the current is externally generated, in-line with the waves, steady in time, non-uniform but slowly varying in the horizontal plane, and has a constant profile in the vertical direction. With these assumptions and some additional statistical approximations Tung and Huang obtained the spectrum of the fluid force as

$$S_{ff} = 4(\rho D)^2\, C_d^2 \sigma_v^4\, [T(\gamma) + |\gamma| P(\gamma)]^2\, S_{vv}(n) + C_m^2\, S_{aa}(n)(\pi \rho D^2/4)^2 \quad (5.18)$$

in which $S_{vv}(n)$ and $S_{aa}(n)$ represent respectively the frequency spectrum of the horizontal components of fluid particle velocity and acceleration; $\gamma = \overline{V}/\sigma_v$; $\overline{V}$, the current velocity; $P(\gamma)$, the error function; and

$$\sigma_v = \left[\int_0^\infty S_{vv}(n)\, dn\right]^{1/2} \quad (5.19)$$

and

$$T(\gamma) = 1/(2\pi)^{1/2} \exp(-\gamma^2/2) \quad (5.20)$$

A sample plot of $S_{ff}(n)$ is reproduced in Fig. 5.7. It is seen that without interactions, the force spectra for positive and negative currents are the same, as would be expected. When there are interactions, the negative current gives rise to an increase in the spectrum over that when interactions are neglected and the situation is reversed under positive current. Evidently, these conclusions are based on the implicit assumption that C_d and C_m remain constant whether there are interactions or not and whether the current is positive or negative. Thus, the conclusions arrived at by Tung and Huang depend, among other assumptions stated, on the selection of proper values of C_d and C_m. As noted earlier, there is not enough experimental data for an intelligent assessment of the proper values of the drag and inertia coefficients.

The conventional method of including the current effects on wave loading is by vectorially superimposing current profiles over the wave velocity field generated in the absence of a current. In doing so, the acceleration to be used in the

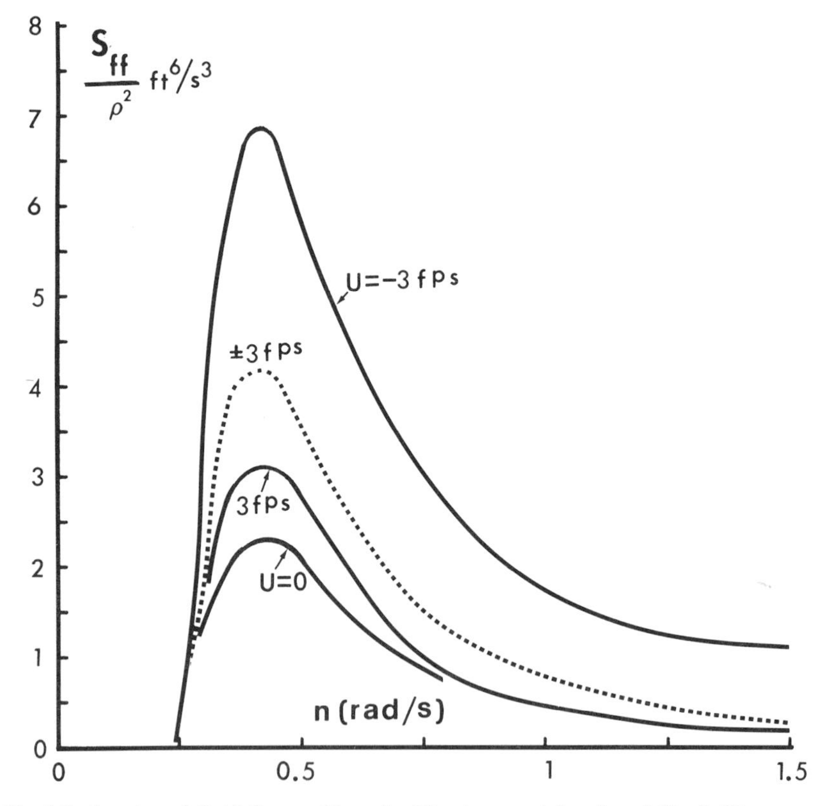

Fig. 5.7. Spectra of fluid force with and without current ($z = 0$ and $D = 1$ ft), - - - no interaction, (Tung and Huang 1976).

calculation of the inertial force should be written as

$$a_x = \frac{Du}{Dt} + \overline{V}\frac{\partial u}{\partial x} = \frac{\partial u}{\partial t} + u\frac{\partial u}{\partial x} + \overline{V}\frac{\partial u}{\partial x} \tag{5.21}$$

Furthermore, it is suggested that the Reynolds number and the Keulegan-Carpenter number be defined as

$$Re = (U_m + \overline{V})D/\nu \quad \text{and} \quad K = (U_m + \overline{V})T/D \tag{5.22}$$

The drag and inertia coefficients to be determined from a systematic experimental investigation may not necessarily correlate with $\overline{V}/U_m$ and Re and K (as defined above). For most practical applications, however, the variations of C_d and C_m with large K and Re (for no current) are rather small.

In summary, much laboratory and field investigations remain to be carried out for a better quantification of the wave-current and wave-current-structure interactions. Small-scale laboratory experiments hint that Morison's equation may not be sufficient to describe the fluid loading through a vectorial superposition of wave and current velocities. For $\overline{V}/U_m$ smaller than about 0.3 and K larger than about 20, the use of Morison's equation, as stated above, together with the drag and inertia coefficients suggested for use in the absence of current will, in general, lead to conservative in-line forces. It is assumed that the presence of the current or wave-current combination does not lead to in-line and/or transverse hydroelastic oscillations. For a model study of wave and current forces on a fixed platform the reader is referred to Sekita (1975).

Wave-current interaction and the scatter component of an incident wave are of some concern in the diffraction analysis of large bodies. It appears that such effects are, in general, small (Hogben 1976) and somewhat obscured by the effect of flow separation (ordinarily ignored) on diffraction analysis. The presence of currents may be more important for large square-shaped bodies than for bodies with no sharp corners.

5.3.6 Effect of Pile Orientation

The effect of body orientation (in the plane of flow) on resistance particularly for bodies of finite length (e.g., a missile at an angle of attack) has been the subject of extensive investigation in steady flows. It has not been possible to correlate the in-plane normal force and the out-of-plane transverse force with a single Reynolds number. Evidently, for a zero angle of attack (flow parallel to the axis of the body), the appropriate Reynolds number is based on the length measured along the body. For a 90-degree angle of attack (flow normal to the axis of the body), the appropriate Reynolds number is based on the diameter of the body. Between the two flow situations, it is not possible to define a simple characteristic length and hence a universal Reynolds number which will correlate the force-transfer coefficients for all flow regimes.

Relf and Powell (1917) tested smooth and multistranded cables in air flow up to yaw angles of 70 degrees. The normal Reynolds number Re_n, based on the velocity component normal to the cable axis, was approximately 10^4. They have concluded that the normal force acting on the cables was directly proportional to $\cos^2 \phi$. A close examination of their drag data on smooth cylinders at Re = 8130 shows a small but consistent increase in drag coefficient with increasing yaw angle as noted by Smith, Moon, and Kao (1972). Increasing the yaw angle at constant free-stream Reynolds number Re results in a reduction in Re_n, and on the basis of unyawed cylinder data in the Reynolds number range of Relf and Powell's data, one expects to observe a decrease in drag coefficient rather than an increase.

Bursnall and Loftin (1951) measured pressure distributions around a circular cylinder in a wind tunnel for yaw angles in the range $0 < \phi < 60$*. They concluded that for Reynolds numbers in the range $10^4 < Re_n < 5 \times 10^5$, the flow and force characteristics of yawed cylinders could not be determined solely by the component of velocity normal to the cylinder axis and that the critical Reynolds number Re_{nc} strongly depends on the yaw angle (see Figs. 5.8a and 5.8b). The importance of the axial component of velocity is also gleaned from Glenny's (1966) review of the data relevant to yawed struts.

Ericsson and Reding (1979) proposed an effective Reynolds number based on the wetted length to the separation point (taken as the quarter periphery of the elliptic cross-section defined by the inviscid flow streamline about the inclined cylinder) and claimed that this effective Reynolds number provides a satisfactory correlation of the sweep effect. Figure 5.8c, prepared using Ericsson and Reding's effective Reynolds number, shows that the correlation is not any better than those shown in Figs. 5.8a and 5.8b.

Hoerner (1965) proposed the independence-or crossflow principle or the 'cosine law' which states that the normal pressure forces are independent of the tangential velocity for subcritical values of Re_n. Such a principle has been used even for examining the galloping instability of yawed transmission lines or cables of guyed towers (Skarecky 1975). Novak (1975) commented that the independence principle may have at least limited validity for smooth bodies and with rough surface, the yaw wind can produce special effects.

Hanson (1966) examined vortex shedding from yawed hot wires in air flow at relatively low Reynolds numbers ($40 < Re_n < 150$) and for angles of yaw $0 < \phi < 72^\circ$ and concluded that the shedding frequency is proportional to $\cos \phi$.

---

*Note that ϕ is the angle between the ambient velocity and the normal to the cylinder axis. θ is the angle between the ambient velocity and the cylinder axis (see Figs. 5.8a and 5.8b).

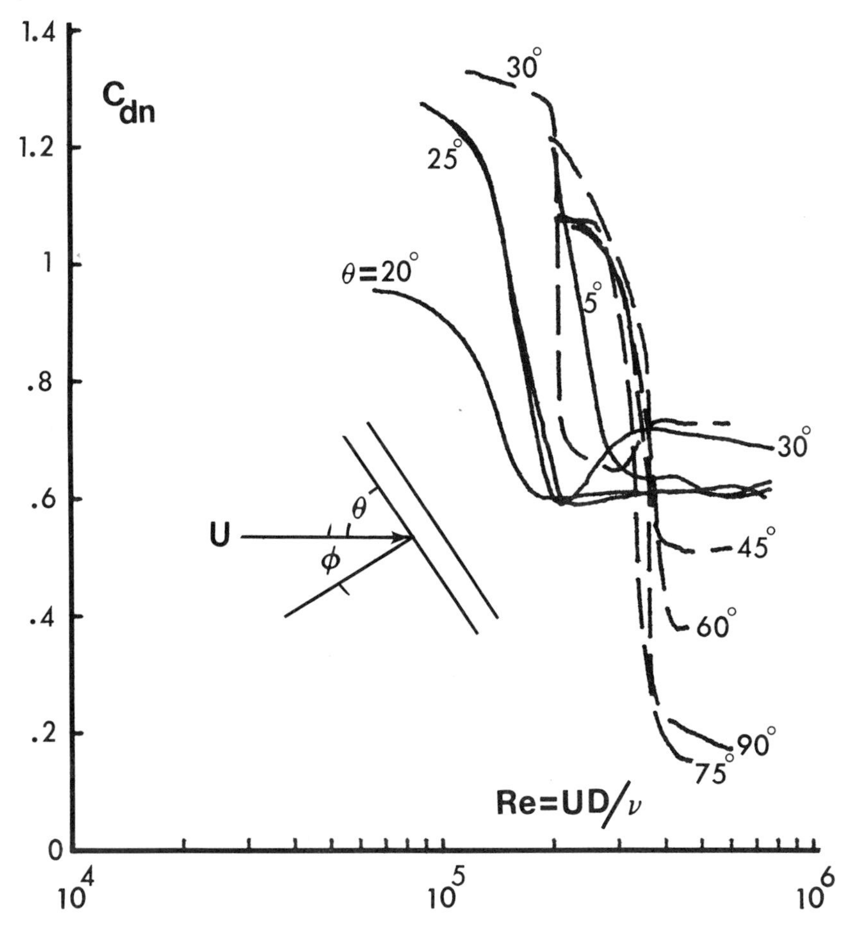

Fig. 5.8a. Normal drag coefficient versus Reynolds number of inclined cylinders (- - - Bursnall and Loftin 1951, —— Trimble 1966).

The Strouhal number S_n based on normal velocity varied with Re_n as in Roshko's (1964) empirical relationship for $0 < \phi < 68°$. For $\phi = 72°$ the results were far removed from the mean curve described by Roshko's equation (see Figs. 3.1a and 3.2).

Van Atta (1968) found a marked deviation of the vortex-shedding frequency from the $\cos \phi$ relation. Furthermore, by varying the tension in the yawed wire, he showed that the effects observed by Hanson (1966) could be attributed to synchronization or lock-in between the natural vortex-shedding frequency and the fourth, fifth, and sixth harmonics of the wire. In doing so, Van Atta has shown that yawed cylinders undergo synchronized oscillations in the range

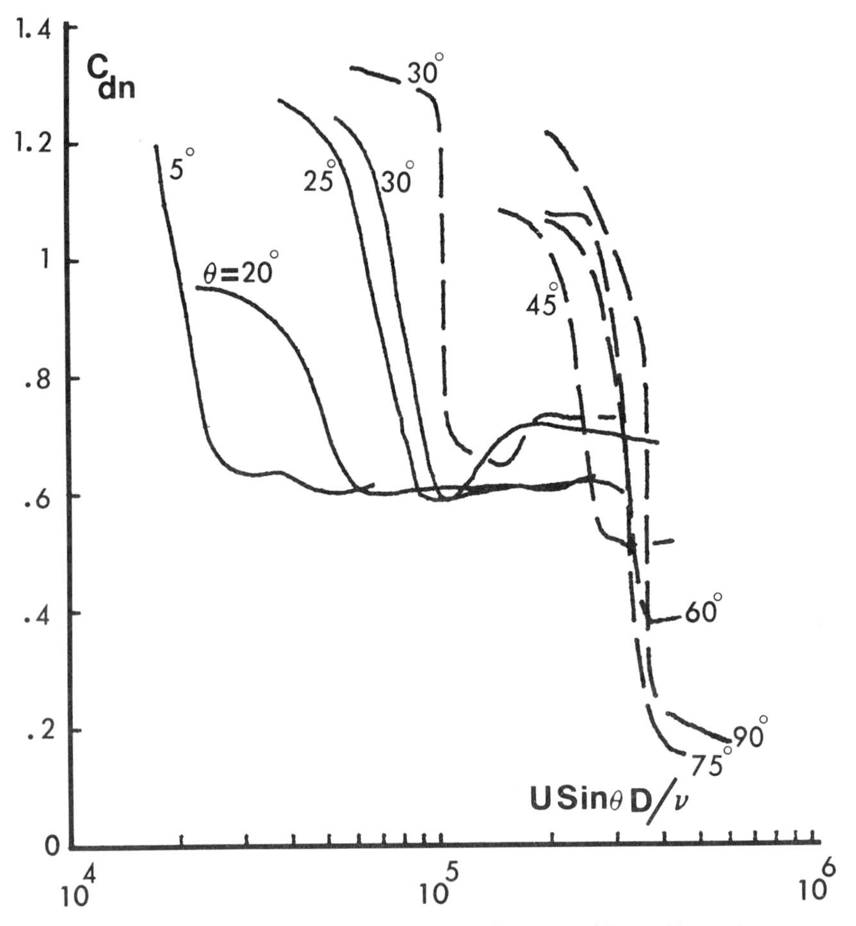

Fig. 5.8b. Replot of Fig. 5.8a with 'Normal Component' Reynolds number.

$5.9 < U \cos \phi / f_n D < 6.4$ (in this connection see also King 1977 and Ramberg 1978).

Chiu (1966) verified Hanson's and Van Atta's results for the higher Reynolds number range $3{,}900 < Re_n < 21{,}200$ and for $0 < \phi < 60°$. Friehe and Schwartz (1968) demonstrated that the length-to-diameter ratio could produce deviations from the cosine law. Smith et al. (1972) conducted experiments with yawed cylinders in the Reynolds number range $2{,}000 < Re_n < 10{,}000$ and found that transition from laminar to turbulent motion in the separated region is promoted as the cylinder is yawed and that the vortex shedding frequency obeys the independence principle (see also Springston 1967).

The studies cited above have been conducted with steady flows and are suffi-

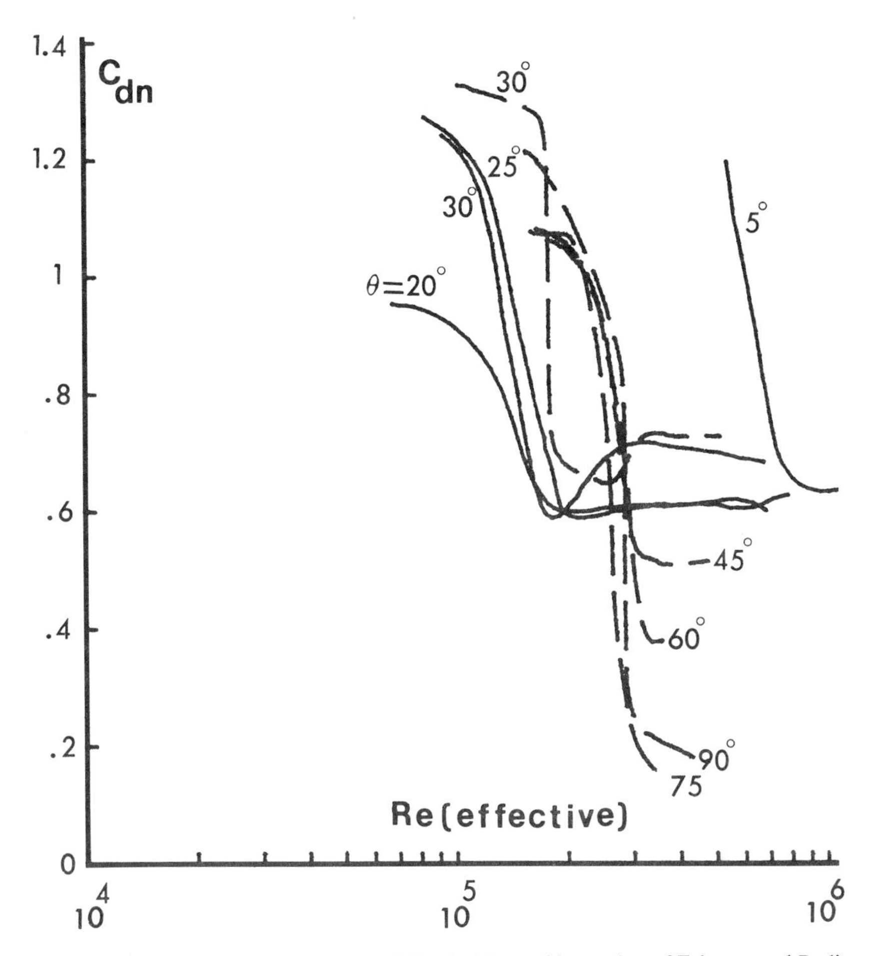

Fig. 5.8c. Replot of Fig. 5.8a with the 'Effective' Reynolds number of Ericsson and Reding (1979).

cient to illustrate the complexity of the problem. The time-dependent flows, in general, and the wave motion, in particular, about oblique cylinders present even more complex problems. The use of the independence principle or the assumption that shedding frequency is proportional to $\cos \phi$ and that the normal component of drag is proportional to $\cos^2 \phi$ may be a gross simplification of the behavior of flow in the near wake. Under these circumstances only experiments can lead to some understanding of the problem and to the evolution of approximate calculation methods.

In what follows, we will deal with waves about oblique cylinders. It will become evident that neither the method of decomposition of velocities and/or forces nor the drag and inertia coefficients appropriate to each method of force decomposition are clear. Furthermore, there are not enough systematic experiments either with waves or with harmonic flows to guide the analysis.

Assuming that Morison's equation may be applied to a cylindrical member oriented in a random manner with respect to the mudline, Eq. (5.9) may be written as

$$\begin{Bmatrix} F_x \\ F_y \\ F_z \end{Bmatrix} = 0.5\ \rho C_d D |W_n| \begin{Bmatrix} u_{nx} \\ u_{ny} \\ u_{nz} \end{Bmatrix} + 0.25\ \pi \rho C_m D^2 \begin{Bmatrix} \dot{u}_{nx} \\ \dot{u}_{ny} \\ \dot{u}_{nz} \end{Bmatrix} \tag{5.23}$$

where C_d and C_m are assumed to be known.

The velocity and acceleration vectors may be defined with respect to a spherical coordinate system as follows (for details see Chakrabarti et al., 1975). Let $\mathbf{e}$ be the unit vector along the cylinder. Then

$$\mathbf{e} = e_x \mathbf{i} + e_y \mathbf{j} + e_z \mathbf{k} \tag{5.24}$$

where $\mathbf{i}, \mathbf{j}, \mathbf{k}$ represent the unit vectors in the x, y, z directions and e_x, e_y, e_z are given by (see Fig. 5.9)

$$e_x = \sin\phi \cos\psi, \qquad e_y = \cos\phi, \qquad e_z = \sin\phi \sin\psi \tag{5.25}$$

The velocity vector *normal* to the pipe is then given by

$$\mathbf{W_n} = \mathbf{i}u_{nx} + \mathbf{j}u_{ny} + \mathbf{k}u_{nz} = \mathbf{e} \times [(\mathbf{i}u + \mathbf{j}v) \times \mathbf{e}] \tag{5.26}$$

which yields

$$\begin{aligned} u_{nx} &= u - e_x(e_x u + e_y v) \\ u_{ny} &= v - e_y(e_x u + e_y v) \\ u_{nz} &= -e_z(e_x u + e_y v) \end{aligned} \tag{5.27}$$

where u and v represent the horizontal and vertical components of the water-particle velocity. u and v as well as $\dot{u}$ and $\dot{v}$ may be obtained from a suitable wave theory such as Stokes 5th or the Stream Function. It should be emphasized that in general the total velocity and acceleration vectors are not collinear.

The absolute value of $\mathbf{W}_n$ is given by

$$|W_n| = [W \cdot W]^{1/2} = [u^2 + v^2 - (e_x u + e_y v)^2]^{1/2} \tag{5.28}$$

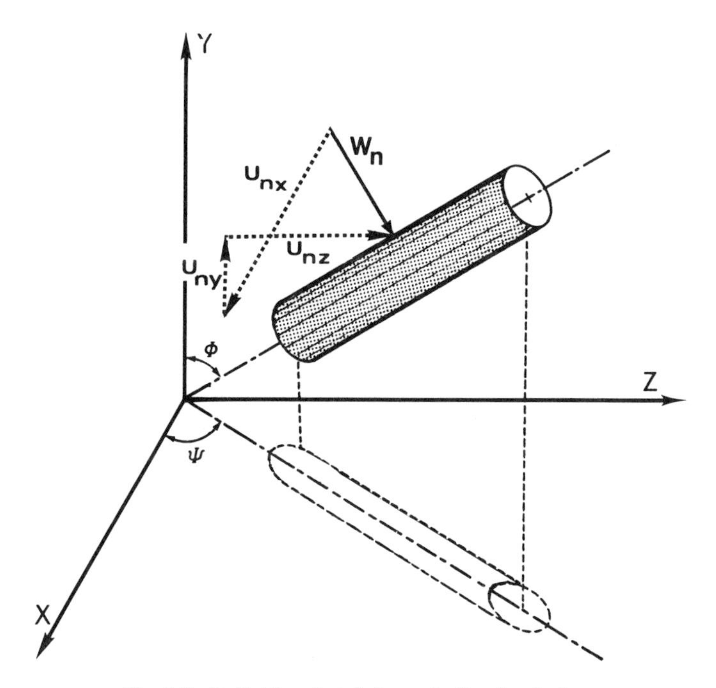

Fig. 5.9. Definition sketch for an inclined cylinder.

As noted earlier, the foregoing development assumes the validity of the 'independence principle.' Thus, consistency requires that the values of C_d and C_m recommended in Section 5.3.3 be used in Eq. (5.23)*. The Reynolds number and the Keulegan-Carpenter number should be defined as

$$Re = |W_n| D/\nu, \quad K = |W_n| T/D \tag{5.29}$$

Chakrabarti et al. (1977) conducted a series of experiments with a small inclined tube in a wave channel and evaluated C_d, C_m, and C_L. The Keulegan-Carpenter number ranged from zero to 16 and β was about 1600. They were unable to show the dependence of the force-transfer coefficients on the Reynolds number because of the limited range of the Reynolds number (less than about 25,000). Although not noted by Chakrabarti et al., a comparison of their results with those obtained by Sarpkaya (1976a), with a harmonically oscillating flow normal to smooth cylinders, shows that the force-transfer coefficients at the corresponding K and β values are nearly identical. This comparison, however

*The critical Reynolds number may be lower for yawed cylinders as seen in Figs. 5.8a through 5.8c.

limited, tends to show that the force-transfer coefficients obtained with cylinders normal to the flow may be used together with Eq. (5.23) to predict the forces acting on inclined tubes.

Wade and Dwyer (1976) examined four methods for calculating wave forces acting on tubular members. Each method is within procedures generally accepted by the industry. Horizontal and vertical wave-induced water-particle kinematic vectors were used in each of the wave force methods on two deep water platforms to compare the horizontal base shear and over-turning moments.

The first method assumes that the resultant drag pressure acts on an area projected on a plane normal to the total water-particle velocity and the resultant inertial pressure acts on an area projected on a plane normal to the total water-particle acceleration. If the total velocity and total acceleration vectors were normal to the axis of the tube this method would have been identical to that described above.

The second method assumes that the resultant drag and inertial pressures may be resolved into normal and tangential components and the tangential pressure components may be ignored. This method is not in conformity with the independence principle since the cosine of the angle rather than its square is multiplied with the square of the velocity.

The third method assumes that the resultant velocity and acceleration may be decomposed into normal and tangential components and the tangential kinematics may be ignored. This method is in conformity with the independence principle and with the method recommended in this section.

Finally, the fourth method assumes that the resultant velocities and accelerations act normal to the members. An area correction factor is applied to pressures when the yaw angle is greater than 60 degrees. This method relies heavily on the data shown in Figs. 5.8a and 5.8b. The said data have been obtained in the transition region with steady flows. Furthermore the assumption that the resultant velocities and accelerations act normal to the members is not appealing.

Evidently, the results of the comparison depend on the proper selection of the drag and inertia coefficients. Wade and Dwyer assumed identical drag and inertia coefficients in applying each method ($C_d = 0.6$, $C_m = 2.0$ for one test structure and $C_d = 0.6$ and $C_m = 1.4$ for the other test structure) and found that the last three methods yielded from 10 percent to 12 percent variation in the base shear relative to the first method. They have also found that the exclusion of the vertical water-particle kinematics had only a small effect, less than 2 percent, upon the structure's base shear and over-turning moment. Such comparisons, however valuable, are not sufficient to assess the validity of one method over the others since the base shear and over-turning moment represent the sum of forces and moments over many members at various angles of inclination.

In summary, considerable additional work is required in order to acquire some understanding of the wave forces on oblique members and, hopefully, to establish uniformly accurate and acceptable design criteria. Until then the use of Eq. (5.23) together with the drag and inertia coefficients applicable to vertical cylinders (see Section 5.3.3) is recommended. The designer should keep in mind the fact that such oblique members seldom occur in isolation. The effect of the wake of one member on the others and the alterations in the velocity field may give rise to significant variations in the resultant forces from those predicted for idealized situations through the use of semi-empirical methods.

5.3.7 Interference Effects

A body's resistance to flow is strongly affected by what surrounds it. When two bodies are in close proximity, not only the flow about the downstream body but also that about the upstream body may be influenced. For example, a simple splitter plate can cause dramatic changes in flow about the cylinder upstream of the plate. Such a situation may arise in any multimember structure. Examples include condenser and boiler tubes in heat transfer, a variety of columns in pressure suppression pools of nuclear reactors, risers and other tubular structures in offshore engineering, turbine and compressor blades in mechanical or aerospace engineering, and high-rise buildings and transmission lines in civil engineering. The quantification of the interference effects in terms of the pressure distribution, lift and drag forces, vortex shedding frequency, and the dynamic response of the members of the array in terms of the governing flow and structural parameters constitute the essence of the problem.

Contrary to common belief, considerable work has been done on flow interference particularly between circular cylinders in various arrangements and between cylinders and a rigid or free surface. Most of the investigations have been prompted by the need to solve problems of immediate practical interest. Obviously, there are infinite numbers of possible arrangements of two or more bodies or cylinders positioned at right or oblique angles to the approaching flow direction. Consequently, there is a danger, however undesirable, of ad hoc testing leading to a proliferation of undigested and uncorrelated data. Such data and observations provide phenomenological explanations, but the intrinsic nature of the flow patterns remains a mystery.

The subject of flow interference may be classified in many categories: separated and unseparated flows, steady or time-dependent flows, partial or mutual interference in all types of flows, etc. In the following, we will first discuss interference effects in separated steady flows. Then we will take up the case of unseparated potential flows. Finally, relatively few studies on interference in wavy flows will be presented.

A careful review of flow interference between two circular cylinders in various arrangements has been presented by Zdravkovich (1977) where an extensive list of references may be found. Numerous studies for the tandem arrangement (one cylinder behind the other) have shown that the changes in drag, lift, and vortex shedding are not necessarily continuous. In fact, the occurrence of a fairly abrupt change in one or all flow characteristics at a critical spacing is one of the fundamental observations of flow interference in cylinder arrays.

It has been shown experimentally that there is strong interference between two cylinders in tandem for spacing ratios L/D smaller than about 3.5, where L is the distance between the centers of the two cylinders. At a spacing ratio of about 3.5 there is a sudden change of the flow pattern in the gap. The critical spacing appears to depend somewhat on the Reynolds number. It is below 3.5 at $Re = 5.8 \times 10^4$, equal to 3.5 for $Re = 8.3 \times 10^4$, and slightly larger than 3.5 at $Re = 1.1 \times 10^5$. There seems to be no data for Re larger than about 2×10^5. At critical spacing the discontinuous change of the flow patterns cause the following: a jump in drag coefficient of the upstream cylinder, the commencement of the vortex shedding, and a drop in the base pressure. For the downstream cylinder, the base and side pressure coefficients drop, vortex shedding frequency jumps, and the gap pressure and the drag coefficient increase suddenly.

Drag coefficient data show (Zdravkovich 1977) that the upstream cylinder takes the brunt of the burden (see Fig. 5.10). The main feature for all Reynolds numbers (smaller than about 2×10^5) is that, beyond the critical spacing, the downstream cylinder has no effect on the upstream one. In other words, there is mutual interference for spacings less than critical and partial interference for spacings larger than critical.

The drag coefficient of the downstream cylinder shows a strong dependence on the Reynolds number. At high subcritical Reynolds numbers, the wake turbulence from the upstream cylinder induces a supercritical flow around the downstream cylinder and hence the drag remains small even at large spacing.

Wardlaw and Cooper (1973) systematically measured drag forces on stranded cables in tandem. However, not much is known about the effect of roughness on interference and additional work is needed.

The side-by-side as well as the staggered arrangement has again been investigated by a number of people, (see Zdravkovich 1977 for extensive references). The most striking qualitative feature of such arrangements is the bistable nature of the flow. For spacing ratios from 1.1 to 2.2, the wakes of the two cylinders interfere and are alternatingly entrained by each other. This gives rise to changes in the base pressures of both cylinders from one steady value to another. The additional feature of the biased flow is that the gap flow biased to one side produces a resultant force on the cylinder which is deflected relative to the free stream direction. Consequently, there is a component of the force acting perpendicular to the free stream direction which may be called a lift force.

The results for the staggered arrangement show that the upstream and downstream cylinders may be subjected to significantly different lift and drag forces.

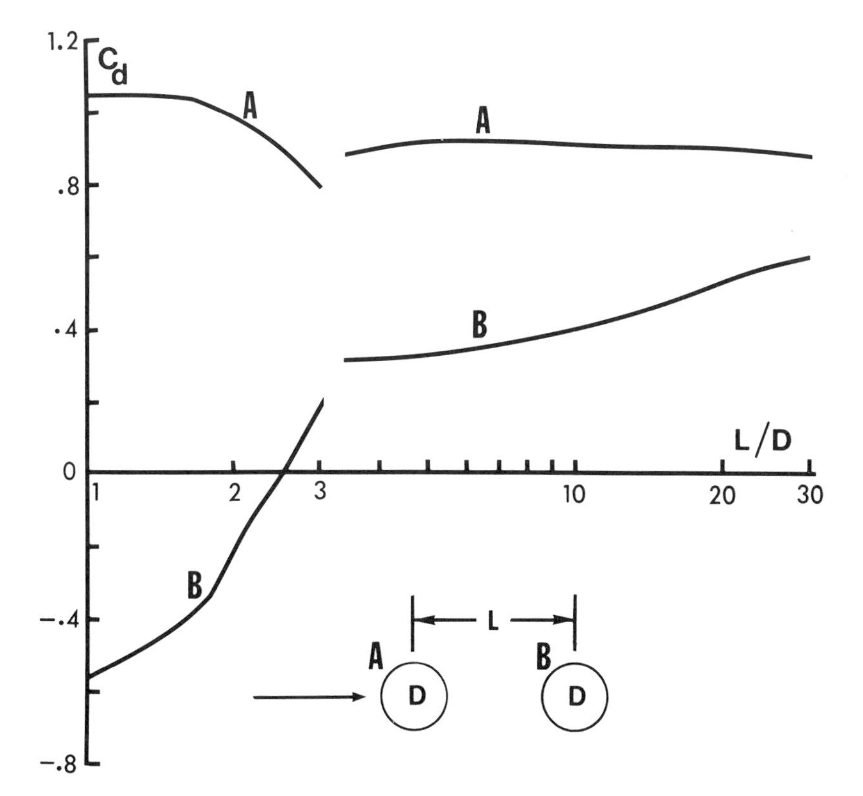

Fig. 5.10. Drag coefficient for tandem cylinders at subcritical Reynolds numbers (a representative plot) (Zdravkovich 1977).

Depending on their relative positions, the cylinders may experience negligible or strong lift and reduced or enhanced drag force.

There is very little data on wake interaction (Fuji and Gomi 1976) and vortex shedding frequencies (Fitz-Hugh 1973) and on lift and drag forces in the range of supercritical Reynolds numbers. Quadflieg (1977) reported some data for the side-by-side arrangement at $Re = 1.51 \times 10^5$. He also used the potential flow theory and the vortex street model to predict the base pressure. A somewhat similar potential flow analysis, based on Rosenhead's (1929) analysis of the wake of a cylinder in a channel, has been presented by Weihs (1977) for the side-by-side arrangement. Weihs found that there is a range of flow parameters where the drag force on each cylinder in the row is less than that of a single cylinder in unbounded flow. He noted that this may serve as a drag reduction criterion for barriers, etc. Experiments of Biermann and Herrnstein (1933) have shown that the interference drag coefficients (difference between the drag coefficient measured on one of the cylinders and the drag coefficient of the single cylinder at the same Reynolds number) may be negative for spacing ratios from 1.2 to 2. However, unlike the underlying assumptions of Weihs' analysis, the flow in this region is bistable and may give rise to severe vibrations. Thus, it turns out that such a critical arrangement must be avoided.

An experimental investigation of the effects of spacing, orientation, and Reynolds number on the drag of each cylinder in a group of two and three cylinders was reported by Dalton and Szabo (1976). They have found that the middle and downstream cylinder drag coefficients are smaller and noticeably more dependent on orientation than is the upstream cylinder. The drag coefficient for the upstream cylinder remained nearly constant. In other words, there is a strong mutual interference between the middle and downstream cylinder but only a partial interaction between the upstream and downstream cylinders. The unsteady wake behind a group of three parallel cylinders has also been investigated by Mujumdar and Douglas (1970).

The steady flow about multitube arrangements has been extensively studied in connection with heat exchangers (for a detailed discussion and some photogaphs see Knudsen and Katz 1958). The purpose of these earlier investigations was to increase the heat transfer and reduce the pressure drop or the energy loss. In general no information has been obtained on the effect of flow interference or shielding on the forces acting on the tubes and on the undesirable hydroelastic oscillations of the tubes. For extensive references and a detailed discussion of the effect of tube spacing in heat exchangers, the reader is referred to Chen (1977).

The aerodynamic behavior of bodies in the wake of other bodies has been examined, among others, by Mair and Maull (1971). They have determined the force acting on one cylinder or square block in a group of similar bodies as a function of the flow direction and have shown that over a small range of flow angle relative to the array the force acting on a single element can change by almost 100 percent. For example, if the mean flow is at an angle of 26 degrees to the group and fluctuates ±2 degrees about the mean, this can produce an oscillating force coefficient on the rear cylinders varying from 0.67 to 1.17, as noted by Mair and Maull. Even though this conclusion is valid only for steady flow about a particular configuration, it shows that one must be as much concerned with the change in transverse forces and vortex shedding frequencies of each and every member of the array as with the in-line or drag forces.

Additional work with tube arrays in steady flows has been reported by a number of people (Laird et al., 1959; Ross 1959; Hoerner 1965; Crua 1967; Hammeke et al., 1967; Chen 1972; Sachs 1972; Arita et al., 1973, and ESDU. 1975).

The experiments of Laird et al., at subcritical Reynolds numbers, showed that the force on the shielded cylinder could vary quite widely, though the mean force is reduced. Ross (1959) conducted large-scale wave tank tests for the case of one cylinder on each side of the test cylinder (side-by-side arrangement) in the range of critical Reynolds numbers. His results appear to indicate that the wave force increases significantly only when the spacing between two cylinders is less than about one diameter.

Hoerner (1965) suggested that the interaction effects in steady subcritical flow are negligible for the side-by-side arrangement of the cylinders if they are more than three diameters apart. For the tandem arrangement, the same effect appeared to be negligible if the cylinders were more than four diameters apart.

Chen (1972) reported average lift coefficients for a number of arrangements and Reynolds numbers. He found that if the cylinders are 3 diameters apart, there is practically no lift force since the wake does not develop sufficiently. For cylinders at relatively smaller spacing, Chen found that the lift on a cylinder in the wake of another can be very large and does not tend to its single-cylinder value until the space between the two cylinders is increased to about 20 diameters. The above is an example of partial interference since the lift on the first or the front cylinder remains practically constant.

Arita et al., (1973) considered a six-pile arrangement and carried out extensive measurements in a calm sea. The outstanding features of their data may be summarized as follows: For the said arrangement, the interference effects on the drag coefficient strongly depend on the Reynolds number. At subcritical Reynolds numbers, the middle cylinders experience the least resistance. At supercritical Reynolds numbers, the drag coefficient of the cylinders in the front row is reduced somewhat. All other cylinders experience the same low drag coefficient which is considerably lower than that of the front row. It should be emphasized that the foregoing is only for a particular combination of six cylinders. Generalizations to other combinations should be done with extreme care. Furthermore, all of the works cited above dealt with steady flows, mostly in the subcritical regime.

For unseparated two-dimensional flow about a group of cylinders the potential flow theory may be used to determine the inertia coefficient for each cylinder through the use of the method of images and complex variables (see e.g., Robertson 1965). The image method will be described here only briefly. For additional details the reader is referred to Dalton and Helfinstine (1971) and Yamamoto (1976) and Yamamoto and Nath (1976).

The complex potential for N-number of cylinders moving arbitrarily in a time-dependent uniform flow of velocity U(t) may be written in terms of an infinite series of doublets as

$$W(z) = -\overline{U}_z + \sum_{k=1}^{N} \sum_{j=1}^{S} B_{kj}/(z - c_{kj}) \tag{5.30}$$

in which $\overline{U}(z)$ represents the conjugate of the complex velocity; W, the complex velocity potential; z, the complex displacement vector; B_{kj} and c_{kj}, the strength and position of the j-th doublet inside the k-th cylinder; and S, the total number of doublets inside the k-th cylinder.

The strength and position of the doublets may be obtained through the use of the circle theorem (see e.g., Robertson 1965) as

$$B_{kj} = -\bar{B}_{\ell m}[a_k/(\bar{c}_{\ell m} - \bar{c}_{k\ell})]^2 \tag{5.31}$$

and

$$c_{kj} = c_{k\ell} + a_k^2/(\bar{c}_{\ell m} - \bar{c}_{k\ell}) \tag{5.32}$$

where a_k is the radius of the k-th cylinder.

The force acting on a stationary cylinder in a time-dependent flow is given by the generalized Blasius theorem (see Robertson 1965) as

$$\bar{F}_k = -i\rho \frac{\partial}{\partial t} \int \overline{W}\, d\bar{z} + \frac{1}{2} i\rho \int \left(\frac{dW}{dz}\right)^2 dz \tag{5.33}$$

The evaluation of the integrals through the use of Eq. (5.30) and the residue theorem yields

$$F_k = \pi\rho a_k^2 \left(\dot{V}_k - 2 \sum_{j=1}^{S} \dot{B}_{kj}\right) + 4\pi\rho \sum_{\substack{\ell=1 \\ \ell \neq k}}^{N} \sum_{j=1}^{S} \sum_{m=1}^{S} \bar{B}_{kj} \bar{B}_{\ell m}/(\bar{c}_{kj} - \bar{c}_{\ell m})^3 \tag{5.34}$$

where V_k is the velocity of the k-th cylinder. Equation (5.34) may be written as

$$F_k = \pi\rho a_k^2 \dot{U} + \pi\rho a_k^2 C_a \dot{U} + \rho a_k C_c |U| U \tag{5.35}$$

in which the first term represents the force due to the pressure gradient to accelerate the fluid; the second term, the force due to the added mass; and the last term, the convective or interference force which would have been zero had there been no interference effects. In general, U and $\dot{U}$ are not collinear and the resultant force may be decomposed or interpreted in various ways.

Yamamoto (1976) used the method of images to calculate the potential flow about a group of randomly arranged cylinders of different diameters. Yamamoto and Nath (1976) conducted a numerical experiment on a 4 by 4 cylinder array using the aforementioned analytical results. They found that the inertial force on a cylinder in the array can be as high as three times that of an isolated cylinder and a large convective force can result on the cylinder.

Dalton and Helfinstine (1971) gave an expression for the C_m values on circular cylinders, arbitrarily spaced in groups in a uniform flow. The C_m values were found to vary significantly with cylinder spacing and configuration. Two general rules emerge from their calculations. For a tandem arrangement (the direction of flow along the line joining the axes of the cylinders), the inertia coefficient *decreases* with decreasing spacing and attains its minimum value when the cylinders touch (the combined body becomes more or less streamlined). For a side-by-side arrangement (the flow normal to the line joining the axes of the cylinders), the inertia coefficient *increases* with decreasing spacing and attains its maximum value when the cylinders touch (the combined body becomes more or less like a vertical streamlined block). Similar conclusions hold true for two spheres (Sarpkaya 1960).

Spring and Monkmeyer (1974) developed a means of calculating the pressures and forces on a cluster of vertical circular cylinders through the use of the method first employed by MacCamy and Fuchs (1954). Their results are more general than those of Dalton and Helfinstine. Spring and Monkmeyer have shown that the force on a given cylinder is significantly affected by the presence of neighboring cylinders. The inertia coefficient may range from 1.19 to 3.38, departing significantly from the often assumed value of 2.0. They have also shown that the force perpendicular to the direction of wave advance may be quite significant when the cylinders are close together, rising in one case to 69 percent of the force component in the direction of wave advance. A similar analysis has been carried out by Chakrabarti (1978a) through the use of the linear potential theory including the wave diffraction effects.

The analyses and the works cited above do not deal with the effects of separation and vortex shedding. Consequently, the results are more appropriate to the determination of earthquake forces on large bodies rather than to the evaluation of the inertial component of the force in the drag/inertia dominated regime. In fact, there is every reason to believe that the inertial force acting on a cylinder in an array is as much affected by vortex shedding as the drag component of the force. Thus, it is recommended that special arrays, requiring a careful evaluation of the drag and inertial forces, be subjected to tests simulating the prototype conditions.

Gibson and Wang (1977) carried out two different experiments to determine the added mass of a series of tube bundles. The bundles consisted of tubes of uniform diameter d, arranged either in a square configuration or a circular configuration. In the first series of experiments, they towed the model of pile cluster under linear acceleration. In the second series, they have vibrated the model at its own natural frequency. For both cases, they have calculated the added mass through the use of the measured force and acceleration and plotted them as a function of the 'solidification ratio' defined by $\Sigma d/\pi D$ where D is the pitch diameter of the bundle. Their results have shown that the added mass increases sharply after the solidification ratio reaches the value of 0.4 to 0.5. Beyond this value, the volume enclosed rather than the volume displaced by the structure becomes important. This result is disputed by Chakrabarti (1978b) and the results of both series of tests are no more applicable to separated wavy or oscillatory flows about tube bundles than those predicted from the potential theory with or without diffraction effects (see also Dalton 1979 for the potential flow calculation of the inertia coefficients for riser type tube bundles). The interference effects between large cylinders in waves (Isaacson 1978) in the diffraction regime are discussed in Chapter 6.

Relatively few studies have been carried out with oscillating tube bundles (Tanida et al., 1973; Bushnell 1977; Sarpkaya 1979; Sarpkaya et al., 1980; Chakrabarti 1980b) where the separation of flow is important.

Tanida et al., (1973) reported an experimental study with two cylinders oscillating in still water. The results, however important for in-line oscillations, are of little relevance to the subject under consideration. Bushnell (1977) investigated the interference effects on the drag and transverse forces acting on a single member of two cylinder configurations through the use of a pulsating water tunnel. He did not evaluate the drag and inertia coefficients through the use of a suitable method, e.g., Fourier averaging. Instead, Bushnell picked out the maximum force values which occurred in each half-cycle and averaged them over ten consecutive values so as to obtain a mean maximum force for each flow direction. The drag and transverse force coefficients were obtained by normalizing the force by $0.5\,\rho LDU_m^2$. The results have shown that the presence of neighbouring cylinders significantly affect the forces on an individual cylinder of an array and the interference effect increases with increasing relative flow displacement. The maximum drag force on shielded cylinders was reduced relative to an exposed cylinder by up to 50 percent. Bushnell has suggested, on the basis of the foregoing, that a design using a high Reynolds number single-cylinder drag coefficient applied throughout the array would have an extra margin of safety against maximum drag loading due to interference effects. He found that the transverse force could be 3 to 4 times larger for interior array positions than that of a single cylinder. Consequently, the total force on each member of the array and the frequency and amplitude of the oscillation of this force become extremely important. In fact, such a cylinder array supported at regular intervals may exhibit very complex dynamic behavior. Some of the segments between supports may undergo in-line oscillation whereas the others may undergo violent transverse oscillations. Furthermore, the behavior of each tube in the bundle-segment may be significantly different.

Sarpkaya (1979) determined the drag and inertia coefficients for various multiple-tube riser configurations. Each configuration consisted of a number of outer pipes of diameter D_0 (uniformly spaced on a circle of diameter D_p) and one central pipe of diameter D_c. The arrays have been subjected to harmonically oscillating flow in a U-shaped water tunnel. The analysis of the in-line force was based on Morison's equation written as

$$F = -0.5\ \rho C_d U_m^2 L\ \Sigma D_i\ |\cos \omega t|\ \cos \omega t + 0.25\ \pi \rho L C_m\ \Sigma D_i^2\ U_m \omega \sin \omega t \quad (5.36)$$

Denoting $D_a = \Sigma D_i$ and $D_e = \Sigma D_i^2$ and inserting in Eq. (5.36), one has

$$F/(0.5\ \rho D_e L U_m^2) = -C_d(D_a/D_e)\ |\cos \omega t|\ \cos \omega t + \pi^2/(U_m T/D_e) C_m \sin \omega t \quad (5.37)$$

The Fourier averages of C_d and C_m are given by

$$C_d = -(3/4)(D_e/D_a) \int_0^{2\pi} [F_m \cos \omega t/(\rho L D_e U_m^2)]\ d\omega t \quad (5.38)$$

and

$$C_m = (2 U_m T/\pi^3 D_e) \int_0^{2\pi} [F_m \sin \omega t/(\rho L D_e U_m^2)]\ d\omega t \quad (5.39)$$

in which F_m represents the measured in-line force; T, the period of flow oscillation; and U_m, the maximum velocity in a cycle ($U = -U_m \cos \omega t$).

The experimental results for one particular array are shown in Fig. 5.11 as a function of the Keulegan-Carpenter number defined by $K = U_m T/D_e$. The drag coefficient decreases gradually with increasing K and reaches an almost constant value for K larger than about 90. The inertia coefficient increases with increasing K and reaches a terminal value of about 6. The average inertia coefficient defined by

$$C_m^* = (\Sigma C_{mi} D_i^2)/\Sigma D_i^2 \quad (5.40)$$

was found to be 2.17, for the particular array shown in Fig. 5.11, through the use of the potential flow theory. The comparison of this value with that obtained experimentally shows that as K approaches zero the experimental value of C_m approaches C_m^*. As K increases, $(C_m - C_m^*)$ increases, showing that some fluid mass is entrapped within the array and that neither the potential flow theory nor the diffraction analysis can adequately describe the behavior of the complex separated flow through the bundle.

The data for two different values of Re/K show that the force coefficients are independent of the Reynolds number within the range of Re and K values shown in Fig. 5.11.

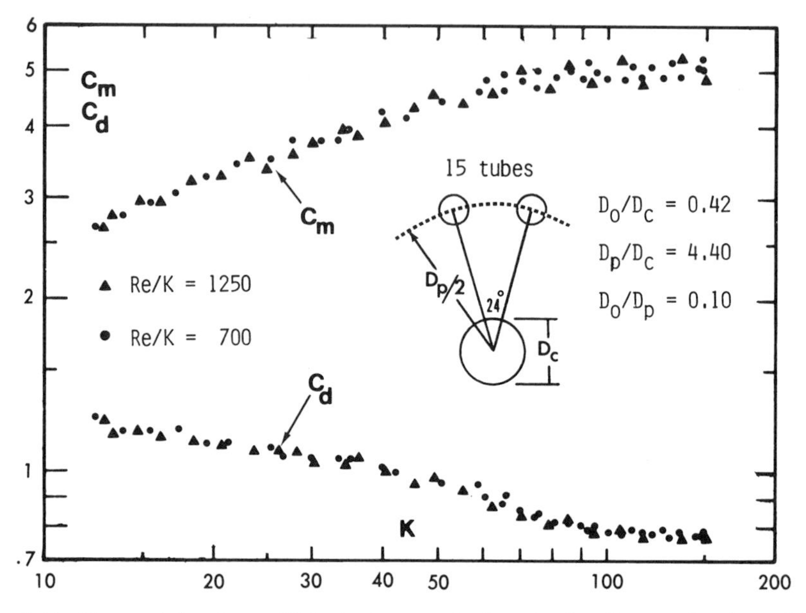

Fig. 5.11. Drag and inertia coefficients for a particular tube bundle as a function of K (Sarpkaya 1979).

The reason for the dependence of C_d and C_m on K is thought to be the dependence of the interaction of the wakes of the outer and inner pipes. The vortices in the wake of a given cylinder lose about 70 percent of their strength within 10 cylinder diameters. Thus, for small values of K, the vortices generated by a small tube at the center front of the bundle arrive at the central tube as weak vortices. Consequently, each tube behaves more or less as if it were independently subjected to a turbulent harmonic flow. As K increases, not only the turbulence level but also the interaction between the wakes of the various cylinders increases. There is a certain amplitude of oscillation beyond which neither the interaction of the wakes nor the increase of the turbulence level affects the overall force acting on the bundle. A comparison of the total drag force acting on the bundle with the sum of the drag forces acting on each cylinder in isolation in harmonic flow (at the corresponding K and Re values appropriate to each tube) shows that the former is about 10 percent smaller. For the hydrodynamic loading of risers see also Løken et al., (1979).

Sarpkaya et al., (1980) determined the lift, drag, and inertia coefficients for a pair of cylinders subjected to harmonic flow. The line joining the centers of the cylinders was rotated at suitable steps relative to the flow direction. The spacing between the cylinder centers was varied from 1.5 diameters to 3.5 diameters (see Figs. 5.12a through 5.12c). The results have been presented in terms of

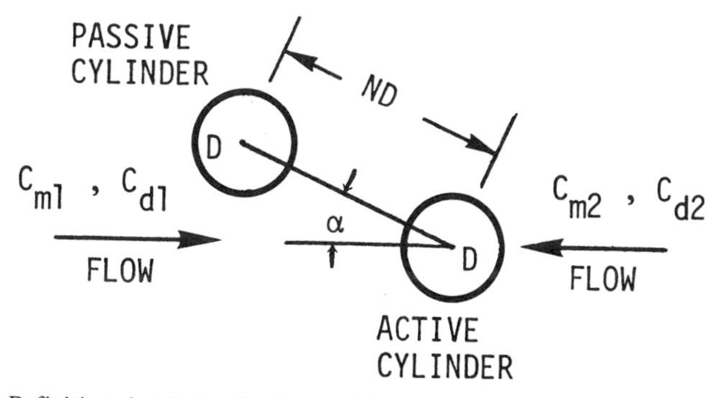

Fig. 5.12a. Definition sketch for the drag and inertia coefficients for two cylinders in harmonic flow (Sarpkaya et al. 1980).

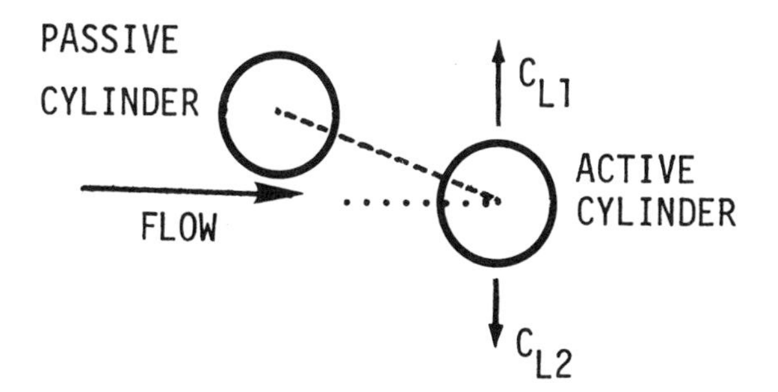

Fig. 5.12b. Definition sketch for the lift coefficients of a shielded cylinder in harmonic flow.

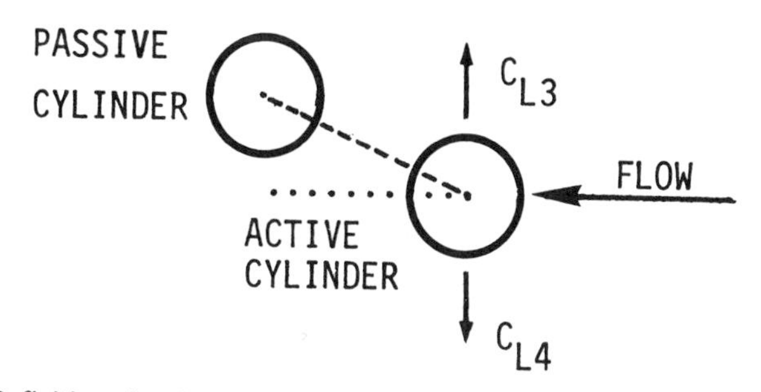

Fig. 5.12c. Definition sketch for the lift coefficients for the front cylinder in harmonic flow.

two drag, two inertia, and four lift coefficients because of the fact that the force acting on a given cylinder is not independent of the direction of the flow.

The results have shown that the drag and inertia coefficients for the tandem arrangement depend on both K and the relative spacing of the two cylinders. As the amplitude of flow oscillation becomes comparable or smaller than the gap between the two cylinders, the drag and inertia coefficients gradually approach those corresponding to an isolated cylinder. The drag and inertia coefficients for the side-by-side arrangement exhibit a similar behavior. For a relative spacing larger than about 2.5 diameters, the cylinders behave as if they were independent.

The quantification of the lift force for an arbitrary arrangement requires four different lift coefficients, depending on the direction of the flow and the direction of the force. The results have shown that the lift force acting on the active cylinder may triple at angles of inclination of about 30 degrees. This result is of great practical significance since the cylinders are subjected to rapidly varying unsteady hydrodynamic loads, both in magnitude and direction, especially for smaller spacings.

Chakrabarti (1979, 1980b) determined the drag and inertia coefficients for the side-by-side arrangement of two, three, and five tubes placed vertically in a wave tank. The spacing between the adjacent tubes were equal and varied from four diameters to a tenth of a diameter. The force coefficients were found to depend on K and the relative spacing. Any relationship with the Reynolds number could not be established primarily because of the small range of Reynolds number covered by the test.

The results have shown that C_d and C_m increase dramatically for a relative spacing smaller than about 1.3. A change in relative spacing from 1.3 to 1.1 for a five-tube arrangement can double the drag coefficient and increase the inertia coefficient by a factor of about three. For relative spacings larger than about 2, C_d and C_m do not significantly differ from those for a single cylinder at the corresponding K and Re values.

It follows from the foregoing that the question of flow interference in closely spaced tube bundles is a complex one and only reliable experiments can guide the designer. For cylinder arrays simulating platform legs, etc., three facts must be taken into consideration. Firstly, the length of the ideal, finite, vortex street on either side of the cylinder cannot exceed about half the wave height or the amplitude of flow oscillations. Thus, one may assume that the interference may be negligible if the spacing between the piles is greater than about half the wave height. Secondly, in a wake comprised of turbulent vortices, the strength of the vortices decays rapidly and in a distance of about ten diameters the vortex strength decreases to about 10 percent of that generated in the boundary layer over a shedding period. Thirdly, the velocity at a point is not constant and varies from zero to U_m during a half cycle. Thus, the use of half wave height as a criterion in determining the interference-free spacing may be too conservative. For example, for a 100-ft wave passing by a 2-ft pile, the H/2 criterion would give a distance of 50 ft whereas, the vortices must have almost completely dissipated within a distance of 20 to 30 ft. Thus, it is suggested that one should compare H/2 and 10-15D distances and use the smaller of the two as the appropriate distance in neglecting the interference effects. In rare cases, however, the flow about the upstream cylinder may be in the range of high subcritical Reynolds numbers. Then the wake turbulence from the upstream cylinder induces a supercritical or post-supercritical flow around the downstream one and hence the drag remains small even at spacings as large as H/2. Thus, the consequences of this type of flow interference are identical to those of an ambient turbulent flow about a single body at sufficiently high subcritical Reynolds numbers.

5.3.8 Pipe Lines and Wall-Proximity Effects

Pipe lines are used for several purposes: to convey oil and gas for power generation, sea-water for desalination, sewage for disposal at sea, and for the protection of communication cables.

There are numerous problems associated with the design and installation of pipelines in deep-water offshore. The subject is too broad to cover entirely, so the purpose of this section is to deal with the prediction of hydrodynamic forces. A comprehensive review of wave forces on submerged pipelines with nearly 100 references has been presented by Davis and Ciani (1976).

The proximity of a cylinder near a plane boundary gives rise to various hydrodynamic and environmental problems. A boundary layer is established near a bottom plane within which the horizontal water particle velocity varies from zero at the wall to the free-stream velocity at some elevation above the wall. The boundary layer has an important effect on cylinder lift and drag. However, in tests made near a bottom boundary, no one has measured the boundary-layer thickness and related it directly to the measured cylinder forces.

Secondly, the flow asymmetry created when a cylinder is placed near a plane generates a lift force normal to the flow which is velocity dependent and which can be of considerable magnitude for small gaps. This lift force which acts in a downward direction (in an unseparated flow), is, in general, different from that due to separation and vortex shedding. However, the two types of lift forces cannot be separated experimentally. Furthermore, there seems to be no need to do so. When the cylinder touches the boundary, a net force exists away from the wall. Thus, a cylinder that is not restrained on a bottom or freed by scour may become unstable, i.e., it can be alternately raised and lowered by the lift force due to flow asymmetry and vortex shedding about the cylinder.

In unsteady oscillatory flow, such as that induced by waves, considerable complexity is added to the flow phenomena around horizontal cylinders. The acceleration of flow gives rise to inertial forces. The force coefficients depend not only on the Reynolds number and the relative gap between the cylinder and the plane boundary but also on the Keulegan-Carpenter number and the time-dependent laminar or turbulent boundary-layer characteristics. Additional complexities arise from the variation in the amplitude and frequency of the waves, orientation of the pipe and of the waves and/or currents, temperature and marine life at the pipe location which determine the kind and type of soft and rigid excrescences on the pipe, the scour and deposition of sediment around the pipe, and the hydroelastic oscillations of the line. Experimental studies which examine these effects in detail do not exist. In fact, the existing data in steady and oscillatory flows about such pipes consist of horizontal and vertical force measurements which, with an applicable wave theory, are used to calculate the force coefficients. There is considerable disagreement as to what are the appropriate force coefficients for use in design primarily because of the multitude of variables affecting the flow.

In summary of the foregoing, no analysis exists for accurately describing the separated wave flow about cylinders with or without bottom effects. Furthermore, measurements of water particle kinematics; vortex shedding; drag, inertia, and lift forces; and the bottom boundary layer are rare in the range of parameters of practical importance.

The special case of unseparated flow about a single or N-number of cylinders situated near a plane boundary has been studied *through the use of the potential theory* by several investigators (see e.g., Yamamoto et al., 1974, and Yamamoto and Nath 1976b). The results show that C_m reaches its maximum value of $\pi^2/3 = 3.29$ for $e/D = 0$, where e is the gap between the cylinder and the wall (see also Kennered 1967). Secondly, both the lift and inertia coefficients nearly reach their ideal values in an infinite fluid ($C_m = 2$ and $C_L = 0$) for $e/D \simeq 1$. In other words, for e/D of about one, the wall-proximity effect *on unseparated flow* about a cylinder is negligible. Thirdly, there is a significant difference between the case of a cylinder touching the boundary and that of a cylinder slightly away from the boundary. When the cylinder touches the boundary a net force exists away from the wall. This is because the maximum velocity and the minimum pressure occur on the cylinder at a point farthest from the plane. However, if even a very small gap exists between the cylinder and the wall then a large net force exists toward the wall. This is merely a consequence of the inability of the potential theory to deal with the consequences of separation. In an ideal case, the smaller the gap, the larger the velocity through it. Thus, the net force is directed towards the wall. It is worth noting that the case of $e = 0$ cannot be obtained from the case of $e \neq 0$ by letting $e \rightarrow 0$. For $e = 0$, $C_L = \pi(\pi^2 + 3)/9 = 4.49$, (see von Müller 1929). For $e \rightarrow 0$, $C_L = -\infty$. In both cases, $C_m = \pi^2/3$.

The foregoing results are based on the potential flow theory and as such they are more appropriate to the determination of fluid loading due to vibrations generated by earthquakes. In real flows, the wall-proximity and separation effects can play significant roles in both the in-line and transverse forces.

Few studies have been carried out with steady flows over cylinders in the vicinity of a wall. Beattie et al., (1971) measured the pressure distribution over smooth and rough cylinders in the range $8 \times 10^4 < Re < 2 \times 10^6$. They used the fluid velocity at the top of the cylinder as the reference velocity in calculating the force coefficients. Their results showed as much as 35 percent scatter.

Wilson and Caldwell (1971) conducted experiments with two parallel cylinders at Reynolds numbers below 8×10^4. They have obtained $C_d = 1.1$ at $Re = 5.7 \times 10^4$ and $C_d = 1.6$ at $Re = 3.3 \times 10^4$.

A series of careful experiments were conducted by Göktun (1975) in a wind and also water tunnel in the range of $9 \times 10^4 < Re < 25 \times 10^4$ for relative gaps of $e/D = 0.1$, 0.125, 0.25, 0.5, 1.0, 1.5, 2, and 2.66. Göktun measured the surface pressures and calculated the lift and drag coefficients. The variation of the drag coefficient with the gap size exhibited an interesting and unexpected trend. The drag was a minimum when the cylinder was resting on the boundary and was a maximum for $e/D = 0.5$. Furthermore, the drag coefficient showed rapid variations as e/D was increased from 0.125 to 0.25.

Bearman and Zdravkovich (1978) investigated experimentally the flow around a circular cylinder placed at various heights above a plane boundary. Distributions of mean pressure around the cylinder and along the plate were measured at a Reynolds number, based on cylinder diameter, of 4.5×10^4. Spectral analysis of hot-wire signals demonstrated that regular vortex shedding was suppressed for all relative gaps of e/D less than about 0.3. For $e/D > 0.3$ the Strouhal number was found to be remarkably constant even though the drag, base pressure and separation position change. Bearman and Zdravkovich concluded that the only influence of the plate on vortex shedding is to make it more highly tuned process as the gap is reduced down to about $e/D = 0.3$.

The foregoing investigations, however significant, are not directly applicable to the determination of wave forces on pipelines. Numerous studies have been conducted with waves or oscillatory flows on the determination of force coefficients for cylinders near a plane boundary (see DSIR 1961, Johansson and Reinius 1963, Johnson 1970, Wilson and Caldwell 1970, Beattie et al., 1971, Priest 1971, Al-Kazily 1972, Grace 1973, Littlejohns 1974, A.S.E.C. 1975, Davis and Ciani 1976, Grace and Nicinski 1976, Herbich 1976, Sarpkaya 1976b, Yamamoto and Nath 1976a, Bowie 1977, Sarpkaya 1977, Grace 1978, Grace 1979, Grace et al., 1979, Graham 1979, Layton and Scott 1979, Wright and Yamamoto 1979, Graham and Machemehl 1980, Sarpkaya and Rajabi 1980).

Grace and Nicinski (1976) used as 17.5-ft long, 16-inch diameter steel pipe, supported 3 inches above a rigid block placed on the ocean floor at a depth 37 feet below the mean level. They have determined the lift, drag, and inertia coefficients through the use of the force and wave records. The data exhibited considerable scatter primarily due to the difficulty of quantifying the complex wave conditions.

Sarpkaya (1977) measured the in-line and transverse forces on cylinders placed at various distances from the bottom of the U-shaped water tunnel. The drag and inertia coefficients C_d and C_m for the in-line force have been calculated through the use of the Fourier analysis and the method of least squares. The lift coefficient C_L has been expressed in the usual manner by normalizing the amplitude of the first harmonic of the lift force by 0.5 ρLDU_m^2.

Sarpkaya has shown that C_d, C_m, and C_L are functions of K, Re, e/D, and the depth of penetration of the viscous wave or the boundary layer thickness, i.e.,

$$[C_d, C_m, C_L] = f_i(K, Re, e/D, \delta/D) \tag{5.41}$$

In these experiments δ/D was approximately 0.073 and the boundary layer effects have been ignored for $e/D > 0.1$. Flow visualization experiments with dye have shown that for very small values of e/D, ($e/D \simeq 0.1$), a jet-like flow exists between the cylinder and the plate. The flow separating from the top of the cylinder contains high frequency oscillations but does not curl up into vortices immediately behind the cylinder. In fact, the immediate wake is essentially free from large vortices. These observations have shown that a plate does not have to be placed near the rear stagnation point of a cylinder, in the form of a splitter plate, in order to interfere with the vortex shedding process. It is apparent that the gap blocks the flow and gives rise to earlier separation over the top of the cylinder. This in turn increases the transverse force as well as the in-line force. No attempt is made here to simplify an extremely complex separated flow situation. The emerging experimental facts are that the wall proximity ceases to affect the flow and the force-transfer coefficients for e/D values larger than about 0.5 and that for smaller values of e/D the frequency of oscillations in the two shear layers are decoupled. The interruption of the regular vortex shedding, earlier separation, and the effect of the vortices shed in the previous cycles lead as a whole to larger force-transfer coefficients for decreasing values of e/D. The increase of the inertia coefficient in separated flow under consideration follows the same trend as in inviscid, steady, unseparated flow as far as the effect of wall-proximity is concerned. For large values of K, the drag component of the in-line force dominates and the increase of the inertia coefficient is not of practical importance since the total inertial force is relatively insignificant.

Figures 5.13 through 5.16 show the drag and inertia coefficients for two representative values of K, namely, K = 40 and K = 100, for e/D = 0.1, 0.2, 0.5, 1.0, and the free cylinder. Evidently, the effect of the wall-proximity is to increase both the drag and inertia coefficients for e/D values smaller than about 0.5. For larger values of e/D, the effect of wall-proximity is practically negligible as evidenced by the comparison of the force coefficients with those obtained for cylinders at larger wall distances.

The transverse force in a given cycle is comprised of two parts. One part is toward the wall (expressed in terms of the lift coefficient C_{LT}) and the other

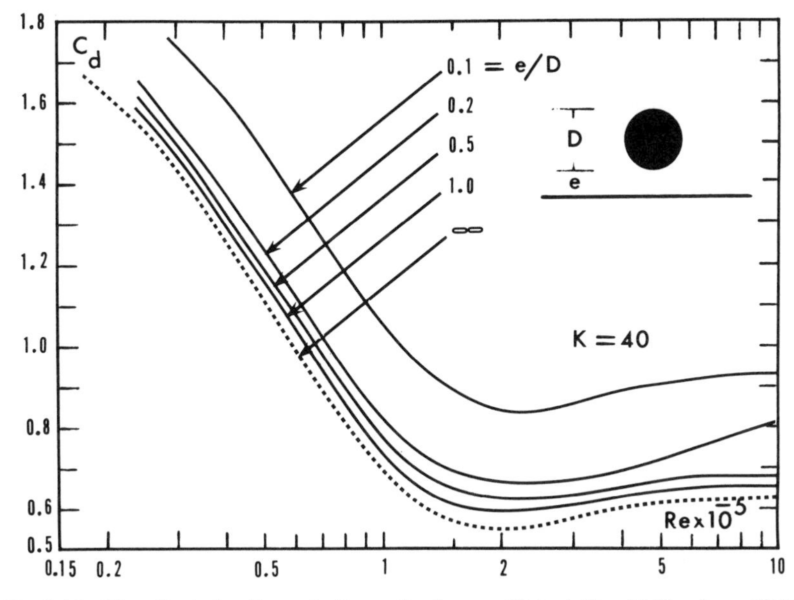

Fig. 5.13. The effect of wall-proximity on the drag coefficient, K = 40 (Sarpkaya 1977).

part is away from the wall (expressed in terms of a lift coefficient C_{LA}), (see Figs. 5.17 and 5.18). The former occurs during the periods of flow where the velocity and hence the separation effects are relatively small. Evidently, had there been no separation the lift force would have been always toward the wall.

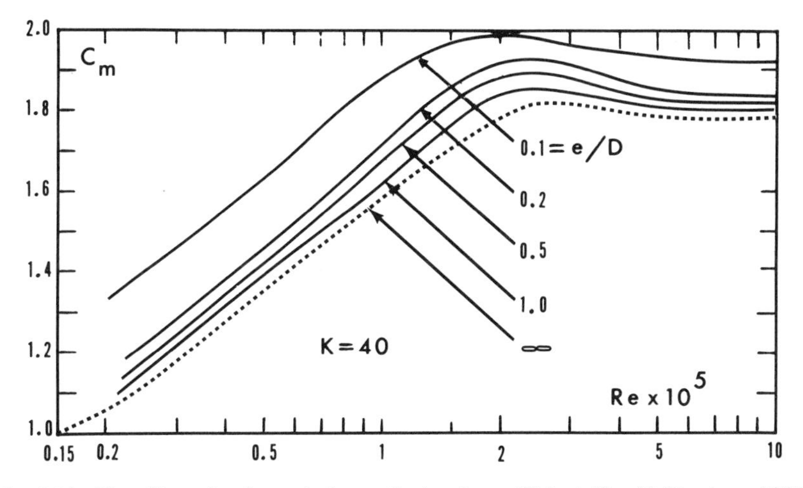

Fig. 5.14. The effect of wall proximity on the inertia coefficient, K = 40 (Sarpkaya 1977).

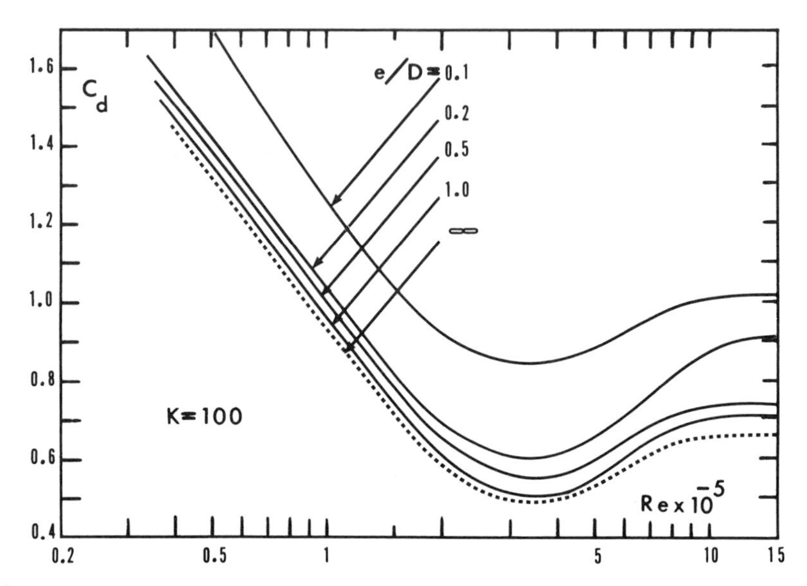

Fig. 5.15. The effect of wall proximity on the drag coefficient, K = 100 (Sarpkaya 1977).

Nevertheless, the separation effects and the effect of the vortices shed in the remainder of the cycle are not entirely eliminated even during the periods of low velocity and the transverse force toward the wall is relatively small and fairly independent of e/D.

The transverse force away from the wall reaches its maximum during the

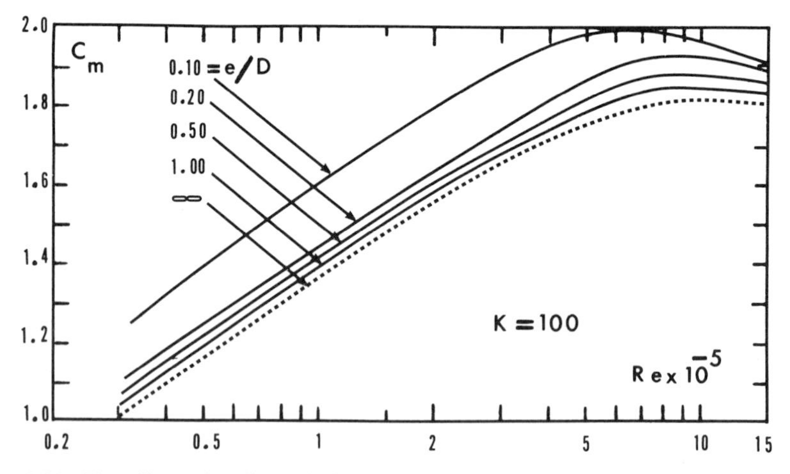

Fig. 5.16. The effect of wall proximity on the inertia coefficient of a circular cylinder, K = 100 (Sarpkaya 1977).

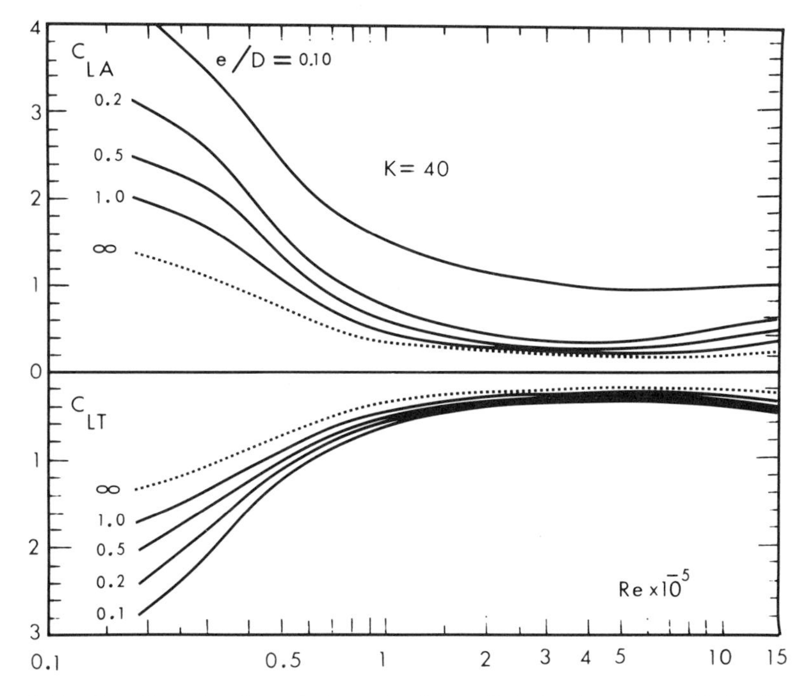

Fig. 5.17. The effect of wall proximity on the lift coefficients of a circular cylinder, K = 40 (Sarpkaya 1977).

periods of large velocities and separation. The position of the separation point, high frequency oscillations in the upper shear layer, and the subsequent formation of the vortices are strongly influenced by the relative spacing of the cylinder for e/D values smaller than about 0.5. Hence, C_{LA} varies considerably with e/D in the range of $e/D < 0.5$ and is significantly larger than the free-cylinder lift coefficients. It is obvious that the shear layer emanating from the lower side of the cylinder is not as free as that emanating from the top side of the cylinder because of the boundedness of the wall jet between the cylinder and the plane wall. Thus, the lift force toward the wall is not as much affected by the variations in e/D. It is also evident that the transverse force toward the wall is about 90 degrees out of phase with that away from the wall.

For cylinders near a plane wall ($e \neq 0$) it may be concluded that:

1. The drag and inertia coefficients for the in-line force acting on the cylinder are increased by the presence of the wall. This increase is most evident in the range of e/D values smaller than about 0.5. Both coefficients depend on the Reynolds number, Keulegan-Carpenter number, and e/D. The effect of the boundary layer or the penetration depth of the viscous wave is small provided that the boundary layer remains laminar. For turbulent oscillatory boundary

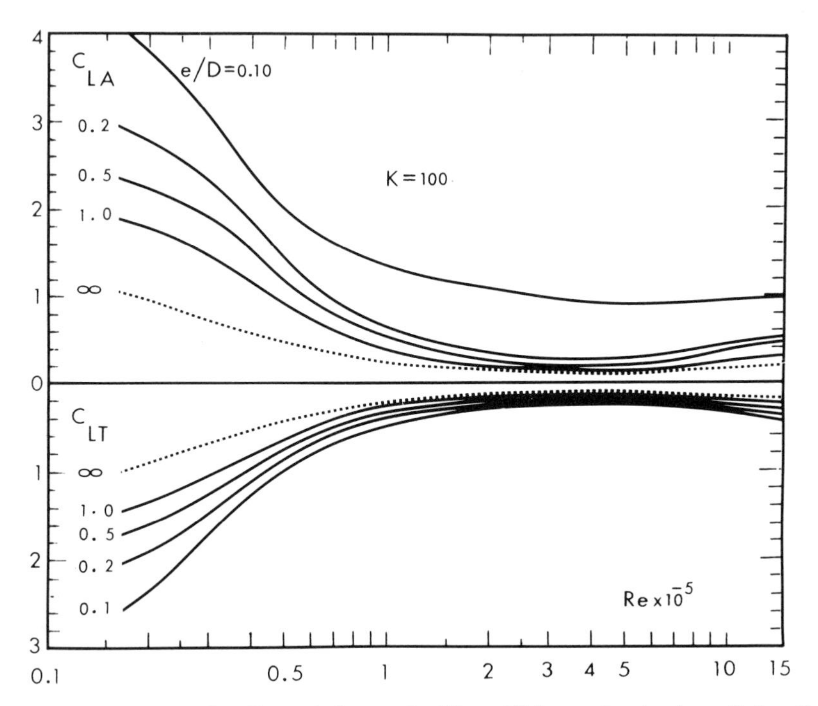

Fig. 5.18. The effect of wall proximity on the lift coefficients of a circular cylinder, K = 100 (Sarpkaya 1977).

layers the characteristics of the wall jet and separation over the cylinder may be significantly affected.

2. The proximity of the wall helps to decouple the frequency of oscillations in the top and bottom shear layers. This decoupling effect prevents the occurrence of regular vortex shedding for small values of e/D.

3. The transverse force toward the wall is relatively small and fairly independent of e/D. It occurs during the periods of low velocity or high acceleration. The transverse force away from the wall is quite large and dependent on e/D particularly for e/D smaller than about 0.5. The two forces are about 90 degrees out of phase.

4. The use of Morison's equation to decompose the in-line force into two components is quite sound. The lumping of the entire in-line force into a single coefficient is not justified and obscures the mechanics of the flow.

As noted earlier, the case of e = 0 (no gap) differs significantly from that with a gap as small as e/D = 0.1. Sarpkaya and Rajabi (1980) conducted a series of experiments with smooth and rough cylinders and determined the drag, inertia, and the lift coefficients. The small gap between the cylinder and the plane boundary (bottom of the U-shaped tunnel) was sealed with a very thin plastic

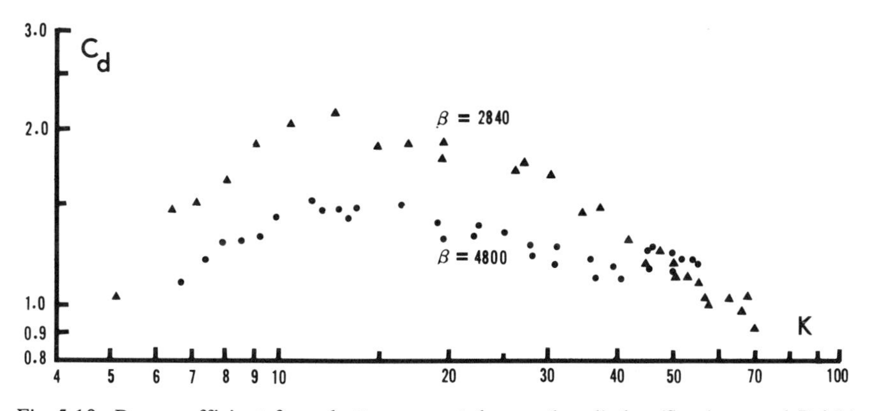

Fig. 5.19. Drag coefficient for a bottom-mounted smooth cylinder (Sarpkaya and Rajabi 1980).

wrapping sheet, attached both to the plane boundary and the bottom of the cylinder. Furthermore, the gaps between the tunnel walls and the cylinder ends were sealed with foamy material. The drag and inertia coefficients for smooth cylinders are shown in Figs. 5.19 and 5.20 for two values of the frequency parameter. Clearly, C_d can reach very high values relative to the case of $e/D \neq 0$ and is a function of the Reynolds number for a given K. The inertia coefficient does not appear to depend on Re and increases with increasing K. For very small values of K where the separation effects are negligible, C_m approaches its theoretical potential flow value of 3.29. No generalizations can be made regarding the relative theoretical and experimental values of C_m. For a cylinder sufficiently away from a boundary, C_m is always smaller than its theoretical value of 2. In the present case ($e/D = 0$), the theoretical value of C_m is smaller than the experimental value, at least within the range of Reynolds numbers encountered.

The drag and inertia coefficients for the rough cylinders ($k/D = 1/100$) are

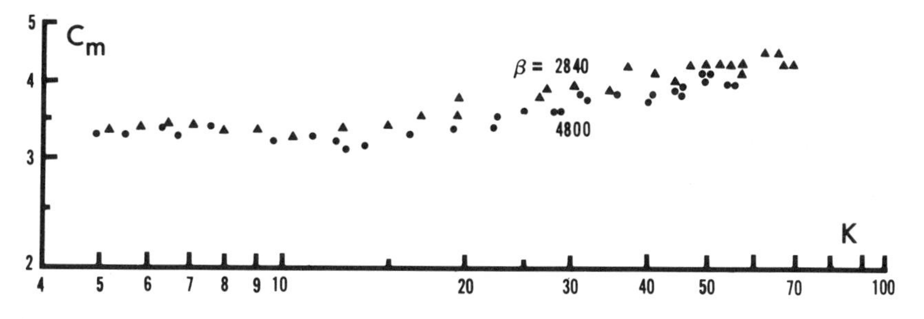

Fig. 5.20. Inertia coefficient for a bottom-mounted smooth cylinder (Sarpkaya and Rajabi 1980).

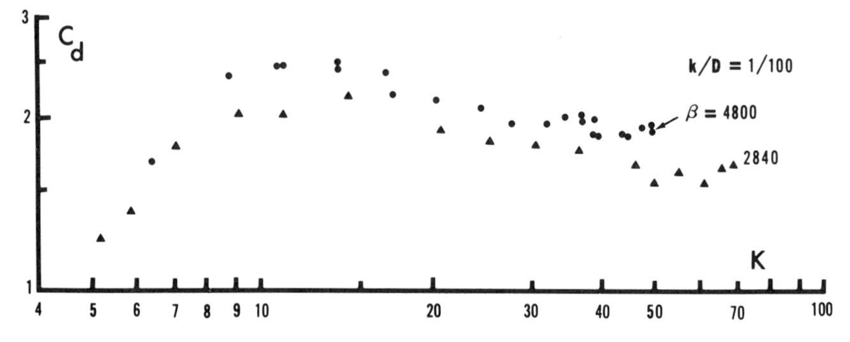

Fig. 5.21. Drag coefficient for a bottom-mounted rough cylinder in harmonic flow (Sarpkaya and Rajabi 1980).

shown in Figs. 5.21 and 5.22. The effect of roughness on C_d is quite significant. Once again, the inertia coefficient is very little affected by the Reynolds number or by roughness.

The maximum and minimum values of the lift coefficient (the lift force is always away from the wall) are shown in Figs. 5.23 and 5.24 for the smooth cylinders. The lift coefficient reaches very high values at relatively small K values. The potential flow value of C_L for $e/D = 0$ is given by von Muller (1929) as $C_L = \pi(\pi^2 + 3)/9 = 4.493$. Clearly, the experimental values for the smooth cylinders are relatively larger than the theoretical value, at about $K = 7$. The effect of separation at this value of K is such as to increase the lift. It is expected that the lift coefficient will reduce to about 4.5 as K approaches zero.

The maximum and minimum values of C_L for rough cylinders are shown in Figs. 5.25 and 5.26. The maximum value of C_L at about $K = 7$ is slightly smaller than that corresponding to the smooth cylinder. Otherwise, there is very little difference between the lift coefficients for the smooth and rough cylinders. The ratio of the lift-force frequency to the flow-oscillation frequency was found to remain constant at a value of 2 for both the smooth and rough cylinders. This shows that it is the separation of flow over the cylinder at each half cycle of

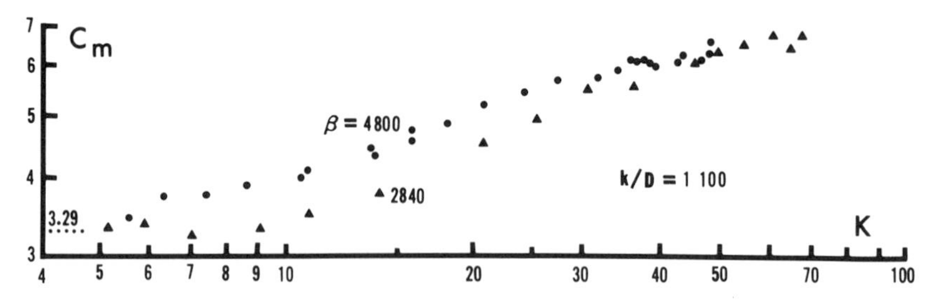

Fig. 5.22. Inertia coefficient for a bottom-mounted rough cylinder in harmonic flow (Sarpkaya and Rajabi 1980).

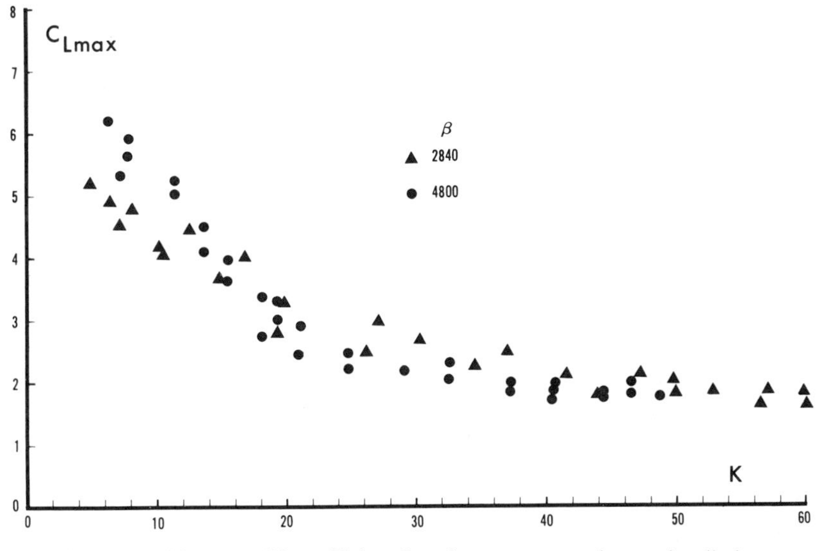

Fig. 5.23. Maximum lift coefficient for a bottom-mounted smooth cylinder.

flow and not the subsequent shedding of vortices that determines the fluctuations of the lift force for the type of flow-cylinder combination considered herein.

It appears that separation over a bottom-mounted cylinder occurs at smaller K values than that for a cylinder away from the wall. In general the effects of separation even at small K values are quite profound and the potential flow values of the inertia coefficient and the lift coefficient tend to underestimate the forces acting on a bottom-mounted cylinder. Evidently, the potential flow

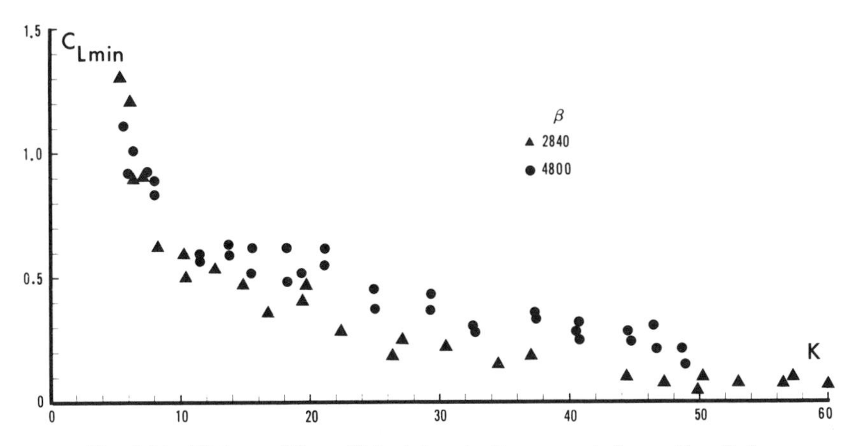

Fig. 5.24. Minimum lift coefficient for a bottom-mounted smooth cylinder.

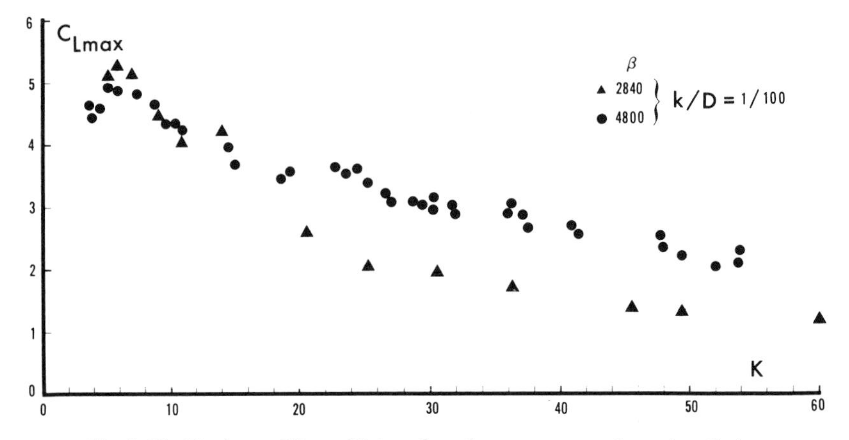

Fig. 5.25. Maximum lift coefficient for a bottom-mounted rough cylinder.

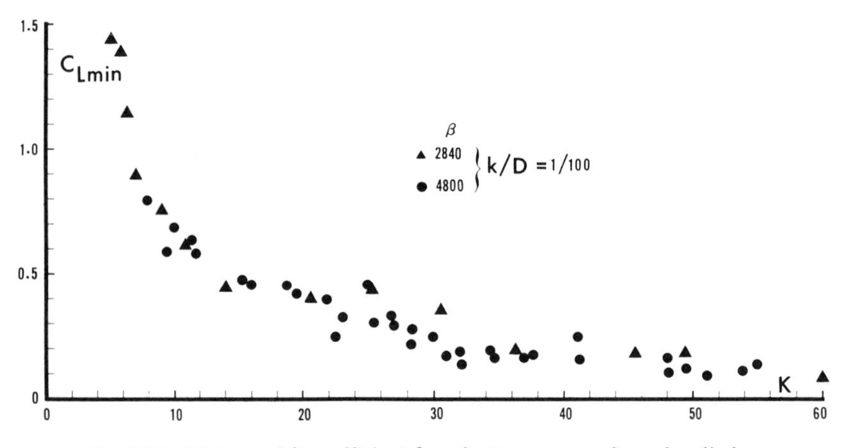

Fig. 5.26. Minimum lift coefficient for a bottom-mounted rough cylinder.

theory gives no clues about the drag force and the calculations must be based on the results obtained experimentally.

5.4 MARINE RISERS

5.4.1 Introduction

A riser is a conductor pipe which connects the wellhead at the seabed to a fixed platform, or a floating platform, or a vssel. It may be categorized as either a production riser or a drilling riser (see Fig. 5.27). The flexible pipe extends from the ball joint at the top of the blow-out-preventer (BOP stack) to the slip joint barrel (telescopic joint) beneath the vessel. The slip joint allows the riser to change its length as the vessel heaves and moves laterally. The additional compo-

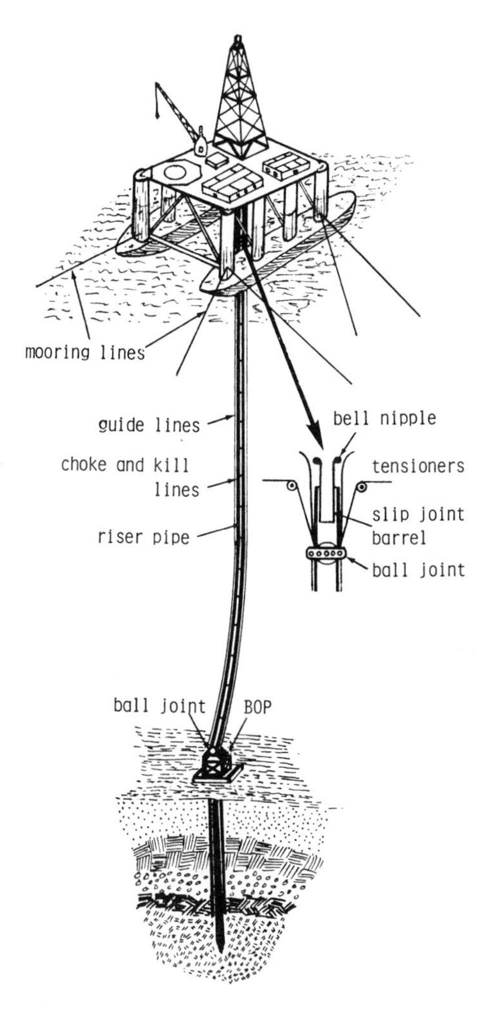

Fig. 5.27. Schematic of a drilling riser.

nents of the riser consist of guide lines, choke-and-kill lines, buoyancy devices, and the tensioner system. The maximum angle, from the vertical, at the lower ball joint is kept below about 8 degrees. The top part of the riser pipe may or may not be equipped with another ball joint, situated just below the slip-joint barrel. The choke and kill lines allow the circulation of drilling mud if and when the BOP is closed.

The riser must sustain production or drilling operations (carry drilling mud, drilling tools, etc.) while undergoing motions (mainly in a plane normal to the riser) due to vessel motion and the waves and currents. The analysis, design, and operation of such a pipe turn out to be rather difficult primarily because of the uncertainties associated with the specification of the environmental loads.

Clearly, the wind, wave, and current conditions must be translated into the prediction of the vessel motions. This can be done, with some approximations, for moored rigs where the motion of the rig is not interfered with by an external power source (thrusters). For dynamically-positioned drill rigs, the problem is more complex since the position of the rig depends on the response of the rig to the waves, currents, and the stationkeeping thrusters. Some rigs have both a dynamic positioning system and a mooring system. A dynamic positioning system can maintain the vessel excursions within a reasonably specified domain relative to the wellhead through the use of an acoustic system (hydrophones, responders, etc.), a computer, and thrusters. Rigs which have both the mooring system and the dynamic positioner may be more energy efficient (less fuel) due to the infrequent use, as opposed to the continuous use, of the thrusters.

It is extremely important that the riser pipe be kept under tension (as constant as possible) at all times in order to prevent damage to the ball joints, connectors, and the pipe itself. The desired tension may be provided by the top tensioners on board the vessel (a hydraulic system, a compressor to pressurize the hydraulic pistons, and tension cables which are attached to the riser just below the upper ball joint). In this case all tensioning loads are borne by the drilling vessel. Additional disadvantages of the use of the top tensioning system only are that the upper parts of the riser are subjected to disproportionately high loads on connectors, heave travel is limited, the wear is excessive, and there is increased danger of loss of tension. It is because of these reasons that additional buoyancy is provided along the riser either by syntactic-foam straps (Watkins and Howard 1976) or open bottom steel cans connected to the riser at suitable points. Each can is provided with compressed air (adjustable buoyancy) by a compressor on board the ship. These devices help to reduce the top tension requirements and tensile stresses. However, they increase the projected area of the riser and hence the fluid forces, and the cost of transportation, operation, and maintenance of the riser system. Air cans are used in smaller depths and provide a relatively cheaper, more rugged, and controllable means for additional buoyancy. Syntactic foam straps do not provide a controllable buoyancy but have some advantages over the cans (no on-board compressor for this purpose, no air lines, etc.).

Most of the foregoing falls in the general category of highly specialized equipment manufactured by a handful of companies from which additional information may be obtained.

5.4.2 Methods of Riser Analysis

The riser analysis has attracted considerable attention partly due to the fact that the riser is and will continue to be a vital link between the floating platform and the subsea bore hole and partly because the analysis poses numerous challenging problems. A number of computer codes has been developed in recent years. In spite of this, however, the issue is far from resolved and each design requires specific analysis with due consideration of the environment, characteristics of the platform or rig, purpose of the riser, water depth, manufacturer of the specific components, etc.

The analysis requires the consideration of wave and current forces due to most severe as well as nominal sea states, water depth, the rig motion, preliminary dimensions of the riser and its components, equation of motion, and suitably defined boundary conditions. The waves and currents may give rise to in-line as well as transverse forces and motions which are not necessarily planar due to the omnidirectionality of the surface and internal waves and the currents throughout the length of the riser.

The analysis may be static or dynamic. The static analysis is concerned with the maximum riser response in a vertical plane and does not take into consideration the time-varying effects of waves, vessel motion and the inertia of the system. The vessel motion (one of the most dominant factors as far as the bending stresses are concerned) and the waves and currents are advanced at suitable time intervals and the maximum values of the critical parameters (position, stress, tension, etc.) are calculated over a suitable period (e.g., wave period).

There are various static analysis methods: finite difference formulations (Bathe et al. 1974, NESC 1966), finite element formulations (Gosse 1969), direct integration using a fourth-order Runge-Kutta method (Burke 1973), and assumed deflection shapes of an elastic catenary (Jones 1975) or power series (Fischer and Ludwig 1966).

The dynamic analysis does take into consideration the relative fluid velocities and the fluid and riser accelerations and yields a time history of the response. The analysis may be restricted to a planar motion (i.e., the riser is assumed to displace in the lateral direction only in a vertical plane held stationary in time) or biplanar motion (i.e., the riser is assumed to displace in an x-y plane normal to the riser). The motion in the z-direction (along the vertical) is not of importance. One must note that the riser may move in various directions in the x-y plane at various elevations due to the omnidirectionality of the waves and currents. Such an analysis is extremely complex due to the fact that it requires information about the variation of the kinematics of the waves and currents as a function of x, y, z, and t, in addition to the rig motion.

The analyses cited above deal primarily with the determination of the riser tension, bending stress, ball joint angle, and the response mode.

5.4.3 Equations of Motion

The assumption of small angle, large deflection theory for a tubular beam column under varying tension and external loads leads to (see e.g., NESC 1966, Morgan

1977)

$$M(z)\frac{\partial^2 y}{\partial t^2} + \frac{\partial^2}{\partial z^2}\left[EI(z)\frac{\partial^2 y}{\partial z^2}\right] - \frac{\partial}{\partial z}\left[T(z,t)\frac{\partial y}{\partial z}\right] = F(x,y,z,t) \qquad (5.42)$$

where $M(z)$ represents the mass of the riser per unit length (i.e., the mass of the riser pipe, flow lines, connectors, buoyancy material, and mud in the pipe). *M(z) does not include the added mass.* $E(z)$ and $I(z)$ represent, respectively, the modulus of elasticity and the moment of inertia of the riser pipe; $T(z, t)$, the effective axial tension in the riser; and $F(x, y, z, t)$, the fluid force, expressed in terms of the modified Morison equation.

Equation (5.42) may be expressed as (see e.g., Young et al., 1977)

$$\begin{aligned} &M(z)\ddot{y} + [EI(z)y'']'' - [T'(z,t) + \gamma_i A_i - \gamma_o A_o]\,y' \\ &- [T(z,t) + p_o A_o - p_i A_i]\,y'' = 0.5\rho_o D_o C_d(z)\,(u_{wy} + V_{cy} - \dot{y})|u_{wy} + V_{cy} - \dot{y}| \\ &\qquad + 0.25\rho_o \pi D_o^2 [C_m(z)\,(\dot{u}_{wy} + \dot{V}_{cy}) - (C_m - 1)\ddot{y}] \qquad (5.43) \end{aligned}$$

in which dots represent derivative with respect to time and primes, with respect to z. γ_i and γ_o represent the specific weight of the inner fluid (mud) and the outer fluid (water); p_o and p_i, the outer and inner pressures on the riser at elevation z (measured upward from the BOP); A_o and A_i, the areas of the outer and inner riser cross sections; ρ_o, density of water; D_o, outer diameter of the riser (including the buoyancy foam. It may also represent an effective diameter representing the pipe and the choke and kill lines or other secondary conductors); $C_d(z)$ and $C_m(z)$, the drag and inertia coefficients which depend on the local Keulegan-Carpenter number, Reynolds number, and the relative roughness (for a first order of approximation); u_{wy}, the wave-induced horizontal velocity at z (i.e., at the point defined by y and z); and V_{cy}, the current velocity. It is assumed that the wave propagates in the y-direction only. Note that the outer pressure p_o increases tension and the inner pressure p_i has the opposite effect.

For a biplanar motion, one also has

$$\begin{aligned} &M(z)\ddot{x} + [EI(z)x'']'' - [T' + \gamma_i A_i - \gamma_o A_o]\,x' - [T + p_o A_o - p_i A_i]\,x'' = \\ &\quad 0.5\rho_o D_o C_d(z)\,(V_{cx} - \dot{x})|V_{cx} - \dot{x}| + 0.25\rho_o \pi D_o^2 [C_m(z)\dot{V}_{cx} - (C_m - 1)\ddot{x}] \end{aligned} \qquad (5.44)$$

In Eqs. (5.43) and 5.44), the variation of the current velocity with time may be ignored since such information is not likely to be obtained with any reasonable degree of accuracy and since it may be negligible during a few wave cycles.

Equation (5.43), and in a similar fashion Eq. (5.44), may be further expanded as

$$[M(z) + (C_m - 1)\frac{\pi}{4}\rho_o D_o^2]\ddot{y} + [EI(z)y'']'' - [T' + 0.25\pi(\gamma_i D_i^2 - \gamma_o D_o^2)]y'$$

$$- \{T(z,t) + 0.25\pi[\gamma_o(h - z)D_o^2 - \gamma_i(h - z)D_i^2]\}y'' = 0.5\rho_o D_o C_d(z)$$

$$\cdot (u_{wy} + V_{cy} - \dot{y})|u_{wy} + V_{cy} - \dot{y}| + 0.25\rho_o \pi D_o^2 C_m(z)(\dot{u}_{wy} + \dot{V}_{cy}) \quad (5.45)$$

In Eq. (5.45) the terms in the brackets containing T' and T may be modified to include the effect of the applied internal riser pressure (diverter pressure), the weight of connectors and other lines, and the weight of the buoyancy material (see Young et al. 1977).

Equation (5.45) needs four boundary conditions. Two of these are provided by assuming that the ball joint at $x = 0$ is fixed and has a rotational stiffness K_{bj}. Thus, one has

$$y(0,t) = 0$$

$$y''(0,t) = K_{bj}y'(0,t) \text{ at all times} \quad (5.46)$$

The two additional boundary conditions are specified at $x = h$, depending on the particular problems: The top of the riser may be assumed to be attached to an elastic support with both lateral and rotational stiffness. The top end may be assumed to be fixed laterally but allowed to rotate with a specified rotational stiffness. Finally, the top tension may be specified assuming that the deflection and rotation of the top end are determined from the vessel motion through a separate analysis. Other boundary conditions in lieu of the foregoing may be specified depending on the type of the riser and the rig. In any case, these boundary conditions are built into the numerical solution process with particular starting values.

5.4.4 Methods of Solution for the Dynamic Analysis

Three basic methods of solution are used for the dynamic response analysis: deterministic time-history analysis, a steady-state or frequency-domain analysis, and a nondeterministic random vibration analysis.

The time-domain solutions include the finite difference (Bennet and Metcalf 1977, Maison and Lea 1977, NESC 1966, Sexton and Agbezuge 1976), and finite element method (Gardner and Kotch 1976, Gosse 1969). The finite difference method converts the equation of motion into a set of nonlinear ordinary differential equations. The time-domain solution is quite flexible and can accommodate variations in riser dimension, buoyancy, boundary conditions, and external time-varying loads and/or motions.

The frequency domain solution is obtained by assuming steady-state wave loadings and vessel motions and reducing the equations of motion to an ordinary differential equation and numerically integrating it [Burke 1973, Kirk et al. 1979 (which includes the normal mode solution of the variable tension), Spanos and Chen (1980) (with the discrete multi-degree-of-freedom model)].

The nonlinear drag is often linearized (see e.g., Krolikowski and Gay 1980) through the use of an equivalent energy principle (see Chapter 8) even though it is not necessary to do so. The frequency-domain model of Young et al. (1977) and Kirk et al. (1979) retains the nonlinear form of the drag force. However, Young et al. assumed linear superposition for random sea states. This may not be permissible in view of the nonlinear nature of the fluid forces, as pointed out by Kirk et al.

The advantages of the frequency-domain analysis are that one can directly apply a frequency-domain definition of the environment or ship motion to the riser and generate, within a relatively short computer run, a response spectrum suitable for subsequent fatigue life estimation. The disadvantages include the unknown effect of drag linearization (if and when it is done) and the sensitivity of the method to minor changes in wave spectra.

The nondeterministic random vibration analysis of Tucker and Murtha (1973) inputs the random wave spectrum to the riser model and obtains the riser response in the form of a spectrum. The difficulties associated with the inclusion of the vessel motion and the nonlinearized form of the drag force limit the application of this method.

Young et al. (1977) assuming small amplitude wave theory, approximated u_{wy} by

$$u_{wy} = (\pi H/T)\, e^{i\omega t}\, e^{-(h-z)\omega^2/g} = U_o(z)\, e^{i\omega t} \tag{5.47}$$

in which H and T represent, respectively, the wave height and period. Assuming

$$y = y_o(z)\, e^{i\omega t} \tag{5.48}$$

Young et al. reduced Eq. (5.45) to

$$\begin{aligned} 0.5\rho_o D_o C_d\, [U_o(z) - i\omega y_o(z)]\, |U_o(z) - i\omega y_o(z)| + 0.25\rho_o \pi D_o^2 i\omega C_m U_o(z) = \\ -\omega^2 [M(z) + (C_m - 1)\frac{\pi}{4}\rho_o D_o^2]\, y_o(z) + [EI(z) y_o''(z)]'' \\ - [T'(z,t) + \gamma_i A_i - \gamma_o A_o]\, y_o'(z) - [T(z,t) + p_o A_o - p_i A_i]\, y_o''(z) \end{aligned} \tag{5.49}$$

Equation (5.49) has been solved iteratively through the use of a finite difference representation of the complex derivatives. The loads were due to lateral forces from regular as well as random waves and currents. Young et al.'s computer program allows for a choice of either the displacement or the force boundary conditions at either end of the riser (ball joints, built-in ends, fully-suspended riser, etc.). Their results have shown that the vessel motion is the dominant source of stress for either long or short risers. However, the fundamental frequencies of short risers may coincide with the shorter wave periods and may give rise to resonance and significant stresses. The importance of the vessel motion on the bending stress has been shown by others also (e.g., Kirk et al. 1979).

Spanos and Chen (1980) used their linearized multi-degree-of-freedom model to conduct a variety of parameter studies regarding the magnitude of the maximum bending stress and bottom angle. Their results, for the specific cases considered by them, have shown that the maximum bending stress occurs at locations close to either end of the riser [similar results have been reported by Burke (1974) and Gardner and Kotch (1976)] ; and that the maximum bending stress and the maximum bottom angle depend nonlinearly on the normalized tension. They have also shown that the bending stiffness cannot be neglected in the analysis of shorter risers.

As noted by Morgan and Peret (1976), transverse periodic forces may also arise due to vortex shedding. The vortex-induced oscillations can give rise to resonance and lead to considerable increases in both the in-line and transverse forces. As noted in detail in Chapter 8, the vortex-induced oscillations are of great practical importance. No riser analysis has so far considered the complex interaction between the in-line motions of the riser and the transverse oscillations. Young et al. (1977) stated that the vortex shedding due to current is included in their analysis but gave no details of how the effects of the transverse motion are coupled with the in-line motion. It is presumed that the projected area (D_o) is suitably increased (see Chapter 8) to account for the transverse oscillations.

The results obtained with the finite element method for a 1200-ft riser are shown in Fig. 5.28 through 5.31. Wave heights of $H = 60$ ft and $H = 35$ ft were considered. The periods were assumed to be 15.5 seconds. In addition, a steady current with a linearly varying velocity (1.1 ft/sec at the bottom and 2.1 ft/sec at the surface) was assumed. The mean vessel offset was taken 50 ft. These figures give a general idea about the variations of the riser displacement, angular deflection, bending stress, and the force per unit length. It must be emphasized that these figures are only for the purpose of illustration and should not be used as a guide for the design of risers in similar depths and environmental states.

5.4.5 Sources of Uncertainty and Recommendations

The advantages and disadvantages of the various methods of solution must be judged within the limits of uncertainty of the numerous assumptions and input parameters. The uncertainty associated with the solutions cannot be easily assessed or minimized for a number of reasons.

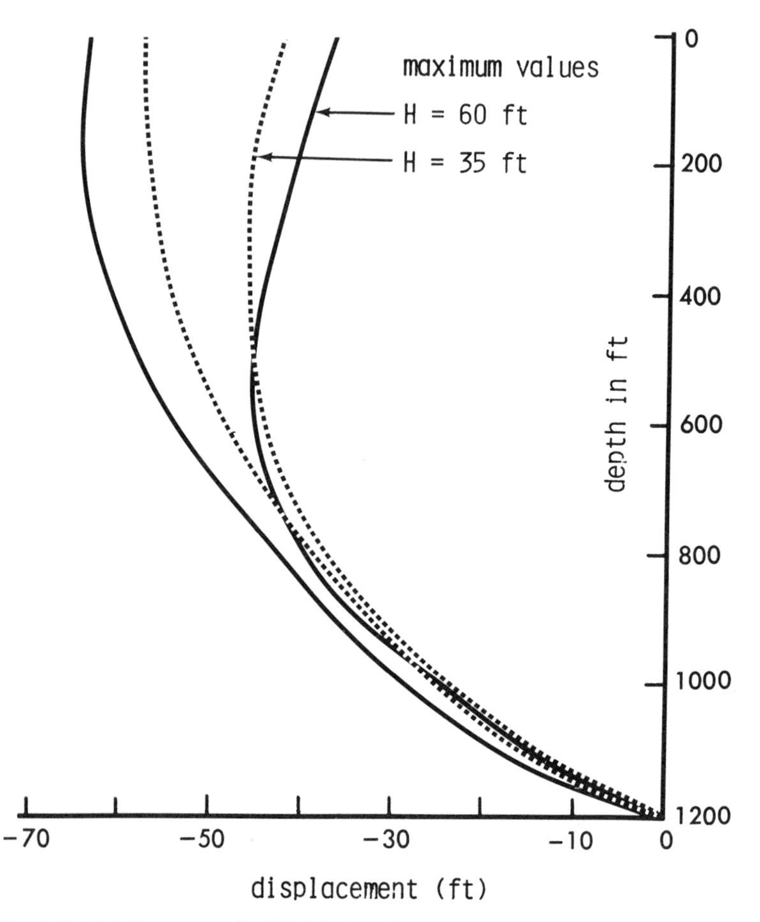

Fig. 5.28. Displacement of a 1200-ft riser due to waves, currents, and rig motion.

(a) Three-dimensional nature of the waves and currents and the riser motion may not necessarily be represented by a planar analysis;

(b) The drag and inertia coefficients for the case under consideration are unknown and the linear superposition of the wave, current, and riser velocities in the 'velocity-square-dependent' drag part of the Morison equation is yet an unjustified assumption. The validity of the use of the drag and inertia coefficients obtained under stationary conditions for wave-current-riser motion combination has not yet been demonstrated. It is because of this reason that the relatively poor definition of the fluid loading constitutes the weakest link in a series of steps toward a reliable riser-response prediction;

(c) The vortex motion along the riser is not perfectly correlated and there are no theoretical means to establish the correct correlation length as a function of

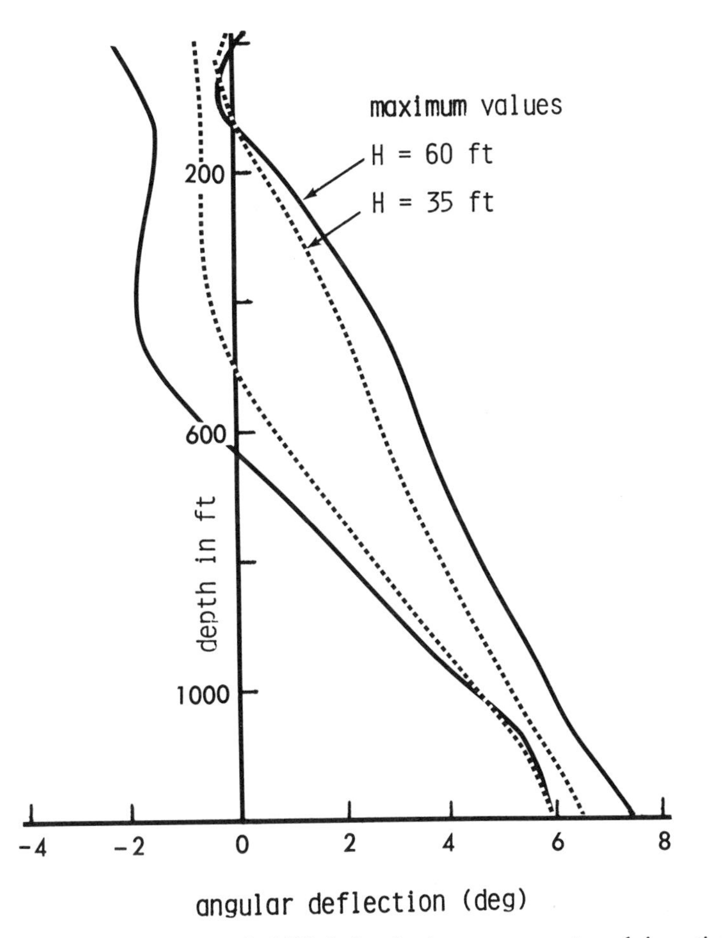

Fig. 5.29. Angular deflection of a 1200-ft riser due to waves, currents, and rig motion.

the wave, current, and riser motion, and the Reynolds and Keulegan-Carpenter numbers. This fact does, in turn, further decrease the reliability of the use of the drag and inertia coefficients obtained under controlled conditions with stationary cylinders;

(d) The effect of structural elements such as the buoyancy devices and the choke and kill lines is unknown and cannot be established theoretically;

(e) The mitigating effects of the ocean environment on the riser loading and response have not been incorporated into the equation of motion (only indirectly through the use of spectra and a number of hidden and intentional safety factors). To illustrate the concept of reduced wave force due to a directional sea, consider two waves of equal amplitude propagating at right angles to each other.

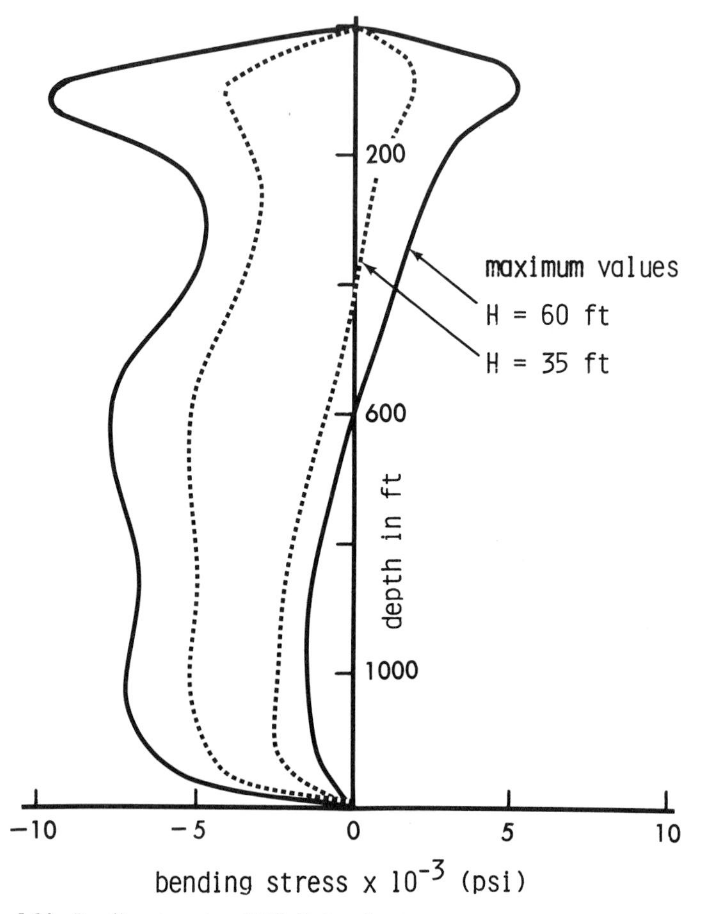

Fig. 5.30. Bending stress in a 1200-ft riser due to waves, currents, and rig motion.

At a location where the crests are in phase the water surface displacement add linearly (for the linear wave theory). On the other hand, the horizontal components of the velocities, being vector quantities, add vectorially. The resultant velocity would, therefore, be 1.414 times larger than that caused by each of the individual waves rather than twice if the waves were collinear.* If the force system is inertia dominant, waves propagating at 90 degrees relative to each other will cause forces which are 0.707 times as large as for the same wave height for collinear waves. If the force is drag dominant, the said forces will be 0.5 times as large as for the same wave height for collinear waves (see Section 7.6.4).

*This example was brought to our attention by Prof. Robert Dean.

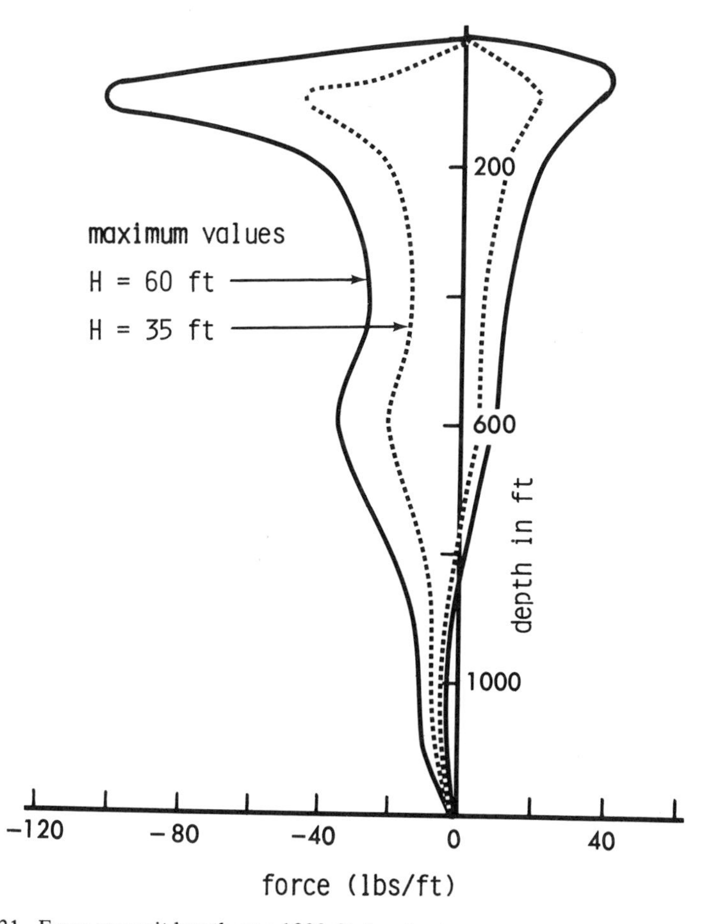

Fig. 5.31. Force per unit length on a 1200-ft riser due to waves, currents, and rig motion.

The problem of determining the force reduction can be discussed in theoretical terms only for the idealized situations (drag or inertia dominant regimes). For the more general case where the force is in the drag/inertia controlled regime and where the energy of the sea state is spread over a continuum of frequencies and directions, no generalizations can be made.

It is evident from the foregoing that the riser dynamics can be described in some detail only for a specific riser system and operating conditions. The variables such as water depth, available tension, vessel motion, stationkeeping system, buoyancy devices, riser joint weight, the number of flow lines attached to the riser, the sea state, and the force-transfer coefficients will have to be described for each individual riser, as close as possible to the real world operating conditions, for an approximate analysis of the system response.

The foregoing reasons as well as many others created by the environment, man, and the elements interfacing the two make it mandatory that measurements on prototype risers be carried out in order to help minimize the uncertainties cited earlier and to put the riser analysis on firmer foundations. This in turn requires that the prototype measurements be carried out (see e.g., Egeland and Solli 1980) with extreme care and advance planning and compared with analytical predictions.

In developing a test program, it is necessary to identify the parameters to be measured, types of tests to be carried out, and the types of instrumentation to be used in making the measurements. At a minimum, the test program should include the test requirements and objectives, a computer simulation of the riser response to guide the design of the tests, a careful selection of the instrumentation and the computer facilities, list of test conditions, test procedures, parameters to be measured, data quality assurance procedures, and methods of data evaluation.

In general, one should measure the characteristics of the sea state (so as to enable one to determine the three-dimensional sea spectrum), the vessel motion (pitch, yaw, heave, roll, surge, and sway, vertical and horizontal accelerations, etc.), subsea parameters (the x- and y-components of the force on instrumented sections of the risers), pressure distributions, choke and kill line loads, tension, bending stress at numerous points, temperature of mud and water, riser position (through the use of a hydrophone-responder system), accelerations of the riser section at various points (for the purpose of independently obtaining the velocity and the displacement of a particular point on the riser), relative velocity of the fluid (either with magnetic velocimeters or perforated balls attached to the riser, at a suitable distance), current profile (using available profilers), etc. The force measurements may be made on clean sections of the riser as well as on other sections fitted with choke and kill lines or buoyancy devices.

The information obtained from such a detailed field study will allow one to verify the methods of calculation of flow kinematics through the use of sea-state measurements, and the calculation of tension, bending stress, riser position, vortex shedding, resonance, ball-joint angles, effective diameter, and practically all other parameters (used either as input or calculated as output) and will enable one to validate and/or upgrade the existing riser models.

5.5 WAVE IMPACT LOADS

5.5.1 Introduction

Information about the forces acting on bluff bodies subjected to wave slamming is of significant importance in ocean engineering and naval architecture. The design of structures that must survive in a wave environment depends on a knowledge of the forces that occur at impact, as well as on the dynamic response of the system. Two typical examples include the structural members of offshore drilling platforms at the splash zone and the often encountered slamming of ships.

The general problem of hydrodynamic impact has been studied extensively, motivated in part by its importance in ordnance and missile technology. Extensive mathematical models have been developed for cases of simple geometry, such as spheres and wedges (Szebehely 1959). These models have been well supported by experiment. Unfortunately, the special case of wave impact has not been studied extensively.

Kaplan and Silbert (1976) developed a solution for the forces acting on a cylinder from the instant of impact to full immersion. Miller (1977) presented the results of a series of wave-tank experiments to establish the magnitude of the wave-force slamming coefficient for a horizontal circular cylinder. He found an average slamming coefficient of $C_s = 3.6$ where C_s is defined by

$$C_s = 2F/(\rho DLU_m^2) \tag{5.50}$$

Faltinsen et al. (1977) investigated the load acting on rigid horizontal circular cylinders (with end plates and length-to-diameter ratios of about 1) that were forced with constant velocity through an initially calm free surface. They found that the slamming coefficient ranged from 4.1 to 6.4. They also conducted experiments with flexible horizontal cylinders and found that the analytically predicted values were always lower (50 to 90 percent) than those found experimentally.

Sarpkaya (1978) conducted slamming experiments with harmonically oscillating flow impacting a horizontal cylinder and found that: (a) the dynamic response of the system is as important as the impact force (i.e., one cannot be determined without accounting for the other); (b) the initial value of the slamming coefficient is essentially equal to its theoretical value of π; (c) the system response may be amplified or attenuated, depending on its dynamic characteristics; (d) the buoyancy-corrected normalized force in the drag-dominated region reaches a maximum at a relative fluid displacement of about 1.75; and (e) roughness increases the rise time of the force and tends to decrease the amplification factor.

5.5.2 Theoretical Analysis

The general case of hydrodynamic impact usually is described by using incompressible potential flow theory. The compressibility of water and air and the cushioning effect of air (air boundary layer, depression of the water surface just before impact, etc.) are ignored. For a moving body with mass M, and velocity v_0, impacting a quiescent surface, the system momentum is Mv_0. Neglecting nonconservative forces, the momentum of the system is unchanged during penetration. However, the mass of the system increases because of the fluid set in motion near the body. Also known as added mass, m results in reducing the velocity. Thus, the system momentum after penetration is $(M + m)v = Mv_0$. The impact force at any instant is a function of m and $\dot{m}$. Therefore, the solution requires knowledge of the added mass and its time derivative. The determination of the added mass is not a simple matter and the results depend on the assumptions made (Moran 1965). The primary source of difficulty is the mathematical singularities encountered at the spray root (Chou 1946, Fabula 1957, Fabula and Ruggles 1955, Karman 1929, and Schnitzer and Hathaway 1953). Spray formation, air cushioning, and the flexibility of the impacting body (dynamic response) are the major sources of error in experiments also. The following analysis is based on the added mass calculated by Taylor (1930) which ignores the spray root problem.

Kaplan and Silbert (1976) have shown that the force acting on a horizontal cylinder by a wave system that propagates normal to it is equal to the sum of the buoyant force and the time-rate of change of momentum. Thus, one has

$$F/L = \rho g A_i + (m + \rho A_i)\ddot{\eta} + \frac{\partial m}{\partial z}\dot{\eta}^2 \tag{5.51}$$

in which F represents the force acting on the cylinder; L, the length of the cylinder; ρ, the density of fluid; g, gravitational acceleration; A_i, the immersed area; m, the added mass per unit length; η, the instantaneous height of the wave surface above the mean water level; and z, the instantaneous depth of immersion (see Fig. 5.32). The first and second derivatives of η with respect to time are denoted by $\dot{\eta}$ and $\ddot{\eta}$. The added mass is given by Taylor (1930) as

$$m = 0.5\,\rho r^2\left[\frac{2\pi^3}{3}\,\frac{(1-\cos\theta)}{(2\pi-\theta)^2} + \frac{\pi}{3}(1-\cos\theta) + (\sin\theta-\theta)\right] \tag{5.52}$$

in which r represents the radius of the cylinder, and θ is defined as shown in Fig. 5.32.

The motion of the free surface is related to the maximum amplitude by

$$\eta = A\sin 2\pi t/T \tag{5.53}$$

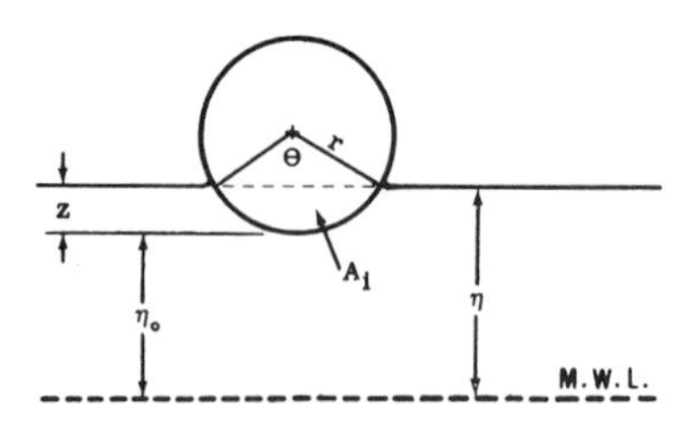

Fig. 5.32. Definition sketch for wave slamming.

where A and T represent the amplitude and period of the free surface in a harmonically oscillating flow. Equation (5.52) also can be written in the form of a slamming coefficient C_s, as defined by Eq. (5.50),

$$C_s = \overline{A}_i(gr/U_m^2) - (\overline{m} + \overline{A}_i)(r/A)\sin 2\pi t/T + \frac{\partial \overline{m}}{\partial \overline{z}} \cos^2 2\pi t/T \qquad (5.54)$$

where

$$\overline{A}_i = A_i/r^2, \quad \overline{m} = m/\rho r^2, \quad \overline{z} = z/r, \quad U_m = 2\pi A/T \qquad (5.55)$$

It is easy to show that at the instant of impact Eq. (5.54) reduces to

$$C_s = C_s^0 = (\partial \overline{m}/\partial \overline{z})_{t=0} = \pi \qquad (5.56)$$

Thus, Eq. (5.54) indicates that C_s, and consequently the impact force, is of an impulsive nature beginning with a finite value at the instant of impact. Since viscous forces are neglected, one would expect the solution to deviate from the actual situation as the cylinder becomes more fully immersed. Where this becomes the case can only be determined experimentally. A comparison of the experimental and theoretical results is shown in Fig. 5.33. Evidently, the predictions of the theoretical model are not valid beyond very small values of z/D. However, this is not of major concern since the largest impact force occurs at the instant of impact.

It is not realistic to assume that the impact force rises from zero to π instantaneously. Several factors, specifically the compressibility of air between the cylinder and water surface, entrapped gases in the water, surface irregularities, and water droplets on the surface of the cylinder would account for some finite rise time. For inclined cylinders the rise time may be quite significant. Nonthe-less, the rise time can be expected to be short, i.e., in the order of milliseconds. The exact nature of the rise is an interesting question for further study.

The realization that the impact force is of an impulsive nature requires consideration of the fact that this force does not act on a perfectly rigid body, but rather on a cylinder which is supported elastically. The response of such a system approaches that of a rigid body only if its natural frequency approaches infinity. Additionally, the response of the system to an impulsive force is heavily dependent on the exact nature of the force itself as well as on the system natural frequency. Sarpkaya (1978) has shown that the slamming coefficient C_s^0 may lie between 0.5π and 1.7π, depending on the rise time and the natural frequency of the elastically mounted cylinder. The significance of the foregoing is that the values of C_s determined experimentally from the measured reaction forces at the supports of a cylinder may show wide scatter depending on the

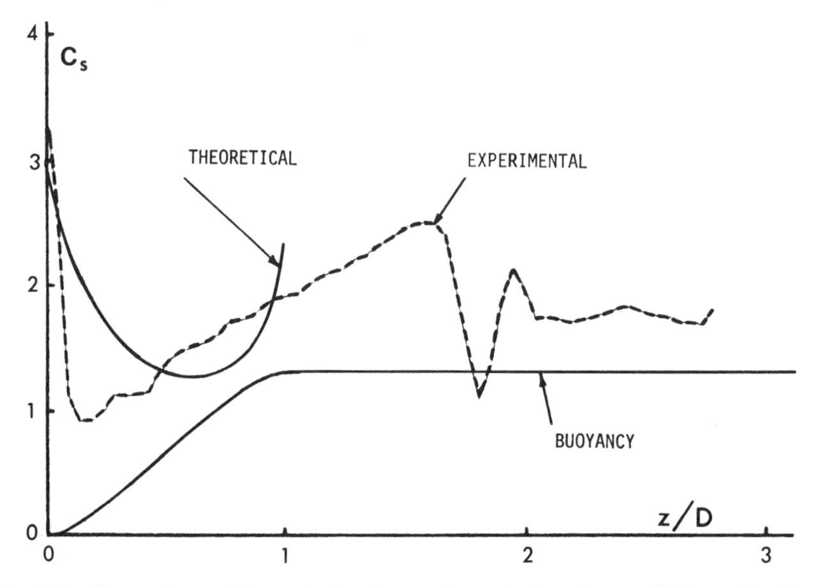

Fig. 5.33. Comparison of theoretical and experimental slamming coefficients (Sarpkaya 1978).

dynamic response of the cylinder and the rise time. Additionally, if the surface is not perfectly plane, rise time may vary from experiment to experiment resulting in an apparent non-repeatibility. This is particularly true for the conditions in the ocean environment. Evidently, controlled laboratory experiments help to establish the ideal value of the impact coefficient and to explain the reasons for the observed scatter in the data. The determination of the impact-force magnification factor for the cylinder in the ocean environment must necessarily consider the random nature of the disturbances at the wave surface, the orientation of the structural member relative to a given wave, currents, and three-dimensional nature of the waves, spray, etc. There are no deterministic means to predict the reaction forces acting on the supports of a member due to wave impact even when the structural characteristics of the member (damping, natural frequency, etc.) and the ideal value of the impact force are known. This is so because the rise time depends on all the nondeterministic conditions just cited. It is on the basis of the foregoing that the following design recommendations are made: (a) use $C_s^0 = 5.5$ *if a dynamic response analysis is not to be carried out*, and (b) use $C_s^0 = 3.2$ if the impact-force magnification is to be determined through a dynamic response analysis.

Kaplan (1979) described some of the problems associated with the measurement of high frequency impact forces on offshore structures and reported C_s values which ranged from 1.88 to 5.11, with just as many values above the theoretical value of π as below, and a mean value of $C_s = 2.98$. While this type of comparison is not quite precise, the general range of magnitudes may be considered to lie within the range of values reported by others, as noted by Kaplan.

Finally, it should be noted that the velocity-square dependent part of the drag may be added to the force expression given by Eq. (5.51) as

$$F/L = \rho g A_i + (m + \rho A_i)\ddot{\eta} + \frac{\partial m}{\partial z}\dot{\eta}^2 + 0.5\,\rho\,\dot{\eta}|\dot{\eta}|\,[2r \sin(\theta/2) C_d(\theta)] \tag{5.57}$$

where $[2r \sin(\theta/2)]$ represents the upper boundary of the immersed sectional area, with a maximum value equal to the cylinder diameter; and $C_d(\theta)$, the drag coefficient of that section for vertical flow, which in general is expected to vary with the degree of immersion (Kaplan 1979). Clearly, the determination of the drag coefficient for such a complex, time-dependent free surface flow is not possible. In general, the contribution of the velocity-square-dependent drag is quite small and may be ignored in view of the rather uncertain nature of the determination of the rise time, dynamic characteristics of the structural member, and the slamming coefficient.

Recently, Campbell and Weynberg (1980) proposed a time history of C_s as

$$C_s = 5.15/(1 + 19\,Ut/D) + 0.55\,Ut/D \tag{5.58}$$

where U is the slamming velocity; t, the time; and D, the diameter of the cylinder. This proposal is in conformity with our recommendation of $C_s^0 = 5.5$ if a dynamic response analysis is not to be carried out. It must be emphasized that the force traces of Campbell and Weynberg contained a dominant response oscillation at approximately 550 Hz and the initial part of the slam load history was difficult to determine from force measurements. Equation (5.58) was based partly on the hyperbolic curves fitted to the *faired force traces* and partly on the integrated pressure distributions (cross faired). Miller's (1980) most recent work has shown that for most structures the current practice of applying a constant slamming coefficient of 3.5 is conservative when estimating extreme stresses. For fatigue estimation this appears to be the case irrespective of the member geometry.

5.6 REFERENCES

Aagaard, P. M. and Dean, R. G. 1969. Wave Forces: Data Analysis and Engineering Calculation Method. OTC Paper No. 1008, Houston, TX.

Al-Kazily, M. F. 1972. Forces on Submerged Pipelines Induced by Water Waves. Ph.D. Thesis, College of Engineering, Univ. of Calif., Berkeley.

American Petroleum Institute (API). 1977. API Recommended Practice for Planning, Designing and Constructing Fixed Offshore Platforms. API RP 2A, API Production Department, 300 Corrigan Tower Bldg., Dallas, TX.

American Science and Engineering Company. 1974. Forces Acting on Unburied Offshore Pipelines. A report prepared for the pipeline research committee of the American Gas Association, Project PR-91-68, (revised April 1975).

Arita, Y., Fujita, H., and Tagaya, K. 1973. A Study on the Force of Current Acting on a Multitubular Column Structure. OTC Paper No. 1815, Houston, TX.

Bathe, K. J., Ozdemir, H., and Wilson, E. L. 1974. Static and Dynamic Geometric and Material Nonlinear Analysis. Report No. UCSESM 74-4, Structural Engineering Lab., Univ. of Calif., Berkeley.

Bea, R. G. and Lai, N. W. 1978. Hydrodynamic Loadings on Offshore Platforms. OTC Paper No. 3064, Houston, TX.

Bearman, P. W. and Zdravkovich, M. M. 1978. Flow Around a Circular Cylinder Near a Plane Boundary. *JFM*, Vol. 89, pt. 1, pp. 33–47.

Beattie, J. F., Brown, L. P., and Webb, B. 1971. Lift and Drag Forces on a Submerged Circular Cylinder. OTC Paper No. 1358, Houston, TX.

Bendat, J. S., Richman, G., Osborne, A. R., and Silbert, M. N. 1979. Cross-Spectral and Coherence Analysis of OTS Data. *Civil Engineering in the Oceans IV, ASCE*, pp. 774–790.

Bennet, B. E. and Metcalf, M. F. 1977. Nonlinear Dynamic Analysis of Coupled Axial and Lateral Motions of Marine Risers. OTC Paper No. 2776, Houston, TX.

Biermann, D. and Herrnstein, W. H., Jr. 1933. The Interference Between Struts in Various Combinations. NACA Tech. Rep. 468.

Bishop, J. R. 1978. The Mean Square Value of Wave Force Based on the Morison Equation. National Maritime Institute Report NMI-R-40.

Bishop, J. R. 1979. R.M.S. Force Coefficients Derived from Christchurch Bay Wave Force Data. National Maritime Institute Report NMI-R-62.

Borgman, L. E. 1958. Computation of the Ocean-Wave Forces on Inclined Cylinders. *Trans. Amer. Geophysical Union*, Vol. 39, pp. 885–888.

Borgman, L. E. 1965. The Spectral Density of Ocean Wave Forces. *Coastal Engineering*, Santa Barbara Specialty Conf., ASCE.

Borgman, L. E. 1967. Spectral Analysis of Ocean Wave Forces on Piling. *Jour. Waterways etc. Div., ASCE*, Vol. 93, WW2, pp. 129–156.

Borgman, L. E. 1969a. Directional Spectra Models for Design Use. OTC Paper No. 1069, Houston, TX.

Borgman, L. E. 1969b. Ocean Wave Simulation for Engineering Design. *Jour. Waterways etc. Div., ASCE*, Vol. 95, WW4, pp. 557–583.

Borgman, L. E. and Yfantis, E. 1979. Three-Dimensional Character of Waves and Forces. *Civil Engineering in the Oceans IV, ASCE*, pp. 791–804.

Bowie, G. L. 1977. Forces Extended by Waves on a Pipeline at or Near the Ocean Bottom. U.S. Army Corps of Engineers, Coastal Engineering Research Center, Technical Paper No. 77-11, Ft. Belvoir, VA.

Brannon, H. R., Loftin, T. D., and Whitfield, J. H. 1974. Deepwater Platform Design. OTC Paper No. 2120, Houston, TX.

Bretschneider, C. L. 1967. Probability Distribution of Wave Forces. *Jour. Waterways etc. Div., ASCE*, Vol. 13, WW2, pp. 5–26.

British Ship Research Association. 1976. A Critical Evaluation of the Data on Wave Force Coefficients. British Ship Research Assoc., Wallsend upon Tyne, Contract Report W-278.

Brown, L. J. and Borgman, L. E. 1966. Tables of the Statistical Distribution of Ocean Wave Forces and Methods for the Estimation of C_d and C_m. Wave Research Report HEL 9-7, Univ. of Calif., Berkeley.

Burke, B. G. 1973. Analysis of Marine Risers for Deep Water. Proceedings of OTC, Vol. 1, pp. 449–464.

Burke, B. G. 1974. An Analysis of Marine Risers for Deep Water. *Journal of Petroleum Technology*, pp. 455–465.

Burnett, S. J. 1979. OSFLAG Project 4: The Effects of Marine Growth on Wave and Current Loading. Offshore Research Focus, CIRIA, U.K., ISSN: 0309-4189, pp. 8–9.

Bursnall, W. J. and Loftin, L. K. 1951. Experimental Investigation of the Pressure Distribution about a Yawed Circular Cylinder in the Critical Reynolds Number Range. NACA Tech. Note 2463.

Bushnell, M. J. 1977. Forces on Cylinder Arrays in Oscillating Flow. OTC Paper No. 2903, Houston, TX.

Campbell, I. M. C. and Weynberg, P. A. 1980. Measurement of Parameters Affecting Slamming. Wolfson Unit for Marine Technology and Industrial Aerodynamics, Univ. of Southampton, Report No. 440.

Chakrabarti, S. K. 1978a. Wave Forces on Multiple Vertical Cylinders. *Jour. Waterways etc. Div., ASCE*, WW2, pp. 147–161.

Chakrabarti, S. K. 1978b. Discussion of "Added Mass of Pile Group" (R. J. Gibson and H. Wang), *Jour. Waterways etc. Div., ASCE*, WW2, pp. 256–258.

Chakrabarti, S. K. 1979. Wave Forces on Vertical Array of Tubes. *Civil Engineering in the Oceans IV, ASCE*, pp. 241–259.

Chakrabarti, S. K. 1980a. Inline Forces on Fixed Vertical Cylinder in Waves. *Jour. Waterways etc. Div., ASCE*, WW2, pp. 145–155.

Chakrabarti, S. K. 1980b. Hydrodynamic Coefficients for a Vertical Tube in an Array. (To appear in Applied Ocean Research).

Chakrabarti, S. K., Tam, W. A., and Wolbert, A. L. 1975. Wave Forces on a Randomly Oriented Tube. OTC Paper No. 2190, Houston, TX.

Chakrabarti, S. K., Wolbert, A. L., and Tam, W. A. 1977. Wave Forces on Inclined Tubes. *Coastal Engineering*, Vol. 1, pp. 149–165.

Chappelear, J. E. 1961. Direct Numerical Calculation of Wave Properties. *Jour. of Geophysical Research*, Vol. 66.

Chen, Y. N. 1972. Fluctuating Lift Forces of the Karman Vortex Streets on Single Circular Cylinders and in Tube Bundles, Part 3—Lift Forces in Tube Bundles. *Jour. Engrg. for Industry, Trans. ASME*, Vol. 24, p. 623.

Chen, Y. N. 1977. The Sensitive Tube Spacing Region of Tube Bank Heat Exchangers for Fluid-Elastic Coupling in Cross Flow. ASME Series PVP-PB-026, (eds. M. K. Au-Yang and S. J. Brown).

Chiu, W. S. 1966. The Boundary-Layer Formation and Vortex Shedding on Yawed Cylinders. Washington State Univ., College of Engrg., Bull. 299.

Chou, P. Y. 1946. On Impact of Spheres Upon Water. *Water Entry and Underwater Ballistics of Projectiles*. Chap. 8, Calif. Inst. of Tech., OSRD Report 2251.

Crua, A. 1967. Druckverlustmessunger an Glattrohrbundeln. Sulger Brothers Report No. 1387.

Dalrymple, R. A. and Dean, R. G. 1975. Waves of Maximum Height on Uniform Currents. *Jour. Waterways etc. Div., ASCE*, Vol. 101, WW3, pp. 317–328.

Dalton, C. 1980. Inertia Coefficients for Riser Configurations. ASME Paper No. 80-Pet-21.

Dalton, C. and Helfinstine, R. A. 1971. Potential Flow Past a Group of Circular Cylinders. *Jour. Basic Engineering, ASME*, pp. 636–642.

Dalton, C. and Nash, J. M. 1976. Wave Slam on Horizontal Members of an Offshore Platform. OTC Paper No. 2500, Houston, TX.

Dalton, C. and Szabo, J. M. 1976. Drag on a Group of Cylinders. ASME Paper No. 76-Pet-42.

Davis, D. A. and Ciani, J. B. 1976. Wave Forces on Submerged Pipelines—A Review with Design Aids. Civil Engineering Lab., NCBC, Port Hueneme, CA., Tech. Report No. R-844.

Dean, R. G. 1965. Stream Function Representation of Nonlinear Ocean Waves. *Jour. of Geophysical Research*, Vol. 70, No. 18, p. 4561.

Dean. R. G. 1970. Relative Validities of Water Wave Theories. *Jour. Waterways etc. Div., ASCE*, Vol. 96, WW1, pp. 105–119.

Dean, R. G. 1976. Methodology for Evaluating Suitability of Wave and Force Data for Determining Drag and Inertia Forces. *BOSS '76*, Trondheim, Vol. 2, pp. 40–64.

Dean, R. G., Lo, J-M., and Johansson, P. I. 1979. Rare Wave Kinematics versus Design Practice. *Civil Engineering in the Oceans IV, ASCE*, pp. 1030–1049.

Department of Energy (U.K.). 1974. Guidance on the Design and Construction of Offshore Installations. London.

Det Norske Veritas. 1974. Rules for the Design, Construction and Inspection of Fixed Offshore Structures. DNV, Oslo.

Engineering Sciences Data Unit, ESDU. 1975. Fluid Forces on Lattice Structures. Data Sheet 75011. [See also Data Sheets 71012 (1971), 70013 (1970), and 70014 (1970)].

Ericsson, L. E. and Reding, J. P. 1979. Vortex-Induced Asymmetric Loads on Slender Vehicles. Lockheed Missiles and Space Company, Inc., Report No. LMSC-D-630807.

Evans, D. J. 1969. Analysis of Wave Force Data. OTC Paper No. 1005, Houston, TX.

Fabula, A. G. 1957. Ellipse Fitting Approximation of Two Dimensional Normal Symmetric Impact of Rigid Bodies on Water. *Proc. 5th Midwestern Conf. on Fluid Mechanics*, Univ. of Michigan.

Fabula, A. G. and Ruggles, I. D. 1955. Vertical Broadside Water Impact of Circular Cylinder. Growing Circular and Approximations. U.S. Naval Ordnance Test Station, China Lake, Calif.

Faltinsen, O., Kjaerland, O., Nøttveit, and Vinje, T. 1977. Water Impact Loads and Dynamic Response of Horizontal Circular Cylinders in Offshore Structures. OTC Paper No. 2741, Houston, TX.

Fischer, W. and Ludwig, M. 1966. Design of Floating Vessel Drilling Risers. *Jour. of Petroleum Technology*, Vol. 272.

Fitz-Hugh, J. S. 1973. Flow-Induced Vibration in Heat Exchangers. Oxford Univ. Report RS-57, AERE-P-7238.

Friehe, C. A. and Schwartz, W. H. 1968. Deviations from the Cosine Law for Yawed Cylindrical Anemometer Sensors. *Trans. ASME, Ser. E, Jour. Applied Mechs.*, Vol. 35, No. 4, pp. 655–662.

Fuji, S. and Gomi, M. 1976. A Note on the Two-Dimensional Cylinder Wake. *Jour. Fluids Engrg., Trans. ASME*, Vol. 98, Ser. 1, No. 2, pp. 318–320.

Galler, S. R. 1969. Boring and Fouling. *Handbook of Ocean and Underwater Engineering*, McGraw-Hill, New York, pp. 7.12 to 7.19.

Gardner, T. N. and Kotch, M. A. 1976. Dynamic Analysis of Risers and Caissons by the Finite Element Method. OTC Paper No. 2651, Houston, TX.

Gaston, J. D. and Ohmart, R. D. 1979. Effects of Surface Roughness on Drag Coefficients. *Civil Engineering in the Oceans IV, ASCE*, pp. 611–621.

Geminder, R. E. and Pomonick, G. M. 1979. The Ocean Test Structure Measurement System. *Civil Engineering in the Oceans IV, ASCE*, pp. 1010–1029.

Gibson, R. J. and Wang, H. 1977. Added Mass of Pile Group. *Jour. Waterways etc. Div., ASCE*, WW2, pp. 215–223.

Giesecke, J. and Hafner, E. 1977. Submarine Pipelines and Their Hydrodynamic Loads. *Wasserwirtsch.*, Vol. 67, No. 10, pp. 299–305 (in German).

Glenny, D. E. 1966. A Review of Flow Around Circular Cylinders, Stranded Cylinders and Struts Inclined to the Flow Directions. Australian Department of Supply, Mechanical Engineering Note 284.

Goktun, S. 1975. The Drag and Lift Characteristics of a Cylinder Placed Near a Plane Surface. MS Thesis submitted to the Naval Postgraduate School, Monterey, CA.

Gosse, C. G. 1969. The Marine Riser–A Procedure for Analysis. OTC Paper No. 1080, Houston, TX.

Grace, R. A. 1973. Available Data for the Design of Unburied Submarine Pipelines to Withstand Wave Action. *First Australian Conference on Coastal Engineering*, Sydney, pp. 59–66.

Grace, R. A. 1979. *Marine Outfall Systems–Planning, Design and Construction*. Prentice-Hall, Englewood Cliffs, NJ.

Grace, R. A. and Nicinski, S. A. 1976. Wave Force Coefficients from Pipeline Research in the Ocean. OTC Paper No. 2676, Houston, TX.

Grace, R. A., Castiel, J., Shak, A. T., and Zee, G. T. Y. 1979. Hawaii Ocean Test Pipe Project: Force Coefficients. *Civil Engineering in the Oceans IV, ASCE*, Vol. 1, pp. 99–110.

Graham, D. S. 1979. A Bibliography of Force Coefficients Literature Germane to Unburied Pipelines in the Ocean. Dept. of Civil Engineering, The Univ. of Florida, Miscellaneous Report.

Graham, D. S. and Machemehl, J. L. 1980. Approximate Force Coefficients for Ocean Pipelines. ASME Paper No. 80-Pet-61.

Hammeke, R., Heinecke, E., and Scholz, F. 1967. Warmeubergangs und Druklustmessungen an Queranstromten Glattrohrbundeln, Insbesondere bei Hohen Reynoldsgablen. *International J. Heat Mass Transfer*, Vol. 10, pp. 427–435.

Hanson, A. R. 1966. Vortex-Shedding from Yawed Cylinders. *AIAA Journal*, Vol. 4, pp. 738–740.

Haring, R. E., Olsen, O. A., and Johansson, P. I. 1979. Total Wave Force and Moment versus Design Practice. *Civil Engineering in the Oceans IV, ASCE*, pp. 805–819.

Haring, R. E. and Spencer, L. P. 1979. The Ocean Test Structure Data Base. *Civil Engineering in the Oceans IV, ASCE*, pp. 669–683.

Heaf, N. J. 1979. The Effect of Marine Growth on the Performance of Fixed Offshore Platforms in the North Sea. OTC Paper No. 3386, Houston, TX.

Heideman, J. C., Olsen, O. A., and Johansson, P. I. 1979. Local Wave Force Coefficients. *Civil Engineering in the Oceans IV, ASCE*, pp. 684–699.

Herbich, J. B. 1976. Scour Around Model Pipelines due to Wave Action. *Proceedings of the 15th Conf. on Coastal Engrg., ASCE*, Honolulu, HI.

Hoerner, S. F. 1965. Fluid-Dynamic Drag. 3rd Ed. Book published by the Author, New Jersey.

Hogben, N. 1976. Wave Loads on Structures. *BOSS '76*, Trondheim, Vol. 1, pp. 187–219.

Hogben, N., Miller, B. L., Searle, J. W., and Ward, G. 1977. Estimation of Fluid Loading on Offshore Structures. *Proc. Institution of Civil Engrs.*, Vol. 63, part 2, pp. 515–562.

Houghton, D. R. 1968. Mechanisms of Marine Fouling. *Proceed. First International Biodeterioration Symposium*, Southampton, U.K.

Houghton, D. R. 1970. Foul Play on the Ship's Bottom. *New Scientist*, 3 Dec.

Hudspeth, R. T., Dalrymple, R. A., and Dean, R. G. 1974. Comparison of Wave Forces Computed by Linear and Stream Function Methods. OTC Paper No. 2037, Houston, TX.

Inman, D. L. and Bowen, A. J. 1963. Flume Experiments on Sand Transport by Waves and Currents. *Proc. 8th Conf. Coastal Engrg.*, Mexico City, Council on Wave Research, pp. 137–150.

Isaacson, M. 1978. Interference Effects Between Large Cylinders in Waves. OTC Paper No. 3067, Houston, TX.

Isaacson, M. 1979. Nonlinear Inertia Forces on Bodies. *Jour. Waterways etc. Div., ASCE*, No. WW3, pp. 213-227.

Johansson, B. and Reinius, E. 1963. Wave Forces Acting on a Pipe at the Bottom of the Sea. IAHR Congress, Paper No. 1.7, London.

Johnson, R. E. 1970. Regression Model of Wave Forces on Ocean Outfalls. *Jour. Waterways etc. Div., ASCE*, Vol. 96, pp. 284-305.

Jones, M. R. 1975. Problems Affecting the Design of Drilling Risers. SPE Paper No. 5268 (Prepared for SPE London).

Jonsson, I. G. 1976. The Dynamics of Waves on Currents over a Weakly Varying Bed. *IUTAM Symposium on Surface Gravity Waves on Water of Varying Depth*, Canberra, Australia, pp. 1-12.

Kaplan, P. 1979. Impact Forces on Horizontal Members. *Civil Engineering in the Oceans IV, ASCE*, Vol. II, pp. 716-731.

Kaplan, P. and Silbert, M. N. 1976. Impact Forces on Platform Horizontal Members in the Splash Zone. OTC Paper No. 2498, Houston, TX.

Karman, Von, T. L. and Wattendorf, F. 1929. The Impact on Seaplane Floats During Landing. NACA TN-321.

Kennered, Z. H. 1967. Irrotational Flow of Frictionless Fluids. Mostly of Invariable Density. David Taylor Model Basin Report No. 2299.

Kim, Y. Y. and Hibbard, H. C. 1975. Analysis of Simultaneous Wave Force and Water Particle Velocity Measurements. OTC Paper No. 2192, Houston, TX.

King, R. 1977. Vortex-Excited Oscillations of Yawed Circular Cylinders. *ASME Jour. Fluids Engrg.*, Vol. 99, pp. 495-502.

Kirk, C. L., Etok, E. U., and Cooper, M. T. 1979. Dynamic and Static Analysis of a Marine Riser. *Applied Ocean Research*, Vol. 1, No. 3, pp. 125-135.

Knudsen, J. G. and Katz, D. L. 1958. *Fluid Dynamics and Heat Transfer*. McGraw-Hill Book Co., New York.

Koterayama, W. 1979. Wave Forces Exerted on Submerged Circular Cylinders Fixed in Deep Water. Reports of Research Institute of Applied Mechanics, Kyushu Univ., Vol. 27, No. 84, pp. 25-46.

Krolikowski, L. P. and Gay, T. A. 1980. An Improved Linearization Technique for Frequency Domain Riser Analysis. OTC Paper No. 3777, Houston, TX.

Laird, A. D. K., Johnson, C. A., and Walker, R. W. 1959. Water Forces on Accelerated Cylinders. *Jour. Waterways etc. Div., ASCE*, Vol. 85, No. WW1, pp. 99-119.

Lambrakos, K. F. and Brannon, H. R. 1974. Wave Force Calculations for Stokes and Non-Stokes Waves. OTC Paper No. 2039, Houston, TX.

Layton, J. A. and Scott, J. L. 1979. Stabilization Requirements for Submarine Pipelines Subjected to Ocean Forces. *Civil Engineering in the Oceans IV, ASCE*, Vol. 1, pp. 60-76.

Le Méhauté, B., Divoky, D., and Lin, A. 1968. Shallow Water Waves: A Comparison of Theory and Experiment. *Proceedings of the 11th Coastal Engineering Conference*, pp. 86-107.

Littlejohns, P. S. G. 1974. Current Induced Forces on Submarine Pipelines. Hydraulic Research Station, Wallingford, INT-138.

Løken, A. E., Torset, O. P., Mathiassen, S., and Arnesen, T. 1979. Aspects of Hydrodynamic Loading in Design of Production Risers. OTC Paper No. 3538, Houston, TX.

Longuet-Higgins, M. S. 1976. Recent Development in the Study of Breaking Waves. *15th Annual Conference on Coastal Engineering*, ASCE, Honolulu, HI., pp. 441-460.

MacCamy, R. C. and Fuchs, R. A. 1954. Wave Force on Piles: A Diffraction Theory. U.S. Army Coastal Engrg. Res. Center, Tech. Memo. No. 69.

Mair, W. A. and Maull, D. J. 1971. Aerodynamic Behavior of Bodies in the Wake of Other Bodies. *Phil Trans. Roy. Soc., London*, Vol. 269A, pp. 425-437.

Maison, J. R. and Lea, J. F. 1977. Sensitivity Analysis of Parameters Affecting Riser Performance. OTC Paper No. 2918, Houston, TX.

Marshall, P. W. 1976. Failure Modes for Offshore Platforms–Fatique. *BOSS '76*, Vol. 2, pp. 234–248.

Matten, R. B. 1977. The Influence of Surface Roughness on the Drag of Circular Cylinders in Waves. OTC Paper No. 2902, Houston, TX.

Maull, D. J. and Norman, S. G. 1979. A Horizontal Circular Cylinders Under Waves. *Mechanics of Wave-Induced Forces on Cylinders* (ed. T. L. Shaw), Pitman, London, pp. 359–378.

Michel, W. H. 1967. Sea Spectra Simplified. *Society of Naval Architects and Marine Engineers*, April 1967 Meeting of the Gulf Section.

Miller, B. L. 1977a. The Hydrodynamic Drag of Roughened Circular Cylinders. *Jour. Roy. Inst. Naval Architects, RINA*, Vol. 119, pp. 55–70.

Miller, B. L. 1977b. Wave Slamming Loads on Horizontal Circular Elements of Offshore Structures. *Jour. Roy. Inst. Naval Arch., RINA*, Paper No. 5.

Miller, B. L. 1980. Wave Slamming on Offshore Structures. National Maritime Institute Report No. NMI-R81.

Mobarek, I. 1965. Directional Spectra of Laboratory Wind Waves. *Jour. Waterways etc. Div., ASCE*, Vol. 91, No. WW3.

Moe, G. and Crandall, S. H. 1977. Extremes of Morison-Type Wave Loading on a Single Pile. ASME Paper No. 77-DET-82.

Moran, J. P. 1965. On the Hydrodynamic Theory of Water-Exit and Entry. Therm. Adv. Res. Rep. TAR-TR-6501.

Morgan, G. W. 1977. Force Systems Acting on Arbitrarily Directed Tubular Members in the Sea. ASME Paper No. 77-Pet-38.

Morgan, G. W. and Peret, J. W. 1976. *Applied Mechanics of Marine Riser Systems.* Petroleum Engineer Publishing Co., Dallas.

Mujumdar, A. S. and Douglas, W. J. M. 1970. The Unsteady Wake Behind a Group of Three Parallel Cylinders. ASME Paper No. 70-Pet-8.

Munk, W. H. 1949. The Solitary Wave and its Application to Surf Problems. *Ann. N. Y. Academy Science*, Vol. 51.

National Engineering and Science Company. 1966. Structural Dynamic Analysis of the Riser and Drill String for Project Mohole. NESC Report 5234, pts. 1 and 2.

Novak, M. 1975. Discussion of "Yaw Effects on Galloping Instability" by R. Skarecky, *Engineering Mechs. Div., ASCE*, EM4, p. 745.

Ohmart, R. D. and Gratz, R. L. 1979. Drag Coefficients from Hurricane Wave Data. *Civil Engineering in the Oceans IV, ASCE*, pp. 260–272.

Pearcey, H. H. 1979. Some Observations on Fundamental Features of Wave-Induced Viscous Flows Past Cylinders. *Mechanics of Wave-Induced Forces on Cylinders* (ed. T. L. Shaw), Pitman, London, pp. 1–54.

Pearcey, H. H. and Bishop, J. R. 1979. Wave Loading in the Drag and Drag-Inertia Regimes; Routes to Design Data. *BOSS '79*, London, Paper No. 23.

Peregrine, D. H. 1976. Interaction of Water Waves and Currents. *Advances in Applied Mechanics*, Vol. 16, Academic Press, New York, pp. 9–117.

Pierson, W. J. and Holmes, P. 1965. Irregular Wave Forces on a Pile. *Jour. Waterways etc. Div., ASCE*, Vol. 91, WW4, pp. 1–10.

Priest, M. S. 1971. Wave Forces on Exposed Pipelines on the Ocean Bed. OTC Paper No. 1383, Houston, TX.

Quadflieg, H. 1977. Vortex Induced Load on the Cylinders Pair at High Reynolds Numbers (in German). *Forschung im Ingenieurwesen*, Vol. 43, No. 1, pp. 9–18.

Ralston, D. O. and Herbich, J. B. 1968. The Effects of Waves and Currents on Submerged Pipelines. Coastal and Ocean Engineering Div. Report No. 101-C.O.E., Texas A&M Univ.

Ramberg, S. E. 1978. The Influence of Yaw Angle Upon Vortex Wakes of Stationary and Vibrating Cylinders. Memorandum Report 3822, Naval Research Lab., Washington, D.C.

Ramberg, S. E. and Niedzwecki, J. M. 1979. Some Uncertainties and Errors in Wave Force Computations. OTC Paper No. 3597, Houston, TX.

Rance, P. J. 1969. Wave Forces on Cylindrical Members of Structures. Hydraulics Res., Hydraulic Research Station, Wallingford, pp. 14–17.

Reid, R. O. 1958. Correlation of Water Level Variations with Wave Forces on a Vertical Pile for Non-Periodic Waves. *Proc. 6th Coastal Engrg. Conf.*, Gainsville, Council for Wave Research, Berkeley, CA., pp. 749–786.

Reid, R. O. and Bretschneider, C. L. 1953. Surface Waves and Offshore Structures. Texas A&M Research Foundation Report.

Relf, E. H. and Powell, C. H. 1917. Tests on Smooth and Stranded Wires Inclined to the Wind Direction and a Comparison of the Results on Stranded Wires in Air and Water. British A.R.C. R and M Report No. 307.

Robertson, J. M. 1965. *Hydrodynamics in Theory and Application*. Prentice-Hall, Englewood Cliffs, NJ.

Rosenhead, L. 1929. The Karman Street of Vortices in a Channel of Finite Breadth. *Phil Trans. Roy. Soc., London*, Vol. 228A, pp. 275–329.

Roshko, A. 1964. On the Drag and Shedding Frequencies of Two-Dimensional Bluff Bodies. NACA Tech. Note TN-3169.

Ross, C. W. 1959. Large-Scale Tests of Wave Forces on Piling (Preliminary Report). U.S. Corps of Engineers Beach Erosion Board, Tech. Memo. 111.

Ruhl, J. A. 1976. Offshore Platforms: Observed Behavior and Comparisons with Theory. OTC Paper No. 2553, Houston, TX.

Sarpkaya, T. 1955. Oscillatory Gravity Waves in Flowing Water. *Trans. ASCE*, Vol. 122, pp. 564–586.

Sarpkaya, T. 1960. Added Mass of Lenses and Parallel Plates. *Jour. Engrg. Mechs. Div., ASCE*, Vol. 86, No. EM3, pp. 141–152.

Sarpkaya, T. 1976a. In-Line and Transverse Forces on Smooth and Sand-Roughened Cylinders in Oscillatory Flow at High Reynolds Numbers. Naval Postgraduate School Technical Report No. NPS-69SL76062, Monterey, CA.

Sarpkaya, T. 1976b. Forces on Cylinders Near a Plane Boundary in a Sinusoidally Oscillating Fluid. *Jour. of Fluids Engrg., ASME*, Vol. 98, Ser. 1, No. 3, pp. 499–505.

Sarpkaya, T. 1977. In-Line and Transverse Forces on Cylinders Near a Wall in Oscillatory Flow at High Reynolds Numbers. OTC Paper No. 2898, Houston, TX.

Sarpkaya, T. 1978. Wave Impact Loads on Cylinders. OTC Paper No. 3065, Houston, TX.

Sarpkaya, T. 1979. Hydrodynamic Forces on Various Multiple-Tube Riser Configurations. OTC Paper No. 3539, Houston, TX.

Sarpkaya, T., Cinar, M., and Ozkaynak, S. 1980. Hydrodynamic Interference of Two Cylinders in Harmonic Flow. OTC Paper No. 3775, Houston, TX.

Sarpkaya, T. and Rajabi, F. 1980. Hydrodynamic Drag on Bottom-Mounted Smooth and Rough Cylinders in Periodic Flow. OTC Paper No. 3761, Houston, TX.

Schnitzer, E. and Hathaway, M. E. 1953. Estimation of Hydrodynamic Impact Loads and Pressure Distributions on Bodies Approximating Elliptical Cylinders with Special Reference to Water Landings of Helicopters. NACA TN-2889.

Sekita, K. 1975. Laboratory Experiments on Wave and Current Forces Acting on a Fixed Platform. OTC Paper No. 2191, Houston, TX.

Sexton, R. M. and Agbezuge, L. K. 1976. Random Wave and Vessel Motion Effects on Drilling Riser Dynamics. OTC Paper No. 2650, Houston, TX.

Skarecky, R. 1975. Yaw Effects on Galloping Instability. *Engrg. Mechs. Div., ASCE*, Proc. Paper No. 11759.

Skjelbreia, L. and Hendrickson, J. A. 1961. Fifth Order Gravity Wave Theory. *Proceedings of the 7th Conference on Coastal Engineering*, Tokyo, Council on Wave Research, Richmond, Calif., Vol. 1, Cpt. 10, pp. 184–196.

Smith, R. A., Moon, W. T. and Kao, T. W. 1972. Experiments on Flow About a Yawed Circular Cylinder. ASME Paper No. 72-FE-2.

Spanos, P-T. D. and Chen, T. W. 1980. Vibrations of Marine Riser Systems. ASME Paper No. 80-Pet-69.

Spring, B. H. and Monkmeyer, P. L. 1974. Interaction of Plane Waves with Vertical Cylinders. *Proceedings of the 14th Coastal Engrg. Conf., ASCE*, Vol. III, pp. 1828–1847.

Springston, G. B. 1967. Generalized Hydrodynamic Loading Functions for Bare and Faired Cables in Two-Dimensional Steady-State Cable Configurations. NSRDC Report 2424.

St. Denis, M. 1973. Some Cautions on the Employment of the Spectral Technique to Describe the Waves of the Sea and the Response Thereto of Ocean Systems. OTC Paper No. 1819, Houston, TX.

Susbielles, G. G., Van Den Bunt, J. R., Deleuil, G., and Michel, D. 1971. Wave Forces on Pile Sections due to Irregular and Regular Waves. OTC Paper No. 1379, Houston, TX.

Szebehely, V. G. 1959. Hydrodynamic Impact. *Applied Mechanics Reviews*, Vol. 12, No. 5, pp. 297–300.

Tanida, Y., Okajima, A., and Watanabe, Y. 1973. Stability of a Circular Cylinder Oscillating in Uniform Flow or in a Wake. *JFM*, Vol. 61, pt. 4, pp. 769–784.

Taylor, J. L. 1930. Some Hydrodynamical Inertia Coefficients. *Phil. Magazine*, Vol. 9, Ser. 7, pp. 161–183.

Thomas, G. P. 1979. Water Wave-Current Interactions: A Review. *Mechanics of Wave-Induced Forces on Cylinders* (ed. T. L. Shaw), Pitman, London, pp. 179–203.

Thrasher, L. W. and Aagaard, P. M. 1969. Measured Wave Force Data on Offshore Platforms. OTC Paper No. 1007, Houston, TX.

Trimble, T. H. 1966. Normal and Tangential Force on Cylinders at Small Inclinations to the Flow. (Unpublished ARL work, see Glenny 1966).

Tucker, T. C. and Murtha, J. P. 1973. Nondeterministic Analysis of a Marine Riser. OTC Paper No. 1770, Houston, TX.

Tung, C. C. and Huang, N. E. 1976. Interactions Between Waves and Currents and Their Influence on Fluid Forces. *BOSS '76*, Trondheim, Vol. I, pp. 129–143.

Van Atta, C. W. 1968. Experiments in Vortex Shedding from Yawed Circular Cylinders. *AIAA Journal*, Vol. 6, No. 5, pp. 931–933.

van Koten, H. 1976. Fatigue Analysis of Marine Structures. *BOSS '76*, Trondheim, Vol. 1, pp. 653–678.

von Müller, W. 1929. Systeme von Doppelquellen in der Ebener Strömung, Insbesondere die Strömung um Zwei Dreiszylinder. *ZAMM*, Vol. 9, No. 3, pp. 200–213.

Wade, B. G. and Dwyer, M. 1976. On the Application of Morison's Equation to Fixed Offshore Platforms. OTC Paper No. 2723, Houston, TX.

Wardlaw, R. L. and Cooper, K. R. 1973. A Wind Tunnel Investigation of the Steady Aerodynamic Forces on Smooth and Stranded Twin Bundled Power Conductors for the Aluminum Company of America. National Aeronautics Establishment, Canada, LTR-LA-117.

Watkins, L. W. and Howard, M. J. 1976. Buoyancy Materials for Offshore Riser Pipe. OTC Paper No. 2654, Houston, TX.

Wheatley, J. H. W. 1976. A British Offshore Research Facility. *Trans. Roy. Inst. of Naval Architects, RINA*, Vol. 118, pp.

Wheeler, J. D. 1969. Method of Calculating Forces Produced by Irregular Waves. OTC Paper No. 1006, Houston, TX.

Weihs, D. 1977. Resistance of a Closely Spaced Row of Cylinders. *Jour. Applied Mechs., Trans. ASME*, Vol. 44, Ser. E., No. 1, pp. 177–178.

Wiegel, R. L. 1969. Waves and Their Effects on Pile Supported Structures. Report HEL-9-15, Hydraulic Engineering Lab., Univ. of Calif., Berkeley.

Wiegel, R. L., Beebe, K. E., and Moon, J. 1957. Ocean Wave Forces on Circular Cylindrical Piles. *Jour. Hydraulics Div., ASCE*, Vol. 83, HY2, pp. 1199.1–1199.36.

Wilson, B. W. 1965. Analysis of Wave Forces on a 30-inch Diameter Pile Under Confused Sea Conditions. Coastal Engrg. Res. Center, U.S. Army Tech. Memo. 15.

Wilson, J. F. and Caldwell, H. M. 1971. Force and Stability Measurements of Submerged Pipelines. *Trans. ASME, Jour. of Engineering for Industry*, pp. 1290–1298.

Wright, J. C. and Yamamoto, T. 1979. Wave Forces on Cylinders Near Plane Boundaries. *Jour. Waterways etc. Div., ASCE*, Vol. 105, No. WW1, pp. 1–13.

Yamamoto, T. 1976. Hydrodynamic Forces on Multiple Circular Cylinders. *Jour. Hydraulics Div., ASCE*, Vol. 102, No. HY9, pp. 1193–1210.

Yamamoto, T. and Nath, J. H. 1976a. High Reynolds Number Oscillating Flow by Cylinders. *Proceed. 15th Coastal Engineering Conf.*, Honolulu, HI, pp. 2321–2340.

Yamamoto, T. and Nath, J. H. 1976b. Forces on Many Cylinders Near a Plane Boundary. ASCE National Water Resources and Ocean Engineering Convention, Preprint No. 2633.

Yamamoto, T. and Nath, J. H. 1976c. Hydrodynamic Forces on Groups of Cylinders. OTC Paper No. 2499, Houston, TX.

Yamamoto, T., Nath, J. H., and Slotta, L. S. 1974. Wave Forces on Cylinders Near Plane Boundary. *Jour. Waterways etc. Div., ASCE*, Vol. 100, No. WW4, pp. 345–359.

Young, R. D., Fisher, E. A., Luke, R., and Fowler, J. R. 1977. Dynamic Analysis as an Aid to the Design of Marine Risers. ASME Paper No. 77-Pet-82.

Zdravkovich, M. M. 1977. Review of Flow Interference Between Two Circular Cylinders in Various Arrangements. *Jour. of Fluids Engineering, Trans. ASME*, Vol. 99, Ser. 1, No. 4, pp. 618–633.

6
Wave Forces on Large Bodies

6.1 THE DIFFRACTION REGIME

6.1.1 Introduction

The Morison equation is based on the assumption that the kinematics of the undisturbed flow in the region near the structure do not change in the incident wave direction. Since flow velocities and accelerations do in fact vary with a wavelength L, the assumption implicit in the use of the Morison equation is that the ratio D/L is small, where D here denotes a characteristic horizontal dimension of the structure, equivalent say to the diameter of a cylinder.

Offshore structures of large horizontal dimensions have now found extensive application and an alternative approach valid over a wide range of D/L becomes necessary. A brief classification of such structures was made in Chapter 1, and it may be seen that many modern offshore structures including gravity platforms, storage tanks, berths and so on, are included in this range. For example, a typical gravity platform may have a base section with a diameter of the order of 300 ft and columns with diameters of the order of 60 ft. Bruun (1976) has presented a general description of gravity structures in the context of North Sea operations, and Hogben (1976) has indicated the wave force regimes applicable to different categories of North Sea structures.

When a body spans a significant fraction of a wavelength, the incident waves generally undergo significant scattering or diffraction and wave force calculations should then take such scattering into account. This situation characterizes the diffraction regime of wave-structure interaction and is generally considered to occur when the structure spans more than about a fifth of the incident wave length. This is in contrast to the interaction of waves with a slender structural element, in which case flow separation dominates the loading behaviour but beyond the immediate vicinity of the element the wave train remains relatively unaffected. These two regimes of wave-structure interaction give rise to two distinct approaches by which wave force problems are treated. The first of these has been described in detail in Chapters 3 and 5 and the second approach concerning the diffraction regime is treated in the present chapter.

Fortunately the influence of wave diffraction is invariably beneficial in that it generally leads to a reduction in the wave loads that would otherwise be predicted were diffraction effects to be neglected. For if the variation of incident wave kinematics in the x direction were disregarded and if, in consequence, the inertia coefficient C_m corresponded to that of a slender structure (D/L small), then the resulting predictions would generally overestimate the actual forces exerted. In physical terms, when the particle accelerations of the undisturbed incident wave are a maximum over one portion of the structure they are not so over the rest of it. It will be seen that we may continue to employ an effective inertia coefficient, the value of which is generally found to decrease as the wavelength is reduced.

A further advantage that arises when wave diffraction is important (i.e. D/L is not too small) is that fluid particle displacements relative to D may themselves become sufficiently small for the effects of flow separation to be minimal or localized. In the specific case of a vertical circular cylinder, linear wave theory gives the Keulegan-Carpenter number at the still water level as

$$K = \frac{\pi H/L}{(D/L) \tanh (kd)} \tag{6.1}$$

and the maximum wave steepness is given approximately as

$$H/L = 0.14 \tanh (kd) \tag{6.2}$$

over a range in depths (but in shallow water H/d is about 0.83). This implies that the value of K is limited approximately to

$$K < 0.44/(D/L) \tag{6.3}$$

Thus when $D/L > 0.2$ (for wave diffraction to be important), K will not exceed about 2.2 and will usually be less than say about 1. Appreciable flow separation should therefore not occur (Chapter 3) and the effects of viscosity will be confined to the boundary layers on the body surface. It is then usually appropriate to treat the flow as irrotational and so attempt to solve the problem on the basis of potential flow theory. However, in certain situations viscous effects remain an important consideration even though the flow is unseparated. This is particularly the case for resonant structural vibrations, where even the small degree of viscous damping associated with a boundary layer flow (for example, see Rosenhead 1963) may have a marked effect.

The above discussion refers to a vertical circular cylinder. When the body contains sharp corners, as for example does a square caisson, flow separation will inevitably occur and viscous effects may no longer be dismissed without further consideration. Additional research is required to advance present understanding of these effects, but provided the effective Keulegan-Carpenter number remains small, the effects of flow separation and vortex shedding will probably be localized within regions near the corners and the overall effects, such as the total force exerted on the structure, will hopefully not be greatly affected. (See also Section 6.8.5.) In any event, if a workable approach is to be adopted it will be necessary to disregard the effects of flow separation in order to proceed with diffraction considerations alone.

6.1.2 Wave Force Regimes

A dimensional analysis relating to the wave force on a fixed body is a convenient means of indicating the conditions under which diffraction or various other effects may assume importance. Any time-invariant force F may thereby be expressed in the form

$$\frac{F}{\rho g H D^2} = f\left(\frac{d}{L}, \frac{H}{L}, \frac{D}{L}, Re\right) \tag{6.4}$$

where Re is a characteristic Reynolds number. The wave depth parameter d/L and the wave steepness H/L are sufficient to describe the incident wave train and have been discussed in Chapters 4 and 9. The body size to wave length ratio D/L is termed a *diffraction parameter* since wave diffraction or scattering becomes important when this is sufficiently large. A value $D/L = 0.2$ is often quoted as a lower limit above which diffraction effects should be taken into account. In view of the earlier discussion concerning flow separation effects, the characteristic Reynolds number may generally be omitted when diffraction is important. If, in addition, the wave steepness is small, a linearizing approximation may be made such that the force varies linearly with wave height, and H/L may then be omitted from the right-hand side of Eq. (6.4). Thus for the linear diffraction problem Eq. (6.4) reduces to

$$\frac{F}{\rho g H D^2} = f\left(\frac{d}{L}, \frac{D}{L}\right) \tag{6.5}$$

Note that the assumption of small wave steepness is invariably made in diffraction analysis and corresponds to the application of linear wave theory. The wave loads associated with other (nonlinear) wave theories under diffraction conditions are much more difficult to deal with and are discussed in Section 6.8.

Equation (6.5) may be written in the alternative form

$$\frac{F}{\rho g H D^2} = f\left(\frac{D}{L}, \frac{D}{d}\right) \tag{6.6}$$

which has the advantage that the force coefficient will vary only with D/L for a given structure located in water of a given depth.

When the ratio D/L is relatively small, flow separation rather than wave diffraction becomes important, and a characteristic Keulegan-Carpenter number K rather than D/L has more physical significance and would then be preferred in Eq. (6.4). For a vertical circular cylinder, the relationship between K (taken at the still water level and based on linear wave theory) and D/L is given by Eq. (6.1), and for waves of maximum steepness is approximately independent of depth (see Eqs. (6.2) and (6.3)). Isaacson (1979a) has presented this relationship in the form shown in Fig. 6.1 in order to provide a convenient indication of conditions under which (a) diffraction, (b) flow separation, and (c) nonlinear effects may be important. (In the figure $a = D/2$ so that $ka(=\pi D/L)$ is simply an alternative to D/L.) The figure indicates, for example, that when both K and D/L are sufficiently small neither flow separation nor wave diffraction should dominate the loading and the force is then inertial. And for steep waves nonlinear effects may be important in either the flow separation or diffraction regimes.

6.1.3 Linear Diffraction Problem

We consider now the linear diffraction problem which arises when the wave height is assumed sufficiently small for linear wave theory to apply. The problem reduces to the determination of the velocity potential ϕ which satisfies the Laplace equation

$$\nabla^2 \phi = 0 \tag{6.7}$$

within the fluid region. This is subject to the boundary conditions (linearized where appropriate):

$$\frac{\partial^2 \phi}{\partial t^2} + g\frac{\partial \phi}{\partial z} = 0 \qquad \text{at } z = 0 \tag{6.8}$$

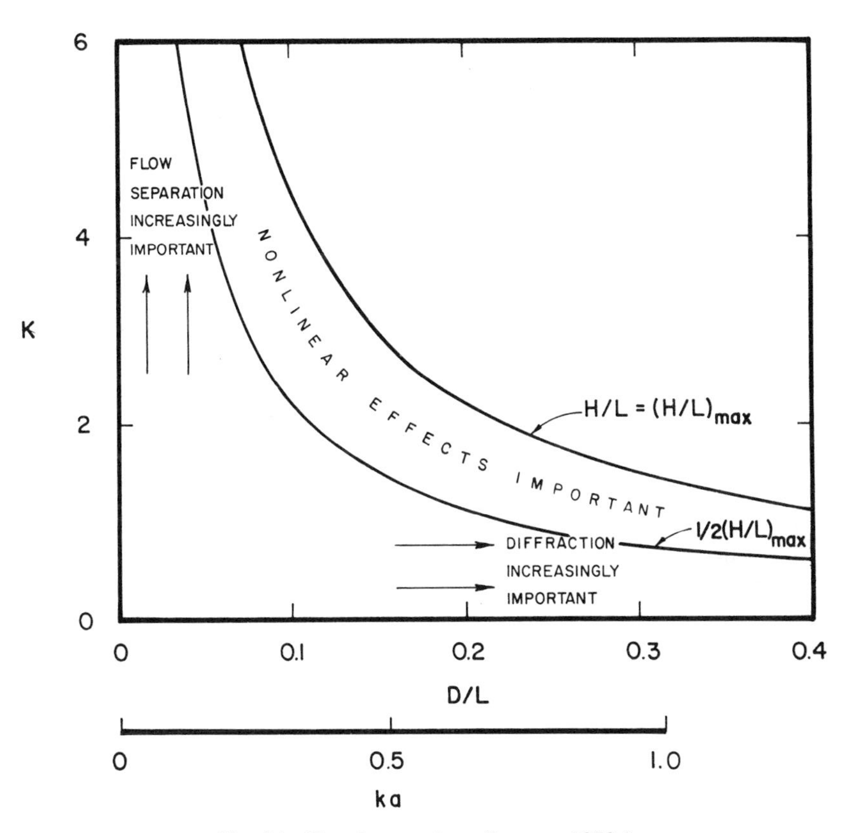

Fig. 6.1. Wave force regimes (Isaacson 1979a).

$$\eta = -\frac{1}{g}\left(\frac{\partial \phi}{\partial t}\right)_{z=0} \tag{6.9}$$

$$\frac{\partial \phi}{\partial z} = 0 \qquad \text{at } z = -d \tag{6.10}$$

$$\frac{\partial \phi}{\partial n} = 0 \qquad \text{at the body surface} \tag{6.11}$$

together with the Sommerfeld radiation condition to be discussed shortly. Here n denotes distance in a direction normal to the body surface. The above conditions have been discussed in Chapter 4 and it will be seen that Eqs. (6.8) and (6.9) derive from the linearized kinematic and dynamic free-surface boundary conditions, while Eqs. (6.10) and (6.11) correspond to the kinematic boundary conditions at the seabed and at the body surface respectively.

The radiation condition may be introduced after first representing the velocity potential as the sum of "incident wave" and "scattered wave" potentials:

$$\phi = \phi_w + \phi_s \tag{6.12}$$

and then requiring that at large distances from the structure ϕ_s corresponds to an outgoing wave. Formally this condition is given as

$$\underset{r \to \infty}{\text{Limit}}\ r^{1/2} \left[\frac{\partial \phi_s}{\partial r} - ik\phi_s \right] = 0 \tag{6.13}$$

where r is the radial ordinate.· A short description of the radiation condition was given in Section 4.9.5, and it is simply repeated here that such a condition is necessary if we are to obtain a unique solution for ϕ_s. For further discussions reference may also be made to Sommerfeld (1949) and to Stoker (1957).

The expression of the velocity potential and associated variables in the form of Eq. (6.12) and involving a separation into undisturbed incident wave and scattered wave components constitutes the basis of diffraction theory. The incident wave potential itself satisfies Eqs. (6.7)-(6.10) and is specified in complex form as

$$\phi_w = A \frac{\cosh(k(z+d))}{\cosh(kd)} e^{i(kx-\omega t)} \tag{6.14}$$

where $A = -igH/2\omega$. Because all the equations of the problem are linear, ϕ_s also satisfies Eqs. (6.7)-(6.10), as well as the radiation condition Eq. (6.13). The body surface boundary condition may be written as

$$\frac{\partial \phi_s}{\partial n} = -\frac{\partial \phi_w}{\partial n} \quad \text{at the body surface} \tag{6.15}$$

and provides for the dependence of ϕ_s on ϕ_w. Thus Eqs. (6.7)-(6.10) applied to ϕ_s, together with Eqs. (6.13) and (6.15), define the problem in terms of ϕ_s. Once ϕ_s and consequently ϕ are determined, the pressure throughout the fluid may be evaluated by the linearized Bernoulli equation

$$p = -\rho g z - \rho \frac{\partial \phi}{\partial t} \tag{6.16}$$

Appropriate integrations of the pressure acting on the body surface may then be carried out to obtain sectional forces, total forces and overturning moments as required.

6.1.4 Diffraction and Effective Inertia Coefficients

For reference purposes, it is sometimes convenient to consider also that component of the total force on the body due to the pressure field of the undisturbed incident waves alone: that is the force which would exist on the body if it were 'transparent' to the wave motion. This force is termed the *Froude-Krylov force* F_k and is relatively simple to calculate since it involves only ϕ_w which is known. It is subsequently convenient to define a *diffraction coefficient* as the ratio of the actual force amplitude to the Froude-Krylov force amplitude. Thus horizontal and vertical force diffraction coefficients, denoted C_h and C_v respectively, may be separately defined in this way with, for example,

$$C_h = F_T/F_k \tag{6.17}$$

in which F_T and F_k are the horizontal total force and horizontal Froude-Krylov force amplitudes respectively.

The slight advantage of this approach is that the determination of the Froude-Krylov force is straightforward and, with F_k known, it may then be possible to rely on approximate values of C_h and C_v to estimate the total force amplitudes on the structure. However, it is emphasized that the technique does not itself provide a solution since, for example, C_h is initially itself as much an unknown as is F_T. In addition, a minor disadvantage is that the Froude-Krylov force may be zero (and the diffraction coefficient will therefore be infinite) at certain wavelengths, but this does not generally impose a serious problem since it is relatively uncommon for a practical range of wave frequencies.

An alternative approach sometimes employed is to consider the force to be given by the inertia component of the Morison equation, even though its phase differs from that of the fluid acceleration of the incident waves at a reference point. In this way an *effective inertia coefficient* C_m may be used to describe the force amplitude, and a phase angle δ to define its phase relative to the incident wave train. Thus a sectional force F_z would be given as

$$F_z = \rho S C_m \left(\frac{\partial u}{\partial t}\right)_m \cos(\omega t - \delta) \tag{6.18}$$

where S is the sectional area of the structure, $(\partial u/\partial t)_m$ denotes the amplitude (maximum) of fluid acceleration at the reference point (say the section center) and δ is the phase angle by which the force lags behind the incident wave crest passing through $x = 0$. (That is, when the force is wholly inertial $\delta = -\pi/2$.)

In the present chapter the diffraction solution for a vertical circular cylinder extending from the seabed and piercing the free surface will first be developed (Section 6.2). This is both of direct practical application and also forms a fundamental reference for many other practical situations. Attention is then given to a numerical approach pertaining to bodies of arbitrary geometry.

There are various extensions and associated problems to be considered beyond this fundamental approach. Of these, the extension of the linear diffraction theory to random waves is straightforward in principle and will be discussed in Chapter 7. Consideration will also be given to extensions which take into account special body configurations (Section 6.4), the effects of body motions (Section 6.6), steep (nonlinear) incident waves (Section 6.8), and underlying currents (Section 6.9). Finally, two fundamental examples of wave force calculations under diffraction conditions will be given (Section 6.10). Reference may also be made to available reviews concerning wave diffraction problems and associated wave forces, which include those by Newman (1972), Hogben (1974a), Hogben et al. (1977), Mei (1978), Garrison (1978) and Isaacson (1979a).

6.2 VERTICAL CIRCULAR CYLINDER

An analytic solution to the linear wave diffraction problem is available for an isolated vertical circular cylinder extending from the seabed and piercing the free surface. This was treated initially by Havelock (1940) for the deep water range, then by Omer and Hall (1949) for the shallow water range and subsequently by MacCamy and Fuchs (1954) for general depths. Omer and Hall were concerned with predicting the wave runup around a circular island and presented a comparison of their prediction with observed tsunami runup around the island of Kauai. The work of MacCamy and Fuchs is widely referred to in the wave force literature, not only because their study was the first pertaining to arbitrary depths, but also because emphasis was given to the wave-induced loads on the cylinder.

It is convenient in the present section to work with a cylindrical coordinate system (r, θ, z) as indicated in Fig. 6.2, and Eqs. (6.7)-(6.15) which define our problem may be transformed accordingly. The Laplace equation in cylindrical coordinates becomes

$$\frac{\partial^2 \phi}{\partial r^2} + \frac{1}{r}\frac{\partial \phi}{\partial \theta} + \frac{1}{r^2}\frac{\partial^2 \phi}{\partial \theta^2} + \frac{\partial^2 \phi}{\partial z^2} = 0 \tag{6.19}$$

while the body surface boundary condition assumes the form

$$\frac{\partial \phi}{\partial r} = \frac{\partial \phi_w}{\partial r} + \frac{\partial \phi_s}{\partial r} = 0 \quad \text{at } r = a \tag{6.20}$$

where a is the cylinder radius. The incident wave potential ϕ_w is known and is given by Eq. (6.14). In the present problem it is re-expressed in the cylindrical

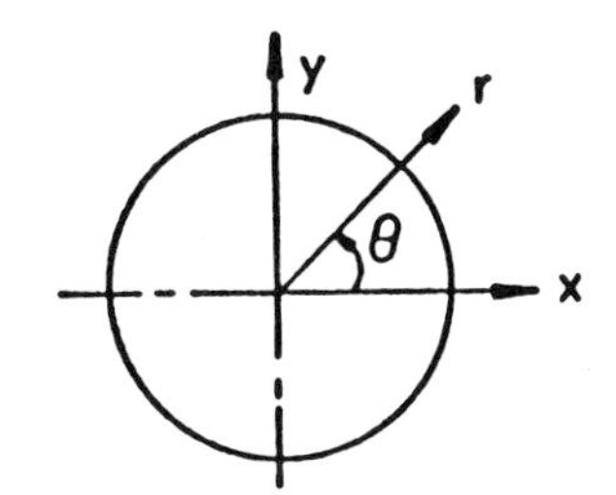

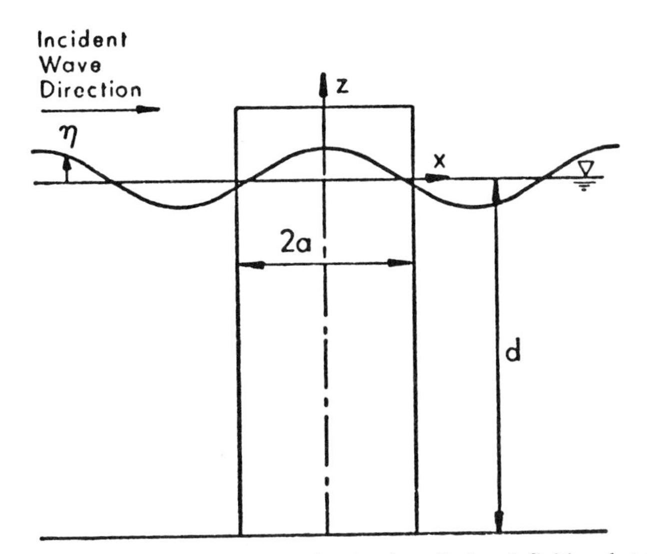

Fig. 6.2. Wave diffraction around a circular cylinder: definition sketch.

coordinate system by making use of the identity

$$e^{ikx} = e^{ikr\cos\theta} = \cos(kr\cos\theta) + i\sin(kr\cos\theta) \qquad (6.21)$$

The right-hand side of Eq. (6.21) may be expanded as an infinite series of Bessel functions (e.g. Abramowitz and Stegun 1965), and the incident velocity potential may accordingly be written as

$$\phi_w = A\frac{\cosh(k(z+d))}{\cosh(kd)}\left[\sum_{m=0}^{\infty} \beta_m J_m(kr)\cos(m\theta)\right] e^{-i\omega t} \qquad (6.22)$$

where

$$\beta_m = \begin{cases} 1 & \text{for } m = 0 \\ 2i^m & \text{for } m \geqslant 1 \end{cases} \qquad (6.23)$$

and $J_m(kr)$ is the Bessel function of the first kind of order m and argument kr. Clearly the even Fourier series in θ is expected on account of the symmetry about the x axis.

We now seek an expression for ϕ_s which satisfies the Laplace equation and the various boundary conditions previously outlined. This is given as a series similar to Eq. (6.22) but based on the Hankel function of the first kind with initially unknown (complex) coefficients

$$\phi_s = A \frac{\cosh(k(z+d))}{\cosh(kd)} \left[\sum_{m=0}^{\infty} \beta_m B_m H_m^{(1)}(kr) \cos(m\theta) \right] e^{-i\omega t} \tag{6.24}$$

Such an expression directly satisfies the bottom and free-surface boundary conditions together with the Laplace equation, in the same way as does the expansion for ϕ_w, Eq. (6.22). Note that the Laplace equation reduces to a Bessel equation describing the Fourier coefficients' dependence on r, and the general solution of this is given by a linear combination of the Hankel functions of the first and second kinds. Of these only $H_m^{(1)}(kr)$ satisfies the far field condition: for large values of the argument kr, the asymptotic form of $H_m^{(1)}(kr)$ is

$$H_m^{(1)}(kr) \to (2/\pi kr)^{1/2} \exp\left[i(kr - (2m+1)\pi/4)\right] \tag{6.25}$$

which is readily seen to satisfy the radiation condition, Eq. (6.13). It remains only to utilize the boundary condition on the cylinder surface to determine the unknown coefficients B_m. Substituting Eqs. (6.22) and (6.24) into Eq. (6.20) we obtain

$$B_m = -J'_m(ka)/H_m^{(1)\prime}(ka) \tag{6.26}$$

where a prime denotes differentiation with respect to the argument. The complete solution is thus given as

$$\phi = A \frac{\cosh(k(z+d))}{\cosh(kd)} \left[\sum_{m=0}^{\infty} \beta_m \left(J_m(kr) - \frac{J'_m(ka)}{H_m^{(1)\prime}(ka)} H_m^{(1)}(kr) \right) \cos(m\theta) \right] e^{-i\omega t} \tag{6.27}$$

Once ϕ has been determined, all the variables of interest may be obtained directly and are listed in Table 6.1. In particular, use is made of the identity (Abramowitz and Stegun 1965)

$$J_m(ka) - \frac{J'_m(ka)}{H_m^{(1)\prime}(ka)} H_m^{(1)}(ka) = \frac{2i}{\pi ka H_m^{(1)\prime}(ka)} \tag{6.28}$$

Table 6.1 Results of diffraction theory for a vertical circular cylinder.

$$\frac{\phi}{gH/\omega} = -\frac{1}{2}\frac{\cosh(ks)}{\cosh(kd)}\left\{\sum_{m=0}^{\infty} i\beta_m\left[J_m(kr) - \frac{J'_m(ka)}{H_m^{(1)'}(ka)}H_m^{(1)}(kr)\right]\cos(m\theta)\right\}e^{-i\omega t}$$

$$\left(\frac{\eta}{H}\right)_{r=a} = \left\{\sum_{m=0}^{\infty}\frac{i\beta_m\cos(m\theta)}{\pi kaH_m^{(1)'}(ka)}\right\}e^{-i\omega t}$$

$$\left(\frac{p}{\rho gH}\right)_{r=a} = -\frac{z}{H} + \frac{\cosh(ks)}{\cosh(kd)}\sum_{m=0}^{\infty}\frac{i\beta_m\cos(m\theta)}{\pi kaH_m^{(1)'}(ka)}e^{-i\omega t}$$

$$\frac{\partial F/\partial s}{\rho gHa} = 2\frac{A(ka)}{ka}\frac{\cosh(ks)}{\cosh(kd)}\cos(\omega t - \delta)$$

$$\frac{F}{\rho gHad} = 2\frac{A(ka)}{ka}\frac{\tanh(kd)}{kd}\cos(\omega t - \delta)$$

$$\frac{M}{\rho gHad^2} = 2\frac{A(ka)}{ka}\left[\frac{kd\sinh(kd) + 1 - \cosh(kd)}{(kd)^2\cosh(kd)}\right]\cos(\omega t - \delta)$$

$$\frac{F_k}{\rho gHad} = -\pi J_1(ka)\frac{\tanh(kd)}{kd}\sin(\omega t)$$

$$C_h = \frac{2A(ka)}{kaJ_1(ka)}, \qquad C_m = \frac{4A(ka)}{\pi(ka)^2}$$

where the real parts of complex expressions are understood, and

$$s = z + d,$$
$$\beta_0 = 1, \beta_m = 2i^m \text{ for } m \geqslant 1,$$
$$A(ka) = [J_1'^2(ka) + Y_1'^2(ka)]^{-1/2},$$
$$\delta = -\tan^{-1}[Y_1'(ka)/J_1'(ka)].$$

The free surface elevation η is obtained by virtue of Eq. (6.9), the runup R around the cylinder is the maximum value of η at r = a, and the pressure p acting on the cylinder periphery is given as

$$p = -\rho gz - \rho\left(\frac{\partial\phi}{\partial t}\right)_{r=a} \tag{6.29}$$

The sectional force acts in the x direction since the flow is symmetric about the x axis, and is found by integrating the pressure around the cylinder as

$$F_z = -\int_0^{2\pi} p(\theta, a)\, a\cos\theta\, d\theta \tag{6.30}$$

The total force F and overturning moment M may then be evaluated from the expressions

$$F = \int_{-d}^{0} F_z \, dz \tag{6.31}$$

$$M = \int_{-d}^{0} F_z(z + d) \, dz \tag{6.32}$$

The Froude-Krylov force F_k, which has been already discussed, is given in the present case as

$$F_k = \rho a \int_{-d}^{0} \int_{0}^{2\pi} \left(\frac{\partial \phi_w}{\partial t}\right)_{r=a} \cos\theta \, d\theta \, dz \tag{6.33}$$

This integration is relatively straightforward and the result is included in the table. Finally the effective inertia coefficient C_m and corresponding phase angle δ may be obtained by using Eq. (6.18) to describe the sectional force already calculated, and expressions for C_m and δ are also set out in the table. It is easily shown that the sectional force, total force and overturning moment are given in terms of C_m and δ as

$$F_z = \frac{\pi}{8} \rho g H k D^2 \frac{\cosh(k(z+d))}{\cosh(kd)} C_m \cos(\omega t - \delta) \tag{6.34a}$$

$$F = \frac{\pi}{8} \rho g H D^2 \tanh(kd) \, C_m \cos(\omega t - \delta) \tag{6.34b}$$

$$M = \frac{\pi}{8} \rho g H \frac{D^2}{k} \left[\frac{kd \sinh(kd) + 1 - \cosh(kd)}{\cosh(kd)}\right] C_m \cos(\omega t - \delta) \tag{6.34c}$$

The variations of $A(ka)$, C_h, C_m and δ with ka are presented in Fig. 6.3 and may be used in conjunction with Table 6.1 to determine the loading for any given conditions.

Mogridge and Jamieson (1976b) have emphasized the use of appropriate curves in calculating as simply as possible the force and moment maxima and corresponding phase angle δ. Equations (6.34b) and (6.34c) can be expressed as

$$F_{max} = \frac{\pi^2 \rho H L D^2}{4T^2} C_m(ka) \tag{6.35}$$

$$M_{max} = \rho g H L D^2 C_m(ka) f(kd) \tag{6.36}$$

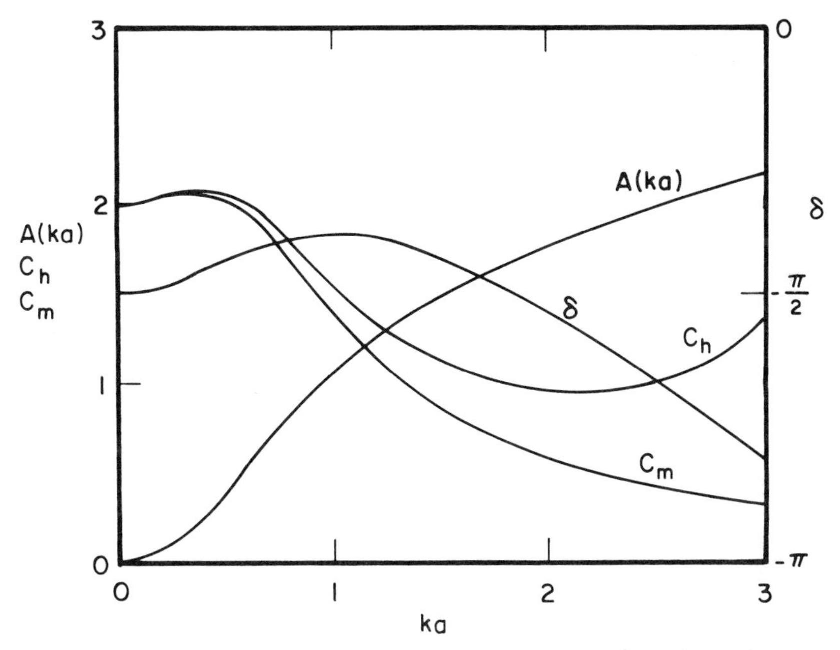

Fig. 6.3. Wave diffraction around a circular cylinder: variations of A(ka), C_h, C_m and δ with ka.

where f(kd) = [kd tanh (kd) + sech (kd) - 1]/16. Mogridge and Jamieson have provided convenient graphs of C_m and δ versus D/L and f(kd) versus d/L, and F_{max}, M_{max} and δ may be obtained from these very simply. Equivalently, it is often useful to refer to a prepared table showing these variations and including other parameters of possible interest such as the maximum runup.

As a simple example, a design wave given by H = 20 ft, T = 6 sec and d = 60 ft and a cylinder of diameter D = 40 ft correspond to the parameters kd and ka being given by linear wave theory as 2.11 and 0.70 respectively. Consequently $C_m = 1.80$, $\delta = 73.5°$, and thus the total force and overturning moment maxima are 1,370 kips and 51,710 kip-ft respectively and these occur 1.2 sec before an incident wave crest crosses the cylinder axis.

Runup profiles $R(\theta)/H$ around the cylinder for various values of ka are given in Fig. 6.4 and the maximum runup R_M/H as a function of ka is given in Fig. 6.5. For the example considered above the maximum runup is accordingly 16.6 ft. Of course the diffraction parameter ka used here is equivalent to D/L (ka = πD/L) which is sometimes used instead and which was adopted in the more general discussion in Section 6.1.

When ka is not too large, series expressions for the Bessel functions may be employed to provide an approximate solution valid over a restricted range of ka, and in the limit ka → 0 the predicted force becomes wholly inertial with

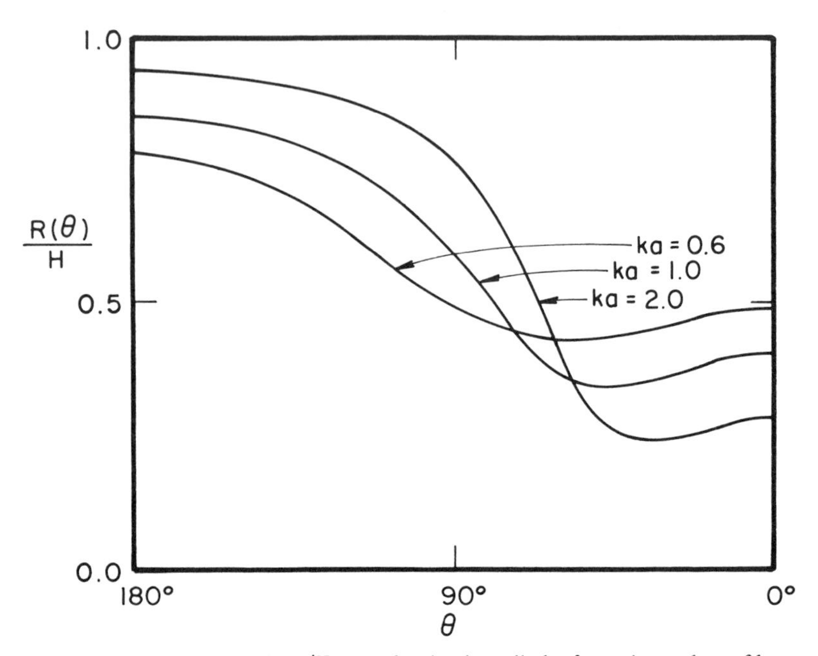

Fig. 6.4. Runup profiles $R(\theta)/H$ around a circular cylinder for various values of ka.

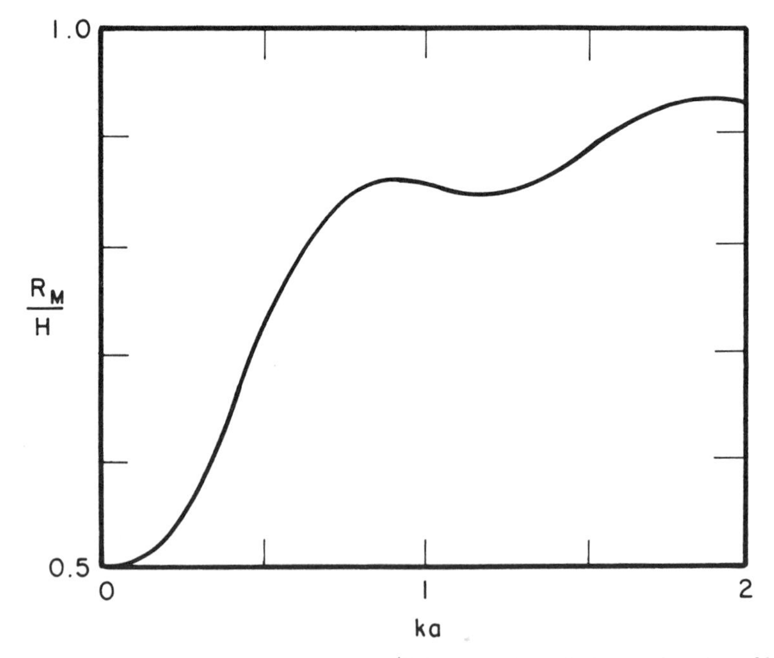

Fig. 6.5. Maximum dimensionless runup R_M/H for a circular cylinder as a function of ka.

$C_m = 2.0$. Such a procedure is really only useful when applied to the runup expression which involves a series. Isaacson (1978a) gives the approximate expression for the runup profile as

$$R(\theta)/H = \frac{1}{2}\{1 + (ka)^2 [2\cos^2\theta + \ln(ka/2) + (2\gamma - 1)/2]\}^{1/2} \quad (6.37)$$

where $\gamma = 0.5772\ldots$ is Euler's constant. The maximum runup is therefore given as

$$R_M/H = \frac{1}{2}\{1 + (ka)^2 [\ln(ka/2) + (2\gamma + 3)/2]\}^{1/2} \quad (6.38)$$

These differ from simpler expressions originally proposed by MacCamy and Fuchs (1954).

It is appropriate at this point to consider various comparisons that have been carried out between experiment and the solution described. Several researchers have conducted experiments aimed at providing a comparison between experimental results and the closed-form solution for a vertical circular cylinder. These include studies by Chakrabarti and Tam (1973), Hogben and Standing (1975), Yamaguchi and Tsuchiya (1974), Raman and Venkatanarasaiah (1976) and Mogridge and Jamieson (1975). Agreement has generally been reasonably good

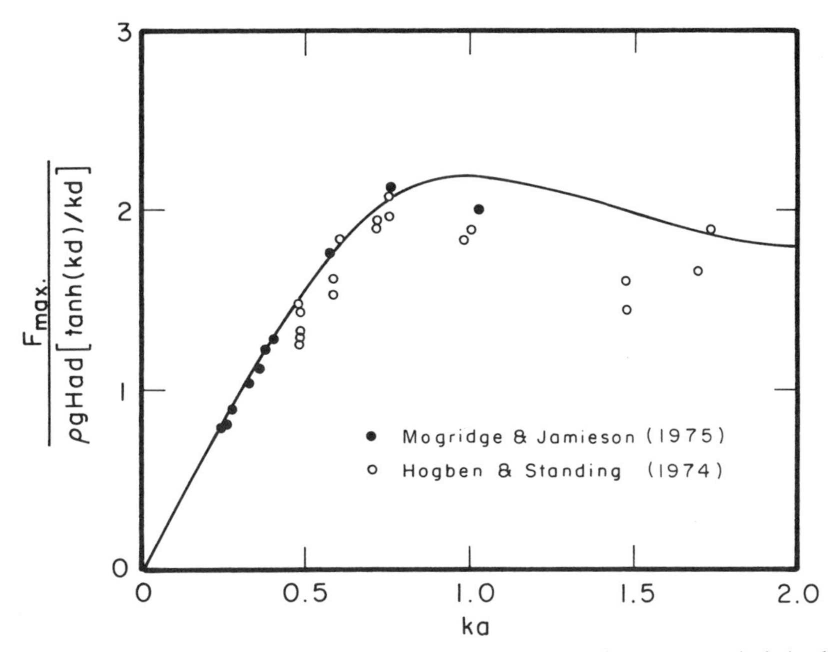

Fig. 6.6. Comparison of the closed-form solution for the wave force on a vertical circular cylinder with the experimental results of Hogben and Standing (1974) and Mogridge and Jamieson (1975).

although the tests reported were conducted outside the shallow water range and with wave steepnesses H/L less than about 0.08, where effects of wave nonlinearity are not expected to be significant. Nevertheless, the experiments do cover the more important practical range of conditions and it follows that the theorectical predictions may be taken as acceptable for design purposes. In order to illustrate the nature of the correlations obtained, the results of Hogben and Standing (1975) and Mogridge and Jamieson (1975), together with the closed-form solution of MacCamy and Fuchs (1954), are shown in Fig. 6.6.

Another measure of comparison between experiment and prediction for the cylinder relates to the runup around the cylinder, and such experiments relating to the diffraction range have been carried out by Chakrabarti and Tam (1975), and by Hallermeier (1976). In this case, however, the agreement with the theoretical solution has not been as close as it is in the comparisons relating to the total force, and it appears that the linear theory generally underestimates the maximum runup. (See also Section 6.8.3.)

6.3 BODIES OF ARBITRARY GEOMETRY

The preceeding theory for a vertical circular cylinder provides an important step in understanding the effects of wave diffraction on large bodies. The solution obtained has a wide range of applications, but it eventually becomes necessary to take up the case of bodies of arbitrary geometry in order to deal with the variety and complexity of design configurations encountered in modern offshore structures. Such a treatment must necessarily be based on a numerical approach and the more common "wave source" or "integral equation" method involves developing a surface integral equation and solving this by a discretization procedure. The method has now become firmly established in design practice and is reviewed in the present section. More detailed descriptions are provided by Garrison and Chow (1972), Hogben and Standing (1974), Hogben, Osborne and Standing (1974) and Garrison (1978). Attention is also drawn to a review paper by Hogben et al. (1977) in which there are tabulated some available diffraction programs based on the wave source method about to be described. These include programs by Lebreton and Cormault (1969), Garrison (Garrison and Rao 1971, Garrison and Chow 1972 and Garrison 1974a, b), van Oortmerssen (1972), Hogben and Standing (1974) and Faltinsen and Michelsen (1974).

The wave source approach described here is a particular case of the method of integral equations, in which an integral equation is set up to define the diffraction problem. For example, alternative integral equations describing a particular two-dimensional problem are outlined in Section 6.4.2; and a separate integral equation for bodies of arbitrary shape is given in Section 6.4.5. An entirely different approach based on the finite element method has been developed more recently and is outlined in Section 6.5.

6.3.1 Theoretical Formulation

Because the problem is linear and the flow is harmonic in time, the dependence on time may be separated and the velocity potential ϕ may be expressed as

$$\phi(x, y, z, t) = \text{Re}\,\{\phi'(x, y, z)e^{-i\omega t}\} \tag{6.39}$$

Here ϕ' is generally complex, and as already described is conveniently expressed as the sum of incident wave and scattered wave components

$$\phi' = \phi'_w + \phi'_s \tag{6.40}$$

In describing the wave source approach, we first note that a fundamental result of potential theory is that the velocity potential of the scattered waves ϕ_s may be represented as due to a continuous distribution of point wave sources over the immersed body surface. This result is described by Lamb (1945) (see also Wehausen and Laitone 1960). If the potential of the fluid due to a point source of unit strength located at the point $\boldsymbol{\xi} = (\xi, \eta, \zeta)$ is known, then on account of the linearity of the problem this may be amplified to any required strength and then superposed with any number of other wave sources. The velocity potential due to the whole (continuous) distribution of sources over the body surface is then given as

$$\phi'_s(\mathbf{x}) = \frac{1}{4\pi}\int_S f(\boldsymbol{\xi})G(\mathbf{x},\boldsymbol{\xi})\,dS \tag{6.41}$$

Here $\mathbf{x}$ represents the point (x, y, z), $G(\mathbf{x},\boldsymbol{\xi})$ is the Green's function of a point wave source of unit strength located at the point $\boldsymbol{\xi} = (\xi, \eta, \zeta)$; $f(\boldsymbol{\xi})$ is the source strength distribution function, and dS is a differential area on the immersed body surface. The Green's function, which is singular at the source point $\boldsymbol{\xi}$, must itself satisfy the Laplace equation, the bottom and linearized free-surface boundary conditions, together with the radiation condition. Such a Green's function was developed by John (1950) and may be expressed either in terms of an integral or as an infinite series. The former expression is

$$G(\mathbf{x},\boldsymbol{\xi}) = \frac{1}{R} + \frac{1}{R'}$$

$$+\, 2 \,{-}\!\!\!\!\!\!\int_0^\infty \frac{(\mu+\nu)e^{-\mu d}\cosh(\mu(\zeta+d))\cosh(\mu(z+d))}{\mu\sinh(\mu d) - \nu\cosh(\mu d)} J_0(\mu r)\,d\mu$$

$$-\, iC_0\cosh(k(\zeta+d))\cosh(k(z+d))J_0(kr) \tag{6.42}$$

where

$$R = [(x - \xi)^2 + (y - \eta)^2 + (z - \zeta)^2]^{1/2}$$

$$R' = [(x - \xi)^2 + (y - \eta)^2 + (z + 2d + \zeta)^2]^{1/2}$$

$$r = [(x - \xi)^2 + (y - \eta)^2]^{1/2}$$

$$\nu = k \tanh (kd) = \omega^2/g$$

$$C_0 = \frac{2\pi(\nu^2 - k^2)}{(k^2 - \nu^2)d + \nu}$$

and f denotes the Cauchy principal value of the integral referred to. The second (and equivalent) form of the Green's function involves a series representation

$$G(\mathbf{x}, \boldsymbol{\xi}) = -iC_0 \cosh (k(\zeta + d)) \cosh (k(z + d)) H_0^{(1)}(kr)$$

$$+ 4 \sum_{m=1}^{\infty} C_m \cos (\mu_m(\zeta + d)) \cos (\mu_m(z + d)) K_0(\mu_m r) \quad (6.43)$$

where K_0 denotes the modified Bessel function of the second kind and order zero,

$$C_m = \frac{\mu_m^2 + \nu^2}{(\mu_m^2 + \nu^2)d - \nu}$$

and μ_m are the real positive roots of the equation

$$\mu_m \tan (\mu_m d) + \nu = 0 \quad (6.44)$$

taken in ascending order. It next becomes necessary to determine the source strength distribution function $f(\boldsymbol{\xi})$. This may be affected by ensuring that the flow satisfies the one remaining boundary condition: that on the body surface and as given by Eq. (6.15). Substituting Eq. (6.41) into Eq. (6.15), and bearing in mind that the effect of the singularity in the normal velocity induced by a source at its own location requires careful evaluation, the boundary condition may be shown (Wehausen and Laitone 1960) to assume the form

$$-\frac{1}{2} f(\mathbf{x}) + \frac{1}{4\pi} \int_S f(\boldsymbol{\xi}) \frac{\partial G}{\partial n} (\mathbf{x}, \boldsymbol{\xi})\, dS = -\frac{\partial \phi'_w}{\partial n} (\mathbf{x}) \quad (6.45)$$

Here n is the distance normal to the surface at the point **x** where the boundary condition is applied (and which must itself lie on the body surface), and the surface integral is taken over the points $\boldsymbol{\xi}$. Equation (6.45) is a Fredholm integral equation of the second kind which must be solved for $f(\boldsymbol{\xi})$.

Milgram and Halkyard (1971) have suggested that numerical iteration is not in general a suitable process for obtaining the solution to Eq. (6.45) since convergence is difficult to achieve. A more suitable approach involves a numerical process of 'discretization' in which the submerged body surface is divided up into a finite number of small elements of area or facets. The boundary condition is then satisfied at the center of each facet, with the source strength distribution taken to be uniform over each facet. In this way the integral equation is reduced to a finite set of linear algebraic equations involving the source strength f_i over the i-th facet. These equations may be written as

$$\sum_{j=1}^{N} B_{ij} f_j = b_i \qquad \text{for } i = 1, 2, \ldots . N \tag{6.46}$$

where N is the number of facets. These equations correspond physically to the influence of all the facets j = 1 to N and the incident wave potential (b_i term) on the boundary condition taken at the i-th facet, and this condition is applied in turn to all the facets i = 1 to N. Fig. 6.7 shows part of a structure divided into facets and indicates a particular combination of i and j. The coefficients b_i and

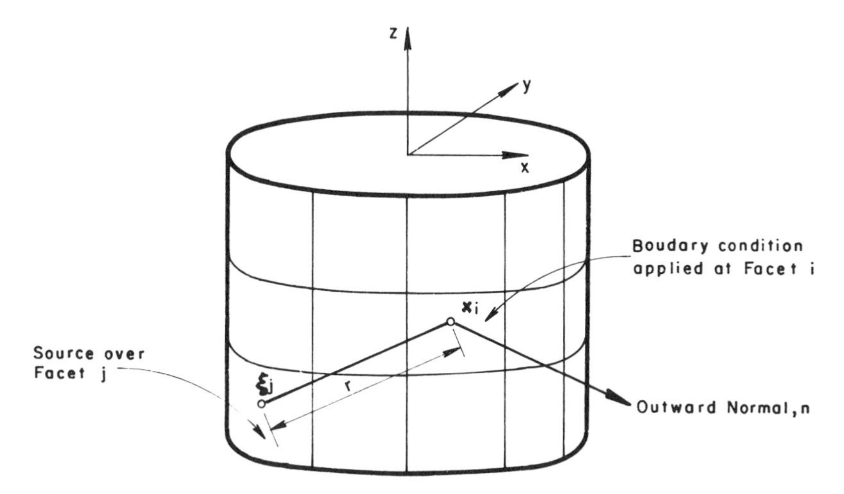

Fig. 6.7. Illustration of boundary condition applied at facet i due to a wave source distribution over facet j.

B_{ij} are given respectively by

$$b_i = -2 \frac{\partial \phi'_w}{\partial n} (\mathbf{x}_i) \tag{6.47}$$

and

$$B_{ij} = -\delta_{ij} + \frac{1}{2\pi} \int_{\Delta S_j} \frac{\partial G}{\partial n} (\mathbf{x}_i, \boldsymbol{\xi})\, dS \tag{6.48}$$

In the above δ_{ij} is the Kronecker delta ($\delta_{ij} = 0$ for $i \neq j$, $\delta_{ii} = 1$), the point $\mathbf{x}_i = (x_i, y_i, z_i)$ is the centroid of the i-th facet of area ΔS_i and n is measured normal to the surface at the point $\mathbf{x}_i$. By taking $\partial G/\partial n$ to be constant over the facet and equal to its value at the facet controid, the expression for B_{ij} approximates to

$$B_{ij} = -\delta_{ij} + \frac{\Delta S_j}{2\pi} \frac{\partial G}{\partial n} (\mathbf{x}_i, \boldsymbol{\xi}_j) \tag{6.49}$$

and $\partial G/\partial n$ may be evaluated in the form

$$\frac{\partial G}{\partial n} = \frac{\partial G}{\partial x} n_x + \frac{\partial G}{\partial y} n_y + \frac{\partial G}{\partial z} n_z \tag{6.50}$$

in which n_x, n_y, n_z are the components of the unit normal vector defining the facet orientation. When $i = j$, the term δ_{ij} is unity and accounts for the velocity at the facet center due to the uniform source distribution on that facet itself, and the last term in Eq. (6.48) may therefore be omitted.

The column vector b_i may be expressed in a reasonably compact form by resort to the expression for ϕ_w, Eq. (6.14), and is given as

$$b_i = -2 \frac{\partial \phi'_w}{\partial n} = -2kA\, e^{ikx} \left[i n_x \frac{\cosh (k(z + d))}{\cosh (kd)} + n_z \frac{\sinh (k(z + d))}{\cosh (kd)} \right] \tag{6.51}$$

This makes possible the evaluation of the matrix B_{ij} and the column vector b_i; and the source strength distribution function f_j may then be obtained by a complex matrix inversion procedure. The potential ϕ'_s around the body surface may subsequently be evaluated by using a discrete version of Eq. (6.41) which is

$$\phi'_s (\mathbf{x}_i) = \sum_{j=1}^{N} A_{ij} f_j \tag{6.52}$$

where

$$A_{ij} = \frac{1}{4\pi} \int_{\Delta S_j} G(\mathbf{x}_i, \boldsymbol{\xi})\, dS \tag{6.53}$$

When $i \neq j$, G is taken to be constant over the j-th facet and equal to it's value at the facet centroid, and thus A_{ij} approximates to

$$A_{ij} = \frac{\Delta S_j}{4\pi} G(\mathbf{x}_i, \boldsymbol{\xi}_j) \tag{6.54a}$$

When $i = j$, a singularity occurs in Eq. (6.53) but the integrand remains capable of evaluation for, if we consider only the singular term in the Green's function (which is dominant), we obtain

$$A_{ii} = \frac{1}{4\pi} \int_{\Delta S_i} \frac{1}{R}\, dS \tag{6.54b}$$

which may be integrated for a given facet shape. For example, if the facet is circular, integration yields

$$A_{ii} = \tfrac{1}{2} \sqrt{\Delta S_i/\pi} \tag{6.55}$$

Hogben and Standing (1974) have provided expressions for evaluating the integral in the case of plane polygonal facets.

Once $(\phi_w')_i$ and $(\phi_s')_i$ have been determined, it becomes possible to obtain the variables of interest by developments based on the following fundamental relationships

$$p = -\rho g z + i\omega\rho\phi' e^{-i\omega t} \tag{6.56}$$

$$\eta = \left(\frac{i\omega\phi'}{g}\right)_{z=0} e^{-i\omega t} \tag{6.57}$$

$$\mathbf{F} = -\int_S p\mathbf{n}\, dS \tag{6.58}$$

$$\mathbf{M} = -\int_S p(\mathbf{r} \times \mathbf{n})\, dS \tag{6.59}$$

Here **F** and **M** are the force and overturning moment vectors acting on the body, **n** is the normal vector at a point on the immersed body surface and taken as positive out of the body (see Fig. 6.7), **r** is the position vector of that point, and the integration is taken over the immersed body surface. In the numerical approach being described, the surface integrals are replaced by summations involving the values of ϕ' and also the unit normal vector components n_x, n_y, n_z at the center of each facet. Note that where more than one body is considered, the loads on any single body will be obtained when the summations are carried over the facets of that body only.

It is mentioned that it is possible to achieve a reduction in computational effort whenever the body contains a vertical plane of symmetry at an arbitrary orientation. This derives from replacing the integral equation for the whole body surface by two integral equations each involving half the body surface. Reference may be made to Hogben, Osborne and Standing (1974) for details of this development.

One difficulty of the wave source method that is sometimes encountered is that it breaks down at certain 'irregular' wave frequencies (see John 1950) which generally correspond to wave lengths of the order of or less than the characteristic size of the body. This behavior has been described by Murphy (1978) and is an inherent property of the source distribution representation Eq. (6.41) rather than of the numerical process of discretization: at these frequencies there is no unique solution to the integral equation and the problem cannot then be solved by employing the integral formula Eq. (6.41). Murphy illustrates the breakdown for the case of the surface-piercing circular cylinder for which a closed-form solution is known (MacCamy and Fuchs 1954). He shows that solutions are not possible using the wave source representation for those wave lengths corresponding to $J_m(ka) = 0$ for any m: that is $ka = 2.40, 3.83, \ldots$. The difficulty is not too serious because the relatively short wave lengths corresponding to the irregular frequencies are not usually critical in design.

This completes in outline the method by which the loading may be determined on the basis of a source distribution representation of the body surface. An alternative approach based on the finite element method will be mentioned in Section 6.5, but in the meanwhile we proceed to consider certain computational aspects that arise in connection with the technique so far described.

6.3.2 Computational Considerations

The method just described is reliable when the facet sizes are small and when a correspondingly large number of facets are employed. In practice, computer storage and costs limit the number of facets to be used to typically of order a hundred. This implies that complex structures may only be represented in a somewhat coarse fashion and attention must therefore be given to the most suitable way to select the facet geometry in order to ensure the reliability of results obtained. Many of the guidelines proposed were initially developed in the context of uniform steady flows past complex shapes (Hess and Smith 1967) but have been restated to accomodate the present context by Hogben and Standing 1974) (see also Hogben, Osborne and Standing 1974). These guidelines are based on experience rather than on rigid principles. The usual representation of the body surface is by plane polygonal facets and this kind of partitioning is schematized in Fig. 6.8.

To ensure that the body is divided up sufficiently finely, facet diameters should be less than about $\frac{1}{8}$ of the incident wave length. This implies that loads due to short waves on a large body may not be reliably computed as any accurate prediction would involve an impracticably large number of facets. However, loads due to such short waves are not usually critical in engineering design and this apparent limitation on the method is usually not too restrictive. There do, however, exist alternative approaches available for certain classes of body geometry and these will be outlined in Section 6.4.

Facet diameters should be less than the (smallest) local radius of curvature of the body surface in order to represent that surface reasonably well. It follows that facets should be concentrated in areas of high curvature. Some smoothing off of sharp corners may be adopted to conform to this. Note, however, that if any smoothing process leads to a reduction in the effective volume of the structure, a corresponding reduction in the computed force may be expected in the

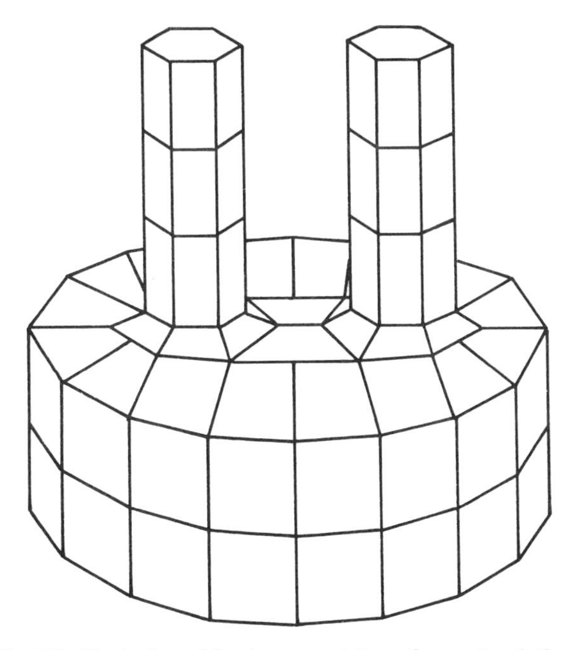

Fig. 6.8. Illustration of facet representation of a gravity platform.

inertia range. Furthermore, Mattioli and Tinti (1979) have pointed out possible inaccuracies of the mid-point approximation in which a Green's function or its derivative is treated as a constant over a facet (for example, such that Eq. (6.48) is simplified to Eq. (6.49)). In the context of a two-dimensional flow (Sections 6.4.2, 6.4.3) they illustrate the inaccuracies of this approximation for neighboring facets inclined to each other. Finally, neighboring facet diameters should not differ by more than about 50%, since an excessively large facet surrounded by smaller facets will lead to computational inefficiency in that the precision provided by the smaller facets will be lost.

A thin wall (as, for example, is contained in an open well-type structure) would require an excessively large number of facets distributed on either side of the wall to ensure that the facet diameter is less than the wall thickness. Thus the method should generally be applied to closed and reasonably blunt bodies. The modeling of a thin wall by a distribution of wave doublets may be possible and is reviewed in Section 6.4.5.

The Green's function for a point wave source is given in integral and series forms by Eqs. (6.42) and (6.43) respectively. In Eqs. (6.49) and (6.54a) it is necessary to evaluate $\partial G/\partial n$ and G at the point $\mathbf{x}_i$ due to a source located at $\boldsymbol{\xi}_j$. Garrison and Chow (1972) recommend that when $kr > 0.1$, the series form converges sufficiently rapidly for it to be employed, whereas when $kr < 0.1$ the integral form is more convenient.

Hogben and Standing (1974) present guidelines useful in evaluating the Green's function by either method. In particular, they discuss the difficulty associated with the singularity in the integrand when $\mu = k$ and describe how this difficulty may be overcome. A second problem discussed is that the Green's function expressions given are not well-behaved in deep water, and in these circumstances they should be replaced by an alternative expression before evaluation: they give a deep water Green's function which is similar to one given by John (1950).

The matrix equation, Eq. (6.46), may be solved by use of a standard complex matrix inversion subroutine when available. Hogben and Standing (1974, see also the discussion of their paper) have pointed out an approach which involves expressing the complex matrix equation in real and imaginary parts. Each set may then be separately solved, a procedure which affords some saving in computational effort. Note also that the matrix diagonal corresponds to the self interaction of sources and as this is generally a dominant effect it follows that an iterative technique, such as the Gauss-Seidel method, may also be directly employed.

6.3.3 Comparisons with Experiment

Several comparisons have been made between experimental results and numerical predictions based on computer programs applicable to bodies of arbitrary geometry. Hogben et al. (1977) have tabulated a number of published comparisons of computer predictions with experiment. Comparisons with simpler body shapes have been made by Lebreton and Cormault (1969) (truncated vertical circular cylinders), Hogben and Standing (1974, 1975) (truncated vertical circular and square section columns) and Mogridge and Jamieson (1976a) (square section column).

Hogben and Standing's comparisons of computed and measured forces and moments for circular and square columns for both h/d = 0.7 (where h is the column height) and for the surface piercing case are reproduced in Fig. 6.9. For the circular column, computer predictions generally form an upper bound to the measured loads, whereas for the square surface-piercing column, measured forces generally exceed the computer predictions. This excess, however, becomes less significant at smaller values of ka which lie in the more practical range.

Published comparisons with offshore structures of more general shape include those of Boreel (1974) (square section column on a pyramidal base), van Oort-

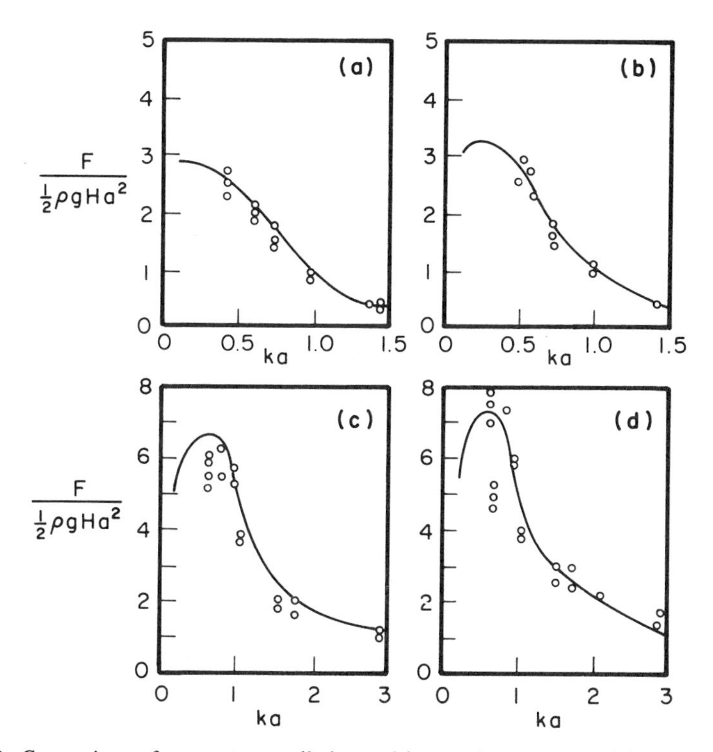

Fig. 6.9. Comparison of computer predictions with experiment reported by Hogben and Standing (1975) for the horizontal wave forces on circular and square columns. (a) circular, h/d = 0.7, (b) square, h/d = 0.7, (c) circular, surface-piercing, (d) square, surface-piercing. (πa^2 = sectional area.)

merssen (1972) (square section column on a pyramidal base), Faltinsen and Michelsen (1974) (floating box), Garrison and Rao (1971) (submerged hemisphere), Garrison and Chow (1972) (submerged tanks), Garrison et al. (1974), Garrison and Stacey (1977) (Condeep platforms), Loken and Olsen (1976) (submerged box with and without column, and floating box), Apelt and MacKnight (1976) (square columns on a submerged box) and Skjelbreia (1979) (Ekofisk and Ninian platforms). Brogen et al. (1974) have compared predictions for an axisymmetric bell shaped storage tank with full scale pressure measurements. This list is by no means exhaustive, but an examination of these comparisons does indicate the reasonable agreement that has generally been found between numerical prediction and experiment.

For example, Garrison et al. (1974) describe a comparison for two different Condeep platforms based on 1/120 scale models and subjected to design waves of height equivalents ranging from 21 to 29 m (69 to 95 ft) and periods from 14 to 17 sec. Fig. 6.10 reproduces their comparison for the horizontal forces and overturning moments on one of these models. Apelt and Macknight (1976) compared measurements using a 1/100 scale model of a berthing structure with predictions based on Garrison's computer program. Their comparison of mea-

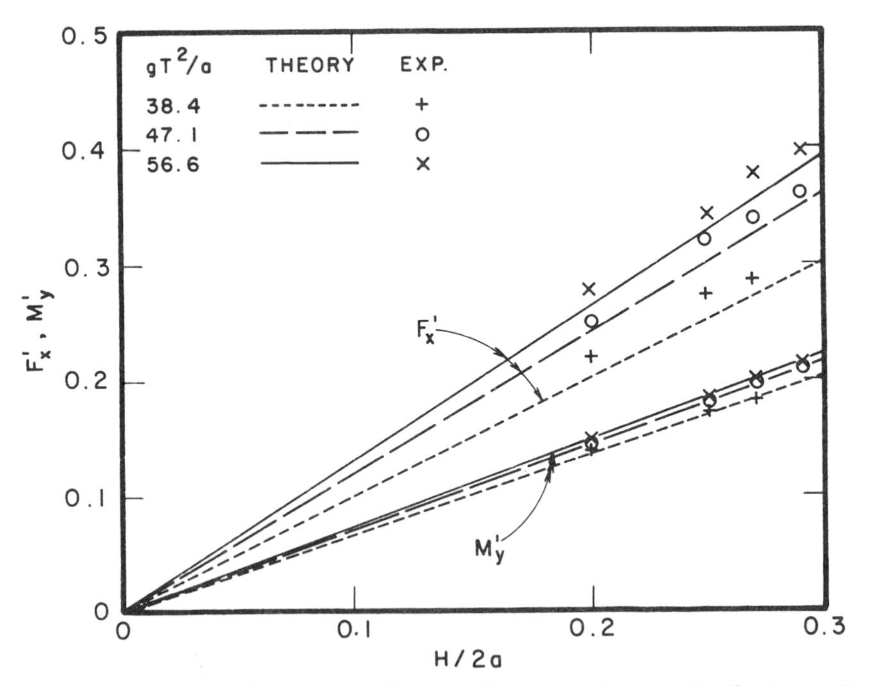

Fig. 6.10. Comparison of computer predictions with experiment reported by Garrison et al. (1974) for the horizontal wave force and base moment on a typical gravity platform. ($F'_x = F_x/\rho g a^3$, $M'_y = M_y/\rho g a^4$, a = characteristic dimension.)

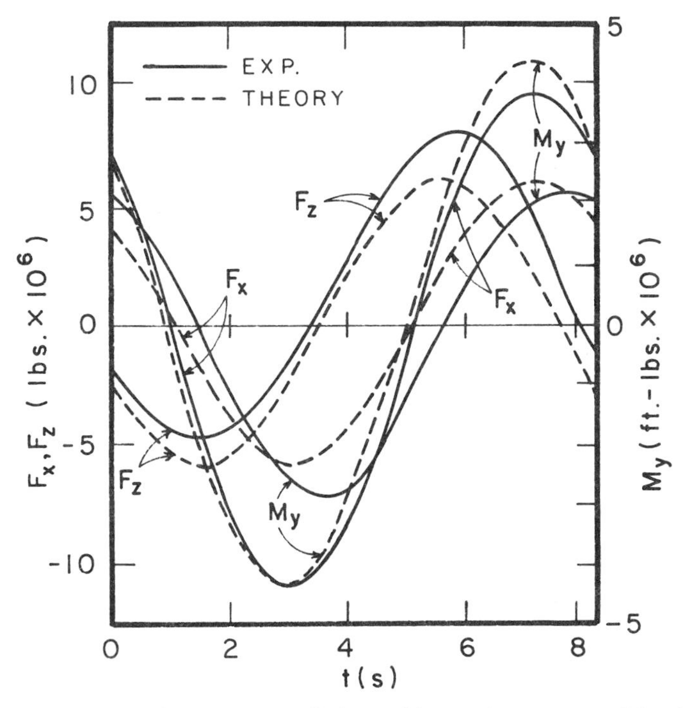

Fig. 6.11. Comparison of computer predictions with experiment reported by Apelt and Macknight (1976) for the wave forces and moment on an offshore berth.

sured and predicted time-varying horizontal and vertical forces and overturning moments are reproduced in Fig. 6.11. The design waves used in the experiment and in the computer calculations are not quite the same, but differences in the comparison are mainly due to the effects of wave nonlinearities and flow separation from the square columns of the structure.

6.4 ALTERNATIVE METHODS APPLICABLE TO PARTICULAR CONFIGURATIONS

The approach described in Section 6.3 not only requires sophisticated computer programming, but it is also costly to run. Alternative methods, which, although generally more economical, are restricted to particular configurations have now to be considered and we take up the problem for

(a) vertical axisymmetric bodies,
(b) horizontal cylinders of arbitrary section and subject to normally incident waves (vertical plane problem),

(c) vertical cylinders of arbitrary section (horizontal plane problem), and
(d) various vertical circular cylindrical bodies.

Because of the shape restrictions considered, problems (a), (b) and (c) are two-dimensional rather than three-dimensional in the sense that the bodies possess only a two-dimensional irregularity. One can take advantage of this such that in the wave source method one or two line integral equations need be solved in place of a surface integral equation. These three configurations, together with indications of their division into segments, are shown in Fig. 6.12. All bodies

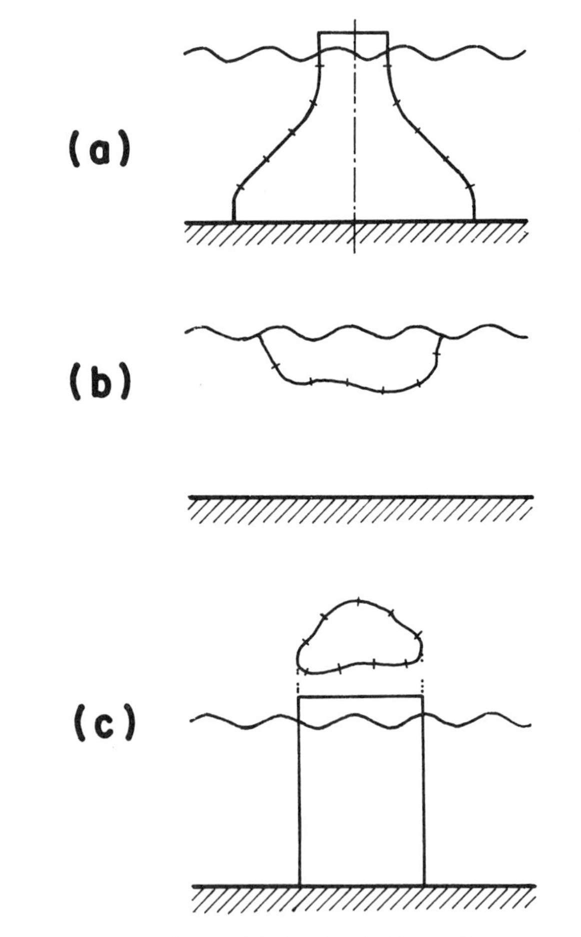

Fig. 6.12. Segment representations of bodies treated by solving line integral equations. (a) vertical axisymmetric bodies, (b) horizontal cylinders of arbitrary section, (c) vertical cylinders of arbitrary section.

considered in this section are taken to be fixed and the question of bodies undergoing motions will be deferred until Section 6.6. We also defer until Section 6.5 any consideration of the finite or hybrid element method.

Mention should first be made of the approach involving diffraction coefficients as outlined in Section 6.1.4. Chakrabarti (1973) has provided expressions for the Froude-Krylov force of various submerged bodies of symmetry and has conducted a literature survey to obtain values of vertical and horizontal force diffraction coefficients. The bodies considered include a hemisphere, sphere, horizontal half-cylinder, horizontal cylinder, rectangular block, vertical cylinder and a horizontal circular plate. Chakrabarti has provided specific values for the diffraction coefficients, but he has emphasized that these are only really valid in the inertia range rather than in the true diffraction range. This approach will not, then, be generally successful in overcoming the underlying problem where diffraction is significant, for the difficulty in evaluating the diffraction coefficients will still persist.

6.4.1 Vertical Axisymmetric Bodies

The case of vertical axisymmetric bodies (Fig. 6.12a) has been treated by Black (1975a) and subsequently by Fenton (1978) using a symmetric Green's function, and also by Bai (1977a) using the finite element method (Section 6.5). Fenton has given particular attention to the treatment of singularities in the governing integral equation involving the symmetric Green's function—a feature apparently not accounted for by Black—and he reduces the relevant line integral equations to forms possessing series which are rapidly convergent.

The reduction in computational effort arises essentially by making an extension of the analysis to each circular element of the body, thus reducing the surface integral equation to a series of line integral equations. Of these only two need be solved in order to determine the forces and moments on the body. We employ a cylindrical coordinate system as shown in Fig. 6.2, and the series form of the Green's function, Eq. (6.43), is expressed as a Fourier series in θ. This is given by Fenton (1978) as

$$G = \sum_{j=0}^{\infty}\left(\sum_{m=0}^{\infty} G_{jm}\right)(2 - \delta_{jo}) \cos(j(\theta - \Theta)) \tag{6.60}$$

with

$$G_{jo} = -iC_o \cosh(k(z+d)) \cosh(k(\zeta + d)) J_j\begin{pmatrix}kR\\kr\end{pmatrix} H_j^{(1)}\begin{pmatrix}kr\\kR\end{pmatrix}$$

$$G_{jm} = 4C_m \cos(\mu_m(z+d)) \cos(\mu_m(\zeta + d)) K_j\begin{pmatrix}\mu_m r\\\mu_m R\end{pmatrix} I_j\begin{pmatrix}\mu_m R\\\mu_m r\end{pmatrix} \quad \text{for } m \geqslant 1$$

Here δ_{j0} is the Kronecker delta and the notation used in Eq. (6.43) is applicable. Also (R, Θ, ζ) correspond to the point $\boldsymbol{\xi} = (\xi, \eta, \zeta)$ on the body surface, I_j denotes the modified Bessel function of the first kind and of order j, the upper value of the alternative argument is used if $r \geqslant R$ and the lower otherwise.

Since the flow is symmetric about the x-axis, the source distribution function $f(R, \Theta, \zeta)$ may be written as an even Fourier series in Θ

$$f(s, \Theta) = \sum_{\ell=0}^{\infty} f_\ell(s) \cos(\ell\Theta) \tag{6.61}$$

where the coordinate $s(R, \zeta)$ specifies a point as measured along the body contour in the x-z plane. With reference to the integral equation specifying the boundary condition, Eq. (6.45), the term involving the incident potential ϕ_w' may likewise be expressed as an even Fourier series, say

$$\frac{\partial \phi_w'}{\partial n} = \sum_{\ell=0}^{\infty} \phi_\ell^{(w)} \cos(\ell\theta) \tag{6.62}$$

where $\phi_\ell^{(w)}$ will be known. Equations (6.60)-(6.62) may now be substituted into Eq. (6.45). Because dS may be expressed as $dS = R\, d\Theta\, ds$ in this case, the surface integral in Eq. (6.45) can now be integrated with respect to Θ to give an even Fourier series in θ with each coefficient involving a line integral over the contour s. All of the Fourier coefficients for each value of ℓ can now be equated to obtain an infinite number of line integral equations, with each independent of θ. Thus

$$-f_\ell(s) + \int_S f_\ell(S) R \left(\sum_{m=0}^{\infty} \frac{\partial G_{\ell m}}{\partial n} \right) dS = -2\phi_\ell^{(w)} \qquad \text{for } \ell = 0, 1, 2, \ldots \tag{6.63}$$

Adopting an equivalent approach to that described already for the three-dimensional case allows each such equation to be approximated as a matrix equation based on the division of the body contour into a finite number of small segments. However, a complication arises because the matrix coefficients may be singular at some points. Fenton pays particular attention to the treatment of such singularities and is able to give each matrix coefficient as a lengthy expression which is non-singular and rapidly convergent. Fenton evaluates the source strengths using a Gauss-Seidel iteration procedure, since the matrix diagonals correspond to self-interaction terms and are strong. The scattered velocity potential ϕ_s' around the body surface may then be determined in terms of source strengths. Once again, the matrix coefficients used to obtain the scattered potential must be derived carefully in order to avoid any singular behavior. Finally, expressions for the forces and moments on the body may now be developed in terms of the velocity potential. Of these components, the vertical force involves only the case $\ell = 0$, the horizontal force and overturning moment (in the x-z plane) involve only the case $\ell = 1$ and the three remaining orthogonal components of force and moment are clearly zero. Thus only the two line integral equations corresponding to $\ell = 0$ and $\ell = 1$ need be solved to determine the wave loads on the body (but not the detailed pressure distribution around the body).

For axisymmetric bodies, Fenton's method involving these two line integral equations clearly requires much less computer effort than the more general three-dimensional Green's function method involving a surface integral equation. This is particularly so for large diameter axisymmetric structures of small height, or those which are located in shallow water. The method is also useful in dealing with various reference configurations including truncated cylinders, compound cylinders, conical structures, spheres and hemispheres. Results based on this method for a hemisphere resting on the seabed are given in Fig. 6.13. This shows the horizontal and vertical force coefficients as functions of the diffraction parameter ka for various values of the depth to radius ratio d/a. (The moment about the axis $x = 0$, $z = -d$ is always zero since the fluid force on any surface element passes through the center $x = 0$, $y = 0$, $z = -d$.) A further example of the method's application to conical structures is given in Section 6.10.3.

6.4.2 Vertical Plane Flows

We turn now to consider a two-dimensional wave motion in the x-z plane past an infinite horizontal cylinder whose axis is parallel to the y axis (see Fig. 6.12b). Research effort has been directed primarily towards two aspects of the problem: the evaluation of wave reflection and transmission coefficients associated with the body's behavior as a breakwater; and the treatment of moored and freely floating bodies undergoing motion. The former case has been mentioned in Section 4.9.4 and the latter will be considered in Section 6.6. We limit our attention for the present to wave forces acting on a fixed body.

Since the motion is two-dimensional, the complex potential may be employed as a convenient means of expressing the problem. A two-dimensional wave-source method may be developed in analogy to the method for three-dimensional bodies described in Section 6.3. The governing equations are essentially the same, except that now the scattered waves do not decay at infinity (because directional spreading is absent), and thus the radiation condition adopts a slightly different form. The radiation condition requires that the scattered waves are outward travelling and thus ϕ_s is proportional to $\exp(-i(kx + \omega t))$ as $x \to -\infty$

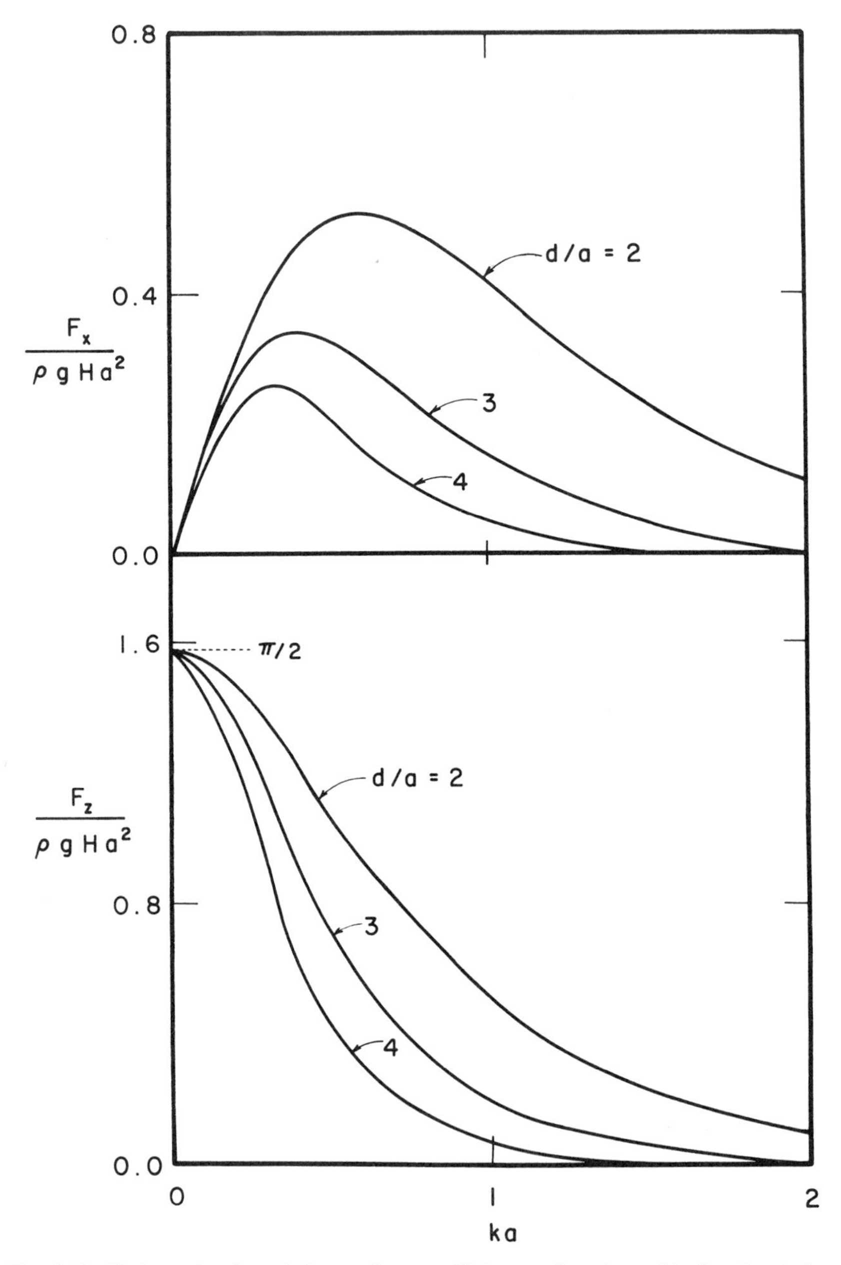

Fig. 6.13. Horizontal and vertical wave force coefficients as functions of ka for a hemisphere resting on the seabed, with d/a = 2, 3 and 4.

and to $\exp(i(kx - \omega t))$ as $x \to \infty$. The condition can therefore be expressed, as in Eq. (4.225), in the form

$$\frac{\partial \phi}{\partial r} = \frac{i\omega}{c} \phi \qquad \text{for } r \to \infty \tag{6.64}$$

The scattered potential ϕ_s is represented by a distribution of sources around the body contour, with the source-strength distribution denoted $f(\boldsymbol{\xi})$, where $\boldsymbol{\xi} = (\xi, \zeta)$ is the source location. In analogy with Eq. (6.42), the Green's function for a two-dimensional wave source in the x, z plane is given by John (1950) as

$$G(\mathbf{x}, \boldsymbol{\xi}) = \ell n\,(RR'/d^2)\ .$$

$$-2 \int_0^\infty \left[\frac{(\mu + \nu) \exp(-\mu d) \cosh(\mu(\zeta + d)) \cosh(\mu(z + d)) \cos(\mu r)}{\mu(\mu \sinh(\mu d) - \nu \cosh(\mu d))} + \frac{\exp(-\mu d)}{\mu} \right] d\mu \tag{6.65}$$

where $\mathbf{x} = (x, z)$ and the notation following Eq. (6.42) applies ($(y - \eta) = 0$ here). John also gave a corresponding series form of the Green's function and also a limiting expression for deep water. Wehausen and Laitone (1960) provide a similar Green's function based on a (directly related) vortex distribution for the deep water case.

The body surface boundary condition now results in a line integral equation analogous to the surface integral equation previously given, Eq. (6.45). This is

$$-\frac{1}{2} f(\mathbf{x}) + \frac{1}{4\pi} \int_S f(\boldsymbol{\xi}) \frac{\partial G}{\partial n} (\mathbf{x}, \boldsymbol{\xi})\, dS = -\frac{\partial \phi_w'}{\partial n} (\mathbf{x}) \tag{6.66}$$

This is solved for the source strength distribution and thus the scattered potential and consequently the wave forces may be calculated.

In most published calculations, a related integral equation method, rather than this wave source method itself, has been used. Such an approach can also be applied more generally to the three-dimensional case: as a consequence of Green's theorem, the velocity potential ϕ' may be expressed for this three-dimensional case as

$$\lambda \phi'(\mathbf{x}) = \int_S \left[\phi'(\boldsymbol{\xi}) \frac{\partial G}{\partial n} (\mathbf{x}, \boldsymbol{\xi}) - G(\mathbf{x}, \boldsymbol{\xi}) \frac{\partial \phi'}{\partial n} (\boldsymbol{\xi}) \right] dS \tag{6.67}$$

where $\mathbf{x}$ represents a general point, $\boldsymbol{\xi}$ represents a point on the body surface S, and G is a Green's function before. λ depends on the location of $\mathbf{x}$ and is given as

$$\lambda = \begin{cases} 0 & \text{for } \mathbf{x} \text{ outside S} \quad (a) \\ \beta & \text{for } \mathbf{x} = \mathbf{x}_+ \text{ on S} \quad (b) \\ 4\pi & \text{for } \mathbf{x} \text{ within S} \quad (c) \end{cases}$$

where $\mathbf{x}_+$ denotes a point on S approached from the fluid side, and β is the solid angle on the fluid side of S which S makes at $\mathbf{x}_+$. The body surface boundary condition can be applied by making use of result (b). By using Eq. (6.67) to describe the scattered potential ϕ_s', applying the boundary condition $\partial\phi_s'/\partial n = -\partial\phi_w'/\partial n$, and putting $\beta = 2\pi$ since S is usually smooth, Eq. (6.67) may be reduced to

$$\phi_s'(\mathbf{x}_+) = \frac{1}{2\pi}\int_S \left[\phi_s'(\boldsymbol{\xi})\frac{\partial G}{\partial n}(\mathbf{x}_+, \boldsymbol{\xi}) + G(\mathbf{x}_+, \boldsymbol{\xi})\frac{\partial \phi_w'}{\partial n}(\boldsymbol{\xi})\right] dS \qquad (6.68)$$

This integral equation can be solved to obtain $\phi_s'(\boldsymbol{\xi})$ and consequently the force on the body may be calculated. This approach may also be applied to two-dimensional problems simply by adopting the appropriate two-dimensional Green's function, and by taking λ in Eq. (6.67) to be 0, $-\beta$, and -2π for the cases (a), (b) and (c) respectively. Thus for a smooth surface S, the factor $\frac{1}{2}\pi$ in Eq. (6.68) would be replaced by $-1/\pi$.

Yet another related integral equation method has also been used for vertical plane problems (e.g. Ijima, Chou and Yoshida 1976, Yamamoto and Yoshida 1978, and Finnigan and Yamamoto 1979). This method avoids the use of a complicated Green's function which has to satisfy various boundary conditions, as has been adopted in the other methods so far described. Green's second identity may be applied to a closed contour containing a fluid region, and in the present context this contour S is chosen as indicated in Fig. 6.14. A corresponding Green's function now need satisfy only the equation of motion but not any boundary conditions, and so may take the particularly simple form $G = \ln(r)$. Green's second identity thus gives the velocity potential ϕ' at any point $\mathbf{x}$, itself on the boundary, as

$$\phi'(\mathbf{x}) = -\frac{1}{\pi}\int_S \left(\phi'(\boldsymbol{\xi})\frac{\partial \ln(r)}{\partial n} - \ln(r)\frac{\partial \phi'}{\partial n}(\boldsymbol{\xi})\right) dS \qquad (6.69)$$

where r is the distance between the points $\mathbf{x}$ and $\boldsymbol{\xi}$, dS is an element of length at the point $\boldsymbol{\xi}$ on the boundary, and n is the normal into the fluid region at the point $\boldsymbol{\xi}$ as before.

In a numerical solution to Eq. (6.69), the boundary S is divided into N segments as indicated in Fig. 6.14, and Eq. (6.69) can thereby by approximated to a matrix equation. This provides N equations relating the 2N segment values of

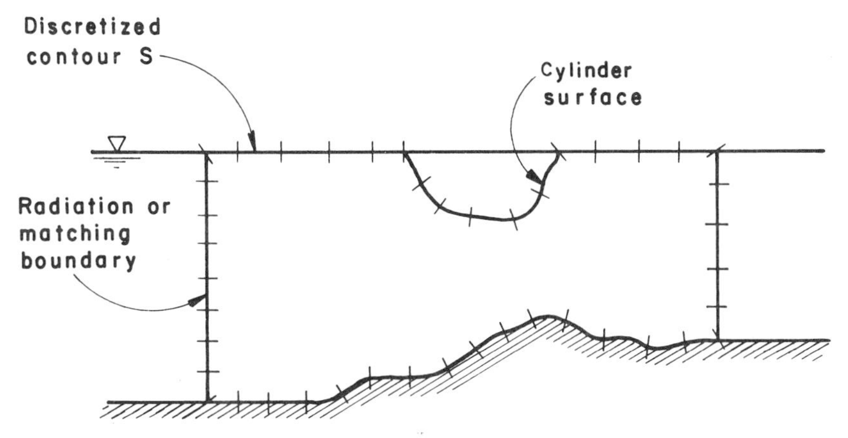

Fig. 6.14. Segment representation of fluid boundaries for a vertical plane problem.

ϕ' and $\partial\phi'/\partial n$. The remaining N relations needed to solve for ϕ' and $\partial\phi'/\partial n$ are obtained from the various boundary conditions around S. If the radiation boundaries are sufficiently far from the body, the radiation condition may be applied directly. Otherwise the boundaries may be chosen somewhat closer to the body, but now the radiation condition is replaced by a matching condition with the flow outside S taken to include evanescent terms involving an infinite series in $\cos(\mu_m(z + d))$, where μ_m is defined in Eq. (6.44).

The advantages of the method are that it may incorporate a variable depth in the vicinity of the body, and that the very simple logarithmic Green's function is used. The main disadvantage is that a relatively long boundary needs to be discretized and that the matching conditions may be slightly awkward. Even so, the method has been used successfully for various vertical plane problems, including those involving moored floating breakwaters (Yamamoto and Yoshida 1978) and permeable breakwaters (Finnigan and Yamamoto 1979).

Mention is made in passing of a related problem in which incident waves approach an infinitely long horizontal cylinder obliquely, say at an angle α to the x axis. The incident wave potential may now be written as

$$\phi_w = A \frac{\cosh(k(z + d))}{\cosh(kd)} \exp[i(kx\cos\alpha + ky\sin\alpha - \omega t)] \tag{6.70}$$

where $A = -igH/2\omega$ as before. The scattered potential may be written in the form

$$\phi_s = \phi_s'(x, z)\exp[i(ky\sin\alpha - \omega t)] \tag{6.71}$$

The problem thus reduces to a vertical plane problem for ϕ_s' with the equation of motion now being given as

$$\frac{\partial^2 \phi_s'}{\partial x^2} + \frac{\partial^2 \phi_s'}{\partial z^2} - k^2 \sin^2 \alpha \, \phi_s' = 0 \tag{6.72}$$

This two-dimensional problem may be solved by similar techniques to those already mentioned, and has been treated, for example, by Bai (1975) using the finite element method.

Specific analytical solutions to vertical plane problems have been presented by Dean and Ursell (1959) for a semi-immersed circular cylinder, and by Dean (1948) and Ursell (1950) for a fully immersed circular cylinder. Ogilvie (1963) has studied the latter case and has presented the first order oscillatory and second order steady forces on the cylinder. Fig. 6.15 shows the first order wave force as a function of ka for a circular cylinder in deep water whose axis is a distance h = 2a and 3a below the still water level. (In deep water, the magnitude of the resultant is constant so that horizontal and vertical force amplitudes are equal.)

More generally, the wave forces on fixed cylinders have often been calculated by solving the corresponding radiation problem in which the body is made to oscillate in otherwise still water, and then utilizing the Haskind relations to

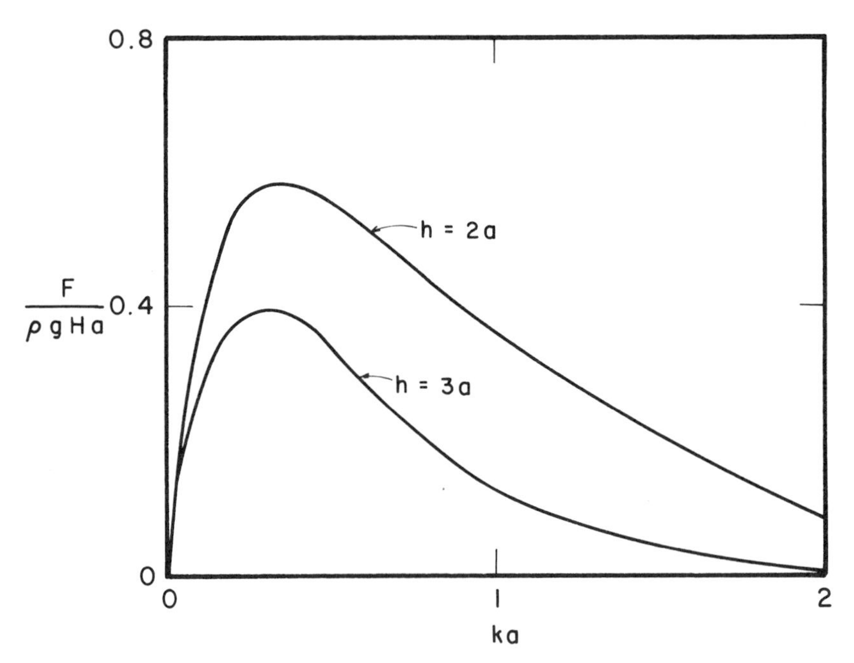

Fig. 6.15. Wave force coefficient as a function of ka for a horizontal circular cylinder in deep water, with h = 2a and 3a.

obtain the wave forces on the stationary body. Section 6.6 describes this approach in more detail, and the presentation of further results for selected bodies is therefore deferred to that section. Some studies explicitly containing wave force results for vertical plane problems include those by Black, Mei and Bray (1971) for a rectangular cylinder either at the free surface or the seabed, Maeda (1974) for fully submerged circular and rectangular cylinders, and Naftzger and Chakrabarti (1979) for a semi-immersed and fully immersed circular cylinder and a half cylinder resting on the seabed.

6.4.3 Vertical Cylinders of Arbitrary Section

This case (Fig. 6.12c) is equivalent to the problem of resonance in harbors of arbitrary shape and constant depth, the fundamental difference being that the solid boundary contour is now closed and the forces on the body, rather than wave elevations, become of primary interest. The situation is much simpler to treat than the general three-dimensional problem since the surface integral equation of the three-dimensional problem is reduced to a single line integral equation taken over the horizontal section of the body or bodies. The corresponding Green's function is particularly simple and may be evaluated rapidly. Such an approach was used by Hwang and Tuck (1970) in the context of harbor resonance, and has been applied by Isaacson (1978d) to wave force calculations. Ijima, Chou and Yumura (1974) and Harms (1979) have used the related integral equation method based on the use of Eq. (6.68), for studying wave diffraction around offshore breakwaters and islands.

The theory of the wave source method may be developed by a separation of variables approach in which the scattered potential ϕ_s is taken to have a hyperbolic cosine variation with depth,

$$\phi_s = A \frac{\cosh(k(z+d))}{\cosh(kd)} \phi_s'(x,y)\, e^{-i\omega t} \tag{6.73}$$

where $A = -igH/2\omega$ as before. This directly satisfies the seabed and linearized free-surface boundary conditions. The Laplace equation for ϕ_s reduces to the Helmholtz equation for ϕ_s',

$$\frac{\partial^2 \phi_s'}{\partial x^2} + \frac{\partial^2 \phi_s'}{\partial y^2} + k^2 \phi_s' = 0 \tag{6.74}$$

The Green's function for the present case must satisfy the Helmholtz equation and the radiation condition, and may be written as

$$G(\mathbf{x}, \boldsymbol{\xi}) = i\pi H_0^{(1)}(kr) \tag{6.75}$$

where $\mathbf{x} = (x, y)$ and $\boldsymbol{\xi} = (\xi, \eta)$ in this case. The body surface boundary condition corresponding to Eq. (6.45) is then given as

$$-\frac{1}{2} f(\mathbf{x}) + \frac{1}{4\pi} \int_S f(\boldsymbol{\xi}) \frac{\partial G}{\partial n} (\mathbf{x}, \boldsymbol{\xi}) \, dS = -\frac{\partial \phi_w'}{\partial n} (\mathbf{x}) \tag{6.76}$$

where $\phi_w' = e^{ikx}$ from Eq. (6.14), and dS now denotes a horizontal differential length along the body surface. A numerical procedure wholly equivalent to that previously outlined for three-dimensional bodies is now carried out, except that we are now dealing with a two-dimensional context.

Once ϕ_s' is known, the variables of interest may readily be determined. Since the velocity potential has a hyperbolic cosine variation with depth, the sectional force, total force and overturning moment are given respectively as

$$F_z = \frac{\rho g H}{k} \frac{\cosh(k(z+d))}{\cosh(kd)} \chi e^{-i\omega t} \tag{6.77}$$

$$F = \frac{\rho g H d}{k} \frac{\tanh(kd)}{kd} \chi e^{-i\omega t} \tag{6.78}$$

$$M = \frac{\rho g H d^2}{k} \left[\frac{kd \sinh(kd) + 1 - \cosh(kd)}{(kd)^2 \cosh(kd)} \right] \chi e^{-i\omega t} \tag{6.79}$$

where χ is a vector in the horizontal plane given as

$$\chi = -\frac{k}{2} \int_S (\phi_w' + \phi_s') \mathbf{n} \, dS \tag{6.80}$$

Equivalently, the results may be expressed in terms of the effective inertia coefficient C_m and corresponding phase angle δ such that Eqs. (6.34a)-(6.34c) are applicable with $\pi D^2/4$ now replaced by the cylinder's sectional area b^2. We then have simply $C_m = 2|\chi|/(kb)^2$ and $\delta = \text{Arg}(\chi)$.

There is a considerable reduction in computational effort when this approach can be used in place of the three-dimensional wave source method. Computer storage and time are proportional at least to the square of the number of unknowns and thus if N segments are employed, the saving when compared to the use of N^2 facets is approximately of the order of $1/N^2$. Furthermore, the Green's function used is particularly simple, being given by a standard computer subroutine, and its evaluation when compared to that of the point-source Green's function gives additional saving. More significantly, the line-integral equation method requires much less preparation both to program as well as to set up for a chosen configuration.

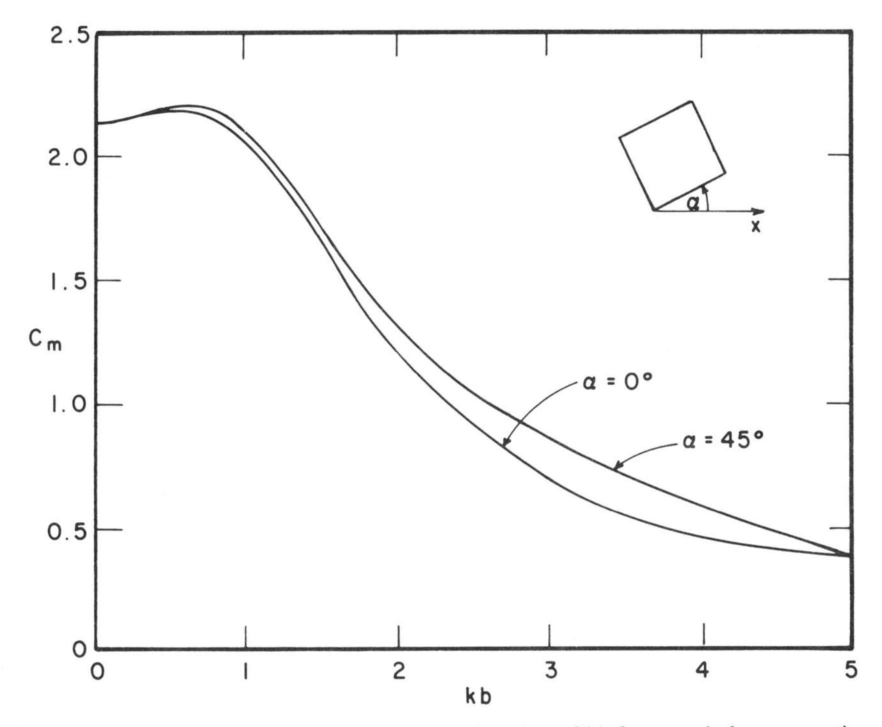

Fig. 6.16. Effective inertia coefficient C_m as a function of kb for a vertical square-section cylinder at orientations $\alpha = 0°$ and 45° (Isaacson 1978d).

Using this approach, Isaacson (1978b,d, 1979b,d) obtained results for an isolated circular cylinder, a pair of circular cylinders, an isolated square column, a pair of square columns, and an isolated rectangular cylinder. Results for a pair of circular cylinders are described in Section 6.7 in the context of the interference effect of one cylinder on its neighbor. In the case of a square column of side b, the variations of effective inertia coefficient C_m with the diffraction parameter kb are reproduced in Fig. 6.16 for the two symmetrical orientations $\alpha = 0°$ and $\alpha = 45°$, where α defines the square's orientation as shown in the figure. A comparison of the computer predictions with the experimental measurements of Mogridge and Jamieson (1976a) and Isaacson (1979b) is given Fig. 6.17.

Computer predictions for the case of a rectangular cylinder of length ℓ and breadth b are shown in Fig. 6.18 in terms of the effective inertia coefficient C_m (see Eqs. (6.34) with $\pi D^2/4$ replaced now by $b\ell$). The figure shows C_m as a function of the diffraction parameter kb for cylinders of various length to breadth ratios oriented so as to give the largest possible forces (see the inset to Fig. 6.18). In the limit $kb \rightarrow 0$, C_m tends towards the true inertia coefficients

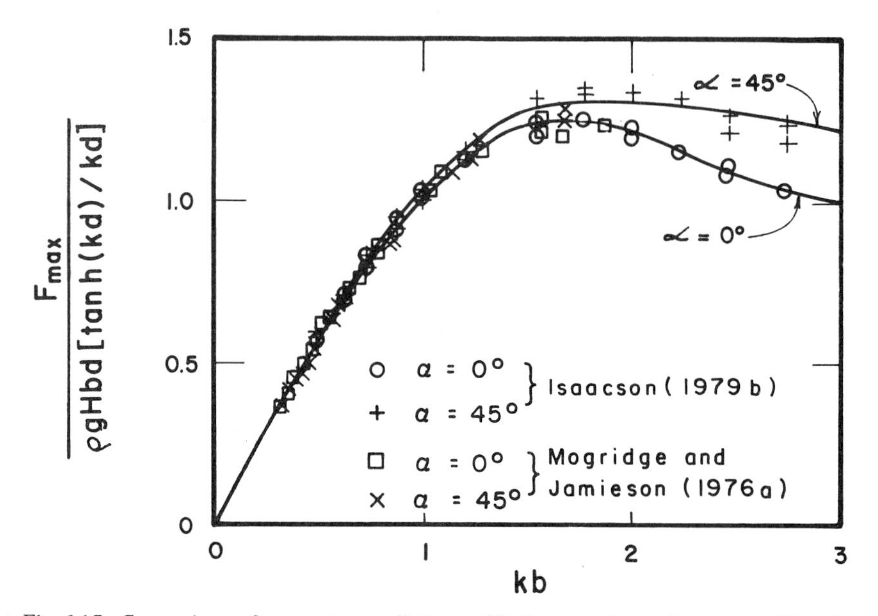

Fig. 6.17. Comparison of computer predictions with the experimental results of Mogridge and Jamieson (1976a) and Isaacson (1979b) for the wave force on a square-section cylinder at orientation $\alpha = 0°$ and 45°.

given previously by Brater, McNown and Stair (1958). Comparisons with experimental measurements for two particular ℓ/b ratios and for various orientations are reproduced in Fig. 6.19.

For completeness, brief mention is made of a limiting case of the geometry considered here which corresponds to a vertical seawall subjected to normally or obliquely incident waves. The normal or oblique reflection of a small amplitude wave train was mentioned in Section 4.9.4, and it is a simple matter to develop expressions for the wave loads on the wall. In the case of normally incident waves the pressure distribution over the wall corresponds to that at an antinode of a standing wave train (see Fig. 4.34) and is contained in Table 4.8. This can be integrated over depth to obtain expressions for the force F′ per unit width and overturning moment M′ per unit width of wall. The corresponding formulae are

$$p = -\rho gz + \rho gH \frac{\cosh(ks)}{\cosh(kd)} \cos(\omega t) \tag{6.81}$$

$$F' = \frac{\rho gd^2}{2} + \rho gHd \frac{\tanh(kd)}{kd} \cos(\omega t) \tag{6.82}$$

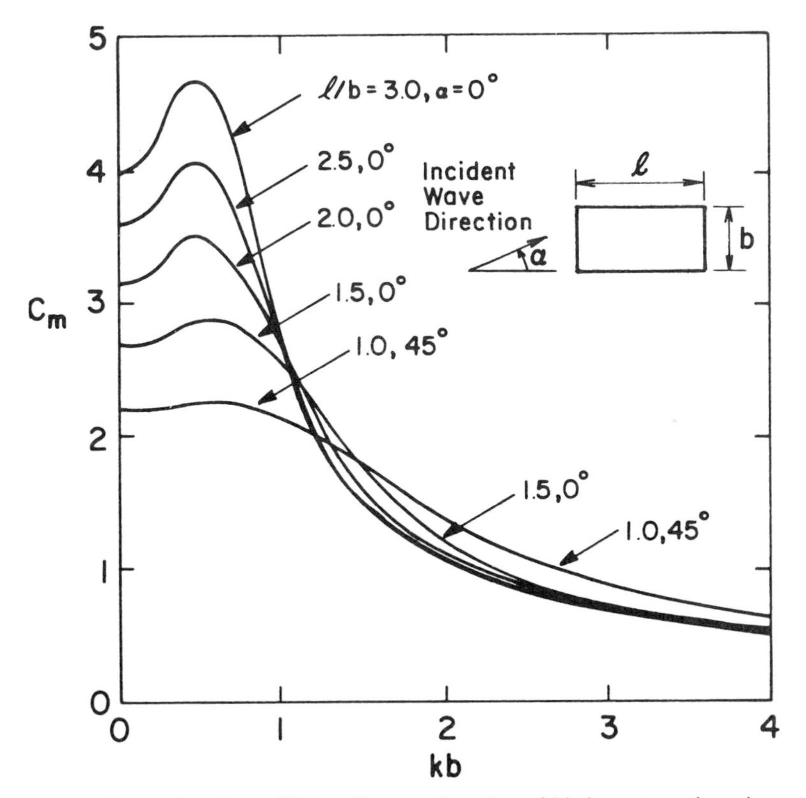

Fig. 6.18. Effective inertia coefficient C_m as a function of kb for rectangular caissons of length to breadth ratios ℓ/b = 1.0, 1.5, 2.0, 2.5 and 3.0 and at orientations α corresponding to maximum forces (Isaacson 1979d).

$$M' = \frac{\rho g d^3}{6} + \rho g H d^2 \left[\frac{kd \sinh(kd) + 1 - \cosh(kd)}{(kd)^2 \cosh(kd)}\right] \cos(\omega t) \qquad (6.83)$$

This situation is of considerable importance in the design of coastal structures. Extensions to it, to account for the effects of finite amplitude waves or of breaking waves, or to cases involving sloping walls, breakwaters, etc. are not described here, and for these the reader is referred to the coastal engineering literature (e.g. U.S. Army Shore Protection Manual 1973, Silvester 1974, Horikawa 1978).

6.4.4 Vertical Circular Cylinders

Various kinds of vertical circular cylindrical bodies have been analysed by alternatives to the general three-dimensional wave source and finite element methods,

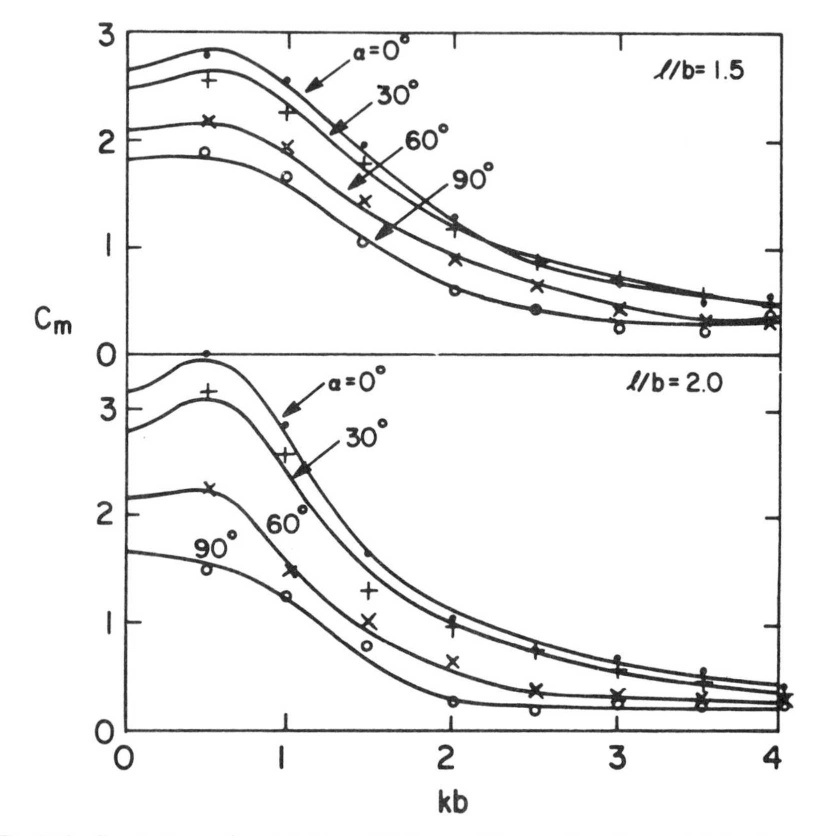

Fig. 6.19. Comparison of computer predictions with experiment reported by Isaacson (1979d) for the effective inertia coefficient C_m of rectangular caissons, with $\ell/b = 1.5$ and 2.0, and orientations $\alpha = 0°, 30°, 60°$ and $90°$.

and some such bodies are indicated in Fig. 6.20. Most of the configurations shown (all but case (b)) are particular examples of axisymmetric bodies and accordingly the axisymmetric Green's function method (Section 6.4.1) is applicable. Also the two-dimensional Green's function method of Section 6.4.3 is valid for the configurations of Fig. 6.20a and 6.20b. The finite element method in either a two-dimensional or an axisymmetric form may be applied to all the cases listed. We now consider the additional methods that have been used for one or other of the configurations shown.

The most simple case is of course the isolated cylinder extending from the seabed up to the free surface (Fig. 6.20a), and the closed-form solution following MacCamy and Fuchs (1954) has already been outlined in detail in Section 6.2. The situation depicted in Fig. 6.20b comprising of two or more surface-piercing cylinders has been treated by Spring and Monkmeyer (1974, 1975) by

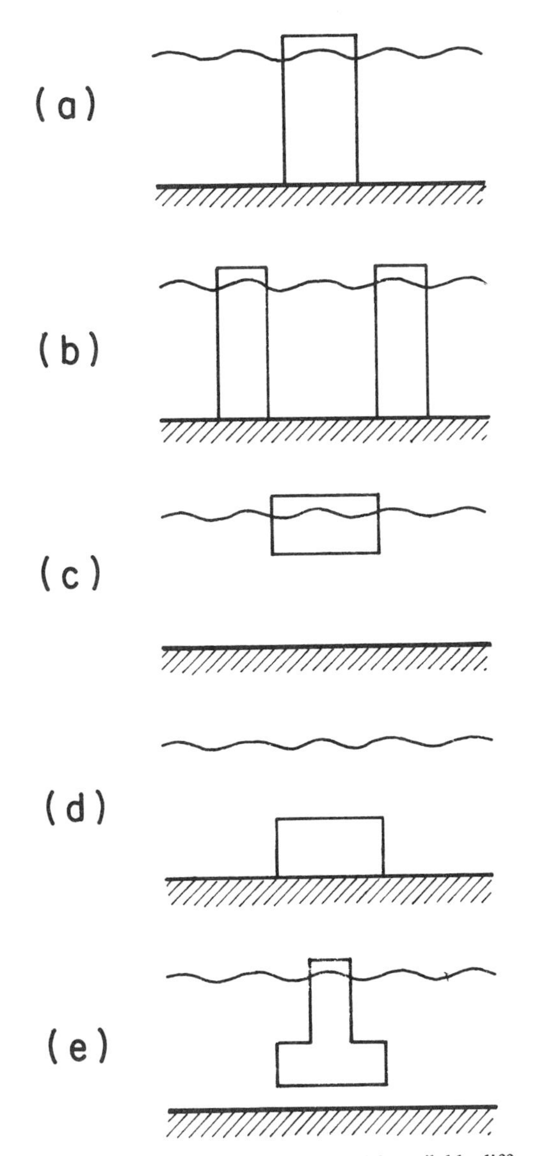

Fig. 6.20. Some vertical circular cylindrical bodies with available diffraction solutions.

using a Fourier series representation for the scattered velocity potential due to each cylinder. The cylinder surface boundary condition is used to set up a matrix equation which is then solved to obtain the initially unknown Fourier coefficients. Chakrabarti (1978a) has followed this approach and has presented results for three and four equally-spaced cylinders. Results for a pair of cylinders are given in Section 6.7.

Black, Mei and Bray (1971) have calculated the wave forces on a truncated cylinder which either extends to the free surface or rests on the seabed (Fig. 6.20c,d). They applied a variational method to the corresponding radiation problem and then used the Haskind relations to determine the forces on the stationary body. They simultaneously treated the corresponding two-dimensional vertical-plane cases as already mentioned in Section 6.4.2. Results for a truncated cylinder resting on the seabed are given in Section 6.10.2.

Garrett (1971) obtained the wave forces on a truncated cylinder extending down from the free surface (a semi-immersed circular dock, Fig. 6.20c). The solution is obtained by adopting different Fourier expansions for the velocity potential in the fluid regions below the dock $r < a$ and exterior to the dock $r > a$; and then matching these at $r = a$ to provide for a continuous velocity potential and radial velocity along the boundary $r = a$. A suitable integration over depth and taken at $r = a$ gives rise to a matrix equation for the unknown Fourier coefficients describing the outer velocity potential. This is solved for a finite number of terms, and in turn the coefficients describing the inner velocity potential may then be calculated. The results of Black, Mei and Bray (1971) for this case are in close agreement with Garrett's own results. As an example of the results obtained, Fig. 6.21 shows the amplitudes of horizontal force F_x, vertical force F_z, and overturning moment M_y, all expressed by dimensionless coefficients, as functions of the diffraction parameter ka for a truncated cylinder with a draft $h = 0.5a$ and in water of depth $d = 1.5a$.

Gran (1973) subsequently attempted an approximate solution for the complementary case of a submerged truncated cylinder resting on the seabed (Fig. 6.20d). In his solution the assumed expressions for velocity potential in the regions $r < a$ and $r > a$ are matched at $r = a$ on the basis of a continuous free surface and continuous radial fluid flux. These matching conditions are less stringent than those of Garrett and permit relatively simple velocity potential expressions to be employed: in both the inner and outer regions the velocity potential is assumed to have hyperbolic cosine variations with depth at appropriate wave numbers. The velocity potential itself is matched only at the still water level and so generally exhibits a discontinuity below the free surface. Gran's findings, while of considerable interest, do show noticeable departures from the more accurate results of Black, Mei and Bray (1971). Isaacson (1979e) has since described a rigorous extension of Garrett's method for this body and his results appear to be as accurate as those of Black, Mei and Bray (1971).

Kokkinowrachos and Wilckens (1974) extended Garrett's method to the still more general case of a compound cylindrical structure which comprises of a circular column resting on a circular base and which may itself be either above or resting on the seabed (Fig. 6.20e). As in Gran's approach, the matching con-

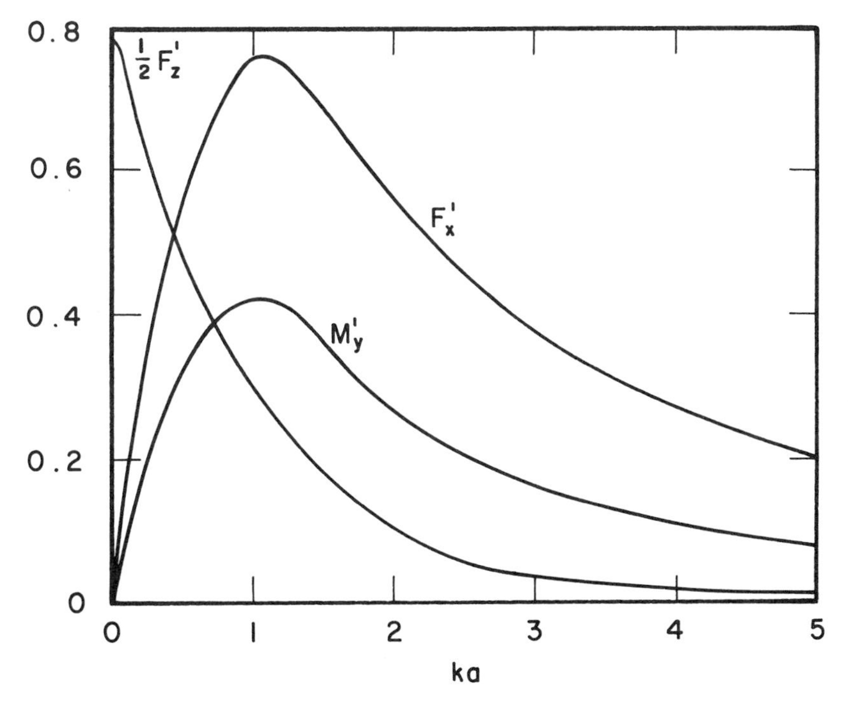

Fig. 6.21. Wave force and moment coefficients as functions of ka for a truncated circular cylinder at the free surface (circular dock), with h/a = 0.5 and d/a = 1.5. ($F'_x = F_x/\rho gHa^2$, $F'_z = F_z/\rho gHa^2$, $M'_y = M_y/\rho gHa^3$, h = draft. M_y is taken about x = 0, z = 0.)

dition of a continuous velocity potential was applied only at the still water level (corresponding to a continuous free surface) rather than over a range of depths, and accordingly the velocity potential in the region above the base is assumed to have a hyperbolic cosine variation with depth.

Once again Isaacson (1979e) has developed a rigorous application of Garrett's method for such a body (in the particular case of the base resting on the seabed), and has presented preliminary results of the forces and moments on either cylindrical component and on the whole structure. Isaacson indicates how the MacCamy and Fuchs solution for a single surface-piercing cylinder may be retrieved as various limiting cases of the compound cylinder geometry. The results are of importance in estimating the interference between a column and base of a gravity structure and are presented in Section 6.7.

Black (1975b) has reviewed the three approaches: (a) Black, Mei and Bray's (1971) variational method, (b) the method used by Garrett (1971) and (c) the axisymmetric Green's function method of Black (1975a). He found that for a case he considered, the relative computer run times were respectively 1, 4 and 6, whereas each method may be applied to a successively wider class of body. In contrast, the general three-dimensional Green's function method (Section 6.3) had a run time of 300. Black concludes that for the more restricted configurations for which they are valid, Garrett's method and the variational method both provide high accuracy without excessive computational effort.

6.4.5 Wave Doublet Representation

Certain "well-type" structures contain a thin, impermeable wall exposed on either side to the wave flow, and a question arises concerning the suitability of a singularity distribution to adequately represent this situation. Since a wave source involves a velocity discontinuity at its location, the wave source representation of a body surface can correctly model the flow on one (the fluid) side of the surface only: that is, if the normal velocity is predicted to be zero on one side of the boundary (wall) then it will not be on the other, and a single sheet of distributed sources cannot correctly reproduce the flow on either side of a thin wall. Since a doublet (dipole) does not exhibit such a velocity discontinuity at its location, it has been proposed (e.g. Hogben, Osborne and Standing 1974) that a wave doublet distribution might be used to reproduce correctly the flow on either side of a wall.

The generalized procedure is not too different in principle from the wave source method and has been outlined briefly by Naftzger and Chakrabarti (1975) and by Mei (1978). It has also been used by Chakrabarti and Naftzger (1976) for axisymmetric shells and by Isaacson (1978c) for the horizontal-plane case involving harbor resonance. The velocity potential ϕ_s' due to a doublet distribution is given as

$$\phi_s'(\mathrm{x}) = \frac{1}{4\pi}\int_S \mu(\boldsymbol{\xi})\,\frac{\partial G}{\partial n'}(\mathrm{x},\boldsymbol{\xi})\,dS \tag{6.84}$$

where G is the Green's function of a wave source as before, μ is the doublet strength distribution function and n' is the direction normal to the surface at the point $\boldsymbol{\xi}$ where the doublet is located. The boundary condition of zero fluid velocity normal to the body surface thus becomes

$$\frac{1}{4\pi}\int_S \mu(\boldsymbol{\xi})\,\frac{\partial^2 G}{\partial n\,\partial n'}(\mathrm{x},\boldsymbol{\xi})\,dS = -\frac{\partial \phi_w'}{\partial n}(\mathrm{x}) \tag{6.85}$$

where n' is measured from the point $\boldsymbol{\xi}$ and n from the point $\mathbf{x}$ at which the boundary condition is applied, as indicated in Fig. 6.22. Equation (6.85) is a Fredholm integral equation of the first kind. The above formulation is equivalent

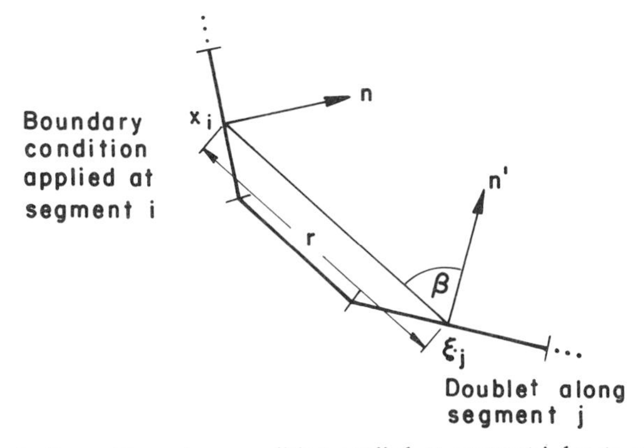

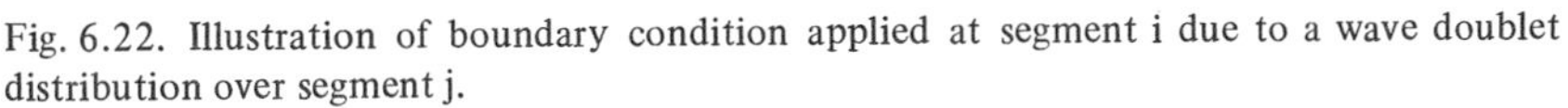
Fig. 6.22. Illustration of boundary condition applied at segment i due to a wave doublet distribution over segment j.

to adopting a Green's function G' for a wave doublet. This may be developed by considering the combined potential of two sources of opposite strength made to approach each other along the doublet axis n' and we thus have $G' = \partial G/\partial n'$. As before, the body is represented by a finite number of facets so that the discretized version of Eq. (6.85) above becomes

$$\sum_{j=1}^{N} B_{ij}\mu_i = b_i \qquad \text{for } i = 1, 2, \ldots N \tag{6.86}$$

where b_i is the known column vector as before, Eq. (6.51), N is the number of facets and

$$B_{ij} = \frac{1}{2\pi} \int_{\Delta S_j} \frac{\partial G'}{\partial n} (x_i, \boldsymbol{\xi})\, dS \tag{6.87}$$

This integral may generally be approximated by calculating $\partial G'/\partial n$ at the segment center (but see the end of this section for a further comment). The net force on an element exposed to fluid on both its faces is due to the pressure discontinuity across it, and only the doublet distribution over the element in question contributes to this discontinuity. The discontinuity in potential is simply μ so that the net force on the j-th element is

$$\Delta F_j = i\omega\rho\mu_j\Delta S_j e^{-i\omega t} \tag{6.88}$$

acting normal to the surface.

Isaacson (1978c) has treated the horizontal plane case applied to situations involving harbor resonance. In this case the Green's function for the doublet is given simply as

$$G' = i\pi \cos \beta H_1^{(1)}(kr), \tag{6.89}$$

where β is the angle at the doublet location (ξ, η) which the general point (x, y) makes with the doublet axis n' (which is itself taken normal to the surface being modeled—see Fig. 6.22).

When the segments i and j are quite far apart such that the variation of $\partial G/\partial n$ over the segment j is sufficiently slight, it is possible to replace the integral expression in Eq. (6.87) by

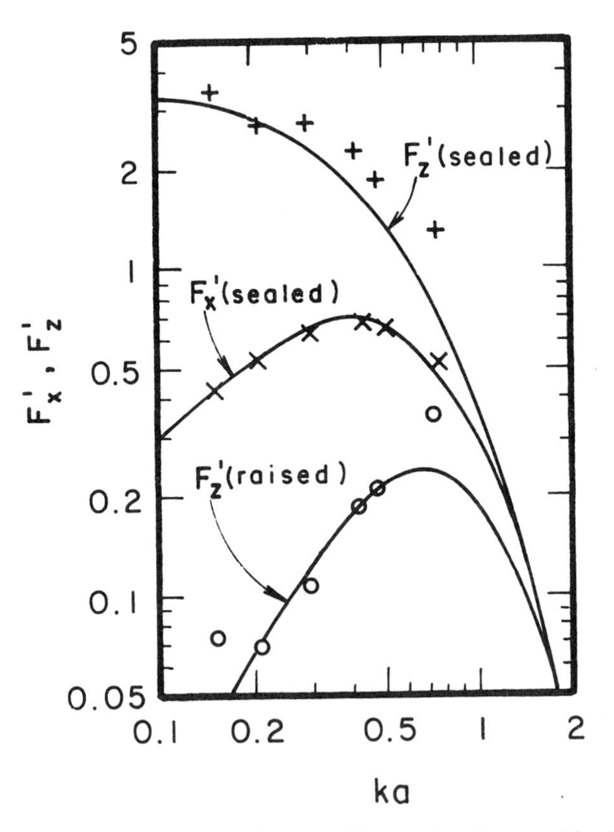

Fig. 6.23. Comparison of computer predictions with experiment reported by Naftzger and Chakrabarti (1975) for the wave forces on a hollow hemisphere near the seabed. ($F_x' = F_x/\frac{1}{2}\,\rho g H a^2$, $F_z' = F_z/\frac{1}{2}\,\rho g H a^2$.)

$$B_{ij} = \frac{\Delta S_j}{2\pi} \frac{\partial G'}{\partial n} (x_i, \xi_j) \tag{6.90}$$

However, a numerical difficulty which is not apparent in the wave source method does arise. Referring to Fig. 6.22, the normal velocity on segment j + 1 induced by a doublet located at ξ_j is not small, as it would be if instead a source were at ξ_j. Consequently for such close segment combinations a mid-point approximation may be unsuitable and a numerical integration becomes necessary. Even so, the method has been found capable of yielding reliable results, as indicated in the studies referred to. In particular, Naftzger and Chakrabarti (1975) used a wave doublet representation to calculate the forces on a hemispherical shell slightly raised above the seabed such that the shell interior is exposed to the fluid motion. A comparison between their computer predictions and experiment for both this "raised" case as well as the case of a sealed shell resting on the seabed is reproduced in Fig. 6.23. The markedly lower vertical forces predicted for the raised case are reproduced by experiment and are associated with the fluid pressure acting within the shell. Chakrabarti and Naftzger (1976) subsequently extended this kind of comparison to a hollow axisymmetric body designed for oil storage.

6.5 FINITE ELEMENT METHODS

The finite element method, which has already been referred to several times, has found increasing use in treating many wave diffraction problems. The general method has been described in detail in the text by Zienkiewicz (1977), and has been reviewed in the context of fluid flow problems by Shen (1977). Surveys by Mei (1978), Zienkiewicz, Bettess and Kelly (1978) and Brebbia and Walker (1979) describe the application of the finite or "hybrid" element method to wave diffraction problems.

From the calculus of variations, the solution to the boundary value problem is taken as the potential which minimizes a certain functional, which itself depends on the various governing equations of the problem. In the present context, the functional may be considered in physical terms to be related to the total energy of the fluid and the work done on it across its boundaries.

Because the free surface boundary conditions have been linearized, all the boundary conditions may be written in the general form

$$\frac{\partial \phi_s'}{\partial n} + \alpha \phi_s' + \beta = 0 \tag{6.91}$$

where α and β are constants which adopt values on the various boundaries as follows

Free surface, S_s ($z = 0$): $\alpha = -\omega^2/g$, $\beta = 0$,

Seabed, $z = -d$: $\alpha = 0$, $\beta = 0$,

Body surface, S_b: $\alpha = 0$, $\beta = \dfrac{\partial \phi_w'}{\partial n}$,

Radiation boundary, S_r: $\alpha = -i\omega/c$, $\beta = 0$.

In general three-dimensional problems governed by the Laplace equation, the functional Π may be expressed as

$$\Pi = \int_\Omega \frac{1}{2}\left[\left(\frac{\partial \phi_s'}{\partial x}\right)^2 + \left(\frac{\partial \phi_s'}{\partial y}\right)^2 + \left(\frac{\partial \phi_s'}{\partial z}\right)^2\right] d\Omega + \int_S \left(\tfrac{1}{2}\,\alpha\phi_s'^2 + \beta\phi_s'\right) dS \quad (6.92)$$

where Ω is the fluid domain, and S the entire fluid boundary. For vertical plane problems the dependence on y in Eq. (6.92) is simply omitted. In the case of generalized horizontal plane problems in which the depth may vary gradually, the governing equation of motion may be approximated (Berkhoff 1972, see also Section 4.9.5) to

$$\frac{\partial}{\partial x}\left(cc_G \frac{\partial \phi'}{\partial x}\right) + \frac{\partial}{\partial y}\left(cc_G \frac{\partial \phi'}{\partial y}\right) + \omega^2 \frac{c_G}{c}\,\phi' = 0 \quad (6.93)$$

where the celerity c and group velocity c_G are now variable, and $\phi' = \phi'(x, y)$ is a two-dimensional potential representing the flow. (Equation (6.93) is an extension to the Helmholtz equation, Eq. (6.74), which is itself valid for constant depth.) The functional corresponding to Eq. (6.93) and the associated boundary conditions is

$$\Pi = \int_\Omega \frac{1}{2}\left\{cc_G\left[\left(\frac{\partial \phi'}{\partial x}\right)^2 + \left(\frac{\partial \phi'}{\partial y}\right)^2 - \omega^2 \frac{c_G}{c}\,\phi'^2\right]\right\} d\Omega + \int_S \left(\tfrac{1}{2}\,\alpha\phi'^2 + \beta\phi'\right) dS \quad (6.94)$$

In the finite element method the fluid domain is divided into a finite number of regions or elements. Figure 6.24 illustrates this for vertical plane and horizontal plane problems corresponding to Figs. 6.12b and 6.12c. Each element has nodes on its boundaries (and possibly within the element), as indicated for some of the elements in the figure, and at the boundary nodes the values of the variable ϕ coincide with corresponding nodal values of the adjacent elements. The potential within any single element can be expressed in terms of the corresponding nodal values ϕ_e of that element and a column vector of interpolation functions (N). In standard matrix notation we have $\phi_s' = (N)^T(\phi_e)$.

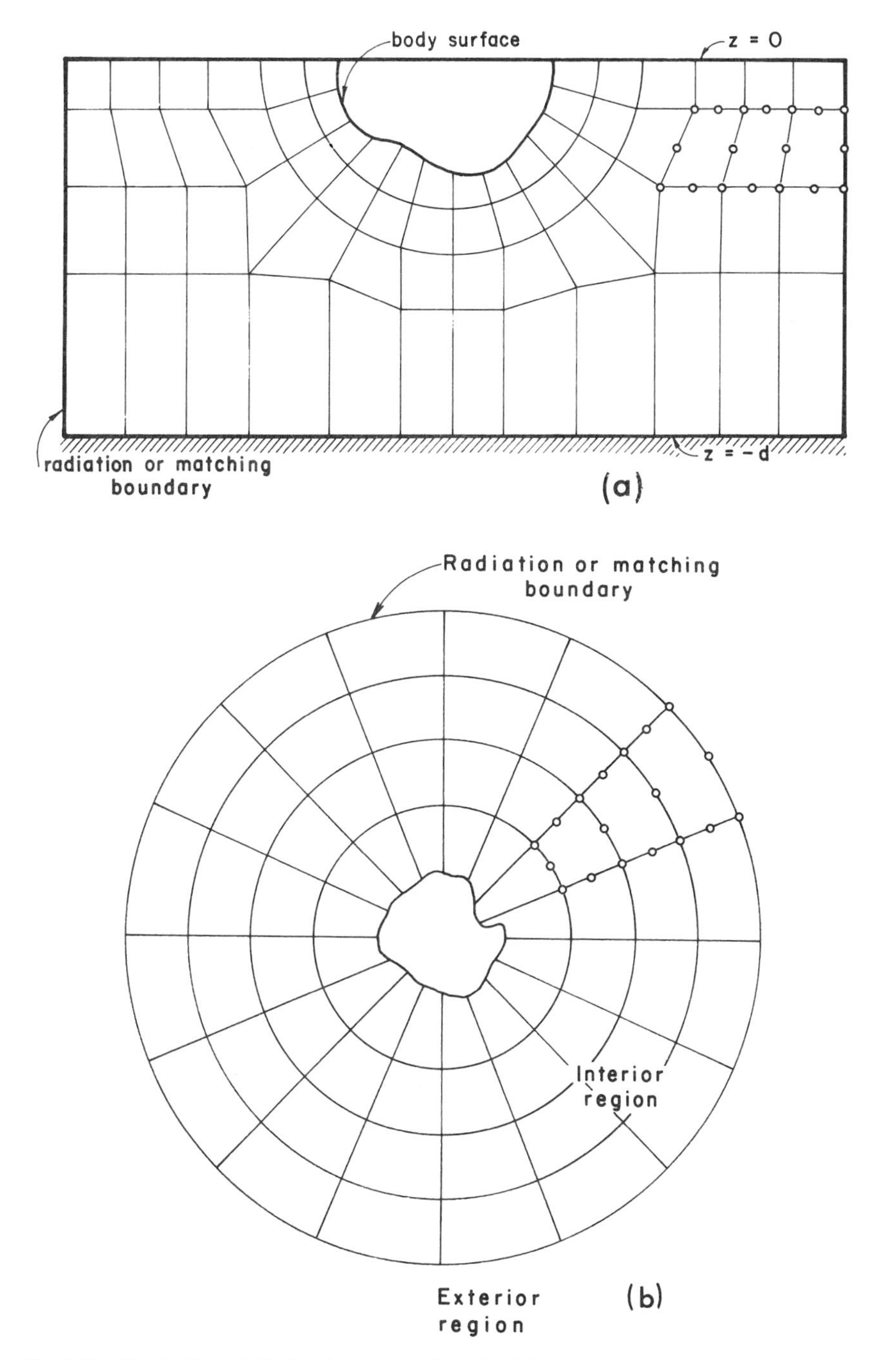

Fig. 6.24. Illustration of finite element meshes for (a) a vertical plane problem, and (b) a horizontal plane problem. Element node and radiation boundary representations are indicated.

The functional is now minimized with respect to the nodal values (ϕ_e) for a single element. For the three-dimensional case, the corresponding element equation may be derived from Eq. (6.92) and is

$$\frac{\partial \Pi}{\partial(\phi_e)} = \int_{\Omega_e} \left[\left(\frac{\partial N}{\partial x}\right)\left(\frac{\partial N}{\partial x}\right)^T + \left(\frac{\partial N}{\partial y}\right)\left(\frac{\partial N}{\partial y}\right)^T + \left(\frac{\partial N}{\partial z}\right)\left(\frac{\partial N}{\partial z}\right)^T \right] (\phi_e)\, d\Omega + \int_{S_e} \left[a(N)(N)^T(\phi_e) + q(N)\right] dS = 0 \tag{6.95}$$

where Ω_e and S_e denote that corresponding integrations are taken over a single element only. All such element equations are assembled to produce a set of global equations pertaining to the fluid region. These may be written in matrix form as

$$[A](\phi_e) = (P) \tag{6.96}$$

The matrix [A] and the column vector (P) are defined for the three-dimensional case as

$$[A] = -\frac{\omega^2}{g} \int_{S_s} (N)(N)^T\, dS - \frac{i\omega}{c} \int_{S_r} (N)(N)^T\, dS + \int_{\Omega} \left[\left(\frac{\partial N}{\partial x}\right)\left(\frac{\partial N}{\partial x}\right)^T + \left(\frac{\partial N}{\partial y}\right)\left(\frac{\partial N}{\partial y}\right)^T + \left(\frac{\partial N}{\partial z}\right)\left(\frac{\partial N}{\partial z}\right)^T \right] d\Omega \tag{6.97a}$$

$$(P) = -\int_{S_b} \frac{\partial \phi'_w}{\partial n} (N)\, dS \tag{6.97b}$$

where S_b denotes the body surface, S_r the radiation boundary and S_s the free surface taken along $z = 0$. The matrix [A] is symmetric and banded, with its half-bandwidth corresponding to the interaction between each node and its immediate neighbours. Equation (6.96) can therefore be solved quite efficiently. Corresponding sets of equations may be developed for various two-dimensional problems.

There is some choice in the element shapes and interpolation functions to be used, and the reader is referred to Zienkiewicz (1977) for a complete discussion of these. In two-dimensional problems, straight-edged or isoparametric triangular or quadrangular elements may be used, with nodes at each of the corners and also along each edge as indicated in Fig. 6.24. And in three dimensions, corresponding tetrahedral or brick-shaped elements are usually used. The interpolation functions are often chosen to describe a quadratic variation in velocity potential across any element. The fluid velocity then has a linear variation across an element, and while the velocity potential is continuous across element boundaries, the velocity itself is not.

One important factor of wave diffraction problems concerns modeling the infinite extent of the ocean. Four methods have been used with the finite element method to ensure that the radiation condition is satisfied. These are as follows:

(i) finite distance radiation boundary,
(ii) analytical series solution for exterior region,
(iii) boundary integral solution for exterior region,
(iv) "infinite" elements.

The first such approach is the simplest and most direct: "radiation" boundaries are taken to lie at some reasonably large but finite distance from the body, and the radiation condition is applied directly at these boundaries. This method has been found to give surprisingly accurate results. In the second and third methods listed above, sometimes termed *hybrid element methods*, the fluid region is divided as sketched in Fig. 6.24b, into an interior region in the vicinity of the body, and an exterior region extending to infinity. A finite element analysis is used only in the interior region and this is matched to an alternative representation of the exterior region. When the matching boundary forms a circular cylinder for three-dimensional or horizontal plane problems, or forms a plane $x = \text{constant}$ for vertical plane problems, the potential in the exterior region may readily be expressed as an analytical series with unknown coefficients (Chen and Mei 1974). This corresponds to the second method listed above. Alternatively, the third method involves expressing the potential in the exterior region in terms of a singularity distribution over the matching boundary, and thus the matching boundary may now possess a more general shape. The fourth method involves "infinite" elements in which the outermost elements themselves extend to infinity, and possess exponentially decaying interpolation functions which ensure that the radiation condition is satisfied (Bettess 1977). Hara, Zienkiewicz and Bettess (1979) have reviewed the alternative methods outlined above and present a comparison of results based on the alternative approaches.

The finite element methods generally compare reasonably well with the integral equation methods already described in Sections 6.3 and 6.4. The finite element method generally requires greater preparation in setting up a particular configuration, but this is offset by the fact that it may be more flexible, for instance in being able to accommodate variable depths in the region near the body. Both approaches can give accurate results and both involve approximately the same order of computer effort. For a two-dimensional problem, this may be indicated as follows. If N sources are used to describe a body contour in the wave source method, then a matrix equation of rank N must be solved, corresponding to the interaction of each source with all the others. In contrast, if M elements are used to describe the corresponding fluid region in the finite element method, then a matrix equation of rank M is to be solved, with M typically larger than N (corresponding to an area rather than a contour being discretized). But now the matrix equation is symmetric and banded and the overall order of computer effort is not too different. Similar comments apply to three-dimensional problems: now a finite fluid volume must be discretized in the finite element method, whereas a surface is discretized in the wave source method. In this case the hybrid element methods, or the use of infinite elements, are essential in order to avoid too large a matrix rank.

Calculations of wave forces using the finite element method for axisymmetric bodies or two-dimensional bodies in the x-z plane have invariably covered the more general case which includes body motions (see Section 6.6). Particular treatments for axisymmetric bodies include studies by Bai and Yeung (1974), Allouard, Chenot and Coudert (1975), and Bai (1977a); and for two-dimensional horizontal cylinders include Newton (1975) and Newton, Chenault and Smith (1974). Bai (1975) has considered an extension to the vertical-plane case corresponding to waves obliquely approaching an infinitely long horizontal cylinder.

Applications of the finite element method to problems in the horizontal plane but in which the depth may be variable, have been carried out mainly in the context of harbor resonance and include studies by Bettess and Zienkiewicz (1977), Berkhoff (1972), Sakai and Tsukioka (1975, 1977), Chen and Mei (1974), Mei and Chen (1975), and Houston (1978). These have generally employed a "hybrid element" approach in dealing with the radiation condition. Three-dimensional problems have been treated more recently by Yue, Chen and Mei (1978) and by Zienkiewicz, Bettess and Kelly (1978). As an example, Fig. 6.25 shows a comparison of wave force results reported by Zienkiewicz, Bettess and Kelly (1978) for the compound column shown. Their results using both two-dimensional elements (i.e. involving a variable depth), and three-dimensional elements are compared to those of Hogben and Standing (1974) based on the source distribution method. Relatively good agreement is obtained, even though quite a course mesh of three-dimensional elements was used.

At this point mention is made in passing of the finite difference technique which has occasionally been employed in wave force calculations. Raichlen and his co-workers (e.g. Raichlen and Naheer 1976) have used it extensively for the related harbor resonance problem. Chan and Hirt (1974), Nicols and Hirt (1976) and Miner et al. (1979) have used finite difference methods for various vertical

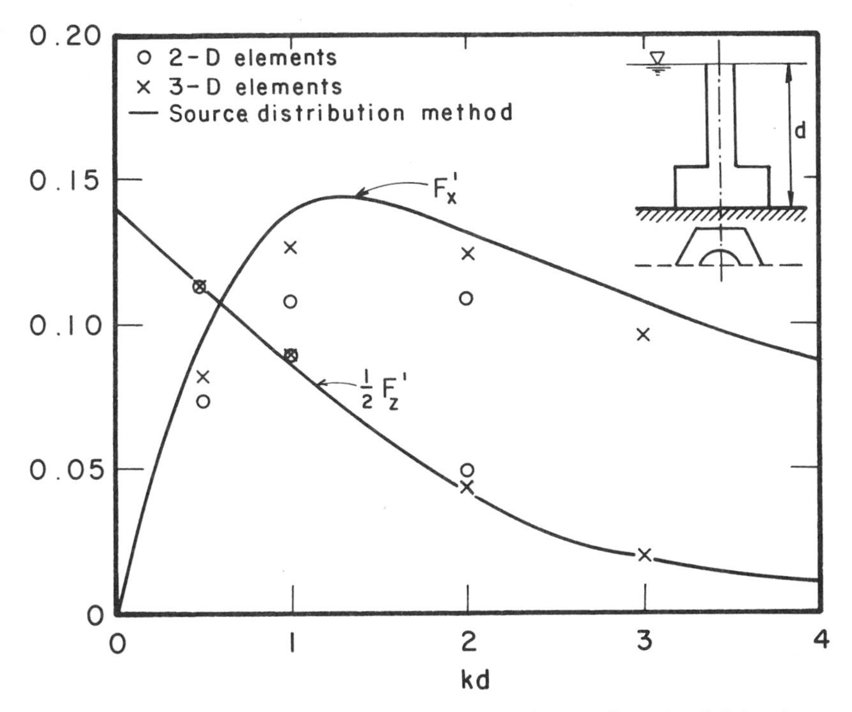

Fig. 6.25. Comparison of predictions based on two- and three-dimensional finite element representations and the source distribution method (Hogben and Standing 1974) reported by Zienkiewicz, Bettess and Kelly (1978) for the horizontal and vertical wave forces on the compound column shown. ($F'_x = F_x/\frac{1}{2}\rho g H d^2$, $F'_z = F_z/\frac{1}{2}\rho g H d^2$.)

plane problems. It appears that the finite difference methods do not have the power of the corresponding finite element methods and have not been developed as extensively.

6.6 FLOATING BODIES

A rigid floating body may have a motion with six degrees of freedom: three translational and three rotational. In the terminology of naval architecture, the translational motions in the x, y and z directions are referred to respectively as *surge*, *sway* and *heave*; and the rotational motions about the x, y and z directions respectively as *roll*, *pitch* and *yaw*. Here the x coordinate is taken to lie along the longitudinal axis of the body and z is the vertical coordinate as before. These motions are depicted in Fig. 6.26, and as shown in the figure the subscripts 1 to 6 are used here to denote these six modes of motion in the above order.

The case of a floating body has been treated extensively in the naval architec-

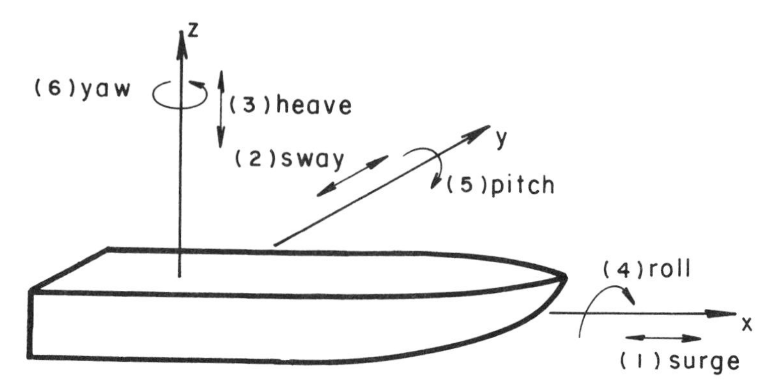

Fig. 6.26. Definition sketch of the six components of motion of a floating body.

ture literature and reference may be made to Wehausen (1971) and Newman (1977) for the theoretical approaches generally used. Developments based on the assumption of a slender body such as a ship are beyond the scope of this text but have been outlined by Newman (1970, 1977). And descriptions of "ship waves" caused by a ship travelling in otherwise still water and of the associated resistance to motion are not given here but are available in several texts (e.g. Stoker 1957, and Newman 1977). We consider here the interaction of a wave train with a floating body. The assumptions made previously for a fixed body in the diffraction range are retained: that is the fluid is incompressible, the fluid motion is irrotational and the waves are of small amplitude. As a consequence of the latter assumption, the body may be taken to oscillate harmonically with a small amplitude motion. The six modes of oscillation may then be written as

$$\alpha_j = a_j e^{-i\omega t} \qquad \text{for } j = 1, 2, \ldots 6 \tag{6.98}$$

where α_j is a displacement for $j = 1, 2, 3$ and a rotation for $j = 4, 5, 6$ and a_j is the corresponding complex amplitude.

The situation may most conveniently be considered in terms of a combination of two fundamental and somewhat related problems: the *radiation problem* of a body forced to oscillate in otherwise still water; and the *scattering problem* of an incident wave train interacting with a fixed body. Because of the linearity of the situation these two motions may be superposed, with the wave forces of the scattering problem providing the forcing function in the radiation problem.

The forced oscillation of a body in otherwise still water will generate waves moving outward in all directions and which may be represented by a "forced" velocity potential ϕ_f. Since the problem is linear, the velocity potential describing the flow field in the presence of incident waves may now be extended from Eq. (6.12) and represented as a sum of three separate components

$$\phi = \phi_w + \phi_s + \phi_f \tag{6.99}$$

As before ϕ_w represents the potential of the incident waves and ϕ_s that of the scattered waves, while the additional component ϕ_f represents the velocity potential of the waves generated by the body motion. These three components separately satisfy the Laplace equation, together with the bottom and free surface boundary conditions; and ϕ_s and ϕ_f must also satisfy the radiation condition. The boundary condition at the body surface must now account for the velocity of the body itself and is given by

$$\frac{\partial \phi_w}{\partial n} + \frac{\partial \phi_s}{\partial n} + \frac{\partial \phi_f}{\partial n} = V_n \qquad \text{at } S_o \tag{6.100}$$

where V_n is the velocity of the body surface in the direction n normal to itself. Since the motions are small, this condition is applied at the equilibrium surface S_o taken at the rest position, rather than at the instantaneous position. Eq. (6.100) may be broken down into the two equations

$$\frac{\partial \phi_w}{\partial n} + \frac{\partial \phi_s}{\partial n} = 0 \qquad \text{at } S_o \tag{6.101}$$

as in the fixed body case, together with

$$\frac{\partial \phi_f}{\partial n} = V_n \qquad \text{at } S_o \tag{6.102}$$

It is now seen that the problem defining ϕ_s is identical to that for the fixed body case and ϕ_s may thus be determined in exactly the same manner as described in earlier sections. However, it will be seen that ϕ_s is often not needed explicitly since it will be possible to express the wave forces directly in terms of ϕ_f.

The problem defining ϕ_f is slightly different on account of the different body surface condition which is now applicable: Eq. (6.102) in place of Eq. (6.101). V_n at any point $\mathbf{x} = (x, y, z)$ on the body is made up of six components associated with each mode of motion and each proportional to the corresponding velocity

$$V_n = \sum_{j=1}^{6} \frac{\partial \alpha_j}{\partial t} n_j = \sum_{j=1}^{6} - i\omega a_j n_j e^{-i\omega t} \tag{6.103}$$

where n_j is given as

$$\left.\begin{aligned} n_1 &= n_x, n_2 = n_y, n_3 = n_z, \\ n_4 &= zn_y - yn_z, \\ n_5 &= xn_z - zn_x, \\ n_6 &= yn_x - xn_y. \end{aligned}\right\} \tag{6.104}$$

n_x, n_y and n_z are the direction cosines of the normal to the surface at the point **x** (taken as directed outward from the body as before). In order to apply the boundary condition Eq. (6.102), and in view of the decomposite structure of Eq. (6.103), it is convenient to decompose also ϕ_f into six components associated with each degree of freedom and each proportional to the displacement amplitudes a_j. That is we put

$$\phi_f = \sum_{j=1}^{6} a_j \phi_j^{(f)} e^{-i\omega t} \tag{6.105}$$

where the six coefficients $\phi_j^{(f)}$ are generally complex. This representation enables the boundary condition Eq. (6.102) to be written in terms of $\phi_j^{(f)}$ and independent of a_j as

$$\frac{\partial \phi_j^{(f)}}{\partial n} = -i\omega n_j \text{ at } S_0, \qquad \text{for } j = 1, \ldots 6 \tag{6.106}$$

The right-hand sides of these equations are known and the six functions $\phi_j^{(f)}$ may now be found in the same manner as is the scattered potential ϕ_s. The only difference is that the body surface boundary condition now involves the body shape at equilibrium in place of the incident potential ϕ_w. It follows that, for example, either a source distribution representation or the finite element method may be employed to solve for $\phi_j^{(f)}$.

The forces and moments on the body due to the fluid pressure acting on the submerged body surface are made up of components due to the hydrostatic pressure and relating to the instantaneous position of the body, together with components due to the hydrodynamic pressure acting on the body. The equations of motion may therefore be written in matrix form as

$$m_{ij} \frac{\partial^2 \alpha_j}{\partial t^2} = -c_{ij}\alpha_j + \rho \int_{S_0} \frac{\partial \phi}{\partial t} n_i \, dS \tag{6.107}$$

where S_0 is the equilibrium body surface, and n_i is given by Eq. (6.104). The mass matrix components m_{ij} and hydrostatic stiffness matrix components c_{ij} are simply derived for a given body configuration and density distribution. The mass matrix components include the body mass, the mass moments of inertia and mass products of inertia as appropriate. The hydrostatic stiffness matrix components are conveniently expressed in terms of the body's waterline profile and the locations of the centers of gravity and buoyancy. Expressions for m_{ij} and c_{ij} are given by Wehausen (1971), Salvesen, Tuck and Faltinsen (1970), and Newman (1977). When extraneous forces act on the body (as in cases such as a moored body or an energy absorbing device) the equations of motion may be modified accordingly.

It remains to calculate the hydrodynamic forces and moments given by the last term of Eq. (6.107). These are conveniently expressed in terms of components $F^{(f)}$ due to the forced potential, and components $F^{(e)}$ due to the incident and scattered potentials. The latter is termed the *exciting force*, and since ϕ_w and ϕ_s are identical in the fixed body case the exciting force is identical to the wave-induced force acting in the fixed body case. (The component of the exciting force associated with ϕ_w is the Froude-Krylov force encountered previously.) We shall now separately consider the two force components $F^{(f)}$ and $F^{(e)}$ below.

Added-Mass and Damping Coefficients

There are six components $F_i^{(f)}$ corresponding to each mode of motion, and each of these may be written as

$$F_i^{(f)} = \rho \int_{S_0} \frac{\partial \phi_f}{\partial t} n_i \, dS = -i\omega\rho \sum_{j=1}^{6} \left(\int_{S_0} \phi_j^{(f)} n_i \, dS \right) \alpha_j \tag{6.108}$$

These are conveniently decomposed into components in phase with the velocity and the acceleration of each mode and we put

$$F_i^{(f)} = - \sum_{j=1}^{6} \left(\mu_{ij} \frac{\partial^2 \alpha_j}{\partial t^2} + \lambda_{ij} \frac{\partial \alpha_j}{\partial t} \right) \quad \text{for } i = i, 2, \ldots 6 \tag{6.109}$$

where the coefficients μ_{ij} and λ_{ij} are taken as real. These are termed the *added-mass* and *damping coefficients* respectively since they assume corresponding roles in the equations of motion. Thus, by writing the last term of Eq. (6.107) as $F_i^{(f)} + F_i^{(e)}$, using Eq. (6.109) and rearranging terms, the equations of motion may be written in the form

$$(m_{ij} + \mu_{ij}) \frac{\partial^2 \alpha_j}{\partial t^2} + \lambda_{ij} \frac{\partial \alpha_j}{\partial t} + c_{ij}\alpha_j = F_i^{(e)} \tag{6.110}$$

It can be seen from this that the added-mass and damping coefficients fullfill their expected roles, and that the exciting force can be considered as the forcing function of the motion. It is emphasized that this equation relates to an unrestrained floating body. The added-mass coefficients μ_{ij} are analogous to, but not the same as, those for a body accelerating in an unbounded fluid. The damping coefficients λ_{ij} are associated with a net outward flux of energy in the radiated waves, and thus represent only damping due to the (radiating) fluid motion. In cases where structural damping or viscous damping are important, these would need to be included in additional terms alongside the λ_{ij} terms. Likewise $F_i^{(e)}$ represents only the force due to the wave field, and if external forces are present these would need to be included alongside $F_i^{(e)}$. It is also stressed that μ_{ij} and λ_{ij} are not dimensionless coefficients but do possess appropriate dimensions. Equating the right-hand sides of Eqs. (6.108) and (6.109), and bearing in mind the boundary condition Eq. (6.106), the added-mass and damping coefficients may be defined explicitly as

$$\mu_{ij} = \frac{\rho}{\omega} \int_{S_0} \mathrm{Im}[\phi_j^{(f)}]\, n_i \, dS = \frac{i\rho}{\omega^2} \int_{S_0} \mathrm{Im}[\phi_j^{(f)}] \frac{\partial \phi_i^{(f)}}{\partial n} dS \tag{6.111}$$

$$\lambda_{ij} = -\rho \int_{S_0} \mathrm{Re}[\phi_j^{(f)}]\, n_i \, dS = -\frac{i\rho}{\omega} \int_{S_0} \mathrm{Re}[\phi_j^{(f)}] \frac{\partial \phi_i^{(f)}}{\partial n} dS \tag{6.112}$$

where Re[ ] and Im[ ] denote real and imaginary parts respectively. Note that in accordance with Eq. (6.106), $\partial\phi_i^{(f)}/\partial n$ is always imaginary over S_0 and therefore μ_{ij} and λ_{ij} are indeed real.

Now a consequence of Green's theorem is that

$$\int_{S_0} \left[\phi_i \frac{\partial \phi_j}{\partial n} - \phi_j \frac{\partial \phi_i}{\partial n} \right] dS = 0 \qquad \text{for } i, j = 1, 2, \ldots 6 \tag{6.113}$$

By applying this to Eqs. (6.111) and (6.112), it is seen that the coefficients are symmetric: $\mu_{ij} = \mu_{ji}$, $\lambda_{ij} = \lambda_{ji}$. It should finally be emphasized also that both the added-mass and the damping coefficients are frequency dependent.

Exciting Force

The forces and moments associated with ϕ_w and ϕ_s comprise the exciting force $F^{(e)}$ on the body, and this is identical to what it would be if the body were fixed. The exciting force could be determined in the same manner as was adopted in the fixed body case, but the calculation does not now require ϕ_s to be determined explicitly. For by an application of Green's theorem it is possible to express the exciting force directly in terms of the incident and forced potentials. Such expressions are the so-called *Haskind relations* which Newman (1962) has emphasized and which play an important role in this area of analysis. They may be given in the form

$$F_i^{(e)} = \rho \int_S \left[\phi_w \frac{\partial \phi_i^{(f)}}{\partial n} - \phi_i^{(f)} \frac{\partial \phi_w}{\partial n} \right] dS \tag{6.114}$$

Thus once the forced potential components $\phi_j^{(f)}$ have been calculated to obtain the added-mass and the damping coefficients, then it may be simpler to use Eq. (6.114) directly rather than first to calculate ϕ_s'. (In fact, in the wave source method the matrix equations used to determine $\phi_j^{(f)}$ and ϕ_s' differ only by their right-hand sides, and thus the additional effort needed to calculate ϕ_s' is small.)

The above approach can be developed further. On the basis of Green's theorem, the above integral can be expressed in terms of S_∞, a control surface in the fluid at a large distance from the body, as

$$F_i^{(e)} = -\rho \int_{S_\infty} \left[\phi_w \frac{\partial \phi_i^{(f)}}{\partial n} - \phi_i^{(f)} \frac{\partial \phi_w}{\partial n} \right] dS \tag{6.115}$$

It follows that an asymptotic form of $\phi_j^{(f)}$ may be employed, and considerations based on the principle of energy conservation may then be used to relate the damping coefficients of the body to the asymptotic behaviour of ϕ_f at S_∞, and subsequently to the exciting force. The damping coefficients and exciting forces are related by means of the equation

$$\lambda_{ii} = \frac{k}{2\pi\rho g c_G} \int_0^{2\pi} \left| \frac{X_i(\theta)}{H} \right|^2 d\theta \tag{6.116}$$

where X_i is the complex amplitude of the exciting force, H is the incident wave height, c_G is the group velocity of the waves, and the incident waves make an angle θ with the x axis.

This approach is particularly useful in the case of an axisymmetric body where the amplitudes (but not the phases) of the exciting force may be expressed explicitly in terms of the damping coefficients. Taking the incident wave direction to lie along the x axis for notational convenience, the force amplitudes are easily seen to be given for the finite depth case as follows

$$F_i = \left\{ K_i \frac{\rho g H^2 \lambda_{ii} \omega}{k^2} \left[1 + \frac{2kd}{\sinh(2kd)} \right] \right\}^{1/2}, \ i = 1, 2, \ldots 6 \tag{6.117}$$

with

$$K_1 = K_5 = 1,$$
$$K_2 = K_4 = K_6 = 0,$$
$$K_3 = \tfrac{1}{2}.$$

The above expression for the vertical force F_3 was given by Newman (1962) for the deep water case. In two-dimensional (vertical plane) problems the body is often symmetric about the plane $x = 0$, and the relation corresponding to Eq. (6.116) is then

$$\lambda_{ii} = \frac{2}{\rho g c_G} \left| \frac{X_i}{H} \right|^2 \tag{6.118}$$

Another simplification is possible when the wave length to body length ratio is sufficiently large. Under such conditions the exciting force components can be expressed directly in terms of the added-mass, damping and hydrostatic coefficients. The underlying theory of this *long wave approximation* is described by Newman (1977).

We are now in a position to consider solving the equations of motion for the body oscillation. Equation (6.110) can be written as a complex matrix equation

$$[-\omega^2(m_{ij} + \mu_{ij}) - i\omega\lambda_{ij} + c_{ij}]\, a_j = F_{0i}^{(e)} \tag{6.119}$$

where $F_0^{(e)}$ is the complex amplitude of the exciting force. Since $F_{0i}^{(e)}$ and the coefficients of a_j are now known, the complex amplitudes of motion may be determined to complete the solution to the problem. This will generally involve a complex matrix inversion procedure.

The two-dimensional problem in the x–z plane has been treated in some detail in the literature and may be approached by employing a complex potential representation. Published computations include those by Yu and Ursell (1961) for a semi-immersed circular cylinder, W. D. Kim (1965) for elliptical sections, C. H. Kim (1969) for various ship-like sections, Vugts (1968) for various shapes including a semi-immersed circular section and different rectangular sections, Adee and Martin (1974), Maeda (1974), Bai and Yeung (1974), Newton (1975) and Bai (1977b). The last three papers cited employ the finite element method. Comparisons with experiment include those reported by Porter (1960) for forced oscillations and by Vugts (1968) for both forced oscillations and freely floating cylinders. Added-mass and damping coefficients calculated by Vugts (1968) for the semi-immersed circular section and for one particular rectangular section in heave and sway are reproduced in Fig. 6.27. Corresponding vertical

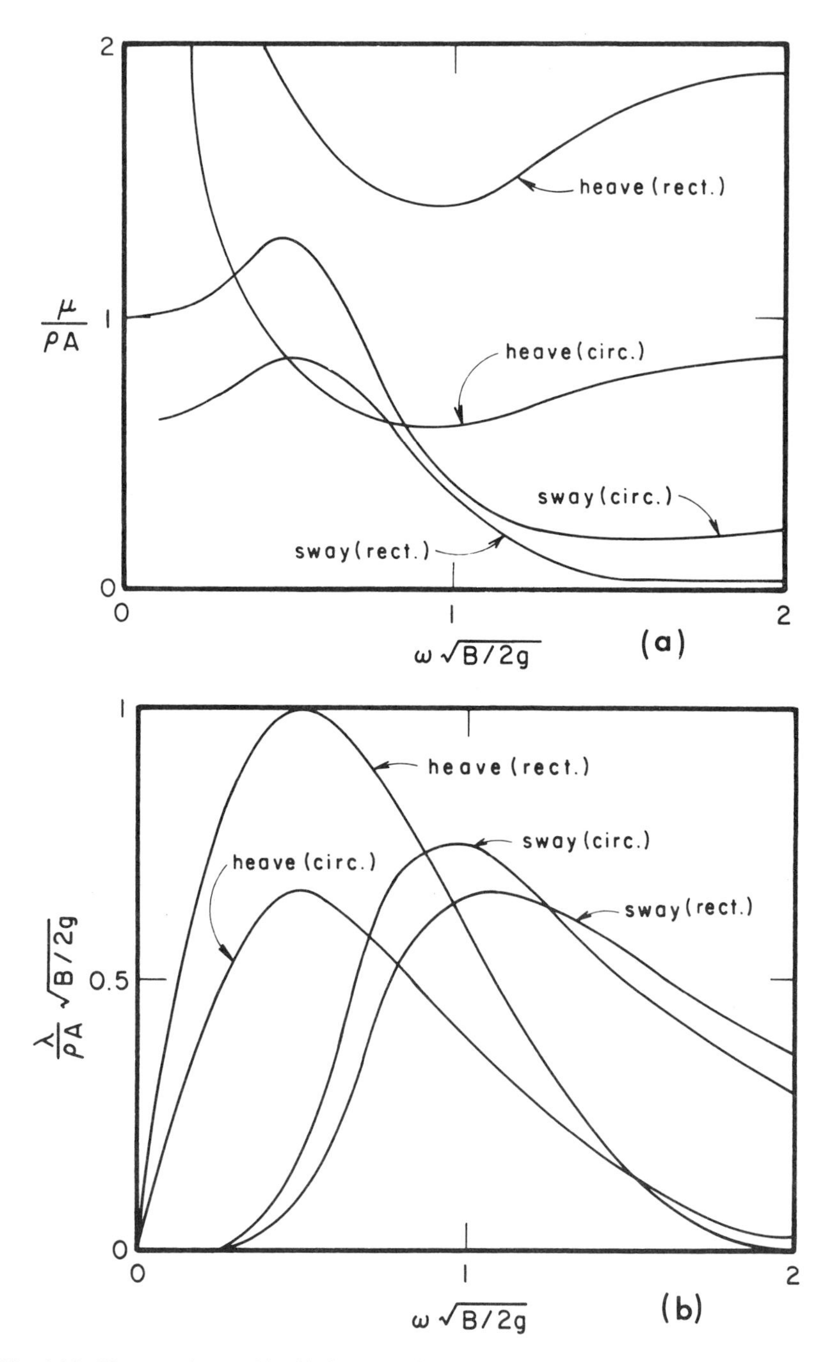

Fig. 6.27. Heave and sway (a) added-mass and (b) damping coefficients for a semi-immersed circular section and a rectangular section (width/draft = 4) in deep water (Vugts 1968). (A = section area, B = section width.)

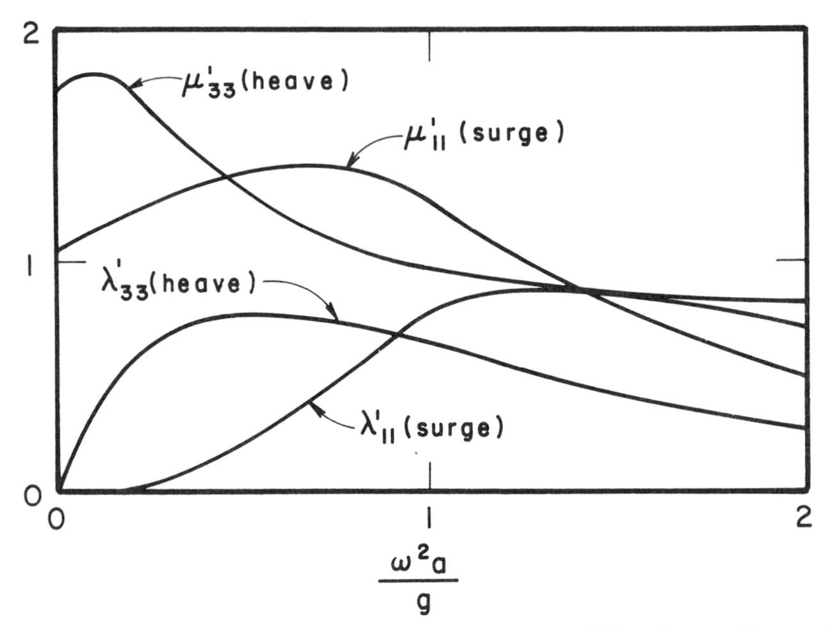

Fig. 6.28. Heave and surge (sway) added-mass and damping coefficients for a semi-immersed sphere in deep water. ($\mu' = \mu/\rho a^3$, $\lambda' = \lambda/\rho a^3 \omega$.)

and horizontal wave force amplitudes for these bodies when stationary are directly related to the damping coefficients by virtue of Eq. (6.118). Vugts also compared his theoretical results with experimental measurements, and the agreement obtained was generally good. One feature often found in such situations is that viscous damping is relatively important for roll motions, especially for a body with sharp corners or a keel (which result in flow separation), or for conditions close to resonance. In order to provide a better comparison with measurements, it is possible (e.g. Salvesen, Tuck and Faltinsen 1970) to include the effects of viscous damping in the equations of motion.

Computer programs for three-dimensional floating bodies of general shape have been described by Garrison (1974a, b, 1975) and by Faltinsen and Michelsen (1974), and have been based on the wave source distribution method. This approach had earlier been used by Kim (1965, 1966) to obtain results for a series of floating ellipsoids, including the case of a sphere.

A floating sphere is a fundamental situation of the axisymmetric problem, and was originally treated by Havelock (1955) for the case of heave motions in deep water. More complete results for a semi-immersed sphere have been presented by Kim (1966) for deep water, and by Garrison (1974a) for finite depths. The added-mass and damping coefficients for heave and surge (sway) motions in deep water are reproduced in Fig. 6.28. The horizontal and vertical exciting force amplitudes may be expressed by Eq. (6.117) and so can be obtained from the figure. The coefficients for roll, pitch and yaw, and the exciting moment components are clearly zero for this case.

Another fundamental axisymmetric body is a floating truncated circular cylinder and numerical results for this case have been presented by Garrison (1978). Comparisons between computer predictions and experimental measurements for three-dimensional floating bodies have been reported by Faltinsen and Michelsen (1974) and Loken and Olsen (1976) and have generally been found to be quite favorable.

6.7 INTERFERENCE EFFECTS

The investigation of the interference effects between large neighboring structures or structural components is necessary in order to access the conditions under which each component may be treated independently, thus permitting reduced computational effort in the analysis performed. Typical examples of effects resulting from interference arise between neighboring offshore storage tanks, between individual columns of a gravity platform, and between the columns and base of a platform.

When the structural members are slender, there is generally a significant interaction between them as the wake of one influences the forces on another. This aspect has been treated in Chapter 5, and we now take up the case of large structures for which the interference is associated with the disturbances created to the incident wave train. Any such interaction may in principle be deduced by the application of a general wave diffraction program extended to the bodies in question, but in practice the modeling of more than one body may be overly ambitious, and due to inadequate facet representation may introduce even greater errors than does the interference effect itself.

It is therefore useful to consider certain reference cases which can be examined by more efficient methods. These include neighboring vertical cylinders, coaxial vertical circular cylinders of different radii, and the vertical plane case of neighboring horizontal circular cylinders, all three cases being sketched in Fig. 6.29.

The interference effect between neighboring vertical circular cylinders has been considered by Lebreton and Cormault (1969) using a diffraction program applicable to bodies of arbitrary geometry, and also by Spring and Monkmeyer (1974, 1975), Ohkusu (1974), Chakrabarti (1978a) and Isaacson (1978b). The approach of Spring and Monkmeyer and of Chakrabarti uses a Fourier series representation for the scattered velocity potential of each cylinder and has been mentioned in Section 6.4.4. Ohkusu (1974) considered cylinders extending from the free surface to an arbitrary depth above the seabed and used an extension to Garrett's (1971) method to deal with three cylinders equally spaced from each other.

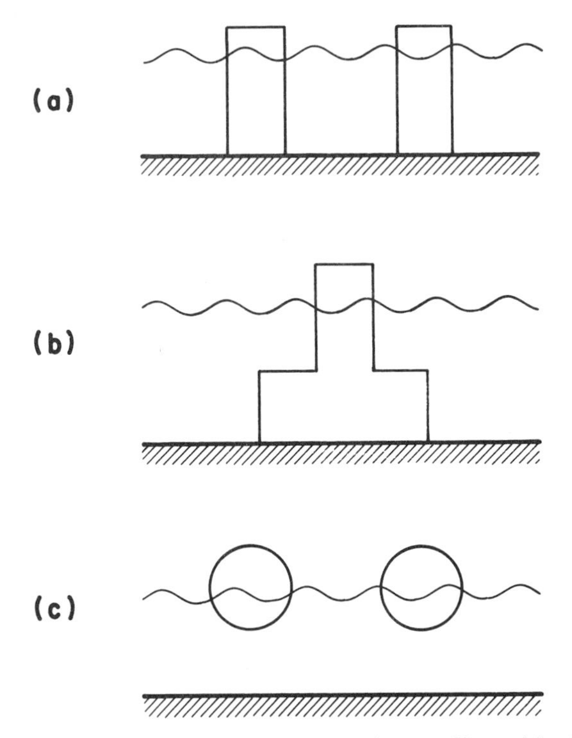

Fig. 6.29. Reference configurations for assessing interference effects. (a) neighboring vertical cylinders, (b) compound cylinder, (c) neighboring horizontal cylinders.

Isaacson (1978b) employed the two-dimensional source method of Section 6.4.3 to investigate the ratio R of the maximum x-ward force on a circular cylinder in the presence of a neighbor to that on the cylinder when isolated. The geometry of a cylinder pair is indicated in Fig. 6.29a and is specified by the cylinder radius a, the distance ℓ between centers and the orientation angle α which the neighbor's center makes with the x-axis at the first cylinder's center. The ratio R may be written in functional form as $R = f(ka, \ell/a, \alpha)$, a relationship which was investigated both numerically and by experiment. The numerical results for the case $\ell/a = 3$ and for various values of α are reproduced in Fig. 6.30. The fluctuations on the curves shown in the figure are predominant for $\alpha = 0°$ and are associated with the effect of the upstream cylinder lying in the standing wave system due to the other. Even for $\ell/a = 10$ the in-line force on a cylinder may be increased by as much as about 25%, while for lower values of ℓ/a this increase is even higher. When $\alpha = 90°$ the interference effect is usually relatively slight, although there is then a notable force acting transverse to the x direction. Chakrabarti (1979) has found good agreement between Isaacson's

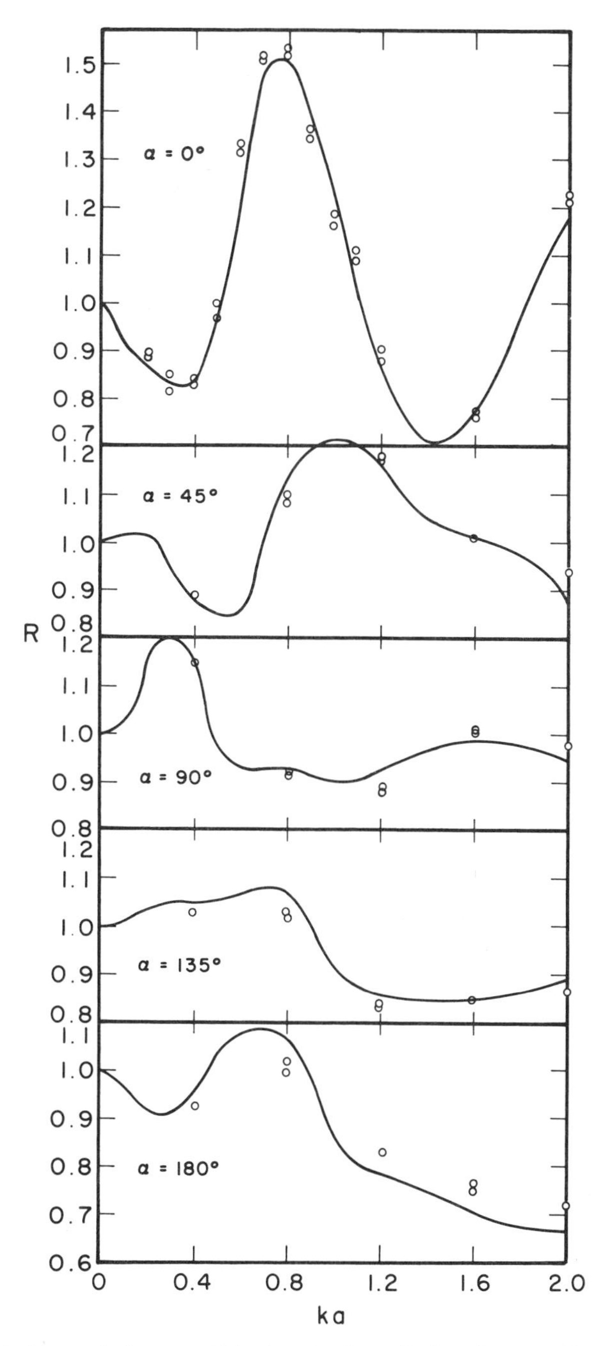

Fig. 6.30. Interference between neighboring circular cylinders shown as the force ratio R as a function of ka for various orientations and for a spacing $\ell/a = 3$ (Isaacson 1978b).

predictions for this problem of neighboring cylinders and those based on the Fourier expansion method of Spring and Monkmeyer (1974).

Isaacson (1979b) subsequently considered the interference effects between a pair of square cylinders using the same numerical procedure and experiments. In this case, vortices form at the corners of the squares, but in spite of this the numerical predictions are found to be in reasonable agreement with experiment. It is useful to provide a typical application demonstrating the significance of these kind of results and Isaacson (1979b) quotes one such example: for a design wave given as H = 4 m (13.1 ft), T = 6 sec, d = 18 m (59.1 ft), and a square caisson of side b = 15 m (49.2 ft) and with sides parallel to the incident wave crests, the effective inertia coefficient is given as $C_m = 1.5$ and the maximum force is then 6415 kN (300 kips). When a second caisson of the same size is located directly downstream with the two centers 3b apart, then the maximum force is increased to as much as 10650 kN (493 kips).

Van Oortmerssen (1979) has also reported on the interference effect between neighboring vertical cylinders. He treated the case of a pair of floating bodies (one cylinder's section was circular and the other was almost square) so that the interference effect involves not only the exciting force as in the fixed body case, but also the various hydrodynamic coefficients.

Many gravity platforms comprise of a relatively large base lying in the diffraction range, together with a number of smaller columns lying in the inertia range. It may therefore be possible to achieve a reasonable approximation by considering wave diffraction around the large component alone, and then applying the resulting velocities and accelerations to the Morison equation in order to estimate the forces on the smaller components. Mention of this approach is made by Hogben and Standing (1974), Hogben et al. (1977), Garrison and Stacey (1977) and Garrison (1978). The limits of such an approximation for the reference case shown in Fig. 6.29b are desirable, and can conveniently be predicted theoretically by using the extension to Garrett's theory as indicated by Isaacson (1979e) (Section 6.4.4). However only preliminary results have been presented and more comprehensive results for this configuration are needed in order to establish suitable guidelines. Fig. 6.31 reproduces some of the results that were presented which dealt with the case $h/d = 0.5$, $a/h = 1.0$, where h is the base cylinder height. The figure shows the amplitude of the horizontal force acting on the base only as a function of the diffraction parameter ka for different values of upper to lower cylinder radius ratio b/a. This shows the manner in which the force on the base increases from that on a single truncated cylinder ($b/a = 0$) to that on a portion of a single surface-piercing cylinder ($b/a = 1$).

The case of neighboring horizontal cylinders subjected to wave action (Fig. 6.29c) has primary application to marine vessels such as semisubmersibles and catamarans. Relevant results thus concern not only the wave-induced forces, but also the added-mass and damping coefficients (Section 6.6) and the ampli-

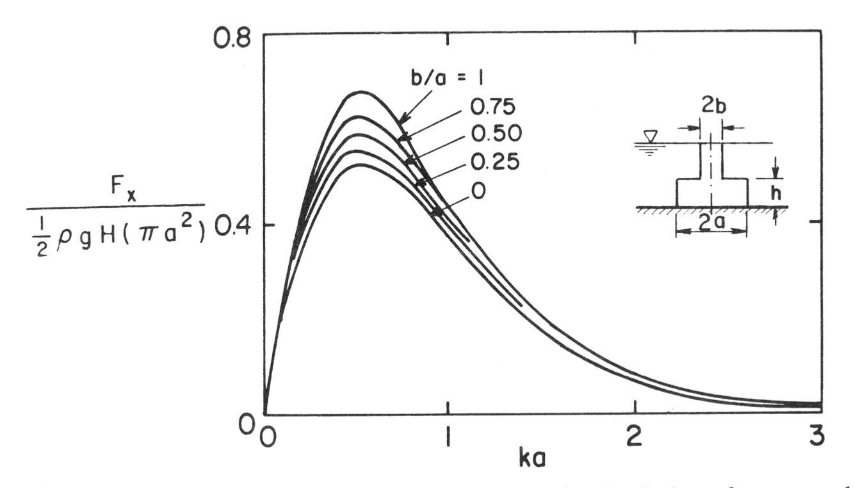

Fig. 6.31. Horizontal wave force coefficient as a function of ka for the base of a compound cylinder, with h/d = 0.5, a/h = 1.0 and various values of b/a (Isaacson 1979e).

tudes of motion. Some pertinent studies include those by Wang and Wahab (1971), Maeda (1974) and Ohkusu (1974) who have considered pairs of semi-immersed cylinders of various sections.

6.8 NONLINEAR WAVE EFFECTS

One of the important limitations of the diffraction methods outlined in preceding sections is that they are based on small amplitude wave theory and the associated assumption of linearity. The severe wave heights encountered in practice have led to some consideration being given to extensions to deal with steep (nonlinear) waves.

Comments on the possible effect of wave nonlinearities on design wave loads on typical gravity platforms have been made by Hogben and Standing (1975), Garrison and Stacey (1977) and Garrison (1978). Hogben and Standing compared linear and Stokes fifth order theory predictions of the inertia force on a column and concluded that the difference for typical North Sea design wave conditions is not large. However, one important effect of wave nonlinearity is that for a given wave period nonlinear wave theory predicts a different (longer) wave length than does linear theory. Garrison and Stacey have pointed out that the higher order components of a nonlinear wave are expected to have little effect on a typical well-submerged caisson on the seabed and they suggest using the first order component of Stokes 5th order theory (which has the appropriate wave length) in place of linear theory itself in the diffraction calculation. The columns of a typical structure lie in the inertia range and so can be calculated by a nonlinear wave theory on the basis of the Morison equation as was described in Chapter 5. Although such an approach lacks mathematical rigour, its use is justified in that it does produce empirically satisfactory results and serves as a basis for a practical design procedure.

Another effect of wave nonlinearities on surface-piercing structures is that forces calculated by integrating pressures up to the still water level on the one hand, or up to the instantaneous free surface on the other, may differ noticeably from each other. Again, Hogben and Standing (1975) have illustrated this difference for the limiting case of inertia force predictions. Formally, this difference is a second order quantity, and is thus of the same order as other second order force contributions which are neglected in the linear diffraction theory. This effect is discussed further in Section 6.8.1 which follows.

Linear diffraction theory may be more unreliable for the relatively steep waves encountered in shallower water or for large structures extending up to the free surface, and a more serious investigation of nonlinear effects then becomes necessary. The important features of the nonlinear problem can formally be investigated by extending the diffraction theory to a second approximation on the basis of the Stokes expansion procedure in a manner analogous to the derivation of Stokes second order wave theory (Section 4.3). This general approach is now described.

6.8.1 Second Order Diffraction Theory

As was indicated in Section 4.3, the nonlinearities of the problem arise on account of the two free-surface boundary conditions. The general (nonlinear) diffraction problem is defined in the same way as the linear problem, Eqs. (6.7)-(6.11), except that the complete nonlinear free-surface boundary conditions now apply in place of Eqs. (6.8) and (6.9). Thus ϕ_2 must satisfy the Laplace equation within the fluid region, and the relevant boundary conditions are given by Eqs. (6.10) and (6.11) as well as the two free-surface conditions, Eqs. (4.3) and (4.4) extended to three dimensions,

$$\frac{\partial \eta}{\partial t} + \frac{\partial \phi}{\partial x}\frac{\partial \eta}{\partial x} + \frac{\partial \phi}{\partial y}\frac{\partial \eta}{\partial y} - \frac{\partial \phi}{\partial z} = 0 \qquad \text{at } z = \eta \qquad (6.120)$$

$$\frac{\partial \phi}{\partial t} + \frac{1}{2}\left[\left(\frac{\partial \phi}{\partial x}\right)^2 + \left(\frac{\partial \phi}{\partial y}\right)^2 + \left(\frac{\partial \phi}{\partial z}\right)^2\right] + g\eta = f(t) \qquad \text{at } z = \eta \qquad (6.121)$$

And finally a suitable radiation condition is applicable.

According to the expansion method, ϕ is expressed as a power series in a perturbation parameter ϵ,

$$\phi = \epsilon\phi_1 + \epsilon^2\phi_2 + \ldots \tag{6.122}$$

This is substituted into the governing equations, and a Taylor series expansion about z = 0 is used to express the free surface conditions directly at z = 0 (see Eqs. (4.50) and (4.51)). Terms with like powers of ϵ are then collected to provide the governing equations of each successive approximation. Terms to order ϵ define the first order (linear) problem, Eqs. (6.7)-(6.11). Terms to order ϵ^2 define the second order problem expressing ϕ_2 in terms of the first order quantities. The governing equations are

$$\frac{\partial^2\phi_2}{\partial x^2} + \frac{\partial^2\phi_2}{\partial y^2} + \frac{\partial^2\phi_2}{\partial z^2} = 0 \quad \text{within the fluid} \tag{6.123}$$

$$\frac{\partial^2\phi_2}{\partial t^2} + g\frac{\partial\phi_2}{\partial z} = -\eta_1 \frac{\partial}{\partial z}\left[\frac{\partial^2\phi_1}{\partial t^2} + g\frac{\partial\phi_1}{\partial z}\right] - \frac{\partial}{\partial t}(\nabla\phi_1)^2 \quad \text{at } z = 0 \tag{6.124}$$

$$\eta_2 = -\frac{1}{g}\left[\frac{\partial\phi_2}{\partial t} + \eta_1\frac{\partial^2\phi_1}{\partial z\partial t} + \frac{1}{2}(\nabla\phi_1)^2\right] \quad \text{at } z = 0 \tag{6.125}$$

$$\frac{\partial\phi_2}{\partial z} = 0 \quad \text{at } z = -d \tag{6.126}$$

$$\frac{\partial\phi_2}{\partial n} = 0 \quad \text{at the body surface} \tag{6.127}$$

together with a radiation condition corresponding to Eq. (6.13).

The above equations for ϕ_2 are identical to those defining the first order potential ϕ_1, except that Eq. (6.124) has a nonzero right-hand side (cf. Eq. (6.8)). In physical terms, this difference is equivalent to solving a first order problem in which an applied pressure distribution equal to $\rho(\nabla\phi_1)^2$ is applied over the whole free surface z = 0 (Lighthill 1979). Isaacson (1977b) has pointed out that in the common case of a body which penetrates the free surface with a curved waterline contour, the second order radial velocity must possess a discontinuity adjacent to the body surface at z = 0. However, this does itself not imply that a rigorous solution to the second-order problem is not possible (e.g. Isaacson 1979f).

As with the linear diffraction problem, it may now be convenient to express ϕ_2 by components associated with the incident waves and scattered waves

$$\phi_2 = \phi_2^{(w)} + \phi_2^{(s)} \tag{6.128}$$

Of these, $\phi_2^{(w)}$ is known and is given in Table 4.3. Eq. (6.128) can then be substituted into Eqs. (6.123)-(6.127) to define the problem in terms of $\phi_2^{(s)}$.

Provided that ϕ_2 is known (although we shall see that it need not actually be determined), the forces on the body may be expressed in terms of ϕ_1 and ϕ_2 in the following way. We consider the general case of a surface-piercing body, and shall assume for simplicity that the body surface penetrates the free surface vertically. The force on the body can thus be expressed in vector form as

$$\mathbf{F} = -\int_S p\mathbf{n}\, dS = -\int_{S_0} p\mathbf{n}\, dS - \int_W \left[\int_0^{\eta(W)} p\, dz\right] \mathbf{n}\, dW \qquad (6.129)$$

where S is the instantaneous wetted body surface (which continuously changes), S_0 is the wetted body surface below the still water level $z = 0$, and W is the waterline curve at $z = 0$. In the above, an element dS of the intermittently wet portion of the surface has been expressed as dz dW, since the body surface in this vicinity is vertical.

Since the pressure variation near the free surface is hydrostatic to the first order, we may write

$$\int_0^{\eta} p\, dz = \int_0^{\eta} \{\rho g(\eta - z) + 0[\epsilon^3]\}\, dz$$

$$= \epsilon^2(\tfrac{1}{2}\rho g \eta_1^2) + 0[\epsilon^3] \qquad (6.130)$$

And the Bernoulli equation gives the pressure on the body surface to the second order as

$$p = -\rho g z + \epsilon\left\{-\rho\frac{\partial\phi_1}{\partial t}\right\} + \epsilon^2\left\{-\rho\frac{\partial\phi_2}{\partial t} - \frac{1}{2}\rho(\nabla\phi_1)^2\right\} + 0[\epsilon^3] \qquad (6.131)$$

Substituting Eq. (6.130) and (6.131) into Eq. (6.129), the total force on the body can be expressed in terms of ϕ_1 and ϕ_2 as

$$\mathbf{F} = \epsilon\mathbf{F}_1 + \epsilon^2\mathbf{F}_2 + 0[\epsilon^3] \qquad (6.132)$$

with

$$\mathbf{F}_1 = \int_{S_0} \rho\frac{\partial\phi_1}{\partial t}\, \mathbf{n}\, dS \qquad (6.133)$$

$$\mathbf{F}_2 = \int_{S_o} \rho \frac{\partial \phi_2}{\partial t} \mathbf{n}\, dS + \int_{S_o} \frac{1}{2} \rho\, (\nabla \phi_1)^2\, \mathbf{n}\, dS - \int_{w} \frac{\rho}{2g} \left(\frac{\partial \phi_1}{\partial t} \right)^2 \mathbf{n}\, dW \tag{6.134}$$

$\epsilon \mathbf{F}_1$ represents the first order force which is obtained by solving the linear diffraction problem. The second order force $\epsilon^2 \mathbf{F}_2$ is seen to contain three terms and these have been described in a physical context by Lighthill (1979). The first is due to the second order velocity potential, the second is due to the lowest order dynamic pressure, and the third is due to the effect of carrying out pressure integrations up to $z = 0$ rather than to $z = \eta$. This third term is a horizontal force which acts at the free surface $z = 0$.

It is the calculation of the first of these terms, involving ϕ_2 which presents the major difficulty. However, Lighthill (1979) has shown that it may be expressed as

$$F_q = -\rho \int_{S_f} w(\nabla \phi_1)^2\, dS_f \tag{6.135}$$

Here w is the vertical velocity at $z = 0$ associated with the flow due to the body oscillating in otherwise still water with unit amplitude in the direction of F_q and at a frequency 2ω, and S_f is the entire plane $z = 0$ exterior to the body. The above results can be derived by applying a reciprocity principle, or equivalently Green's theorem, between the second order diffraction problem which includes the pressure distribution $(\nabla \phi_1)^2$ acting over S_f, and the above radiation problem. This development can readily be extended to include the moment acting on the body. In the general three-dimensional case, the solution to the second order problem would involve numerical integrations over S_f, and it is expected that solutions for various body configurations should shortly become available.

The surface-piercing vertical circular cylinder constitutes a fundamental reference situation, and some effort has been directed towards solving this problem to a second or higher order approximation (Chakrabarti 1972, 1975, Yamaguchi and Tsuchiya 1974, Raman and Venkatanarasaiah 1976, and Kurata and Ijima 1979). Chakrabarti's (1972) fifth order solution does not satisfy the free-surface boundary conditions. The other (second order) solutions are complicated and may contain errors (e.g. Chakrabarti 1978b) so that the validity of any one solution has as yet not been confirmed. Kurata and Ijima's (1979) approach involves the use of the finite element method. In this, a Taylor series expansion is used to transform the integral over the free surface, contained in the corresponding functional, to one over the plane $z = 0$.

In spite of the considerable efforts taken towards developing a solution to the second order diffraction problem based on Stokes expansion procedure as outlined in the foregoing development, it is emphasized that the corresponding solution should be valid over relatively restricted conditions as in the case of an undisturbed wave train. For example, nonlinear effects are expected to be particularly significant for steep, shallower water waves, which are precisely the conditions under which the second order theory would become invalid. (See Fig. 4.5 and the discussion accompanying that figure.)

6.8.2 Drift Forces

In addition to the oscillatory wave forces acting on a body, steady second order forces are also present. These drift forces are an order smaller than the oscillatory forces, being proportional to the square of wave heights, but nevertheless they may be of considerable importance in certain circumstances. This is particularly so for random wave loading, since the drift force will then vary slowly with time, and this low frequency variation may excite a low frequency resonance in a body's mooring system, possibly causing it to fail.

The drift force may be demonstrated by examining the general expression for the second order wave force on a body, given by Eq. (6.134). This shows that the time-averages of the second and third terms of the second order force are generally non-zero, and, furthermore, that the second order potential is not needed to calculate the drift force.

Maruo (1960) has developed formulae for the drift force on a floating body in deep water in terms of a far-field approximation to the velocity potential. Newman (1967) extended these formulae to the drift (yaw) moment. In the two-dimensional case of a vertical plane flow, which may be used to approximate a ship in beam seas, Maruo's formula for the drift force is particularly simple, involving only the reflected wave height. For the vertical plane flow Longuet-Higgins (1977) has given a more general formula, which is valid for arbitrary depths and applies also to case of energy absorption or dissipation. This is:

$$F_D = \frac{1}{16}\rho g \left[1 + \frac{2kd}{\sinh(2kd)}\right](H_i^2 + H_r^2 - H_t^2) \tag{6.136}$$

where H_i, H_r and H_t are the incident, reflected and transmitted wave heights respectively. When no energy is dissipated, we have $H_i^2 = H_r^2 + H_t^2$, and consequently Eq. (6.136) simplifies to

$$F_D = \frac{1}{8}\rho g \left[1 + \frac{2kd}{\sinh(2kd)}\right]H_r^2 \tag{6.137}$$

For three-dimensional problems various authors have calculated drift forces for ship-like vessels on the basis of the slender body approximation or strip theory (e.g. Newman 1977). For more arbitrary body shapes, the three-dimensional wave source method described in Section 6.3 may be extended to calculate these forces. This has been carried out by Faltinsen and Michelsen (1974) on the basis of Newman's (1967) formulae, and by Garrison (1974b, 1975) by using the time-averaged terms of Eq. (6.134). Numerical predictions are generally found to agree reasonably well with experiments. Finally, extensions to calculating drift forces due to random waves are described by Hsu and Blenkarn (1970), Newman (1974) and Faltinsen and Loken (1979).

6.8.3 Cnoidal Wave Diffraction

As an alternative line of investigation into nonlinearity effects, Isaacson (1977a) has considered the diffraction of cnoidal waves around a large vertical circular cylinder, and has developed a first approximation for the loads on the cylinder. In this method the Friedrich's shallow wave expansion procedure (Section 4.4.1) is applied, with all variables transformed to dimensionless forms which incorporate an appropriate stretching of the z ordinate. The incident wave potential ϕ_w, given by the first approximation to cnoidal wave theory, is then expanded as a Fourier series in time t and angular coordinate θ. The cylinder surface boundary condition can then be used to develop an expression for the scattered potential. The total potential is thereby given as

$$\frac{k\phi}{\sqrt{gd}} = \frac{H}{d}\sum_{n=1}^{\infty}\frac{A_n}{in}\left\{\sum_{m=0}^{\infty}\beta_m\left[J_m(nkr) - \frac{J_m'(nka)}{H_m^{(1)\prime}(nka)}H_m^{(1)}(nkr)\right]\cos(m\theta)\right\}e^{-in\omega t} + 0\left[\left(\frac{H}{d}\right)^2\right] \quad (6.138)$$

where the cylindrical coordinate system of Fig. 6.2 is applicable, and A_n are the Fourier coefficients obtained by expanding $cn^2 q$ as in Eq. (4.88) and are given by Eq. (4.90). The total force F and overturning moment M are eventually given as:

$$\frac{F}{\rho gHad} = \frac{2M}{\rho gHad^2} = \frac{4}{ka}\left[\sum_{n=1}^{\infty}\frac{A_n}{nH_1^{(1)\prime}(nka)}\right]e^{-i\omega t} \quad (6.139)$$

A_n depends on the Ursell number $U = HL^2/d^3$, which has been encountered in Chapter 4 as essentially a nonlinearity parameter appropriate for shallow wave conditions. Hence the dimensionless force depends not only on the diffraction

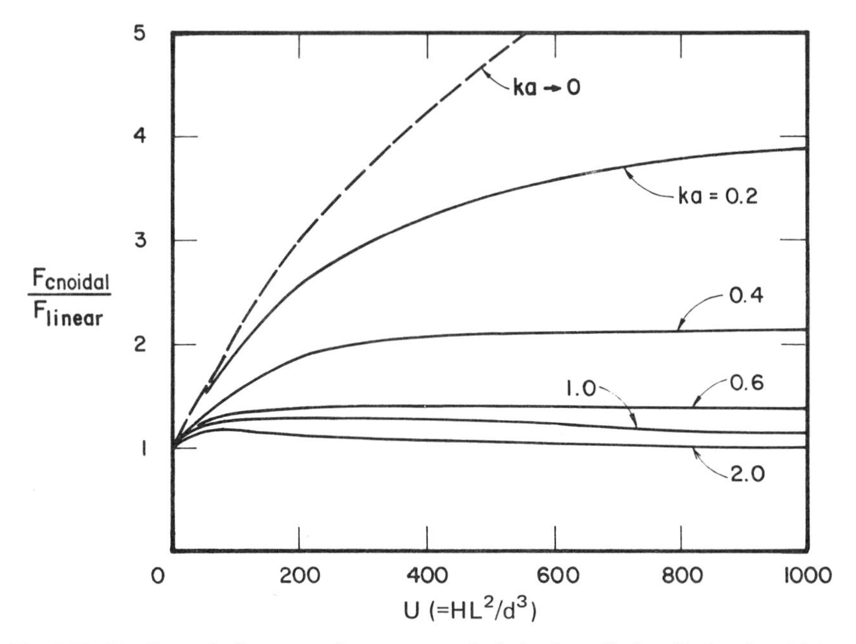

Fig. 6.32. Nonlinear shallow wave forces on a vertical circular cylinder. Ratio of cnoidal to shallow sinusoidal wave theory force predictions as a function of Ursell number U for various values of ka (Isaacson 1977a).

parameter ka, as with linear theory, but also on the Ursell number. The results are summarized in Fig. 6.32, which shows the ratio of the predicted force to that given by shallow water sinusoidal wave theory, this ratio being expressed as a function of Ursell number U for various values of the diffraction parameter ka. The figure may conveniently be used to provide an estimate of the cnoidal diffraction theory results by simply applying the appropriate factor to the forces predicted by shallow water sinusoidal theory. The theory based on cnoidal wave diffraction is however only valid in the shallow water range (say $kd < 0.6$) and is available only for the one reference case. It cannot therefore have widespread application, although it is useful in emphasizing the severe nonlinear effects that are possible in shallow water. An example Isaacson provides is indicative of this: for a water depth of 20 ft, a design wave of height 8 ft and length 460 ft, and a cylinder of radius 30 ft, which correspond to $ka = 0.41$, $kd = 0.273$ and $U = 212$, the cnoidal diffraction theory predicts a maximum force which is about 80% greater than that given by linear wave theory. This relatively high prediction can be considered to be due to the high fluid accelerations associated with the sharp crests of the wave train.

Comprehensive wave force experiments with fairly steep waves in the shallow water range, where the above theory may be expected to apply, have not been carried out. However, Isaacson (1978a) subsequently measured the wave runup on a cylinder under such conditions and compared the results with both sinusoidal and cnoidal wave diffraction theory predictions. He found that the cnoidal theory predictions approached the measured profiles more closely, although the maximum runup around the cylinder periphery was still significantly higher than the predictions of either theory.

6.8.4 Nonlinear Inertia Forces

As part of our consideration of nonlinear wave effects it is instructive to investigate also the limiting case of wave diffraction as the diffraction parameter becomes sufficiently small for the force to be taken as wholly inertial. That is, the body size to wave length ratio is now taken to be relatively small, although flow separation effects continue to be neglected (see Fig. 6.1).

The inertia force as applied to the Morison equation is usually taken as proportional either to the local or to the total fluid acceleration at a point in nonlinear waves (see Section 5.3.1). Even though these two accelerations may differ considerably for a typical design wave, no formal justification exists for adopting one or the other formulation. Indeed, the conventional expression for the inertia force, Eq. (1.1), is derived for an unsteady but otherwise uniform flow past a body, whereas nonlinear wave motion constitutes the more general case of an unsteady and nonuniform flow. Isaacson (1979c) has examined the theoretical formulation of the inertia force for this more general case, and considers the theoretical influence of nonlinear convective acceleration terms on the inertia force. He derives a generalized force expression which depends on all the added-mass coefficients and cross-coefficients of the body. In the usual case of a body with zero added-mass cross-coefficients, the expression for the inertia force in the x direction, F_x, reduces to:

$$F_x = (\rho V + m_{11})\left(\frac{\partial u}{\partial t} + u\frac{\partial u}{\partial x}\right) + (\rho V + m_{22})v\frac{\partial v}{\partial x} + (\rho V + m_{33})w\frac{\partial w}{\partial x} \tag{6.140}$$

where V is the body volume, and m_{11}, m_{22} and m_{33} are the added masses of the body in the x, y and z directions respectively. This equation is found to predict wave forces on cylinders and spheres which are generally lower than those obtained by taking the inertia force as proportional to the local acceleration. However, Isaacson (1979c) emphasizes that such results should not be interpreted in terms of the practical application of the Morison equation (whose coefficients are in any case empirical), but rather they should contribute towards understanding the inertia regime limit of the nonlinear diffraction problem.

Lighthill (1979) has considered the inertia force limit to the second order diffraction problem for the particular case of a surface-piercing circular cylinder in deep water. This represents a particular solution to the formulation given by

Eq. (6.134) and the result may be expressed in the form

$$\frac{F}{\rho g H D^2} = -\frac{\pi}{4}\left[\sin(\omega t) + \frac{5\pi}{8}\left(\frac{H}{L}\right)\sin(2\omega t)\right] \tag{6.141}$$

It is found that 80% of the second order force is due to the third term of Eq. (6.134), associated with reducing pressure integrations up to $z = 0$ rather than $z = \eta$. Consequently the empirical modification to linear diffraction theory whereby pressure integrations are extended to the instantaneous free surface may produce realistic force estimates. In any case, for waves of maximum steepness $H/L = 0.141$, this formula predicts a maximum force which is about 12% greater than the linear theory prediction, and thus for typical design wave steepnesses the increase would be nearer, say, 5%.

6.8.5 Vortex Shedding Effects

At this point it is mentioned that in the case of a sharp-edged body, vortex-induced forces, arising as a consequence of flow separation, will generally vary nonlinearly with wave height and may be just as important as nonlinear wave effects themselves. It is difficult to account for vortex shedding and wave diffraction simultaneously, but a rough estimate of vortex shedding effects may be made by considering lower values of the diffraction parameter where the wave force is mainly inertial, and thus the usual unseparated flow inertia force is taken to represent diffraction theory predictions. Graham (1980) (see also Bearman and Graham 1979) has described the vortex-induced forces on cylinders in a two-dimensional sinusoidal flow at low Keulegan-Carpenter numbers (i.e. in the inertia force regime). Their results have been used by Isaacson (1979d) to develop a rough estimate of vortex shedding effects on sharp-edged cylinders in waves. It was found that vortex shedding should in general not cause any noticeable increase in maximum wave forces, except at relatively low values of the diffraction parameter (in which case the Morison equation approach would be used in preference to a diffraction analysis). Indeed, wave force experiments with square and rectangular section vertical cylinders (Hogben and Standing 1975, Mogridge and Jamieson 1976a and Isaacson 1979b,d) have indicated that the influence of vortex shedding on the total force is generally not noticeable. On the other hand, relatively high local pressures near the edges may occur.

6.9 EFFECTS OF CURRENTS

The effect of currents on the loading on large bodies has been summarized by Hogben and Standing (1975) who mention three possible modes of influence on the wave force as now described. The incident wave motion may itself be altered in the presence of a current as described in Section 4.9.3. It will be recalled that on the basis of linear wave theory, the waves propagating in the presence of a current undergo a modification of wave speed according to the magnitude of the current.

Secondly, although drag forces are not taken into account in the diffraction approach, they are nevertheless liable to be significant in the presence of an appreciable current, this effect being predicted on the basis of the Morison equation. In order to assess the magnitude of such forces, the case of a surface-piercing circular cylinder may be considered in the light of the Morison equation alone so that the inertia force is taken to represent the limiting prediction of diffraction theory. According to this approach the ratio of maximum drag to maximum inertia force on a structural element in the presence of a current U is given as $(C_d K/\pi^2 C_m)(1 + U/U_w)^2$, where K is the Keulegan-Carpenter parameter at the section and U_w is the amplitude of the orbital velocity associated with the wave motion. Thus for the reference situation of a surface-piercing vertical circular cylinder, the ratio of maximum drag to maximum inertia force, F_D/F_I, is given by linear wave theory as

$$\frac{F_D}{F_I} = \frac{C_d K}{\pi^2 C_m}\left[\left(\frac{2kd + \sinh(2kd)}{2\sinh(2kd)}\right) + 2\left(\frac{U}{U_w}\right) + \frac{kd}{\tanh(kd)}\left(\frac{U}{U_w}\right)^2\right] \tag{6.142}$$

where K and U_w in this formula corresponds to values at the still water level. This ratio is presented as a function of relative current magnitude U/U_w in Fig. 6.33 for the particular case $C_d = 1$, $C_m = 2$, and $K = 1$. This gives some indication of the extent to which a current may increase the drag force, and of how linear diffraction theory predictions which ignore flow separation effects may be unreliable when the current is relatively strong. It will be appreciated, however, that the effects of the current on the wake structure of the flow and consequently on the force coefficients C_d and C_m have been neglected. The approach has also been considered by Dalrymple (1975) and has been taken up rather more fully in Chapter 5.

With regard to the third mode of influence, we note that when a surface-piercing structure is subjected to a steady current, a surface wave pattern is set up which gives rise to an additional force—the so-called 'wave-making resistance'—acting on the body. This aspect may be treated on a potential flow basis involving a different form of Green's function to that already encountered here for wave motion. For a circular cylinder piercing the free surface, the solution may be obtained by employing a distribution of sources or doublets over the body surface, but any such solution is not unique (Kotik and Morgan 1969). The question has been considered by Hogben (1974b) in relation to large offshore

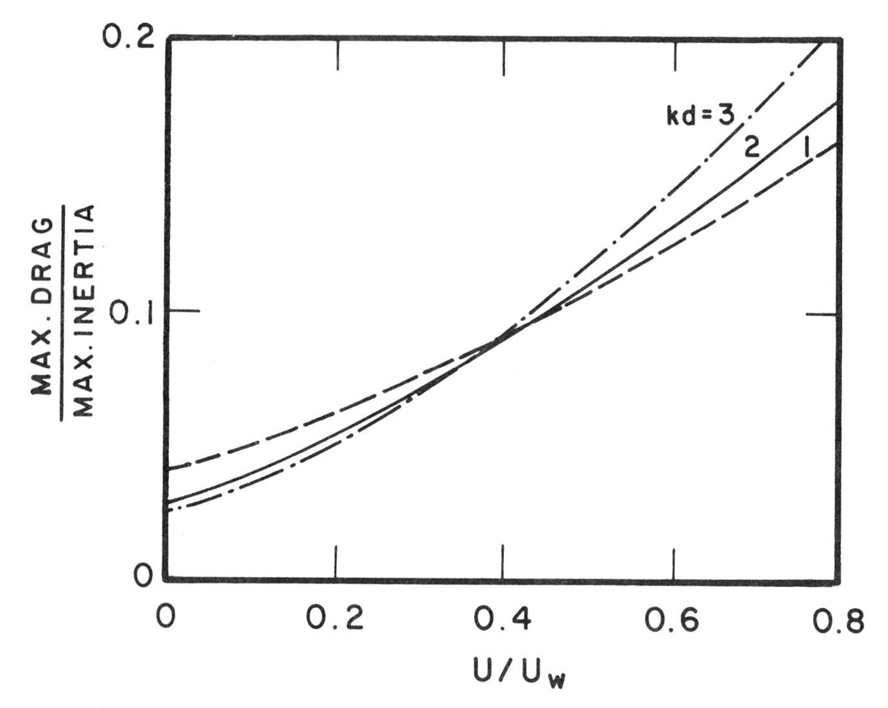

Fig. 6.33. Ratio of drag to inertia force amplitudes for a vertical circular cylinder as a function of relative current U/U_W for various values of depth parameter kd and for K = 1 (Isaacson 1979a).

structures, who concludes that for typical large structures the wave-making resistance is negligible compared to the drag force (see also Hogben 1976).

6.10 APPLICATIONS OF DIFFRACTION THEORY

6.10.1 Reliability of Linear Diffraction Results

In the practical application of any calculations based on linear diffraction theory, it is essential to consider the relevance and validity of the predictions made. There are two aspects to be considered in assessing the reliability of any particular prediction. The first concerns the accuracy of the numerical prediction with respect to the "true" solution of the boundary value problem as posed. That is, what inaccuracies may arise on account of the particular numerical method or solution technique adopted? The accuracy of any more general numerical method can be assessed theoretically by carrying out comparisons with other solutions known to be accurate (e.g. MacCamy and Fuchs 1954 for a surface-piercing cylinder, Garrett 1971 for a floating dock, and Black, Mei and Bray

1971, or Isaacson 1979e, for a truncated cylinder on the seabed). Indeed this requirement demonstrates the need for highly accurate results for selected fundamental configurations.

A second question concerns the relevance of the stated boundary value problem with respect to the intended application. Some possible effects which may be important in a particular situation, but which may not be accounted for in the linear diffraction calculations, might include those due to flow separation, underlying currents, wave nonlinearities, viscous damping, interference between neighboring bodies, etc. To a certain extent comparisons with the results of model tests may be used to assess the importance of some of these effects. Thus selected comparisons with experiment have been mentioned in the preceding sections. In particular, Section 6.2 includes a description of comparisons with the closed-form solution for a surface-piercing circular cylinder, and Section 6.3.3 describes some comparisons based on three-dimensional source computer programs for bodies of more general shape. Programs and methods developed for simpler geometries, as outlined in Section 6.4, essentially duplicate the results of the more general programs, but with reduced effort or greater accuracy or both.

A useful preliminary to a complete diffraction calculation is to obtain initially a rapid, ready estimate of results, invoking whatever approximations may be made: this will permit an appreciation of any particular features to be considered in a more accurate calculation. This might then be followed by the application of an appropriate computer program and/or a series of physical model tests.

At this point it may be useful to provide additional results and discussion for two fundamental configurations which are of practical importance, and we now consider in turn a truncated circular cylinder and a surface-piercing conical structure.

6.10.2 Truncated Cylinder

We initially consider a truncated circular cylinder resting on the seabed (Fig. 6.20d) as a fundamental example whose results are frequently useful in providing ready force estimates for more complex geometries. Alternative calculation methods for this case have been described in Section 6.4.4, and comparisons between experiment and predictions based on a three-dimensional wave source program have been reported by Hogben and Standing (1975) (see Figure 6.9).

For this case, force coefficients depend on three parameters, say ka, h/d and a/d, where h is the truncated cylinder height. The horizontal force, vertical force and overturning moment coefficients are shown in Fig. 6.34 as a function of the diffraction parameter ka for various values of h/d and a/d = 0.5. The calculations used have been based on the extension to Garrett's (1971) method described by Isaacson (1979e).

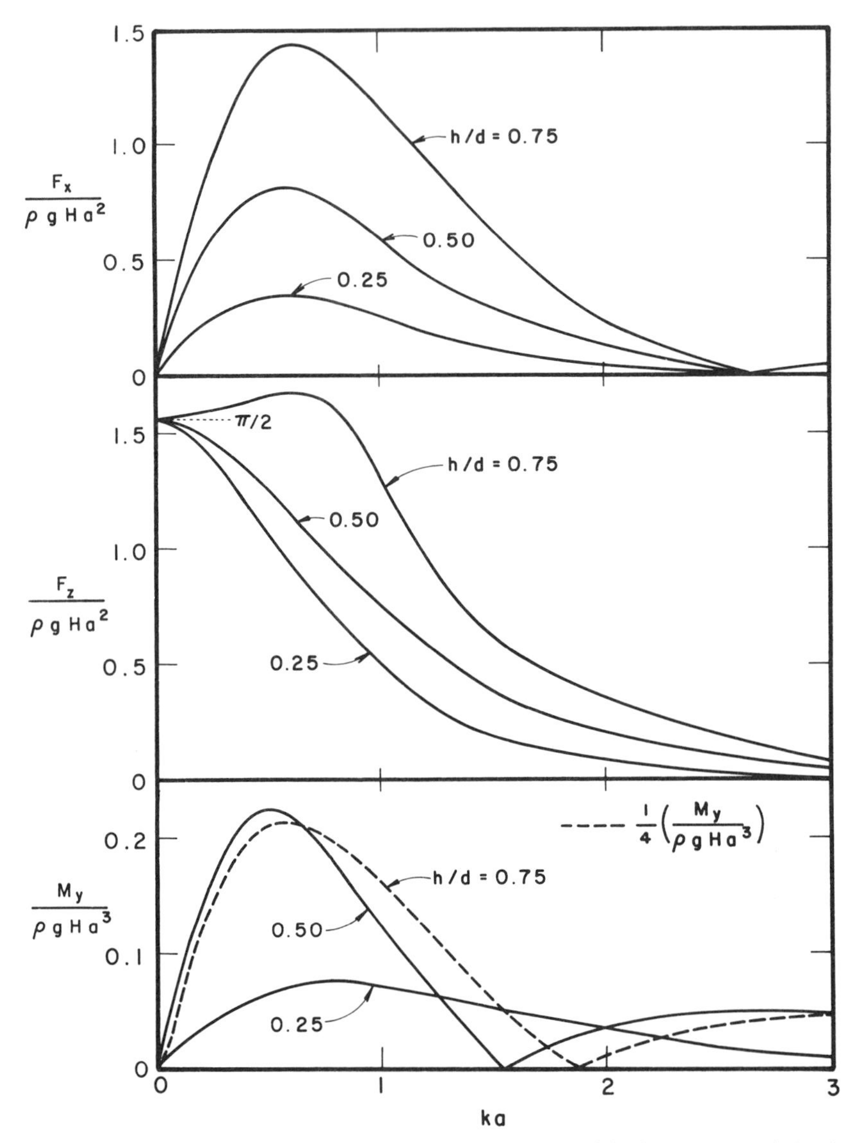

Fig. 6.34. Wave force and moment coefficients as functions of ka for a truncated circular cylinder resting on the seabed, with a/d = 0.5 and with h/d = 0.25, 0.50 and 0.75. (M_y is taken about x = 0, z = −d.)

Hogben and Standing (1975) presented simple formulae to approximate the maximum forces on the cylinder. These are given in terms of the horizontal force, vertical force and overturning moment diffraction coefficients, C_h, C_v and C_y respectively, as follows

$$\left.\begin{aligned} C_h &= 1 + 0.75\left(\frac{h}{2a}\right)^{1/3} [1 - 0.3(ka)^2] \\ C_v &= \begin{cases} 1 + 0.74(ka)^2\left(\frac{h}{2a}\right) & \text{for } \alpha < 1 \\ 1 + 0.5(ka) & \text{for } \alpha > 1 \end{cases} \\ C_y &= 1.9 - 0.35(ka) \end{aligned}\right\} \tag{6.143}$$

where $\alpha = 1.48(ka)(h/2a)$. The ranges of validity of these approximations were given as

$$\left.\begin{aligned} &h/d < 0.6 \\ &0.3 < \frac{h}{2a} < 2.3 \quad \text{for } C_h \text{ and } C_v \\ &0.6 < \frac{h}{2a} < 2.3 \quad \text{for } C_y \end{aligned}\right\} \tag{6.144}$$

The Froude-Krylov force amplitudes, which are needed to calculate the actual force amplitudes using Eqs. (6.143), are given by the expressions

$$F_k^{(h)} = \pi \rho g H a d J_1(ka) \frac{\sinh(kh)}{kd \cosh(kd)} \tag{6.145a}$$

$$F_k^{(v)} = \frac{\pi}{2} \rho g H a^2 [J_0(ka) + J_2(ka)] \frac{\cosh(kh)}{\cosh(kd)} \tag{6.145b}$$

$$F_k^{(y)} = \pi \rho g H a d^2 J_1(ka) \left[\frac{kh \sinh(kh) + 1 - \cosh(kh)}{(kd)^2 \cosh(kd)}\right] - \frac{\pi}{4} \rho g H a^3 [J_1(ka) + J_3(ka)] \frac{\cosh(kh)}{\cosh(kd)} \tag{6.145c}$$

One noticeable feature of Fig. 6.34, and one which is typical of gravity platforms with a large submerged base, is that the horizontal force and overturning moment coefficients reach a maximum at certain values of ka: that is, for a given design wave height the force reaches a maximum at a specific wave length or period. In physical terms, this can be considered to be due to the combined effect of lower fluid accelerations as the period increases, and less exposure as the period decreases. Indeed, the simple inertia force amplitude on the cylinder is given by linear theory as:

$$\frac{F}{\rho g H D^2} = \frac{\pi}{4} \frac{\sinh(kh)}{\cosh(kd)} \tag{6.146}$$

which exhibits this feature for $h < d$.

As a specific example of the application of the results describing Fig. 6.34, suppose that in one instance $T = 8$ sec, $d = 100$ ft, $a = 80$ ft and $h = 50$ ft. For $h/d = 0.5$, $d/a = 1.25$ and $ka = 1.59$, the force coefficients are calculated to be

$$\frac{F_x}{\rho g H a^2} = 0.397, \qquad \frac{F_z}{\rho g H a^2} = 0.592, \qquad \frac{M_y}{\rho g H a^3} = 0.155.$$

The force and moment amplitudes may be obtained from these for any chosen wave height.

6.10.3 Conical Structure

As a second fundamental example, we now consider the wave loading on a conical structure as sketched in Fig. 6.35. Such a structure and similar axisymmetric structures are particularly suitable in Arctic environments where ice loading effects are of major importance (e.g. Stenning and Schumann 1979). Published wave force calculations for this situation are relatively unavailable, and the axisymmetric wave source method described in Section 6.4.1 is ideally suited to generate such results—thus, the body's axisymmetry is exploited to provide reduced computer effort or increased accuracy when compared to results based on a three-dimensional approach.

In order to present results in a compact manner, the force (moment) amplitudes F may be expressed in dimensionless form as

$$\frac{F}{\rho g H a^2} = f\left(ka, \frac{a}{d}, \alpha\right) \tag{6.147}$$

where a is the base radius and α the cone angle. A computer program based on Fenton's* method has been used to generate results for the particular case

*A correction to Fenton's expressions is necessary, which may be demonstrated by considering how the closed-form solution for a surface-piercing vertical cylinder is retrieved.

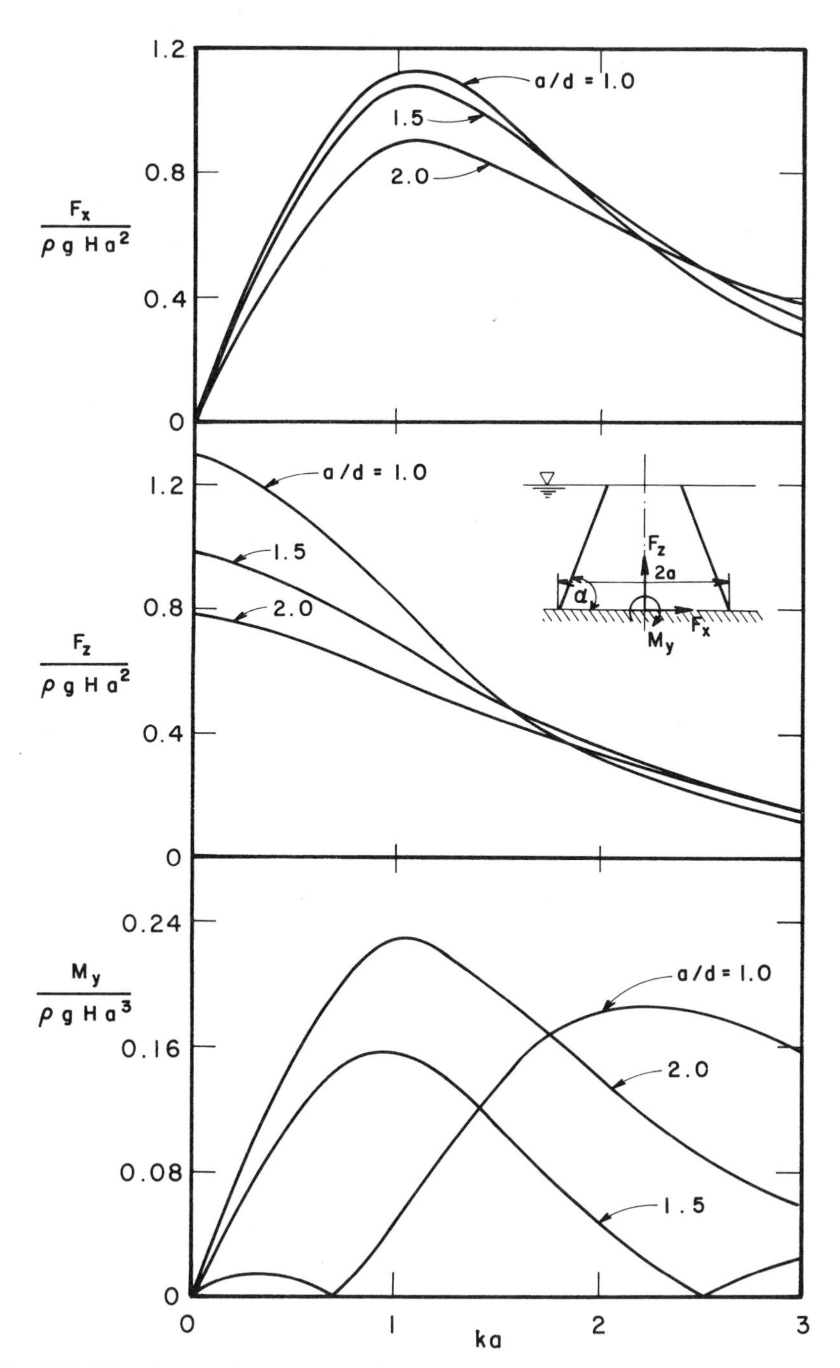

Fig. 6.35. Wave force and moment coefficients as functions of ka for a conical structure as shown, with $\alpha = 60^\circ$ and with a/d = 1.0, 1.5 and 2.0.

$\alpha = 60°$, and these are plotted in Fig. 6.35. The figure shows the horizontal force, vertical force and overturning moment coefficients as functions of the diffraction parameter ka for three different values of a/d.

As an example of the application of such results, we consider the particular case corresponding to H = 30 ft, T = 10 sec, d = 100 ft, a = 100 ft, $\alpha = 60°$. For the parameters ka = 1.39, a/d = 1, the force and moment amplitudes are estimated from Fig. 6.35 to be

$$F_x = 1.057\,\rho g H a^2 = 19{,}800 \text{ kips},$$

$$F_z = 0.595\,\rho g H a^2 = 11{,}150 \text{ kips},$$

$$M_y = 0.117\,\rho g H a^3 = 219{,}530 \text{ kip-ft}.$$

6.11 REFERENCES

Abramowitz, M. and Stegun, I. A. 1965. *Handbook of Mathematical Functions*. Dover, New York.

Adee, B. H. and Martin, W. 1974. Analysis of Floating Breakwater Performance. *Proc. Floating Breakwater Conf.*, Univ. of Rhode Island, pp. 21–40.

Allouard, Y., Chenot, J. L., and Coudert, J. F. 1975. Les Principales Méthodes D'évaluation Des Efforts Provoqués Par La Houle Sur Divers Types D'ouvrages. *La Houille Blanche*, Vol. 30, pp. 439–446.

Apelt, C. J. and MacKnight, A. 1976. Wave Action on Large Off-Shore Structures. *Proc. 15th Coastal Eng. Conf.*, Honolulu, Vol. III, pp. 2228–2247.

Bai, K. J. 1975. Diffraction of Oblique Waves by an Infinite Cylinder. *JFM*, Vol. 68, pp. 513–535.

Bai, K. J. 1977a. Zero-Frequency Hydrodynamic Coefficients of Vertical Axisymmetric Bodies at a Free Surface. *J. Hydronautics*, Vol. 11, pp. 53–57.

Bai, K. J. 1977b. The Added Mass of Two-Dimensional Cylinders Heaving in Water of Finite Depth. *JFM*, Vol. 81, pp. 85–105.

Bai, K. J. and Yeung, R. 1974. Numerical Solutions of Free Surface Problems. *Proc. 10th Symp. Naval Hydrodyn.*, Cambridge, Mass., pp. 609–647.

Bearman, P. W. and Graham, J. M. R. 1979. Hydrodynamic Forces on Cylindrical Bodies in Oscillatory Flow. *Proc. 2nd Int. Conf. on the Behaviour of Offshore Structures, BOSS '79*, London, Vol. I, pp. 309–322.

Berkhoff, J. C. W. 1972. Computation of Combined Refraction-Diffraction. *Proc. 13th Coastal Eng. Conf.*, Vancouver, Vol. I, pp. 471–490.

Bettess, P. 1977. Infinite Elements. *Int. J. Num. Methods in Eng.*, Vol. 11, pp. 53–64.

Bettess, P. and Zienkiewicz, O. C. 1977. Diffraction and Refraction of Surface Waves Using Finite and Infinite Elements. *Int. J. Num. Methods in Eng.*, Vol. 11, pp. 1271–1290.

Black, J. L. 1975a. Wave Forces on Vertical Axisymmetric Bodies. *JFM*, Vol. 67, pp. 369–376.

Black, J. L. 1975b. Wave Forces on Large Bodies, A Survey of Numerical Methods. *Proc. Civil Eng. in the Oceans III*, ASCE, Univ. of Delaware, pp. 924–933.

Black, J. L. Mei, C. C., and Bray, M. C. G. 1971. Radiation and Scattering of Water Waves by Rigid Bodies. *JFM*, Vol. 46, pp. 151–164.

Boreel, L. J. 1974. Wave Action on Large Offshore Structures. *Proc. Inst. Civil Eng. Conf. on Offshore Structures*, London, pp. 7-14.

Brater, E. F., McNown, J. S., and Stair, L. D. 1958. Wave Forces on Submerged Structures. *J. Hyd. Div., ASCE*, Vol. 84, No. HY6, pp. 1833-1-1833-26.

Brebbia, C. A. and Walker, S. 1979. *Dynamic Analysis of Offshore Structures*, Newnes-Butterworths, London.

Brogen, E., Soderstrom, J., Snider, R., and Stevens, J. 1974. Field Data Recovery System—Khazzan Dubai No. 3. *Proc. Offshore Tech. Conf.*, Houston, Paper No. OTC 1943, pp. 89-98.

Bruun, P. 1976. North Sea Offshore Structures. *Ocean Eng.*, Vol. 3, No. 5, pp. 361-373.

Chakrabarti, S. K. 1972. Nonlinear Wave Forces on Vertical Cylinder. *J. Hyd. Div., ASCE*, Vol. 98, No. HY11, pp. 1895-1906.

Chakrabarti, S. K. 1973. Wave Forces on Submerged Objects of Symmetry. *J. Waterways Harbors and Coastal Eng. Div., ASCE*, Vol. 99, No. WW2, pp. 147-164.

Chakrabarti, S. K. 1975. Second-Order Wave Force on Large Vertical Cylinder. *J. Waterways Harbors and Coastal Eng. Div., ASCE*, Vol. 101, No. WW3, pp. 311-317.

Chakrabarti, S. K. 1978a. Wave Forces on Multiple Vertical Cylinders. *J. Waterway Port Coastal and Ocean Div., ASCE*, Vol. 104, No. WW2, pp. 147-161.

Chakrabarti, S. K. 1978b. Comments on Second-Order Wave Force on Large-Diameter Vertical Cylinder. *J. Ship Res.*, Vol. 22, pp. 266-268.

Chakrabarti, S. K. 1979. Discussion of 'Vertical Cylinders of Arbitrary Section in Waves' by M. de St. Q. Isaacson. *J. Waterway Port Coastal and Ocean Div., ASCE*, Vol. 105, No. WW2, pp. 208-210.

Chakrabarti, S. K. and Naftzger, R. A. 1976. Wave Interaction with a Submerged Open-Bottom Structure. *Proc. Offshore Tech. Conf.*, Houston, Paper No. OTC 2534, Vol. II, pp. 109-123.

Chakrabarti, S. K. and Tam, W. A. 1973. Gross and Local Wave Loads on a Large Vertical Cylinder—Theory and Experiment. *Proc. Offshore Tech. Conf.*, Houston, Paper No. OTC 1818, Vol. I, pp. 813-826.

Chakrabarti, S. K. and Tam, W. A. 1975. Wave Height Distribution Around Vertical Cylinder. *J. Waterways Harbors and Coastal Eng. Div., ASCE*, Vol. 101, No. WW2, pp. 225-230.

Chan, R. K. and Hirt, C. W. 1974. Two-dimensional Calculations of the Motion of Floating Bodies. *Proc. 10th Symp. Naval Hydrodyn.*, Cambridge, Mass., pp. 667-683.

Chen, H. S. and Mei, C. C. 1974. Oscillations and Wave Forces in a Man-Made Harbor in the Open Sea. *Proc. 10th Symp. Naval Hydrodyn.*, Cambridge, Mass., pp. 573-596.

Dalrymple, R. A. 1975. Waves and Wave Forces in the Presence of Currents. *Proc. Civil Eng. in the Oceans III*, ASCE, Univ. of Delaware, pp. 999-1018.

Dean, R. G. and Ursell, F. 1959. Interaction of a Fixed, Semi-Immersed Circular Cylinder with a Train of Surface Waves. *Hydrodynamics Lab.*, M.I.T., Tech. Rept. No. 37, Mass.

Dean, W. R. 1948. On the Reflection of Surface Waves by a Submerged Circular Cylinder. *Proc. Camb. Phil. Soc.*, Vol. 44, pp. 483-491.

Faltinsen, O. M. and Loken, A. E. 1979. Slow Drift Oscillations of a Ship in Irregular Waves. *Applied Ocean Res.*, Vol. 1, No. 1, pp. 21-31.

Faltinsen, O. M. and Michelsen, F. C. 1974. Motions of Large Structures in Waves at Zero Froude Number. *Proc. Int. Symp. on the Dynamics of Marine Vehicles and Structures in Waves*, Univ. College, London, pp. 91-106.

Fenton, J. D. 1978. Wave Forces on Vertical Bodies of Revolution. *JFM*, Vol. 85, pp. 241-255.

Finnigan, T. D. and Yamamoto, T. 1979. Analysis of Semi-submerged Porous Breakwaters. *Proc. Civil Engineering in the Oceans IV*, ASCE, San Francisco, Vol. I, pp. 380-397.

Garrett, C. J. R. 1971. Wave Forces on a Circular Dock. *JFM*, Vol. 46, pp. 129–139.

Garrison, C. J. 1974a. Hydrodynamics of Large Objects in the Sea, Part I–Hydrodynamic Analysis. *J. Hydronautics*, Vol. 8, pp. 5–12.

Garrison, C. J. 1974b. Dynamic Response of Floating Bodies. *Proc. Offshore Tech. Conf.*, Houston, Paper No. OTC 2067, Vol. II, pp. 365–377.

Garrison, C. J. 1975. Hydrodynamics of Large Objects in the Sea, Part II–Motions of Free-Floating Bodies. *J. Hydronautics*, Vol. 9, pp. 58–63.

Garrison, C. J. 1978. Hydrodynamic Loading of Large Offshore Structures. Three-Dimensional Source Distribution Methods. In *Numerical Methods in Offshore Engineering*, eds. O. C. Zienkiewicz, R. W. Lewis, and K. G. Stagg, Wiley, Chichester, England, pp. 97–140.

Garrison, C. J. and Chow, P. Y. 1972. Wave Forces on Submerged Bodies. *J. Waterways Harbors and Coastal Eng. Div., ASCE*, Vol. 98, No. WW3, pp. 375–392.

Garrison, C. J. and Rao, V. S. 1971. Interaction of Waves with Submerged Objects. *J. Waterways Harbors and Coastal Eng. Div., ASCE*, Vol. 97, No. WW2, pp. 259–277.

Garrison, C. J. and Stacey, R. 1977. Wave Loads on North Sea Gravity Platforms: A Comparison of Theory and Experiment. *Proc. Offshore Tech. Conf.*, Houston, Paper No. OTC 2794, Vol. I, pp. 513–524.

Garrison, C. J., Torum, A., Iverson, C., Leivseth, S., and Ebbesmeyer, C. C. 1974. Wave Forces on Large Volume Structures–A Comparison Between Theory and Model Tests. *Proc. Offshore Tech. Conf.*, Houston, Paper No. OTC 2137, Vol. II, pp. 1061–1070.

Graham, J. M. R. 1980. The Forces on Sharp-Edged Cylinders in Oscillatory Flow at Low Keulegan-Carpenter Numbers. *JFM*, Vol. 97, pp. 331–346.

Gran, S. 1973. Wave Forces on Submerged Cylinders. *Proc. Offshore Tech. Conf.*, Houston, Paper No. OTC 1817, Vol. I, pp. 801–812.

Hallermeier, R. J. 1976. Nonlinear Flow of Wave Crests Past a Thin Pile. *J. Waterways Harbors and Coastal Eng. Div., ASCE*, Vol. 102, No. WW4, pp. 365–377.

Hara, H., Zienkiewicz, O. C., and Bettess, P. 1979. Application of Finite Elements to Determination of Wave Effects on Offshore Structures. *Proc. 2nd Int. Conf. on the Behaviour of Off-Shore Structures, BOSS '79*, London, Vol. I, pp. 383–390.

Harms, V. W. 1979. Diffraction of Water Waves by Isolated Structures. *J. Waterway Port Coastal and Ocean Div., ASCE*, Vol. 105, No. WW2, pp. 131–147.

Havelock, T. H. 1940. The Pressure of Water Waves Upon a Fixed Obstacle. *Proc. Royal Soc.*, London, Ser. A, Vol. 963, pp. 175–190.

Havelock, T. H. 1955. Waves Due to a Floating Sphere Making Periodic Heavy Oscillations. *Proc. Royal Soc.*, London, Ser. A, Vol. 231, pp. 1–7.

Hess, J. L. and Smith, A. N. O. 1967. Calculation of Potential Flow About Arbitrary Bodies. *Prog. Aeronaut. Sci.*, Vol. 8, pp. 1–138.

Hogben, N. 1974a. Fluid Loading on Offshore Structures, A State-of-the-Art Appraisal: Wave Loads. *Maritime Tech. Monograph*, No. 1, RINA.

Hogben, N. 1974b. Wave Resistance of Surface Piercing Vertical Cylinders in Uniform Currents. *National Physical Laboratory, Ship Div.*, London, Report No. 183.

Hogben, N. 1976. Wave Loads on Structures. *Proc. Conf. Behaviour of Off-Shore Structures, BOSS '76*, Trohnheim, Vol. I, pp. 187–219.

Hogben, N., Miller, B. L., Searle, J. W., and Ward, G. 1977. Estimation of Fluid Loading on Offshore Structures. *Proc. Inst. Civil Eng.*, Vol. 63, pp. 515–562.

Hogben, N., Osborne, J., and Standing, R. G. 1974. Wave Loading on Offshore Structures–Theory and Experiment. *Proc. Symp. Ocean Eng.*, National Physical Laboratory, London, RINA, pp. 19–36.

Hogben, N. and Standing, R. G. 1974. Wave Loads on Large Bodies. *Proc. Int. Symp. on the Dynamics of Marine Vehicles and Structures in Waves*, Univ. College, London, pp. 258–277.

Hogben, N. and Standing, R. G. 1975. Experience in Computing Wave Loads on Large

Bodies. *Proc. Offshore Tech. Conf.*, Houston, Paper No. OTC 2189, Vol. II, pp. 413-431.

Horikawa, K. 1978. *Coastal Engineering, An Introduction to Ocean Engineering.* Wiley, New York.

Houston, J. R. 1978. Interaction of Tsunamis with the Hawaiian Islands Calculated by a Finite-Element Numerical Model. *J. Phys. Oceanography*, Vol. 8, pp. 93-102.

Hsu, F. H. and Blenkarn, K. A. 1970. Analysis of Peak Mooring Forces Caused by Slow Vessel Drift Oscillations in Random Seas. *Offshore Tech. Conf.*, Houston, Paper No. OTC 1159, Vol. I, pp. 135-146.

Hwang, L. S. and Tuck, E. O. 1970. On the Oscillations of Harbours of Arbitrary Shape. *JFM*, Vol. 42, pp. 447-464.

Ijima, T., Chou, C. R., and Yoshida, A. 1976. Method of Analysis for Two-Dimensional Water Wave Problems. *Proc. 15th Coastal Eng. Conf.*, Honolulu, Vol. III, pp. 2717-2736.

Ijima, T., Chou, C. R., and Yumura, Y. 1974. Wave Scattering by Permeable and Impermeable Breakwater of Arbitrary Shape. *Proc. 14th Conf. Coastal Eng.*, Copenhagen, Vol. III, pp. 1886-1905.

Isaacson, M. de St. Q. 1977a. Shallow Wave Diffraction Around Large Cylinder. *J. Waterway Port Coastal and Ocean Div., ASCE*, Vol. 103, No. WW1, pp. 69-82.

Isaacson, M. de St. Q. 1977b. Nonlinear Wave Forces on Large Offshore Structures. *J. Waterway Port Coastal and Ocean Div., ASCE*, Vol. 103, No. WW1, pp. 100-104.

Isaacson, M. de St. Q. 1978a. Wave Runup Around Large Circular Cylinder. *J. Waterway Port Coastal and Ocean Div., ASCE*, Vol. 104, No. WW1, pp. 69-79.

Isaacson, M. de St. Q. 1978b. Interference Effects Between Large Cylinders in Waves. *Proc. Offshore Tech. Conf.*, Houston, Paper No. OTC 3069, Vol. I, pp. 185-192. Also, *J. Petroleum Tech.*, Vol. 31, No. 4, pp. 505-512.

Isaacson, M. de St. Q. 1978c. Long Wave Resonance in Harbours of Arbitrary Shape. *Int. Symp. on Long Waves in the Ocean*, National Research Council Canada, Ottawa, Manuscript Rept. Ser., No. 53, pp. 145-149.

Isaacson, M. de St. Q. 1978d. Vertical Cylinders of Arbritrary Section in Waves. *J. Waterway Port Coastal and Ocean Div., ASCE*, Vol. 104, No. WW4, pp. 309-324.

Isaacson, M. de St. Q. 1979a. Wave Induced Forces in the Diffraction Regime. In *Mechanics of Wave-Induced Forces on Cylinders*, ed. T. L. Shaw, Pitman, London, pp. 68-89.

Isaacson, M. de St. Q. 1979b. Wave Forces on Large Square Cylinders. In *Mechanics of Wave-Induced Forces on Cylinders*, ed. T. L. Shaw, Pitman, London, pp. 609-622.

Isaacson, M. de St. Q. 1979c. Nonlinear Inertia Forces on Bodies. *J. Waterway Port Coastal and Ocean Div., ASCE*, Vol. 105, No. WW3, pp. 213-227.

Isaacson, M. de St. Q. 1979d. Wave Forces on Rectangular Caissons. *Proc. Civil Engineering in the Oceans IV*, ASCE, San Francisco, Vol. I, pp. 161-171.

Isaacson, M. de St. Q. 1979e. Wave Forces on Compound Cylinders. *Proc. Civil Engineering in the Oceans IV*, ASCE, San Francisco, Vol. I, pp. 518-530.

Isaacson, M. de St. Q. 1979f. Closure to 'Wave Runup Around Large Circular Cylinder'. *J. Waterway Port Coastal and Ocean Div., ASCE*, Vol. 105, No. WW4, pp. 477-478.

John, F. 1950. On the Motion of Floating Bodies, II. *Comm. Pure and Applied Mathematics*, Vol. 3, pp. 45-101.

Kim, C. H. 1969. Hydrodynamic Forces and Moments for Heaving, Swaying and Rolling Cylinders on Water of Finite Depth. *J. Ship Research*, Vol. 13, pp. 137-154.

Kim, W. D. 1965. On the Harmonic Oscillations of a Rigid Body on a Free Surface. *JFM*, Vol. 21, pp. 427-451.

Kim, W. D. 1966. On a Free Floating Ship in Waves. *J. Ship Research*, Vol. 10, pp. 182-191, 200.

Kokkinowrachos, K. and Wilckens, H. 1974. Hydrodynamic Analysis of Cylinder Offshore

Oil Storage Tanks. *Proc. Offshore Tech. Conf.*, Houston, Paper No. OTC 1944, Vol. I, pp. 99–112.

Kotik, J. and Morgan, R. 1969. The Uniqueness Problem for Wave Resistance Calculated From Singularity Distributions which are Exact at Zero Froude Numer. *J. Ship Research*, Vol. 13, pp. 61–68.

Kurata, K. and Ijima, T. 1979. Finite Element Analysis of Wave Forces on Structures. *Proc. Civil Engineering in the Oceans IV*, ASCE, San Francisco, Vol. I, pp. 172–186.

Lamb, H. 1945. *Hydrodynamics*. 6th ed., Dover, New York; also Cambridge University Press, 1932.

Lebreton, J. C. and Cormault, P. 1969. Wave Action on Slightly Immersed Structures, Some Theoretical and Experimental Considerations. *Proc. Symp. Research on Wave Action*, Delft.

Lighthill, J. 1979. Waves and Hydrodynamic Loading. *Proc. 2nd Int. Conf. on the Behaviour of Off-Shore Structures, BOSS '79*, London, Vol. I, pp. 1–40.

Loken, A. E. and Olsen, O. A. 1976. Diffraction Theory and Statistical Methods to Predict Wave Induced Motions and Loads for Large Structures. *Proc. Offshore Tech. Conf.*, Houston, Paper No. OTC 2502, Vol. I, pp. 797–820.

Longuet-Higgins, M. S. 1977. The Mean Forces Exerted by Waves on Floating or Submerged Bodies with Applications to Sand Bars and Wave Power Machines. *Proc. Roy. Soc.*, London, Ser. A, Vol. 352, pp. 463–480.

MacCamy, R. C. and Fuchs, R. A. 1954. Wave Forces on Piles: A Diffraction Theory. U.S. Army Corps of Engineers, *Beach Erosion Board*, Tech. Memo No. 69.

Maeda, H. 1974. Hydrodynamical Forces on a Cross Section of a Stationary Structure. *Proc. Int. Symp. on the Dynamics of Marine Vehicles and Structures in Waves*, Univ. College, London, pp. 80–90.

Maruo, H. 1960. The Drift of a Body Floating on Waves. *J. Ship Research*, Vol. 4, pp. 1-10.

Mattioli, F. and Tinti, S. 1979. Discretization of Harbor Resonance Problem. *J. Waterway Port Coastal and Ocean Div., ASCE*, Vol. 105, No. WW4, pp. 464–469.

Mei, C. C. 1978. Numerical Methods in Water-Wave Diffraction and Radiation. *Ann. Rev. Fluid Mech.*, Vol. 10, pp. 393–416.

Mei, C. C. and Chen, H. S. 1975. Hybrid-Element Method for Water Waves. *Proc. Symp. on Modeling Techniques*, San Francisco, pp. 63–81.

Milgram, J. H. and Halkyard, J. E. 1971. Wave Forces on Large Objects in the Sea. *J. Ship Research*, Vol. 15, pp. 115-124.

Miner, E. W., Griffin, O. M., Ramberg, S. E., and Fritts, M. J. 1979. Numerical Calculation of Wave Effects on Structures. *Proc. Civil Engineering in the Oceans IV*, ASCE, San Francisco, Vol. I, pp. 17-27.

Mogridge, G. R. and Jamieson, W. W. 1975. Wave Forces on a Circular Caisson: Theory and Experiment. *Can. J. Civil Eng.*, Vol. 2, pp. 540–548.

Mogridge, G. R. and Jamieson, W. W. 1976a. Wave Forces on Square Caissons. *Proc. 15th Coastal Eng. Conf.*, Honolulu, Vol. III, pp. 2271–2289.

Mogridge, G. R. and Jamieson, W. W. 1976b. Wave Loads on Large Circular Cylinders: A Design Method. *Hydraulics Lab., Nat. Research Council Canada*, Report No. MH-111, Ottawa.

Murphy, J. E. 1978. Integral Equation Failure in Wave Calculations. *J. Waterway Port Coastal and Ocean Div., ASCE*, Vol. 104, No. WW4, pp. 330–334.

Naftzger, R. A. and Chakrabarti, S. K. 1975. Wave Forces on a Submerged Hemispherical Shell. *Proc. Civil Eng. in the Oceans III*, ASCE, Univ. of Delaware, pp. 959–978.

Naftzger, R. A. and Chakrabarti, S. K. 1979. Scattering of Waves by Two-Dimensional Circular Obstacles in Finite Water Depths. *J. Ship Research*, Vol. 23, pp. 32–42.

Newman, J. N. 1962. The Exciting Forces on Fixed Bodies in Waves. *J. Ship Research*, Vol. 6, pp. 10–17.

Newman, J. N. 1967. The Drift Force and Moment on Ships in Waves. *J. Ship Research*, Vol. 11, pp. 51–60.

Newman, J. N. 1970. Applications of Slender Body Theory in Ship Hydrodynamics. *Ann. Rev. Fluid Mech.*, Vol. 2, pp. 67–94.

Newman, J. N. 1972. Diffraction of Water Waves. *App. Mechanics Rev.*, Vol. 25, pp. 1–7.

Newman, J. N. 1974. Second Order Slowly Varying Forces on Vessels in Irregular Waves. *Proc. Int. Symp. on the Dynamics of Marine Vehicles and Structures in Waves*, Univ. College, London, pp. 182–186.

Newman, J. N. 1977. *Marine Hydrodynamics*. M.I.T. Press, Mass.

Newton, R. E. 1975. Finite Element Analysis of Two-Dimensional Added Mass and Damping. In *Finite Elements in Fluids–Vol. I, Viscous Flow and Hydrodynamics*, eds. R. H. Gallagher, J. T. Oden, C. Taylor and O. C. Zienkiewicz, John Wiley, New York, pp. 219–232.

Newton, R. E., Chenault, D. W. H., and Smith, D. A. 1974. Finite Element Solution for Added Mass and Damping. In *Finite Element Methods in Flow Problems*, eds. J. T. Oden, et al., UAH Press, Huntsville, Alabama, pp. 159–170.

Nicols, B. D. and Hirt, C. W. 1976. Numerical Calculation of Wave Forces on Structures. *Proc. 15th Coastal Eng. Conf.*, Honolulu, Vol. III, pp. 2254–2270.

Ogilvie, T. F. 1963. First and Second Order Forces on a Cylinder Submerged Under a Free Surface. *JFM*, Vol. 16, pp. 451–472.

Ohkusu, M. 1974. Hydrodynamic Forces on Multiple Cylinders in Waves. *Proc. Int. Symp. on the Dynamics of Marine Vehicles and Structures in Waves*, Univ. College, London, pp. 107–112.

Omer, G. C. and Hall, H. H. 1949. The Scattering of a Tsunami by a Cylindrical Island. *J. Seismological Soc. of Amer.*, Vol. 39, No. 4, pp. 257–260.

Porter, W. R. 1960. Pressure Distribution, Added-Mass and Damping Coefficients for Cylinders Oscillating in a Free Surface. Contract No. N-ONR-222(30), Series No. 82, Issue No. 16, *Inst. Eng. Research*, Univ. California, Berkeley, California.

Raichlen, F. and Naheer, E. 1976. Wave Induced Oscillations of Harbors with Variable Depth. *Proc. 15th Coastal Eng. Conf.*, Honolulu, Vol. IV, pp. 3536–3556.

Raman, H. and Venkatanarasaiah, P. 1976. Forces Due to Nonlinear Waves on Vertical Cylinders. *J. Waterways Harbors and Coastal Eng. Div., ASCE*, Vol. 102, No. WW3, pp. 301–316.

Rosenhead, L. (ed.) 1963. *Laminar Boundary Layers*. Oxford Univ. Press, Oxford.

Sakai, F. and Tsukioka, K. 1975. Application of the Finite Element Method to Surface Wave Analysis–The 2nd Report: The Analysis of Harbor Oscillations. *Coastal Eng. in Japan*, Vol. 18, pp. 45–52.

Sakai, F. and Tsukioka, K. 1977. Finite Element Simulation of Surface Wave Problems. *Finite Elements in Water Resources*, eds. W. G. Gray, G. F. Finder and C. A. Brebbia, Pentech Press, London, England, pp. 4.3–4.18.

Salvesen, N., Tuck, E. O., and Faltinsen, O. 1970. Ship Motions and Sea Loads. *Trans. SNAME*, Vol. 78, pp. 250–279.

Shen, S. F. 1977. Finite Element Methods in Fluid Mechanics. *Ann. Rev. Fluid Mech.*, Vol. 9, pp. 421–445.

Silvester, R. 1974. *Coastal Engineering 1, 2*. Elsevier, Amsterdam.

Skjelbreia, L. 1979. Wave Forces on Large North Sea Gravity Structures. *Proc. Civil Engineering in the Oceans IV*, ASCE, Vol. I, pp. 137–160.

Sommerfeld, A. 1949. *Partial Differential Equations in Physics*. Academic Press, New York.

Spring, B. H. and Monkmeyer, P. L. 1974. Interaction of Plane Waves with Vertical Cylinders. *Proc. 14th Coastal Eng. Conf.*, Copenhagen, Vol. III, pp. 1828–1847.

Spring, B. H. and Monkmeyer, P. L. 1975. Interaction of Plane Waves with a Row of Cylinders. *Proc. Civil Eng. in the Oceans III*, ASCE, Univ. of Delaware, Vol. III, pp. 979–998.

Stenning, D. G. and Schumann, C. G. 1979. Arctic Production Monocone. *Offshore Tech. Conf.*, Houston, Paper No. OTC 3630, Vol. IV, pp. 2357–2365.

Stoker, J. J. 1957. *Water Waves*. Interscience, New York.

Ursell, F. 1950. Surface Waves on Deep Water in the Presence of a Submerged Circular Cylinder II. *Proc. Camb. Phil. Soc.*, Vol. 46, pp. 141–152.

U.S. Army, Coastal Engineering Research Center. 1973. *Shore Protection Manual.*

Van Oortmerssen, G. 1972. Some Aspects of Very Large Offshore Structures. *Proc. 9th Symp. Naval Hydrodynamics*, Paris, pp. 957–1001.

Van Oortmerssen, G. 1979. Hydrodynamic Interaction Between Two Structures, Floating in Waves. *Proc. 2nd Int. Conf. on the Behaviour of Off-Shore Structures, BOSS '79*, London, Vol. I, pp. 339–356.

Vugts, J. H. 1968. The Hydrodynamic Coefficients for Swaying, Heaving and Rolling Cylinders in a Free Surface. *Int. Shipbuilding Prog.*, Vol. 15, pp. 251–276.

Wang, S. and Wahab, R. 1971. Heaving Oscillations of Twin Cylinders in a Free Surface. *J. Ship Research*, Vol. 15, pp. 33–48.

Wehausen, J. V. 1971. The Motion of Floating Bodies. *Ann. Rev. Fluid Mech.*, Vol. 3, pp. 237–268.

Wehausen, J. V. and Laitone, E. V. 1960. Surface Waves. In *Handbuch der Physik*, ed. S. Flugge, Springer-Verlag, Berlin, Vol. IX, pp. 446–778.

Yamaguchi, M. and Tsuchiya, Y. 1974. Nonlinear Effect of Waves on Wave Pressure and Wave Force on a Large Cylindrical Pile. (in Japanese) *Proc. Civil Eng. Soc. in Japan*, No. 229, pp. 41–53.

Yamamoto, T. and Yoshida, A. 1978. Elastic Moorings of Floating Breakwaters. *Proc. 7th Int. Harbour Congress*, Antwerp.

Yu, Y. S. and Ursell, F. 1961. Surface Waves Generated by an Oscillating Circular Cylinder on Water of Finite Depth: Theory and Experiment. *JFM*, Vol. 11, pp. 529–551.

Yue, D. K. P., Chen, H. S., and Mei, C. C. 1978. A Hybrid Element Method for Diffraction of Water Waves by Three Dimensional Bodies. *Int. J. Num. Methods in Eng.*, Vol. 12, No. 2, pp. 245–266.

Zienkiewicz, O. C. 1977. *The Finite Element Method*. 3rd ed., McGraw-Hill, London.

Zienkiewicz, O. C., Bettess, P., and Kelly, D. W. 1978. The Finite Element Method for Determining Fluid Loadings on Rigid Structures. Two- and Three-Dimensional Formulations. In *Numerical Methods in Offshore Engineering*, eds. O. C. Zienkiewicz, P. Lewis, and K. G. Stagg, John Wiley, Chichester, England, pp. 141–183.

7
Random Waves and Wave Forces

7.1 SUMMARY OF STATISTICAL CONCEPTS

Ocean waves exhibit a notably random behaviour and attention is now given to the probabilistic aspects of waves and also to the resulting random loads on offshore structures. The subject has been developed to quite a high degree of sophistication and at the present time the response of an offshore structure subjected to random waves forms an integral part of usual design procedure.

One of the first attempts to take into consideration the random characteristics of waves was made by Sverdrup and Munk (1947) who introduced the concept of the *significant wave height*, denoted H_s or $H_{1/3}$. This is defined as the average height of the highest one-third of all the waves for a particular sea state, and is in fact found to be close to the wave height reported on the basis of observation. The concept of the significant wave height remains in current usage and is one of the parameters often used to describe sea conditions. In 1952 Longuet-Higgins applied the statistical theory of random signals, developed earlier in connection with electrical noise by Rice (1944–1945), to the random water surface elevation associated with wave motion, and since then considerable progress has been made in describing ocean wave statistics.

In the present text, we first summarize various statistical properties of a random signal x(t) varying continuously with time t as depicted in Fig. 7.1. This might represent, for example, the free surface elevation $\eta(t)$ at a particular location, a force F(t) on a structural element, and so on. Detailed accounts of the

Fig. 7.1. Realization of a random process x(t).

statistics of random signals are given, for example, by Davenport and Root (1958), and in the context of ocean waves by Kinsman (1965), Price and Bishop (1974), Phillips (1977), and Hallam, Heaf and Wootton (1977).

7.1.1 Probabilistic Properties

We consider now some probabilistic properties which may be used to describe x(t). Initially these are concerned with the spread of values of x and not with the manner in which these values may vary with time.

The probability distribution function or cumulative probability P(x) is the probability that a general value x(t) is less than or equal to the value of x being considered. That is

$$P(x) = \text{Prob}\ [x(t) \leqslant x] \tag{7.1}$$

Clearly this must vary between zero and unity such that $P(x = -\infty) = 0$ and $P(x = \infty) = 1$. The probability density p(x) is defined as

$$p(x) = \frac{dP(x)}{dx} \tag{7.2}$$

so that p(x) dx represents the probability that x(t) lies within the small range x to x + dx.

Two probability distributions which are of particular interest in the study of random waves are the Gaussian (or normal) distribution and the Rayleigh distribution. These are commonly employed to describe the probability distributions of water surface elevation η and of wave height H respectively. Curves showing the corresponding cumulative probabilities and probability densities for both distributions are indicated in Figs. 7.2 and 7.3. The cumulative probability and probability density of the Gaussian distribution are given as

$$P(x) = \frac{1}{\sigma_x\sqrt{2\pi}} \int_{-\infty}^{x} \exp\left[-\frac{(x' - \mu_x)^2}{2\sigma_x^2}\right] dx' \tag{7.3a}$$

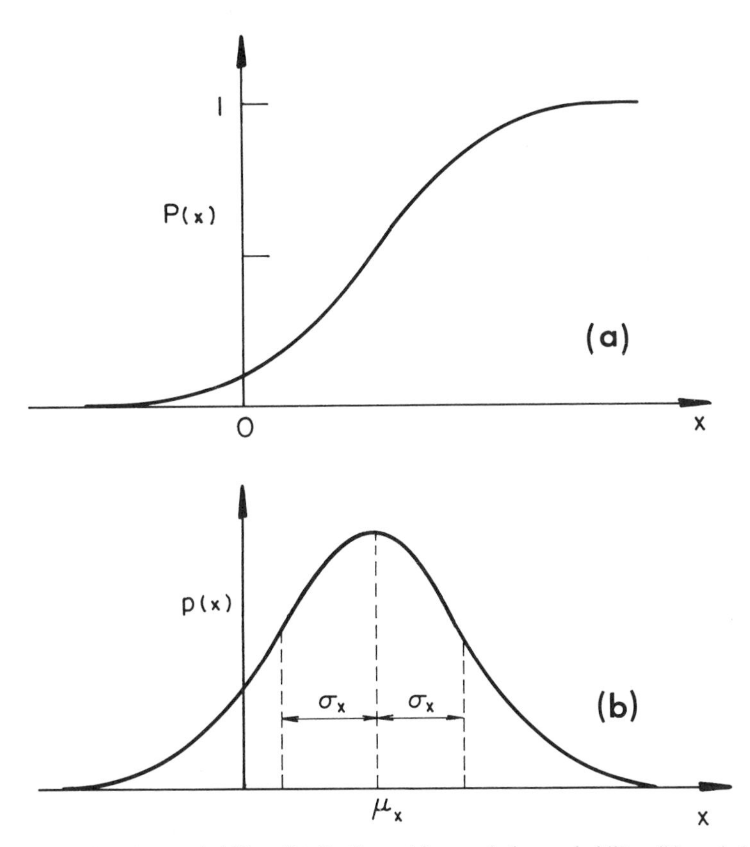

Fig. 7.2. The Gaussian probability distribution. (a) cumulative probability, (b) probability density.

$$p(x) = \frac{1}{\sigma_x \sqrt{2\pi}} \exp\left[-\frac{(x - \mu_x)^2}{2\sigma_x^2}\right] \tag{7.3b}$$

while those of the Rayleigh distribution are given as

$$P(x) = \begin{cases} 1 - \exp\left[-\frac{\pi}{4}\left(\frac{x}{\mu_x}\right)^2\right] & \text{for } x \geqslant 0 \\ 0 & \text{otherwise} \end{cases} \tag{7.4a}$$

$$p(x) = \begin{cases} \frac{\pi x}{2\mu_x^2} \exp\left[-\frac{\pi}{4}\left(\frac{x}{\mu_x}\right)^2\right] & \text{for } x \geqslant 0 \\ 0 & \text{otherwise} \end{cases} \tag{7.4b}$$

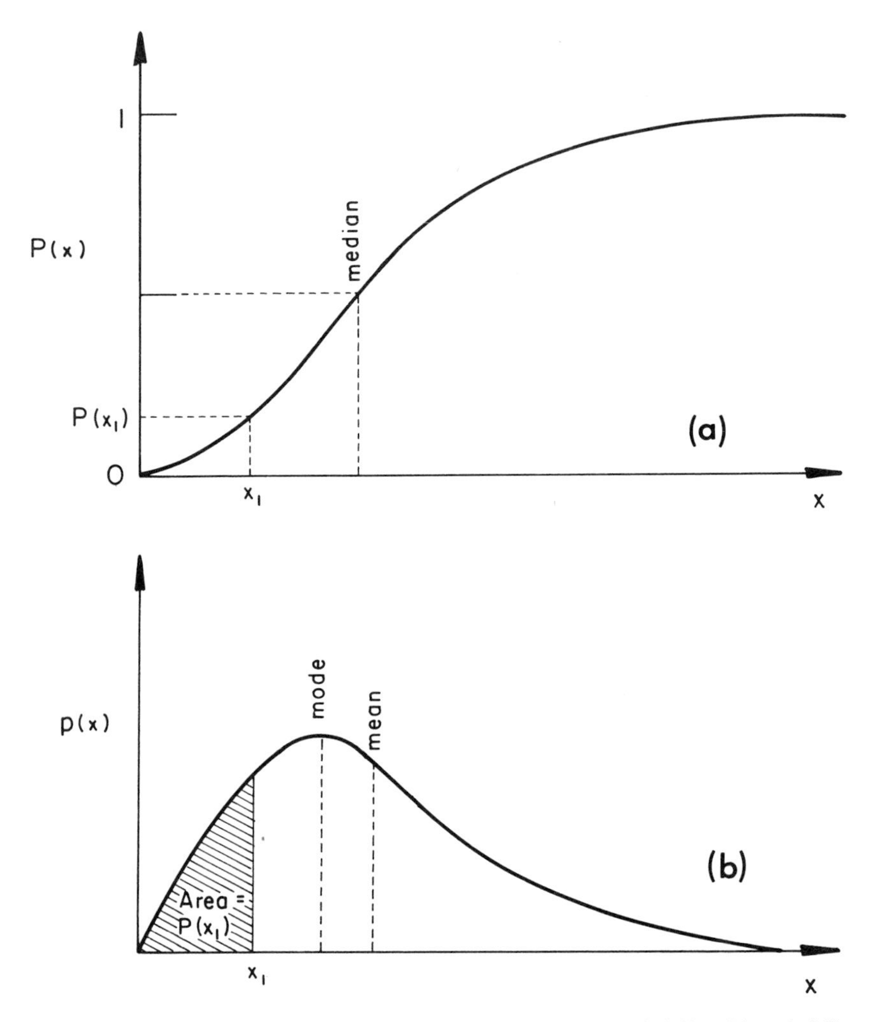

Fig. 7.3. The Rayleigh probability distribution. (a) cumulative probability, (b) probability density.

In the above, the symbols μ_x and σ_x denote the mean and standard deviation respectively of x and are constants of the probability distributions which will be defined shortly. By definition, the probability density curves Figs. 7.2b and 7.3b are equal to the gradients of the corresponding cumulative probability curves Figs. 7.2a and 7.3a. On the other hand the cumulative probability P(x) is expressed in terms of p(x) as

$$P(x) = \int_{-\infty}^{x} p(x')\, dx' \tag{7.5}$$

and graphically represents the area under the p(x) curve lying to the left of the argument x chosen. This is indicated in Fig. 7.3. It follows also that the total area under the p(x) curve must be unity

$$\int_{-\infty}^{\infty} p(x)\, dx = P(x = \infty) = 1 \tag{7.6}$$

And more generally the area under the p(x) curve between two values x_1 and x_2 represents the probability that x lies between these two values

$$\int_{x_1}^{x_2} p(x)\, dx = P(x_2) - P(x_1) = \text{Prob}\ [x_1 \leqslant x(t) \leqslant x_2] \tag{7.7}$$

Various characteristic values of x related to its probability distribution are of interest. The mean or expected value of x, generally denoted $\overline{x}, \mu_x$ or E [x], is defined in terms of the probability density of x as

$$\mu_x = E\,[x] = \int_{-\infty}^{\infty} x p(x)\, dx \tag{7.8}$$

And more generally the expected value of any function g(x) of x is given as

$$E\,[g(x)] = \int_{-\infty}^{\infty} g(x)\, p(x)\, dx \tag{7.9}$$

provided that the integral $\int_{-\infty}^{\infty} |g(x)|\, p(x)\, dx$ converges. Adopting this approach we may now define the n-th moment of x as

$$E\,[x^n] = \int_{-\infty}^{\infty} x^n\, p(x)\, dx \tag{7.10}$$

It is often convenient to take moments about the mean of x rather than about x = 0 and we may thus define the n-th central moment of x as

$$E\,[(x - \mu_x)^n] = \int_{-\infty}^{\infty} (x - \mu_x)^n\, p(x)\, dx \tag{7.11}$$

The mean-square of x corresponds to the second moment of x, $E\,[x^2]$, and the root-mean-square of x is the positive square root of this. The variance of x, denoted σ_x^2, is defined as the second central moment

$$\sigma_x^2 = E[(x - \mu_x)^2] = E[x^2] - \mu_x^2 \tag{7.12}$$

And the standard deviation σ_x is the positive square root of the variance. It follows that in those cases where $\mu_x = 0$ the variance is identically equal to the mean-square of x and the standard deviation is equal to the root-mean-square of x. The standard deviation σ_x has the same dimensions as x itself and is conveniently used to characterize the spread of the values of x about the mean. The mean and standard deviation of x for a Gaussian distribution are indicated in Fig. 7.2b.

The skewness of x is defined as its third central moment $E[(x - \mu_x)^3]$ and provides an indication of the degree of asymmetry of the probability distribution about the mean value. Clearly the skewness is zero for a Gaussian distribution but nonzero for a Rayleigh distribution. Other characteristic values of x that are commonly used in addition to those above include the mode, which is the value of x at which p(x) is a maximum: this is often referred to as the 'most probable' value of x; and the median, which is the value of x at which $P(x) = \frac{1}{2}$: that is general values of x are just as likely to be greater than as to be less than the median. The mean, mode and median of x for the Rayleigh distribution are indicated in Fig. 7.3.

7.1.2 The Random Process

The properties of x(t) outlined so far do not depend upon this being a continuous function of time, and indeed they may be developed when x is a discrete rather than a continuous variable. It is, however, convenient at this point to regard x(t) as varying continuously with time and to make the further assumptions that x(t) is both *stationary* and *ergodic*. These concepts may be most readily understood by reference to an ensemble or set of signals $x_1(t)$, $x_2(t)$, $x_3(t)$, . . . which represent different realizations or measurements of the process x(t) at the given location.

The requirement of stationarity of x is ensured provided that the probability distribution obtained by taking $x_1(t_1)$, $x_2(t_1)$, $x_3(t_1)$, . . . as variables is independent of the instant chosen t_1. The assumption of stationarity implies then that the statistical properties of x(t) are independent of the origin of time measurement and thus there is no 'drift' with time in the statistical behaviour of x(t). In cases where x(t) can be measured at different locations (as can the free-surface elevation $\eta(t)$) then an analogous assumption that x(t) is *homogeneous* may also be made. This implies that the statistics of x(t) do not depend on its location. In practice, x(t) may be assumed stationary and homogeneous for a certain duration and region only. Thus the sea surface may be considered to be stationary during the course of a few hours, whereas more generally it is non-stationary as storms vary in severity over a longer time scale.

The assumption that $x(t)$ is ergodic implies that the measured realization of $x(t)$—say $x_1(t)$—is typical of all other possible realizations such that the variation of the single (measured) function with respect to time $x_1(t_1)$, $x_1(t_2)$, $x_1(t_3)$, . . . may be used to represent the various possible realizations of $x(t)$ at one instant $x_1(t_1)$, $x_2(t_1)$, $x_3(t_1)$, . . . This enables one to interchange the expected value of a function $g(x)$ with the temporal average of $g(x)$. That is, the ergodicity assumption permits us to write

$$E[g(x)] = \underset{T\to\infty}{\text{Limit}} \left\{ \frac{1}{T} \int_{-T/2}^{T/2} g(x)\, dt \right\} \tag{7.13}$$

Thus we have

$$\mu_x = \langle x(t) \rangle = \underset{T\to\infty}{\text{Limit}} \left\{ \frac{1}{T} \int_{-T/2}^{T/2} x(t)\, dt \right\} \tag{7.14}$$

$$\sigma_x^2 = \langle (x(t) - \mu_x)^2 \rangle = \underset{T\to\infty}{\text{Limit}} \left\{ \frac{1}{T} \int_{-T/2}^{T/2} (x(t) - \mu_x)^2\, dt \right\} \tag{7.15}$$

and so on. The angle brackets $\langle\ \rangle$ denote the temporal mean of terms contained between them.

The properties of x encountered so far describe the spread of values of x but do not describe the way that $x(t)$ varies with time. In order to describe the temporal variation of $x(t)$ it is necessary to extend the statistical description of $x(t)$ and introduce further statistical properties. For convenience the continuous signal $x(t)$ is defined in the present chapter to have a zero mean and thus the mean square of x can be taken as its variance σ_x^2.

The *autocorrelation function* $R_x(\tau)$ relates the value of x at time t to its value at a later time $t + \tau$ and so provides an indication of the correlation of the signal with itself for various time lags τ. $R_x(\tau)$ is defined as

$$R_x(\tau) = \langle x(t)x(t+\tau) \rangle = \underset{T\to\infty}{\text{Limit}} \left\{ \frac{1}{T} \int_{-T/2}^{T/2} x(t)x(t+\tau)\, dt \right\} \tag{7.16}$$

In practice the period T used to obtain the temporal average is of course finite but sufficiently large so as to ensure that its value does not effect the estimate of $R_x(\tau)$ obtained. The term autocovariance function is also used for $R_x(\tau)$ in cases where $x(t)$ has a zero mean as considered here. It is convenient to express the autocorrelation function in dimensionless form as the autocorrelation coefficient $\rho_x(\tau)$ given as

$$\rho_x(\tau) = \frac{R_x(\tau)}{\langle x^2(t)\rangle} = \frac{\langle x(t)x(t+\tau)\rangle}{\langle x^2(t)\rangle} \tag{7.17}$$

Clearly a signal is perfectly correlated with itself for zero lag τ and we have $\rho_x(\tau) = 1$ at $\tau = 0$. On the other hand, as $\tau \to \infty$, $x(t)$ and $x(t + \tau)$ for a random signal are uncorrelated, the product $x(t)x(t + \tau)$ is then just as likely to be positive as negative (since $\mu_x = 0$) and in consequence $\rho_x(\tau) \to 0$ as $\tau \to \infty$. It is noted incidently that a further property of $R_x(\tau)$ is that it must be an even function: $R_x(-\tau) = R_x(\tau)$.

The above concepts may readily be extended to two different signals x and y by defining a *cross-correlation function* $R_{xy}(\tau)$ as

$$R_{xy}(\tau) = \langle x(t)y(t+\tau)\rangle = \underset{T\to\infty}{\text{Limit}} \left\{\frac{1}{T}\int_{-T/2}^{T/2} x(t)y(t+\tau)\,dt\right\} \tag{7.18}$$

This is useful in relating signals (e.g. velocities or pressures) at two separate points in a random fluid flow. Two different functions $R_{xy}(\tau)$ and $R_{yx}(\tau)$ may be separately defined for a particular pair of signals and these will be related by $R_{xy}(\tau) = R_{yx}(-\tau)$, while each function itself is not even. The term cross-covariance function is also used for $R_{xy}(\tau)$ in cases where x and y are both defined to have zero means.

7.1.3 Spectral Density

The frequency content of a random signal can conveniently be used to describe the temporal variation of the signal and is of considerable importance in many applications. This may be obtained by a harmonic or spectral analysis in which the spectral density $S_x(f)$ of the signal $x(t)$ is developed. Detailed accounts of spectral analysis are given by Davenport and Root (1958) and in the context of ocean waves by Kinsman (1965), Phillips (1977) and Price and Bishop (1974).

It is well known that a periodic signal may be represented by a Fourier series which contains components at multiples of the fundamental frequency f_0. This may be written in complex form as

$$x(t) = \sum_{n=-\infty}^{\infty} a_n\, e^{i2\pi n f_0 t} \tag{7.19}$$

where a_n are the complex Fourier coefficients given as

$$a_n = \frac{1}{T} \int_{-T/2}^{T/2} x(t)\, e^{-i2\pi n f_0 t}\, dt \tag{7.20}$$

and $T = 1/f_0$ is the period of the signal. If x(t) is real we must have $a_{-n} = a_n^*$, where the asterisk denotes the complex conjugate, and we may prefer the alternative representation

$$x(t) = \text{Re}\left\{\sum_{n=1}^{\infty} A_n\, e^{-i2\pi n f_0 t}\right\} \tag{7.21}$$

with $A_n = 2a_n$. ($a_0 = 0$ for $\bar{x}$ to be zero.) The variance σ_x^2 may then be written as

$$\sigma_x^2 = \sum_{n=1}^{\infty} \tfrac{1}{2} |A_n|^2 \tag{7.22}$$

Thus $\frac{1}{2}|A_n|^2$ represents the contribution to the variance σ_x^2 which is associated with the component frequency nf_0.

By an intuitive extension to this approach, a random signal may be considered as periodic with infinite period such that it contains a continuous range of frequencies rather than discrete harmonics. That is we consider $f_0 \to 0$, while $nf_0 \to f$ with $0 < f < \infty$, and also A_n is now considered to be a random quantity. In this way the variance σ_x^2 of a random signal (with zero mean) may be considered to be made up of contributions associated with all possible frequencies such that the summation of Eq. (7.22) may be replaced by an integral

$$\sigma_x^2 = \int_0^{\infty} S_x(f)\, df \tag{7.23}$$

where $S_x(f)$ is a continuous function of frequency. That is, $S_x(f)\, df$ represents the contribution to the variance (or 'energy') of the signal x(t) due to its content within the frequency range f to f + df. This might be symbolized as

$$S_x(f)\, df = \sum_{f_n}^{f_n + df} \tfrac{1}{2} |A_n|^2 \tag{7.24}$$

$S_x(f)$ is known alternatively as the spectral density, the power spectral density or the energy spectrum of x.

This kind of description can be developed more formally to enable $S_x(f)$ to be defined explicitly: In certain circumstances a non-periodic signal may be represented as a Fourier series by considering the fundamental frequency to be infinitesimal $f_0 \to df$, while $nf_0 \to f$ such that the discrete harmonics tend towards a continuous range of frequencies. The Fourier series of x(t) then becomes a Fourier integral, and we may write

$$x(t) = \int_{-\infty}^{\infty} A(f)\, e^{i2\pi ft}\, df \qquad (7.25)$$

where

$$A(f) = \int_{-\infty}^{\infty} x(t)\, e^{-i2\pi ft}\, dt \qquad (7.26)$$

A(f) is the Fourier transform of x(t), and x(t) and A(f) constitute a Fourier transform pair. Such a representation may be employed provided that $\int_{-\infty}^{\infty} |x(t)|\, dt$ and $\int_{-\infty}^{\infty} x^2(t)\, dt$ exist and are finite. But these conditions do not hold for a stationary random signal. However, they do hold for a related signal $x_T(t)$ given as

$$x_T(t) = \begin{cases} x(t) & \text{for } -T/2 < t < T/2 \\ 0 & \text{otherwise} \end{cases} \qquad (7.27)$$

and the Fourier transform $A_T(f)$ of $x_T(t)$ does therefore exist. It is possible to express the variance of the original signal x(t) in terms of $A_T(f)$ as

$$\sigma_x^2 = \int_0^{\infty} \underset{T\to\infty}{\text{Limit}} \left\{ \frac{2}{T} |A_T(f)|^2 \right\} df \qquad (7.28)$$

such that the integrand itself is finite. Comparing this to Eq. (7.23) we may now formally define the spectral density as

$$S_x(f) = \underset{T\to\infty}{\text{Limit}} \left\{ \frac{2}{T} |A_T(f)|^2 \right\} \qquad (7.29)$$

It will often be convenient to consider the spectral density as a function of angular frequency $\omega = 2\pi f$ or period $T = 1/f$, or even of wave number k (which is related to ω by the dispersion relation). In the general case the energy content within corresponding frequency intervals da and db must be the same, $S(a)\,da = S(b)\,db$, and we therefore have, since S must be positive,

$$S(a) = \frac{S(b)}{|da/db|} \tag{7.30}$$

In particular $S(\omega) = S(f)/2\pi$ so that it will not be confusing to adopt either form of spectral density as desired. Note though that the dimensions of spectral density depend not only on those of x, but also on the manner in which the frequency is represented. Thus the units of $S_x(f)$ are X^2/Hz, where X denotes the units of x.

Spectral moments may conveniently be used to characterize a spectral distribution just as probability moments are used to characterize a probability distribution. In general the n-th spectral moment is defined as

$$m_n = \int_0^\infty f^n S_x(f)\,df \tag{7.31}$$

It will be seen that the zeroth spectral moment is identical to the variance, $m_0 = \sigma_x^2$. We see also that the spectral moments $m_n^{(\omega)}$ based on angular frequency ω differ from the moments $m_n^{(f)}$ based on f, these being related by

$$m_n^{(\omega)} = (2\pi)^n\, m_n^{(f)} \tag{7.32}$$

In the present text spectral moments will be based on circular frequency f and the superscript (f) is omitted.

It is found mathematically convenient to consider the formal development of the spectral density in terms of a "two-sided" density $S'_x(f)$ which is a real and even function of frequency in the range $-\infty < f < \infty$. The one-sided spectral density $S_x(f)$ used in the present text is defined for positive frequencies only, $0 < f < \infty$, and has the advantage of avoiding the consideration of negative frequencies. No confusion will arise since these alternative forms are simply related by

$$S_x(f) = \begin{cases} 2S'_x(f) & \text{for } f \geqslant 0 \\ 0 & \text{otherwise} \end{cases} \tag{7.33}$$

The autocorrelation function $R_x(\tau)$ described earlier is directly related to the spectral density such that $R_x(\tau)$ and $S'_x(f)$ form a Fourier transform pair. The one-sided spectral density is consequently related to the autocorrelation function by the pair of equations,

$$S_x(f) = \int_0^\infty 4R_x(\tau)\cos(2\pi f\tau)\,d\tau \tag{7.34}$$

$$R_x(\tau) = \int_0^\infty S_x(f)\cos(2\pi f\tau)\,df \tag{7.35}$$

Equation (7.34) can be used in place of Eq. (7.29) as a basis for defining the spectral density.

Autocorrelation coefficients and spectral densities corresponding to three distinct signals are sketched in Fig. 7.4 in order to illustrate some of their fundamental features. Fig. 7.4a shows a general random signal, while Fig. 7.4b shows a signal possessing a narrow-band spectrum in that its frequency content is concentrated over a narrow range. This signal is of considerable importance since

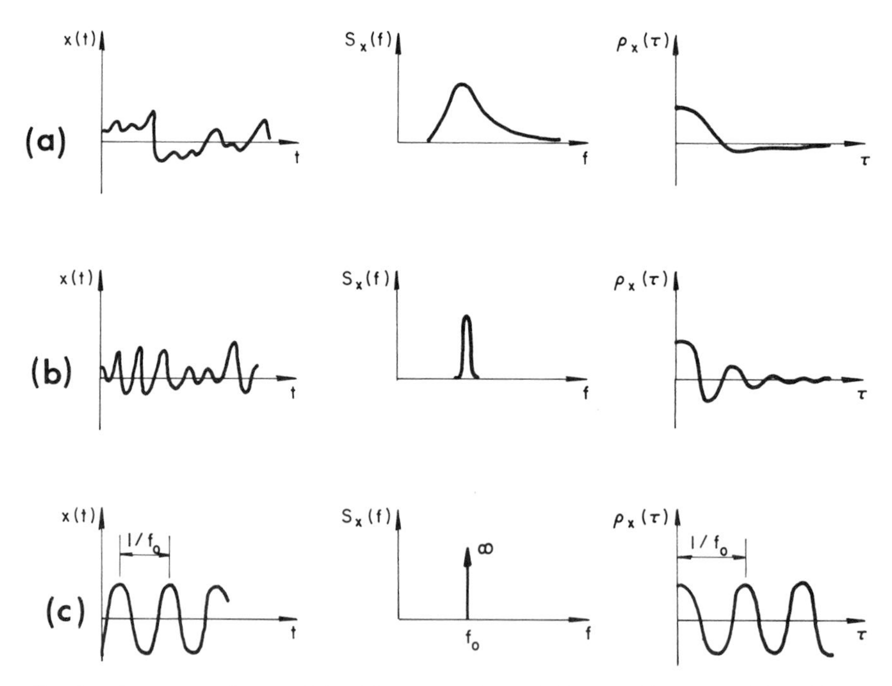

Fig. 7.4. Examples of spectral densities and autocorrelation coefficients. (a) broad-band spectrum, (b) narrow-band spectrum, (c) sinusoidal signal.

useful simplifications to its statistical properties are possible. And Fig. 7.4c shows a perfectly sinusoidal signal which contains a single discrete frequency and therefore its spectral density is infinite at that frequency and zero otherwise in order that Eq. (7.23) is satisfied. We see then that the spectral density essentially enables a signal to be represented in the frequency domain rather than the time domain and most random signal problems are conveniently described in this way.

Just as a cross-correlation function was defined with respect to two signals, we may accordingly define a cross-spectral density $S_{xy}(f)$ with respect to the two signals x and y. The two-sided cross-spectral density and the cross-correlation function form a Fourier transform pair

$$S'_{xy}(f) = \int_{-\infty}^{\infty} R_{xy}(\tau)\, e^{-i2\pi f\tau}\, d\tau \tag{7.36}$$

$$R_{xy}(\tau) = \int_{-\infty}^{\infty} S'_{xy}(f)\, e^{i2\pi f\tau}\, df \tag{7.37}$$

and the one- and two-sided cross-spectral densities continue to be related as in Eq. (7.33). The alternative definitions equivalent to those of Eqs. (7.24) and (7.29) are

$$S_{xy}(f)\, df = \sum_{f_n}^{f_n+df} \tfrac{1}{2} A_n^* B_n \tag{7.38}$$

and

$$S_{xy}(f) = \operatorname*{Limit}_{T\to\infty} \left[\frac{2}{T} A_T^*(f) B_T(f)\right] \tag{7.39}$$

respectively. Here A_n and A_T relate to the signal x, and B_n and B_T to the signal y. For a given pair of signals, two different cross-spectral densities $S'_{xy}(f)$ and $S'_{yx}(f)$ may be separately defined. These are generally complex and are related by

$$S'_{xy}(f) = S'_{yx}(-f) = S'^{*}_{yx}(f) \tag{7.40}$$

Other related quantities may be introduced and are found useful in this area of analysis. These include the co-spectrum given as $\mathrm{Re}\{S'_{xy}(f)\}$, the quadrature-spectrum given as $-\mathrm{Im}\{S'_{xy}(f)\}$, where Re{ } and Im{ } denote the real and imaginary parts respectively, and the coherence function $\gamma_{xy}^2(f)$ given as

$$\gamma_{xy}^2(f) = \frac{|S_{xy}(f)|^2}{S_x(f)S_y(f)} \leqslant 1 \tag{7.41}$$

In the foregoing outline x has so far been taken to vary only with time. When it becomes desirable to consider spatial variations of x then other spectral representations which reflect such a dependence may also be employed. Thus in the case of the free surface elevation $\eta(x, y, t)$, it is in principle possible to employ a three-dimensional wave spectrum $S_\eta(k_x, k_y, f)$ which has as arguments not only the wave frequency f, but also the wave number components (analogous to spatial frequencies) in the two orthogonal directions x and y. Thus $S_\eta(k_x, k_y, f)\, dk_x\, dk_y\, df$ represents the contribution to the variance σ_η^2 associated with waves possessing wave number components and frequencies lying simultaneously within the range k_x to $k_x + dk_x$, k_y to $k_y + dk_y$ and f to f + df. The reader is referred to Kinsman (1965) and to Price and Bishop (1974) for detailed accounts of such possible developments. In ocean engineering practice a directional wave spectrum $S_\eta(\theta, f)$, which has as arguments wave frequency and direction, has found extensive usage and will subsequently be introduced in Section 7.3.1.

7.1.4 Response to Random Loading

We now proceed to consider the relation between an input random signal x(t) (e.g. a force) and the resulting output random signal y(t) (e.g. a displacement) deriving from a fixed-parameter linear system: that is a system in which $y(t) = f[x(t)]$ such that $f(ax_1 + bx_2) = af(x_1) + bf(x_2)$. If the input signal is sinusoidal and written in complex form as

$$x(t) = X\, e^{-i2\pi ft} \tag{7.42}$$

where X is the complex amplitude of x(t), then the response may be written in the same form

$$y(t) = Y\, e^{-i2\pi ft} \tag{7.43}$$

such that

$$Y = H(f)X \tag{7.44}$$

H(f) is generally complex and is known as the system function or receptance. The mean square of y will then be $|H(f)|^2$ times that of x. In the context of harmonic motion, the linearity of the system implies that the output at a given frequency depends only on the input at that frequency.

More generally if x is periodic and represented by a Fourier series then y will also be periodic with each Fourier coefficient equal to the corresponding one of x multiplied by H(f), with f the frequency of each particular harmonic nf_0. Thus if the mean squares of x and y are expressed in the form of Eq. (7.22) then the contributions of the n-th harmonics to the mean squares are related by

$$\tfrac{1}{2}|Y_n|^2 = |H(nf_0)|^2 \tfrac{1}{2}|X_n|^2 \tag{7.45}$$

Finally when x and y are random, then a corresponding relation between their spectral densities may be developed by considering the Fourier integral representation of the related signal x_T or by a technique based on the Duhamel integral for impulsive loading. The relationship is

$$S_y(f) = |H(f)|^2 S_x(f) \tag{7.46}$$

and $|H(f)|^2$ is termed the *transfer function* of the system.

A fundamental example illustrating the above relationship is provided by the response y(t) of a spring-mass-dashpot system to a force x(t). Here the equation of motion is

$$m\frac{d^2y}{dt^2} + \lambda\frac{dy}{dt} + sy = x \tag{7.47}$$

where m is the mass, λ is the damping constant, and s is the spring constant of the system. With x and y given by Eqs. (7.42) and (7.43) we have

$$\begin{aligned} H(f) &= \{-4\pi^2 f^2 m - i2\pi f\lambda + s\}^{-1/2} \\ &= \{s[1 - (f/f_n)^2 - 2i\xi(f/f_n)]\}^{-1/2} \end{aligned} \tag{7.48}$$

where $f_n = (s/m)^{1/2}/2\pi$ and $\xi = \lambda/2(ms)^{1/2}$. These are respectively the natural frequency and the damping ratio of the system. The transfer function may then be written as

$$|H(f)|^2 = \{s[(1 - (f/f_n)^2)^2 + (2\xi(f/f_n))^2]\}^{-1} \tag{7.49}$$

The multiplicative operation underlying Eq. (7.46) is represented graphically for this system in Fig. 7.5. This illustrates how different peaks appearing in both the input spectrum $S_x(f)$ and the transfer function $|H(f)|^2$ may appear in the output spectrum $S_y(f)$. In many cases where the system is lightly damped corresponding to a high and narrow peak in the transfer function, then the output spectrum may be narrow-banded reflecting only that peak.

It is possible to extend this approach to a system with several random input and output signals. In this case the various output spectra and cross-spectra are

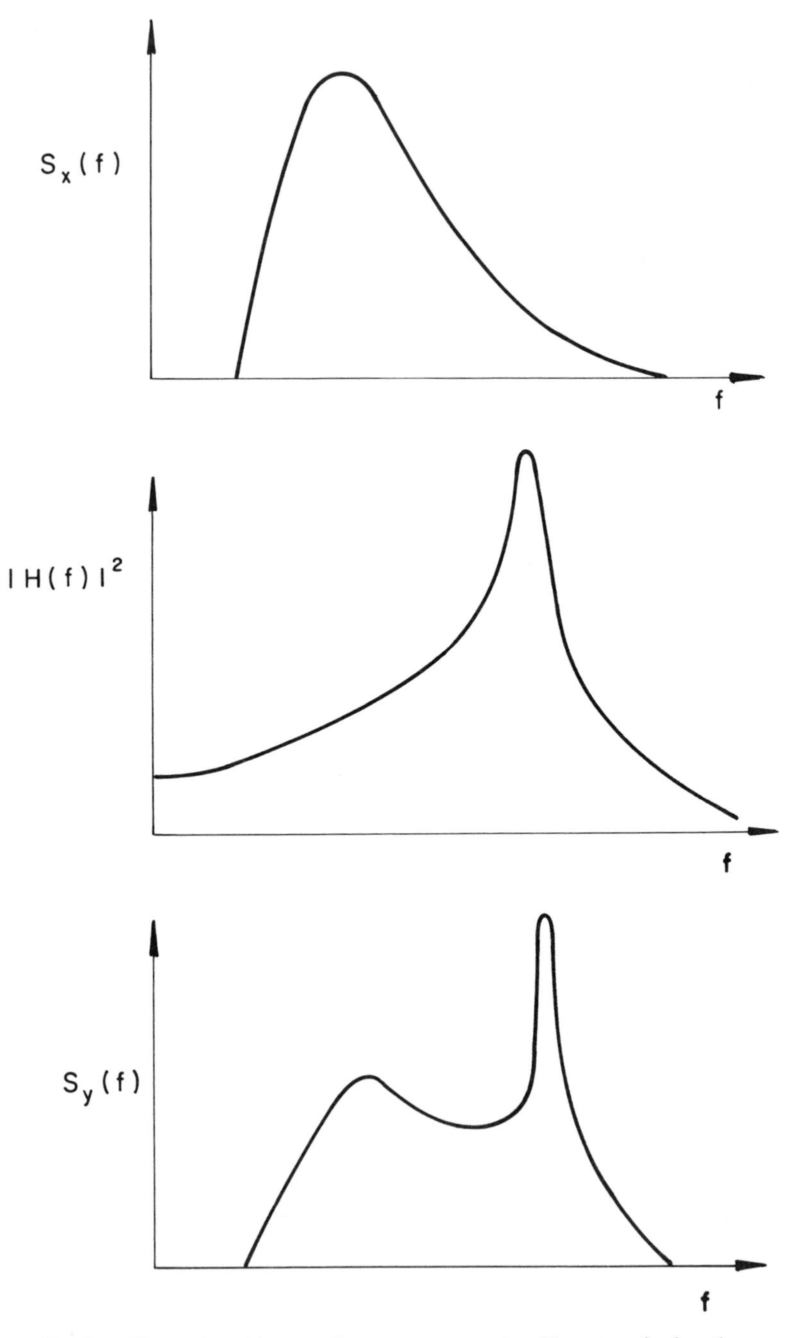

Fig. 7.5. Illustration of input and output spectra related by a transfer function.

conveniently represented in matrix form and are given by the matrix equation

$$[S'_{yy}(f)] = [H^*(f)]\ [S'_{xx}(f)]\ [H^T(f)] \tag{7.50}$$

In the above $[S'_{yy}(f)]$ represents the matrix of the (two-sided) output spectra and cross-spectra with elements $S'_{y_iy_j}(f)$; $[S'_{xx}(f)]$ represents the corresponding input spectra matrix; and $[H^*(f)]$ and $[H^T(f)]$ are respectively the complex conjugate and transpose of the receptance matrix $[H(f)]$. The general element $H_{ij}(f)$ of the receptance matrix corresponds to the receptance relating to the output $y_i(t)$ to the input $x_j(t)$.

This completes a rather concise outline of those statistical concepts needed as background for the remainder of the chapter. The brevity has been intentional and once more the reader is referred to some of the texts already referenced for comprehensive accounts of the properties of random variables.

7.2 PROBABILISTIC PROPERTIES OF OCEAN WAVES

We now focus our attention on the statistical behaviour of ocean waves themselves, making the assumption that the water surface elevation $\eta(t)$ forms a stationary random process. Thus the developments outlined here apply to a particular sea state which may last several hours, rather than to the longer term wave climate at a site. Most of the theoretical results described in this section concern the following properties of the water surface:

(a) the probability distribution of wave heights, amplitudes, or maxima of $\eta(t)$,
(b) characteristic values of this distribution (a),
(c) the probability distribution of wave periods,
(d) the probability distribution of the maximum wave height in a finite length of record, and
(e) characteristic values of this distribution (d).

The different spectra which may be used to represent random waves will be outlined in Section 7.3.

Considerable simplifications are posssible when the waves possess a narrow-band spectrum and this case will be considered in Section 7.2.1, while more general results for spectra of arbitrary shape will be outlined in Section 7.2.2.

7.2.1 Narrow-Band Wave Spectra

Longuet-Higgins (1952) first applied the statistical results of Rice (1944-1945) to ocean waves possessing a narrow-band spectrum and found that the wave heights then possess a Rayleigh distribution. At about the same time, Putz (1952) reported on a statistical analysis of selected wave records and found that the wave height distribution was approximately described by the Gamma distribution function. This is a more general distribution which includes the Rayleigh distribution as a special case, but representation in terms of the Gamma distribution has not found general use and will therefore not be pursued further.

Probability Distribution of H

If ocean waves are characterized by a narrow-band spectrum in that their component frequencies are concentrated over a narrow range, then the resulting variation of $\eta(t)$ corresponds to a regular sinusoid with slowly varying envelope (amplitude) and phase as sketched in Fig. 7.6. Thus the surface elevation, which forms a narrow-band Gaussian process, may conveniently be expressed as

$$\eta(t) = a(t) \cos(2\pi f_0 t + \delta(t)) \tag{7.51}$$

where f_0 is the central frequency, $a(t)$ is the amplitude of the envelope, $\delta(t)$ is a phase angle, and both $a(t)$ and $\delta(t)$ are random quantities whose variations are slow compared to f_0.

It is often reasonable to assume that the phase $\delta(t)$ is uniformly distributed, or equivalently that $\eta(t)$ has a Gaussian distribution—the one implies the other (Rice 1944-1945). Such an assumption, with the associated symmetry about the still water level, is expected to be realistic for small amplitude waves. For steep waves, however, the crests reach heights above the still water level that are greater in magnitude than are the depths of the troughs below it. Furthermore, the breaking process imposes a limit on the height of waves and therefore on $\eta(t)$. In such cases the distribution of $\eta(t)$ is skew and the Gaussian assumption

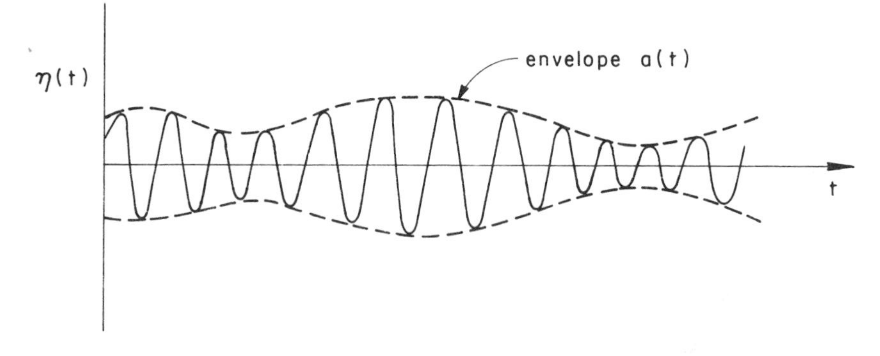

Fig. 7.6. Realization of a narrow-band process $\eta(t)$.

is unrealistic. The effects of wave nonlinearities on the probability distribution of η have been discussed by Longuet-Higgins (1963). Nevertheless, the assumption that $\eta(t)$ is normally distributed is generally a reasonable one and is invariably adopted in engineering problems.

Making this assumption then—or its equivalent that $\delta(t)$ is uniformly distributed—it may be shown that the wave amplitude a(t) is characterized by a Rayleigh distribution

$$P(a) = \begin{cases} 1 - \exp(-a^2/2m_0) & \text{for } a \geqslant 0 \\ 0 & \text{otherwise} \end{cases} \tag{7.52a}$$

$$p(a) = \begin{cases} (a/m_0) \exp(-a^2/2m_0) & \text{for } a \geqslant 0 \\ 0 & \text{otherwise} \end{cases} \tag{7.52b}$$

where m_0 is the zeroth spectral moment which is wholly equivalent to σ_η^2, the variance of $\eta(t)$. This distribution was shown in Fig. 7.3 and of course exhibits the necessary feature that a can never be negative. It is pointed out here that in the context of ocean waves the symbol E is sometimes used to denote $E = 2m_0 = 2\sigma_\eta^2$.

It is commonly the case to prefer working with wave heights H rather than amplitudes a where possible. For a narrow-band motion, and individual wave will have a height very nearly twice its amplitude and the wave height may therefore be expected to possess a Rayleigh distribution also. Following coastal engineering practice, it is convenient to define a root-mean-square wave height as

$$H_{rms} = \left[\frac{1}{N} \sum_{i=1}^{N} H_i^2\right]^{1/2} \tag{7.53}$$

where H_i denotes the height of a particular wave as measured, say, from trough to the preceding crest on a time record, and N is sufficiently large so as not to influence H_{rms}. [Note that although we have adopted the notation of some publications, e.g. U.S. Army Shore Protection Manual (1973), the symbol H_{rms} has sometimes been used to denote σ_η, the root-mean-square of $\eta(t)$, e.g. Draper (1963), Tucker (1963).]

Since any single wave is closely sinusoidal for a narrow-band motion, σ_η^2 may be expressed as the average of integrals taken over each wave. [For a single wave $(\sigma_\eta^2)_i = H_i^2/8$.] In this manner we obtain

$$H_{rms} = 2\sqrt{2}\,\sigma_\eta = 2\sqrt{2m_0} \tag{7.54}$$

and Eq. (7.52) may be written in the form

$$P(H) = \begin{cases} 1 - \exp(-H^2/H_{rms}^2) & \text{for } H \geqslant 0 \\ 0 & \text{otherwise} \end{cases} \tag{7.55a}$$

$$p(H) = \begin{cases} \dfrac{2H}{H_{rms}^2} \exp(-H^2/H_{rms}^2) & \text{for } H \geqslant 0 \\ 0 & \text{otherwise} \end{cases} \tag{7.55b}$$

Characteristic Values of p(H)

The quantities H_{rms} and $\sqrt{m_0}$ are two convenient heights used to characterize the probability distribution of wave heights and they are related as in Eq. (7.54) above. Various other characteristic heights relating to the distribution may be calculated in terms of these and are assembled in Table 7.1. These include the mode ('most probable'), median, mean and significant wave heights. In particular, the average wave height of the highest 1/n-th of all the waves (which is the significant wave height when n = 3) is given by considering first the height H'

Table 7.1 Some Wave Height Relations Based on the Rayleigh Distribution

Characteristic Height		$\frac{H}{H_{rms}}$	$\frac{H}{\sqrt{m_0}}$	$\frac{H}{H_s}$
Standard deviation of free surface,	$\sigma_\eta = \sqrt{m_0}$	$[1/2\sqrt{2} =]$ 0.354	1.0	0.250
Root-mean-square height,	H_{rms}	1.0	$[2\sqrt{2} =]$ 2.828	0.706
Mode,	$\mu(H)$	$[1/\sqrt{2} =]$ 0.707	$[2 =]$ 2.000	0.499
Median height,	$H(P = \frac{1}{2})$	$[(\ln 2)^{1/2} =]$ 0.833	$[(8 \ln 2)^{1/2} =]$ 2.355	0.588
Mean height,	$\overline{H} = H_1$	$[\sqrt{\pi}/2 =]$ 0.886	$[\sqrt{2\pi} =]$ 2.507	0.626
Significant height	$H_s = H_{1/3}$	1.416	4.005	1.0
Average of tenth highest waves,	$H_{1/10}$	1.800	5.091	1.271
Average of hundreth highest waves,	$H_{1/100}$	2.359	6.672	1.666

above which lie the highest 1/n-th of the waves

$$1 - P(H') = \text{Prob}\,(H > H') = \int_{H'}^{\infty} p(H)\,dH = 1/n \tag{7.56}$$

$H_{1/n}$ will then be the average of all waves of height greater than H' and is given by

$$H_{1/n} = \int_{H'}^{\infty} Hp(H)\,dH \Big/ \int_{H'}^{\infty} p(H)\,dH = n \int_{H'}^{\infty} Hp(H)\,dH \tag{7.57}$$

Substituting the Rayleigh distribution Eq. (7.55b) for $p(H)$, H' is given simply as

$$H'/H_{rms} = [\ln(n)]^{1/2} \tag{7.58}$$

and $H_{1/n}$ may then eventually be expressed as

$$\frac{H_{1/n}}{H_{rms}} = [\ln(n)]^{1/2} + \frac{n\sqrt{\pi}}{2}\{1 - \text{erf}\,[(\ln(n))^{1/2}]\} \tag{7.59}$$

where the error function erf ( ) is defined (e.g. Abramowitz and Stegun 1965) as

$$\text{erf}\,(x) = \frac{2}{\sqrt{\pi}} \int_0^x \exp(-t^2)\,dt \tag{7.60}$$

From Eq. (7.59) the significant wave height itself is obtained by putting $n = 3$ and is

$$H_s = 1.416 H_{rms} \simeq 4\sqrt{m_0} = 4\sigma_\eta \tag{7.61}$$

Corresponding values for $H_{1/10}$ and $H_{1/100}$ are included in Table 7.1.

Probability Distribution of T

We now turn to consider the wave period T, measured as the interval between successive zero up-crossings: that is between instants at which $\eta(t) = 0, \partial\eta/\partial t > 0$. T will vary slightly from wave to wave and remains close to $1/f_0$. Rice (1944-1945) gives an approximate expression for the probability density of the time interval between successive zero crossings. Developing this approach, Longuet-Higgins (1958, 1962, 1975) has given the probability density of the wave period

T in the form

$$p(\zeta) = \frac{1}{2}(1 + \zeta^2)^{-3/2} \tag{7.62}$$

where $\zeta = (T - \overline{T})/\nu$ and is a dimensionless representation of the wave period, chosen to have a zero mean. Here $\overline{T}$ is the mean wave period and ν describes the width of the spectrum, being defined as

$$\nu = (m_2/m_0)^{1/2}\,\overline{T} \tag{7.63}$$

where m_2 is based on circular frequency.

The joint wave period–amplitude probability density has also been considered by Longuet-Higgins (1957a, 1975) who derives the result

$$p(\xi, \zeta) = \frac{\xi^2}{\sqrt{2\pi}} \exp\left[-\frac{1}{2}\xi^2(1 + \zeta^2)\right] \tag{7.64}$$

where $\xi = a/\sqrt{m_0}$ is a dimensionless representation of the wave amplitude and ζ is defined above. $p(\xi, \zeta)\, d\xi\, d\zeta$ represents the probability that ξ and ζ simultaneously lie within the small intervals ξ to $\xi + d\xi$ and ζ to $\zeta + d\zeta$ respectively. Contours of this distribution are reproduced in Fig. 7.7, and show for example

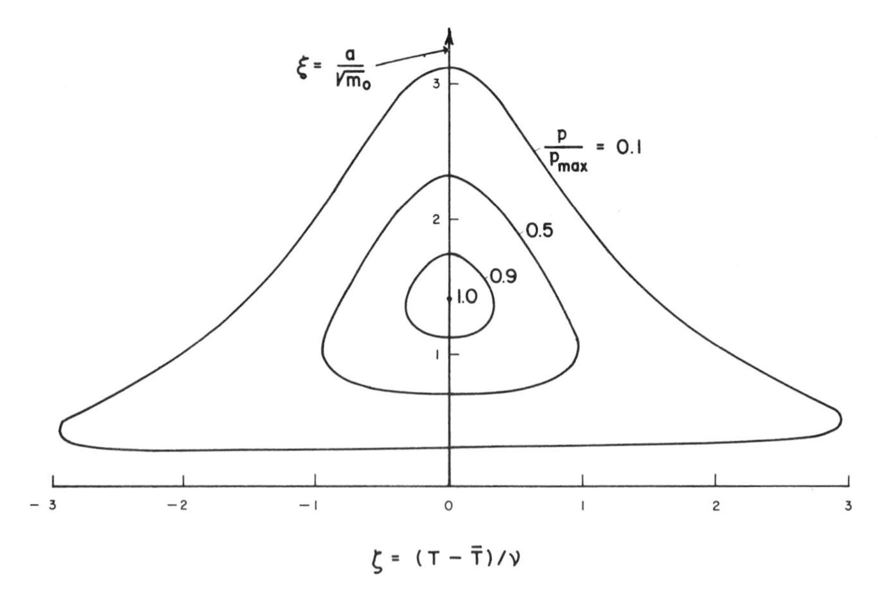

Fig. 7.7. Joint wave height-wave period probability distribution given by Longuet-Higgins (1975).

that the distribution of wave periods is relatively wide for smaller wave amplitudes. Longuet-Higgins compared this distribution to observed wave properties reported by Bretschneider (1959) and found good agreement.

Probability Distribution of a_{max}

Quite apart from the distribution of all the individual wave heights, in practical applications one requires an estimate of the largest wave height that may occur over a given duration. Equivalently, one requires the probability distribution of the largest wave amplitude a_{max} that occurs in a sample of N waves. The probability that the amplitude a of any wave is less than a specified amplitude a_{max} is denoted by $P_1(a_{max})$, where $P_1(\ )$ corresponds to $P(\ )$ in Eq. (7.52a). Thus the probability of every value of a being less than a_{max} in N (independent) waves is $P_1^N(a_{max})$. That is $P_1^N(a_{max})$ represents the probability that the maximum of all the a values is less than a_{max}, and thus $P(a_{max}) = P_1^N(a_{max})$. From this an expression for the probability density $p(a_{max})$ can be developed

$$p(a_{max}) = \frac{dP(a_{max})}{d\,a_{max}} = N\,P_1^{N-1}(a_{max})p_1(a_{max}) \tag{7.65}$$

This expression is quite general. But for the present, we are considering a narrow-band spectrum with $p_1(\ )$ is given by the Rayleigh distribution, Eq. (7.52b), and we then obtain eventually

$$p(a_{max}/\sqrt{m_0}) = N(a_{max}/\sqrt{m_0})\exp(-a_{max}^2/2m_0)\,[1 - \exp(-a_{max}^2/2m_0)]^{N-1} \tag{7.66}$$

Characteristic Values of $p(a_{max})$

Now that the probability distribution of a_{max} has been obtained, various representative values of a_{max} can be calculated. These include the mean or expected value, and the mode or most probable value.

The expected value of a_{max} can be expressed by definition as indicated in Eq. (7.8) and reduces to the integral

$$\frac{E(a_{max})}{\sqrt{m_0}} = \frac{1}{\sqrt{2}}\int_0^\infty [1 - (1 - \exp(-\theta))^N]\,\theta^{-1/2}\,d\theta \tag{7.67}$$

For large values of N an asymptotic approximation to this is

$$\frac{E(a_{max})}{\sqrt{m_0}} = \sqrt{2}\left\{[\ln(N)]^{1/2} + \frac{1}{2}\gamma\,[\ln(N)]^{-1/2} + 0[(\ln(N))^{-3/2}]\right\} \tag{7.68}$$

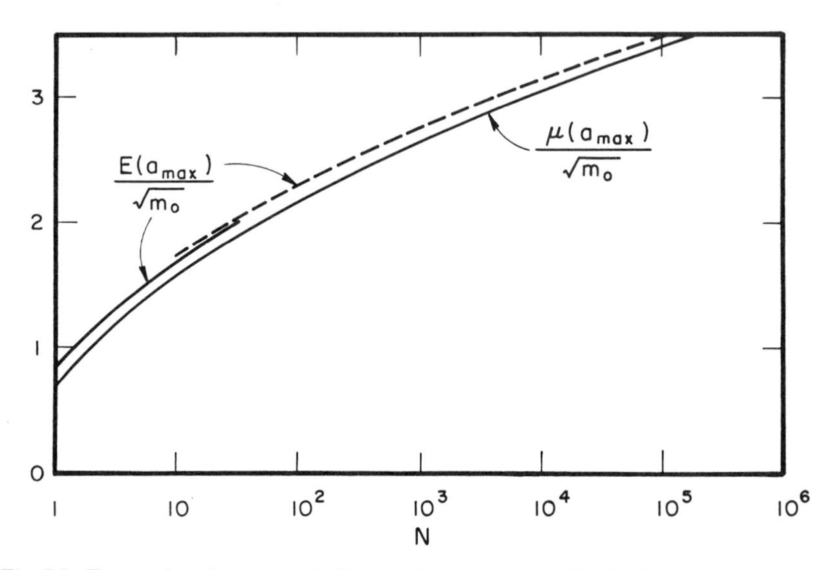

Fig. 7.8. Expected and most probable maximum wave amplitudes in narrow-band wave motion as a function of the number of waves. – – –, asymptotic formula.

where γ is Euler's constant (=0.5772 . . .). This differs from the complete expression in Eq. (7.67) by less than 3% when $N \geqslant 50$. The variation of $E(a_{max})/\sqrt{m_0}$, as given by both the exact and asymptotic expressions above, is shown in Fig. 7.8.

The mode of a_{max}, denoted here $\mu(a_{max})$, may be obtained by putting $dp(a_{max})/da_{max} = 0$. This gives $\mu(a_{max})$ implicitly, and an approximate explicit expression for $\mu(a_{max})$ can then be obtained as

$$\frac{\mu(a_{max})}{\sqrt{m_0}} = [2 \ln (N)]^{1/2} + 0[(\ln (N)^{-3/2}]. \tag{7.69}$$

This relation is included in Fig. 7.8.

Finally, it may be preferable to calculate the value $(a_{max})_\alpha$ which should be exceeded with only a specified small probability α, rather than use some best estimate of a_{max}. This is simply obtained by setting $P(a_{max}) = 1 - \alpha$, and for small α approximates to

$$\frac{(a_{max})_\alpha}{\sqrt{m_0}} = [2 \ln (N/\alpha)]^{1/2} \tag{7.70}$$

As an extension of the foregoing results, the 'expected' and 'most probable' wave heights over a given number of waves (or time interval) can be expressed in terms of the significant wave height by taking $\sqrt{m_0} = H_s/4$ and $H_{max} = 2a_{max}$.

For example, in a run of 500 waves the expected value of the maximum wave height $E(H_{max}) = 7.378\sqrt{m_0} = 1.845\ H_s$; and the most probable value $\mu(H_{max}) = 7.051\sqrt{m_0} = 1.763\ H_s$. These compare closely with the value $1.8\ H_s$ for the maximum wave height of a given sea state proposed some time ago by Wiegel (1949).

When N is sufficiently large for only the leading term of Eq. (7.68) to be significant, then both the expected and most probable values coincide and are given by the very simple formula

$$\frac{H_{max}}{H_s} = \left(\frac{\ln(N)}{2}\right)^{1/2} \tag{7.71}$$

And since a narrow-band wave motion is being considered, we have $N = f_0 t$ where t is the specified duration of the record. Thus if a narrow-band stationary wave train has a significant height $H_s = 10$ ft and a central frequency $f_0 = 0.1$ Hz, then the maximum height expected over a 20 minute period is given as $H_{max} = 10[\frac{1}{2}\ln(0.1 \times 1200)]^{1/2}$ ft $= 15.5$ ft.

A number of studies have compared measured wave height distributions with the Rayleigh distribution (e.g. Forristall 1978, Nolte and Hsu 1979). The Rayleigh distribution is found to overpredict the heights of larger waves. This may be due to the limiting effects of wave breaking, and to the assumptions of a linear, Gaussian, narrow-band free surface and of independent consecutive waves.

7.2.2 Arbitrary Wave Spectra

The results considered so far relate to the special case of a narrow-band spectrum. Cartwright and Longuet-Higgins (1956) extended the treatment, once more on the basis of Rice's (1944–1945) work, to an investigation of the maxima of $\eta(t)$ with no restriction now being placed on the width of the wave spectrum. In this case it is difficult to deal directly with the wave heights themselves and attention is directed instead to the distribution of the crest elevations, that is to the maxima of $\eta(t)$ denoted by a.

We continue to assume uniformly distributed random phase angles in any harmonic representation of $\eta(t)$ and this generally implies that $\eta(t)$ itself has a Gaussian distribution. It may be shown that the probability density of a together with various characteristic values of a, depend upon only one additional parameter ϵ termed the *spectral width parameter* and defined in terms of the spectral moments, Eq. (7.31), as

$$\epsilon = \left(1 - \frac{m_2^2}{m_0 m_4}\right)^{1/2} \tag{7.72}$$

ϵ takes values between 0 and 1, and for a narrow-band spectrum $\epsilon \to 0$. By setting $\epsilon = 0$ in the various results that are obtained, the corresponding results for a narrow-band spectrum are retrieved. As a simple example of the calculation of ϵ directly from Eq. (7.72), it can readily be shown that for a spectrum given as $S(f) = \text{constant}$ for $f < f_0$, zero otherwise, the spectral width parameter is independent of f_0 and is $\epsilon = \frac{2}{3}$.

Probability Distribution of a

The probability distribution of the crest elevations a is found to be given as

$$p(\xi) = (2\pi)^{-1/2}\,\epsilon \exp\left(-\tfrac{1}{2}\,\xi^2/\epsilon^2\right) + \tfrac{1}{2}(1-\epsilon^2)^{1/2}\,\xi \exp\left(-\tfrac{1}{2}\,\xi^2\right)\left[1 + \operatorname{erf}\left(\xi(1-\epsilon^2)^{1/2}/\epsilon\sqrt{2}\right)\right] \quad (7.73)$$

where the error function erf ( ) is defined in Eq. (7.60) and $\xi = a/\sqrt{m_0}$ is a dimensionless representation of a based on the characteristic length $\sqrt{m_0}$, and has already been used. This probability density is reproduced in Fig. 7.9 for various values of ϵ. As we should expect, the curve for $\epsilon = 0$ reduces to the Rayleigh distribution obtained already for a narrow-band spectrum. On the other hand, as $\epsilon \to 1$ the distribution becomes Gaussian so that there are then as many crests below the still water level as above it corresponding to a high frequency content entailed in a wide spectrum.

Characteristic Values of p(a)

Having described the probability distribution of a, we are now able to determine a number of its characteristic values. These include the mean of a, denoted $\bar{a}$, the root-mean-square value a_{rms}, defined in the same way as was H_{rms} in Eq.

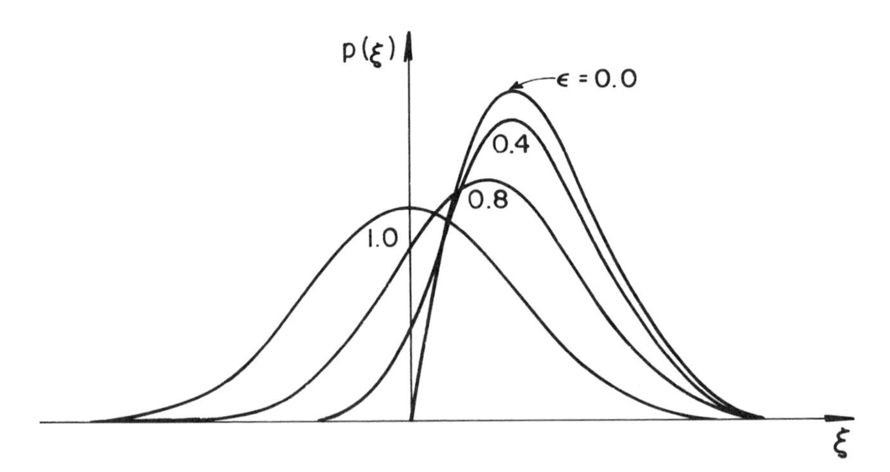

Fig. 7.9. Probability densities of the maxima $\xi = a/\sqrt{m_0}$ for various values of the spectral width parameter ϵ.

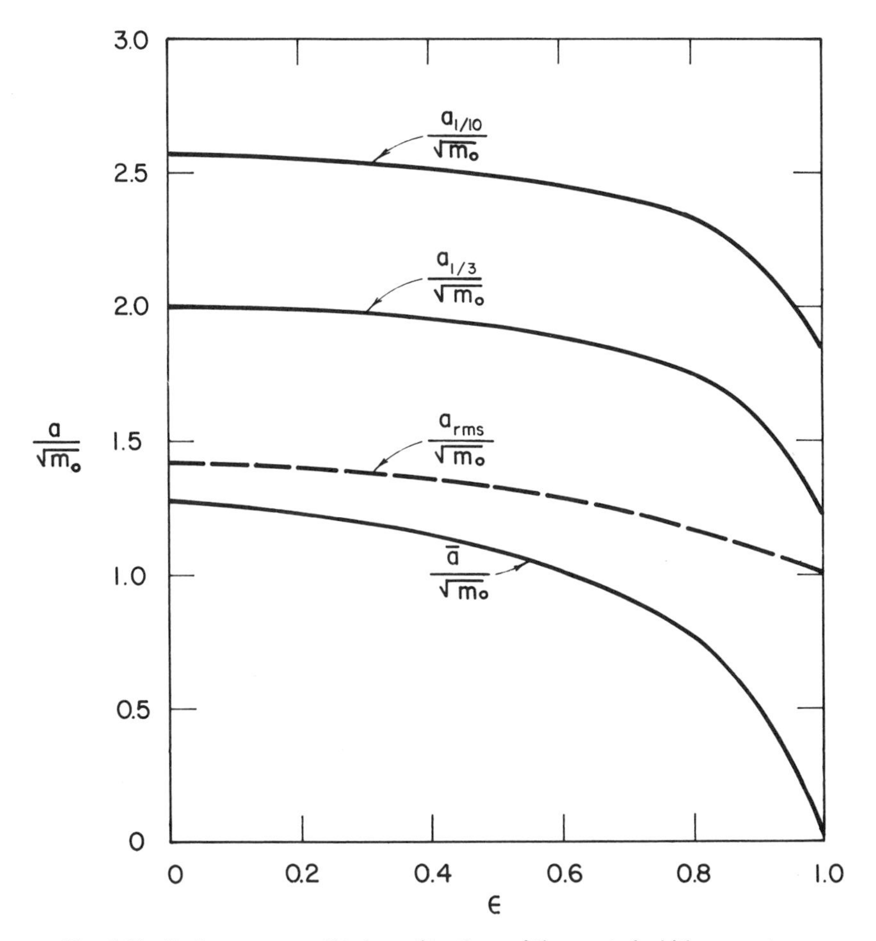

Fig. 7.10. Various wave amplitudes as functions of the spectral width parameter ϵ.

(7.53), and the average elevation $a_{1/n}$ of the highest fraction 1/n of the wave crests. The mean and root-mean-square are given respectively as

$$\bar{a}/\sqrt{m_0} = (\pi/2)^{1/2}(1 - \epsilon^2)^{1/2} \tag{7.74}$$

$$a_{rms}/\sqrt{m_0} = (2 - \epsilon^2)^{1/2} \tag{7.75}$$

The average $a_{1/n}$ of the highest fraction 1/n of all the a values may be found by applying the probability density Eq. (7.73) to the relationships already described: Eqs. (7.56) and (7.57). For a given n, $a_{1/n}/\sqrt{m_0}$ depends only on ϵ and has been evaluated numerically by Cartwright and Longuet-Higgins (1956). The variations of $a_{1/3}/\sqrt{m_0}$, $a_{1/10}/\sqrt{m_0}$, $\bar{a}/\sqrt{m_0}$ and $a_{rms}/\sqrt{m_0}$ with ϵ are reproduced in Fig. 7.10.

Estimates of ϵ and $\sqrt{m_0}$

The spectral width parameter ϵ may in principle be calculated from Eq. (7.72) once the wave spectrum is specified. However, it may conveniently be estimated from a wave record in a fairly straight-forward manner by virtue of some additional results. The proportion r of negative maxima (that is, maxima below the still water level) out of all the maxima depends only on ϵ is given as

$$r = \frac{1}{2}\left[1 - (1 - \epsilon^2)^{1/2}\right] \tag{7.76a}$$

Thus for a specified length of record, the parameter ϵ may be simply estimated by counting the numbers of positive and negative maxima and applying the result

$$\epsilon = \left[1 - (1 - 2r)^2\right]^{1/2} \tag{7.76b}$$

An alternative and equally simple method depends on estimating the crest period T_c and the zero-crossing period T_z. These are defined respectively as the average period between successive crests and the average period between successive zero up-crossings. These periods may again be estimated directly from a given length of record. They are given in terms of the spectral moments as

$$T_c = (m_2/m_4)^{1/2} \tag{7.77}$$

$$T_z = (m_0/m_2)^{1/2} \tag{7.78}$$

(where the spectral moments m_n are based on circular frequency, see Eq. (7.32)). It follows from Eqs. (7.72), (7.77) and (7.78) that an estimate of ϵ may be obtained by applying the relationship

$$\epsilon = \left[1 - (T_c/T_z)^2\right]^{1/2} \tag{7.79}$$

The second parameter used to define the probability density p(a) and the characteristic values of a is $\sqrt{m_0}$. Tucker (1963) has proposed a simple method for estimating $\sqrt{m_0}$ which is based on the use of measurements of the extreme crests and troughs in a wave record. The theoretical basis for the method was given earlier by Cartwright (1958). If H_1 is the vertical distance between the lowest trough and the highest crest (which generally does not correspond to the same wave), and H_2 is the vertical distance between the second lowest trough and the second highest crest, then $\sqrt{m_0}$ is given from that record in terms of either H_1 or H_2 as

$$\sqrt{m_0} = H_1(8\theta)^{-1/2}(1 + 0.289\theta^{-1} - 0.247\theta^{-2})^{-1} \tag{7.80a}$$

$$\sqrt{m_0} = H_1(8\theta)^{-1/2}(1 - 0.211\theta^{-1} - 0.103\theta^{-2})^{-1} \tag{7.80b}$$

where $\theta = \ln(N_z)$ and N_z is the number of zero up-crossings in the record.

As a simple example of the foregoing results, we consider a 6-minute wave record containing 45 crests and 40 zero up-crossings, and a vertical distance H_1 measured as 20 ft. We may obtain directly $T_c = 8$ sec, $T_z = 9$ sec and from Eq. (7.79) the spectral width parameter is then estimated as $\epsilon = 0.458$. Also $\theta = \ln(N_z) = 3.689$ and therefore from Eq. (7.80a) the parameter $\sqrt{m_0}$ is estimated to be $\sqrt{m_0} = 3.47$ ft. With values of ϵ and $\sqrt{m_0}$ now obtained, other characteristic values of a may be calculated. From Eqs. (7.74) and (7.75) and from Fig. 7.10 we thus obtain $\bar{a} = 3.87$ ft, $a_{rms} = 4.65$ ft and $a_{1/3} = 6.73$ ft.

Probability Distribution of T

The probability distribution of wave periods for an arbitrary spectrum is much more difficult to deal with. An expression for the probability density of the duration between successive zero crossings was obtained by Rice (1944-1945) and developed also by Longuet-Higgins (1958, 1962). On the basis of experimental data, Bretschneider (1959) has suggested that wave lengths may be characterized by a Rayleigh distribution, and consequently in deep water so also are the squares of wave periods. The distribution of wave periods in deep water is thus expressed as

$$P(T) = \begin{cases} 1 - \exp\left[-0.675(T/\overline{T})^4\right] & \text{for } T \geqslant 0 \\ 0 & \text{otherwise} \end{cases} \tag{7.81a}$$

$$p(T) = \begin{cases} \dfrac{2.7T^3}{\overline{T}^4} \exp\left[-0.675(T/\overline{T})^4\right] & \text{for } T \geqslant 0 \\ 0 & \text{otherwise} \end{cases} \tag{7.81b}$$

where $\overline{T}$ is the mean period. The coefficient 0.675 approximates $[\Gamma(\frac{5}{4})]^4$, Γ being the Gamma function, and arises when the mean of T^2 appearing in the Rayleigh distribution is replaced by the square of $\overline{T}$. Bretschneider (1959) also suggested that wave heights and periods may be considered as independent random variables [but see Longuet-Higgins (1975) results for the narrow-band case].

The distribution of breaking waves is a feature of associated interest. In deep water a wave breaks when its height and period are related by $H/gT^2 = 0.027$ (see Fig. 4.39). Nath and Ramsey (1974) employed Bretschneider's (1959) proposed probability distributions of H and T, Eqs. (7.55) and (7.81) respectively, to obtain such results as the probability distribution of breaking wave heights, and the distribution of the maximum heights of breaking waves. Houmb and Overvik (1976) worked instead with Longuet-Higgins' (1975) joint probability distribution of H and T, Eq. (7.64), and obtained corresponding results.

Characteristic Values of $p(a_{max})$

We proceed now to consider the largest maximum, a_{max}, that occurs in a specified finite length of record just as we have done for the case of a narrow-band spectrum. The record is taken to contain N maxima, of which N^+ are positive maxima (occuring above the still water level). A general expression for the probability distribution of a_{max} was developed in Eq. (7.65), but now the parent distribution $p_1(\ )$ is given by Eq. (7.73) in place of Eq. (7.52). Various representative values of a_{max} which are of interest are, as before, the expected value $E(a_{max})$, the mode or most probable value $\mu(a_{max})$ and the value $(a_{max})_\alpha$ which should be exceeded with only a specified small probability α.

The expected value of a_{max} was derived by Cartwright and Longuet-Higgins (1956) and is given as

$$\frac{E(a_{max})}{\sqrt{m_0}} = \sqrt{2}\ \{\Lambda^{1/2} + \tfrac{1}{2}\gamma\Lambda^{-1/2} + 0[\Lambda^{-3/2}]\} \tag{7.82}$$

where $\Lambda = \ln((1 - \epsilon^2)^{1/2} N)$ and $\gamma = 0.5772\ldots$ is Euler's constant.

Ochi (1973) has given an approximation to the mode of a_{max} which is valid when $\epsilon < 0.9$. This is

$$\frac{\mu(a_{max})}{\sqrt{m_0}} = \left\{2 \ln\left[\frac{2(1 - \epsilon^2)^{1/2} N^+}{1 + (1 - \epsilon^2)^{1/2}}\right]\right\}^{1/2} \tag{7.83}$$

Ochi has also presented an approximation to $(a_{max})_\alpha$ which is valid when α is small and again when $\epsilon < 0.9$. The result obtained is

$$\frac{(a_{max})_\alpha}{\sqrt{m_0}} = \left\{2 \ln\left[\left(\frac{(1 - \epsilon^2)^{1/2}}{1 + (1 - \epsilon^2)^{1/2}}\right)\left(\frac{2N^+}{\alpha}\right)\right]\right\}^{1/2} \tag{7.84}$$

For a narrow-band spectrum the various results above reduce to those already given in Eqs. (7.68)-(7.70) by taking $\epsilon \to 0$ and $N^+ = N$. For the general case $\epsilon > 0$, it is convenient to express the above results directly in terms of the specified duration of record t. By virtue of Eqs. (7.76a) and (7.77), N and N^+ are given approximately as

$$N = t(m_4/m_2)^{1/2} \tag{7.85a}$$

$$N^+ = \tfrac{1}{2} N[1 + (1 - \epsilon^2)^{1/2}] \tag{7.85b}$$

Consequently, the three characteristic values of a_{max} may be expressed as

$$\frac{E(a_{max})}{\sqrt{m_0}} = [2 \ln(\tau)]^{1/2} + \frac{1}{2}\gamma\,[\ln(\tau)]^{-1/2} \tag{7.86}$$

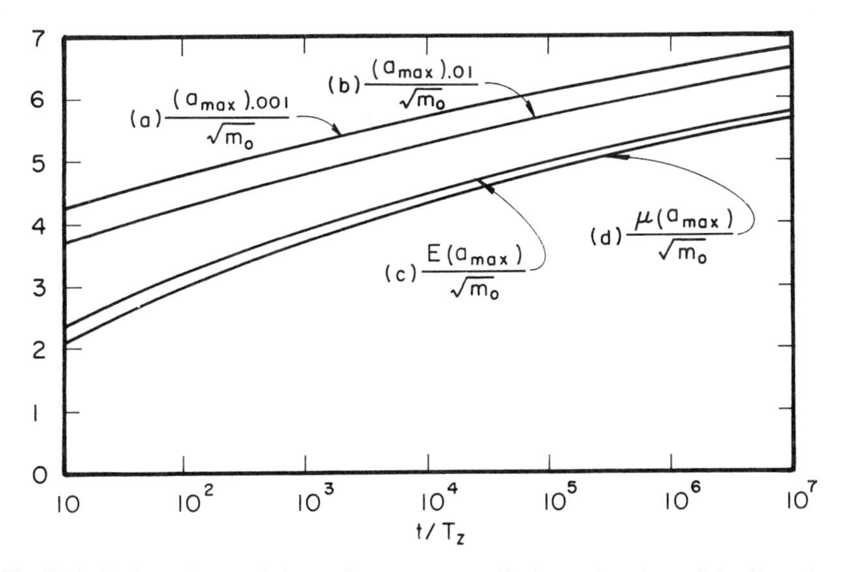

Fig. 7.11. Various characteristic maximum wave amplitudes as functions of the dimensionless duration τ. (a), (b), maximum amplitudes which should not be exceeded with probabilities of $\alpha = 0.001$ and 0.01 respectively; (c), expected maximum amplitude; (d), most probable maximum amplitude.

$$\frac{\mu(a_{max})}{\sqrt{m_0}} = [2 \ln (\tau)]^{1/2} \tag{7.87}$$

$$\frac{(a_{max})_\alpha}{\sqrt{m_0}} = [2 \ln (\tau/\alpha)]^{1/2} \qquad \text{for small } \alpha \tag{7.88}$$

where $\tau = t(m_2/m_0)^{1/2} = t/T_z$. These values are shown plotted in Fig. 7.11 as functions of the dimensionless duration parameter τ. It is seen that the above three equations are identical in form to those developed previously for narrow-band motion and Fig. 7.11 may conveniently be used for that case. For the narrow-band case τ can also be expressed as $f_0 t$.

Wave Groups

A knowledge of the spectral density and probability distribution of wave maxima may not always provide an adequate description of random wave behavior. A further phenomenon that is considered significant in certain problems and that has recently received some attention concerns the properties of a wave group. This is defined as a succession of consecutive waves with heights larger than a specified value. Contributions to the development of this topic include studies by Goda (1970), Nagai (1973), Nolte and Hsu (1973), Rye (1974), Ewing

(1973) and Siefert (1976), and have been reviewed by Goda (1976). For wave group analysis Goda (1970) proposed the use of a dimensionless spectral peakedness parameter Q defined as

$$Q = \frac{2}{m_0^2} \int_0^\infty f S^2(f)\, df \tag{7.89}$$

This is a measure of the narrowness of the spectrum (e.g. $Q = 2f_0/\Delta f$ for a reference rectangular spectral shape defined as S = constant for $|f - f_0| < \Delta f/2$, zero otherwise), and is considered more meaningful than the spectral width parameter ϵ in the study of wave groups.

A primary parameter to be investigated is the run length, that is the number of consecutive waves whose height is greater than a specified value, and this has been expressed in terms of the spectral moments and the specified envelope amplitude. Goda (1976) indicates that prospective applications of wave group analysis include studies of long period oscillations of moored ships, ship capsizing tests and investigations into the severity of wave action on breakwaters.

7.3 SPECTRAL PROPERTIES OF OCEAN WAVES

Attention has so far been confined to the temporal variation of η at a single point, and no consideration has been given to the variation of η with space in the vicinity of the point of interest. It has already been mentioned in Section 7.1.3 that the more general case accounting for such spatial variations may be represented by a three-dimensional wave spectrum $S_\eta(k_x, k_y, f)$, which has as arguments not only frequency f, but also the wave number components k_x and k_y. Various two-dimensional spectra, which are more applicable for general use may also be employed, and we consider now the alternatives which may be possible. Provided no confusion arises it will be convenient to omit the subscript η to the spectral density S( ) in the various descriptions that will be set out in this section.

7.3.1 Alternative Spectral Representations

Alternative approaches to developing a one-dimensional frequency spectrum were mentioned in Section 7.1.3 and any of these may be extended to two- or three-dimensional spectra. We consider here the approach based on Eqs. (7.19)-(7.24). Thus a random sea may be represented as an infinite sum of sinusoidal components travelling with different wave numbers k, frequencies f and directions θ (referred to a base direction, say that of the predominant wind). The free surface elevation is thus written as

$$\eta = \sum_{n=1}^{\infty} A_n \exp \{i[k_n \cos(\theta_n)x + k_n \sin(\theta_n)y - 2\pi f_n t]\} \tag{7.90}$$

where the complex amplitudes A_n are random quantities as in Eq. (7.21).

As a generalization of the previous approach, we may adopt a spectrum $S(k, f, \theta)$, such that $S(k, f, \theta)\, dk\, df\, d\theta$ represents the contribution to the variance σ_η^2 due to component waves with wave numbers between k and k + dk, frequencies between f and f + df and directions between θ and $\theta + d\theta$. That is, Eq. (7.24) is now replaced by

$$S(k, f, \theta)\, dk\, df\, d\theta = \sum_{k_n}^{k_n+dk} \sum_{f_n}^{f_n+df} \sum_{\theta_n}^{\theta_n+d\theta} \tfrac{1}{2}|A_n|^2 \tag{7.91}$$

and the variance itself, given previously by Eq. (7.23), may now be written as

$$\sigma_\eta^2 = \int_0^\infty \int_0^\infty \int_{-\pi}^{\pi} S(k, f, \theta)\, dk\, df\, d\theta \tag{7.92}$$

This three-dimensional spectrum is based on a polar representation of the wave number vector **k**. A Cartesian representation involving the wave number components $k_x = k \cos\theta$ and $k_y = k \sin\theta$ is equivalent and the corresponding spectra will be related (on the basis of an extension to Eq. (7.30)) by

$$S(k_x, k_y, f) = \frac{1}{k} S(k, f, \theta) \tag{7.93}$$

A three-dimensional spectrum is too unwieldy for general use, and to a certain extent is redundant since it may be possible to exploit the linear dispersion relationship between wave numbers and frequencies to effect a reduction from three arguments to two. It is emphasized, however, that this reduction will only be an approximation, particularly so for the higher frequencies (shorter waves), on account of nonlinear interactions between various wave components (e.g. Phillips 1977). In any event, a two-dimensional spectrum may always be adopted, whether or not it completely describes the corresponding three-dimensional spectrum.

A two-dimensional directional spectrum $S(f, \theta)$ may be developed from Eq. (7.90) by considering $S(f, \theta)\, df\, d\theta$ to represent the contribution to the variance σ_η^2 due to waves with frequencies between f and f + df, and directions between θ and $\theta + d\theta$, regardless of their wave numbers. That is

$$S(f, \theta)\, df\, d\theta = \sum_{f_n}^{f_n+df} \sum_{\theta_n}^{\theta_n+d\theta} \tfrac{1}{2}|A_n|^2 \tag{7.94}$$

and

$$\sigma_\eta^2 = \int_0^\infty \int_{-\pi}^{\pi} S(f, \theta)\, df\, d\theta \tag{7.95}$$

The directional spectrum can be obtained from the three-dimensional spectrum by the following integration,

$$S(f, \theta) = \int_0^\infty S(k, f, \theta)\, dk \tag{7.96}$$

The directional spectrum $S(f, \theta)$ has found general application in engineering and some expressions for it which have been proposed are given in Section 7.3.4. A sketch of a directional spectrum is shown in Fig. 7.12.

A directional wave number spectrum $S(k, \theta)$ may be developed in the same way, with the relationships corresponding to Eqs. (7.94)-(7.96) now being

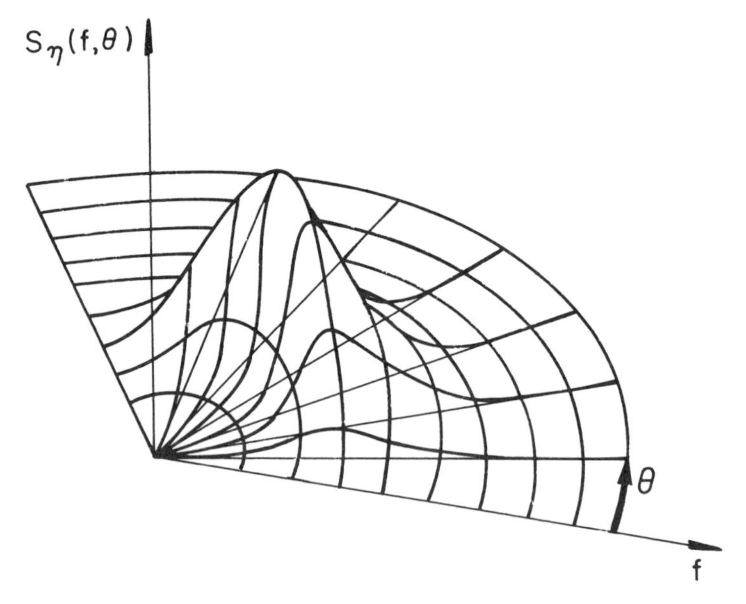

Fig. 7.12. Illustration of a directional wave spectrum.

$$S(k, \theta)\, dk\, d\theta = \sum_{k_n}^{k_n+dk} \sum_{\theta_n}^{\theta_n+d\theta} \tfrac{1}{2}|A_n|^2 \tag{7.97}$$

$$\sigma_\eta^2 = \int_0^\infty \int_{-\pi}^{\pi} S(k, \theta)\, dk\, d\theta \tag{7.98}$$

and

$$S(k, \theta) = \int_0^\infty S(k, f, \theta)\, df \tag{7.99}$$

According to Eq. (7.30), the relationship between the directional wave number and frequency spectra involves the group velocity $c_G = d\omega/dk = 2\pi(df/dk)$ and must be

$$S(k, \theta) = \frac{1}{2\pi} c_G S(f, \theta) \tag{7.100}$$

One-dimensional frequency and wave number spectra may be obtained by integrating the corresponding directional spectra over θ. Thus

$$S(f) = \int_{-\pi}^{\pi} S(f, \theta)\, d\theta \tag{7.101}$$

$$S(k) = \int_{-\pi}^{\pi} S(k, \theta)\, d\theta \tag{7.102}$$

and the variance is given in terms of these as

$$\sigma_\eta^2 = \int_0^\infty S(f)\, df = \int_0^\infty S(k)\, dk \tag{7.103}$$

Again, provided the linear dispersion relationship is assumed valid, the relationship between these is given by Eq. (7.100) with the argument θ absent.

Further spectral representations are also possible, such as spectra involving the gradient of η and cross-spectra relating to η measured at two separate points, and so on.

7.3.2 Transformation of Wave Spectra

In certain applications it is the rate of energy transfer rather than the energy density itself which is of importance. The average rate of energy transfer per unit width, or energy flux, of a uni-directional random wave train is given as

$$P = \rho g \int_0^\infty c_G S(f)\, df \tag{7.104}$$

where c_G is the group velocity. Thus the transformation of a unidirectional spectrum during wave shoaling is given by

$$c_G S(f) = \text{constant} \tag{7.105}$$

(Wave refraction and energy dissipation are absent, and energy transfer between frequencies does not occur on account of the linearity assumption.)

Equation (7.104) is also of interest when considering the vertical plane interaction of a random wave train with an energy extracting device or other fixed or floating body, and may be used to set up an energy balance equation relating the incident, reflected and transmitted wave energy fluxes and the rate of energy dissipation. (The measurement of incident and reflected wave spectra from a composite wave record is described, for example, by Thornton and Calhoun (1972) and by Goda and Suzuki (1976).)

A description of the transformation of a random wave train due to refraction may be obtained by an extension of the regular wave refraction relationships, and was originally developed by Longuet-Higgins (1956, 1957b). The energy flux between neighboring orthogonals of any component wave train remains constant, as expressed by Eq. (4.181). In the random wave case, $S(f)\, df\, d\theta$ is used in place of H^2 to represent the energy density, and therefore Eq. (4.181) relating conditions at any location (unsubscripted variables) to those at the corresponding deep water location (variables with subscript 0) now becomes

$$\begin{aligned} S(f, \theta)\, df\, d\theta\, c_G b &= S_0(f_0, \theta_0)\, df_0\, d\theta_0\, c_{G_0} b_0 \\ &= \text{constant} \end{aligned} \tag{7.106}$$

Because of linearity, no energy transfer between frequencies occurs, $df_0 = df$. Also it may be shown (Longuet-Higgins 1957b) that

$$bk\, d\theta = b_0 k_0\, d\theta_0 = \text{constant along an orthogonal} \tag{7.107}$$

Substituting Eq. (7.107) into Eq. (7.106), and bearing in mind that $k/k_0 = c_0/c$

since f is constant, we obtain

$$c c_G S(f, \theta) = c_0 c_{G0} S_0 (f, \theta_0)$$

$$= \text{constant along an orthogonal} \qquad (7.108)$$

Equivalently, this relationship may be expressed in terms of the wave number spectrum in the form derived by Longuet-Higgins (1957b). By virtue of Eqs. (7.93) and (7.100), this is simply

$$S(k_x, k_y) = S_0 (k_{x0}, k_{y0}) = \text{constant along an orthogonal} \qquad (7.109)$$

More generally, energy is dissipated because of friction at the seabed and the wave field may be non-stationary, so that the directional spectrum depends not only on location, but also on time. The energy balance equation for this general case is

$$\frac{\partial S}{\partial t} + \nabla(S c_G) = -F \qquad (7.110)$$

where F represents an energy dissipation function, and the gradient operator ∇ refers to the arguments f, θ, x and y. Applications of Eq. (7.110), in which non-stationarity and/or energy dissipation effects are retained, have been described by Karlsson (1969), Collins (1972), Tang and Ou (1972) and Cardone, Pierson and Ward (1976).

One effect of the increased wave heights associated with shoaling is that the nonlinear transfer of energy between different frequency components becomes significant. Eventually, at wave breaking the spectral peak associated with the breaking wave will diminish, with the corresponding energy being transferred to higher frequencies and into turbulence. The treatment of nonlinear interactions between different frequency components, and of a random wave train described to the second order, is an important consideration, particularly for the wave generation process. Contributions to this topic are described in the text by Phillips (1977) and further mention of it is made in Section 7.6.5.

7.3.3 Some Proposed Frequency Spectra

It will now be of interest to list some of the better known one-dimensional frequency spectra that have been employed to describe ocean waves. Many of these were developed in terms of a reference wind speed U as a parameter. Of those spectra given here, the Bretschneider and Pierson-Moskowitz spectra—which are essentially of the same form—are perhaps the most commonly used at present, while the JONSWAP spectrum, which is an extension of the Pierson-Moskowitz spectrum to account for a much sharper spectral peak, is more recent and involves additional parameters. The measurement of wave spectra from wave recordings is mentioned in Section 7.4.2. For conformity, all the spectra are presented here in terms of circular frequency f.

Darbyshire spectrum (*1952*). This was one of the earliest to be used and may be written

$$S(f) = \begin{cases} A \exp\left[-\dfrac{10.79\,(f - f_0)}{(f - f_0 + 0.0422)^{1/2}}\right] & \text{for } f - f_0 > -0.0422 \\ 0 & \text{otherwise} \end{cases} \tag{7.111}$$

Here A and f_0 are functions of wind speed U and are given as

$$A = 1.169 \times 10^{-5}\, U^4$$

$$f_0 = 1/(1.94\, U^{1/2} + 2.5 \times 10^{-7}\, U^4)$$

where f_0 is in Hz, U is in m/sec and A is in m^2/Hz. f_0 is the peak frequency, that is the frequency at which S(f) is a maximum. This spectrum is applicable to fully-developed conditions and so depends only on the characteristic wind speed U.

Neumann spectrum (*1953*). This is given as

$$S(f) = \frac{Kg^2}{f^6} \exp\left(-\frac{B}{f^2}\right) \tag{7.112}$$

where

$$B = g^2/2\pi^2 U^2$$

$$K = 2 \times 10^{-5}\ \text{Hz}$$

The peak frequency f_0 can readily be shown to be

$$f_0 = g/\pi\sqrt{6}\, U = (B/3)^{1/2}$$

And once again this spectrum requires only U and so relates to fully-developed conditions.

Bretschneider spectrum (*1959*). This spectrum is given in terms of the significant wave height H_s and peak frequency f_0 rather than the wind speed itself. Thus H_s and f_0 are themselves first obtained from hindcasting relations (see Section 7.4.1) in terms of the wind speed U, the fetch F and the duration t and so the spectrum therefore applies to developing seas. It may be written as

$$S(f) = \frac{5H_s^2}{16f_0} \frac{1}{(f/f_0)^5} \exp\left[-\frac{5}{4}\left(\frac{f}{f_0}\right)^{-4}\right] \tag{7.113}$$

The peak frequency f_0 is empirically related to the significant wave period T_s. The Bretschneider spectrum is designed so as to ensure that the area m_0 under the spectrum corresponds to $H_s/16$, as should be the case on the assumption of a Rayleigh distribution of wave heights.

Pierson-Moskowitz spectrum (*1964*). This is written as

$$S(f) = \frac{\alpha g^2}{(2\pi)^4 f^5} \exp\left(-\frac{B}{f^4}\right) \tag{7.114}$$

where $\alpha = 8.1 \times 10^{-3}$ and is termed the Phillips' constant and $B = 0.74\ (g/2\pi U)^4$. As with the Neumann spectrum, this depends only on the wind speed U and so refers to fully-developed conditions.

It will be noted that both the Bretschneider and Pierson-Moskowitz spectra may be written in the general form

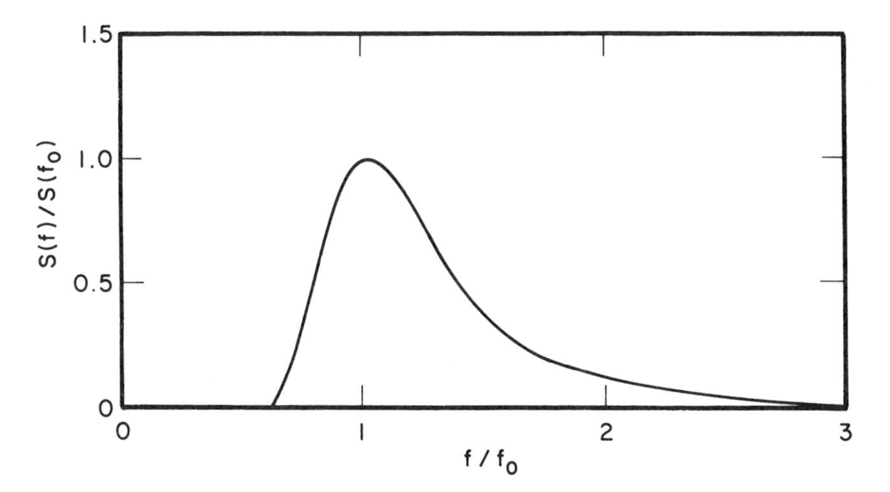

Fig. 7.13. Dimensionless form of the Bretschneider and Pierson-Moskowitz spectra.

$$S(f) = \frac{A}{f^5} \exp\left(-\frac{B}{f^4}\right) \tag{7.115}$$

and differ only in the magnitudes assigned to A and B. For the Pierson-Moskowitz spectrum A is a constant and B depends only on U, while for the Bretschneider spectrum $A = 5H_s^2 f_0^4/16$ and $B = 5f_0^4/4$ and both H_s and f_0 depend on U, F and t. Equation (7.115) may be written in a normalized form in terms of the peak frequency f_0 and peak spectral density $S(f_0)$ as

$$\frac{S(f)}{S(f_0)} = e^{5/4}\left(\frac{f}{f_0}\right)^{-5} \exp\left[-\frac{5}{4}\left(\frac{f}{f_0}\right)^{-4}\right] \tag{7.116}$$

and this profile is shown in Fig. 7.13. The moments of the spectrum Eq. (7.115) are given as

$$m_n = AB^{(n/4-1)}\Gamma(1 - n/4)/4 \qquad \text{for } n \leqslant 4 \tag{7.117}$$

where Γ is the Gamma function, and thus in particular

$$m_0 = A/4B, \quad m_2 = \sqrt{\pi}\, A/4\sqrt{B} \quad \text{and} \quad m_4 = \infty.$$

It therefore follows that the spectral width parameter ϵ is formally unity. In practice the ripples with very low energy content which are contained in this spectral form are disregarded (i.e. a high frequency cut-off is imposed) so that ϵ is generally less than unity and may be obtained independently from a wave record.

JONSWAP spectrum. This derives from the Joint North Sea Wave Project (Hasslemann et al. 1973) and constitutes a modification to the Pierson-Moskowitz spectrum to account for the effect of fetch restrictions and to provide for a much more sharply peaked spectrum. It is given as

$$S(f) = \frac{\alpha g^2}{(2\pi)^4 f^5} \exp\left[-\frac{5}{4}\left(\frac{f}{f_0}\right)^{-4}\right]\gamma^a \tag{7.118}$$

where

$$a = \exp\left[-(f - f_0)^2/2\sigma^2 f_0^2\right]$$

$$\sigma = \begin{cases} \sigma_a = 0.07 & \text{for } f \leqslant f_0 \\ \sigma_b = 0.09 & \text{for } f > f_0 \end{cases}$$

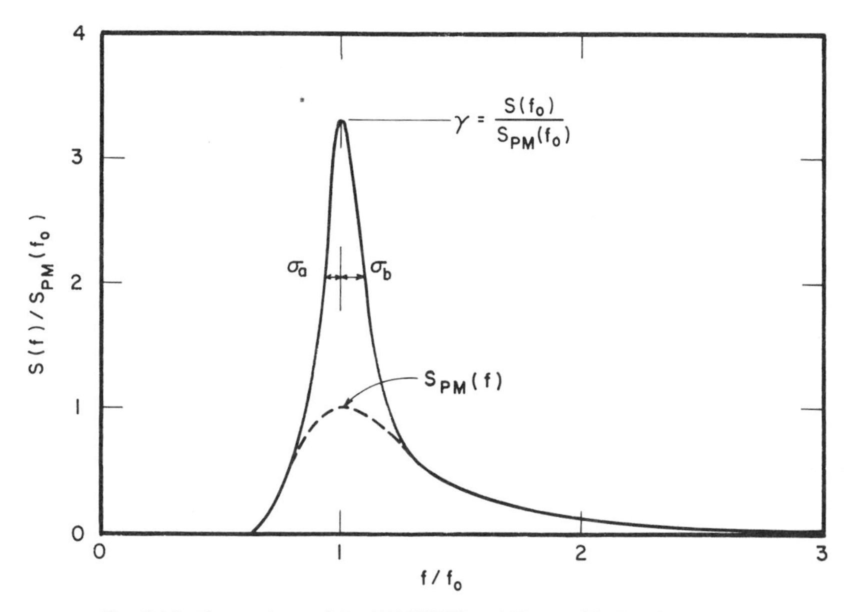

Fig. 7.14. Comparison of the JONSWAP and Pierson-Moskowitz spectra.

Here f_0 is the peak frequency at which S(f) is a maximum and is found to be related to the fetch parameter by $f_0 = 2.84(gF/U^2)^{-0.33}$; σ_a and σ_b relate respectively to the widths of the left and right sides of the spectral peak; α is equivalent to the Phillips' constant but is now taken to depend on the fetch parameter: $\alpha = 0.066(gF/U^2)^{-0.22}$; and finally γ is the ratio of the maximum spectral density to that of the corresponding Pierson-Moskowitz spectrum. This was found to have a mean value of about 3.3, corresponding to a much more sharply peaked spectrum than predicted by the Pierson-Moskowitz formula. The profiles of both these spectra are compared in Fig. 7.14.

The spectra mentioned so far represent only a concise selection of the various frequency spectra which have been proposed. Others include the Scott (1965) spectrum, Mitsuyasu's (1971) modification to the Pierson-Moskowitz spectrum to account for fetch limitations, Mitsuyasu's (1972) own proposed spectrum, the ISSC spectrum and so on. Ochi and Wang (1976) (see also Ochi and Hubble 1976) have proposed a further spectrum which depends on six parameters, and which exhibits two peaks, one associated with underlying swell and the other with locally generated waves. This spectrum is the sum of two terms, each specified by characteristic frequency, height and shape parameters.

It is appropriate at this point to mention that the wave spectrum is sometimes presented as a period spectrum S(T) rather than as a frequency spectrum S(f) or $S(\omega)$. These presentations are, however, related to one another in the form of

Eq. (7.30). Thus

$$S(T) = f^2 S(f) = (\omega^2/2\pi) S(\omega) \tag{7.119}$$

In particular the Bretschneider period spectrum (Bretschneider 1961) may be written in the form

$$S(T) = AT^3 \exp(-BT^4) \tag{7.120}$$

where A and B are identical to those of Eq. (7.115), $A = 5H_s^2 f_0^4/16$ and $B = 5f_0^4/4$. Note though that the spectral moments of a period spectrum take on a different significance from those of a frequency spectrum; in particular the spectral width parameter for the Bretschneider period spectrum is given as $\epsilon = 0.463$.

Now that a variety of proposed spectral formulae have been outlined, it is appropriate at this point to summarize a selection of wave periods and frequencies that are commonly adopted to characterize a random wave motion and which may generally be determined from a specified spectrum. A list of the more common quantities together with a description of each is given in Table 7.2. All of these except the peak period T_p, modal period T_0 and the mean frequency $\bar{f}$ have already been encountered. It is emphasized that the list is not exhaustive and characteristic quantities based on other definitions have also been used. These include the half spectrum period T_h defined as

Table 7.2 Some Characteristic Wave Periods and Frequencies

Period or Frequency	Symbol	Description	Equation
Crest period	T_c	Average period between successive crests	$T_c = \sqrt{m_2/m_4}$
Zero-crossing period	T_z	Average period between successive zero up-crossings	$T_z = \sqrt{m_0/m_2}$
Peak frequency	f_0	Frequency at which S(f) is a maximum	$\dfrac{dS(f)}{df} = 0 \quad \text{at } f = f_0$
Peak period	T_p	Period at which S(f) is a maximum	$T_p = 1/f_0$
Modal period	T_0	Period at which S(T) is a maximum	$\dfrac{dS(T)}{dT} = 0 \quad \text{at } T = T_0$
Mean frequency	$\bar{f}$	Zero first moment of the frequency spectrum about the mean	$\bar{f} = m_1/m_0$
Significant period	T_s	Average period of the highest one-third of the waves	$T_s = \sqrt[4]{4/5}\, f_0^{-1}$ (suggested by Bretschneider 1977)

$$\int_0^{T_h} S(T)\, dT = \tfrac{1}{2} \int_0^{\infty} S(T)\, dT \tag{7.121}$$

and the wave-group period (Thompson 1972) which may be obtained from wave sequences appearing in a record.

Once the wave spectrum has been defined, either by a named functional form or by measurement, then the relationships between the various alternative characteristics will generally be known. Thus, in the case of the Bretschneider spectrum, the periods and frequencies listed in Table 7.2 are interrelated as follows:

$T_c = 0$ (corresponding to non-zero high frequency content);

$T_z = 0.710\, f_0^{-1}$; $T_p = f_0^{-1}$; $T_0 = 0.880\, f_0^{-1}$;

$\bar{f} = 1.296\, f_0$; $T_s = 0.946\, f_0^{-1}$.

Note here that the significant wave period T_s is not theoretically defined in terms of the spectrum, but use is made of the empirical relation given in Table 7.2 which is due to Bretschneider. On the basis of experimental measurement, Goda (1974) recommends that the relationship $T_s = 1/1.05\, f_0$ may be used, a formula which is virtually the same as Bretschneider's own.

Before concluding the present discussion, mention is made of the concept of the *Envelope of Spectra* as described by Bretschneider (1975) and Bretschneider and Tamaye (1976). This is a curve which envelopes all the spectra occuring at a certain location at different times during the wave generation process, and is sketched in Fig. 7.15. Thus the higher frequency component spectra of low energy occur earlier in the generation process and are transformed into the lower frequency spectra with greater energy. The envelope can be represented by a formula such as Eq. (7.115), but it is emphasized that the envelope is not itself a spectrum as such and that the area under the envelope is greater than that under any component spectrum. Such a concept accounts for the high frequency content present during the early stages of wave generation which may be important in some applications but which are disregarded when considering only a fully-developed sea state.

7.3.4 Some Proposed Directional Spectra

We now consider briefly some forms of the directional spectrum $S(f, \theta)$ that have been proposed. It is convenient to separate the one-dimensional spectrum $S(f)$ and write

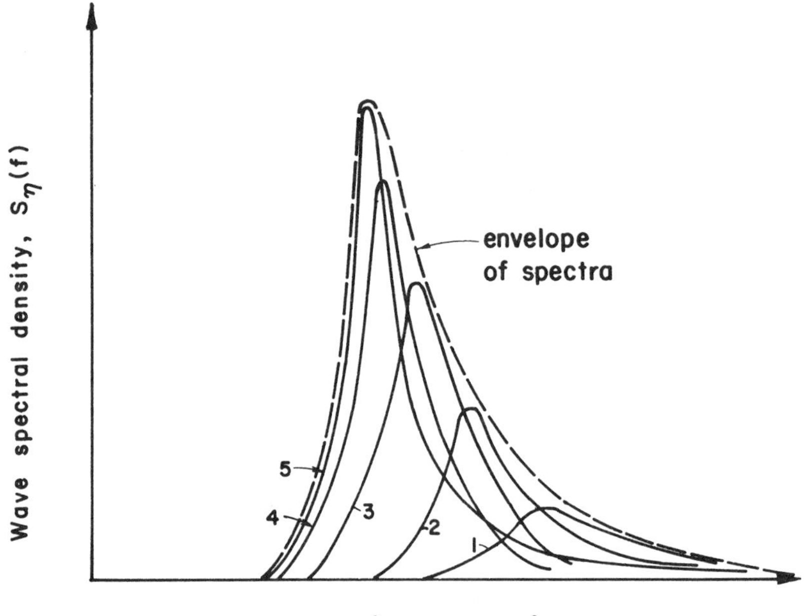

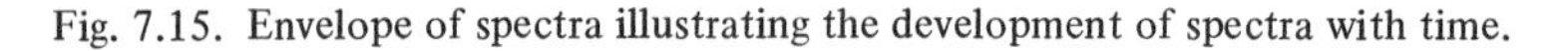

Fig. 7.15. Envelope of spectra illustrating the development of spectra with time.

$$S(f, \theta) = S(f)G(f, \theta) \tag{7.122}$$

where $G(f, \theta)$ is a *directional spreading function* which is not necessarily independent of f. It follows from Eqs. (7.101) and (7.122) that $G(f, \theta)$ must satisfy

$$\int_{-\pi}^{\pi} G(f, \theta)\, d\theta = 1 \tag{7.123}$$

Thus, in the special case of uni-directional random waves propagating in the direction $\bar{\theta}$, G would be given in terms of the Dirac delta function δ as

$$G(\theta) = \frac{1}{2\pi}\, \delta(\theta - \bar{\theta}) \tag{7.124}$$

Various semi-empirical expressions for $G(f, \theta)$ have been proposed. As the most obvious simplification it is convenient to consider G to be independent of frequency, and a selection of formulae which have been proposed for $G(\theta)$ and which may include one or more parameters are now given.

Cosine-squared. This was proposed by St. Denis and Pierson (1953) and is given as

$$G(\theta) = \begin{cases} \dfrac{2}{\pi} \cos^2 \theta & \text{for } |\theta| < \pi/2 \\ 0 & \text{otherwise} \end{cases} \tag{7.125}$$

$G(\theta)$ is a maximum along the direction $\theta = 0$.

Cosine-power.

$$G(\theta) = C(s) \cos^{2s} [\tfrac{1}{2} (\theta - \bar{\theta})] \tag{7.126}$$

where $C(s)$ is a normalizing function needed to ensure that Eq. (7.123) is satisfied, and is given by

$$C(s) = \frac{2^{2s-1}}{\pi} \frac{\Gamma^2(s+1)}{\Gamma(2s+1)}$$

$$= \frac{1}{2\sqrt{\pi}} \frac{\Gamma(s+1)}{\Gamma(s+\frac{1}{2})} \tag{7.127}$$

and Γ is the Gamma function. $\bar{\theta}$ is the direction about which the spectrum is centered. Longuet-Higgins, Cartwright and Smith (1961) found this function to provide a reasonable fit to their measured spectra. Figure 7.16 shows the directional spreading function $G(\theta)$ given by Eq. (7.126) for different values of the parameter s. It is seen that s describes the degree of spread about the direction $\bar{\theta}$, with a high value of s corresponding to a more concentrated spectrum.

An alternative cosine-power function which has also been used (e.g. Borgman 1969a) is

$$G(\theta) = \begin{cases} C'(s) \cos^{2s} (\theta - \bar{\theta}) & \text{for } |\theta - \bar{\theta}| < \pi/2 \\ 0 & \text{otherwise} \end{cases} \tag{7.128}$$

In this case the normalizing function $C'(s)$ must be given as

$$C'(s) = \frac{1}{\sqrt{\pi}} \frac{\Gamma(s+1)}{\Gamma(s+\frac{1}{2})} = 2\, C(s) \tag{7.129}$$

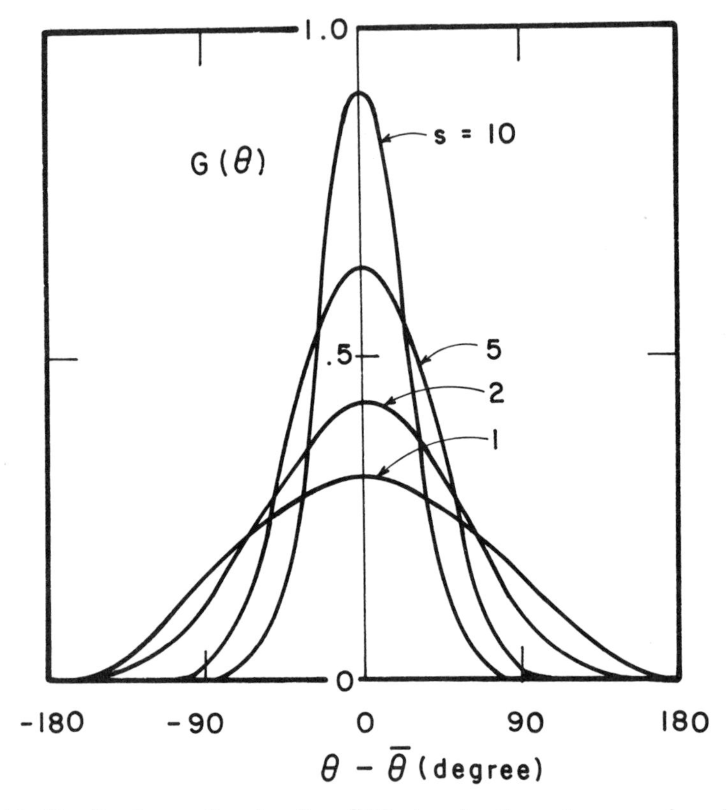

Fig. 7.16. Directional spreading function G(θ) given by the cosine-power formula Eq. (7.127) for different values of the parameter s.

Finite Fourier Series

In principle, the directional spreading function can always be represented by an infinite Fourier series

$$G(\theta) = \frac{1}{2\pi} + \sum_{n=1}^{\infty} \left[a_n \cos(n\theta) + b_n \sin(n\theta)\right] \tag{7.130}$$

(The term $1/2\pi$ ensures that Eq. (7.123) is satisfied.) However, in any measurement of the directional spreading function, only the first few coefficients a_n and b_n up to say $n = N$, will generally be obtained. The corresponding truncated series will not necessarily provide the most desirable fit to the infinite series since, for example, negative values of G may then be predicted. It will generally be more appropriate to combine the coefficients obtained with a set of weighting coefficients w_n in order to provide a better approximation to the infinite series. That is one may take

$$G(\theta) = \frac{1}{2\pi} + \sum_{n=1}^{N} w_n \left[a_n \cos(n\theta) + b_n \sin(n\theta)\right] \qquad (7.131)$$

Longuet-Higgins, Cartwright and Smith (1961) developed this approach to estimate the directional distribution from measurements using a buoy.

A further drawback of the Fourier series representation is that an unduly large number of coefficients would be needed to reproduce G = 0 over most of the range $\pi/2 < |\theta| < \pi$, which is expected. This can be overcome by adjusting the period of the Fourier series from 2π to some other prescribed direction range $\Delta\theta$ (e.g. $\Delta\theta = \pi$) so that a truncated Fourier series representation might then be (Borgman 1977)

$$G(\theta) = \begin{cases} \dfrac{1}{\Delta\theta} + \displaystyle\sum_{n=1}^{N} \left[a_n \cos\left(\dfrac{2\pi}{\Delta\theta} n\theta\right) + b_n \sin\left(\dfrac{2\pi}{\Delta\theta} n\theta\right)\right] & \text{for } |\theta| < \Delta\theta/2 \\ 0 & \text{otherwise} \end{cases} \qquad (7.132)$$

The above selection of spreading functions dependent on θ only is not complete and several other formulations have been proposed. These include the von Mises or circular spreading function, the wrapped-around Gaussian function (Borgman 1969a) and so on.

The various spreading functions listed above may be generalized so as to become functions of θ and f (see Eq. (7.122)) by taking $\bar{\theta}$, s or any other parameters involved to be functions of frequency. A further directional spreading function which depends on both θ and f was proposed by Cote et al. (1962) in connection with the Stereo-Wave Observation Project (SWOP). This is

$$G(f, \theta) = \begin{cases} \dfrac{1}{\pi} \left[1 + (0.5 + 0.32a) \cos(2\theta) + 0.32a \cos(4\theta)\right] & \text{for } |\theta| < \dfrac{\pi}{2} \\ 0 & \text{otherwise} \end{cases} \qquad (7.133)$$

where

$$a = \exp\left[-\frac{1}{2}\left(\frac{2\pi f U}{g}\right)^4\right]$$

Mitsuyasu et al. (1975) (see also Mitsuyasu and Mizuno 1976) have reported on extensive measurements of directional spectra and proposed that as an idealization Eq. (7.126) may be used with $\bar{\theta} = 0$ for all frequencies and they suggest how s varies with frequency and with fetch. This variation exhibits a maximum in s (a small degree of spreading) at frequencies near the peak frequency.

7.4 ESTIMATION OF SHORT-TERM WAVE STATISTICS

Various procedures may be used to specify design wave conditions and these depend in part upon the applications intended. In a great many cases it is customary to select suitable characteristics of a single wave, say the significant wave, and subsequently to apply wave theory to the characteristics selected. This approach has attained widespread usage in coastal engineering practice: swell is often narrow-banded and the selection of a large representative wave is physically appealing. Furthermore, we shall see that this approach constitutes a useful preliminary to obtaining a wave spectrum. A second representation of design conditions, then, is by resort to the wave spectrum (either one-dimensional or directional) and this may be used to simulate either operational or extreme conditions.

As a further preliminary comment, we should note that there are two complementary methods for estimating design conditions. These involve, firstly, the application of meteorological or wind data, and, secondly, the collection of data provided by wave-recording instruments and relating to water surface elevations. The former, employing past wind records, is termed *wave hindcasting* and is first considered.

7.4.1 Estimates Based on Wind Data

At a location where no information on the wave climate is directly available, the wave characteristics may be estimated by application of available wind data. Advances in understanding the processes of wave generation have been made by Jeffreys (1924), Sverdrup and Munk (1947), Phillips (1957), Miles (1957-62), (see the review by Phillips (1967) for the state of knowledge at this time), Barnett (1968), Ewing (1971), Hasselmann et al. (1976), Sanders (1976) and others. An important effect concerns the nonlinearities inherent in a developing sea state, whereby energy is transferred between different frequency components by means of resonant wave interactions (Hasselmann 1962-1963 and Longuet-Higgins 1969). Reviews of wave generation processes are given also in the texts by Kinsman (1965), Neumann and Pierson (1966) and Phillips (1977). Details of all the various contributions are outside the scope of the present work, and we limit our concern to a simplified consideration of the available methods of using wind data as a basis for predicting overall wave characteristics.

Two kinds of approach have been used to determine wave characteristics from a known wind field. In one of those, termed the "significant wave" method or SMB method, after Sverdrup and Munk (1947) and Bretschneider (1958), the significant wave height H_s and significant wave period T_s are determined directly in terms of representative wind speed U, fetch F and duration t over which the wind acts. The significant wave period T_s is the average period of the highest one-third of the waves in any particular sea state. Bretschneider (1959) extended this approach to the prediction of a wave spectrum, by expressing the spectrum directly in terms H_s and T_s.

In the alternative approach, the wave spectrum is expressed directly in terms of the wind characteristics, and several such spectra were encountered in Section 7.3.3. This is sometimes termed the PNJ method after Pierson, Neumann and James (1955) who adopted it. Where required, a significant height and characteristic period may then be estimated from the spectrum and either method, SMB or PNJ, may therefore be used to obtain equivalent results. Following Sverdrup and Munk (1947) and subsequently Bretschneider (1958), we proceed by investigating the significant wave height H_s and significant wave period T_s which are associated with the pertinent variables already mentioned, U, F and t. The water depth d is omitted since we shall not consider wave generation in shallow water—see Bretschneider (1966) for a review of hindcasting procedures applied to both deep and shallow water. We may express H_s and T_s in functional form as

$$\left.\begin{aligned} H_s &= f_1(U, F, t, g) \\ T_s &= f_2(U, F, t, g) \end{aligned}\right\} \tag{7.134}$$

The wave speed if required may be expressed in terms of T_s as $c_0 = gT_s/2\pi$. A dimensional analysis applied to Eq. (7.134) gives

$$\left.\begin{aligned} gH_s/U^2 &= f_1(gF/U^2, gt/U) \\ gT_s/U &= f_2(gF/U^2, gt/U) \end{aligned}\right\} \tag{7.135}$$

Under given conditions it will be either the fetch or the wind duration that imposes a limit on H_s and T_s and these are termed the *fetch-limited* and *duration-limited* cases respectively. Thus if gt/U is sufficiently large so as not to influence H_s and T_s, but rather it is gF/U^2 that controls them, then we have fetch-limited waves,

$$\left.\begin{aligned} gH_s/U^2 &= f_1(gF/U^2) \\ gT_s/U &= f_2(gF/U^2) \end{aligned}\right\} \tag{7.136}$$

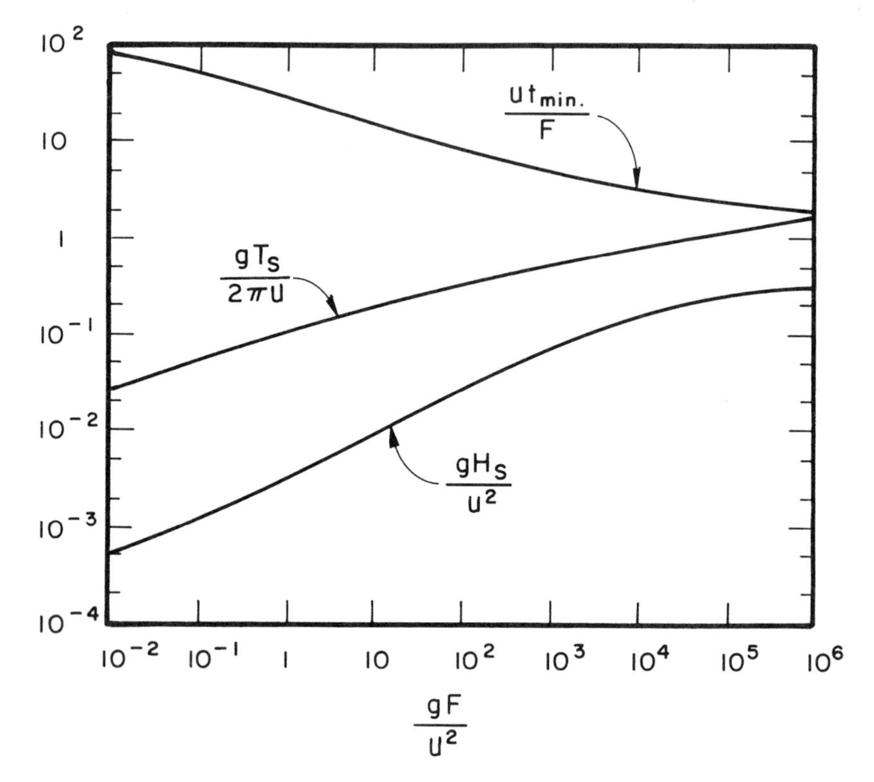

Fig. 7.17. Significant wave height, wave period and minimum duration as functions of fetch plotted in dimensionless form.

And conversely for duration-limited waves,

$$\left.\begin{aligned} gH_s/U^2 &= f_1(gt/U) \\ gT_s/U &= f_2(gt/U) \end{aligned}\right\} \tag{7.137}$$

The condition for the waves to be fetched-limited is that $t > t_{min}$ where t_{min} is the minimum necessary duration and $gt_{min}/U = f(gF/U^2)$. When both the fetch and duration are sufficiently large for H_s and T_s to reach limiting values, these will become dependent only on the wind speed U and the condition of a fully arisen sea or a fully developed sea then exists.

Hindcasting curves used to obtain H_s and T_s have been given by Wiegel (1964) and by Bretschneider (1958) with subsequent revisions (Bretschneider 1970, 1973) and are also reproduced in the U.S. Army Shore Protection Manual (1973). It is emphasized that these curves are based on data which necessarily exhibit considerable scatter. The curves of Bretschneider (1973) are given by the empirical equations

$$gH_s/U^2 = 0.283 \tanh [0.0125 (gF/U^2)^{0.42}] \tag{7.138}$$

$$c_0/U = gT_s/2\pi U = 1.2 \tanh [0.077(gF/U^2)^{0.25}] \tag{7.139}$$

$$gt_{min}/U = 6.5882 \exp \{[0.0161\Lambda^2 - 0.3692\Lambda + 2.2024]^{1/2} + 0.8798\Lambda\} \tag{7.140}$$

where $\Lambda = \ln (gF/U^2)$. These relations are plotted in Fig. 7.17 and are reproduced in a convenient dimensional form in the U.S. Army Shore Protection Manual (1973). When using Fig. 7.17, it is intended that if the duration t occuring exceeds t_{min}, then the waves are fetched-limited and gF/U^2 determines H_s and T_s; on the other hand if the duration t is less than t_{min} for the given gF/U^2, then the waves are duration-limited and a (smaller) fetch corresponding to $t = t_{min}$ determines H_s and T_s. For example, when a wind field is specified as U = 30 mph, F = 40 miles and t = 3 hrs, the hindcasting curves indicate that t_{min} = 6.2 hrs and consequently the waves are duration-limited and are described by H_s = 4.7 ft and T_s = 4.7 sec.

Useful simplifications to the hindcasting relations can be made under certain conditions. Thus in the case of a fully arisen sea Eqs. (7.138) and (7.139) reduce respectively to

$$\left.\begin{aligned} gH_s/U^2 &= 0.283 \\ gT_s/2\pi U &= 1.2 \end{aligned}\right\} \tag{7.141}$$

The fully arisen sea occurs when gF/U^2 is sufficiently large (which is generally difficult to attain in practice). However, the conditions for the predicted wave height to be 90% and 99% of the fully arisen sea wave height are respectively gF/U^2 = 85400 and 346000.

For short fetches and high wind speeds, gF/U^2 is relatively small and approximations to Bretschneider's formulae then give

$$\left.\begin{aligned} gH_s/U^2 &= 0.0035375 (gF/U^2)^{0.42} \\ gT_s/2\pi U &= 0.0924 (gF/U^2)^{0.25} \end{aligned}\right\} \tag{7.142}$$

This simplification is valid when gF/U^2 is less than about 2000. Alternative formulae to those mentioned above have been proposed by various authors and have been reviewed, for example, by Silvester (1974).

Bretschneider (1977) has since modified the significant wave period formula to predict slightly lower periods for high wind speeds than had previously been forecast. The new relation is given in terms of a peak frequency f_0 which corresponds to a maximum in the spectral density curve and it may be expressed in dimensionless form as

$$g/f_0 U = 7.62 \tanh\left\{\left[\frac{1}{2} \ln\left(\frac{1 + 3.546 gH_s/U^2}{1 - 3.546 gH_s/U^2}\right)\right]^{0.6}\right\} \tag{7.143}$$

The significant wave period T_s may be retrieved from this by Bretschneider's proposed relation

$$T_s = \sqrt[4]{4/5}\, f_0^{-1} \tag{7.144}$$

Note that in enclosed regions the limited fetch width restricts the wave generation processes and wave heights will be lower than would otherwise be the case. This effect may be taken into account by estimating an "effective fetch" in the manner described in the U.S. Army Shore Protection Manual (1973); but see also Seymour (1977) for a discussion of restricted fetch widths.

Darbyshire and Draper (1963) have also provided hindcasting curves corresponding to those of Bretschneider, and which were derived from wave records obtained in the Atlantic ocean and neighbouring coastal areas. A noticeable difference was found between the records obtained in the open ocean and those in coastal locations, and consequently two corresponding sets of curves were proposed.

In the case of wave generation by a hurricane, the fetch itself is continuously changing and an alternative graphical procedure proposed by Wilson (1955, 1961) may then be used. This involves the application of families of curves for different wind speeds U, plotted on a graph in which F, H_s, T_s, and t lie along the positive abscissa and ordinate and negative abscissa and ordinate respectively. The procedure can provide the time variation of significant wave height and period from a description of the hurricane wind field.

A second approach to wave hindcasting involves the prediction of a wave spectrum directly in terms of the prescribed wind field, and indeed many of the wave spectra listed in Section 7.3.3 are given in terms of a characteristic wind speed. Most of them refer to fully developed conditions, but some of these (e.g. the Pierson-Moskowitz and JONSWAP spectra) have been adapted to account for limited fetch or duration, or for temporal and spatial variations in the wind field. Significant contributions to such models include those by Pierson, Tick and Baer (1966), Inoue (1967), Barnett (1968), Ewing (1971), Cardone, Pierson and Ward (1976), Hasselmann et al. (1976) and Sanders (1976).

In particular, Cardone, Pierson and Ward (1976) described a procedure for determining the directional wave spectrum from historical meteorological data. A hurricane wind-field model is used to provide a description of the wind field, which is then used to determine directional spectra at successive time steps for an array of points within the region of interest.

7.4.2 Estimates Based on Wave Data

In order to complement the prediction methods based on wind data, and to obtain more direct information on wave characteristics, it is often desirable to install a wave recorder at a site. This will provide a continuous record of the surface elevation $\eta(t)$ and should be used over a reasonably long period (a few years) so as to obtain data representative of the various conditions occuring. In order to avoid the accumulation of too unwieldy a mass of data, waves are recorded intermittently as indicated in Fig. 7.18. The recording interval is the time that lapses between successive starts of the recording instrument and could, for example, be 3 or 6 hours. The recording period is the duration of a single continuous recording and may be say 10 or 20 minutes.

We now consider the question of how best to analyse a wave record in order to expose the corresponding spectrum $S_\eta(f)$ and other quantities that may be of interest. For any individual record the zero-crossing period T_z, the crest period T_c, and the vertical distance from the lowest trough to the highest crest can be measured without first obtaining a spectrum. On the basis of Eqs. (7.79) and (7.80) these values may then be used to estimate respectively the spectral width parameter ϵ and $\sqrt{m_0}$. The two parameters $\sqrt{m_0}$ and T_z are sufficient to determine the wave amplitude $a_{1/n}$. These steps have already been indicated in a previous example. It should be mentioned that the significant wave height H_s is often estimated from $H_s = 4\sqrt{m_0}$ and taken as independent of ϵ.

The above estimates do not require a spectral analysis to be carried out. But the spectrum is often of primary interest in itself, besides being of use in the calculation of additional parameters. The procedure for deriving the frequency spectrum from a given signal is common to many branches of engineering. The

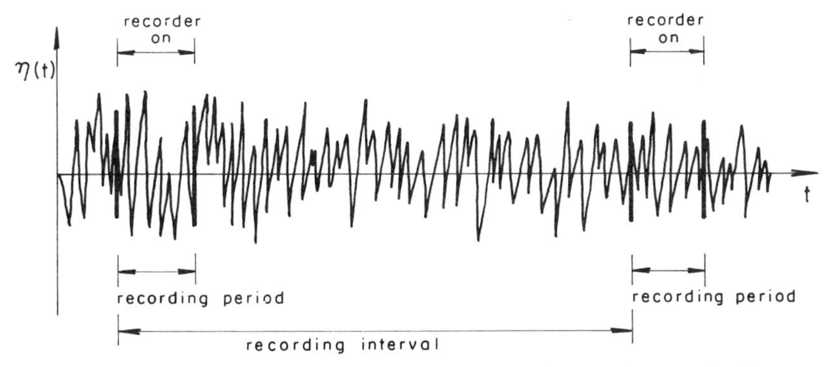

Fig. 7.18. Recording interval and recording period for a random signal $\eta(t)$.

earlier approach was based on first computing the autocorrelation function and then taking the cosine transform of this as indicated in Eq. (7.34). This method is documented in detail by Blackman and Tukey (1959) and in the context of ocean waves by Kinsman (1965).

Since about 1965, however, this approach has been superceded by the Fast Fourier Transform technique which was introduced by Cooley and Tukey (1965) and which uses the Fast Fourier Transform computer algorithm to determine the spectrum directly from the data, a much more efficient procedure than the previous one mentioned above. Bendat and Piersol (1971) provide a detailed discussion of the techniques involved. Reference may also be made to Borgman (1972b), Harris (1972, 1974) and Wilson, Chakrabarti and Snider (1974) among others for discussions on the computational factors involved as applied to ocean waves. Numerous examples of wave spectra derived from measured recordings and the critical evaluation of associated parameters appear in the literature (e.g. Proceedings of the Coastal Engineering Conferences).

If a directional wave spectrum is to be determined, it is clear that indications of the spatial variations of η in the vicinity of the point of interest will be necessary. Various methods have been developed to obtain the necessary information. These are often based on the use of an array of wave probes (e.g. Barber 1961) and have also involved the use of a disc shaped or "cloverleaf" buoy which can record spatial gradients of η (Longuet-Higgins, Cartwright and Smith 1961). Remote measurements may be made by various optical or radio backscatter techniques. A review of various methods of measurement and of procedures for the analysis of directional wave spectra has been presented by Panicker (1974), who also provides an extensive bibliography on the subject. In addition the symposium Ocean Wave Measurement and Analysis (1974) itself also provides useful additional references. More general outlines of the development and uses of directional spectra are given in the texts by Kinsman (1965), Price and Bishop (1974) and Phillips (1977).

Reverting to a consideration of recordings $\eta(t)$ at a single point, it is clearly desirable to obtain wave spectra for each of a large number of recording periods which should ideally extend over several years. Pertinent data may be extracted from these spectra and presented in various ways. Draper (1966) has attempted to unify the forms of presentation and some of the more usual ones are listed here. Corresponding figures are given in Fig. 7.19. In these, the significant wave height H_s and peak period T_p for each recording period are conveniently obtained directly from each corresponding spectrum, i.e. by taking $H_s = 4\sqrt{m_0}$ and by the definition $T_p = 1/f_0$.

Scatter diagram. This shows the number of occurences of various significant wave height ranges corresponding to various peak period T_p or zero-crossing period T_z ranges. An example is given in Fig. 7.19a.

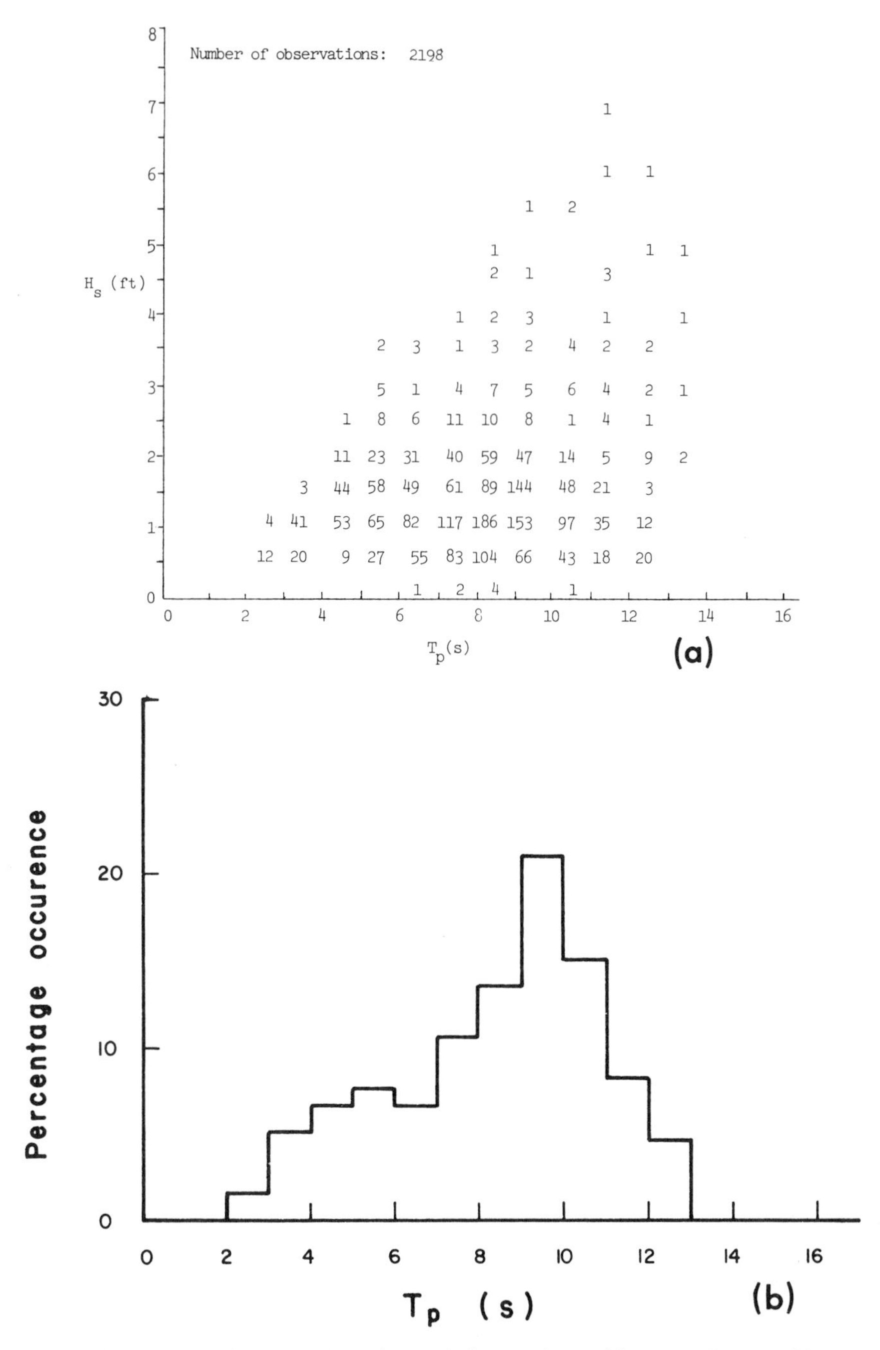

Fig. 7.19. Methods of presentation of recorded wave data. (a) scatter diagram, (b) wave period histogram, (c) wave height exceedence diagram, (d) significant wave height vs. time plot.

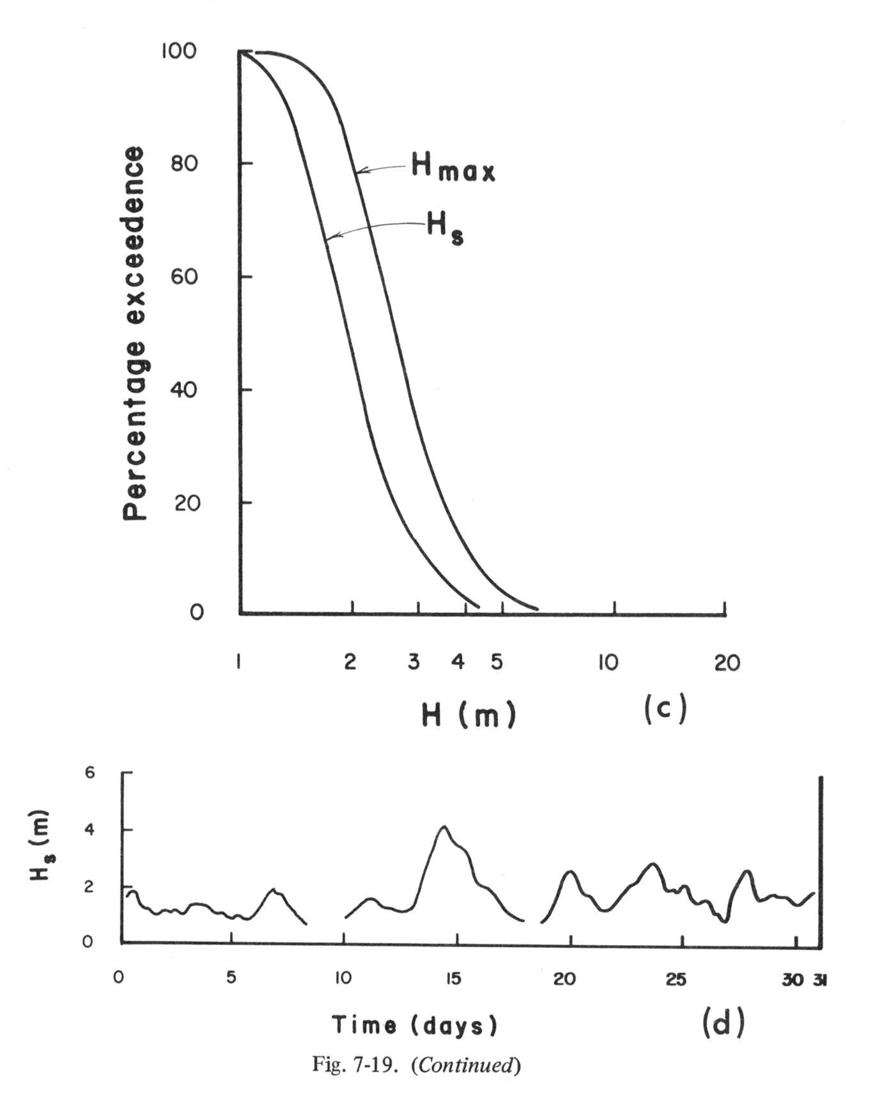

Fig. 7-19. (*Continued*)

Wave period histogram. This shows the percentage of records for which the peak (or zero-crossing) period falls within each of the ranges indicated (Fig. 7.19b).

Wave height exceedence diagram. This shows the percentage of wave records for which the significant wave height exceeds a range of different values. The corresponding curve for the maximum wave height, (i.e. the most probable or expected maximum wave height occuring over a recording period or recording interval on the basis of Longuet-Higgins' (1952) results) is often superposed on this plot. An example is given in Fig. 7.19c.

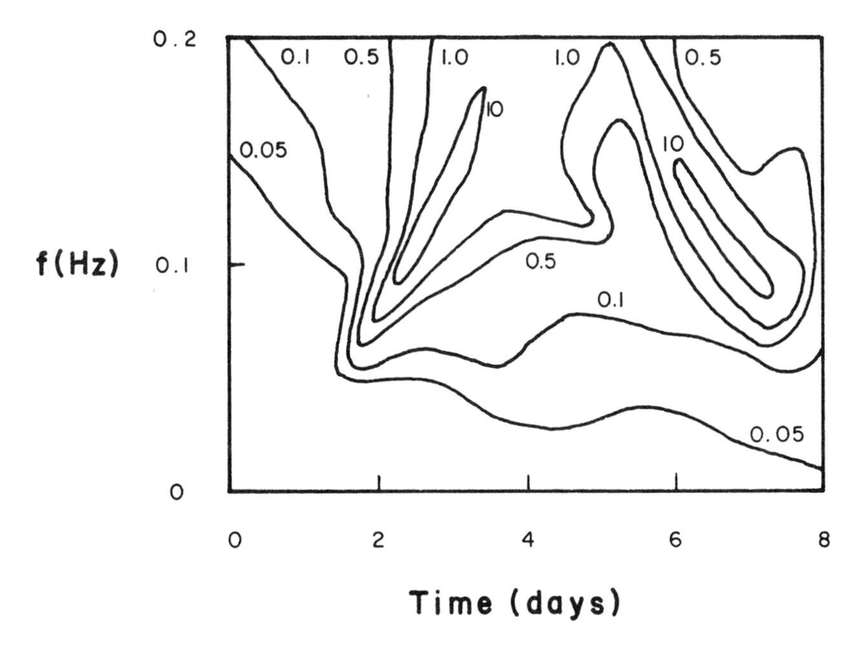

Fig. 7.20. Wave spectra shown as a function of time as presented by Draper (1976).

Significant wave height vs. time plot. This is a continuous trace obtained by linear interpolation between adjacent recording periods. Gaps in the plot might arise due to missing or faulty records. An example is given in Fig. 7.19d. Such data may in turn be used to prepare 'Cumulative Persistence of Storms' or 'Persistence of Calms' diagrams. These indicate respectively the number and duration of occasions when wave heights are above or below a threshold value.

Draper (1976) has also proposed a convenient form for presenting wave spectrum data which involves plotting contours of spectral density on a graph with time in days and wave frequency forming the two axes. This approach is illustrated in Fig. 7.20 and conveniently shows the growth of a wave field (contours aligned towards the upper left) or the arrival of swell from a different generating area (contours aligned towards the upper right). Notice the relationship between this kind of presentation and that used in Fig. 7.15.

7.5 ESTIMATION OF EXTREME WAVES

An important step often encountered in design is the estimation of an extreme design wave on the basis of recorded or hindcast wave data. This generally involves selecting and fitting a suitable probability distribution to wave height data, and extrapolating this to locate a suitable design wave, such as the so-called "50-year wave". This is a characteristically large wave height that might be expected with a certain small probability during the lifetime of the structure.

Analogies to this kind of procedure lie in the prediction of extreme winds, earthquakes and floods.

It is emphasized that the pertinent time scale is now very long such that the assumption of a continuously stationary variable must be abandoned. Since the various statistical results obtained so far have in fact been based upon that assumption, they can be expected to apply only to relatively short durations such as occur during the course of a (possibly extreme) storm rather than to all time leading up to an extreme event (storm).

The selection of a design wave from a series of wave records is usually carried out in the following stages:

(i) Data consisting of wave heights and periods are collected over a long time (say a few years) at the site of interest. Alternatively, a hindcasting technique may be used to provide wave height data over a much longer time span (say 50 years or more).
(ii) A plotting formula is used to reduce the data to a set of points describing the probability distribution of wave heights.
(iii) These points are plotted on an extreme value probability paper corresponding to a chosen probability distribution function.
(iv) A straight line is fitted, often by eye, through the points to represent a trend.

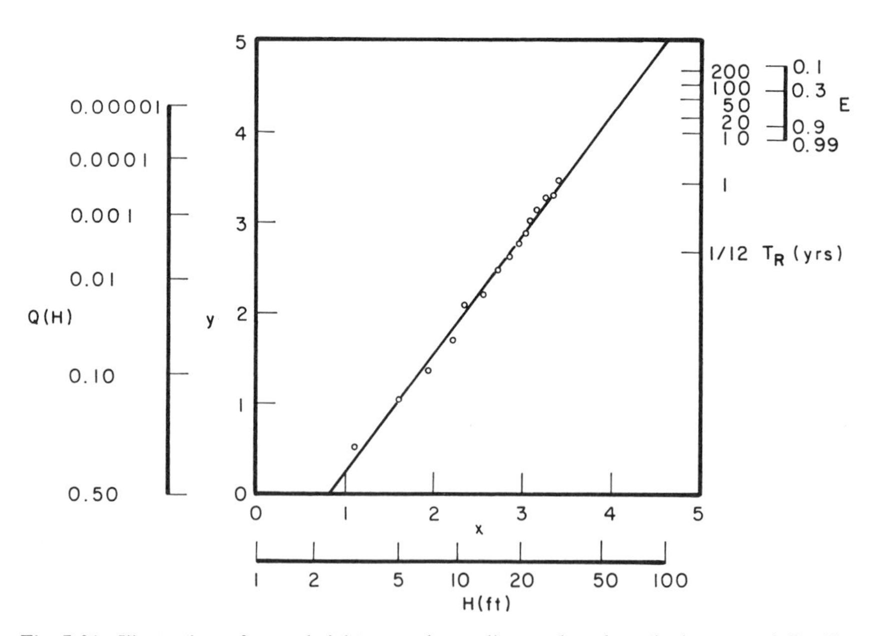

Fig. 7.21. Illustration of wave height exceedence diagram based on the log-normal distribution.

(v) The line is then extrapolated to locate a design value corresponding to a chosen return period T_R, or a chosen encounter probability E.

A wave height exceedence diagram indicating this procedure is sketched in Fig. 7.21. These steps are now described in some detail, since a reliable prediction of extreme conditions is of considerable importance in offshore design.

7.5.1 Collection of Data Sample

Various methods have been used to assemble a sample of wave statistics to which the extrapolation procedure is to be applied. The individual data points may be defined in quite different ways, such as the significant wave height or expected or most probable maximum wave height during a recording interval, individual wave heights, the most probable maximum wave height in an individual storm, and so on. Alternative choices of the data points have been reviewed and compared by Nolte (1973).

In the analogous procedure of estimating extreme wind speeds, available data generally extends over a period of many years (say 20) and each data point (annual maximum hourly speed) is itself an extreme value. Such an approach is occasionally possible also for extreme wave prediction, but generally wave records for a chosen site will not generally extend over more than a very few years.

A common method of collecting wave data consists of recording the free-surface elevation intermittently as described in Section 7.4.2. Any data point or statistic of interest associated with each recording interval may then be derived and assembled, for example as in a scatter diagram.

As an alternative procedure, such data may be used to obtain the long-term probability distribution of *individual* wave heights $P_L(H)$. This distribution $P_L(H)$ may be developed in terms of the long-term distribution of sea states, characterized by H_s and T (as in a scatter diagram), and the Rayleigh distribution $R(H|H_s, T)$ which describes the short-term distribution of individual wave heights H corresponding to a particular H_s and T (sea state). Thus the Rayleigh distribution of individual wave heights is weighted by the probability of occurence of each sea state and by the variability of the numbers of waves in each sea state. Battjes (1970) derives an expression for $P_L(H)$ to be

$$P_L(H) = \frac{1}{\overline{T^{-1}}} \int_0^\infty \int_0^\infty \frac{1}{T} R(H|H_s, T)\, p(H_s, T)\, dH_s\, dT \qquad (7.145)$$

where $p(H_s, T)$ is the bivariate probability density of H_s and T. $\overline{T^{-1}}$ is the long-term average number of waves per unit time and is defined as

$$\overline{T^{-1}} = \int_0^\infty \int_0^\infty \frac{1}{T} P(H_s, T)\, dH_s\, dT \tag{7.146}$$

Instead of direct wave measurement, the wave height data may be obtained by hindcasting from meteorological data. Such an alternative is generally used for hurricane generated waves (e.g. Cardone, Pierson and Ward 1976). A statistic that is generally used in this context is the most probable maximum height occuring in an individual storm greater than a certain intensity (e.g. Petrauskas and Aagaard 1971 and Jahns and Wheeler 1973). Nolte (1973) has illustrated how unreliable a design wave estimate can be when based on a single year's wave records, and emphasizes the importance of an adequate data base for reliable predictions to be made. The alternative procedure involving hindcasting has the considerable advantage that data deriving from a relatively long time span may be reconstructed.

7.5.2 Plotting Formulae

In order to plot the data one must first assign a value of P(H) to each value in the sample. To do this, the data is ordered according to height and the suffix m is used to denote its rank, with $m = 1$ corresponding to the largest value, and $m = N$ to the smallest value in a sample containing N wave heights.

A simple estimate of the exceedence $Q(H) = 1 - P(H)$ for each of the N heights is then given as

$$Q(H_m) = 1 - P(H_m) = \frac{m}{N + 1} \tag{7.147}$$

In those cases where N is large (such as when a scatter diagram or an individual wave height distribution is used), Q(H) may be calculated for fewer heights selected to be the larger heights in the sample, or selected at convenient intervals (e.g. as pertaining to a scatter diagram). In this case the formula Eq. (7.147) still applies, but takes the following form

$$Q(H) = 1 - P(H) = \frac{\text{Number of height values} \geqslant H}{N + 1} \tag{7.148}$$

This plotting formula, Eq. (7.147), is not unique, and in fact has been demonstrated (e.g. Kimball 1960, Gringorten 1963a) to introduce a bias peculiar to the distribution being estimated. An unbiased formula depends on the particular distribution being used, and also on the specific parameters involved. A more general plotting formula may be written in the form

$$Q(H_m) = 1 - P(H_m) = \frac{m - A}{N + B} \tag{7.149}$$

where A and B are to be determined for the particular distribution used. (See Gringorten (1963a) and Petrauskas and Aagaard (1971) respectively for the Type I and III_L distributions which will shortly be described.) Even so, most authors have adopted Eq. (7.147) as a plotting formula for reducing the data, because of its simplicity, but in spite of it resulting in a possible departure from any "true" underlying distribution (Gringorten 1963a).

7.5.3 Extreme Value Probability Distributions

The process of extrapolation plays a fundamental role in this area of analysis and it is essential therefore to fit empirically a convenient probability distribution which describes the available data as closely as possible.

Several probability distributions have been used or proposed to describe extreme wave statistics. These include the log-normal distribution, and the Extremal Types I, II and III (both lower and upper bound) probability distributions. Although these all have a theoretical base, they are used here essentially as an empirical fit to the data. In describing the distributions it is convenient to adopt the following notation for the parameters used to define any specific distribution: α is a *shape parameter* which determines the basic shape of a particular distribution; θ is a *scale parameter* which controls the degree of spread along the abscissa (variate axis); and ϵ is a *location parameter* which locates the position of the density function along the abscissa. In the special case of the Type III distribution to be described, ϵ locates one end of the density function. Table 7.3 gives the expressions for P(H) defining the different distributions considered here, and also includes expressions for their means and variances. Sketches of each of the corresponding probability densities, plotted with linear scales, are shown in Fig. 7.22.

It is common practice to plot any set of measured data such that the selected distribution lies on a straight line, since this aids visualization of the extrapolation procedure, and permits eye fitting where appropriate. Thus a particular probability paper is first selected, or equivalently corresponding scales are constructed, in order to meet this requirement. The linear ordinate scale y is related to the cumulative probability P( ), and the linear abscissa scale x is related to the variate H according to the relationships given in Table 7.4. There now exists a linear relationship $y = ax + b$, with slope a and intercept b which are given in terms of the parameters α, θ and ϵ as indicated in the table. In the light of these comments, each distribution is now briefly mentioned in turn.

Table 7.3 Selected Extreme Value Probability Distributions.

Distribution	Range	Cumulative Probability P(H)	Mean	Variance
Lognormal	$0 < H < \infty$ $-\infty < \theta < \infty$ $0 < \alpha < \infty$	$\frac{1}{\sqrt{2\pi}} \int_0^H \frac{1}{\alpha h} \exp\left[-\frac{1}{2}\left(\frac{\ln(h) - \theta}{\alpha}\right)^2\right] dh$	$\exp\left(\theta + \frac{\alpha^2}{2}\right)$	$\exp(2\theta + \alpha^2)\,[\exp(\alpha^2) - 1]$
Type I	$-\infty < H < \infty$ $-\infty < \epsilon < \infty$ $0 < \theta < \infty$	$\exp\left\{-\exp\left[-\left(\frac{H - \epsilon}{\theta}\right)\right]\right\}$	$\epsilon + \gamma\theta$ $(\simeq \epsilon + 0.58\theta)$	$\frac{\pi^2}{6}\theta^2$ $(\simeq 1.64\theta^2)$
Type II	$0 < H < \infty$ $0 < \theta < \infty$ $0 < \alpha < \infty$	$\exp\left[-\left(\frac{H}{\theta}\right)^{-\alpha}\right]$	$\theta\,\Gamma\left(1 - \frac{1}{\alpha}\right)$	$\theta^2\left[\Gamma\left(1 - \frac{2}{\alpha}\right) - \Gamma^2\left(1 - \frac{1}{\alpha}\right)\right]$
Type III_L (Lower Bound)	$\epsilon < H < \infty$ $0 < \theta < \infty$ $0 < \alpha < \infty$	$1 - \exp\left[-\left(\frac{H - \epsilon}{\theta}\right)^{\alpha}\right]$	$\epsilon + \theta\,\Gamma\left(1 + \frac{1}{\alpha}\right)$	$\theta^2\left[\Gamma\left(1 + \frac{2}{\alpha}\right) - \Gamma^2\left(1 + \frac{1}{\alpha}\right)\right]$
Type III_U (Upper Bound)	$-\infty < H < \epsilon$ $0 < \theta < \infty$ $0 < \alpha < \infty$	$\exp\left[-\left(\frac{\epsilon - H}{\theta}\right)^{\alpha}\right]$	$\epsilon - \theta\,\Gamma\left(1 + \frac{1}{\alpha}\right)$	$\theta^2\left[\Gamma\left(1 + \frac{2}{\alpha}\right) - \Gamma^2\left(1 + \frac{1}{\alpha}\right)\right]$

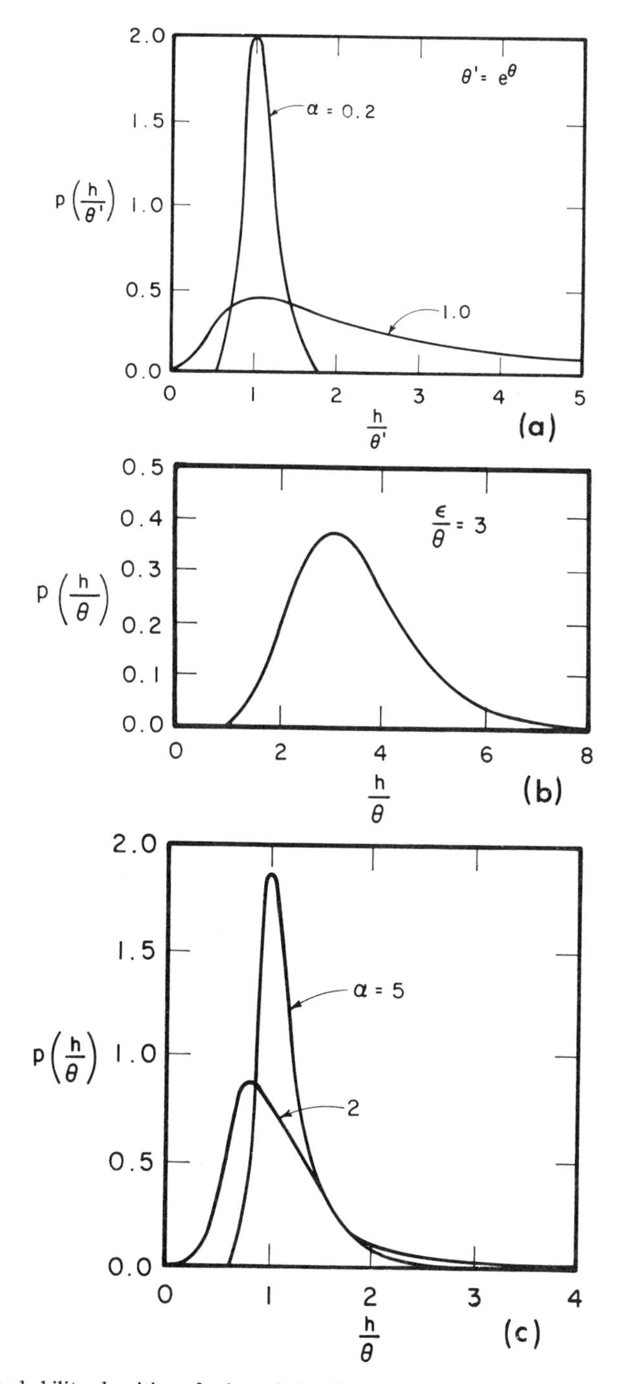

Fig. 7.22. Probability densities of selected distributions plotted with linear scales. (a) log-normal, (b) Type I, (c) Type II, (d) Type III_L, (e) Type III_U.

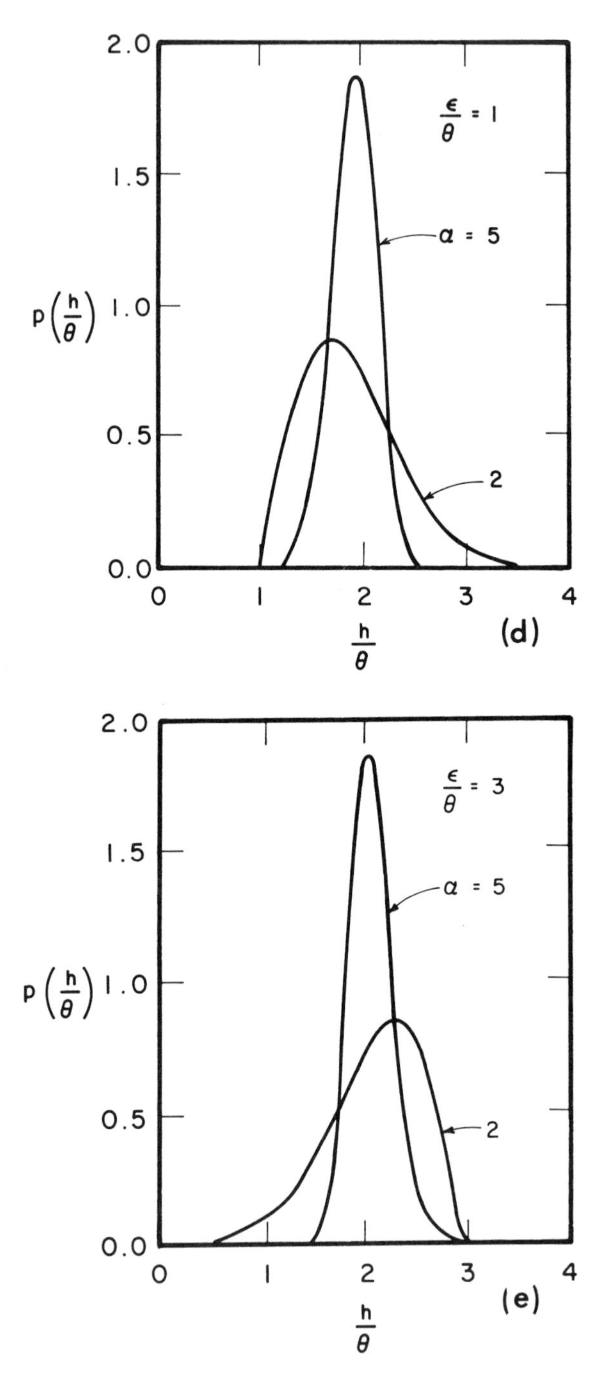

Fig. 7-22. *(Continued)*

Table 7.4 Scale Relationships of Selected Extreme Value Probability Distributions.

Distribution	Abscissa Scale x	Ordinate Scale y	Slope a	Intercept b
Lognormal	$\ln(H)$	$P(H) = \frac{1}{\sqrt{2\pi}} \int_0^y e^{-t^2/2}\, dt$	$1/\alpha$	$-\theta/\alpha$
Type I	H	$-\ln[-\ln(P(H))]$	$1/\theta$	$-\epsilon/\theta$
Type II	$\ln(H)$	$-\ln[-\ln(P(H))]$	α	$-\alpha \ln \theta$
Type III_L	$\ln(H - \epsilon)$	$\ln[-\ln(Q(H))]$	α	$-\alpha \ln \theta$
(Lower Bound)	H	$[-\ln(Q(H))]^{1/\alpha}$	$1/\theta$	$-\epsilon/\theta$
Type III_U	$-\ln(\epsilon - H)$	$-\ln[-\ln(P(H))]$	α	$\alpha \ln \theta$
(Upper Bound)	H	$-[-\ln(P(H))]^{1/\alpha}$	$1/\theta$	$-\epsilon/\theta$

Log-Normal Distribution

The log-normal distribution, corresponding to ln (H) possessing a normal distribution, was the first distribution to be fitted to long term wave height data (Jasper 1956) and has commonly been used in extreme wave prediction following the work of Draper (1963). (See, for example, Draper 1966, Draper and Driver 1971, and Dattari 1973.) The corresponding log-normal probability paper is readily available, and is easy to construct since the ordinate scale, relating y to P as given in Table 7.4, is listed in probability tables (e.g. Abramowitz and Stegun 1965).

Extremal Type I Distribution

(Also termed the Gumbel or the Fisher-Tippett I distribution.)

Gumbel (1958) developed this distribution to a considerable degree in the context of flood prediction. It is also frequently used in extreme wind prediction, and has been discussed in the context of extreme wave prediction by St. Denis (1969, 1973). The corresponding extremal ('Gumbel') probability paper is easily constructed, with the ordinate given simply as $y = -\ln(-\ln(P))$ and the abscissa scale given directly as H.

Extremal Type II Distribution

(Also termed the Fretchet or the Fisher-Tippett II distribution.)

The Type II distribution is used in the prediction of extreme wind speeds, following the work of Thom (1954). It's application to extreme wave prediction, deriving from the use of wave hindcast formulae to wind speed distributions has subsequently been proposed (Thom 1971, 1973b, and reviewed by Yang, Tayfun and Fallah 1975), and will be discussed shortly. The distribution corresponds to ln (H) possessing a Type I distribution, and thus plots as a straight line on log-extremal paper (sometimes termed 'Weibull' paper), in which the ordinate is scaled as with extremal paper, $y = -\ln(-\ln(P))$, while ln (H) in place of H is plotted linearly along the abscissa.

Extremal Type III Distribution

(Also termed the Weibull distribution.)

There are two alternative forms which have been adopted or proposed for extreme wave prediction. These are lower-bound and upper-bound distributions, denoted here by Types III_L and III_U, which depend on whether the location parameter ϵ is used to describe a lower-bound (Type III_L) or an upper-bound (Type III_U) to the wave heights. Thus in either case ϵ represents a limiting value of the variate H beyond which no occurence is possible.

In the Type III_L distribution, ϵ may be considered to correspond to a small wave height representing low level background wave activity which is always present, or a lower bound to the wave heights included in the data sample. For particular parameter values, this distribution reduces to the Rayleigh distribution (when $\alpha = 2$, $\epsilon = 0$) or to the Exponential distribution (when $\alpha = 1$). The Type III_L distribution has been used quite successfully for extreme wave prediction (e.g. Battjes 1970 and Petrauskas and Aagaard 1971).

The Type III_U distribution has not generally been adopted for extreme wave prediction, although its use has been advocated by St. Denis (1973) and by Borgman (1975). The upper height limit entailed in the Type III_U distribution makes it particularly attractive, for example, at locations where height may be limited by water depth, fetch, or other such feature, or in any case where a finite maximum wave height corresponding to $P(H) = 1$ is preferred.

It will be seen that the Type III distributions depend on three parameters, rather than two as with the other distributions considered, and so provide added flexibility in attempting to fit data. Against this is the fact that the theoretical variation on log-extremal paper does not fall on a straight line. In order to obtain a straight line plot it becomes necessary to adopt scales which depend on one or other of the parameters to be estimated, as indicated in Table 7.4. Scales providing a straight line plot thus depend on making an estimate of α or ϵ before plotting the data. However, as an alternative and simpler approach, chosen values of ϵ (Battjes 1970) or α (Petrauskas and Aagaard 1971), rather than best estimates of these parameters, may be adopted. This simplification enables data to be plotted directly, and the Type III_L distribution has generally been used for wave prediction in this way.

It is seen from Table 7.3 that both Types I and III_U distributions admit negative values of H, and also that the absence of an upper limit of H in the log-normal, Types I, II and III_L distributions indicates that any value of H, however large, may be encountered provided that P(H) is sufficiently close to unity. Both these objections introduce no practical difficulty, and are somewhat academic since a good empirical fit to measured data is all that is required.

7.5.4 Methods of Parameter Estimation

Having a selected one distribution as a likely model, it remains to estimate the parameter values which will provide the best empirical fit between the distribution and the data.

The most straightforward approach is to plot the individual data points on the selected probability paper and then draw a straight line through these by eye. (In the case of the Type III distributions, this would apply only if α or ϵ were chosen in advance.) Alternatively, the best-fit line may be derived by three possible methods:

(a) the method of moments,
(b) the method of least squares,
(c) the method of maximum likelihood.

The method of moments operates by equating the first two or three moments of the distribution to those of the data, and thereby relationships may be established between the estimated parameters and the sample mean, variance and skewness. (Table 7.3 gives expressions for the distribution means and variances in terms of the distribution parameters).

The least squares method in its most basic form is directly applicable to the log-normal, Type I and Type II distributions, since the scales are known *a priori* and only two parameters are needed in each case. The method gives the slope a and intercept b of the best-fit line $y = ax + b$ in terms of the coordinates (x_m, y_m) of all the data points. The corresponding estimated values of the parameters of the distribution, if required, may then be obtained from the slope and intercept by the expressions given in Table 7.4.

In the case of the Type III distributions, either the ordinate or abscissa is dependent upon one of the parameters to be estimated, and thus without prior knowledge of this parameter the data cannot even be plotted unless ϵ or α is chosen to take some prescribed or trial value. In any case, the least-squares procedure can readily be extended to provide estimates of the three unknowns and involves an iteration procedure which entails no serious difficulty.

The method of maximum likelihood attempts to provide estimated parameters which would give the data sample the highest probability of being observed in its particular form. The method results in estimated parameters which are unbiased and have a relatively small variance. However, the solution usually requires lengthy iterative manipulation and the method is not generally used in the present context.

7.5.5 Confidence Intervals

Once a distribution has been fitted to a set of data by one or other method described, it becomes desirable to appraise the closeness of fit of the data points to the fitted distribution. The scatter of the data may best be described in terms of confidence limits on either side of the fitted line. Thus, curves drawn on either side of the best-fit line provide a series of confidence bands which indicate the confidence attached to any particular data point. This is sketched in Fig. 7.23, which shows, for example, the 50% and 90% confidence bands, within which data is expected to lie with probabilities of 50% and 90% respectively.

A procedure for constructing confidence bands was developed by Gumbel (1958), and has been described by St. Denis (1969) in the context of wave prediction. The method is simple to use, but is based on an approximation which is really only suitable for data near the center of the ordered sample, and becomes invalid for the highest points in the sample (which is of greater concern in the present case). Furthermore, the method described applies only to the Type I distribution.

The more complete derivation of confidence bands, as applied to any chosen distribution, is described by Borgman (1961) and Gringorten (1963b). For a

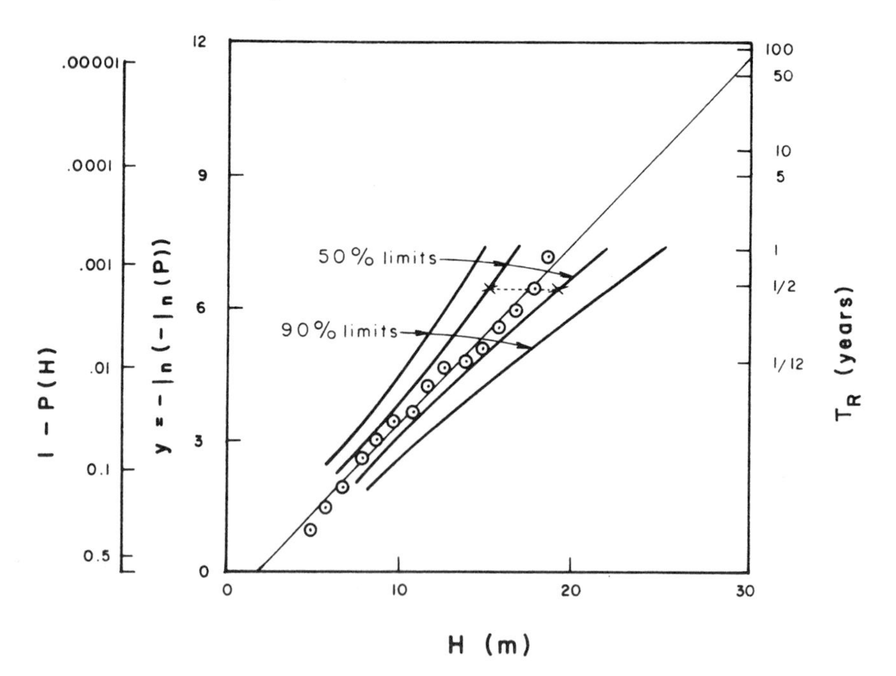

Fig. 7.23. Wave height exceedence diagram illustrating the use of confidence bands. x, example of a pair of calculated height limits used to derive confidence bands.

given sample size N, rank m and chosen confidence probability level, the formulation provides a pair of height limits for the m-th statistic. These values may then be plotted on either side of the m-th data point, and a faired line drawn through equivalent limits for the remaining points. This procedure is indicated in Fig. 7.23.

An alternative approach has been used by Petrauskas and Aagaard (1971) and involves Monte Carlo simulation to generate random sets of data deriving from the best-fit distribution that has been obtained. The spread of this simulated data can then be used to describe the confidence or uncertainty attached to any chosen value. The major advantage of this approach is that it can be used for any values of height, including predicted values, and thus enables confidence limits to be attached to any chosen design wave height.

7.5.6 Design Wave Selection

The particular probability distribution that has been fitted to the wave height data may now be used to select some design wave (or design sea state). This may be chosen on the basis of a *return period* T_R or *encounter probability* E, and furthermore the wave period associated with the chosen design wave height still needs to be established.

The return period, or recurrence interval, T_R, is the average time interval between successive events of the design wave being equalled or exceeded, and is directly related to the probability exceedence Q(H) as

$$\frac{T_R}{r} = \frac{1}{Q(H)} = \frac{1}{1 - P(H)} \tag{7.150}$$

Here r is the recording interval associated with each data point. (Thus in the case where each data point corresponds to an individual storm, r would be the average time interval between such storms.) From this formula, return periods may be scaled along the ordinate as indicated in Fig. 7.21. Thus a prescribed return period has an associated value of P(H), and the corresponding wave height may then be determined from the (extrapolated) best-fit line that has been plotted. This design value will be defined in the same way as were the individual data points (e.g. the significant height over a recording interval, or the most probable maximum height in an individual storm), and should be equalled or exceeded on the average once over the duration of any return period. Other characteristic heights, such as the most probable maximum height, or the height corresponding to a chosen probability value, may then be obtained in terms of the height obtained by the use of statistical formulae applicable to short-term (stationary) variations.

Nolte (1973) has indicated that in many cases a correction to the return period is necessary, because its intended meaning is distorted due to the occurrence of more than one data point in a single storm; that is, the occurrence of large valued data points will tend to group together, and the average duration between successive storms containing a design wave, which is really the intended requirement, will thus be larger than the true return period. The corresponding "grouping correction factor" is relatively important when an individual wave model is used (since more data points will tend to occur in groups), and is unnecessary when a storm model is used.

In many cases design is carried out with wave heights corresponding to some prescribed return period, e.g. 50 or 100 years. However, it is appropriate in the context of offshore design to consider also the encounter probability E. This is the probability that the design wave is equalled or exceeded during a prescribed period L, say the design lifetime of a structure; and indeed it may be preferable instead to select T_R to correspond to prescribed values of E and L. The relationship between these quantities (Borgman 1963) is

$$E = 1 - (1 - r/T_R)^{L/r} \tag{7.151}$$

When $T^2/Lr \gg 1$ (i.e. generally when $T/r \gg 1$), a suitable approximation to E which is independent of the recording interval is

$$E \simeq 1 - \exp(-L/T_R) \tag{7.152}$$

Thus, for example, the return period giving rise to an encounter probability of 0.1 for a design lifetime L = 50 years is 475 years; and again, the probabilities of a 50 year design wave being exceeded within 10, 50 and 100 years are 0.181, 0.632 and 0.865 respectively.

If the design lifetime of the structure is prescribed, the encounter probability E may be scaled along the ordinate alongside, or in place of, T_R, as sketched in Fig. 7.21.

There are various approaches to estimating the extreme wave period associated with the design wave height which has been obtained. The first is to repeat the entire procedure using wave periods instead of wave heights as statistics. By using the same return period as for wave heights, one may obtain a predicted value of the corresponding characteristic wave period (e.g. zero-crossing or peak period). Draper (1963) has suggested that this value of period may be used with the predicted height.

An alternative approach involves using the predicted wave height to set a lower limit to the wave period (due to wave breaking), and then using a series of values of wave period above this lower limit to find the worst possible effect on the structure. Battjes (1970) has indicated that limitations on wave steepness due to breaking imply that periods should lie approximately within the range

$2\pi H_s/gT^2 \leqslant \frac{1}{16}$. Thus a lower wave period limit T_L for a given significant wave height may be set as

$$T_L = \left(\frac{32\pi H_s}{g}\right)^{1/2} \tag{7.153}$$

Research has recently been directed (e.g. Houmb and Overvik 1976) towards fitting a bivariate probability distribution to long-term data of both wave heights and periods. This should account for the wave period limits due to wave breaking, and should eventually yield different combinations of height and period corresponding to known probabilities of exceedence.

The expected value of the maximum height occuring in relation to say the root-mean-square or significant wave height depends on the duration of the extreme storm considered and is given approximately by Eq. (7.71) when the number of waves N is reasonably large. Draper (1972) has emphasized, however, that the storm duration specified has less effect upon the prediction of extreme values than might, at first sight, appear to be the case. This is because a short duration leads to a relatively high root-mean-square or significant height, but the ratio of the expected maximum height to this will be smaller than would otherwise be the case. As an example of this effect, Draper (1972) has considered a 12-hour storm duration which is found to give higher values as deduced by his method than do other durations. In this way, Draper (1972) has provided 50-year design wave conditions for the North Sea.

Thom (1971, 1973b) has proposed a method for transforming data deriving from extreme wind distributions into extreme significant wave height distributions, and thus in turn into estimates of the extreme wave height and period corresponding to a given return period. Extratropical and tropical storms are distinguished and a mixed Frechet distribution is used to describe the wind velocity V as follows

$$P(V) = (1 - m)\exp\left[-\left(\frac{V}{\theta_{EV}}\right)^{-\alpha_{EV}}\right] + m\exp\left[-\left(\frac{V}{\theta_{TV}}\right)^{-\alpha_{TV}}\right] \tag{7.154}$$

Here the subscripts E and T refer to extratropical and tropical conditions respectively and m is a tropical cyclone mixing parameter which indicates the relative frequency of tropical cyclones: thus m = 0 and m = 1 correspond to the occurence of only extratropical and only tropical storms respectively, and in either case a regular Frechet distribution is then applicable. In a separate paper, Thom (1973a) indicated a method by which the mixing parameter m could conveniently be estimated. Finally, θ and α above are respectively the scale and shape parameters of each component Frechet distribution. Thom (1973b) assembled wind data and indicated how these wind distribution parameters can be estimated.

In extending these results to extreme wave prediction, Thom assumes that a mixed Frechet distribution is applicable also to the significant wave height (subscript H). This assumed distribution therefore requires the determination of two scale parameters, θ_{EH} and θ_{TH}, and two shape parameters, α_{EH} and α_{TH}, in terms of those parameters relating to the wind distribution, and Thom has proposed corresponding formulae.

Once the mixing parameter m and the wave distribution parameters have been obtained, the distribution of significant wave heights is fully defined and some extreme height can then be found in the usual way.

For the extreme sea state corrresponding to this significant height, H, Thom suggests that the maximum height occurring H_m and the corresponding period $T(H_m)$ can be based on Wiegel's (1949, 1964) formulae

$$\left.\begin{aligned} H_m &= 1.8H \\ T(H_m) &= T(H) = 15.6(H/g)^{1/2} \end{aligned}\right\} \tag{7.155}$$

Yang, Tayfun and Hsiao (1974) have subsequently proposed modifications to some of the steps involved for the case of tropical storms. The whole procedure is also summarized briefly by Yang, Tayfun and Fallah (1975).

7.6 RANDOM WAVE FORCES

Discussion has so far centered on the statistical nature of the waves themselves and we turn now to consider the consequential loading on a structural element. This is approached on the assumption that the Morison equation is continuously valid and we shall therefore initially require a statistical description of the fluid velocities and accelerations.

It is assumed as before that the free surface elevation has a Gaussian probability distribution and that the results of linear wave theory are valid. It may then be shown (Borgman 1967a) that the horizontal particle velocity u and acceleration a are statistically independent and that each possesses a Gaussian probability distribution. Their spectral densities may readily be expressed in terms of the wave spectrum: The horizontal velocity u and acceleration a of a component wave train are given in complex form by linear theory as follows

$$u(t) = H_u(f)\,\eta(t) \tag{7.156}$$

$$a(t) = H_a(f)\,\eta(t) \tag{7.157}$$

where

$$H_u(f) = 2\pi f \frac{\cosh(k(z+d))}{\sinh(kd)} \tag{7.158}$$

$$H_a(f) = -i4\pi^2 f^2 \frac{\cosh(k(z+d))}{\sinh(kd)} \tag{7.159}$$

$$\eta(t) = \frac{H}{2} \exp(-i2\pi ft) \tag{7.160}$$

H_u and H_a are complex receptances and the wave number k itself influences their frequency dependence through the linear dispersion relationship. Equations (7.156) and (7.157) are analogous to Eq. (7.44) relating the input and output of a linear system, and consequently the corresponding velocity and acceleration spectra are given as

$$S_u(f) = |H_u(f)|^2 S_\eta(f) \tag{7.161}$$

$$S_a(f) = |H_a(f)|^2 S_\eta(f) \tag{7.162}$$

An illustration of the horizontal velocity and acceleration spectra at a particular point which derive from a particular wave spectrum is given in Fig. 7.24. Corresponding formulae can likewise be developed for vertical velocity and acceleration spectra. It is emphasized, once more, that the above relationships are based on the assumption that linear wave theory is valid.

Now that the probabilistic and spectral properties of u and a are defined, the Morison equation may be used to obtain the statistical properties of the force on a pile. The force per unit length, denoted here as F, may be written as

$$F(t) = K_d u(t)|u(t)| + K_i a(t) \tag{7.163}$$

where K_d and K_i are assumed constants and are given as

$$K_d = \tfrac{1}{2}\rho D C_d,\ K_i = \rho(\pi D^2/4)C_m \tag{7.164}$$

Here ρ is the fluid density, C_d and C_m are the drag and inertia coefficients repectively and D is the cylinder diameter. (For a noncircular section D may be taken as a suitable characteristic length.) The assumption that the coefficients C_d and C_m are constants is admittedly questionable but is invariably made for purposes of developing a practical approach.

There are two aspects to describing the wave force. One concerns the probabilistic properties of the force and force maxima analogous to those obtained for

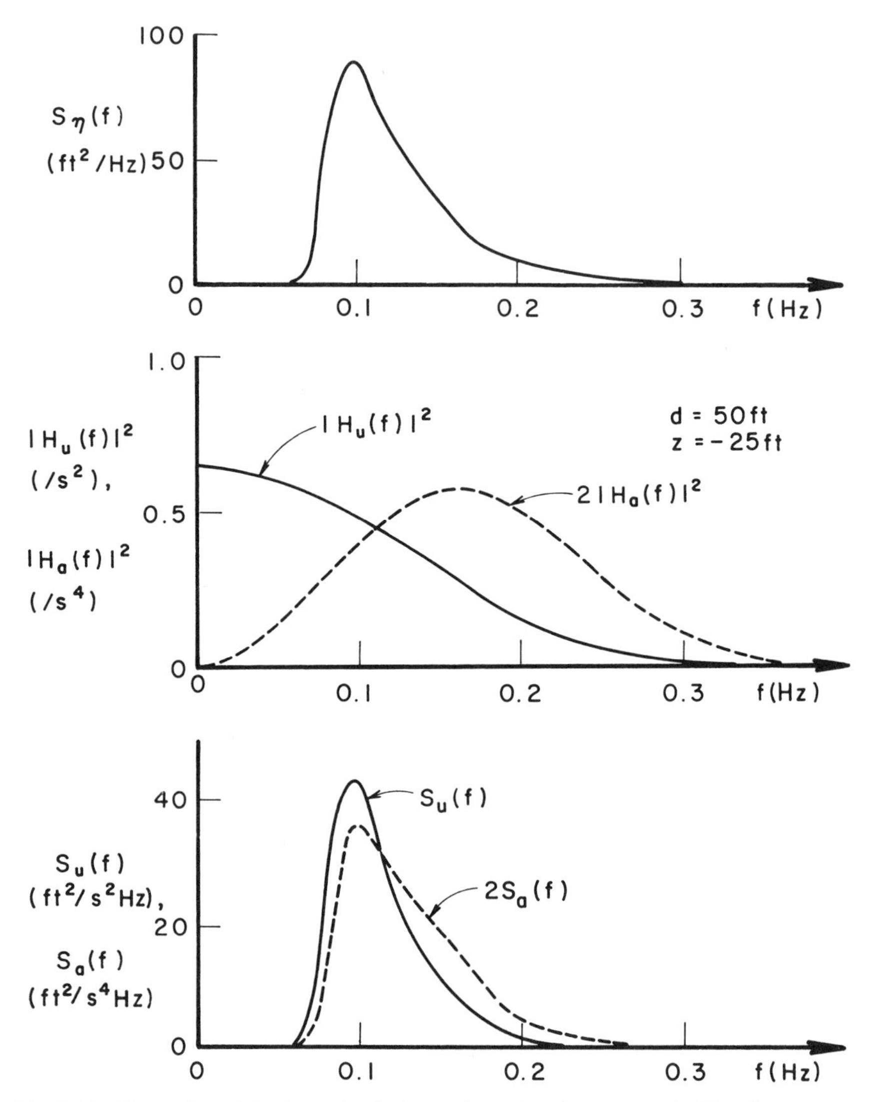

Fig. 7.24. Illustration of horizontal velocity and acceleration spectra deriving from a prescribed wave spectrum.

the free surface elevation and maxima in Section 7.2. The second is the determination of the force spectrum in terms of the prescribed wave spectrum. And in the same way as with the free surface elevation, certain simplifications are possible if the wave spectrum is narrow-banded: the consequences of this assumption will be considered separately in Section 7.6.2.

7.6.1 Morison Equation with an Arbitrary Wave Spectrum

But first we treat the case when the wave spectrum has an arbitrary shape, and investigate the probability distributions of the force maxima and the spectral density of the force. For a review of the results outlined here see Borgman (1972a).

Probability Density of F

The probability density of the sectional force F was derived by Pierson and Holmes (1965) and may be expressed (Tickell 1977) as

$$p(F) = (2\pi K_i \sigma_u \sigma_a)^{-1} \int_{-\infty}^{\infty} \exp\left[-\frac{1}{2}\left(\frac{u^2}{\sigma_u^2} + \frac{a^2}{\sigma_a^2}\right)\right] du \qquad (7.165)$$

Here σ_u and σ_a are the deviations of u and a respectively and may be obtained once the spectra $S_u(f)$ and $S_a(f)$ have been calculated using Eqs. (7.161) and (7.162). Borgman (1967a) expressed this distribution in an alternative form

$$p(\xi) = (\alpha/8\pi)^{1/2} \exp(-\xi^2/2) \{\exp[(\alpha + \xi)^2/4] \, U(0, \alpha + \xi) + \exp[(\alpha - \xi)^2/4] \, U(0, \alpha - \xi)\} \qquad (7.166)$$

where $\xi = F/K_i\sigma_a$ is a dimensionless representation of the force F, U(0, x) is the parabolic cylinder function (e.g. Abramowitz and Stegun 1965) which may be expressed in terms of modified Bessel functions if required, and $\alpha = K_i\sigma_a/2K_d\sigma_u^2$. We see that α is the ratio of a characteristic inertia to a characteristic drag force and thus indicates the relative magnitude of inertia to drag forces.

The above distribution is symmetric with zero mean and with standard deviation $\sigma_\xi = (1 + 3/(4\alpha^2))^{1/2}$. It is made up of a linear combination of a Gaussian inertia component and a non-Gaussian drag component: for large values of α, when the force is predominantly inertial, the distribution becomes Gaussian, while for smaller values of α the distribution then has a longer tail than does the Gaussian distribution. That is a Gaussian distribution would underestimate ξ for a given $P(\xi)$ near 1.

Probability Distribution of $\hat{F}$

The probability density of force maxima $p(\hat{F})$ is more difficult to develop but various expressions have been obtained by Tung (1974), Tickell (1977) and Moe and Crandall (1977). The latter authors derive an expression for the extreme rate density $\mu(\hat{F})$ which can be used to describe the probability density $p(\hat{F})$ through the relationship $p(\hat{F}) = \mu(\hat{F})/\int_0^\infty \mu(F)\,dF$. Their expression is asymptotically valid for large $\hat{F}$ and also takes account of a small current U (such that

$U \ll \sigma_u$). A relatively simple expression for $p(\hat{F})$ is possible when the wave spectrum is narrow-banded and this will be considered shortly.

Spectral Density of F

The spectral density $S_F(f)$ of the force may be obtained by first considering its autocorrelation function $R_F(\tau)$. Borgman (1967b) showed that this could be written in terms of those of the velocity and acceleration as

$$R_F(\tau) = K_d^2 \sigma_u^4 G(R_u(\tau)/\sigma_u^2) + K_i^2 R_a(\tau) \quad (7.167)$$

where

$$G(r) = [(4r^2 + 2) \sin^{-1}(r) + 6r(1 - r)^{1/2}]/\pi \quad (7.168)$$

This function may be expressed as a power series in r,

$$G(r) = \frac{1}{\pi}\left\{8r + \frac{4}{3}r^3 + \frac{1}{15}r^5 + 0[r^7]\right\} \quad (7.169)$$

In principle the spectral density $S_F(f)$ may be found by taking the Fourier transform of $R_F(\tau)$. However this is not straightforward since powers of $R_u(\tau)$ appear in the complete expression and the component drag force spectrum then involves a series of self-convolutions of the velocity spectrum. A linearized approximation based on the first term only of Eq. (7.169) is therefore introduced. Thus, Eq. (7.167) is written in the approximate form

$$R_F(\tau) = (8/\pi) K_d^2 \sigma_u^2 R_u(\tau) + K_i^2 R_a(\tau) \quad (7.170)$$

It then follows that

$$S_F(f) = (8/\pi) K_d^2 \sigma_u^2 S_u(f) + K_i^2 S_a(f) \quad (7.171)$$

$S_u(f)$ and $S_a(f)$ have already been expressed in terms of $S_\eta(f)$, Eqs. (7.161) and (7.162), so the spectral density $S_F(f)$ may finally be written directly in terms of $S_\eta(f)$ as

$$S_F(f) = S_\eta(f)\left\{4\pi^2 f^2 \frac{\cosh^2(k(z+d))}{\sinh^2(kd)}\left[\frac{8}{\pi} K_d^2 \sigma_u^2 + 4\pi^2 f^2 K_i^2\right]\right\} \quad (7.172)$$

The steps involved in the calculation of the force spectrum as described above are indicated schematically in Fig. 7.25.

The linearized approximation made in obtaining Eq. (7.170) is equivalent to linearizing the Morison equation itself into the form

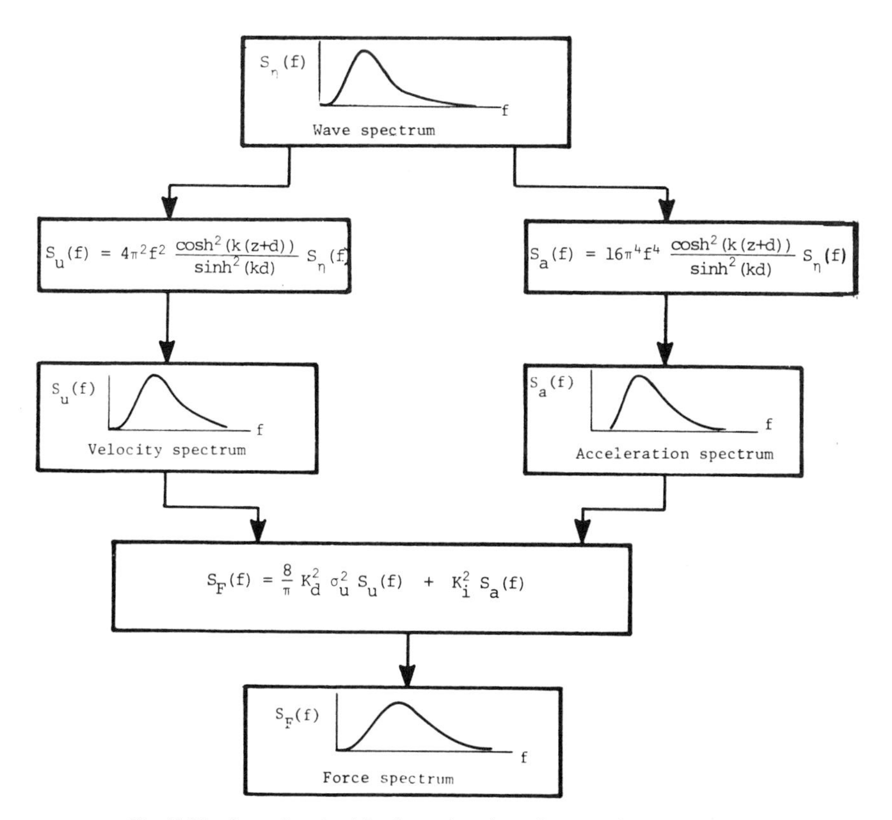

Fig. 7.25. Steps involved in the estimation of a wave force spectrum.

$$F = (8/\pi)^{1/2} K_d \sigma_u u + K_i a \qquad (7.173)$$

This linearization of the drag force ensures that the velocity component at one frequency effects only the drag component at that same frequency and thus a transfer function can be used in the usual way as in Eq. (7.171). This is in contrast to the complete Morison equation in which a sinusoidal velocity results in a drag force which contains components at odd multiples of the fundamental frequency.

The drag force in Eq. (7.173) can be considered as a single term approximation to a complete power series representation of $u|u|$ (Borgman 1969b). Borgman suggests that the linearized approximation is reasonable over the range $|u| < 2\sigma_u$. However, as Tickell (1977) has emphasized, the corresponding Gaussian distribution significantly underestimates forces corresponding to small given exceedences. Thus the linearized approximation should not be used for purposes of developing the probability distribution of F or $\hat{F}$ as already described.

It is noted that if drag forces are negligible over the range of frequencies contained in the wave spectrum, then the response is in any case linear and is relatively straightforward to calculate. Furthermore, when the waves so considered also lie in the deep water range over the complete spectrum, then the total force on a single vertical circular cylinder subjected to a sinusoidal wave train within this frequency range is

$$F(t) = -\frac{\pi}{8}\rho g D^2 C_m H \sin(\omega t) \tag{7.174}$$

Consequently the force spectrum is given simply as

$$S_F(f) = \left(\frac{\pi}{4}\rho g D^2 C_m\right)^2 S_\eta(f) \tag{7.175}$$

and the transfer function is then a constant independent of frequency.

The preceding analysis largely relates to the sectional force F on a pile. In order to obtain the total force or overturning moment acting on a structural element or a pile, the integration of this sectional force over the length of the member is necessary. Borgman (1967b) has presented the analysis for an array of N vertical cylinders. For the case of a single cylinder extending from the seabed and piercing the free surface, the total force and overturning moment spectra are eventually given as

$$S_{F_T}(f) = S_\eta(f)\left\{\frac{8}{\pi}\left[\frac{2\pi f K_d}{\sinh(kd)}\int_{-d}^{0}\sigma_u(z)\cosh(k(z+d))\,dz\right]^2 + \frac{(2\pi f)^4 K_i^2}{k^2}\right\} \tag{7.176}$$

$$S_M(f) = S_\eta(f)\left\{\frac{8}{\pi}\left[\frac{2\pi f K_d}{\sinh(kd)}\int_{-d}^{0}(z+d)\sigma_u(z)\cosh(k(z+d))\,dz\right]^2 + \left[\frac{(2\pi f)^2 K_i\,[(kd)\sinh(kd) + 1 - \cosh(kd)]}{k^2\sinh(kd)}\right]^2\right\} \tag{7.177}$$

The integrals in these two equations contain $\sigma_u(z)$ and so can be evaluated once $S_\eta(f)$ is known.

In the more general case of a structure with many members of various lengths, sections and orientations, the distributed loading may conveniently be approximated by "lumped" loads in which sectional forces (multiplied by suitable lengths) predicted by the Morison equation, Eq. (7.171), are taken to act at a discrete set of nodes on the structure. This approach has been explored by Burrows (1979).

It is emphasized that the force spectrum derived here by virtue of the Morison equation is valid only for a stationary structure since no consideration has been given to any possible motion of the structure itself. In practical situations dynamic excitation of the structure is very possible, particularly in deeper water, and consequent resonant behavior may be of crucial importance. The Morison equation may be extended to account for structural motions in terms of the *relative* fluid velocity and acceleration. The mass, (structural) damping and stiffness properties of any structural element would be known, and those fluid force components associated with the structural motion may conveniently be considered as added mass and hydrodynamic damping terms, while the force calculated to act on the structure if stationary is now treated as the excitation of the modified dynamic system. Along these lines, a matrix equation of motion may be set up and solved for the motions of the various degrees of freedom of the structure. The dynamic excitation of a structure, and the calculation procedures involved—both under regular and random wave loading—are discussed in Chapter 8.

The influence of a steady current flowing in the direction of a random wave train has been considered by Tung and Huang (1973a, b). In an earlier paper (Huang et al., 1972) it had been established that the wave spectrum is modified by the presence of a current U to

$$S_U(\omega) = \frac{4}{\alpha(1+\alpha)^2} S_0(\omega) \tag{7.178}$$

where $\alpha = 1 + 4U\omega/g$ and $S_0(\omega)$ is the wave spectrum in the absence of the current. This formula indicates that increased values of S occur for a countercurrent (U negative) and that this increase is greater at higher frequencies.

Tung and Huang used the Pierson-Moskowitz spectrum to describe the waves in terms of wind speed, and the Morison equation with constant coefficients ($C_d = 0.5$, $C_m = 1.4$), to obtain various statistical properties including the probability distribution of the force maxima. They find that the presence of even a moderate current can appreciably affect the statistical properties of the force. Some of their results have been reviewed in a more recent paper (Tung and Huang 1976), (see Section 5.3.5 and Fig. 5.7).

7.6.2 Morison Equation with a Narrow-band Spectrum

When the incident waves have a narrow-band spectrum, various simplifications may be made in the statistical representation of the wave forces (Borgman 1965) and we now consider some of these. The surface elevation η is suitably represented as an amplitude-modulated sinusoid given by Eq. (7.51), and consequently the velocity u and acceleration a may be expressed in an equivalent manner. The Morison equation may then be written as

$$F = A \cos(2\pi f_0 t + \delta_u) |\cos(2\pi f_0 t + \delta_u)| - B \sin(2\pi f_0 t + \delta_a) \quad (7.179)$$

where δ_u and δ_a are slowly varying phase angles, f_0 is the central wave frequency and A and B are also slowly varying functions of time. They may be expressed directly in terms of the wave height H as

$$A = K_d \left(\frac{\pi H}{T} \frac{\cosh(k(z+d))}{\sinh(kd)} \right)^2 \quad (7.180)$$

$$B = K_i \left(\frac{2\pi^2 H}{T^2} \frac{\cosh(k(z+d))}{\sinh(kd)} \right) \quad (7.181)$$

Any particular wave is closely sinusoidal and the maximum force $\hat{F}$ for that wave is given as

$$\hat{F} = \begin{cases} A + B^2/4A & \text{for } 2A > B \\ B & \text{for } 2A \leqslant B \end{cases} \quad (7.182)$$

Probability Distribution of $\hat{F}$

It is then a relatively straightforward procedure to obtain the probability density of the envelope $\hat{F}$ since H is known to possess a Rayleigh distribution and the above equation expresses $\hat{F}$ directly in terms of H. The cumulative probability distribution must involve a parameter of the wave height distribution, say H_{rms}, and the constant parts of A and B, i.e. A/H^2 and B/H. The cumulative probability is conveniently expressed in the form

$$P(\zeta) = \begin{cases} 1 - \exp\left[-\frac{1}{2}\zeta_c(2\zeta - \zeta_c)\right] & \text{for } \zeta > \zeta_c \\ 1 - \exp(-\zeta^2/2) & \text{for } 0 < \zeta < \zeta_c \\ 0 & \text{otherwise} \end{cases} \quad (7.183)$$

And consequently the probability density is

$$p(\zeta) = \begin{cases} \zeta_c \exp\left[-\frac{1}{2}\zeta_c(2\zeta - \zeta_c)\right] & \text{for } \zeta > \zeta_c \\ \zeta \exp(-\zeta^2/2) & \text{for } 0 < \zeta < \zeta_c \\ 0 & \text{otherwise} \end{cases} \quad (7.184)$$

In the above ζ is a dimensionless representation of the force maxima chosen here such that the component Rayleigh distribution $(0 < \zeta < \zeta_c)$ has a mode of unity: $\zeta = (\sqrt{2}\, H/BH_{rms})\hat{F}$; while ζ_c, the value at which the form of $p(\zeta)$ changes, is given as $\zeta_c = BH/\sqrt{2}\, AH_{rms}$. The probability density $p(\zeta)$ is shown

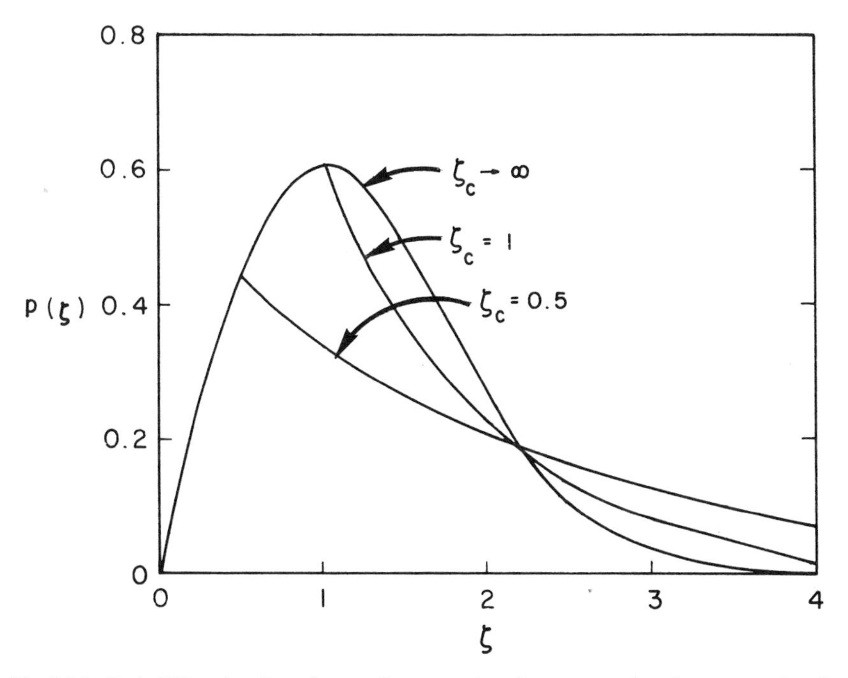

Fig. 7.26. Probability density of wave force maxima in a narrow-band wave motion for various values of the parameter ζ_c.

in Fig. 7.26 for various values of ζ_c. The figure indicates how the force maxima are inertial at most levels and most of the distribution follows the Rayleigh distribution. But when ζ_c is small then drag begins to affect the force maxima above a much lower force level corresponding to a discontinuity in the density gradient. It is mentioned here that Moe and Crandall (1977) have also obtained this result and extended it to account for the presence of a small current $U \ll \sigma_u$.

For the sectional force considered above, A and B are given by Eqs. (7.180)-(7.181) and thus ζ and ζ_c can be expressed directly in terms of the constants of the problem. When the total force or overturning moment on the pile are required, corresponding results may be obtained by using the appropriate expressions for A and B in the definitions of ζ and ζ_c. Thus if the total force maxima on the pile are of interest, we would put

$$\zeta = \left(\frac{8\sqrt{2}}{\pi\rho g D^2 C_m H_{rms} \tanh(kd)}\right)\hat{F} \tag{7.185}$$

$$\zeta_c = \left(\frac{4\pi C_m D}{\sqrt{2}\, C_d H_{rms}}\right)\left(\frac{\sinh^2(kd)}{2kd + \sinh(2kd)}\right) \tag{7.186}$$

Spectral Density of F

The force spectrum on the basis of the complete Morison equation would involve the same procedure as already described for spectra of arbitrary shape. In this case, however, with the wave spectrum being narrow-banded at frequency f_0, the force spectrum will contain narrow bands at the odd harmonics f_0, $3f_0$, $5f_0$, . . . , each with rapidly diminishing energy content. As before it is generally appropriate to linearize the drag force so that the force spectrum obtained will also be narrow-banded at the frequency f_0.

A second situation with a narrow-band motion is also of practical interest. This arises when a flexible pile is subjected to wave motion with an arbitrary spectrum such that the pile's response itself possesses a narrow-band spectrum. The transfer function for a spring-mass-dashpot system is described in Section 7.1.4 and is given by Eq. (7.49). An example of its effect on the response spectrum has already been indicated in Fig. 7.5. When the damping ratio is small, the response–which may be a displacement, force, bending moment or stress–will generally be characterized by a narrow-band spectrum centred at f_n, the natural frequency of the system. A similar approach to that already described for narrow-band motion may be made in order to determine the various statistical properties of the response. (See Chapter 8.)

7.6.3 Estimates of Force Coefficients

It is appropriate here to reiterate the assumptions and approximations made at various stages in estimating the different random force properties described. These have already been discussed in detail in Chapter 5. The three fundamental approximations that have been made in the analysis so far are that

a) the surface elevation is Gaussian and linear wave theory is valid;
b) in deriving the force spectrum a linearized form of the Morison equation is used in which $u|u|$ is replaced by a term proportional to u; however, the derived probabilistic properties of the force have not invoked this assumption, nor does a dynamic analysis require it (see Section 8.3); and thirdly,
c) the coefficients C_d and C_m are assumed to be known constants.

The first, Gaussian approximation is found by experience to be reasonable provided the waves are not near-breaking and the limiting height and appreciable wave steepness do not therefore significantly distort the symmetrical probability distribution of $\eta(t)$. Longuet-Higgins (1963) has investigated the probability distribution of finite amplitude random waves and has given a correction to the Gaussian distribution to account for the wave nonlinearities. The assumed validity of linear wave theory is more dubious and does not, for example, adequately describe force variations between the still water level and the free surface itself.

The second assumption concerning the linearization of the Morison equation has been discussed in some detail by Borgman (1967b, 1969b, 1972a) and appears reasonable in obtaining the shape of the force spectrum. We note here merely that in the case of a regular sinusoidal motion the contribution to the mean-square force arising from components at the fundamental frequency and at the third and fifth harmonics are 96.1%, 3.8% and 0.1% respectively. However, this assumption also implies that the force is Gaussian, and force maxima corresponding to a given probability of occurrence are then considerably underestimated. Thus this assumption is not applied in developing the probabilistic properties of the force.

The third assumption that C_d and C_m are constants is perhaps the most difficult to justify. Moreover, discrepancies arising from this assumption tend to mask the lesser discrepencies that derive from assumptions (a) and (b) above. It has already been indicated that in regular two-dimensional flow C_d and C_m vary markedly with the parameters characterizing that flow and the question immediately arises therefore as to what values should be assigned to these coefficients for a random wave motion.

Let us now go on to examine what methods are in fact used to estimate C_d and C_m from force measurements, particularly in a random sea (see also Section 5.3.1). Once again Borgman (1972a) has provided a valuable review of the various methods employed. The simplest approach, used by Wiegel, Beebe and Moon (1957) in field tests off the California Coast, is to assume that linear wave theory applies and to use measured values of sectional forces at specific instants when either the drag or inertia is assumed zero (that is at a wave trough or crest or when $\eta = 0$). In this way drag and inertia coefficients can be obtained for individual waves. This method can easily be extended in two directions. Firstly, instead of alloting instantaneous values to the coefficients (i.e. to C_d when inertia is zero and C_m when drag is zero), they can be averaged over a complete wave cycle as indicated in Section 3.8.2. Secondly, an alternative wave theory such as the stream function theory may be used in place of linear wave theory to evaluate fluid velocities and accelerations to be used in the Morison equation. However, even adopting these extensions to the method, pairs of values of C_d and C_m will be obtained for individual waves and there will generally be significant scatter in their values.

An approach of this nature has been used by Dean and Aagaard (1970), Thrasher and Aagaard (1970), Wheeler (1970) and Evans (1970) in analyzing data from field tests in the Gulf of Mexico, and average values of C_d and C_m (or values which serve as suitable upper bounds) may be extracted from the data provided.

Wheeler's (1970) approach involved a comparison of predicted maximum forces and measured maximum forces regardless of any phase difference between them. Dean Aagaard (1970) first grouped the force data into several Reynolds number ranges and for each of these obtained C_d and C_m values by minimising the differences between measured and predicted forces. More recently, data from the "Ocean Test Structure" located in the Gulf of Mexico during 1976–1978 have been used to provide C_d and C_m values by averaging over individual half wave cycles and by analyzing short drag-dominant and inertia-dominant segments (Heideman et al. 1979).

Several statistical methods for estimating C_d and C_m have been described by Brown and Borgman (1967) and reviewed by Borgman (1972a). In the method of moments, the second and fourth moments of the measured force distribution are equated to those of the theoretical distribution function. Brown and Borgman (1967) provide various tables to enable this procedure to be applied.

Another approach is to obtain estimates of C_d and C_m which provide the best least-squares fit of the theoretical autocorrelation function and spectral density of wave force to the measured functions. A further alternative is to treat C_d and C_m as frequency dependent rather than as being unique constants. The spectra and cross-spectra of the free-surface elevation and wave force may then be used to provide estimates of C_d and C_m at each frequency (Brown and Borgman 1967). This approach has also been considered by Schüeller and Shah (1972). Finally, C_d and C_m may themselves be treated as random variables possessing their own probability distributions. Under such an approach, the task of establishing the spectral density and probability distribution of wave force needs to be completely reconsidered, and this approach has not become established in practical applications.

7.6.4 Effects of Wave Directionality

The preceding outline of random wave force calculation procedures has been based on the application of one-dimensional frequency spectra used to describe the incident wave motion, and no attention has been given to the effects of the directional spread of waves. Fortunately, this spread tends to result in forces which are somewhat less than those predicted on the basis of a one-dimensional spectrum. This may be illustrated simply by considering two regular wave trains propagating at an angle to each other: the resulting fluid horizontal velocities and accelerations (which add vectorially), and thus the resulting forces acting on an element, are generally less than twice those of a single wave train. Forristall et al. (1978) have compared measured particle velocities in a storm with predictions based on uni- and multi-directional waves, and emphasize the importance of accounting for directional spreading.

The influence of directional spreading has been investigated by Dean (1977) on the basis of the linearized Morison equation, and also by Huntington (Huntington and Thompson 1976, Huntington 1979 and Huntington and Gilbert 1979) in the context of wave loading on a large cylinder in the diffraction regime (Section 7.6.5).

The horizontal force F in the direction of a component wave train may be written as some linear function of the corresponding free surface elevation η,

$$F(t) = H(f)\eta(t) \tag{7.187}$$

Thus the force component F_ϕ in any direction ϕ due to a component wave train propagating in the direction θ is

$$F_\phi(t) = H(f) \cos(\theta - \phi)\eta(t) \tag{7.188}$$

Equation (7.188) is analogous to Eq. (7.44) relating the input and output of a linear system, and thus may be extended to a relationship between the corresponding spectra. The value of F_ϕ at any one frequency f is due to the various wave components at that same frequency but propagating in the different possible directions. We therefore have

$$S_\phi(f) = \int_{-\pi}^{\pi} |H(f)|^2 \cos^2(\theta - \phi) S_\eta(f, \theta)\, d\theta \tag{7.189}$$

where $S_\phi(f)$ is the spectral density of F_ϕ. The rms value of F_ϕ may now be obtained by an integration over frequency f. This may be compared to the corresponding rms value of the force which would be obtained if directional spreading were ignored, and we may thus define a force reduction factor R_F as the ratio of these two characteristic forces. Thus

$$R_F^2 = \frac{\displaystyle\int_0^\infty \int_{-\pi}^{\pi} |H(f)|^2 \cos^2(\theta - \phi) S_\eta(f, \theta)\, df\, d\theta}{\displaystyle\int_0^\infty |H(f)|^2 S_\eta(f)\, df} \tag{7.190}$$

In the case where the directional spreading function $G(f, \theta)$ is taken to be independent of frequency, the integration over θ is independent of $S_\eta(f)$ and can be separated out to yield

$$R_F = \left[\int_{-\pi}^{\pi} G(\theta) \cos^2(\theta - \phi)\, d\theta\right]^{1/2} \tag{7.191}$$

Furthermore, if the principal wave direction $\overline{\theta}$ is a constant independent of frequency, then the principal force direction for which R_F is a maximum will be $\phi = \overline{\theta}$.

Dean (1977) considered the case of a single cylinder when the directional spreading function is given as a cosine-power function as in Eq. (7.128). In this case

$$R_F = \sqrt{\frac{C'(s)}{C'(s)+1}} \tag{7.192}$$

where $C'(s)$ is defined in Eq. (7.129). However, when drag forces are included, the expression for $H(f)$ contains the rms of horizontal velocity σ_u which itself is influenced by the degree of directional spreading. Equation (7.192) is therefore valid only for the inertia force, and more generally we have

$$R_F = \begin{cases} \sqrt{\dfrac{C'(s)}{C'(s)+1}} & \text{for inertia predominant} \\[2ex] \dfrac{C'(s)}{C'(s)+1} & \text{for drag predominant} \end{cases} \tag{7.193}$$

For the particular case $s = 1$ corresponding to a cosine-squared spreading function, Eq. (7.125), R_F reduces to

$$R_F = \begin{cases} 0.866 & \text{for inertia predominant} \\ 0.75 & \text{for drag predominant} \end{cases} \tag{7.194}$$

which shows that the force may be reduced by as much as 25%.

Dean (1977) describes an extension to this approach to deal with the case when the principal wave direction $\bar{\theta}$ itself varies with frequency, as may occur near the center of a hurricane. When this effect is taken into account, a further force reduction is predicted.

The above discussion concerns the force component in some fixed direction ϕ, and in particular concerns the (principal) direction corresponding to the maximum rms value of the force component. The time-varying resultant force itself will in fact change direction continuously and is more awkward to deal with. In order to describe its probabilistic properties, Huntington and Thompson (1976) considered the force components in-line and transverse to the principal wave direction, given by Eq. (7.188) with $\phi = 0$ and $\pi/2$ respectively when $\bar{\theta} = 0$. Huntington and Gilbert (1979) subsequently described the probabilistic properties of the resultant force, and in particular the force magnitude which should not be exceeded with a small prescribed probability (c.f. Eq. (7.88) which concerns wave amplitudes). This can be defined in terms of the spectra and cross-spectra of the in-line and transverse force components.

7.6.5 Effects of Wave Nonlinearities

An underlying assumption throughout the present discussion of spectral methods, and indeed throughout most of this chapter, has been that of linearity which is strictly associated with waves of infinitesimal height. Wave force predictions for random finite amplitude waves are much more difficult to deal with, and St. Denis (1973) has cautioned against the use of the conventional spectral approach when dealing with more severe sea states. The influence of wave nonlinearities are more readily investigated with respect to regular uni-directional waves, but we here consider briefly the simultaneous effects of wave nonlinearities and randomness.

As an empirical approach to dealing with finite amplitude and random waves, Dean (1977) has proposed a "hybrid method" in which the forces due to nonlinear, random (directional) waves might be represented in terms of those due to nonlinear, regular, unidirectional waves, modified by a coefficient T_L to account for the randomness of the waves but based on linear theory. Thus

$$F(\text{nonlinear, random}) = T_L F(\text{nonlinear, regular}) \tag{7.195}$$

Reference cases considered by Dean include a single pile subjected to unidirectional random waves, directional waves at a single frequency, as well as the general case of a complete directional spectrum. In this latter case the coefficient T_L is identical to the force reduction factor R_F described in the preceding section, see Eqs. (7.190)-(7.194).

The hybrid method of combining nonlinear, regular wave theory predictions with those for linear, random directional waves is a useful and practical approach, but is strictly an intuitive one whose validity must be established. As a more formal approach to the problem, it is possible to extend the random wave analysis to a second order along the lines of a Stokes expansion procedure (Section 4.3). This may readily be appreciated by considering the discretization of a linear wave spectrum at frequencies and directions taken at intervals Δf and $\Delta\theta$ respectively. Thus Eq. (7.90) applies to the first order free surface elevation with complex amplitudes

$$A_n = \sqrt{2S(f_n, \theta_n)\Delta f\, \Delta\theta} \tag{7.196}$$

and in practice the number of terms N will be finite. Second order components can be developed in terms of these first order amplitudes by the application of the two nonlinear free surface boundary conditions taken to second order, Eqs. (4.54) and (4.55). Thus, for example, additional frequencies will be realized at the sum and difference frequencies of any two linear components.

Investigations into second order effects in random waves have been carried out by Tick (1959, 1961), Hasselmann (1962-1963) and Longuet-Higgins

(1963), and more recently by Hudspeth (1975), Hudspeth and Chen (1979) and Sharma and Dean (1979). Hudspeth and Chen (1979) compared the statistical properties of simulated second order random waves with measured data of hurricane-generated waves. Hudspeth (1975) had earlier obtained simulated wave force spectra and probability distributions which compared reasonably well with field measurements made during "Wave Force Project II."

Sharma and Dean (1979) have developed the approach mentioned here to second order random, directional waves for finite depths, and have used it to simulate nonlinear wave and wave force spectra, and surface and wave force profiles. A comparison with the hybrid method already mentioned indicates that the hybrid method does provide an improvement to linear random analysis.

A further second order effect described by Dean (1979) relates to a narrow-band random wave train (for which maximum wave heights coincide with the maximum amplitude of a wave group, see Fig. 7.6). Dean points out that (Longuet-Higgins and Stewart 1964) the wave group associated with any maximum wave height forces a second order coupled wave system which is about 180° out of phase with the group's envelope, and which has a wave length near that of the group. These secondary waves give rise to a second order current which must extend to relatively large depths (on account of the large secondary wave length). This current is a maximum when individual wave heights are a maximum and so can modify drag forces quite noticeably.

7.6.6 Random Forces on Large Bodies

The random loads on large offshore structures are considerably simpler to treat than are those on slender elements because no appreciable (nonlinear) drag forces are present. It is recalled from Chapter 6 that on the basis of linear wave theory the force on a large body in a regular unidirectional wave train is proportional to the wave height. It may be written in complex form in terms of the free surface elevation η as

$$F(t) = H(ka)\frac{H}{2}\,e^{-i2\pi ft} = H(ka)\eta(t) \qquad (7.197)$$

where k is the wave number, a is a characteristic dimension of the body and H(ka) is a complex receptance which depends on ka and consequently on wave frequency f. Equation (7.197) is completely analogous to Eq. (7.44) relating the input and output of a linear system. In accordance with the general case outlined in Section 7.1.4, the spectral density of the force $S_F(f)$ is given in terms of the spectral density of the free surface $S_\eta(f)$ (wave spectrum) as

$$S_F(f) = |H(ka)|^2\, S_\eta(f) \qquad (7.198)$$

where $|H(ka)|^2$ is thus the transfer function and the dependence of the transfer function on frequency is obtained through the use of the linear dispersion relation. For a given body and water depth, $|H(ka)|^2$ is a known function of ka and may be obtained by repeated use of the methods outlined in Chapter 6. For example, if the force amplitude F_m on the body has been determined in the form of a coefficient C_F defined as

$$C_F = \frac{F_m}{\rho g H a^2} \tag{7.199}$$

then C_F will depend only on ka (for a given body and water depth) and the transfer function is

$$|H(ka)|^2 = [2\rho g a^2 C_F(ka)]^2 \tag{7.200}$$

Or again, in the case of a vertical circular cylinder of radius a the regular wave results may be expressed in terms of an effective inertia coefficient $C_m(ka)$, see Eq. (6.34), and the transfer function is then

$$|H(ka)|^2 = [\rho g(\pi a^2) \tanh(kd) C_m(ka)]^2 \tag{7.201}$$

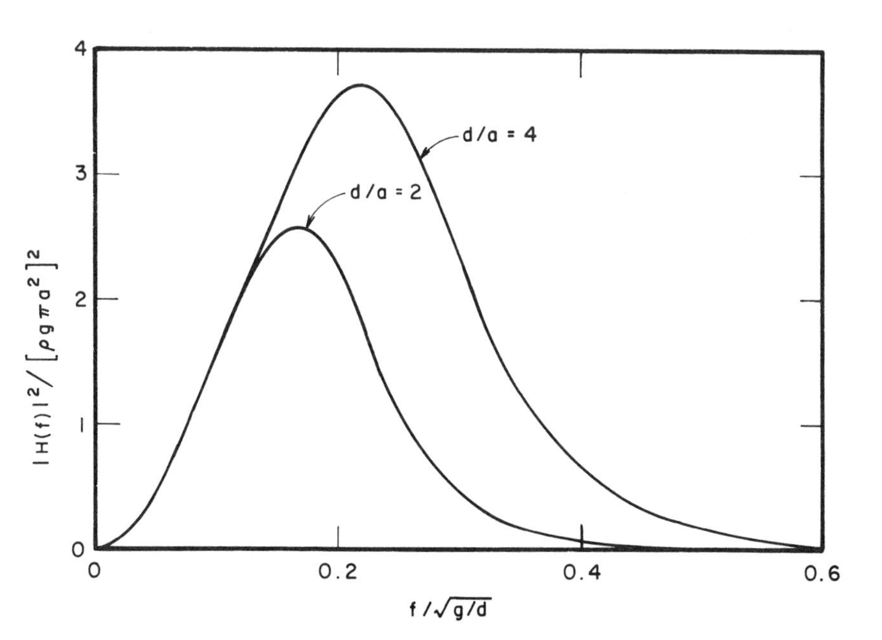

Fig. 7.27. Transfer function relating wave and wave force spectra for a large vertical circular cylinder.

This transfer function is shown in dimensionless form for the two cases d = 2a and d = 4a in Fig. 7.27.

In short-crested seas where a directional wave spectrum $S_\eta(f, \theta)$ is employed, the approach described in Section 7.6.4 may be used. The horizontal and vertical force coefficients C_H and C_V corresponding to C_F in Eq. (7.199) above will now depend both on ka and also on the component wave direction θ measured relative to a given body orientation. Once more the variation of C_H and C_V with ka and θ, and of the corresponding transfer functions $|H_H(f, \theta)|^2$ and $|H_V(f, \theta)|^2$ given by Eq. (7.200), may be derived by repeated use of the regular wave methods described in Chapter 6. The contributions from waves of different directions can be summed to obtain the vertical force, and its spectrum $S_V(f)$ will thereby be given as

$$S_V(f) = \int_{-\pi}^{\pi} |H_V(f, \theta)|^2 S_\eta(f, \theta)\, d\theta$$

$$= \int_{-\pi}^{\pi} [2\rho g a^2 C_V(f, \theta)]^2 S_\eta(f, \theta)\, d\theta \qquad (7.202)$$

The horizontal force acts in varying directions, and it is most appropriate to consider the spectra and cross-spectra of the component forces F_x and F_y in-line and transverse to the principal wave direction as described in Section 7.6.4. This approach has been outlined by Huntington and Thompson (1976), Huntington (1979) and Huntington and Gilbert (1979) for the case of a vertical circular cylinder. They also reported on experiments involving a vertical circular cylinder subjected to multi-directional random waves and have found satisfactory agreement with the theoretical predictions. In this particular case of course there is no vertical force while the horizontal force coefficient C_H is independent of the component wave direction.

7.6.7 Long-term Force Distributions

The present section has been concerned entirely with the short-term statistics of wave forces, which describe force variations during any specified sea state. In offshore design, the long-term statistics extending over the design lifetime of a structure are also needed, as for example in calculations of fatigue damage.

The long-term probability distribution of wave forces may be developed in an analogous way to that of the individual wave heights given by Eq. (7.145). That is, the probability distribution of force during any specific sea state defined by H_s and T, is weighted by the probability of occurence of that sea state. This procedure has been described by Tickell, Burrows and Holmes (1976) and by

Tickell (1979). The long-term distribution of the force F is thus given as

$$P(F) = \int_0^\infty \int_0^\infty P(F|H_s, T)p(H_s, T)\, dH_s\, dT \tag{7.203}$$

where $P(F|H_s, T)$ denotes the cumulative probability of force F for a given sea state defined by H_s and T, and $p(H_s, T)$ is the bivariate probability density of H_s and T. The long-term distribution of force maxima $\hat{F}$ is given by an equation analogous to Eq. (7.145) in which a weighting term is necessary to account for the number of force maxima per unit time, T_p^{-1}, during any sea state. This can be written as

$$P(\hat{F}) = \frac{1}{\overline{T_p^{-1}}} \int_0^\infty \int_0^\infty \frac{1}{T_p} P(\hat{F}|H_s, T)p(H_s, T)\, dH_s\, dT \tag{7.204}$$

where $\overline{T_p^{-1}}$ is the long-term average number of force maxima per unit time and is defined as

$$\overline{T_p^{-1}} = \int_0^\infty \int_0^\infty \frac{1}{T_p} p(H_s, T)\, dH_s\, dT \tag{7.205}$$

Also $P(\hat{F}|H_s, T)$ is the cumulative probability of force maxima for a given sea state defined by H_s and T. In any numerical calculation, such as one employing a scatter diagram, the above formulae may be expressed as summations. A generalization of these expressions is also possible to account for variability in wave directions such that H_s, T and principal wave direction $\overline{\theta}$ would be used to define any particular sea state.

7.7 REFERENCES

Abramowitz, M. and Stegun, I. A. 1965. *Handbook of Mathematical Functions*. Dover, New York.

Battjes, J. A. 1970. Long-Term Wave Height Distribution at Seven Stations Around the British Isles. *National Institute of Oceanography*, Godalming, England, Report No. A44.

Barber, N. F. 1961. The Directional Resolving Power of an Array of Wave Detectors. In *Ocean Wave Spectra*, Prentice-Hall, Englewood Cliffs, N.J., pp. 137–150.

Barnett, T. P. 1968. On the Generation, Dissipation and Prediction of Ocean Wind Waves. *J. Geophys. Res.*, Vol. 73, pp. 513–530.

Bendat, J. S. and Piersol, A. G. 1971. *Random Data: Analysis and Measurement Procedures*, 2nd ed., Wiley, New York.

Blackman, R. B. and Tukey, J. W. 1959. *The Measurement of Power Spectra*. Dover, New York.

Borgman, L. E. 1961. The Frequency Distribution of Near Extremes. *J. Geophys. Res.*, Vol. 66, pp. 3295–3307.

Borgman, L. E. 1963. Risk Criteria. *J. Waterways and Harbors Div., ASCE*, Vol. 89, No. WW3, pp. 1–35.

Borgman, L. E. 1965. Wave Forces on Piling for Narrow-Band Spectra. *J. Waterways and Harbors Div., ASCE*, Vol. 91, No. WW3, pp. 65–90.

Borgman, L. E. 1967a. Random Hydrodynamic Forces on Objects. *Ann. Math Statist.*, Vol. 38, pp. 37–51.

Borgman, L. E. 1967b. Spectral Analysis of Ocean Wave Forces on Piling. *J. Waterways and Harbors Div., ASCE*, Vol. 93, No. WW2, pp. 129–156.

Borgman, L. E. 1969a. Directional Spectra Models for Design Use. *Proc. Offshore Tech. Conf.*, Houston, Paper No. OTC 1069, pp. 721–746.

Borgman, L. E. 1969b. Ocean Wave Simulation for Engineering Design. *J. Waterways and Harbors Div., ASCE*, Vol. 95, No. WW4, pp. 557–583.

Borgman, L. E. 1972a. Statistical Models for Ocean Waves and Wave Forces. *Advances in Hydroscience*, Vol. 8, pp. 123–456.

Borgman, L. E. 1972b. Confidence Intervals for Ocean Wave Spectra. *Proc. 13th Coastal Eng. Conf.*, Vancouver, Vol. I, pp. 237–250.

Borgman, L. E. 1975. Extremal Statistics in Ocean Engineering. *Proc. Civil Engineering in the Oceans III*, ASCE, Univ. of Delaware, pp. 117–133.

Borgman, L. E. 1977. Directional Spectra From Wave Sensors. In *Ocean Wave Climate*, eds. M. D. Earle and A. Malahoff, Plenum Press, New York, pp. 269–300.

Bretschneider, C. L. 1958. Revisions in Wave Forecasting: Deep and Shallow Water. *Proc. 6th Coastal Eng. Conf.*, Miami, pp. 30–67.

Bretschneider, C. L. 1959. Wave Variability and Wave Spectra for Wind-Generated Gravity Waves. U.S. Army Corps of Engineers, *Beach Erosion Board*, Tech. Memo. No. 118.

Bretschneider, C. L. 1961. A One-Dimensional Gravity Wave Spectrum. In *Ocean Wave Spectra*. Prentice-Hall, Englewood Cliffs, N.J., pp. 41–65.

Bretschneider, C. L. 1966. Wave Generation by Wind, Deep and Shallow Water. In *Estuary and Coastline Hydrodynamics*, ed. A. T. Ippen, McGraw-Hill, New York, pp. 133–196.

Bretschneider, C. L. 1970. Forecasting Relations for Wave Generation. *Look Lab/Hawaii*, Vol. 1, No. 3, Univ. of Hawaii.

Bretschneider, C. L. 1973. Prediction of Waves and Currents. *Look Lab/Hawaii*, Vol. 3, No. 1, Univ. of Hawaii.

Bretschneider, C. L. 1975. The Envelope Wave Spectrum. *Proc. 3rd Int. Conf. on Port and Ocean Eng. Under Arctic Conditions*, Univ. of Alaska, Vol. II, pp. 689–703.

Bretschneider, C. L. 1977. On the Determination of the Design Ocean Wave Spectrum. *Look Lab/Hawaii*, Vol. 7, No. 1, Univ. of Hawaii.

Bretschneider, C. L. and Tamaye, E. E. 1976. Hurricane Wind and Wave Forecasting Techniques. *Proc. 15th Coastal Eng. Conf.*, Honolulu, Vol. I, pp. 202–237.

Brown, L. J. and Borgman, L. E. 1967. Tables of the Statistical Distribution of Ocean Wave Forces and Methods of Estimating Drag and Mass Coefficients. *Coastal Engineering Research Center*, U.S. Army Corps of Engineers, Tech. Memo. No. 24.

Burrows, R. 1979. Probabilistic Description of the Response of Offshore Structures to Random Wave Loading. In *Mechanics of Wave-Induced Forces on Cylinders*, ed. T. L. Shaw, Pitman, London, pp. 577–595.

Cardone, V. J., Pierson, W. J., and Ward, E. G. 1976. Hindcasting the Directional Spectra of Hurricane Generated Waves. *J. Petrol. Tech.*, Vol. 28, pp. 385–394.

Cartwright, D. E. 1958. On Estimating the Mean Energy of Sea Waves From the Highest Waves in a Record. *Proc. Roy. Soc.*, Ser. A, Vol. 247, pp. 22–48.

Cartwright, D. E. and Longuet-Higgins, M. S. 1956. The Statistical Distribution of the Maxima of a Random Function. *Proc. Roy. Soc.*, Ser. A, Vol. 237, pp. 212-232.

Collins, J. I. 1972. Prediction of Shallow-Water Spectra. *J. Geophys. Res.*, Vol. 77, pp. 2693-2707.

Cooley, J. W. and Tukey, J. W. 1965. An Algorithm for the Machine Computation of Complex Fourier Series. *Math. Comp.*, Vol. 19, pp. 297-301.

Cote, L. J., Davis, J. O., Marks, W., Mcgough, R. J., Mehr, E., Pierson, W. J., Ropek, J. F., Stephenson, G., and Vetter, R. C. 1960. The Directional Spectrum of a Wind Generated Sea as Determined From Data Obtained by the Stereo Wave Observation Project. *Meteorological Paper*, Vol. 2, No. 6, New York Univ., College of Engineering, pp. 1-88.

Darbyshire, J. 1952. The Generation of Waves by Wind. *Proc. Roy. Soc.*, Ser. A, Vol. 215, pp. 299-328.

Darbyshire M. and Draper, L. 1963. Forecasting Wind Generated Sea Waves. *Engineering*, Vol. 195, pp. 482-484.

Dattari, J. 1973. Waves Off Mangalore Harbor–West Coast of India. *J. Waterways Harbors and Coastal Engineering Div., ASCE*, Vol. 99, No. WW1, pp. 39-58.

Davenport, W. B. and Root, W. L. 1958. *An Introduction to the Theory of Random Signals and Noise*. McGraw-Hill, New York.

Dean, R. G. 1977. Hybrid Method of Computing Wave Loading. *Proc. Offshore Tech. Conf.*, Houston, Paper No. OTC 3029, pp. 483-492.

Dean, R. G. 1979. Kinematics and Forces Due to Wave Groups and Associated Second Order Currents. *Proc. 2nd Int. Conf. on the Behaviour of Off-Shore Structures, BOSS '79*, London, Vol. I, pp. 87-94.

Dean, R. G. and Aagaard, P. M. 1970. Wave Forces: Data Analysis and Engineering Calculation Method. *J. Petrol. Tech.*, Vol. 22, pp. 368-375.

Draper, L. 1963. Derivation of a 'Design Wave' From Instrumental Measurements of Sea Waves. *Proc. Inst. Civil Eng.*, Vol. 26, pp. 291-304.

Draper, L. 1966. The Analysis and Presentation of Wave Data–A Plea for Uniformity. *Proc. 10th Coastal Eng. Conf.*, Tokyo, pp. 1-11.

Draper, L. 1972. Extreme Wave Conditions in British and Adjacent Waters. *Proc. 13th Coastal Eng. Conf.*, Vancouver, Vol. I, pp. 157-165.

Draper, L. 1976. Revisions in Wave Data Presentation. *Proc. 15th Coastal Eng. Conf.*, Honolulu, Vol. I, pp. 3-9.

Draper, L. and Driver, J. S. 1971. Winter Waves in the Northern North Sea at 57°30'N 3°00'E, Recorded by M. V. Famita. *Proc. 1st Int. Conf. on Port and Ocean Engineering Under Artic Conditions*, Trondheim, Vol. II, pp. 966-978.

Evans, D. J. 1970. Analysis of Wave Force Data. *J. Petrol. Tech.*, Vol. 22, pp. 339-346.

Ewing, J. A. 1971. A Numerical Wave Prediction Method for the North Atlantic Ocean. *Deut. Hydrol. Z.*, Vol. 24, pp. 241-261.

Ewing, J. A. 1973. Mean Length of Runs of High Waves. *J. Geophys. Res.*, Vol. 78, pp. 1933-1936.

Forristall, G. Z. 1978. On the Statistical Distribution of Wave Heights in a Storm. *J. Geophys. Res.*, Vol. 83, pp. 2353-2358.

Forristall, G. Z., Ward, E. G., Cardone, V. J. and Borgmann, L. E. 1978. The Directional Spectra and Kinematics of Surface Gravity Waves in Tropical Storm Delia. *J. Phys. Oceanography*, Vol. 8, pp. 888-909.

Goda, Y. 1970. Numerical Experiments on Wave Statistics with Spectral Simulation. *Rept. Port and Harbour Res. Inst.*, Japan, Vol. 9, No. 3, pp. 320-337.

Goda, Y. 1974. Estimation of Wave Statistics From Spectra Information. *Proc. Int. Symp. on Ocean Wave Measurement and Analysis*, ASCE, New Orleans, pp. 320-337.

Goda, Y. 1976. On Wave Groups. *Proc. Behaviour of Off-Shore Structures, BOSS '76*, Trondheim, Vol. I, pp. 115-128.

Goda, Y. and Suzuki, Y. 1976. Estimation of Incident and Reflected Waves in Random Wave Experiments. *Proc. 15th Coastal Eng. Conf.*, Honolulu, Vol. I, pp. 828-845.

Gringorten, I. I. 1963a. A Plotting Rule for Extreme Probability Paper. *J. Geophys. Res.*, Vol. 68, pp. 813-814.

Gringorten, I. I. 1963b. Envelopes for Ordered Observations Applied to Meterological Extremes. *J. Geophys. Res.*, Vol. 68, pp. 815-826.

Gumbel, E. J. 1958. *Statistics of Extremes*. Columbia Univ. Press, New York.

Hallam, M. G., Heaf, N. J., and Wootton, L. R. 1977. Dynamics of Marine Structures: Methods of Calculating the Dynamic Response of Fixed Structures Subject to Wave and Current Action. *CIRIA Underwater Engineering Group*, London, Report UR8.

Harris, D. L. 1972. Characteristics of Wave Records in the Coastal Zone. In *Waves on Beaches and Resulting Sediment Transport*, ed. R. E. Meyer, Academic Press, New York, pp. 1-51.

Harris, D. L. 1974. Finite Spectrum Analyses of Wave Records. *Proc. Int. Symp. on Wave Measurement and Analysis*, ASCE, New Orleans, pp. 107-124.

Hasselmann, K. 1962-1963. On the Non-Linear Energy Transfer in a Gravity-Wave Spectrum. Part I. *JFM*, Vol. 12, pp. 481-500, Part II, *JFM*, Vol. 15, pp. 273-281, Part III, *JFM*, Vol. 15, pp. 385-398.

Hasselmann, K. et al. 1973. Measurements of Wind-Wave Growth and Swell Decay During the Joint North Sea Wave Project. *Deut. Hydrogr. Z.*, Reihe A, No. 12.

Hasselmann, K., Ross, D. B., Müller, P., and Sell, W. 1976. A Parametric Wave Prediction Model. *J. Phys. Oceanography*, Vol. 6, pp. 200-228.

Heideman, J. C., Olsen, O. A. and Johansson, P. I. 1979. Local Wave Force Coefficients. *Proc. Civil Eng. in the Oceans IV*, ASCE, San Francisco, Vol. II, pp. 684-699.

Houmb, O. G. and Overvik, T. 1976. Parameterization of Wave Spectra and Long Term Joint Distribution of Wave Height and Period. *Proc. Behaviour of Off-Shore Structures, BOSS '76*, Trondheim, Vol. I, pp. 144-169.

Huang, N. E., Chen, D. T., Fung, C. C., and Smith, J. R. 1972. Interactions Between Steady Non-Uniform Current and Gravity Waves with Applications for Current Measurements. *J. Phys. Oceanography*, Vol. 2, pp. 420-431.

Hudspeth, R. T. 1975. Wave Force Predictions From Nonlinear Random Sea Simulations. *Proc. Offshore Tech. Conf.*, Houston, Paper No. OTC 2193, pp. 471-485.

Hudspeth, R. T. and Chen, M-C. 1979. Digital Simulation of Nonlinear Random Waves. *J. Waterway Port Coastal and Ocean Div., ASCE*, Vol. 105, No. WW1, pp. 67-85.

Huntington, S. W. 1979. Wave Loading on Large Cylinders in Short-Crested Seas. In *Mechanics of Wave-Induced Forces on Cylinders*, ed. T. L. Shaw, Pitman, London, pp. 636-649.

Huntington, S. W. and Gilbert, G. 1979. Extreme Forces in Short Crested Seas. *Proc. Offshore Tech. Conf.*, Houston, Paper No. OTC 3695, Vol. III, pp. 2075-2084.

Huntington, S. W. and Thompson, D. M. 1976. Forces on a Large Vertical Cylinder in Multi-Directional Random Waves. *Proc. Offshore Tech. Conf.*, Houston, Paper No. 2539, Vol. II, pp. 169-183.

Inoue, T. 1967. On the Growth of the Spectrum of a Wind-Generated Sea According to a Modified Miles-Phillips Mechanism and Its Application to Wave Forecasting. *Tech. Rept. No. 67-5*, Geophysical Sciences Lab., New York Univ.

Jahns, H. O. and Wheeler, J. D. 1973. Long-Term Wave Probabilities Based on Hindcasting of Severe Storms. *J. Petrol. Tech.*, Vol. 25, pp. 473-486.

Jasper, N. H. 1956. Statistical Distribution Patterns of Ocean Waves and of Wave-Induced

Ship Stresses and Motions with Engineering Applications. *Trans. SNAME*, Vol. 64, pp. 375-432.

Jeffreys, H. 1924. On the Formation of Water Waves by Wind. *Proc. Roy. Soc.*, Ser. A, Vol. 107, pp. 189-199.

Karlsson, T. 1969. Refraction of Continuous Ocean Wave Spectra. *J. Waterways and Harbors Div., ASCE*, Vol. 95, No. WW4, pp. 437-448.

Kimball, B. F. 1960. On the Choice of Plotting Positions on Probability Paper. *J. Amer. Stat. Assn.*, Vol. 55, pp. 546-560.

Kinsman, B. 1965. *Wind Waves*. Prentice-Hall, Englewood Cliffs, N.J.

Longuet-Higgins, M. S. 1952. On the Statistical Distribution of the Heights of Sea Waves. *J. Mar. Res.*, Vol. 11, pp. 245-266.

Longuet-Higgins, M. S. 1956. The Refraction of Sea Waves in Shallow Water. *JFM*, Vol. 1, pp. 163-176.

Longuet-Higgins, M. S. 1957a. The Statistical Analysis of a Random, Moving Surface. *Phil. Trans. Roy. Soc.*, Ser. A, Vol. 249, pp. 321-387.

Longuet-Higgins, M. S. 1957b. On the Transformation of a Continuous Spectrum by Refraction. *Proc. Camb. Phil. Soc.*, Vol. 53, pp. 226-229.

Longuet-Higgins, M. S. 1958. On the Interval Between Successive Zeros of a Random Function. *Proc. Roy. Soc.*, Ser. A, Vol. 246, pp. 99-118.

Longuet-Higgins, M. S. 1962. The Distribution of Intervals Between Zeros of a Stationary Random Function. *Phil. Trans. Roy. Soc.*, Ser. A, Vol. 254, pp. 557-599.

Longuet-Higgins, M. S. 1963. The Effect of Non-Linearities on Statistical Distributions in the Theory of Sea Waves. *JFM*, Vol. 17, pp. 459-480.

Longuet-Higgins, M. S. 1969. A Nonlinear Mechanism for the Generation of Sea Waves. *Proc. Roy. Soc.*, Ser. A, Vol. 311, pp. 371-389.

Longuet-Higgins, M. S. 1975. On the Joint Distribution of the Periods and Amplitudes of Sea Waves. *J. Geophys. Res.*, Vol. 80, pp. 2688-2694.

Longuet-Higgins, M. S., Cartwright, D. E., and Smith, N. D. 1961. Observations of the Directional Spectrum of Sea Waves Using the Motions of a Floating Buoy. In *Ocean Wave Spectra*, Prentice-Hall, Englewood Cliffs, N.J., pp. 111-132.

Longuet-Higgins, M. S. and Stewart, R. W. 1964. Radiation Stress in Water Waves: A Physical Discussion, with Applications. *Deep Sea Res.*, Vol. 11, pp. 529-549.

Miles, J. W. 1957-1962. On the Generation of Surface Waves by Shear Flows. *JFM*, Vol. 3, pp. 185-204. Also *JFM*, Vol. 6, pp. 568-582; Vol. 6, pp. 583-598; Vol. 7, pp. 469-478; Vol. 13, pp. 433-448.

Mitsuyasu, H. 1971. On the Form of Fetch-Limited Wave Spectrum. *Coastal Eng. in Japan*, Vol. 14, pp. 7-14.

Mitsuyasu, H. 1972. The One-Dimensional Wave Spectra at Limited Fetch. *Proc. 13th Coastal Eng. Conf.*, Vancouver, Vol. I, pp. 289-298.

Mitsuyasu, H. and Mizuno, S. 1976. Directional Spectra of Ocean Surface Waves. *Proc. 15th Coastal Eng. Conf.*, Honolulu, Vol. I, pp. 329-348.

Mitsuyasu, H. et al. 1975. Observations of the Directional Spectrum of Ocean Waves Using a Cloverleaf Buoy. *J. Phys. Oceanography*, Vol. 5, pp. 750-760.

Moe, G. and Crandall, S. H. 1977. Extremes of Morison-Type Wave Loading on a Single Pile. *Trans. ASME, J. Mech. Design*, Vol. 100, No. 1, pp. 100-104.

Nagai, K. 1973. Runs of the Maxima of the Irregular Sea. *Coastal Eng. in Japan*, Vol. 16, pp. 13-18.

Nath, J. H. and Ramsey, F. L. 1974. Probability Distributions of Breaking Wave Heights. *Proc. Int. Symp. on Ocean Wave Measurement and Analysis*, ASCE, New Orleans, pp. 379-395.

Neumann, G. 1953. On Ocean Wave Spectra and a New Method of Forecasting Wind-Generated Sea. U.S. Army Corps of Engineers, *Beach Erosion Board*, Tech. Memo. No. 43.

Neumann, G. and Pierson, W. J. 1966. *Principles of Physical Oceanography*. Prentice-Hall, Englewood Cliffs, N.J.

Nolte, K. G. 1973. Statistical Methods for Determining Extreme Sea States. *Proc. 2nd Int. Conf. on Port and Ocean Engineering Under Arctic Conditions*, Univ. of Iceland, pp. 705-742.

Nolte, K. G. and Hsu, F. H. 1973. Statistics of Ocean Wave Groups. *Proc. Offshore Tech. Conf.*, Houston, Paper No. OTC 1688, Vol. II, pp. 637-656.

Nolte, K. G. and Hsu, F. H. 1979. Statistics of Larger Waves in a Sea State. *J. Waterway Port Coastal and Ocean Div., ASCE*, Vol. 105, No. WW4, pp. 389-404.

Ocean Wave Measurement and Analysis. 1974. Proceedings, ASCE, New Orleans.

Ochi, M. K. 1973. On Prediction of Extreme Values. *J. Ship Res.*, Vol. 17, No. 1, pp. 29-37.

Ochi, M. K. and Hubble, E. N. 1976. Six-Parameter Wave Spectra. *Proc. 15th Coastal Eng. Conf.*, Honolulu, Vol. I, pp. 301-328.

Ochi, M. K. and Wang, S. 1976. Prediction of Extreme Wave-Induced Loads on Ocean Structures. *Proc. Behaviour of Off-Shore Structures, BOSS '76*, Trondheim, Vol. I, pp. 170-186.

Panicker, N. W. 1974. Review of Techniques for Directional Wave Spectra. *Proc. Int. Symp. on Ocean Wave Measurement and Analysis*, ASCE, New Orleans, pp. 669-688.

Petrauskas, C. and Aagaard, P. 1971. Extrapolation of Historical Storm Data for Estimating Design-Wave Heights. *J. Soc. Petrol. Eng.*, Vol. 11, pp. 23-37.

Phillips, O. M. 1957. On the Generation of Waves by Turbulent Wind. *JFM*, Vol. 2, pp. 417-445.

Phillips, O. M. 1967. The Theory of Wind-Generated Waves. *Advances in Hydroscience*, Vol. 4, pp. 119-149.

Phillips, O. M. 1977. *The Dynamics of the Upper Ocean*. 2nd ed., Camb. Univ. Press.

Pierson, W. J. and Holmes, P. 1965. Irregular Wave Forces on a Pile. *J. Waterways and Harbors Div., ASCE*, Vol. 91, No. WW4, pp. 1-10.

Pierson, W. J. and Moskowitz, L. 1964. A Proposed Spectral Form for Fully Developed Wind Seas Based on the Similarity Theory of S.A. Kitaigorodskii. *J. Geophys. Res.*, Vol. 69, pp. 5181-5190.

Pierson, W. J., Neumann, G., and James, R. W. 1955. Practical Methods for Observing and Forecasting Ocean Waves. *U.S. Hydrogr. Office*, Pub. No. 603.

Pierson, W. J., Tick, L. J., and Baer, L. 1966. Computer Based Procedure for Preparing Global Forecasts and Wind Field Analysis Capable of Using Wave Data Obtained From a Spacecraft. *Proc. 6th Symp. Naval Hydrodynamics*, Washington, D.C.

Price, W. G. and Bishop, R. E. D. 1974. *Probabilistic Theory of Ship Dynamics*. Chapman and Hall, London.

Putz, R. R. 1952. Statistical Distribution for Ocean Waves. *Trans. Am. Geophys. Union*, Vol. 33, No. 5, pp. 685-692.

Rice, S. O. 1944-1945. Mathematical Analysis of Random Noise. *Bell System Tech. J.*, Vol. 23, pp. 282-332; Vol. 24, pp. 46-156; also in *Noise and Stochastic Processes*, ed. N. Wax, Dover, New York, 1954, pp. 133-294.

Rye, H. 1974. Wave Group Formation Among Storm Waves. *Proc. 14th Coastal Eng. Conf.*, Copenhagen, Vol. I, pp. 164-183.

St. Denis, M. 1969. On Wind Generated Waves. In *Topics in Ocean Engineering*, ed. C. L. Bretschneider, Gulf Publishing Co., Texas, Vol. I, pp. 37-41.

St. Denis, M. 1973. Some Cautions on the Employment of the Spectral Technique to Describe the Waves of the Sea and the Response Thereto of Oceanic Systems. *Proc. Offshore Tech. Conf.*, Houston, Paper No. OTC 1819, pp. 827–837.

St. Denis, M. and Pierson, W. J. 1953. On the Motions of Ships in Confused Seas. *Trans. SNAME*, Vol. 61, pp. 280–357.

Sanders, J. W. 1976. A Growth-Stage Scaling Model for the Wind-Driven Sea. *Deut. Hydrogr. Z.*, Vol. 29, pp. 136–161.

Schüeller, G. I. and Shah, H. C. 1972. Probabilistic Approach to Determine Wave Forces on Ocean Pile Structures. *Proc. 13th Coastal Eng. Conf.*, Vancouver, Vol. III, pp. 1683–1702.

Scott, J. R. 1965. A Sea Spectrum for Model Tests and Long-Term Ship Protection. *J. Ship Res.*, Vol. 9, pp. 145–152.

Seymour, R. J. 1977. Estimating Wave Generation on Restricted Fetches. *J. Waterway Port Coastal and Ocean Div., ASCE*, Vol. 103, No. WW2, pp. 251–264.

Sharma, J. N. and Dean, R. G. 1979. Second-Order Directional Seas and Associated Wave Forces. *Proc. Offshore Tech. Conf.*, Houston, Paper No. OTC 3645, Vol. III, pp. 2505–2514.

Siefert, W. 1976. Consecutive High Waves in Coastal Waters. *Proc. 15th Coastal Eng. Conf.*, Honolulu, Vol. I, pp. 171–182.

Silvester, R. 1974. *Coastal Engineering, 1, 2.* Elsevier, Amsterdam.

Sverdrup, H. U. and Munk, W. H. 1947. Wind, Sea and Swell; Theory of Relations for Forecasting. *U.S. Navy Hydrogr. Office*, Pub. No. 601.

Tang, F. L. W. and Ou, S-H. 1972. Researches on the Deformation of Wave Spectra in Intermediate Water Area by Calculation. *Proc. 13th Coastal Eng. Conf.*, Vancouver, Vol. I, pp. 271–288.

Thom, H. C. S. 1954. Frequency of Maximum Wind-Speeds. *J. Struc. Div., ASCE*, Vol. 80.

Thom, H. C. S. 1971. Asymptotic Extreme Value Distributions of Wave Heights in the Open Ocean. *J. Mar. Res.*, Vol. 29, pp. 19–27.

Thom, H. C. S. 1973a. Distribution of Extreme Winds Over Oceans. *J. Waterways Harbors and Coastal Eng. Div., ASCE*, Vol. 99, No. WW1, pp. 1–17.

Thom, H. C. S. 1973b. Extreme Wave Height Distribution Over Oceans. *J. Waterways Harbors and Coastal Eng. Div., ASCE*, Vol. 99, No. WW3, pp. 355–374.

Thompson, W. C. 1972. Period by the Wave-Group Method. *Proc. 13th Coastal Eng. Conf.*, Vancouver, Vol. I, pp. 197–214.

Thornton, E. B. and Calhoun, R. J. 1972. Spectral Resolution of Breakwater Reflected Waves. *J. Waterways Harbors and Coastal Eng. Div., ASCE*, Vol. 98, No. WW4, pp. 443–460.

Thrasher, L. W. and Aagaard, P. M. 1970. Measured Wave Force Data on Offshore Platforms. *J. Petrol. Tech.*, Vol. 22, pp. 339–346.

Tick, L. J. 1959. A Non-Linear Random Model of Gravity Waves I. *J. Maths. and Mech.*, Vol. 8, No. 5, pp. 643–651.

Tick, L. J. 1961. Nonlinear Probability Models for Ocean Waves. In *Ocean Wave Spectra*, Prentice-Hall, Englewood Cliffs, N.J., pp. 163–169

Tickell, R. G. 1977. Continuous Random Wave Loading on Structural Members. *The Structural Engineer*, Vol. 55, No. 6, pp. 209–222.

Tickell, R. G. 1979. The Probabilistic Approach to Wave Loading on Marine Structures. In *Mechanics of Wave-Induced Forces on Cylinders*, ed. T. L. Shaw, Pitman, London, pp. 152–178.

Tickell, R. G., Burrows, R., and Holmes, P. 1976. Long-Term Wave Loading on Offshore Structures. *Proc. Inst. Civil Eng.*, Vol. 61, pp. 145–162.

Tucker, M. J. 1963. Analysis of Records of Sea Waves. *Proc. Inst. Civil Eng.*, Vol. 26, pp. 305–316.

Tung, C. C. 1974. Peak Distribution of Random Wave-Current Force. *J. Eng. Mech. Div., ASCE*, Vol. 100, No. EM5, pp. 873–884.

Tung, C. C. and Huang, N. E. 1973a. Statistical Properties of Wave-Current Force. *J. Waterways Harbors and Coastal Eng. Div., ASCE*, Vol. 99, No. WW3, pp. 341–354.

Tung, C. C. and Huang, N. E. 1973b. Combined Effects of Current and Waves on Fluid Force. *Ocean Eng.*, Vol. 2, pp. 183–193.

Tung, C. C. and Huang, N. E. 1976. Interactions Between Waves and Currents and Their Influence on Fluid Forces. *Proc. Behaviour of Off-Shore Structures, BOSS '76*, Trondheim, Vol. I, pp. 129–143.

U.S. Army, Coastal Engineering Research Center. 1973. *Shore Protection Manual.*

Wheeler, J. D. 1970. Method for Calculating Forces Produced by Irregular Waves. *J. Petrol. Tech.*, Vol. 22, pp. 359–367.

Wiegel, R. L. 1949. An Analysis of Data from Wave Recorders on the Pacific Coast of the United States. *Trans. Am. Geophys. Union*, Vol. 30, pp. 700–704.

Wiegel, R. L. 1964. *Oceangraphical Engineering.* Prentice-Hall, Englewood Cliffs, N.J.

Wiegel, R. L., Beebe, K. E., and Moon, J. 1957. Ocean Wave Forces on Circular Cylindrical Piles. *J. Hyd. Div., ASCE*, Vol. 83, No. HY2, pp. 89–116.

Wilson, B. W. 1955. Graphical Approach to the Forecasting of Waves in Moving Fetches. U.S. Army Corps of Engineers, *Beach Erosion Board*, Tech. Memo. No. 73.

Wilson, B. W. 1961. Deep Water Wave Generation by Moving Wind Systems. *J. Waterways and Harbors Div., ASCE*, Vol. 87, No. WW2, pp. 113–141.

Wilson, B. W., Chakrabarti, S. K., and Snider, R. H. 1974. Spectrum Analysis of Ocean Wave Records. *Proc. Int. Symp. on Ocean Wave Measurement and Analysis*, ASCE, New Orleans, pp. 87–106.

Yang, C. Y., Tayfun, M. A., and Hsiao, G. C. 1974. Extreme Wave Statistics for Delaware Coastal Waters. *Proc. Int. Symp. on Ocean Wave Measurement and Analysis*, ASCE, New Orleans, pp. 352–361.

Yang, C. Y., Tayfun, M. A., and Fallah, M. H. 1975. Extreme Wind, Wave and Tide. *Proc. Civil Engineering in the Oceans III*, ASCE, Univ. of Delaware, pp. 134–141.

8

Dynamic Response of Framed Structures and Vortex-Induced Oscillations

8.1 INTRODUCTION

In the design of offshore structures, particularly in deep waters, dynamic response of the entire structure to surface and internal waves, currents, wind and gust, and earthquake loading and of the individual members to cyclic buoyancy, wave impact, and wave and/or current-induced vortex forces must be considered. A dynamic analysis is particularly important for waves of moderate height since they make the greatest contribution to fatigue damage. The magnification due to dynamic response is greatest when the sea state dominant frequency coincides with the structure natural frequency and the degree of magnification depends on the damping of the system. The first essential step in such an analysis is the formulation of an appropriate mathematical model representing the structure and its foundation by suitably defined mass, damping, and stiffness matrices and the hydrodynamic force by a modified version of Morison's equation. The dynamic response analysis of local members to impact and vortex-induces forces require special consideration.

8.2 BASIC ASSUMPTIONS AND UNCERTAINTIES

Dynamic analysis turns out to be an exceedingly complex problem not so much because of the difficulty of solving the resulting equations (often linearized and uncoupled) but rather because of the limitations of the existing knowledge for an adequate description of the sea state and the interactions of the sea, structure, and the soil. The method of analysis depends on a critical assessment of all the input parameters. It is necessary to achieve a proper balance between computational accuracy, model sophistication, physical realities, and the uncertainties of the input parameters.

The problem reduces to the solution of the dynamic equation of motion given by

$$[M]\{\ddot{x}\} + [C]\{\dot{x}\} + [K]\{x\} = \{P\} \tag{8.1}$$

where $\{x\}$ is the displacement vector; $[M]$, the mass matrix; $[K]$, the stiffness matrix; $[C]$, the damping matrix; and $\{P\}$, the time-dependent load vector. Clearly, the response will be amplified due to resonant effects when the frequency of the loading is close to a natural frequency of the structure. Thus, the static analysis based only on the third and fourth terms of Eq. (8.1) is often followed by a dynamic analysis.

Every term in Eq. (8.1) is subject to varying degrees of approximations.

Mass matrix: The structure is often modeled as lumped nodal masses. The lumping of the system into discrete points is not unique and the choice made may have some effect on the ensuing solutions (natural frequencies, mode shapes, etc.). The mass matrix often includes the frame mass, added mass of water (often assumed to be equal to its ideal potential flow value*), piles mass, fluid inside the piles, superstructure, and the mass of the accumulated excrescences. The lumping of the system into discrete points along one, or two, or more columns may yield different results not only because of the differences in the mass, damping, degrees of freedom, and force distributions but also because of the differences of the wave forces on the structure (i.e., the cancelling effect obtained when the wave length is a certain multiple of the leg spacing). [M] may or may not include rotational inertia depending on the desire of the designer. Clearly, the mass matrix, which at first sight appears to be the most deterministic of all, is not free from ambiguities. The fact that the rate of change of the fluid forces with depth as well as the displacements of the structure are largest (for the first mode) near the free surface, the discretization of mass in the upper parts of the structure deserves special care (see Fig. 8.1).

The discretization depends on the structure and personal judgment. Usually a static analysis is performed applying the distributed mass of the structure to

---

*Added mass of vibrating cylinders in steady and wavy flows is smaller than the ideal value (see Fig. 3.44). Thus, the mass is overestimated. This is desirable since it lowers the natural frequency of the structure.

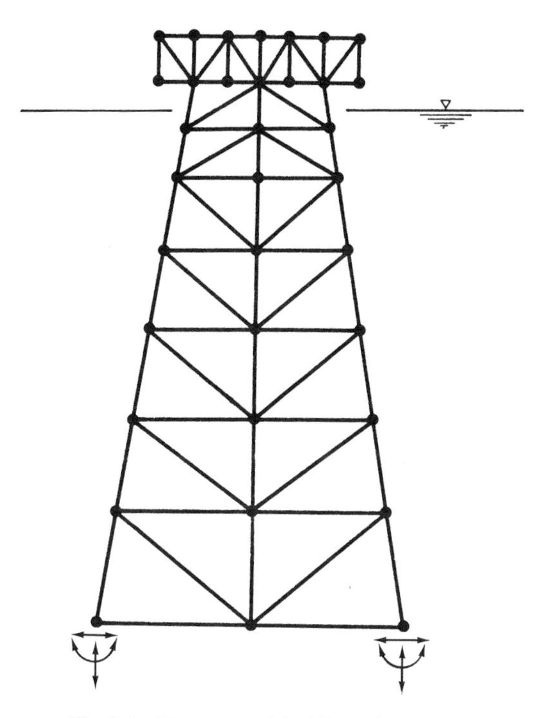

Fig. 8.1. Discrete model of framed structure.

determine the reactions at the mass nodes, restrained by simple pinned joints. These reactions are then taken as the lumped mass values. One may also use Guyan's (1965) energy consistent mass condensation technique but it leads to a full (as opposed to a diagonal) mass matrix (for numerical examples see Hallam et al. 1977).

Damping matrix is subject to large uncertainty. It plays a very important role in the determination of the magnification factor. Ordinarily, [C] would include only the structural damping (often assumed linear viscous), nonlinear interface damping at the joints, and the foundation radiation and foundation damping. These are not easy to determine and often recourse is made to field measurements and past experience (field measurements naturally include the fluid effects). Strictly speaking, there would be no need for the "fluid damping" if one were able to express all the fluid forces resulting from the fluid-structure interaction in {P}. However, it is common practice to linearize the nonlinear drag force and regard part of it as the fluid damping and part as the driving force. This procedure does not violate the basic principles. However, the uncertainty stems partly from the linearization (or from the approximation of the nonlinear drag by another relatively more manageable nonlinear form) and partly from the use of force-transfer coefficients appropriate to static conditions. The force transfer coefficients (C_d, C_m, C_L) under static conditions are not the same under dynamic conditions (see e.g., Figs. 3.43 and 3.44). In general, the force-transfer coefficients depend on K, Re, k/D, x/D, V_c/U_m, proximity effects, member orientation, etc. Existing analyses do not take this into consideration primarily because of lack of reliable data. The drag and inertia coefficients are often assumed to be constant. In the range of $10 < K < 30$, damping calculated through the use of static values of C_d may be overestimated by a factor of 2 (see Fig. 3.43). Furthermore, the lift force is never taken into consideration in the dynamic analysis of the full structure. This is justifiable partly because of phase shifts in lift acting on various members and partly because of the relatively high frequency of vortex shedding. For individual members as well as tube bundles, however, the periodic lift force is extremely important and may give rise to in-line and/or transverse oscillations.

Finally, it should be noted that the damping is important not only near resonance but also away from resonance (near one of the secondary peaks). Thus, considerable research is needed in quantifying $[C]$ in terms of the appropriate parameters for the transient as well as steady-state response of structures (note that the stiffness and damping of soil in the transient state may significantly differ from those in steady state).

Stiffness matrix can be calculated by the normal structural methods used in static design analysis. The uncertainties associated with the stiffness matrix stem primarily from the soil-structure interaction. For a detailed discussion of the approximate methods and values the reader is referred to Matlock and Reese (1970), Matlock (1970), McClelland (1974), and API RP 2A (1977) for the pile foundations, and to Barkan (1962), Hsieh (1962), Whitman (1966), Girard and Picard (1970), Richart (1970), Whitman (1972), Hallam et al. (1977), and Zienkiewicz et al. (1978) for the gravity structures.

Force vector: Its specification requires numerous simplifying assumptions: (a) The sea state and the kinematics of the flow field may be accurately represented by the existing deterministic or stochastic methods; (b) The modified Morison equation (with appropriate relative velocities and accelerations) is valid for all members of the structure irrespective of the amplitude of their motion and of their orientation with respect to a global coordinate system; (c) The force-transfer coefficients determined for static conditions are applicable to yawed cylinders undergoing oscillations either individually and/or as part of the entire structure in multidirectional waves and currents; (d) Interference effects and the modification of the flow and force due to complex assemblages of structural members are negligible; and (e) Morison's equation may be linearized or the full nonlinear term may be replaced by another approximate relation which retains the nonlinear character of the drag force.

There is neither enough data nor sufficient physical understanding of the flow situations to be encountered for a critical assessment of the errors and the mitigating effects introduced by these assumptions. The purpose of the foregoing is partly to point out, albeit indirectly, the numerous research topics and partly to warn the reader that one should not try to read too much into the results and one should not think that the more complex the model, the more accurate the results will be.

8.3 APPRROXIMATE EQUATIONS OF MOTION

Let us first consider a single member of mass M (mass of pipe and the fluid inside), structural damping C, and stiffness K. Assume that this member is normal to the direction of wave propagation. Then the equation of motion, within the limitations of the assumptions cited above, takes the following form [see Eq. (2.55)]

$$\begin{aligned} M\ddot{x} + C\dot{x} + Kx = P(t) &= \rho\forall\ddot{u} + \rho(C_m - 1)\forall(\ddot{u} - \ddot{x}) + 0.5\rho C_d A|\dot{u} - \dot{x}|(\dot{u} - \dot{x}) \\ &= \rho\forall C_m\ddot{u} - \rho(C_m - 1)\forall\ddot{x} + 0.5\rho C_d A|\dot{u} - \dot{x}|(\dot{u} - \dot{x}) \qquad (8.2) \end{aligned}$$

in which u, $\dot{u}$, and $\ddot{u}$ represent, respectively, the displacement, velocity, and the acceleration of the fluid.*

Equation (8.2) is often written as

$$(M + \rho C_a\forall)\ddot{x} + C\dot{x} + Kx = \rho\forall C_m\ddot{u} + 0.5\rho C_d A|\dot{u} - \dot{x}|(\dot{u} - \dot{x}) \qquad (8.3)$$

where $C_a = C_m - 1$, as usual, and $(M + \rho C_a\forall)$ now represents the structural mass, water inside the pipe, and the added mass. The last term in Eq. (8.3), representing the drag force due to the relative velocity of fluid, is nonlinear. In order to apply spectral methods the drag term is often linearized with an equivalent linearization technique (Krylov and Bogoliubov 1947, Lin 1967, Malhotra and Penzien 1970). Defining $\dot{v}_r$ as the relative velocity

$$\dot{v}_r = \dot{u} - \dot{x} \qquad (8.4)$$

one writes the quadratic term as

$$|\dot{v}_r|\dot{v}_r = a_f\dot{v}_r \qquad (8.5)$$

*Fluid dynamicists prefer to denote the x-component of velocity by u. However, practically every paper dealing with dynamic response denotes the said velocity of $\dot{u}$. This practice has been followed here for ease of perusal of other works. Clearly, it eliminates any ambiguity which might result from the meaning of single and double dots on x and u.

where a_f is the linearized drag factor. With the assumption that the excitation is a zero mean ergodic Gaussian process and the system is linear, the expression for a_f reduces to

$$a_f = \sqrt{8/\pi}\, \sigma_{\dot{v}_r} \tag{8.6}$$

where $\sigma_{\dot{v}_r}$ is the root mean square value of the relative velocity.*

Letting $S_{\dot{v}_r \dot{v}_r}$ represent the one-sided spectral density function for $\dot{v}_r$, $\sigma_{\dot{v}_r}$ is determined from

$$\sigma^2_{\dot{v}_r} = m_0 \tag{8.7}$$

where

$$m_\lambda = \int_0^\infty \omega^\lambda S_{\dot{v}_r \dot{v}_r}(\omega)\, d\omega \tag{8.8}$$

Combining Eqs. (8.3), (8.4), and (8.5), one obtains

$$(M + \rho C_a \forall)\ddot{x} + (C + 0.5\rho C_d A a_f)\dot{x} + Kx = \rho \forall C_m \ddot{u} + 0.5\rho C_d A a_f \dot{u} \tag{8.9}$$

where $(0.5 C_d A a_f)$ is now interpreted as the hydrodynamic or fluid-damping coefficient. Evidently, the evaluation of the fluid damping is an iterative process, beginning with $\dot{x} = 0$ on the first iteration.

The linearization introduced into the Eq. (8.9) is effective when the drag term plays a relatively minor role. Otherwise, (e.g., when the critical loading is caused by high-amplitude long-period waves giving rise to small amplification of response) the nonlinearity of the drag term may be retained through the use of the approximate relation (Penzien 1976)

$$|\dot{u} - \dot{x}|(\dot{u} - \dot{x}) = |\dot{u}|\dot{u} - 2(\langle|\dot{u}|\rangle)\dot{x} \tag{8.10}$$

where $\langle|\dot{u}|\rangle$ represents the time average of $|\dot{u}|$. Then Eq. (8.3) becomes

$$(M + \rho C_a \forall)\ddot{x} + (C + \rho C_d A \langle|\dot{u}|\rangle)\dot{x} + Kx = \rho \forall C_m \ddot{u} + 0.5\rho C_d A |\dot{u}|\dot{u} \tag{8.11}$$

*For a more generalized expression for a_f including a steady and constant current V_c, the reader is referred to Wu and Tung (1975). It should be used with care since currents can change the wave form and direction, wake structure, wake bias, and the force-transfer coefficients.

In the foregoing we have dealt only with a single element normal to the direction of flow for the purpose of emphasizing the underlying assumptions while keeping the equations relatively uncluttered. The next question to deal with is the effect of orientation. This matter has been discussed in detail in Section 5.3.6 and it has been pointed out that the appropriate method for the determination of the fluid force is not yet clear because of lack of data and that the "independence principle" may be used with some reservations. In other words, one should use, in the modified Morison equation, the normal component of the velocities. It should be emphasized that this does not necessarily render the dynamic response analysis more accurate relative to the use of other methods, e.g., the assumption that the resultant drag pressure acts on an area projected on a plane normal to the total water particle velocity. In either case one does not know the appropriate force-transfer coefficients. Thus, it is not simply the use of one or the other method but rather the use of the combination of the method of velocity, acceleration, and force decomposition and the force-transfer coefficients that dictate the relative accuracy of the predictions. Many authors have used the simple "projected area" method rather than the normal component of velocity (see e.g., Berge and Penzien 1974, Penzien 1976, Wu 1976, Cronin et al., 1978, Angelides and Connor 1979). The computational simplicity afforded by this approximation is evident: one does not have to resolve the velocity and acceleration vectors (not collinear) into normal and tangential components on each and every element, determine the drag and inertial forces due to these normal components, and then resolve the resultant forces back to the global coordinates.

The total force at a node is obtained by summing the contributions of the members contained in the nodal tributary zone (the region halfway above and below the node). Using the projected area method rather than the cross-flow assumption and introducing the double subscript ij to denote the j-th member in zone i, Angelides and Connor (1979) expressed the fluid force in Eq. (8.2) as

$$P_i(t) = \sum_j \rho(C_{mij} - 1)\frac{\pi}{4} D_{ij}^2 L_{ij}(\ddot{u}_{ij} - \ddot{x}_i) + \rho\frac{\pi}{4} D_{ij}^2 L_{ij}\ddot{u}_{ij}$$

$$+ 0.5\rho C_{dij} D_{ij} L_{ij} |\dot{u}_{ij} - \dot{x}_i|(\dot{u}_{ij} - \dot{x}_i) \quad (8.12)$$

where C_{dij}, C_{mij} are the drag and inertia coefficients for the member ij, and D_{ij} and L_{ij} are the diameter and projected length of the tubular members at node level i.

The drag term is linearized in a manner similar to those given by Eqs. (8.4) through (8.8) and yields

$$\dot{v}_{rij} = \dot{u}_{ij} - \dot{x}_i \quad (8.4a)$$

$$|\dot{v}_{rij}|\dot{v}_{rij} = a_{fij}\dot{v}_{rij} \tag{8.5a}$$

$$a_{fij} = \sqrt{\frac{8}{\pi}}\,\sigma_{\dot{v}_{rij}} \tag{8.6a}$$

$$\sigma^2_{\dot{v}_{rij}} = (m_0)_{ij} \tag{8.7a}$$

$$(m_\lambda)_{ij} = \int_0^\infty \omega^\lambda S_{\dot{v}_{rij}\dot{v}_{rij}}(\omega)\, d\omega \tag{8.8a}$$

Note that the method used by Angelides and Connor (1979) allows for the variation of the force-transfer coefficients as a function of the Keulegan-Carpenter and the Reynolds number, defined by

$$(Re)_{ij} = \sigma_{\dot{v}_{rij}} D_{ij}/\nu \quad \text{and} \quad (K)_{ij} = \sigma_{\dot{v}_{rij}}(T_0)_{ij}/D_{ij} \tag{8.13}$$

where $(T_0)_{ij}$ is defined by

$$(T_0)_{ij} = 2\pi\,[(m_0)_{ij}/(m_2)_{ij}]^{1/2} \tag{8.14}$$

The above expressions, coupled with iteration on $\sigma_{\dot{v}_{rij}}$, allow one to identify the consistent values for the force coefficients and a_{fij}.

Superimposing the upper structure and foundation stiffness at the base node, Eq. (8.2) is generalized to

$$(\mathbf{M} + \mathbf{M_a})\ddot{\mathbf{x}} + (\mathbf{C} + \mathbf{C}_{fd})\dot{\mathbf{x}} + (\mathbf{K} + \mathbf{K}_F)\mathbf{x} = \mathbf{F} \tag{8.15}$$

in which **M**, **C**, **K** represent the conventional mass, viscous damping, and complex stiffness and K_F, the complex foundation stiffness matrix. At node i, one has

$$M_{ai} = \rho\frac{\pi}{4}\sum_j (C_{mij} - 1)D_{ij}^2 L_{ij} \tag{8.16}$$

$$C_{fdi} = 0.5\rho\sum_j C_{dij} D_{ij} L_{ij} a_{fij} \tag{8.17}$$

and

$$F_i = \rho\,\frac{\pi}{4}\sum_j D_{ij}^2 C_{mij}\ddot{u}_{ij} + 0.5\rho\sum_j C_{dij} D_{ij} L_{ij} a_{fij}\dot{u}_{ij} \tag{8.18}$$

Angelides and Connor (1979) determined the quasi-linear pile stiffness coefficients by extending the finite element pile model of Blaney et al. (1976) to evaluate the vertical stiffness of piles, simulate floating piles, and to treat non-linear soil behavior in an approximate way (Angelides 1978). Equation (8.18) was then solved in the frequency domain for the complex nodal displacements through the iteration of the force-transfer coefficients given by Sarpkaya (1976) and through the use of a suitable wave spectrum and the spectral density function for the relative velocity. For additional details of the numerical procedure used, the reader is referred to Angelides (1978) and Angelides and Connor (1979).

The extensive sensitivity analysis carried out by Angelides and Connor of the response to various parameters have shown that (a) the use of K and Re-dependent C_d and C_m yields more conservative values than those based on constant force-transfer coefficients; (b) roughness effects are important only when the force is drag dominated (Angelides and Connor did not allow for the increase of D_{ij} with roughness); (c) the inclusion of the flexibility of the foundation lowers the natural frequency and excites the fundamental mode and increases the r.m.s. value of the displacements, at least for the case studied by Angelides and Connor; and (d) the effect of foundation-stiffness degradation becomes more pronounced with increasing wave heights. Representative results obtained by Angelides and Connor are shown in Figs. 8.2 through 8.5.

It should be noted again that the above conclusions depend to an unknown extent on the numerous assumptions made and are for a specific structure only. Nevertheless, they do give some insight into the dynamic behavior of similar structures.

Penzien and Tseng (1978) used the cross-flow component of the velocities and accelerations and retained the nonlinear character of the drag force through the use of Eq. (8.10). Firstly, they have shown that the normal fluid velocities and accelerations $\dot{\mathbf{u}}_n(s)$ and $\ddot{\mathbf{u}}_n(s)$ and the normal displacement velocities and accelerations $\dot{\mathbf{r}}_{tn}(s)$ and $\ddot{\mathbf{r}}_{tn}(s)$ are related to the instantaneous velocity and acceleration vectors $\dot{\mathbf{u}}(s)$, $\ddot{\mathbf{u}}(s)$ and the displacement velocity and acceleration vectors $\dot{\mathbf{r}}_t(s)$ and $\ddot{\mathbf{r}}_t(s)$ by

$$\begin{aligned} \dot{\mathbf{u}}_n(s) &= \mathbf{N}\dot{\mathbf{u}}(s), & \ddot{\mathbf{u}}_n(s) &= \mathbf{N}\ddot{\mathbf{u}}(s) \\ \dot{\mathbf{r}}_{tn}(s) &= \mathbf{N}\dot{\mathbf{r}}_t(s), & \ddot{\mathbf{r}}_{tn}(s) &= \mathbf{N}\ddot{\mathbf{r}}_t(s) \end{aligned} \tag{8.19}$$

Here s is the variable dimension measured along the member from its end node, and

$$\mathbf{N} = \begin{bmatrix} (1 - s_1^2) & -s_1 s_2 & -s_1 s_3 \\ -s_1 s_2 & (1 - s_2^2) & -s_2 s_3 \\ -s_1 s_3 & -s_2 s_3 & (1 - s_3^2) \end{bmatrix} \tag{8.20}$$

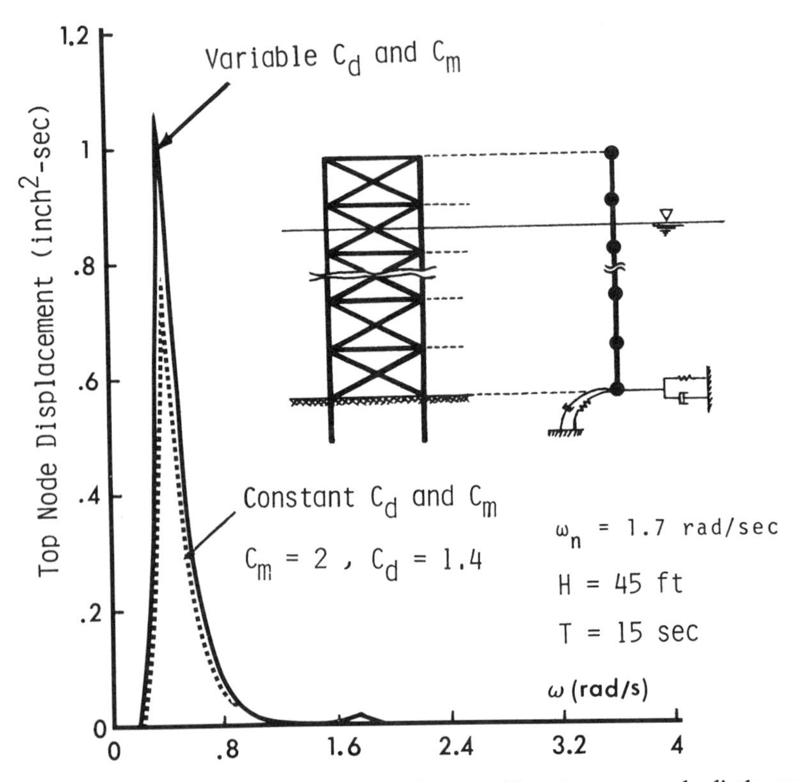

Fig. 8.2. Effect of constant versus variable force coefficient on top node displacement (Angelides and Connor 1979).

where s_1, s_2, and s_3 are the components of unit vectors in the global coordinates. In the foregoing $\ddot{\mathbf{r}}_t$ represents the total acceleration vector of the nodal points and combines the contributions of the nodal acceleration vector $\ddot{\mathbf{r}}$, relative to the moving ground, and the translational ground acceleration $\ddot{\mathbf{r}}_g$, i.e.,

$$\ddot{\mathbf{r}}_t = \ddot{\mathbf{r}} + \ddot{\mathbf{r}}_g = \ddot{\mathbf{r}} + \mathbf{B}\ddot{\mathbf{u}}_g \tag{8.21}$$

where $\ddot{\mathbf{u}}_g$ is the vector representing the translational ground-acceleration time history and $\mathbf{B}$ is a simple matrix relating the direct contribution of ground accelerations to the total nodal acceleration. Also note that

$$\dot{\mathbf{r}}_t(s) = \dot{\mathbf{r}}_g + \dot{\mathbf{r}}(s) \tag{8.22}$$

and

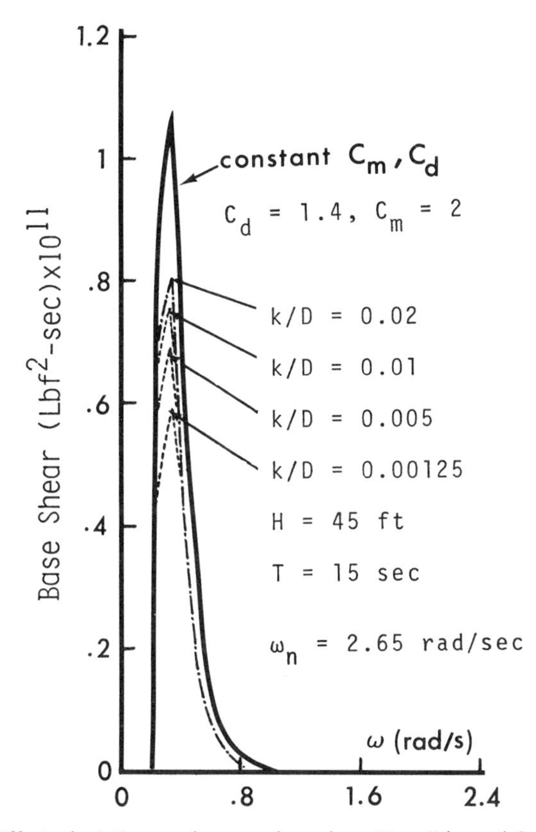

Fig. 8.3. Effect of relative roughness on base shear (Angelides and Connor 1979).

$$\dot{\mathbf{u}}_n(s) - \dot{\mathbf{r}}_{tn}(s) = \dot{\mathbf{u}}_n(s) - \dot{\mathbf{r}}_{gn} - \dot{\mathbf{r}}_n(s) \equiv \dot{\mathbf{q}}_n(s) - \dot{\mathbf{r}}_n(s) = \mathbf{N}[\dot{\mathbf{q}}(s) - \dot{\mathbf{r}}(s)] \tag{8.23}$$

where

$$\dot{\mathbf{q}}_n(s) = \dot{\mathbf{u}}_n(s) - \dot{\mathbf{r}}_{gn} \tag{8.24}$$

The fluid force per unit length along member i at location s becomes [see Eqs. (8.2) and (8.10)]

$$\mathbf{P}(s) = \rho A C_m \mathbf{N}\ddot{\mathbf{u}}(s) - \rho A(C_m - 1)\mathbf{N}\ddot{\mathbf{r}}_t(s) + 0.5\rho D C_d |\mathbf{N}\dot{\mathbf{q}}(s)|\mathbf{N}\dot{\mathbf{q}}(s) - \rho D C_d |\mathbf{N}\dot{\mathbf{q}}(s)|\mathbf{N}\dot{\mathbf{r}}(s) \tag{8.25}$$

Further progress may be made only by rendering $\mathbf{P}(s)$ independent of s, i.e., by

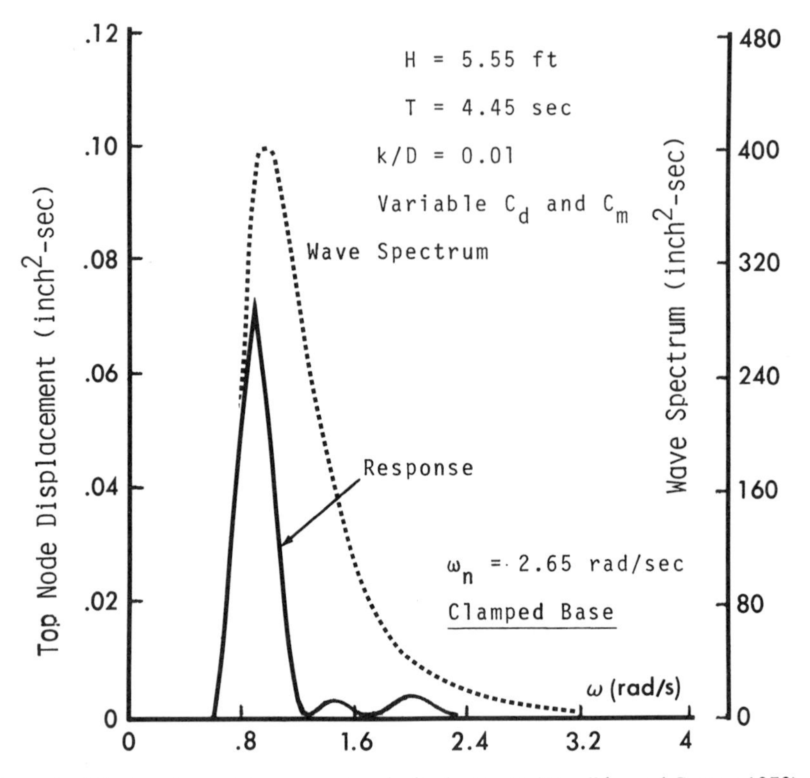

Fig. 8.4. Effect of clamped base on top node displacement (Angelides and Connor 1979).

discretizing $\mathbf{P}(s)$ into a 12-component force vector $\mathbf{P}_i$, corresponding to the 6 nodal displacements at each end of the i-th member. The use of appropriate interpolation functions and finite element procedures enabled Penzien and Tseng to write*

$$\mathbf{P}_i = \rho C_m \mathbf{\forall} \ddot{\mathbf{u}} - \rho(C_m - 1)\mathbf{\forall}\ddot{\mathbf{r}}_t + 0.5\rho C_d |\mathbf{T}\dot{\mathbf{q}}|\mathbf{A}\mathbf{T}\dot{\mathbf{q}} - \rho C_d \langle|\mathbf{T}\dot{\mathbf{q}}|\rangle \mathbf{A} \quad (8.26)$$

in which $\mathbf{\forall} = (\pi D^2/4)L\mathbf{T}$, $\mathbf{A} = DL\mathbf{T}$, and $\mathbf{T}$ is the 12×12 matrix given by

$$\mathbf{T} = \frac{1}{2}\left[\begin{array}{cc|cc} \mathbf{N} & 0 & 0 & 0 \\ 0 & 0 & 0 & 0 \\ \hline 0 & 0 & \mathbf{N} & 0 \\ 0 & 0 & 0 & 0 \end{array}\right] \quad (8.27)$$

*Note that in Penzien and Tseng's notation $C_d = 2K_D$. Throughout this text C_d and C_m represent the commonly defined drag and inertia coefficients.

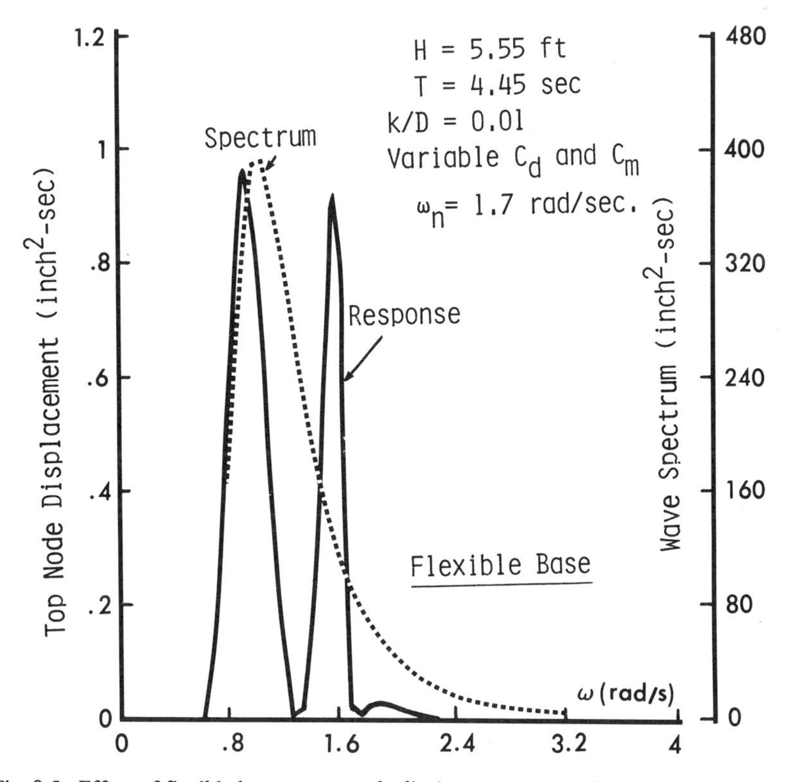

Fig. 8.5. Effect of flexible base on top node displacement (Angelides and Connor 1979).

By assembling $\mathbf{P}_i$ into the n-component vector $\mathbf{P}$, one has

$$\mathbf{P} = \sum_{i=1}^{m} \mathbf{P}_i = \sum_{i=1}^{m} \rho C_{mi} \mathbf{V}_i \ddot{\mathbf{u}}_i - \rho (C_{mi} - 1) \mathbf{V}_i \ddot{\mathbf{r}}_{ti} + 0.5 \rho C_{di} | \mathbf{T}_i \dot{\mathbf{q}}_i | \mathbf{A}_i \dot{\mathbf{q}}_i - \rho C_{di} \langle | \mathbf{T}_i \dot{\mathbf{q}}_i | \rangle \mathbf{A}_i \dot{\mathbf{r}}_i \quad (8.28)$$

where m is the total number of members in the structural system and C_{di} and C_{mi} represent the usual drag and inertia coefficients.

Through the use of the foregoing Penzien and Tseng (1978) reduced Eqs. (8.11) and (8.15) to the following form

$$(\mathbf{M} + \mathbf{M}_a)\ddot{\mathbf{r}} + (\mathbf{C} + \mathbf{C}_{fd})\dot{\mathbf{r}} + \mathbf{K}\mathbf{r} = \sum_{i=1}^{m} \rho C_{mi} \mathbf{V}_i \ddot{\mathbf{u}} - \left[\mathbf{M} + \sum_{i=1}^{m} \rho \mathbf{V}_i (C_{mi} - 1) \right] \mathbf{B} \ddot{u}_g + \mathbf{F}_D \quad (8.29)$$

where

$$M_a = \sum_{i=1}^{m} \rho(C_{mi} - 1)\mathbf{V}_i \tag{8.30}$$

$$C_{fd} = \sum_{i=1}^{m} \rho C_{di} \langle |\mathbf{T}_i \mathbf{q}_i| \rangle \mathbf{A}_i \tag{8.31}$$

and

$$\mathbf{F}_D = \sum_{i=1}^{m} 0.5 \rho C_{di} |\mathbf{T}_i \dot{\mathbf{q}}_i| \mathbf{A}_i \dot{\mathbf{q}}_i \tag{8.32}$$

A comparison of Eqs. (8.15) through (8.18) with Eqs. (8.29) through (8.32) shows that the differences between the two sets of equations stem primarily from the use of the cross-component of velocity, simplified nonlinear form of the drag, and the ground acceleration $\ddot{u}_g$. One set of the equations is not necessarily more accurate than the other because of the many uncertainties and assumptions underlying the development of both sets of equations.

Penzien and Tseng described methods for the modeling of the foundation of the gravity structures and framed structures. For the gravity structures, the foundation impedance are obtained by Veletsos and Wei (1971) and Luco and Westman (1971) assuming a uniform elastic half-space and by Luco (1979) assuming a layered viscoelastic halfspace. For the cap motion of piled foundations, impedance functions are obtained through the use of finite element methods (Lysmer and Kuhlemeyer 1969, Waas and Lysmer 1972, Kansel and Roesset 1975). The detailed discussion of these topics is beyond the scope of this text.

A number of approximate methods has been developed to solve the equations of motion. For a detailed discussion of these methods the reader is referred to Clough and Penzien (1975), Felippa (1976), Hallam et al. (1977), Cronin et al. (1978), Penzien and Tseng (1978), Sigbjörnsson et al. (1978), Wilson (1978), and Angelides and Connor (1979). Numerous computer programs are now available to facilitate the calculation of the dynamic response, stresses, and the fatigue life of offshore structures. The fatigue life is determined through the use of the Palmgren-Miner ratio (Palmgren 1924, Miner 1945, see also Bogdanoff 1978).

It appears from the foregoing that the dynamic-response analysis of a complex structure is in need of a great deal of research. The ambiguities stem primarily from the "fluid" part of the model and from the soil-structure interaction. Future research may help to resolve the effects of linearization of the drag force, variation of the force-transfer coefficients with the amplitude and frequency of oscillations (in addition to Re, K, k/D, and V_c/U_m), proximity and orientation of the members, and of the uncertainties in the description of the sea state. At present, the use of relatively simpler models such as the one developed by Angelides and Connor (1979), together with a careful sensitivity analysis, may provide valuable design information about the end-on, broad-side, and torsional oscillations of framed structures.

8.4 VORTEX-INDUCED OSCILLATIONS

8.4.1 Description and Consequences of the Phenomenon

Numerous experiments have shown that when the vortex-shedding frequency brackets the natural frequency of an elastic or elastically mounted rigid cylinder with a suitable afterbody (capable of giving rise to a transverse force), the cylinder takes control of the shedding in apparent violation of the Strouhal relationship. Then the frequencies of vortex shedding and the body oscillation collapse into a single frequency close to the natural frequency of the body (see Fig. 8.6). This phenomenon is known by various names: lock-in, locking-on, synchronization, hydroelastic or fluid-elastic oscillations, wake capture, self-controlled or self-excited oscillations. A vortex-excited oscillation is actually a forced one having a self-excited character also to some degree due to lift force amplification through nonlinear interactions.

The facts which have emerged from two decades of work on vortex-induced oscillations may be summarized as follows [for extensive reviews see Parkinson (1974), Blevins (1977), Chen (1977), King (1977), Blevins (1979), and Sarpkaya (1979)]:

1. When a body is close to its linear resonance conditions it can undergo sustained oscillations at a frequency close to its natural frequency. That a cylinder should be excited at its natural frequency when the frequency of the exciting force is equal to its natural frequency is not surprising. But that the phenomenon encompass a range of ±25 to 30 percent of the natural vortex-shedding frequency and that the vibration and vortex-shedding frequencies lock together and control the shedding process are surprising.

2. The interaction between the oscillation of the body and the action of the fluid is nonlinear.

3. The vortex shedding does not necessarily result in an alternating transverse force. There must be a suitable afterbody, and hence an alternating lift force, and the shedding frequency should coincide with or be near the natural frequency of the body, or one of its harmonics, for sustained oscillations to occur. Thus, the magnitude and occurrence of sustained oscillations strongly depend on the lift coefficient of the stationary body.

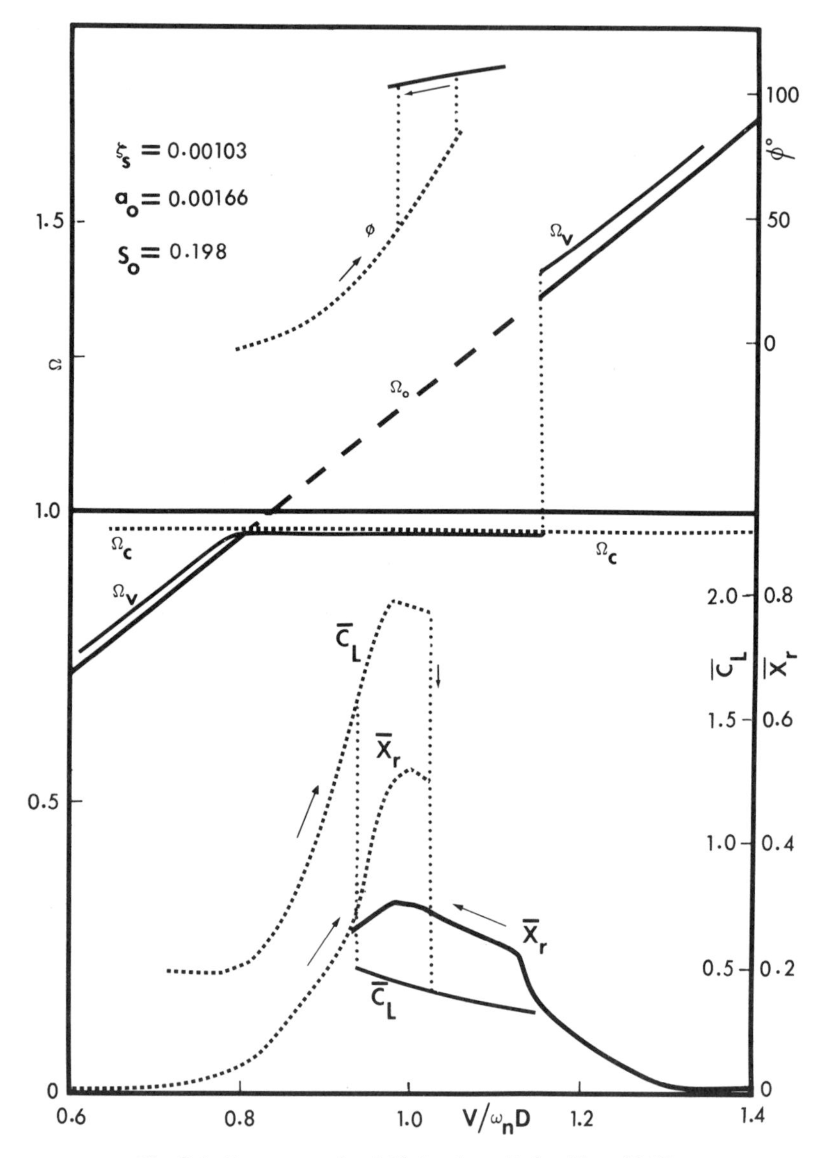

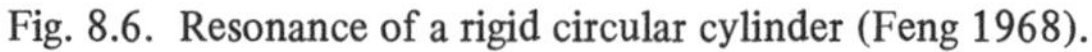

Fig. 8.6. Resonance of a rigid circular cylinder (Feng 1968).

4. The parameters $V_r = V/f_n D$ and $K_r = m\zeta/\rho D^2$ or $R = m\zeta/\rho D^2 C_{L0}$, or $\Delta_r = (2\pi\zeta)(\pi S_0)^2/\rho_r$ are of major importance in determining the amplitude of oscillations and the range of synchronization for a given body. Note that V represents the velocity of the steady flow; f_n, the natural frequency of the cylinder; D, the

diameter of the cylinder; ζ, the damping factor; and C_{L0}, the lift coefficient for the stationary cylinder. The Strouhal number $S_0 = f_0 D/V$ where f_0 is the frequency of vortex shedding for the stationary cylinder in steady flow.

5. The excitation range of cross-flow oscillations in air extends over $4.75 < V_r < 8$ and maximum amplitudes occur in the range $5.5 < V_r < 6.5$. In water, the excitation range of the transverse oscillations can be increased to $4.5 < V_r < 10$ with maximum amplitude falling within the range of $6.5 < V_r < 8$.

6. For a circular cylinder with large L/D, synchronization begins when $f_0 \simeq f_n$ and ends at about $f_0/f_n = 1.4$. The maximum amplitude occurs near the middle of this range. At the end of the lock-in range, vortex-shedding frequency jumps to that governed by the Strouhal relationship, but the cylinder continues to oscillate at $f_c \simeq f_n$, where f_c is the cylinder oscillation frequency. This is true at both ends of the lock-in range and shows that the response is not a simple forced vibration at the exciting natural Strouhal frequency.

7. The correlation length increases rapidly with amplitude. The increase of the correlation length in smooth flow is much larger than in turbulent flow. In smooth flow the correlation length is estimated by numerical extrapolation to increase from about 3.5D to 40D for $Re = 1.9 \times 10^4$ in the range $0.05 < A/D < 0.1$, (A is the amplitude of cylinder oscillation). In turbulent flow, it is again estimated to vary from about 2.5D to 10D in the same A/D range. The rate of increase is steeper than linear but does not show any abrupt change which would indicate a sudden development of the lock-in once a threshold amplitude is achieved.

8. For bodies, with mobile or fixed separation points, undergoing sustained oscillations, the vortex strength is increased. This could be either or both due to increased rate of vorticity flux or due to less destruction of vorticity in the near wake. The growing vortex on a vibrating cylinder seems to roll up more quickly.

9. A hysteresis behavior may exist in the amplitude variation and frequency capture depending on the approach to the resonance range—whether from a low velocity or from a high velocity (see Fig. 8.6). No universal behavior is noted and the reasons are not yet clear. The jump condition (double amplitude response) may originate in the fluid system, and therefore in the lift force, and not in the cylinder elastic system. The jump may be the consequence of a variable structural damping (depending on the interaction of the ambient flow and structure support, etc.) or of a nonlinear spring behavior, or of a sudden change in the shedding of vortices. It is also entirely possible that the jump condition originates in both the fluid and the cylinder elastic system, including blockage effects and the vibrations of the test apparatus.

10. In-line oscillations occur within two adjacent regions. The first is in the range of $1.25 < V_r < 2.5$, maximum amplitudes occurring at $V_r \simeq 2.1$. The second region extends from $V_r \simeq 2.7$ to $V_r \simeq 3.8$ with maximum amplitudes at $V_r \simeq 3.2$. The first instability region is accompanied by symmetric vortex

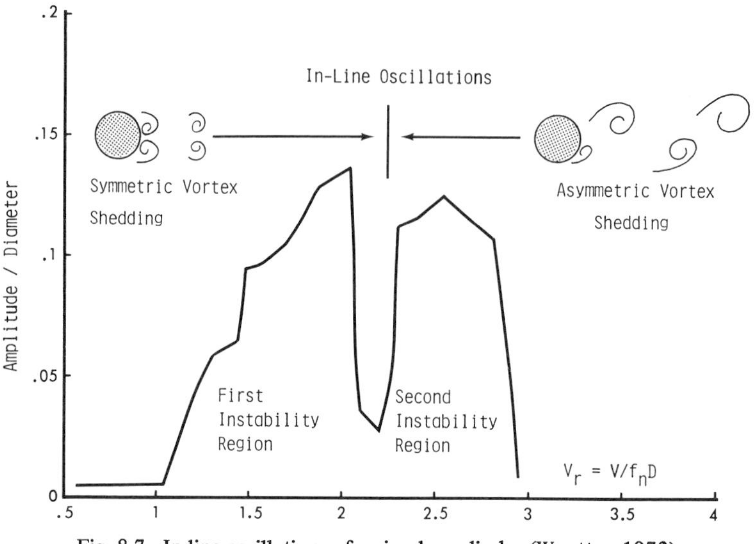

Fig. 8.7. In-line oscillations of a circular cylinder (Wootton 1972).

shedding (as if the flow started impulsively from rest at each cycle) and the second by alternate vortex shedding (see Fig. 8.7).

8.4.2 Wake-Oscillator Models for Transverse Oscillations

Several mathematical models have been proposed in an attempt to simulate and/or explain some of the experimental results just summarized. These models do not include the analysis of the flow field and the fluid-mechanical arguments invoked in their evolution are not altogether convincing. Thus their value should be measured not so much by their capacity to obtain functional relations among significant parameters that lead to the basic understanding of the phenomenon but rather by their ability to produce results which are qualitatively similar to those obtained experimentally.

The most noteworthy among the oscillator models is the one proposed by Hartlen and Currie (1970) where a van der Pol-type soft nonlinear oscillator for the lift force is coupled to the body motion by a linear dependence on cylinder velocity. This model is based partly on a suggestion by Birkhoff and Zarantonello (1953) and by Bishop and Hassan (1963) in connection with their experiments with oscillating cylinders in uniform flow. The pair of equations which result from this concept are

$$\ddot{x}_r + 2\zeta_s\dot{x}_r + x_r = a_0\Omega_0^2 C_L \tag{8.33a}$$

$$\ddot{C}_L - \alpha\Omega_0\dot{C}_L + (\gamma/\Omega_0)\dot{C}_L^3 + \Omega_0^2 C_L = B\dot{x}_r \tag{8.33b}$$

in which*

$$\omega_n = 2\pi f_n = \sqrt{k/m}, \quad \Omega_0 = f_0/f_n, \quad \Omega_c = f_c/f_n, \quad \Omega_v = f_v/f_n,$$

$$\tau = \omega_n t, \quad S_0 = f_0 D/V, \quad x_r = x/D, \quad \bar{x}_r = A/D, \quad \bar{x}_{rm} = (A/D)_{max},$$

$$\dot{x}_r = dx_r/d\tau, \quad C_L = 2F_L/(\rho LDV^2), \quad a_0 = \rho LD^2 f_0^2/(2mS_0^2 \omega_n^2 \Omega_0^2),$$

$$\Delta_r = \zeta_s/a_0 = (2\pi\zeta_s)(\pi S_0)^2/\rho_r, \quad \rho_r = \rho/\rho_s = 2\pi^3 a_0 S_0^2, \quad \omega_0 = 2\pi f_0,$$

$$\omega_v = 2\pi f_v, \quad \omega_c = 2\pi f_c \tag{8.34}$$

The parameters α and γ are the van der Pol coefficients and B is the interaction parameter. Finally, ζ_s represents the material damping factor for the elastic system. There is no other damping imposed on the body.

In Eq. (8.33b), the first and fourth terms generate a simple harmonic oscillator of normalized frequency Ω_0, the second term (the so-called negative lift damping) provides the growth of C_L, and the third term prevents unlimited growth. The fifth term provides the feedback from the body motion to the fluid motion and hence to the lift force. Equation (8.33b) is not the only form which could provide the desired variation in C_L, but it is the simplest.

Of the three parameters (α, γ, and B), two must be chosen to provide the best fit to the data. This is not too unusual and all other models require some retrofitting to experiment. The ratio α/γ is related to the amplitude of the steady-state oscillation of the lift force by $C_{L0} = \sqrt{4\alpha/3\gamma}$.

Experiments have shown that in large-amplitude, steady-state, vortex-induced oscillation, the displacement and the exciting force have nearly sinusoidal forms and oscillate at the same frequency ω_c, close to ω_n, with a phase angle ϕ. Thus,

$$x_r = \bar{x}_r \sin \Omega_c \tau \quad \text{and} \quad C_L = \bar{C}_L \sin(\Omega_c \tau + \phi) \tag{8.35}$$

Substituting Eqs. (8.35) in Eq. (8.33a) and equating to zero the resulting coefficients of $\sin \Omega_c \tau$ and $\cos \Omega_c \tau$, Hartlen and Currie (1970) obtained

$$(1 - \Omega_c^2)\bar{x}_r = a_0 \Omega_0^2 \bar{C}_L \cos\phi \tag{8.36}$$

$$2\zeta \Omega_c \bar{x}_r = a_0 \Omega_0^2 \bar{C}_L \sin\phi \tag{8.37}$$

Likewise, substituting Eqs. (8.35) in Eq. (8.33b), dropping terms arising from

*Additional parameters appearing in Eq. (8.34) will be referred to later.

C_L^3, and using Eqs. (8.36) and (8.37), one has

$$\Omega_0 = \Omega_c \left[1 - \frac{2a_0 B \zeta_s \Omega_c^2}{(1 - \Omega_c^2)^2 + 4\zeta_s^2 \Omega_c^2} \right]^{-1/2} \tag{8.38}$$

which shows that $\Omega_c < \Omega_0$ or $f_c < f_0$.

Hartlen and Currie applied their analysis to a comparison with Jones et al. (1969) data ($Re = 3.6 \times 10^5$ to 1.9×10^7, $\bar{x}_r = 0.08$) with good qualitative results. However, some doubt has been raised (Berger and Wille 1972) as to whether the oscillations in Jones et al., experiments had reached synchronization and whether an amplitude of $\bar{x}_r = 0.08$ was sufficiently greater than the threshold amplitude (Berger 1964, Koopman 1967) necessary to bring about synchronization. Furthermore, Jones et al. data did not show an increase in drag (Sarpkaya 1978, Sarpkaya and Shoaff 1979) in the transcritical range of Reynolds numbers.

There are some important quantitative differences between the predictions of Hartlen and Currie (see Figs. 8.6 and 8.8) and the measurements of Feng (1968):

1. The phase angle between the exciting force and response vary continuously, as does ω_c, $\bar{C}_L$, and $\bar{x}_r$, in Fig. 8.8. There is no oscillation hysteresis.

2. In Feng's data (1968) the cylinder is seen to continue to oscillate at $\omega_c \simeq \omega_n$ on both sides of the region outside the lock-in range. This feature is not presented in Fig. 8.8.

3. The force and response maximums occur in Fig. 8.6 at about the same V_r value. In Fig. 8.8, $\bar{C}_L$ remains constant over a broad range of synchronization, reaching a maximum at a V_r value considerably smaller (relative to the width of the lock-in range) than that for $\bar{x}_r$. The explanation of these discrepancies and the relationship between the particular changes in the wake and/or the elastic system which trigger the hysteresis remained unresolved.

Additional attempts to improve the Hartlen and Currie model have been made by Griffin et al., (1975), Landl (1973), Szechenyi (1975), and Iwan and Blevins (1974). The maximum amplitude at perfect synchronization is given by Iwan and Blevins (1974) as

$$(A/D)_{max} = \frac{0.44\gamma}{\Delta_r + 11.94 S_0^2} \left[0.3 + \frac{4.52 S_0}{\Delta_r + 11.94 S_0^2} \right]^{1/2} \tag{8.39}$$

in which γ is the dimensionless mode shape factor and is given by

$$\gamma = \psi_{max}(y/L) \left[\frac{\int_0^L \psi^2(y)\,dy}{\int_0^L \psi^4(y)\,dy} \right]^{1/2} \tag{8.40}$$

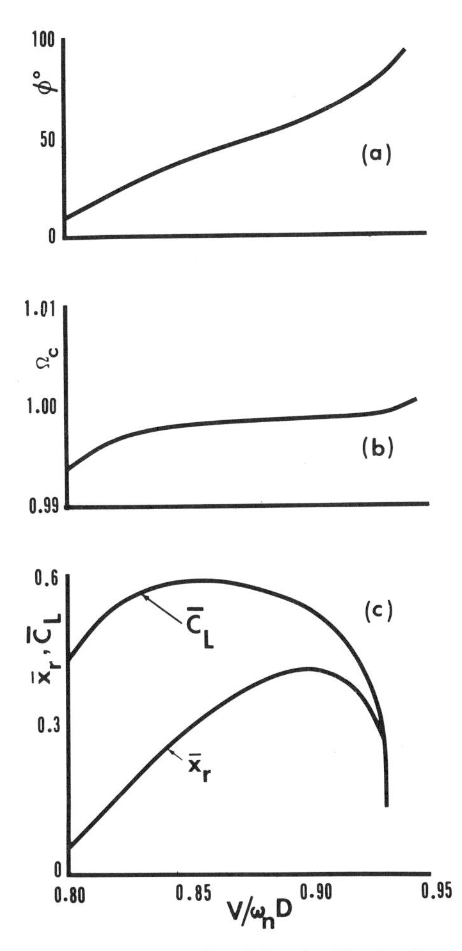

Fig. 8.8. Response characteristics as predicted by the Hartlen-Currie model (Hartlen and Currie 1970).

where $\psi(y)$ is the mode shape at each spanwise point y. Note that $\gamma = 1$ for a spring-supported rigid cylinder, 1.305 for the first mode of a cantilever, and 1.155 for a sinusoidal mode.

The results of Eq. (8.39) are quite sensitive to the Strouhal number, particularly for Δ_r smaller than about 0.4. Griffin et al. (1975) obtained a least-squares fit to the existing data which resulted in

$$(A/D)_{max} = 1.29\gamma/[1 + 0.43\Delta_r^*]^{3.35} \tag{8.41}$$

in which Δ_r^* includes both the structural and still-fluid effects. Sarpkaya (1978)

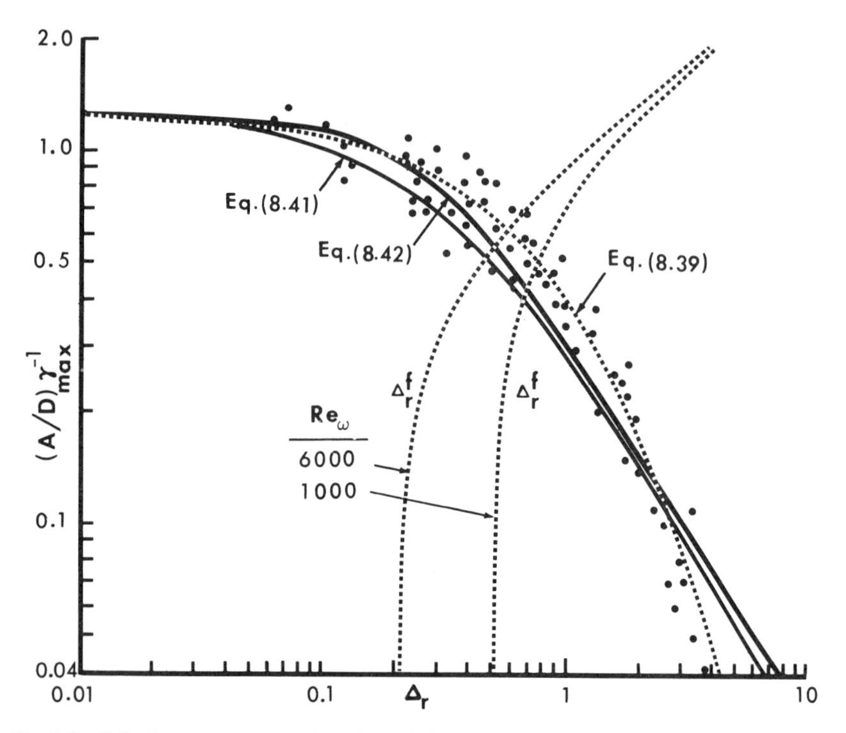

Fig. 8.9. Cylinder response as a function of the response parameter. Full circles represent various data points as compiled by Griffin et al. (1975).

obtained, through a semi-empirical analysis,

$$(A/D)_{max} = 0.31\gamma/[0.062 + \Delta_r^2]^{1/2} \tag{8.42}$$

Equations (8.39), (8.41), and (8.42) yield essentially similar results as shown in Fig. (8.9). For amplitudes of oscillation of order 0.1 diameter or smaller, these equations tend to overpredict the amplitude of the vortex-induced transverse oscillations because of the fact that the vortices are shed with increasingly random spanwise phase.

All of the models cited above correctly predict the self-limiting behavior of the oscillations as Δ_r approaches zero. In fact, Eq. (8.39) yields $(A/D)_{max}$ = 1.37γ; Eq. (8.41), 1.29γ; and Eq. (8.42), 1.25γ.

The effects of cylinder inclination, ambient-flow turbulence, tube proximity, surface roughness, shear, and the time-dependence of the ambient flow on the transverse oscillations of cylinders and other bodies are largely unknown. King (1977b), and Ramberg (1978) have demonstrated that the vortex-induced oscillations of circular cylinders are largely independent of the angle of inclination of the cylinder and that the "independence principle" may be used, i.e., the component of the free stream velocity normal to the cylinder may be incorporated into any and all relations found for the normal incidence flow to describe the flow over a yawed cylinder. Additional experiments at higher Reynolds numbers with direct force measurements are needed.

Body proximity, as in cylindrical arrays in cross-flow, is of great practical importance (Savkar 1976). It appears that the tube-proximity effects considerably reduce the exciting force and the width of the lock-in range (Pettigrew et al. 1978, Grover and Weaver 1978). Furthermore, tube vibrations are often limited to the first few rows and to relatively small Reynolds numbers (Re smaller than about 1,200). There seems to be no lock-in at higher Reynolds numbers.

There are no systematic experiments on the effect of turbulence on vortex-induced vibrations. It is possible that turbulence will increase vorticity diffusion, reduce vortex strengths, and lead to smaller structural response. The effect of flow shear remains largely unexplored (Stansby 1976).

There is an extreme lack of information on the vortex-shedding from and vortex-induced motions of three-dimensional bluff bodies (Rail et al. 1977). Wind and gust induced vibrations of buildings and towers fall in this category (Simui and Scanlan 1978). Considerable work is needed to check the validity of the various modeling and scaling techniques; to determine the influence of the shape of the velocity profile, turbulence intensity and scale, and of the surface roughness on structural response; to develop suitable analytical models; and to devise methods and devices for the reduction of flow excited vibrations.

8.4.3 Flow-Field Models

A number of flow-field models has been proposed (see Parkinson 1974). They are based primarily on the use of one or two concentrated point vortices. The position and the time-dependent strength of these vortices are chosen to give good agreement with some of the measurements. These models are highly empirical and lack the mathematical and phenomenological sophistication necessary to relate the motion of the body to the changes in the boundary layer, separation points, in-line and transverse forces, and the evolution of the wake. The numerical solution of the Navier-Stokes equations for a transversely oscillating cylinder can provide basic information only for laminar wakes where $Re < 300$. Even then the difficulty of describing the time-dependent conditions with a finite-element grid and the computer time needed for a few cycles of oscillations pose extremely difficult problems and confine the Reynolds number range to about 100. The boundary-condition problem may be overcome by introducing a transformation of the Navier-Stokes equations to a non-inertial frame of reference (Hurlbut et al. 1978). Potential flow-field models can provide realistic information on practically every phase of the motion and allow one to conduct

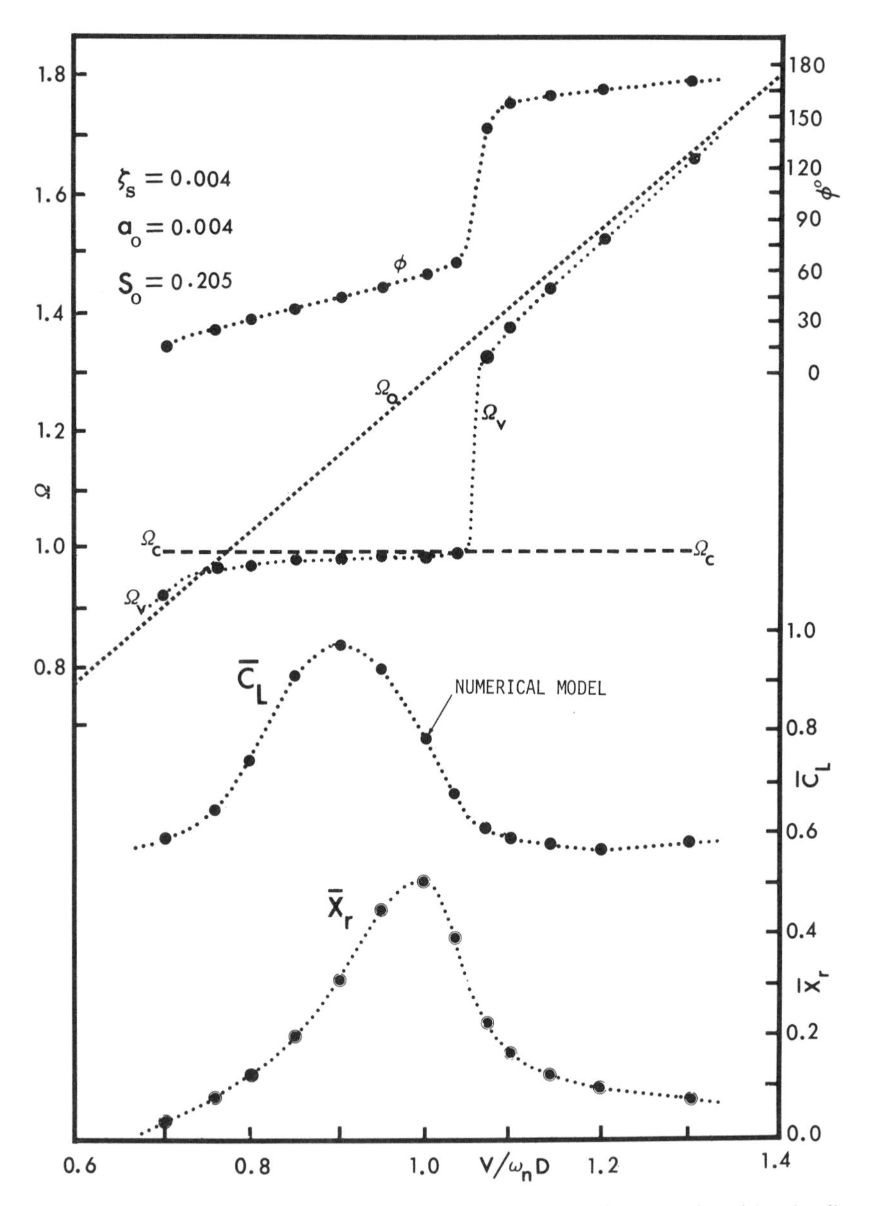

Fig. 8.10. Response characteristics of a freely oscillating cylinder as predicted by the discrete vortex model (Sarpkaya and Shoaff 1979a).

numerical experiments or sensitivity analysis on any part of the flow field or on the variation of any parameter characterizing the elastic system or the flow.

Recently, Sarpkaya and Shoaff (1979a) developed a comprehensive discrete-vortex model based on potential flow and boundary-layer interaction, rediscretization of the shear layers, and the circulation dissipation to determine the characteristics of an impulsively started flow about a circular cylinder. The evolution of the flow from the start to very large times, lift and drag forces, Strouhal number, oscillations of the stagnation and separation points, and the vortex-street characteristics have been calculated and found to be in good agreement with those obtained experimentally. The numerical model was then applied (Sarpkaya and Shoaff 1979) to the flow about a transversely oscillating circular cylinder.

The numerical results obtained with $\zeta_s = 0.004$ and $a_0 = 0.004$ [see Eq. (8.34)] are shown in Fig. 8.10. Other ζ_s and a_0 combinations yielded similar results. Apparently, the results follow essentially the same trends as those in Fig. 8.6. However, no hysteresis effect was found for any combination of ζ_s, a_0, and Ω_0 within the lock-in range. Furthermore, unlike in Fig. 8.6, $\overline{C}_L$ reached its maximum at a $V/\omega_n D$ value smaller than that corresponding to $(\overline{x}_r)_{max}$. The numerical values of $(\overline{x}_r)_{max}$ were found to be slightly larger than those predicted by Eqs. (8.39), (8.41), and (8.42), (for additional details see Sarpkaya and Shoaff 1979b).

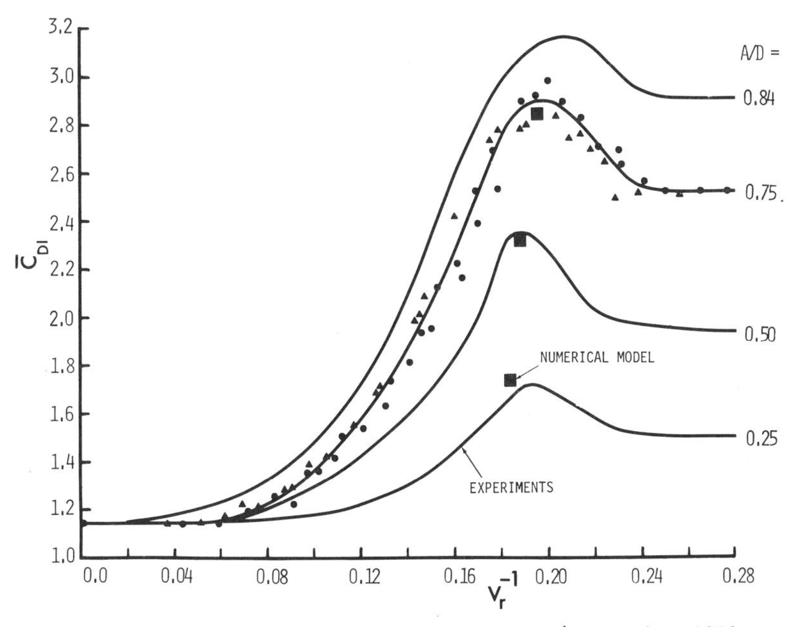

Fig. 8.11. Variation of the in-line drag coefficient with $1/V_r$ (Sarpkaya 1978).

The numerical as well as experimental results (Bishop and Hassan 1963, Griffin et al. 1975, Stansby 1976, King 1977, and Sarpkaya 1978) have shown that the mean in-line force and its amplitude of oscillation increase with increasing amplitude of transverse oscillations. The maximums of the numerically obtained mean values of the in-line drag coefficient C_{DI} are compared in Fig. 8.11 with those obtained experimentally (Sarpkaya 1978). A calculation based on the stationary-cylinder drag coefficient C_{DS} and the apparent projected area for the oscillating cylinder, i.e., $C_{DI} = C_{DS}(1 + 2\overline{x}_{rm})$, yields values that are almost equal to the maximum C_{DI}-values shown in Fig. 8.11. Evidently, the oscillations increase the absolute value of the base pressure as confirmed by experiments (Stansby 1976). It is also evident that the increase of the in-line force due to transverse oscillations must be taken into account in the design of tubular structures.

8.4.4 Transverse Resonant Oscillations in Harmonic Flow

Hydroelastic response of cylinders in time-dependent flows has not been sufficiently explored. Laird (1962) studied the effects of support flexibility by oscillating a vertical cylinder through still water. He found that (1) the forces acting on a flexibly-supported oscillating cylinder can exceed 4.5 times the drag force of the cylinder rigidly-mounted while moving at a uniform velocity equal to the maximum velocity during the oscillation; and that (2) a cylinder, flexible enough to have transverse oscillations with amplitudes more than half the diameter, while performing large amplitude oscillations in water, tends to oscillate transversely at the eddy frequency and to vibrate at twice the eddy frequency in the in-line direction.

Vaicaitis (1976) investigated the response of deep-water piles due to cross-flow forces generated by wind-induced ocean waves. The resulting lift forces were treated as random processes in time-space domain and are assumed to be dependent on fluid velocities and vortex shedding processes. Verley and Every (1977) measured wave-induced stress on similar rigid and flexible vertical cylinders in a wave channel at relatively low Keulegan-Carpenter numbers. Even though they were unable to correlate their data with any suitable parameter governing the motion, they concluded that the vibration is caused by the cylinder's response to eddy shedding and that there is no-fluid-structure interaction. They found that the vibration occurs if V_r is greater than about unity for any natural frequency, wave frequency, and damping. The reduced velocity V_r never reached sufficiently high values in Verley and Every's experiments for the cylinder to undergo sustained oscillations. Sawaragi (1977) investigated the in-line and transverse dynamic response of a cantilevered circular cylinder, with a concentrated mass at its top, in waves of small amplitude. The Reynolds number ranged from 1,500 to 6,200 and the r.m.s. value of the Keulegan-Carpenter number (calculated over the submerged length of the cylinder through the use of the r.m.s. value of the maximum of the horizontal velocities) ranged from 2 to about 20. They have approximated the lift coefficient for a rigid cylinder by a Rayleigh distribution and calculated the dynamic response of the test cylinder. The results are of limited value since the interaction between the synchronization and the force amplification was ignored.

Sarpkaya (1976b) and Rajabi (1979) carried out a series of detailed experiments with spring-mounted smooth and rough cylinders undergoing sustained oscillations in a harmonically oscillating flow in a U-shaped water tunnel. The results have shown that (1) an elastically-mounted cylinder may undergo synchronized oscillations when the reduced velocity $V_r = U_m/f_n D$ is in the range of 5 to 7.5; (b) perfect synchronization, at which the response is maximum, occurs at $V_r \simeq 5.6$; (c) synchronized oscillations occur at an average Strouhal number of 0.16; (d) in the region of synchronous oscillations, the r.m.s. of the lift coefficient and the amplitude of the predominant harmonic of the normalized lift force are amplified by a factor of about 2.35 relative to that for a stationary cylinder in harmonic flow at the corresponding Keulegan-Carpenter number and Reynolds number; (e) the Fourier and spectral analyses of the exciting force and response, in the resonance region, show that the exciting force is well represented by its three most important harmonics and the response, by the predominant harmonic alone. The relative amplitude of the n-th harmonic is given by

$$\overline{X}_n/D = \frac{0.5\left[\frac{\rho L D^2}{M}(C_L^0)_{rms}\right](U_m/\omega_n D)^2 \overline{C}_{Ln}/(C_L^0)_{rms}}{\left(\left\{1-\left(\frac{n\omega_f}{\omega_n}\right)^2\right\}^2 + (2\zeta\omega_f/\omega_n)^2\right)^{1/2}} \tag{8.43}$$

in which $(C_L^0)_{rms}$ represents the r.m.s. value of the lift force coefficient for the corresponding stationary cylinder (see Sarpkaya 1976); and ω_f, the circular frequency of the lift force oscillations. The amplification ratios $C_{Ln}/(C_L^0)_{rms}$ have been evaluated for the fundamental and two harmonics through the use of the Fourier-decomposed exciting force. The results have shown that the said ratio for the predominant harmonics at *perfect synchronization* are given by

$$C_{Lp_1}/(C_L^0)_{rms} = 2.75 \pm 0.15, \quad C_{Lp_2}/(C_L^0)_{rms} = 1.15 \pm 0.10$$
$$C_{Lp_3}/(C_L^0)_{rms} = 0.70 \pm 0.07 \tag{8.44}$$

and

$$[C_L(t)]_{rms}/(C_L^0)_{rms} = 2.80 \pm 0.20 \tag{8.45}$$

in which p_i represents the predominant harmonics.

A critical examination of the relative significance of the second and third harmonics on the cylinder response has shown that the response is largely determined by the predominant harmonic. The second and third harmonics of the relative displacement are an order of magnitude smaller than the predominant harmonic. Assuming that the fundamental harmonic of the exciting force is representable by $(C_L)_{rms}$ [see Eq. (8.45)] and that $\omega_f \simeq \omega_n$ at synchronization, Eq. (8.43) reduces to

$$\overline{X}/D = \frac{(U_m/f_n D)^2 \ [(C_L)_{rms}/(C_L^0)_{rms}]}{16\pi^2 M\zeta/[\rho L D^2 (C_L^0)_{rms}]} \tag{8.46}$$

A comparison of Eqs. (8.44) and (8.45) shows that

$$(C_L)_{rms}/(C_L^0)_{rms} \simeq C_{Lp_1}/(C_L^0)_{rms} \tag{8.47}$$

Thus, it is reasonable to expect that Eq. (8.46) will yield the fundamental harmonic of the relative displacement fairly accurately.

Equation (8.46) shows that a relationship should exist between $\overline{X}/D$ and the response parameter R_p defined by

$$R_p = \frac{M\zeta}{\rho D^2 L (C_L^0)_{rms}} \tag{8.48}$$

for both smooth and rough cylinders. Figure 8.12 shows the relationship between $\overline{X}/D$ and R_p for all smooth and rough test cylinders. The correlation between the two parameters is rather remarkable in view of the fact that only the fundamental harmonic of the exciting force has been used and it was assumed that $\omega_f \simeq \omega_n$. Figure 8.12 also shows that the response of a given body to a given flow must necessarily depend on the starting conditions, i.e., on the dynamics of the same flow past the same body when the latter is held stationary. The use of a response parameter given by $M\zeta/(\rho L D^2)$, as done previously by others, without $(C_L^0)_{rms}$ could not have correlated the response for all smooth and rough cylinders. The generalization of this concept to hydroelastic oscillations in steady flow shows that Eq. (8.42) should be recast as

$$(A/D)_{max} = 9.56\gamma (C_L^0)_{rms}/[1.94 + \Delta_r^2/(C_L^0)_{rms}^2]^{1/2} \tag{8.49}$$

in which $\Delta_r/(C_L^0)_{rms}$ is related to R_p by

$$\Delta_r/(C_L^0)_{rms} = 8\pi^2 S^2 R_p \tag{8.50}$$

in which S is the Strouhal number given by $S = f_f D/U_m$.

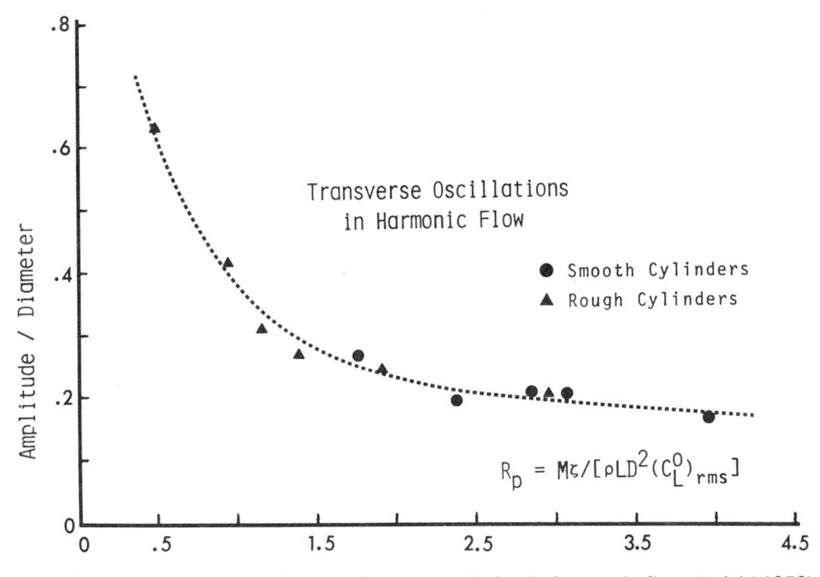

Fig. 8.12. Free transverse oscillations of circular cylinders in harmonic flow (Rajabi 1979).

In summary, it has been shown that the synchronization in harmonic flow occurs over a range of U_m/f_nD values from 5 to 7.5, with perfect synchronization at $U_m/f_nD \simeq 5.6$. The average amplification factor for the synchronous region is about 2.35, and the response is a unique function of the response parameter R_p for all smooth and rough cylinders. The effect of the Keulegan-Carpenter number and the Reynolds number on the cylinder response is imbedded in the dependence of $(C_L^0)_{rms}$ on the starting conditions. Finally, the predominant harmonic of the response is an order of magnitude larger than those of the higher harmonics, which enable one to determine the response through the use of a relatively simple analysis. The modulations in amplitude and phase of the response is quite well predicted both by a generalized solution based on the Fourier decomposition of the exciting force and by a straightforward application of the Duhamel's integral (for details see Rajabi 1979).

8.4.5 Vortex-Induced In-Line Oscillations

This type oscillations in uniform flow have attracted some attention partly because of the damaging vortex-induced oscillations of trashracks in currents and of piles in tidal waters and partly because of the need to determine the wave and current-induced forces on offshore structures (see Section 3.9).

Lock-in occurs when the in-line frequency approaches twice the Strouhal frequency. The amplitude of the alternating force (drag fluctuations) and the

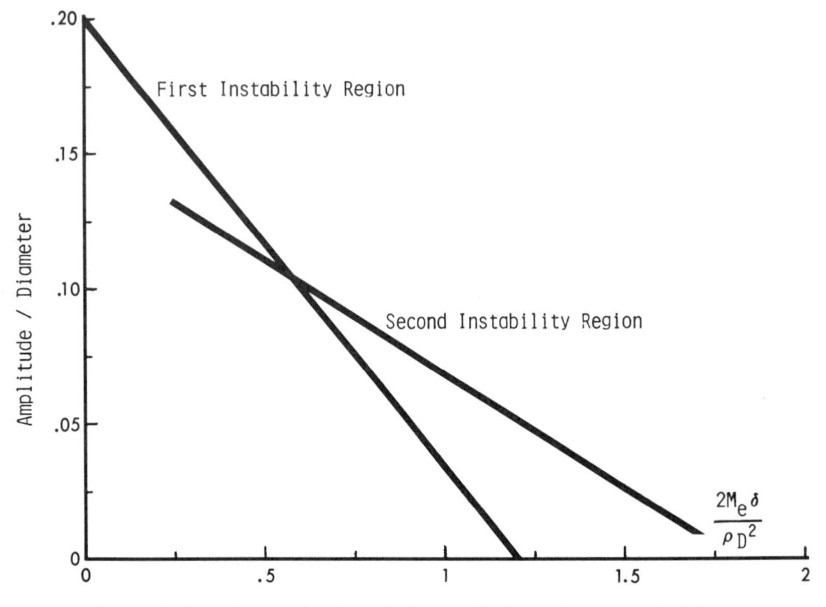

Fig. 8.13. Relative amplitude of in-line oscillations (Hallam et al. 1977).

response of the cylinder are an order of magnitude smaller than those in the transverse direction. The range of observed frequencies and other features of the phenomenon have been noted previously. The most interesting feature of the phenomenon is the occurrence of two regions of instability separated by $V_r \simeq 2$ (see Fig. 8.7). For $V_r < 2$, usually a symmetric vortex shedding is observed (Wootton 1972). At $V_r \simeq 2$, the vortex shedding changes to the commonly observed form of alternate shedding, indicating a radical change in the cause of excitation and dynamic response. These have been discussed in detail in Section 3.9.

The limited data presented by King (1977) show that the amplitude of in-line oscillations in the first and second regions of instability is related to the parameter $(2M_e\delta/\rho D^2)$ as shown in Fig. 8.13. Note that M_e represents the mass of the pipe, the fluid inside, and the added mass; and $\delta = 2\pi\zeta_f$ where ζ_f includes the fluid damping, as determined in still water (see Section 8.4.6).

The fluctuating component of the drag force may be approximated by $C_d' = 0.12 + 3.8\ x/D$ in the first instability region, and by $C_d' = 0.08 + 2.66\ x/D$ in the second instability region (x is the amplitude of the in-line oscillation at a given point z along a pipe, measured from a nodal point). These approximate expressions are of importance in determining the total fluctuating force on a segment of a pipe undergoing in-line oscillations in the first or second region of instability.

It is apparent from the foregoing that the quantification of the in-line oscillations requires great deal of additional research particularly with wavy and harmonic flows. Experiments with a cantilevered cylinder at very low Reynolds numbers and Keulegan-Carpenter numbers ($K < 10$) in wavy flows by Sawaragi (1977) have shown that (a) for $f_w/f_n > 0.9$, (f_w = wave frequency, f_n = natural frequency in water), the displacement of the top of the pile in the direction of in-line force is predominant in comparison with that in the direction of lift force and the locus shows a nearly straight line in the in-line direction; (b) in the range $0.6 < f_w/f_n < 0.9$, the locus draws a figure of ∞; (c) in the range $0.4 < f_w/f_n < 0.6$, the locus shows a nearly double ellipse (the natural frequency of the pile is twice the frequency of the wave so that the pile is resonated by the lift force); and (d) in the range $0.3 < f_w/f_n < 0.4$, the locus shows a long ellipse inclined with respect to the in-line direction. In this case the cantilever may vibrate in the transverse direction with a frequency three times the wave frequency.

The foregoing results, obtained mostly in the inertia-dominated region, cannot be generalized to other situations. They are cited here simply to point out the complex interaction between the in-line and transverse oscillations and the need for further research on vortex induced oscillations of structural members.

8.4.6 Added Mass and Damping

There are some fundamental questions in the current approaches to empirical correlations of the dynamic response and to the modeling of vortex-induced oscillations which have not been adequately dealt with. These questions stem from the use of such concepts as still-water added mass and fluid damping for the case of flow with vortex shedding. As they are presently used, added mass and fluid damping are quantities which are defined only for the zero flow condition and they are, in fact, measured under these conditions (e.g., by plucking excitation in still fluid). There is no theoretical basis for assuming that these quantities should play the same roles in the case of flow with vortex shedding.

Some wake oscillator models require that both the still-fluid added mass and damping be added to the "structural" part of the equation. Certain other models require only that the added mass be added to the structural element. Both of these approaches are fundamentally in error and both require more information for the structural part than should be required. Hence, rather than arbitrarily introducing these quantities in the structural part of the equation describing the model, the "fluid" part of the model should be constructed from the outset so as to include all fluid dynamic effects.

There is no reason in the laws of fluid dynamics that the added mass should be the same for a body oscillating, at its natural frequency, in a fluid otherwise at rest with that oscillating in a fluid in motion. Experiments have shown that the added mass coefficient of a circular cylinder oscillating in still fluid is slightly under unity (Sarpkaya 1976), the inviscid-flow value being exactly 1.0 for A/D less than about unity. This does not prove that it remains constant for all types of fluid motion. In fact, added mass depends on the type of motion of the body or of the fluid about the body; proximity of other bodies, free surface, etc., and time. It is not always possible nor advisable to determine the instantaneous value of the added-mass coefficient. In short, it would be wrong in general to equate the two flow situations where in one case the cylinder oscillates with amplitudes of A/D smaller than unity in still fluid and the other case where the cylinder oscillates at similar amplitudes in a transverse or in-line direction in steady flow. The former does not involve vortex shedding (A/D smaller than about 0.7) whereas the latter is accompanied by complex separation and vortex shedding phenomenon even though the amplitude of oscillation is of comparable magnitude. However, as will be noted shortly, the perfect lock-in state appears to be an exception.

King (1971) conducted a series of experiments in still water (by plucking a cantilevered cylinder) and also in flowing water at the fundamental mode of the flow-excited vibrations. From a comparison of the measured and calculated frequencies (using the ideal value of the added mass) he concluded that the still-water and flow-excited frequencies are virtually identical, i.e., the added mass is unaffected by streaming flow. Apparently, this conclusion is valid only at the resonant condition.

Sarpkaya (1978) has demonstrated experimentally that the added-mass coefficient decreases rapidly with increasing V_r, *becomes nearly equal to unity at perfect synchronization*, and then becomes negative as V_r increases further. Evidently, the use of $C_a = 1$ for off-resonance conditions is not justified.

The problem associated with the determination of the damping ratio of a structure in vacuum ζ_s, in still fluid ζ_{sf}, and in flowing fluid ζ_{ff} (at resonance) are far more complex and inherent in choosing the appropriate damping ratio for all vibrating structures. Experiments in vacuum are difficult to perform and slight changes in support conditions may lead to large differences. Experiments in still fluid are relatively easier but the results are not easy to interpret since ζ_f is a nonlinear function of $Re_\omega = \omega D^2/\nu$ and A/D. For a circular cylinder undergoing harmonic oscillations in still fluid, separation occurs at about A/D = 0.2 and the vortex shedding at about A/D = 0.7. An approximate theoretical analysis of the problem is possible only for very small amplitudes (A/D smaller than about 0.1) (Batchelor 1967, Chen et al. 1976).

Batchelor (1967) has shown that for small values of A/D (less than about 0.05) ζ_f (fluid damping alone) for a circular cylinder is given by

$$\zeta_f = \frac{\rho}{\rho_s} \frac{4\sqrt{2}}{\sqrt{Re_\omega}} \tag{8.51}$$

As A/D increases, the effect of separation and hence the pressure drag increases. Experiments have shown that (Sarpkaya 1976, Verley 1978) $C_{dp} = 1.25\,A/D$ for

A/D smaller than about 1.6 and $Re_\omega < 10^4$. Thus noting that

$$\zeta = \frac{8}{3\pi^2} \frac{\rho A}{\rho_s D} C_{dp} \tag{8.52}$$

which is based on equivalent energy dissipation per cycle, and combining with the viscous-dissipation contributions given by Eq. (8.51), one has

$$\zeta_f = \frac{\rho}{\rho_s}\left[\frac{4\sqrt{2}}{\sqrt{Re_\omega}} + 0.34(A/D)^2\right] \tag{8.53}$$

Then the total damping in still fluid is $\zeta_{sf} = \zeta_s + \zeta_f$.

The relative significance of the roles played by ζ_s, ζ_f, and ζ_{ff} may be further clarified by writing

$$\Delta_r^{sf} = \Delta_r^s + \Delta_r^f = \zeta_s/a_0 + \zeta_f/a_0 \tag{8.54}$$

and combining with Eq. (8.53) to yield

$$\Delta_r^{sf} = \Delta_r^s + 2\pi^3 S_0^2 [4\sqrt{2}\ Re_\omega^{-1/2} + 0.34(A/D)^2] \tag{8.55}$$

The last term in Eq. (8.55), representing Δ_r^f, is plotted in Fig. 8.9 together with the experimental data and the empirical correlations. Evidently, for Re_ω = 6000, A/D less than about 0.2, and Δ_r larger than about 0.5, the contribution of Δ_r^f to Δ_r is small and becomes insignificant as Δ_r increases. Thus, the region of small amplitudes or larger Δ_r's is material-damping dominated. This is of course the region where ζ_{sf} is mostly due to ζ_s, i.e., the use of ζ_{sf} in lieu of ζ_s does not change the correlation in Fig. 8.9. The region where Δ_r is smaller than about 0.2 is dominated by fluid damping. In this region, in particular, ζ_{sf} may not be assumed to be equal to ζ_s. Fortunately, as the experiments show, A/D does not significantly vary with Δ_r for Δ_r less than about 0.2 or 0.3. Thus, the use of ζ_{sf} in lieu of ζ_s does not materially affect the correlation even if ζ_{sf} is several times larger than ζ_s.

A third region, between the two regions cited above, may be identified where the use of ζ_{sf} in lieu of ζ_s leads to larger scatter in the data (see Fig. 8.9) since both ζ_s and ζ_f acquire equal importance. The data reported in the literature are not always comprehensive enough to delineate the effect of ζ_s, let alone to assess the amplitude or frequency dependence of ζ_s.

8.4.7 An Example

Consider a 24-inch diameter, 560-feet long production riser attached to an offshore platform at 70-feet intervals. Assuming the riser to be a simply-supported

Table 8.1 Sample Calculation of In-Line and Transverse Oscillations.

Elevations	Velocities	V/f_nD	Oscillations
-560--490	3.95- 4.25	0.73-0.79	unlikely
-490--420	4.25- 5.75	0.79-1.06	unlikely
-420--350	5.75- 7.50	1.06-1.39	in-line (first kind)
-350--280	7.50-10.00	1.39-1.85	in-line (first kind)
-280--210	10.00-13.35	1.85-2.47	in-line (most severe)
-210--140	13.35-18.25	2.47-3.38	in-line (second kind)
-140- -70	18.25-25.75	3.38-4.77	in-line (second kind)
-70- 0	25.75-32.00	4.77-5.92	possibly transverse

Note: In calculating f_n, one writes $W = M_e g$ where M_e includes the mass of the pipe, the mass of oil, and the added mass of the surrounding water.

beam, the natural frequency is given by

$$f_n = (\pi n^2/2)\,[EIg/L^4W]^{1/2} \tag{8.56}$$

For $n = 1$ and the appropriate values of the remaining parameters one has (for the specific example under consideration) $f_n = 2.7$ Hz. Assume that the wave and current-induced velocities at various elevations are calculated through the use of a suitable wave theory and tabulated as shown in Table 8.1 (wave period is assumed to be $T = 15$ seconds). One can now calculate the reduced velocity $V_r = V/f_nD$ and determine as to whether any part of the riser could undergo in-line or transverse oscillations. Evidently, several segments of the riser may be subjected to in-line oscillations in the first or second region of instability. The top segment of the riser may be subjected to vortex-induced transverse oscillations. The foregoing suggests that the force induced by the in-line and cross-flow oscillations on the riser segments be determined. The analysis may be limited to the design wave only since smaller waves are not likely to cause larger stresses. This is because of the fact that V/f_nD decreases with V whereas the fluctuating force decreases with the square of the velocity. For smaller waves fatigue may be a more serious problem and should be examined in terms of the mean and fluctuating stresses.

For a first order of approximation the force acting on the unit length of the riser may be assumed to be equal to the sum of the forces due to wave induced velocities on the stationary cylinder and those due to in-line oscillations, i.e.,

$$F = 0.5\rho DV^2 C_d + 0.5\rho DV^2 C_d' \sin 2\pi f_n t \tag{8.57}$$

As noted earlier C_d' depends on the amplitude of oscillation. The selection of

C_d is based on the Reynolds number, at least for the case under consideration. Ordinarily, C_d is a function of K and Re. For the riser under consideration the Reynolds number varies from about 5×10^5 to 4×10^6 and the Keulegan-Carpenter number from about 30 to 250. In other words, the force is drag dominated and the drag coefficient may be taken $C_d = 0.6$. The fluctuating drag coefficient C_d' for the first instability region has been approximated by (see Section 8.4.5)

$$C_d' = 0.12 + 3.8x/D \tag{8.58}$$

where x is the amplitude of oscillation at a given point on the segment. Denoting the maximum amplitude of oscillation at the mid-point of the riser segment by X_m, one has

$$x(z)/D = (X_m/D) \sin \pi z/L \tag{8.59}$$

where z is the distance along the segment with the origin at the support of the segment, and L is the length of the segment (here L = 70 feet). Combining the foregoing, the total force acting on the unit length of the segment becomes

$$F = 0.5\rho DC_d V^2 + 0.5\rho DV^2 [0.12 + 3.8(X_m/D) \sin \pi z/L] \sin 2\pi f_n t \tag{8.60}$$

A similar expression may be derived for the second region of the in-line oscillations by replacing C_d' by $C_d' = 0.08 + 2.66x/D$. It should be noted that we have used the strip theory in the development of Eq. (8.60). In other words, we have ignored the effect of three-dimensionality. This assumption is not expected to yield materially different results from those based on some arbitrary correlation lengths, etc.

The further use of Eq. (8.60) requires the calculation of V at suitable intervals along z and some knowledge of the expected maximum amplitude of the mid-point of the riser segment. The latter information is obtained from Fig. 8.13. For the example under consideration, m_s (unit riser mass) = 7.36 slugs, m_a (added mass with $C_a = 1$) = 6.28 slugs, and m_w (oil mass per unit length) = 4.44 slugs. Thus, $M_e = 18.1$ slugs. Assuming δ (in still water) = 0.1, one has

$$K_s = 2M_e\delta/(\rho D^2) = 0.45 \tag{8.61}$$

From Fig. 8.13, one has $X_m/D = 0.12$. Then Eq. (8.60) for the particular values of V = 13 ft/sec, $\rho = 2$ slugs/ft^3, $C_d = 0.6$, $K_s = 0.45$, $X_m/D = 0.12$, and $z/L = 0.5$ (mid point) reduces to

$$F = 203 + 169 \sin 17t (\text{Lbs/ft}) \tag{8.62}$$

Evidently, in-line oscillations can significantly increase the fluid loading.

A more rigorous analysis of the hydroelastic oscillations of the riser under consideration requires the inclusion of the inertial forces and the use of Eq. (8.3). Such an analysis for the case under consideration shows that the force is fairly well represented by Eq. (8.60) primarily because of the fact that the driving force is in the drag-dominated regime and the nonlinearity of the drag is not of major importance.

8.4.8 Suppression Devices

The suppression of vortex-induced oscillations can be accomplished either mechanically or fluid dynamically. The system can be detuned mechanically by increasing the structural stiffness or using mechanical dampers. This detuning generally tries to ensure that the natural frequency of the structure is separated by at least an order of magnitude from the vortex shedding frequency. The installation of dampers or the stiffening of the structure is not always technically or economically possible and a fluid dynamic solution would be needed.

Fluid dynamically, the oscillations of a cylinder can be substantially reduced by introducing disturbances on or near the surface of the cylinder which interact with the vortex-shedding mechanism. This interaction can affect the shedding mechanism in four different ways: (a) minimizing the adverse pressure gradient by influencing the point of separation; (b) interfering with the vortex interaction in the near wake; (c) distrupting the vortex formation length in the wake; and (d) disrupting the coherence of the vortex shedding or the spanwise coherence.

Numerous devices have been proposed in the past 20 years. Their use requires sound judgment, experience, and additional experiments. An excellent review of the suppression devices is presented by Hafen et al., (1976). They have noted that streamlined fairings, vortex generators and studs have been used to influence the boundary layer separation; splitter plates have been used to prevent vortex interaction; "hair" fairings, "fringe" fairings and ribbons have been used to disrupt the vortex formation length; helical strakes, "hair," ribbons, herringbone and twisted pairs of cable have been used to disrupt the spanwise coherence. Figure 8.14 shows some of the representative suppression devices. Clearly, it is not possible to compare the advantages and disadvantages of these various devices. In general they tend to increase the in-line drag. Furthermore, the use of cables and pipes covered with suppression devices may introduce operational difficulties.

Splitter plates increase the base pressure (reduced drag) and can entirely eliminate the vortex formation. The omnidirectionality of the ocean waves and

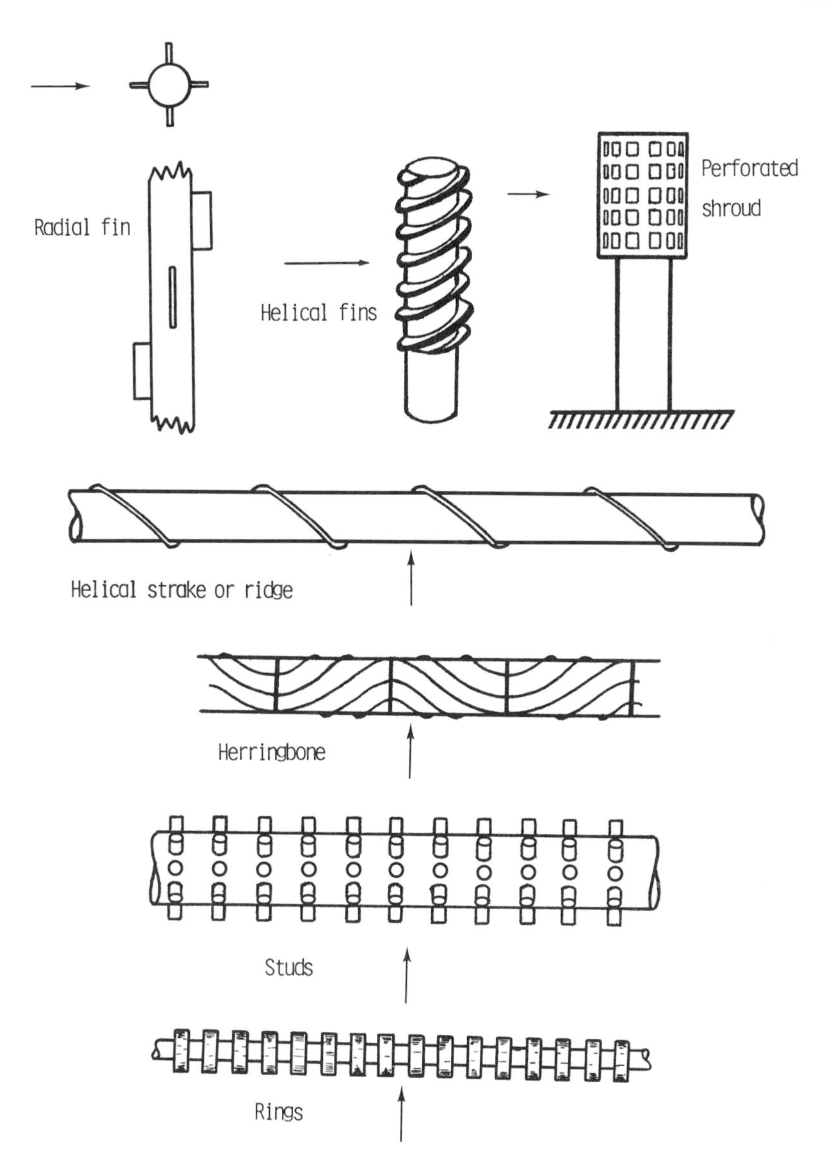

Fig. 8.14a. Strumming suppression devices (Hafen et al. 1976).

currents preclude the use of fixed splitter plates. The use and maintenance of self-aligning splitter plates (weather vane fairings) are rather difficult and costly (for additional discussion see Diggs 1974).

Doolittle (1974) investigated the effect of trailing ribbons, helical wraps, fairing bodies, masses, rings, and collars. He found that the trailing ribbons can

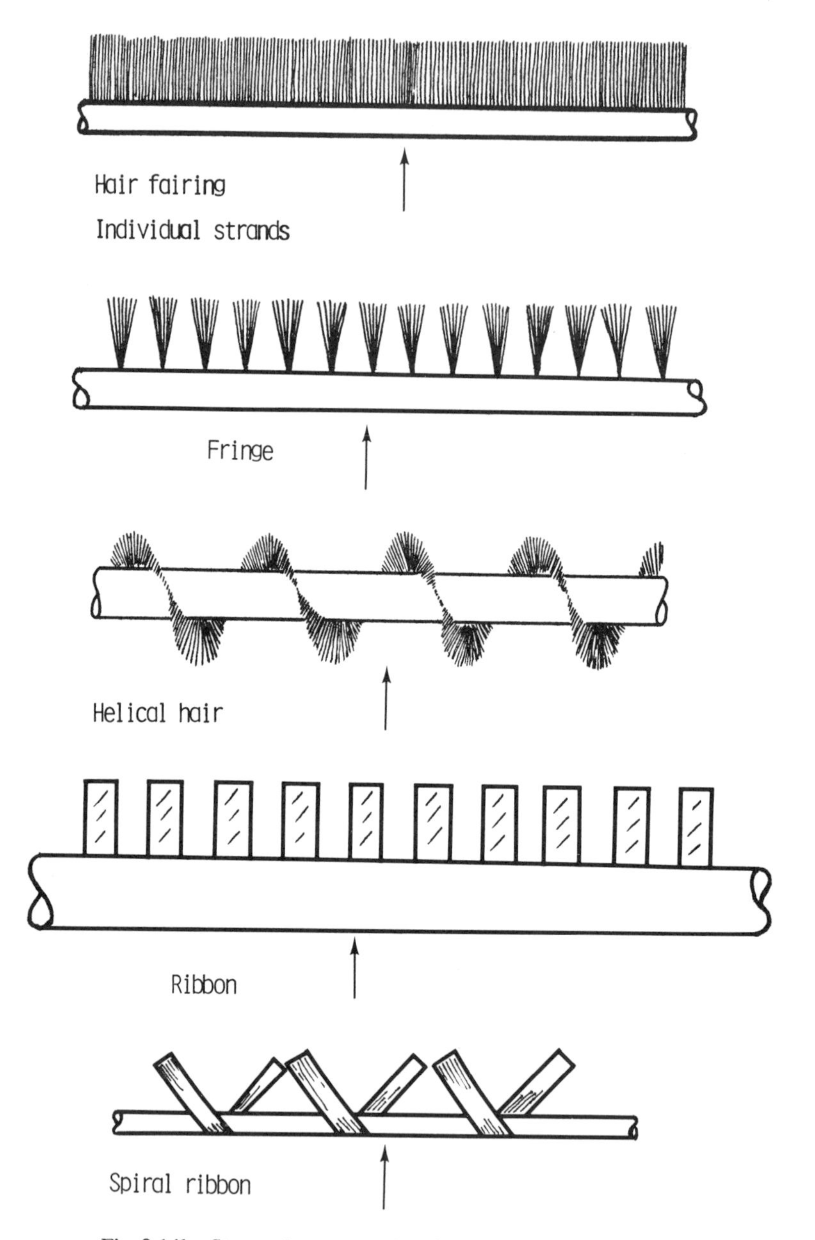

Fig. 8.14b. Strumming suppression devices (Hafen et al. 1976).

reduce strumming as much as 99 percent depending on their length and width, and a single helical wrap, reversing its spiral at midspan, as much as 66 percent.

Fabula and Bedore (1974) carried out extensive experiments with various suppression devices. Their results show a wide range of strum reduction effectiveness dependent on speed and flow angle. Heavy flags with adequate width and lee-side tie-on hair had outstanding performance. Single helical ridges were mostly moderately effective but ineffective for certain conditions. Lighter flags and various ribbon treatments were moderately effective. Boundary-layer trips were only slightly effective and mainly made the vibration more tonal. The hydrodynamic load coefficients varied appreciably with the type of strum reduction device. It is important to note that these results have been obtained with steady flow about cables. Information about the effectiveness of practically every suppression device in time-dependent flows is not available. Thus, special care must be exercised in interpreting the results obtained in steady flows and in using the devices in wavy flows.

Strakes are among the mostly widely used devices. Usually, a three-start helical fin of 5 diameters pitch, each helix protruding about 0.1 diameter from the cylinder surface, prevents in-line and transverse oscillations. The drag coefficient of the straked cylinder in steady flow is nearly independent of Reynolds number and has a value of 1.3 (based on the cylinder diameter).

Shrouds consist of an outer shell, separated from the cylinder by a gap of about 0.1 diameter, with many small rectangular holes. Circular holes are not found to be as effective. Shrouds as well as other devices may be used to cover only part of the pipe. However, final design must be based on a careful experimental investigation through the use of sound modeling techniques.

8.5 REFERENCES

Angelides, D. C. 1978. Stochastic Response of Fixed Offshore Structures in Random Sea. Ph.D. Thesis, Dept. of Civil Engineering, M.I.T.

Angelides, D. C. and Connor, J. J. 1979. A Probabilistic Model for Stiffness Degradation of Steel Jacket Structures. *BOSS '79*, Paper No. 48, Imperial College, London.

American Petroleum Institute (API). 1979. API RP 2A, Recommended Practice for Planning, Designing and Construction of Offshore Fixed Platforms. 10th edition.

Barkan, D. D. 1962. *Dynamics of Bases and Foundations* (translated from Russian). McGraw-Hill, New York.

Batchelor, G. K. 1967. *An Introduction to Fluids Dynamics*. Cambridge University Press, Cambridge, England.

Berge, B. and Penzien, J. 1974. Three-Dimensional Stochastic Response of Offshore Towers to Wave Forces. OTC Paper No. 2050.

Berger, E. 1964. Unterdrueckung der Laminaren Wirbelstroemung und des Turbulenzeinsatzes der Karmanschen Wirbelstrasse im Nachlauf eines Schwingenden Zylinders bei kleinen Reynolds-Zahlen. *Jahrbuch der WGLR*, pp. 164–172.

Berger, E. and Wille, R. 1972. Periodic Flow Phenomena. *Annual Reviews of Fluid Mechanics*, Vol. 4, pp. 313–340.

Birkhoff, G. and Zarantonello, E. H. 1957. *Jets, Wakes, and Cavities*. Academic Press, New York.

Bishop, R. E. D. and Hassan, A. Y. 1963. The Lift and Drag Forces on a Circular Cylinder in a Flowing Fluid. *Proceedings of Royal Society*, London, Series A, Vol. 277, pp. 32-50.

Blaney, G. W., Kausel, E., and Roesset, J. M. 1976. Dynamic Stiffness of Piles. 2nd International Conference on Numerical Methods in Geomechanics, V.P.I., ASCE, pp. 1001-1012.

Blevins, R. D. 1977. *Flow-Induced Vibrations*. Van Nostrand Reinhold Co., New York.

Blevins, R. D. 1979. Flow-Induced Vibration of Nuclear Reactors: A Review. *Progress in Nuclear Energy*, Vol. 4, pp. 25-49.

Bogdanoff, J. L. 1978. A New Cumulative Damage Model–Part I. *ASME Jour. of Applied Mechanics*, Vol. 45, pp. 246-250, (see also *Jour. of Applied Mechanics*, Vol. 47, March 1980, pp. 40-44).

Chen, S. S. 1977. Flow-Induced Vibrations of Circular Cylindrical Structures, Part I: Stationary Fluids and Parallel Flow. *Shock Vibration Digest*, Vol. 9, pp. 25-38.

Chen, S. S., Wambsganss, M. W., and Jendrzejczyk, J. A. 1976. Added Mass and Damping of a Vibrating Rod in Confined Viscous Fluids. *ASME Jour. of Applied Mechanics*, Vol. 43, pp. 325-329.

Clough, R. W. and Penzien, J. 1975. *Dynamics of Structures*. McGraw-Hill.

Cronin, D. J., Godfrey, P. S., Hook, P. M., and Wyatt, T. A. 1978. Spectral Fatigue Analysis for Offshore Structures. In *Numerical Methods in Offshore Engineering* (eds. Zienkiewicz, O. C. et al.), John Wiley and Sons, Chapter 9, pp. 281-316.

Diggs, J. S. 1974. A Survey of Vortex Shedding from Circular Cylinders with Application Toward Towed Arrays. Technical Report No. 122, MAR Inc., Rockville, MD.

Doolittle, R. D. 1974. NSRDC Basin Cable Strum Tests. Unpublished Report, MAR Inc., Rockville, MD.

Fabula, A. G. and Bedore, R. L. 1974. Tow Basin Tests of Cable Strum Reduction (second series). Naval Undersea Center, NUC TN-1379, San Diego, CA.

Felippa, C. A. 1976. Procedures for Computer Analysis of Large Non-linear Structural Systems. *International Symposium on Large Engineering Systems*, Univ. of Manitoba, Winnipeg, Canada.

Feng, C. C. 1968. The Measurement of Vortex-Induced Effects in Flow Past Stationary and Oscillating Circular and D-Section Cylinders. MA-Sc Thesis, University of British Columbia, Vancouver.

Girard, J. and Picard, J. 1970. Etude Experimentale du Dynamique des Massifs de Fondation de Machine. *Annales de L'Institut Technique du Batiment et des Travaux Publics*, No. 273.

Griffin, O. M., Skop, R. A., and Ramberg, S. E. 1975. The Resonant Vortex Excited Vibrations of Structures and Cable Systems. OTC Paper No. 2319.

Grover, L. K. and Weaver, D. S. 1978. Cross-Flow Induced Vibrations in a Tube Bank-Vortex Shedding. *Jour. Sound and Vibration*, pp. 263-276.

Guyan, R. J. 1965. Reduction of Stiffness Mass Matrices. *Jour. American Institute of Aeronautics and Astronautics*, Vol. 3, pp. 380-386.

Hafen, B. E., Meggitt, D. J., and Liu, F. C. 1976. Strumming Suppression–An Annotated Bibliography. Civil Engineering Lab. Technical Report N-1456, Port Hueneme, CA.

Hallam, M. G., Heaf, N. J., and Wootton, L. R. 1977. Dynamics of Marine Structures: Methods of Calculating the Dynamic Response of Fixed Structures Subject to Wave and Current Action. CIRIA Underwater Engineering Group, London.

Hartlen, R. T. and Currie, I. G. 1970. Lift-Oscillator Model of Vortex-Induced Vibration. *Jour. of Engineering Mechanics*, ASCE, Vol. 96, pp. 577-591.

Hsieh, T. K. 1962. Foundation Vibrations. Proc. *Institution of Civil Engineers*, Vol. 22, pp. 211-226.

Hurlbut, S. E., Spaulding, M. L., and White, F. M. 1978. Numerical Solution of the Time-Dependent Navier-Stokes Equations in the Presence of an Oscillating Cylinder. ASME Book No. H00118, pp. 201-206.

Iwan, W. D. and Blevins, R. D. 1974. A Model for the Vortex-Induced Oscillation of Structures. *ASME Jour. of Applied Mechanics*, Vol. 41, pp. 581-586.

Jones, G. W., Cincotta, J. J., and Walker, R. W. 1969. Aerodynamic Forces on a Stationary and Oscillating Circular Cylinder at High Reynolds Numbers. NASA Tr-R-300.

Kausel, E. and Roesset, J. M. 1975. Dynamic Effects of Circular Foundation. *Jour. of Engineering Mechanics Div., ASCE*, Vol. 101, pp. 1625-1651.

King, R. 1971. The Added Mass of Cylinders. The British Hydromechanics Research Association, BHRA, Report TN-1100.

King, R. 1977a. A Review of Vortex-Shedding Research and Its Applications. *Ocean Engineering*, Vol. 4, pp. 141-172.

King, R. 1977b. Vortex Excited Oscillations of Yawed Circular Cylinders. *ASME Journal of Fluids Engineering*, Vol. 99, pp. 495-502.

Koopman, G. H. 1967. The Vortex Wakes of Vibrating Cylinders at Low Reynolds Numbers. *Journal of Fluid Mechanics*, Vol. 28, pp. 501-512.

Kryloff, N. and Bogoliuboff, N. 1943. *Introduction to Nonlinear Mechanics*. Princeton University Press, NJ.

Laird, A. D. K. 1962. Water Forces on Flexible Oscillating Cylinders. *Jour. Waterways etc. Div., ASCE*, Vol. 88, No. WW3, pp. 125-137.

Landl, R. 1973. Theoretical Model for Vortex-Excited Oscillations. *Proceedings of the International Symposium on Vibration Problems in Industry*, Keswick, England.

Lin, Y. K. 1967. *Probabilistic Theory of Structural Dynamics*. McGraw-Hill.

Luco, J. E. 1979. Impedance Functions for a Rigid Foundation on a Layered Medium. *Nuclear Engineering and Design*, Vol. 31, pp. 204-217.

Luco, J. E. and Westman, R. A. 1971. Dynamic Response of Circular Footings. *Jour. Engineering Mechanics Div., ASCE*, Vol.. 97, No. EM5, pp. 1381-1395.

Lysmer, J. and Kuhlemeyer, R. L. 1969. Finite Element Model for Infinite Media. *Jour. Engineering Mechanics Div., ASCE*, Vol. 95, No. EM4, pp. 859-877.

Malhotra, A. K. and Penzien, J. 1970. Nondeterministic Analysis of Offshore Tower Structures. *Jour. Engineering Mechanics Div., ASCE*, Vol. 96, No. EM6, pp. 985-1003, (see also Vol. 97, EM3, 1971, pp. 1028-1029).

Matlock, H. 1970. Correlations for the Design of Laterally Loaded Piles in Soft Clay. Proceedings of OTC, Vol. I, pp. 577-594.

Matlock, H. and Reese, L. C. 1970. Generalized Solution for Laterally Loaded Piles. *Jour. Soil Mech. Foundation Div., ASCE*, Vol. 86, No. SM5, pp. 63-91.

McClelland, B. F. 1974. Design of Deep Penetration Piles for Ocean Structures. *Trans. ASCE, Jour. Geotechnique Div.*, No. GT7, pp. 709-747.

Miner, M. A. 1945. Cumulative Damage in Fatigue. *Trans. ASCE*, Vol. 67, pp. A159-A164.

Palmgren, A. 1924. Die Lebensdauer von Kngellagern. *VDI, Zeit*, Vol. 68, p. 339.

Parkinson, G. V. 1974. Mathematical Models of Flow-Induced Vibrations of Bluff Bodies. In *Flow-Induced Structural Vibrations* (ed. Naudascher, E.), Springer-Verlag, Berlin, pp. 81-127.

Penzien, J. 1976. Structural Dynamics of Fixed Offshore Structures. *BOSS '76*, Vol. 1, Trondheim, Norway.

Penzien, J. and Tseng, S. 1978. Three-Dimensional Dynamic Analysis of Fixed Offshore Platforms. In *Numerical Methods in Offshore Engineering* (eds. O. C. Zienkiewicz et al.), John Wiley and Sons, New York, pp. 221-243.

Pettigrew, M. J. and Gorman, D. J. 1978. Vibration of Heat Exchange Components in Liquid and Two-Phase Cross Flow. International Conference on Vibration in Nuclear Plant, Keswick, U.K., Paper No. 2.3.

Rail, R. D., Hafen, B. E., and Meggitt, D. J. 1977. Flow-Induced Vibrations of Three-Dimensional Bluff Bodies in Cross Flow, an Annotated Bibliography. Technical Note No. N-1493, Civil Engineering Laboratory, Port Hueneme, CA.

Rajabi, F. 1979. Hydroelastic Oscillations of Smooth and Rough Cylinders in Harmonic Flow. Ph.D. Thesis, Naval Postgraduate School, Monterey, CA.

Ramberg, S. E. 1978. The Influence of Yaw Angle Upon Vortex Wakes of Stationary and Vibrating Cylinders. Memorandum Report 3822, Naval Research Laboratory, Washington, DC.

Richart, F. E., Jr., Hall, J. R., Jr., and Woods, R. D. 1970. *Vibrations of Soils and Foundations*. Prentice-Hall, Englewood Cliffs, NJ.

Sarpkaya, T. 1976. Vortex Shedding and Resistance in Harmonic Flow About Smooth and Rough Circular Cylinders at High Reynolds Numbers. Technical Report No. NPS-59SL76021, Naval Postgraduate School, Monterey, CA. (see also OTC Paper No. 2533, 1976).

Sarpkaya, T. 1978. Fluid Forces on Oscillating Cylinders. *Jour. Waterways etc. Div., ASCE*, Vol. 104, No. WW4, pp. 275-290.

Sarpkaya, T. 1979a. Vortex-Induced Oscillations–A Selective Review. *Jour. of Applied Mechanics*, Vol. 46, No. 2, pp. 241-258.

Sarpkaya, T. 1979b. Dynamic Response of Piles to Vortex Shedding in Oscillating Flows. OTC Paper No. 3647, Houston, Texas.

Sarpkaya, T. and Shoaff, R. L. 1979a. A Discrete-Vortex Analysis of Flow About Stationary and Transversely Oscillating Circular Cylinders. Technical Report No. NPS-69SL79011, Naval Postgraduate School, Monterey, CA.

Sarpkaya, T. and Shoaff, R. L. 1979b. Inviscid Model of Two-Dimensional Vortex Shedding by a Circular Cylinder. *AIAA Journal*, Vol. 17, No. 11, pp. 1193-1200.

Savkar, S. D. 1976. A Survey of Flow-Induced Vibrations of Cylindrical Arrays in Cross Flow. ASME Paper No. 76-WA-FE-21.

Sawaragi, T. 1977. Dynamic Behavior of a Circular Pile due to Eddy Shedding in Waves. *Coastal Engineering in Japan*, Vol. 20, pp. 109-120.

Sigbjörnsson, R., Bell, K., and Holand, I. 1978. Dynamic Response of Framed and Gravity Structures to Waves. In *Numerical Methods in Offshore Engineering* (ed. O. C. Zienkiewicz), John Wiley and Sons, New York, pp. 245-280.

Simui, E. and Scanlan, R. H. 1978. *Wind Effects on Structures: An Introduction to Wind Engineering*. John Wiley and Sons, New York.

Stansby, P. K. 1976. The Locking-on of Vortex Shedding to the Cross-Stream Vibration of Circular Cylinders in Uniform and Shear Flows. *Journal of Fluid Mechanics*, Vol. 74, pp. 641-665.

Szechenyi, E. 1975. Modele Mathematique de Mouvement Vibratoire Engendre par un Echappment Tourbillonaire. *La Recherche Aerospatiale*, No. 5, pp. 301-312.

Vaicaitis, R. 1976. Cross-Flow Response of Piles due to Ocean Waves. *Jour. Engineering Mechanics Div., ASCE*, Vol. 102, EM1, pp. 121-134.

Veletsos, A. S. and Wei, Y. T. 1971. Lateral and Rocking Vibrations of Footings. *Jour. Soil Mechanics and Foundations Div., ASCE*, Vol. 97, No. SM9.

Verley, R. L. P. 1978. An Investigation into the Damping of Oscillations of a Cylinder in Still Water. River and Harbor Laboratory Report No. STF-60-A78049, Norwegian Institute of Technology, Trondheim-NTH.

Verley, R. L. P. and Every, M. J. 1977. Wave-Induced Vibrations of Cylindrical Structures. OTC Paper No. 2899, Houston, Texas.

Waas, G. and Lysmer, J. 1972. Shear Waves in Plane Infinite Structures. *Jour. Engineering Mechanics Div., ASCE*, EM1, pp. 85–105.

Whitman, R. V. 1976. Soil-Platform Interaction. *BOSS '76*, Trondheim.

Whitman, R. V. 1972. Analysis of Soil-Structure Interaction–A State of the Art Review. *Proceedings of the Symposium on Structural Dynamics*, Southampton Univ., U.K.

Wilson, E. L. 1978. Numerical Methods for Dynamic Analysis. In *Numerical Methods in Offshore Engineering* (ed. O. C. Zienkiewicz et al.), John Wiley and Sons, New York, pp. 195–220.

Wootton, L. R. 1972. Oscillations of Piles in Marine Structures. CIRIA Underwater Engineering Group Report No. 40.

Wu, S. C. 1976. The Effects of Current on Dynamic Response of Offshore Platforms. OTC Paper No. 2540, Houston, Texas.

Wu, S. C. and Tung, C. C. 1975. Random Response of Offshore Structures to Wave and Current Forces. Sea Grant Publication UNC-SG-75-22, North Carolina State Univ., Raleigh, NC.

Zienkiewicz, O. C., Bettess, P., and Kelly, D. W. 1978. The Finite Element Method for Determining Fluid Loading on Rigid Structures, Two- and Three-Dimensional Formulations. In *Numerical Methods in Offshore Engineering* (ed. O. C. Zienkiewicz et al.), John Wiley and Sons, New York, pp. 141–183.

9 Models and Prototypes

9.1 PRINCIPLES OF MODEL LAWS

This section of this final chapter is relatively straightforward and is presented here only for the sake of completeness. The reader who is familiar with the fundamental principles of dimensional analysis and model laws may proceed directly to Section 9.2.

Small scale physical models of offshore structures or of structural components are a convenient means of predicting full-scale performance. Their use can result in considerable economies in helping avoid disastrous mistakes in prototype design. Model tests are particularly invaluable when analytical methods of prediction are inadequate or unavailable as in some separated flow and dynamic response problems within the general area of fluid-structure interaction. More generally, even when analytical methods are available, model tests can provide a useful check on the predictions made.

Familiarization with the problem at hand is an essential first step in order to clarify the phenomena which are significant and to interpret the reliability of experimental results obtained. Thus, even though a model test will invariably provide some results, the relevance and reliability of these must be carefully assessed.

Model testing is quite expensive and must be carried out with considerable care. Indeed, if a novel experiment is being planned, it is often prudent first to construct a small crude model for observational purposes only in order to detect possible sources of difficulty and to assist in planning the more detailed model tests.

Before describing the modelling of offshore structures and the various problems which may be encountered, we first summarize the principles of model laws and the manner in which they may be derived for a given problem.

9.1.1 Model Laws Via Dimensional Analysis

Model laws provide relationships between variables pertaining to the model and prototype, thus enabling measurements made in the model tests to be used to predict prototype values. One common approach to the planning of model experiments lies in carrying out a dimensional analysis of the situation at hand. This enables the relevant factors influencing a particular problem to be assembled in a manner which is suggestive of its underlying structure and indicative of requirements which the model should satisfy.

It is assumed that the reader is familiar with the background and procedure of dimensional analysis and Pi theorem. If necessary, reference may be made, for example, to Bridgeman (1963), Langhaar (1951), or Sedov (1959). The application of dimensional analysis to establishing model laws is summarized here, being based largely on the account given by Isaacson and Isaacson (1975).

Suppose we wish to investigate the influence of a number of independent variables B, C, D, . . . (such as water depth, body size, fluid viscosity, etc.) on a particular variable A (such as force), then the relationship may be expressed in functional form as

$$A = f(B, C, D, \ldots) \tag{9.1}$$

where A is the dependent variable whose value is required in a prototype situation, and B, C, D, . . . are the independent variables whose values are prescribed. A dimensional analysis of Eq. (9.1) yields

$$\pi_1 = f(\pi_2, \pi_3, \ldots) \tag{9.2}$$

where $\pi_1, \pi_2, \pi_3, \ldots$ are independent dimensionless products (or groups) containing the pertinent variables, and where the dependent variable A appears in the product π_1 only. A model is then constructed in such a manner that

$$(\pi_i)_m = (\pi_i)_p \qquad \text{for } i = 2, 3, \ldots \tag{9.3}$$

in which the subscripts m and p denote values pertaining to model and prototype, respectively. Note that despite this equality between $(\pi_i)_m$ and $(\pi_i)_p$ the values of a variable contained within these products may–and generally will–be different, as follows from the relative disparity in size between the model and the prototype. Provided Eq. (9.3) is made to hold, then $(\pi_1)_m = (\pi_1)_p$ and with the value of $(\pi_1)_m$ available from experimental model investigation, the value of the unknown variable contained within $(\pi_1)_p$ may then be readily determined.

As a trivial but pertinent example, we consider the use of a model to determine the maximum force F on a fixed rigid structure of characteristic size D subjected to a regular two-dimensional wave train propagating in deep water. The force F may be expressed as

$$F = f(\rho, H, L, g, D) \tag{9.4}$$

We have chosen H, L, and g to represent the incident wave motion. The depth is of no significance for the deep water case being considered and, as we shall see later, additional variables such as the wave period and speed are superfluous. We have intentionally omitted fluid viscosity as a pertinent variable (that is a large structure is implicitly assumed) and shall reserve a discussion of its effects to Section 9.2.3. A dimensional analysis of Eq. (9.4) yields, for example,

$$F/(\rho g H D^2) = f(D/L, H/L) \tag{9.5}$$

For given prototype values of H, L, and D, a model is constructed to a suitably reduced scale, taking care to maintain the constancy of the dimensionless parameters D/L and H/L. This constancy entails the equality of the ratios

$$H_m/H_p = L_m/L_p = D_m/D_p \tag{9.6}$$

It then follows from Eq. (9.5) that

$$[F/(\rho g H D^2)]_m = [F/(\rho g H D^2)]_p \tag{9.7}$$

and with F_m determined by measurement, F_p remains the only unknown quantity in Eq. (9.7) above and so may readily be evaluated.

9.1.2 Dynamic Similarity

An alternative manner of regarding the conditions to be met in a model experiment is to introduce the requirement that a certain similarity should be maintained between the model and prototype, and we consider here several aspects of the *physical similarity* that is involved. In particular *geometric similarity* exists when the ratio of corresponding lengths is constant; *kinematic similarity* exists when the ratio of corresponding velocities is constant; and *dynamic similarity* exists when the ratio of corresponding forces is constant. In the latter case, dynamic similarity is generally met only by certain types of force (inertial, viscous, etc.) but not by others (capillary) with the result that we then have *partial dynamic similarity*.

Table 9.1 Common Dimensionless Numbers.

Dimensionless Number	Definition	Force Ratio
Froude number, Fr	$U/\sqrt{gL}$	inertia/gravitational
Reynolds number, Re	UL/ν	inertia/viscous
Weber number, Wb	$U/\sqrt{\sigma/\rho L}$	inertia/capillary
Cauchy number, Ca	$U/\sqrt{E/\rho}$	inertia/elastic
Euler number, Eu	$U/\sqrt{2\Delta p/\rho}$	inertia/pressure force

The condition of dynamic similarity with regard to two relevant forces is equivalent to the requirement that a corresponding dimensionless group remains constant. If, for example, the two forces considered are the inertial and viscous fluid forces, then the similarity requirement can be shown to entail the constancy of the Reynolds number between the model and prototype—as is also required by the approach based on dimensional analysis when viscosity is recognized as a relevant variable. Developing this approach, the forces that may be relevant generally include inertia, gravitational, friction (usually viscous in the present context), capillary, elastic, and pressure forces. The fluid inertia force is common to all fluid dynamic problems and consequently any other relevant force is conveniently introduced as a ratio to this inertial force. Such force ratios are listed in Table 9.1 and are commonly referred to in hydraulic modelling practice. In the case of a fluid-structure interaction problem where the elasticity of the structure is relevant, the quantity E appearing in the Cauchy number is the Young's modulus of the material. (The ratio of inertial to compressive 'elastic' forces in a fluid corresponds to the Mach number).

9.1.3 Scale Factors

The requirement of model laws may conveniently be carried out by resort to scale factors. The scale factor of a quantity f, denoted k_f, is the ratio of the value of f in the model to that in the prototype, i.e., $k_f = f_m/f_p$. The constancy of a particular dimensionless product provides a relationship between various scale factors. Thus, the condition of geometric similarity implies that a single length scale factor k_L applies to all lengths pertaining to the problem and thus, for example, Eq. (9.6) would be satisfied. And again, Eq. (9.7) provides that $k_F = k_\rho k_g k_L^3$.

For problems involving wave motions we inevitably meet the following conditions: (a) $k_\rho = 1$ (water is used in the model); (b) $k_g = 1$; (c) Froude similarity is maintained, i.e., $k_u^2/k_L k_g = 1$. From these three conditions and from relevant defining relations (e.g., moment = force × length) the following relationships may be derived:

$$k_u = k_T = k_L^{1/2}, \quad k_f = k_L^{-1/2}, \quad k_{ac} = 1, \quad k_p = k_L,$$

$$k_{F'} = k_L^2, \quad k_F = k_L^3, \quad k_M = k_L^4 \tag{9.8}$$

in which u represents the velocity; T, time; f, frequency; ac, acceleration; p, pressure or stress; F', sectional force; F, force and M, moment.

9.1.4 Model Laws Via Governing Equations

Model laws may also be derived by an examination of the governing equations of a problem when these are known. These equations–which include defining equations involving the relevant quantities–must hold for both the model and the prototype and can be used to develop appropriate model laws, that is relationships between the different relevant scale factors.

As a simple example, the response x of a spring-mass-dashpot system to an imposed force F may be described by the equation

$$m\ddot{x} + \lambda\dot{x} + sx = F \tag{9.9a}$$

where m is a mass; λ, a damping constant; and s, a spring constant. Related parameters of interest include the natural frequency f_n and the damping ratio ζ, defined respectively as,

$$f_n = (1/2\pi)(s/m)^{1/2}, \quad \zeta = \lambda/[2(ms)^{1/2}] \tag{9.9b}$$

Equations (9.9) must hold for both the model and the prototype. Applying Eq. (9.9a) to the model and introducing appropriate scale factors, $k_m = m_m/m_p$, etc., this may be written as

$$(k_m k_x/k_t^2) m_p \ddot{x}_p + (k_\lambda k_x/k_t)\lambda_p \dot{x}_p + (k_s k_x) s_p x_p = k_F F_p \tag{9.10}$$

By dividing Eq. (9.10) by $k_s k_x$ and comparing the corresponding terms in Eqs. (9.9a) and (9.10), we must have

$$k_m/k_s k_t^2 = k_\lambda/k_s k_t = k_F/k_s k_x = 1 \tag{9.11a}$$

Applying Eq. (9.9b) to model and prototype, one has

$$k_{f_n} = (k_s/k_m)^{1/2}, \quad k_\zeta = k_\lambda/(k_m k_s)^{1/2} = 1 \tag{9.11b}$$

If k_m, k_x, and k_t were prescribed, then the remaining scale factors can be expressed in terms of these as,

$$k_s = k_m/k_t^2, \quad k_\lambda = k_m/k_t, \quad k_F = k_m k_x/k_t^2 \tag{9.12}$$

A complete outline of this 'group theory' approach to obtaining model laws is given by Hansen (1964).

9.2 MODELLING OF WAVES AND WAVE FORCES

The preceding summary has been fairly straightforward and we consider now its application to various problems involving waves and wave forces. General discussions of the modelling of waves and offshore structures have been given by Hallam et al. (1978), Le Méhauté (1976b), Remery (1971), Silvester (1974), and Vugts (1971). Other examples of modelling applications within coastal and ocean engineering include investigations of ice forces, sediment motion (scour), harbor resonance, breakwater stability, and so on. Atkins (1975) described the modelling of ice breaking behavior, and Le Méhauté (1976a, b), O'Brien (1977), and Silvester (1974) have indicated the scope and difficulties of coastal engineering model studies.

9.2.1 Modelling of a Wave Train

We initially examine the most appropriate manner of characterizing a two-dimensional steady wave motion in water of finite depth. Various quantities which may possibly be used to characterize the motion include H, d, T, L, g, ω, k, c, crest particle velocity U_c, maximum velocity at the seabed U_b, etc. However, of these only four independent variables are necessary and sufficient to characterize the motion and all others will necessarily be related to the four selected on the basis of definition or of the laws of physics as reflected in a particular wave theory. It is usually convenient to adopt the wave height H, the still water depth d, and the gravitational acceleration g as three of the independent variables, and for the fourth we choose either L, k, or T, as required. Thus, waves of a given height and period propagating in water of given depth will have a specific wave length, wave speed etc., which a wave theory attempts to predict. (Note, however, that for a given H, L, and d, the wave speed (or T) is not unique and depends on the magnitude of an underlying second-order, uniform current corresponding to alternative assumptions used in defining the wave speed, as noted in Section 4.3.2).

The finite depth case of the example previously considered may be represented by rewriting Eq. (9.4) in the extended form,

$$F = f(\rho, g, H, L, d, D) \tag{9.13}$$

which may be reduced by dimensional analysis to

$$F/(\rho g H D^2) = f(d/L, H/L, D/L) \tag{9.14}$$

In a model test the constancy of the dimensionless parameters d/L and H/L between the model and prototype ensures that the wave train is suitably represented. Alternatively, it may be convenient to include initially both L and T among the list of relevant variables, and subsequently to invoke the dispersion relationship between L and T to obtain eventually the same number of dimensionless groups. For example, one has

$$F/(\rho g H D^2) = f(d/gT^2, H/L, D/L) \tag{9.15}$$

Although both L and T appear in this selection, there is no redundancy in the groups now adopted.

A wide variety of dimensionless parameters may be used to define the wave motion. A selection from among those which have been used in the literature include the relative height H/d, the wave steepness H/L, kH, H/gT^2, the Ursell number HL^2/d^3, the relative-depth parameters kd, d/L, and d/gT^2, and the dimensionless period $T/(d/g)^{1/2}$. The selection of any pair of independent parameters from this list, say H/d and d/gT^2, ensures that the wave motion has been adequately characterized in dimensionless form. Thus, in Chapter 4, H/gT^2 and d/gT^2 were adopted in Figs. 4.16, 4.17, 4.39, etc., and $(d/gT^2)^{1/2}$ and H/d were used in Figs. 4.18 and 4.23.

For waves in deep water the situation is somewhat simplified since d is no longer a relevant variable and only a single group is necessary: say H/L or H/gT^2. Again, in the case of a solitary wave, the wave length or period becomes infinite so that the relative height H/d is sufficient to describe the wave train.

The above concepts may readily be extended to the case of a random wave motion. This may be described by a wave spectrum—either measured or one prescribed in terms of a characteristic wind speed, significant wave height, etc. Either spectrum may be represented by its profile expressed in dimensionless form in terms of a characteristic wave height, say the root-mean-square surface elevation $(m_0)^{1/2}$, and a characteristic wave period or frequency, say the peak frequency f_0. Evidently, in modelling a certain spectrum these two quantities would be scaled as k_L and $k_L^{-1/2}$ respectively [see Eq. (9.8)], while the dimensionless profile of the spectrum [say $f_0 S(f/f_0)/m_0$ versus f/f_0] should be reproduced identically in the model. The extension to the case of directional spectra introduces no further difficulty, i.e., the additional variable θ is dimensionless and must assume identical values in both the model and prototype.

9.2.2 Wave Forces in the Diffraction Regime

The general problem concerning the forces acting on a fixed body of characteristic dimension D, when subjected to a regular wave train is of fundamental importance. This problem has already been adopted as a simple example [Eq. (9.5)], but in the general case viscous effects may be important and the fluid viscosity must specifically be taken into account. Any time-invariant force F, such as the maximum in-line force, may be expressed in functional form as,

$$F = f(\rho, \nu, D, H, L, d, g) \tag{9.16}$$

from which one has, for example,

$$F/(\rho g H D^2) = f(d/L, H/L, D/L, Re) \tag{9.17}$$

where Re is a characteristic Reynolds number which accounts for the consequences of fluid viscosity but which need not be defined explicitly as yet.

The wave-depth parameter d/L and the wave steepness H/L define the incident wave train. When D/L is sufficiently large (say $D/L > 0.2$) wave diffraction or scattering is important. Furthermore, the amplitude of water particle oscillation relative to D is necessarily small and the flow separation effects may then be neglected (see Section 6.1). Consequently, the Reynolds number may be omitted from Eq. (9.17) and the flow may conveniently be taken as irrotational and so treated on the basis of potential theory. In addition, if the wave steepness is small, a linearizing approximation is possible in which F is proportional to H, and thus H/L may also be omitted from Eq. (9.17). Thus, in the usual linear diffraction problem we have

$$F/(\rho g H D^2) = f(d/L, D/L) \tag{9.18a}$$

which may also be expressed as,

$$F/(\rho g H D^2) = f(D/d, D/L) \tag{9.18b}$$

in which D/d is fixed for a given body located in water of given depth.

As an example of the application of the foregoing to modelling, experimental or computer results would provide relationships such as Eq. (9.18) and these can then be used for the prototype. For example, for the prototype values given as H = 15 ft, L = 200 ft, and d = D = 60 ft, the independent ratios D/d and D/L are 1.0 and 0.3, respectively. If $F/(\rho g H D^2)$ has been found to be 1.1 for these values, as determined by experimental or computer results, then it follows that the prototype force would be $F = 3.37 \times 10^6$ Lbs.

Some further simplifications to Eq. (9.18) are possible under different circumstances. When, in addition to the assumptions already made (i.e., flow separation and wave nonlinearity effects are unimportant), the body or bodies consist of vertical surface-piercing cylinders, the velocity potential varies with depth according to the hyperbolic cosine and the situation then further reduces to

$$F/[\rho g H D^2 \tanh(kd)] = f(D/L) \tag{9.19}$$

In deep water d is not a relevant variable and, with the above restrictions on body shape no longer necessary, we have

$$F/(\rho g H D^2) = f(D/L) \tag{9.20}$$

as in the simple example considered earlier. This sequence of relations, then, indicates the varying degree of simplification that may be possible under appropriate conditions.

9.2.3 Wave Forces in the Flow Separation Regime

Let us now consider the more difficult case where D/L in Eq. (9.17) is small. The body will no longer significantly scatter the incident wave field and it follows that D/L has no direct physical significance. However, under such conditions a representative Keulegan-Carpenter number K assumes importance, as discussed in detail in Chapter 3, and is to be preferred in place of D/L. This might conveniently be defined as the value at the still water level as predicted by linear wave theory [see Eq. (6.1)]. In view of the fact that K = f(D/L, H/L, d/L), we emphasize that no reduction in the number of groups is achieved, but rather we have made an alternative selection which has more physical significance in the new situation considered. The counterpart to Eq. (9.17) now becomes,

$$F/(\rho g H D^2) = f(d/L, H/L, K, Re) \tag{9.21}$$

Now that the separation effects are predominant, the influence of the Reynolds number can no longer be dismissed without further consideration. The difficulty of maintaining a high Reynolds number in model tests is a major one and will be discussed shortly. Provided changes in Reynolds number are taken to be admissible, Eq. (9.21) represents a situation that can be reproduced in the laboratory.

Under certain conditions it may be possible to affect a further simplification. For example, the two-dimensional sinusoidal flow normal to the axis of a vertical cylinder can be characterized by U_m and T. Then the time-invariant sectional force F' may be written as

$$F'/(0.5\,\rho D U_m^2) = f(K, Re) \text{ with } K = U_m T/D \text{ and } Re = U_m D/\nu \tag{9.22a}$$

If required, the instantaneous (time-dependent) force F_t' would be written in a corresponding manner as,

$$F'_t/(0.5\ \rho D U_m^2) = f(K, Re, t/T) \tag{9.22b}$$

The primary application of Eqs. (9.22a) and (9.22b) is to piles and other structural elements, provided a 'strip' approach is assumed to be applicable. In this case, the sectional force at any instant and location (depth) is written as Eq. (9.22b), with K and Re values referring to that section. Experimental results with sectional models are thus used to evaluate sectional forces in the wave flow and at any instant the force may then be integrated to obtain the total instantaneous force and the overturning moment as described in Chapter 5.

The flow about a submarine pipeline can also be represented by a two-dimensional oscillatory flow. In this case the separation distance e from the seabed enters as an additional variable, and thus Eq. (9.22a) would be extended to:

$$F'/(0.5\ \rho D U_m^2) = f(K, Re, e/D) \tag{9.23}$$

Experimental results along these lines have been provided by Sarpkaya (1978) and Sarpkaya et al. (1980).

Finally, since we are here taking the flow separation to be significant, the roughness of the body surface may well have a marked influence on the wave forces. Clearly, this implies the inclusion of an additional dimensionless parameter, say a characteristic roughness height to body size ratio k_r/D and, moreover, consideration must also be given to the nature of roughness being modelled. Sarpkaya (1976) has in particular shown that roughness does have a marked influence, particularly at higher Reynolds numbers, and so should not be ignored (see Chapters 3 and 5).

Numerous experimental studies have been carried out to investigate different aspects of the various relationships outlined in the preceding paragraphs, and indeed many of these have been covered in detail in the previous chapters and will not be repeated here.

Effects of Reynolds Number

We now come to one of the most significant sources of difficulty in the modelling of offshore structures. In the flow about slender bodies, flow separation occurs and the consequences of separation must be taken into consideration. It follows that a Reynolds number is now a further dimensionless group affecting the problem. In the present context it may be defined as $U_m D/\nu$, where U_m is some representative particle velocity, e.g., the maximum horizontal velocity at $z = 0$. Evidently, either the Reynolds number or the frequency parameter $\beta = D^2/\nu T$ may be used depending on the experimental procedure (see Chapter 3).

It is a well-known fact that it is difficult to model a free-surface flow so as to ensure constancy of both a Froude and a Reynolds number. For a given length scale factor k_L, the velocity and time scale factors are determined by ensuring the constancy of d/gT^2 (or equivalently any kind of Froude number) and we have $k_T = k_u = k_L^{1/2}$. In order to ensure equal Reynolds numbers we have to hold $k_u k_L / k_\nu = 1$, and it then follows that we require $k_\nu = k_L^{3/2}$. For example, for a $\frac{1}{20}$ scale model of a structure in water we require the kinematic viscosity of the model fluid to be approximately $\nu_m = 10^{-7}$ ft^2/sec, ($k_\nu = \frac{1}{89}$).

Most liquids have viscosities greater than or about the same as that of water. Mercury is a notable exception having a kinematic viscosity some 8.7 times less than that of water. However, the above relationship indicates that a model involving mercury could only be about 4.2 times smaller than one using water so that any theoretical advantage due to reduced model size would be quite overshadowed by the disadvantages of prohibitive cost, experimental inconvenience and difficulties associated with the much greater level of energy to be supplied and dissipated. There is no satisfactory answer to this dilemma and water is in consequence invariably used as the fluid in models involving wave action.

Several approaches which attempt to by-pass this difficulty have been devised. In discussing these we first emphasize that when no flow separation occurs, or when it is localized so as not to influence the overall loads on a structure, then differences in Reynolds number should be unimportant, and the effect may be neglected as was assumed in the diffraction analysis (see Section 9.2.2).

Field tests give rise to relatively high Reynolds numbers and may be found appropriate in certain circumstances, but they have the disadvantage of high cost and also the lack of control on the incident waves. However, field tests are important in assessing the applicability of the laboratory results and the mitigating effects of the ocean environment (see Section 5.3.2).

A different approach, and one which is particularly suitable from a research viewpoint is to restrict the simulation to the two-dimensional oscillatory flow, rather than to represent the entire wave motion. Such a flow has been discussed in detail in Chapters 3 and 5. Gravitational effects are no longer relevant and this provides added flexibility in attempting to maintain a constant Reynolds number. Furthermore, water can be used in the model and the Reynolds number or the frequency parameter can be matched by suitably scaling the oscillation period, i.e., $k_T = k_L^2$ and $k_u = k_L^{-1}$.

Experimental facilities involving a two-dimensional oscillatory flow generated by a piston have been described by Dedow (1966) and by Brebner and Riedel (1973); and ones involving a U-shaped tube, containing fluid in oscillatory motion, have been described by Lundgren and Sorensen (1958), Carstens and Neilson (1961) and by Sarpkaya (1976). In addition, numerous published studies have involved the complementary situation of a body oscillating in otherwise still water. The advantages and disadvantages of these methods have been described by Sarpkaya (1976).

It is known that in steady flow the effects of modest changes in Reynolds number are relatively unimportant provided that the flow remains in the sub-critical or post-supercritical range as the case may be. Aerodynamic modelling techniques often involve the use of trip wires to cause transition to a turbulent boundary layer in order to simulate a post-supercritical flow at relatively low Reynolds numbers. Some attempt has been made to apply this technique to wave motion (Bushnell 1977) but the oscillatory nature of the separated flow and the complex vortex interaction with the structure suggest that such a procedure may be questionable.

9.3 MODELLING OF ELASTIC STRUCTURES

Attention must often be given to the use of an elastic model designed to reproduce the dynamic behavior of a structure. This problem has received relatively widespread attention in aeroelastic modelling of aircraft. More recently, it has gained considerable importance in the field of offshore technology.

When taking account of the dynamic response of an elastic member, a number of additional parameters are needed to characterize its behavior. These include its density (or a characteristic density) ρ_s, modulus of elasticity E, and damping properties, conveniently characterized by the damping ratio ζ or the logarithmic decrement δ.

Dimensional analysis indicates that three additional independent parameters must be considered. These are the density ratio ρ_s/ρ, an elasticity parameter $E/\rho U^2$, and a dimensionless damping parameter, say the structural damping ratio ζ. Thus, in the case of a regular wave train interacting with a fixed (but flexible) body, Eq. (9.21) would be extended to:

$$F/(\rho g H D^2) = f(d/L, H/L, K, Re, \rho_s/\rho, E/\rho U^2, \zeta) \qquad (9.24)$$

Ideally all three additional parameters should of course be held constant between the model and prototype, but in practice this may not be possible.

The density ratio ρ_s/ρ represents the ratio of structural inertia force (or weight) to fluid inertia force. In many cases a suitably averaged density may be scaled instead of the detailed density distribution over the whole structure, i.e., a parameter $m/\rho L^3$ may be used in place of ρ_s/ρ, where m is the mass of the structure and L its characteristic length.

The structural damping ratio ζ should also be held constant, but very often accurate information concerning prototype values is difficult to obtain and the damping ratio is only approximately duplicated. However, it is sometimes possible to accept an altered damping ratio, the intention being that resonances will occur at approximately the correct frequencies although the response amplitudes will be altered. If the model is known to be under-damped then the results should overpredict the response and so lie on the safe side.

The elasticity parameter $E/\rho U^2$ describes the ratio of structural elastic to fluid inertia forces and is directly related to the Cauchy number (see Table 9.1). This parameter is also difficult to hold constant in a fully elastic model. When geometric and Froude similarity are maintained (as required in wave force problems), Eq. (9.8) lists the scale factors of various quantities in terms of the length scale factor k_L. In particular, $k_\rho = 1$, $k_u = k_L^{1/2}$ and therefore the constancy of the elasticity parameter $E/\rho U^2$ indicates that we should have $k_E = k_L$. In practice this is awkward to achieve: k_E varies between about $\frac{1}{4}$ to 1 for most metals and is about $\frac{1}{60}$ for plastics, values which are generally incompatible with a prescribed k_L. The most usual ways of overcoming such a difficulty are by resorting to simpler *sectional* or *linear mode* models, or by incorporating a sectional distortion of the complete model. These are now briefly described in turn.

In many problems the most important effect of E lies in describing the structure's (fundamental) natural frequency f_n, and it is then convenient to adopt the alternative parameter $f_n L/U$ in place of $E/\rho U^2$. The related parameter $U/f_n L$ is the reduced velocity, discussed in detail in Chapter 8. The scale factor for frequency is thus given by as $k_{f_n} = k_L^{-1/2}$. The required natural frequency may be obtained by a model which is itself rigid but which is elastically mounted, rather than by any flexibility in the model itself. This is the approach adopted with a sectional model, which is a rigid elastically mounted model representing only a typical section of a structure. This technique is particularly useful for investigating two-dimensional hydroelastic oscillations as described in Chapter 8. For a single-degree-of-freedom system a linear mode model may be used. With this, a rigid cantilever elastically mounted so as to pivot about its base would be used to simulate a flexible cantilever oscillating in its fundamental mode.

Such models are relatively straightforward to construct. The required natural frequency (stiffness) can be obtained by a spring arrangement fixed externally to the model so as not to obstruct significantly the incident flow. The necessary damping is generally introduced as electromagnetic damping acting on a plate attached to the model, or as external viscous damping.

When a fully elastic model of the entire structure is required, the situation is more difficult but a sectional distortion of the structure may be used to achieve the necessary requirements. As already mentioned, the single most important effect of E is the natural frequency f_n and the reduced velocity may be employed in place of $E/\rho U^2$. When the natural frequency is associated with flexural oscillations—as is often the case—then we generally have that $f_n = C(EI/mL^3)^{1/2}$, where C is a constant and EI is the flexural rigidity of the structure. Noting that $k_{f_n} = k_L^{-1/2}$ (since $f_n L/U$ is constant and $k_u = k_L^{1/2}$) and $k_m = k_L^3$, one has

$$k_{EI} = k_L^5 \tag{9.25}$$

Ideally, we would require $k_E = k_L$ as already indicated; and $k_I = k_L^4$ derives from the definition of I. But since E is taken to influence the problem only through bending, E and I appear only in the combination EI, and thus provided this compound variable is itself correctly scaled according to Eq. (9.25), it may be acceptable that E and I each in isolation may not be.

This condition can generally be achieved by distorting the cross-section of structural members relative to the length scale k_L. Thus, we would have $k_A \neq k_L^2$, where A is a characteristic cross-sectional area of the structure. This is most conveniently carried out by distorting the internal dimensions of pipes or hollow structural elements and leaving the external dimensions undistorted. Flow patterns and added-mass effects should then be correctly reproduced.

Evidently, if the same material is used in both the model and prototype, $k_E = 1$, we would require $k_I = k_L^5$. In the case of a thin hollow cylinder of diameter D and wall thickness t ($\ll$D), we have $k_I \simeq k_L^3 k_t$ and, therefore, we would require $k_t = k_L^2$ implying that the model cylinder would be relatively thin.

But now that the model is distorted the effect on its required mass and density must be reconsidered. The submerged weight to hydrodynamic force ratio should be held constant, (for no distortion and $k_\rho = 1$, this implies that ρ_s/ρ is constant as before). This condition now gives

$$(\Delta\rho gAL/F)_m = (\Delta\rho gAL/F)_p \tag{9.26}$$

where $\Delta\rho = \rho_s - \rho$. Taking $k_F = k_L^3$, and k_A not necessarily equal to k_L^2 we have

$$k_{\Delta\rho} = k_L^2/k_A \tag{9.27}$$

The required mass distribution is obtained by fitting the model to the above formula by the addition of local masses around the structure, but taking care to leave the structural stiffness unaltered.

As an example of the foregoing approach we consider modelling a hollow cylindrical member constructed of 0.5 inch steel plate ($\rho_s/\rho = 7.8$) and whose diameter is 30 inches. The requirements of a $\frac{1}{20}$ scale plastic ($\rho_s/\rho = 1.2$) model might be estimated as follows. Given that $k_L = \frac{1}{20}$, $k_E = \frac{1}{60}$, we have $k_t = k_L^2/k_E = \frac{3}{20}$ (since $k_{EI} = k_L^5$ and $k_I = k_t k_L^3$). Therefore, the tube thickness in the model is $t_m = 0.075$ inch. The tube's mass should be scaled according to Eq. (9.27), $k_{\Delta\rho} = k_L/k_t = \frac{1}{3}$ and, therefore, for the given values of density, $(\rho_s)_m/\rho = 3.27$. The actual density ratio for the plastic tube is 1.2 and thus additional masses should be placed along the tube to increase its mass by a factor of 2.7.

The elastic modelling of offshore structures along the lines outlined above has been described by several authors. Plate and Nath (1968) have described this procedure of distorting internal dimensions as applied to offshore structures. Dawson (1976) has considered the more general case where elastic effects are not limited to bending and analyzed a typical offshore platform with distinct deck and support structure masses. Examples of elastic models of offshore pipelines, with consideration being focussed also on the fluid within the pipe, have been given by Kolkman and van der Wielde (1971), and by Clauss and Kruppa (1974). Hydroelastic oscillations and the basic governing parameters have been discussed in Chapter 8, (see also Griffin et al. 1980, King 1977, and Sarpkaya 1979).

9.4 EXPERIMENTAL TECHNIQUES

In a typical laboratory model test of an offshore structure, we attempt to generate a regular two-dimensional progressive wave train of permanent form and to locate a model at a test section of the tank in order to measure the loads on it. A variety of other phenomena may also be investigated and these may include the scour around the base of the structure, the runup around the structure, the dynamic response and motions of the structure, etc.

The waves may be generated by a wedge-shaped plunger, a paddle or a piston. The plunger is oscillated vertically and is best suited to deep-water wave generation. A paddle is often hinged at the channel bottom and is suitable for intermediate depth to deep water waves. Shallow wave conditions may be obtained if the channel bottom is made to slope upwards gently over a short distance from the paddle so that the depth along the major part of the channel is constant, but less than that at the generation section. A piston made to oscillate horizontally generates a harmonic motion which does not decay with depth and is best suited to shallow wave generation. In many cases the generator motion may be designed to incorporate both translational (piston derived) and rotational (paddle derived) components so as to be more effective than either alone over a wide range of relative depths.

The classical 'wave-maker' theory, which concerns the manner in which a wave train is generated is treated by Havelock (1929), Biesel (1951), and by Ursell et al. (1960). For small amplitude motion, Biesel obtained the relation between the wave height H and paddle or piston amplitude R as

$$H/2R = \begin{cases} 2\sinh^2(kd)/[kd + \sinh(kd)\cosh(kd)] & \text{for a piston} \\ \dfrac{[2\sinh(kd)/kd]\,[1 - \cosh(kd) + kd\sinh(kd)]}{[kd + \sinh(kd)\cosh(kd)]} & \text{for a paddle} \end{cases} \tag{9.28}$$

where $k = 2\pi/L$.

For a given water depth, we are usually interested in the variation of H/2R with $T/(d/g)^{1/2}$ which derives from the above equation. For details of wave generator design, the reader is referred to Gilbert et al. (1971).

In many model investigations, the laboratory generation of random waves is required. Among others, Sorensen (1971) has outlined the application of random wave experiments to harbor design. The technique generally adopted for random wave generation involves an electro-hydraulic servo system, in which a wave paddle is driven by an activator whose motion is controlled by a servo valve. A closed-loop computerized control system may be used such that the output corresponding to the paddle motion is compared to a reference input signal and the activator drive is based on the difference between these two.

The major problem lies in obtaining a suitable reference input signal which will result in the random wave motion possessing the desired statistical properties. The primary requirement here is to obtain a specified wave spectrum, although the probability distributions of water surface elevation or of wave heights will also be of importance. In establishing the spectral density of the input signal, the transfer function between the paddle motion and wave motion must be taken into account. The theoretical relationship between paddle amplitude and wave height, valid for small motions, is given by Eq. (9.28), but in general experimental results are required in this regard. In some cases (Kimura and Iwagaki 1976) the wave generator may comprise of a paddle and a piston that may be operated either separately or simultaneously and which will be more suitable over a wider range of frequencies than if one component only had been used.

Several methods have been used to generate the random input signal possessing the necessary properties. These include an analogue method in which electrical white noise is passed through a set of band pass filters each with slightly different central frequencies and with gains adjusted so that the superposition of all the filter outputs produces a random signal with the required spectrum. A further method (Goda 1970) is to superpose a large number (say 50 or more) of sinusoidal waves with different amplitudes and frequencies in order to simulate a specified wave spectrum. In this case some care is required to avoid a perfectly periodic signal. Digital methods are often used in which a sequence of random numbers generated by a computer are digitally processed (filtered) and transmitted to a digital-to-analogue converter in such a manner that the appropriate random signal is then produced (see e.g., Fryer et al. 1966, Funke 1974, Kimura and Iwagaki 1976, and Webber and Christian 1974).

In many cases a wave filter is placed a short distance from the generator. This serves the dual purpose of filtering out high frequency irregularities of the generated motion and also of absorbing reflected waves in order to reduce re-reflection. This is particularly necessary in tests involving seawalls, breakwaters or harbors, where primary reflections at the test section are an inherent part of the experiment and so further reflection from the wave generator must be avoided.

A reasonable channel length is allowed in order to permit the waves to adopt a freely travelling form rather than that of a forced motion (Biesel 1951). Even so, the generated waves may not be of permanent form particularly in the shallow water range where soliton separation occurs and different components of the wave may be found to travel at different speeds. A beach located at the far end of the channel serves to dissipate as much of the incident wave energy as possible, and so reduce wave reflection to a minimum. The beach may be made of sand or gravel, but hair matting or other artificial material is frequently used. An account of an 'active' absorber is given by Milgram (1970) and a useful discussion of the characteristics of wave absorbers by Le Méhauté (1973). Even though the beach slope is constructed with as gentle a gradient as possible in order to minimize wave reflections, these may yet be significant for longer and higher waves. Greslou and Mahe (1954) and more recently Johnson and Tokikawa (1975) provided data on wave reflections in a laboratory channel under various conditions.

In order to account for the interference between the incident and reflected wave motions, the wave height may be measured by the technique described in Section 4.9.4. We have, in addition, to note that the waves attenuate during their propagation and it is necessary, therefore, to measure the wave height in the vicinity of the test section. On the assumption that the wave attenuation is linear over a short distance, and taking wave reflection into account a simple formula may be developed to obtain the incident wave height from measurement of two consecutive maximum heights H_A and H_C and two adjacent minimum heights H_B and H_D. The incident wave height is given in terms of these (Treloar and Brebner 1970) as

$$H = (H_A + 3H_B + 3H_C + H_D)/8 \tag{9.29}$$

An alternative approach to eliminating wave reflections is to generate only a short wave train and to complete the measurements before reflections reach the test section. In this case care must be taken to ensure that regular conditions are established at the time measurements are taken, since the leading waves of the group generally produce different loads and effects than those of the regular and established wave motion.

A variety of resonant and other spurious wave motions may be induced and require special attention. These include a long period standing wave along the channel length in the fundamental or higher mode; transverse standing waves causing, for example, a wave crest at the channel centerline alternately to rise and dip during its travel. 'Cross waves' of this nature are noticeable when the channel width is greater than the wave length and so need consideration when a 'wave basin' is used. Reference may be made to Mahoney (1972), Barnard and Pritchard (1972), and Madsen (1974). Instabilities in deep water are described by Benjamin and Feir (1967). Instabilities in shallow water, involving soliton separation and the distortion of the wave train from one permanent form are discussed by Goda (1967), Madsen et al. (1970), Zabusky and Galvin (1971), and Galvin (1972), (see also Section 4.9.7).

The channel width may have an influence on the model tests when the ratio of this to the body size is not sufficiently large. It may be possible to perform additional tests corresponding to different channel width to body size ratios, while holding constant the remaining parameters so that any possible effect may be detected. In the case of large bodies, a reasonable indication is provided by results pertaining to the interference between neighboring vertical cylinders aligned with the incident wave crests. These results (see e.g., Spring and Monkmeyer 1975) indicate that such width effects are generally quite small when the cylinders are separated by more than a few diameters. In the case of slender bodies where vortex shedding occurs particular care should be taken to ensure that the motions of the vortices are not constrained by the channel walls.

Wave channels, such as we have been considering, represent one of the more common types of facility employed to investigate wave-structure interactions. Others include U-shaped water tunnels previously mentioned in connection with the study of two-dimensional oscillatory flows. Hogben (1976) has described a *wave dozer* used to generate particularly high waves by incorporating a paddle which travels along the tank during the generation process.

It is now convenient to summarize briefly the procedures generally adopted for measuring the more usual quantities involved. It is relatively straightforward to obtain the depth and wave period by direct measurements. In the case of regular wave motion, the wave height may be obtained with a hook-and-pointer gage which is raised and lowered by measured amounts so that the wave trough and crest elevations may conveniently be determined.

For a varying wave motion, the surface elevation can be continuously recorded by a wave probe which measures a change in water level as a change in the resistance or capacitance of a Wheatstone-bridge circuit. The output signal can be conditioned and recorded or displayed as required. In the case of a random wave motion, the signal is generally recorded on a tape recorder, for subsequent spectral or other analysis, or directly processed by computer.

A pressure, force, or moment can be measured by the use of an appropriate transducer or dynamometer. The overall loading on a structure may be represented by three forces along and three moments about three orthogonal axes. It may thus be convenient to employ a six-component dynamometer in which these can all be measured simultaneously.

To list a few of the factors to-be considered in the design of models of rigid structures for force measurements we should include the following: arrangements should be made for the measured strains to be sufficiently large to provide a reliable signal; the maximum deflection of the model must be sufficiently small that it does not alter the flow and hydrodynamic loading; the natural frequency of the model to wave frequency ratio must be large enough to avoid significant dynamic amplification; and any gaps between a flexible force-measuring element and the neighboring rigid geometry must be small enough to ensure that the required flow is adequately modelled. Data obtained under questionable conditions (surface irregularities, severe vibrations, etc.) do not allow one to assess critically either the force-transfer coefficients or the behavior of the prototype (for a full discussion of such effects see Sarpkaya and Collins 1978).

9.5 EXPERIENCE WITH PROTOTYPES

We conclude this chapter with a few remarks concerning reliability estimates pertaining to prototype structures. Numerous uncertainties associated with the design of offshore platforms such as the random nature of the oceanographic data, foundation characteristics, material properties, construction procedures, methods of analysis, etc. raise a fundamental question concerning the provision of adequate safety margins. The designer is often forced to introduce both intentional as well as unintentional safety factors. Oceanographic data (wave height and period, wind and current velocity, spectral characteristics, etc.) and the material characteristics are intentionally chosen on the conservative side, whereas the mitigating effects of the ocean environment (often resulting from the relatively less organized nature of the flow about the structural members) introduce unintentional safety factors. With regard to wave force predictions, the Morison equation is often written as

$$F = k_1(0.5 k_2 \rho C_d D|U|U + 0.25 k_3 C_m D^2 \, dU/dt) \tag{9.30}$$

where k_1, k_2, and k_3 are coefficients larger than one. This procedure, with its intentional safety factors, is not in conformity with the extreme concern generally expressed regarding the most appropriate values of the drag and inertia coefficients.

The foregoing raises numerous questions such as: how safe is safe, how expensive is safe, what are the consequences of failure in terms of human life, environmental damage, and lost revenue, what are the failure modes, what are the relative influences of various uncertainties on the failure modes?, etc. Admittedly, these questions may be answered only approximately, since the sensitivity and failure analyses are subject to the probabilistic assumptions made regarding the parameters examined.

Planeix et al. (1979) have carried out an extensive sensitivity analysis on the influence of environmental factors on the design of jacketed structures and gravity platforms. For jacketed structures, they have shown that the most important environmental parameters are the design wave height and the force-transfer coefficients. A risk analysis accounting for the hidden safety factors in the choice of these parameters and for the explicit safety factors specified by the design codes may allow for savings of at least 10 percent in steel weight and cost.

For gravity platforms, Planeix et al. (1979) have concluded that the influence of the environmental parameters is most paramount on stability constraints, and that the overall weight and cost are much more sensitive than is the case for jacketed structures.

Generally, wave heights or assumed sea spectra have important effects on all three types of structures: jacketed, gravity, and semi-submersible. Thus, their definition requires relatively more precise information than those of other

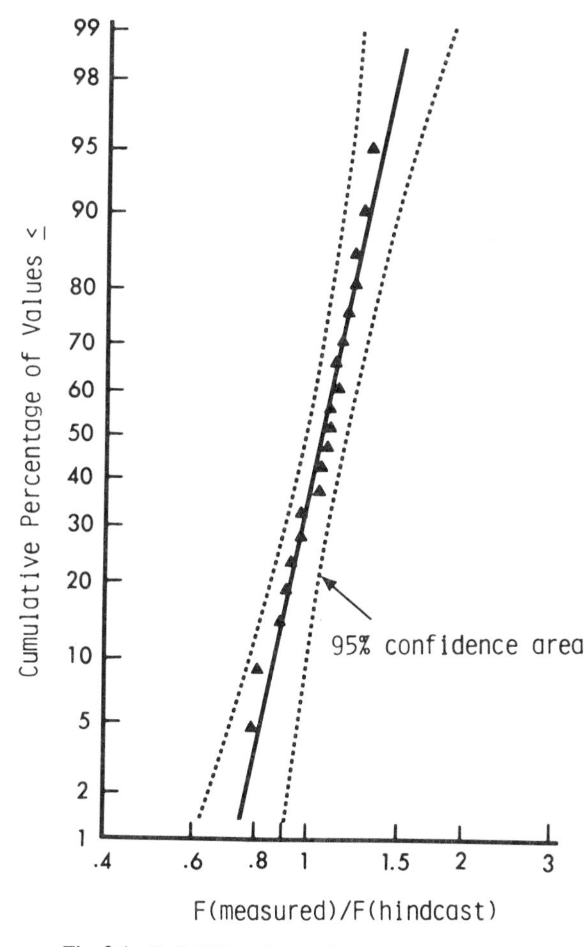

Fig. 9.1. Reliability of wave force hindcast model.

parameters. A better knowledge of the force-transfer coefficients for jacketed structures will certainly allow a better safety and cost balance.

Sensitivity and risk analyses must also deal with such additional factors as the characteristics of piles and foundations, the relationship between the overall forces and moments and those measured on actual platforms.

Bea (1974) and Marshall and Bea (1976) presented a detailed discussion of the environmental loads on and failure of offshore platforms. Bea considered eight major loading categories (extreme condition wave and current forces, current forces, wave fatigue forces, wave dynamic excitation and interaction forces, wind forces, seismic forces, soil forces, and ice forces) and presented precision factors for each loading category in terms of the desired accuracy. The precision factors are given as ratios of 'actual' to 'predicted' effects or conditions (e.g., the precision ratio for the extreme condition wave and current forces, F_a/F_p, ranges from 0.5 to 1.1). In addition, Bea presented a concise discussion of the long-term goals of the desired-state-of-the art factors, in addition to an evaluation or estimation of the present-state-of-the-art factors.

In general, measurements are not available on the total lateral loads developed by hurricane waves and currents on platforms. However, some measurements are available on single isolated piles. Figure 9.1 (after Bea 1974) shows the ratio of measured and hindcast forces on such a pile for the 20 highest wave-current forces measured in the Gulf of Mexico (obtained during hurricane 'Carla'). These data show that the hindcast model makes an unconservatively biased

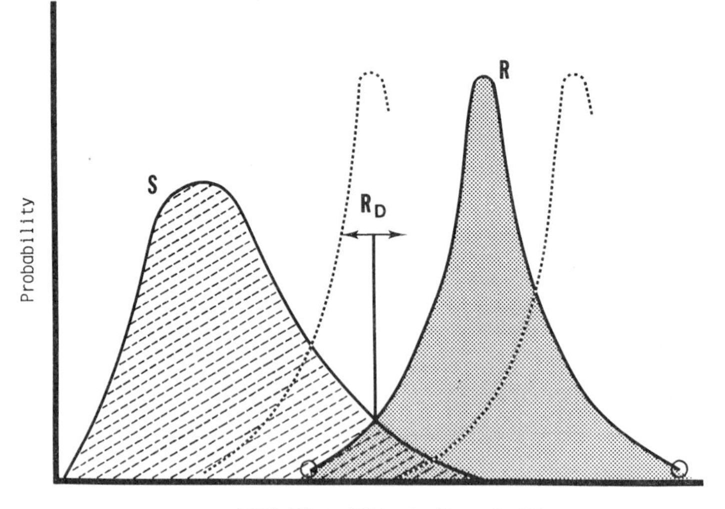

Fig. 9.2. Acceptable reliability based on probabilistic definitions of lifetime loads and ultimate strength of the platform (Reliability is inversely proportional to the overlapp area).

estimate, i.e., F(measured)/F(hindcast) at the 50th percentile is 1.06. However, one must note that the prediction of wave forces is not independent of the prediction of wave heights, and these two procedures should be combined in defining the resultant probability of a platform experiencing various magnitudes of life-time loadings. As shown in Fig. 9.2 (after Bea 1974), a choice must be made to define the desired probability of the platform. This definition of reliability (probability of a failureless platform life-time) will in turn define the required design load.

Industry's experience over the past 40 years has helped to develop a history of many platform-exposure years. This history shows (Marshall and Bea 1976) an annual probability of failure of about 0.6 percent. For platforms in deep water, the historical annual probability of failure is about 1.1 percent. Evidently, the industry has achieved remarkable success through an unparalleled integration of varied disciplines and backgrounds in spite of the uncertainties of the hostile ocean environment. Acquisition of data on environmental forces and conditions will lead to further verification and calibration of offshore engineering models.

9.6 REFERENCES

Atkins, A. G. 1975. Icebreaking Modeling. *Jour. Ship Research*, Vol. 19, No. 1, pp. 40–43.

Barnard, B. J. S. and Pritchard, W. G. 1972. Cross Waves, Part 2. Experiments. *JFM*, Vol. 55, pp. 245-255.

Bea, R. G. 1974. API Review, Environmental Loads on Ocean Structures. Shell Oil Co., Note CE-8, Houston, TX.

Benjamin, T. B. and Feir, J. E. 1967. The Disintegration of Wave Trains on Deep Water. *JFM*, Vol. 27, pp. 417-430.

Biesel, F. 1951. Les Appareils Générateures de Houle en Laboratoire. *La Houille Blanche*, Vol. 6, pp. 147–165.

Brebner, A. and Riedel, P. H. 1973. A New Oscillating Water Tunnel. *Jour. Hydraulic Research*, Vol. 111, pp. 107-121.

Bridgeman, P. W. 1963. *Dimensional Analysis*. Yale Univ. Press, revised edition.

Bushnell, M. J. 1977. Forces on Cylinder Arrays in Oscillating Flow. OTC Paper No. 2903, Houston, TX.

Carstens, M. R. and Neilson, F. M. 1967. Evolution of a Duned Bed in Oscillatory Flow. *Jour. Geophys. Res.*, Vol. 72, pp. 3053-3059.

Clauss, G. and Kruppa, C. 1974. Model Testing Techniques in Offshore Pipelining. OTC Paper No. 1937, Houston, TX.

Dawson, T. H. 1976. Scaling of Fixed Offshore Structures. *Ocean Engineering*, Vol. 3, pp. 421–427.

Dedow, H. R. A. 1966. A Pulsating Water Tunnel for Research in Reversing Flow. *La Houille Blanche*, Vol. 7, p. 837.

Fryer, D. K., Gilbert, G., and Wilkie. 1966. A Wave Spectrum Synthesizer. *Jour. Hydraulic Res.*, Vol. 11, pp. 193-204.

Funke, R. 1974. Random Wave Signal Generation by Minicomputer. *Proc. 14th Coastal Engineering Conf., Copenhagen*, pp. 352-371.

Galvin, C. J. 1972. Wave Breaking in Shallow Water. *Waves and Beaches and Resulting Sediment Transport* (ed. R. E. Meyer), Academic Press, NY, pp. 413-456.

Gilbert, G., Thompson, D. M., and Brewer, A. J. 1971. Design Curves for Regular and Random Wave Generators. *Jour. Hyd. Res.*, Vol. 9, pp. 163-196.

Goda, Y. 1967. Travelling Secondary Wave Crests in Wave Channels. *Port and Harbor Res. Inst., Japan*, Vol. 13, p. 32.

Goda, Y. 1970. Numerical Experiments on Wave Statistics with Spectral Simulation. *Port and Harbor Res. Inst., Japan*, Vol. 9, pp. 7-57.

Greslou, L. and Mahe, Y. 1954. Etude du Coefficient de Reflexion d'une Houle sur un Obstacle Constitue par un Plan Incline. *Proc. 5th Conf. Coastal Engineering, Grenoble, France*, pp. 66-84.

Griffin, O. W., Pattison, J. H., Skop, R. A., Ramberg, S. E., and Meggitt, D. J. 1980. Vortex-Excited Vibrations of Marine Cables. *Jour. Waterway etc. Div., ASCE*, WW2, pp. 183-204.

Hallam, M. G., Heaf, N. J., and Wootton, L. R. 1977. Dynamics of Marine Structures: Methods of Calculating the Dynamic Response of Fixed Structures Subject to Wave and Current Action. *CIRIA Underwater Engineering Group*, London, Report UR-8.

Hansen, A. G. 1964. *Similarity Analyses of Boundary Value Problems in Engineering.* Prentice-Hall, Englewood Cliffs, NJ.

Havelock, T. 1929. Forced Surface Waves on Water. *Phil. Mag.*, ser. F, Vol. 8, pp. 569-576.

Hogben, N. 1976. The 'Wave Dozer': A Travelling Beam Wavemaker. *Proc. 11th ONR Symp. on Naval Hydrodynamics.*

Isaacson, E. and Isaacson, M. 1975. *Dimensional Methods in Engineering and Physics.* Edward Arnold, London.

Johnson, J. W. and Tokikawa, K. 1975. Laboratory Study of Wave Reflection by Slopes with Different Roughness. *Coastal Engrg. in Japan*, Vol. 18, pp. 35-44.

Kimura, A. and Iwagaki, Y. 1976. Random Wave Simulation in a Laboratory Wave Tank. *Proc. 15th Coastal Engrg. Conf.*, Honolulu, pp. 368-387.

King, R. 1977. A Review of Vortex Shedding Research and Its Applications. *Ocean Engineering*, Vol. 5, No. 3, pp. 141-171.

Kolkman, P. A. and van der Weide, J. 1971. Elastic Similarity Models as a Tool for Offshore Engineering Development. *Symp. on 'Offshore Hydrodynamics'*, Wageningen, Vol. VII, pp. 1-15.

Langhaar, H. L. 1951. *Dimensional Analysis and the Theory of Models.* Chapman and Hall, NY.

Le Méhauté, B. 1973. Progressive Wave Absorber. *Jour. Hydraulic Res.*, Vol. 10, pp. 153-169.

Le Méhauté, B. 1976a. Similitude in Coastal Engineering. *Jour. Waterway etc. Div., ASCE*, Vol. 102, WW3, pp. 317-335.

Le Méhauté, B. 1976b. *An Introduction to Hydrodynamics and Water Waves.* Springer-Verlag, Berlin.

Lundgren, H. and Sorensen, T. 1958. A Pulsating Water Tunnel. *Proc. 6th Conf. Coastal Engineering*, Miami, pp. 356-358.

Madsen, O. S. 1974. A Three Dimensional Wave Marker, Its Theory and Application. *Jour. Hydraulics Res.*, Vol. 12, pp. 205-222.

Madsen, O. S., Mei, C. C., and Savage, R. P. 1970. The Evolution of Time-Periodic Waves of Finite Amplitude. *JFM*, Vol. 44, pp. 195-208.

Mahoney, J. J. 1972. Cross Waves, Part 1. Theory. *JFM*, Vol. 55, pp. 229-245.

Marshall, P. W. and Bea, R. G. 1976. Failure Modes of Offshore Platforms. *BOSS '76*, Trondheim, pp. 579-635.

Milgram, J. H. 1970. Active Water Wave Absorbers. *JFM*, Vol. 43, pp. 845-859.

O'Brien, M. P. 1977. Discussion of 'Similitude in Coastal Engineering', by B. Le Méhauté, *Jour. Waterways etc. Div., ASCE*, Vol. 103, WW3, pp. 393-400.

Planeix, J. M., Ciolina, J., Delueil, J., Doris, C. G., and Heas, J. Y. 1979. Are Offshore Structures Over-Designed? Relative Influence of Environmental Parameters. OTC Paper No. 3387, Houston, TX.

Plate, E. J. and Nath, J. H. 1968. Modelling of Structures Subjected to Wind Generated Waves. *Proc. 11th Conf. Coastal Engrg.*, London, pp. 745-760.

Remery, G. F. M. 1971. Model Testing for the Design of Offshore Structures. *Symp. on 'Offshore Hydrodynamics'*, Wageningen, Vol. III, pp. 1-31.

Sarpkaya, T. 1976. Vortex Shedding and Resistance in Harmonic Flow About Smooth and Rough Circular Cylinders at High Reynolds Numbers. Technical Report No. NPS-59SL 76021, Naval Postgraduate School, Monterey, CA.

Sarpkaya, T. 1978. In-line and Transverse Forces on Cylinders Near a Wall in Oscillatory Flow at High Reynolds Numbers. OTC Paper No. 2898, Houston, TX.

Sarpkaya, T. 1979. Vortex Induced Oscillations, A Selective Review. *Journal of Applied Mechanics, ASME*, Vol. 46, pp. 241-258.

Sarpkaya, T. and Collins, N. J. 1978. Discussion of 'Drag and Inertia Forces on a Cylinder in Periodic Flow' (by Garrison, C. J. et al. Jour. Waterways etc. Div., ASCE, Proc. Paper No. 12913), *Jour. Waterway etc. Div., ASCE*, WW1, pp. 96-98.

Sarpkaya, T. and Rajabi, F. 1980. Hydrodynamic Drag on Bottom-Mounted Smooth and Rough Cylinders in Periodic Flow. OTC Paper No. 3761, Houston, TX.

Sedov, L. I. 1959. *Similarity and Dimensional Methods in Mechanics*. Academic Press, New York.

Silvester, R. 1974. *Coastal Engineering*. Vol. II, Elsevier, Amsterdam.

Sorensen, T. 1973. Model Testing with Irregular Waves. *Dock and Harbour Authority*, Vol. 54 (631), pp. 2-5.

Spring, B. H. and Monkmeyer, P. L. 1975. Interaction of Plane Waves with a Row of Cylinders. *Proc. Civil Engineering in the Oceans III*, Univ. of Delaware, pp. 979-998.

Treloar, P. D. and Brebner, A. 1970. Energy Losses under Wave Action. *Proc. 12th Coastal Engrg. Conf.*, Washington, DC, pp. 257-267.

Ursell, F., Dean, R., and Yu, Y. 1960. Forced Small Amplitude Water Waves; A Comparison of Theory and Experiment. *JFM*, Vol. 7, pp. 33-52.

Vugts, J. H. 1971. The Role of Model Tests and Their Correlation with Full Scale Observations. *Symp. on 'Offshore Hydrodynamics'*, Wageningen, Vol. VII, pp. 1-30.

Webber, N. B. and Christian, C. D. 1974. A Programmable Irregular Wave Generator. *Proc. 14th Coastal Engrg. Conf.*, Copenhagen, pp. 340-351.

Author Index

Subject Index